AIDS UPDATE 2005

An Annual Overview of Acquired Immune Deficiency Syndrome

GERALD J. STINE, PH.D.

Department of Biology
University of North Florida, Jacksonville

San Francisco Boston New York
Cape Town Hong Kong London Madrid Mexico City
Montreal Munich Paris Singapore Sydney Tokyo Toronto

**AIDS IS A WAR THAT NO ONE WANTS TO LOSE
BUT NO ONE YET KNOWS HOW TO WIN**

• • • • • •

*This book, as with my other HIV/AIDS college-level textbooks,
is also dedicated to
those who have died of AIDS,
those who have HIV disease,
and to those who must prevent the spread of this plague—*

EVERYONE, EVERYWHERE.

Publisher: *Daryl Fox*
Executive Editor: *Leslie Berriman*
Editorial Assistant: *Blythe Robbins*
Executive Marketing Manager: *Lauren Harp*
Managing Editor: *Deborah Cogan*
Production Supervisor: *Beth Masse*
Senior Manufacturing Buyer: *Stacey Weinburger*
Cover Design Supervisor: *Mark Ong*
Cover Designer: *Armen Kojoyian*
Production Supervision/Composition: *WestWords, Inc.*
Cover Photos: *Getty Images*

ISBN: 0-8053-7310-1

2 3 4 5 6 7 8 9 10–CRS–07 06 05
www.aw-bc.com

AIDS Update 2005

WHY DO I WRITE ABOUT HIV/AIDS?

I began writing about a new disease, later called AIDS, in 1981, shortly after the Centers for Disease Control and Prevention issued the first of its reports. These writings were limited in scope because not much was known at the time. My writing then was for classroom use. Little did I know at the time the passion that I would develop about this disease. As the number of infected people and their deaths continued to rise, fear and discrimination reared their ugly heads because, in some cases, people are what they are, and, in other cases, because of the lack of available nonbiased educational material. At the time, in the mid-1980s, I felt a need to write, to help educate people about this disease, hopefully to answer questions and reduce the blatant discrimination occurring against those who already had the overwhelming burden of HIV disease. So, I began—I created a college-level HIV/AIDS course, taught it for several years, constantly shaping and reshaping the information necessary to help others learn the facts about this pandemic, and to destroy destructive myths. After I felt I had set the record straight in my classroom, I began writing HIV/AIDS college-level textbooks so that information on this pandemic could be shared more broadly.

There are many reasons I have not stopped writing about HIV/AIDS. First, because as Yogi Berra supposedly said, "It ain't over till it's over," and it "ain't" near over yet! Second, AIDS summons up the greatest themes in literature, among them sex, faith, and death—themes that are universal and unexpectedly permanent. Anyone who has lost a loved one to death, untimely or by nature, can read about AIDS and understand the emotional forces involved. Anyone who has taken care of someone who has been ill understands the need for compassion. Anyone who has faced death from a prolonged or life-threatening illness should be able to identify with those who suffer with AIDS. Anyone interested in uncovering acts of human kindness or, conversely, acts of despicable behavior can find them in writings about this disease.

But more than any of these reasons, my writing about AIDS is fueled by a need to do something, anything, to help. For the millions of people who are in pain and dying, I have little to offer except my writing. Although my only known risk factor was a blood transfusion in the mid-1970s, I cannot help but feel lucky that so far neither HIV/AIDS nor any other serious disease threatens my life or the lives of my wife and children. Is there a reason I have been spared to my present age when so many others have died? I write about AIDS, not just because I live but because it is part of our history. The AIDS pandemic has changed my feelings and attitudes toward people. I find myself more sensitive to others. AIDS has changed the world in some way for most everyone, and at least some of the information about this disease—its impact and repercussions—needs telling. So, I write about AIDS because someone needs to tell the story.

INTO THE NEW MILLENNIUM

The war on AIDS is still closer to the beginning than it is to the end, but I have never been more positive about the future for those who are HIV-infected in the United States and other developed nations. The use of anti-HIV drugs in combination has led to a dramatic reduction in HIV-infected babies and AIDS-related deaths, as well as to a significant reduction in new AIDS cases and reductions in opportunistic infections. These events, along with new insight into the biology and pathology of HIV, may lead to a preventive vaccine and provide a shining light against a stark history of the first 24 years of a pandemic first reported in the United States in 1981.

With the creation of the Global Fund for HIV/AIDS, Malaria, and Tuberculosis in 2001, the G8 (group of eight industrial powers) founding of the International Global Fund for HIV/AIDS, and the decision by the world's largest antiretroviral manufacturers to dramatically lower drug costs to developing nations, there is now hope in those countries. However, the drugs are not yet free, and therefore are out of reach for millions of people living in poverty. But hope and help must begin somewhere—someone must first benefit before others can. **Let us pull together so that help, not just hope, will be available to all who need it!**

—Gerald Stine, Ph.D.

The information on HIV/AIDS within this text goes where the pandemic has taken us over the last 24 years. This information is not designed to replace the relationship that exists between you and your doctor or other health adviser. For all medical events/needs, consult with an appropriate physician.

Contents

Preface x

Introduction: Histories of Global Pandemics, World AIDS Conferences, Overview of HIV/AIDS, and the AIDS Quilt 1

1 Discovering AIDS, Naming the Disease 23

AIDS: A Disease or a Syndrome? 23
What Causes AIDS? 24
How Does HIV Cause AIDS? 25
The Centers for Disease Control and Prevention Reports 26
Discovery of the AIDS Virus 27
Defining the Illness: AIDS Surveillance 29
Summary 31
Review Questions 31

2 What Causes AIDS: Origin of the AIDS Virus 32

The Cause of AIDS: The Human Immunodeficiency Virus (HIV) 32
What Do You Believe and Why? 34
Evidence That HIV Causes AIDS 37
Origin of HIV: The AIDS Virus 39
Summary on the Origin of HIV 45
Summary 46
Review Questions 47

3 Biological Characteristics of the AIDS Virus 48

Viruses Need Help 48
Viruses Are Parasites 49
Retroviruses 49
Human Immunodeficiency Virus (HIV): Its Life Cycle 50
HIV Life Cycle 51
Basic Genetic Structure of Retroviral Genomes 55
Genetic Revolution 60
Formation of Genetic Mutations in HIV 60
Formation of Genetic Recombination in HIV 61
Distinct Genotypes (Subtypes/Clades) of HIV-1 Worldwide Based on ENV and GAG Proteins 63
Summary 67
Review Questions 67

4 Anti-HIV Therapy 68

Anti-HIV Therapy 68
Anti-HIV Drugs With FDA Approval 72
The HIV Medicine Chest: HIV/AIDS Drugs Receiving FDA Approval 72
FDA-Approved Nucleoside, Non-Nucleoside Analog Reverse Transcriptase Inhibitors and Protease Inhibitors 75

Use of Non-Nucleoside Analog Reverse Transcriptase Inhibitors 77
FDA-Approved Protease Inhibitors 77
Protease Inhibitors 79
Development and Selection of HIV Drug-Resistant Mutants 80
Development of Drug-Resistant Protease Inhibitors 81
How Combination Drug Therapy Can Reduce the Chance of HIV Drug Resistance 84
HIV Treatment Regimen Failure 86
Viral Load: Its Relationship to HIV Disease and AIDS 88
Medical Complications Associated With Anti-HIV Drug Therapies 90
Two Most Sobering Reports on AIDS-Related Deaths 96
Disclaimer 105
Summary 105
Some AIDS Therapy Information Hotlines 106
Review Questions 107

5 The Immunology of HIV Disease/AIDS 109

The Immune System 109
Human Lymphocytes: T Cells and B Cells 110
Antibodies and HIV Disease 122
Immune System Dysfunction 124
FUSIN or CXCKR-4 (R-4) 128
CC-CKR-5(R-5) Receptor 128
CC-KR-2(R-2) and CC-KR-3(R-3) Receptors 129
Genetic Resistance to HIV Infection 129
T4 Cell Depletion and Immune Suppression 131
Impact of T4 Cell Depletion 134
Role of Monocytes and Macrophages in HIV Infection 134
Summary 137
Review Questions 137

6 Opportunistic Infections and Cancers Associated With HIV Disease/AIDS 139

What Is an Opportunistic Disease? 139
The Prevalence of Opportunistic Diseases 140
Prophylaxis Against Opportunistic Infections 140
Opportunistic Infections in HIV-Infected People 140
TB and HIV 154
Cancer or Malignancy in AIDS Patients 154
Disclaimer 159
Summary 159
Review Questions 160
HIV/AIDS Self-Evaluation/Education Quiz 161

7 A Profile of Biological Indicators for HIV Disease and Progression to AIDS 162

HIV Disease Defined 162
Stages of HIV Disease (Without Drug Therapy) 164
HIV Disease Without Symptoms, With Symptoms, and AIDS 175
Aspects of HIV Infection 176
Production of HIV-Specific Antibodies 176
Prognostic Biological Markers Related to AIDS Progression 179
HIV Infection of the Central Nervous System 185
Neuropathies in HIV Disease/AIDS Patients 188
Pediatric Clinical Signs and Symptoms 188
Summary 189
Review Questions 189

8 Epidemiology and Transmission of the Human Immunodeficiency Virus 191

We Must Stop HIV Transmission Now! 193
Epidemiology of HIV Infection 193
Transmission of Two Strains of HIV (HIV-1/HIV-2) 193
Is HIV Transmitted by Insects? 194
HIV Transmission 195
HIV Transmission in Family/Household Settings 195
Noncasual Transmission 197
United States: Heterosexual IDU 221
Caesarean Section 233
Public Confidence: Acceptance of Current Dogma on Routes of HIV Transmission 234
Conclusion 237
National AIDS Resources 237
Summary 237
Review Questions 238

9 Preventing the Transmission of HIV 240

The AIDS Generation: "I Knew Everything About It, and I Still Got It!" 240
Prevention, Not Treatment, Is the Least Expensive and Most Effective Way To Stop the Spread of HIV/AIDS 240
Global Prevention 241
Advancing HIV Prevention: New Strategies for a Changing Epidemic 243
Preventing the Transmission of HIV 243
Behavioral Change 247
New Rules to an Old Game: Promoting Safer Sex 251
Injection-Drug Use and HIV Transmission: The Twin Epidemics 265
HIV Prevention for Injection-Drug Users 265
Prevention of Blood and Blood Product HIV Transmission 268
Infection Control Procedures 273
Sexual Partner Notification 275
Vaccine Development 278
Types of Experimental HIV Vaccines 281
DNA Vaccine 281
Problems in the Search for HIV Vaccine 282
Human HIV Vaccine Trials 285
Summary 287
Review Questions 288

10 Prevalence of HIV Infections, AIDS Cases, and Deaths Among Select Groups in the United States and AIDS in Other Countries 290

A Word About Data 291
The Millennium: Year 2005 294
Formula for Estimating HIV Infections 296
Behavioral Risk Groups and Statistical Evaluation 296
Estimates of HIV Infection and Future AIDS Cases 308
Newly Infected 310
Issues of Credibility in U.S. Estimates 311
Shape of the HIV Pandemic: United States 311
Shape of AIDS Pandemic: United States 313
Estimates of Deaths and Years of Potential Life Lost Due to AIDS in the United States 315
United States and Global: Who Is Dying of AIDS? 319

Life Expectancy 320
Connected But Separate 320
Workforce: South Africa: Some Examples 323
Education: Teachers and Students: South Africa and Some Countries in Southern Africa 324
Other HIV/AIDS Time Bombs: Asia, India, China, and Russia 325
Summary 329
Review Questions 329

11 Prevalence of HIV Infection and AIDS Cases Among Women, Children, and Teenagers in the United States 331

International Women's Day 332
Women: AIDS and HIV Infections Worldwide 332
Women: HIV-Positive and AIDS Cases—United States 334
Internet 345
Pediatric HIV-Positive and AIDS Cases—United States 346
Orphaned Children Due to HIV Infection and AIDS 347
The Phenomenon of AIDS Orphans 349
Global HIV Infections in Young Adults 355
How Large Is the Teenage and Young Adult Population In the United States? 356
HIV/AIDS Won't Affect Us! 357
Estimate of HIV-Infected and AIDS Cases Among Teenagers and Young Adults 357
Summary 363
Internet 363
Other Useful Sources 363
Review Questions 363

12 Testing for Human Immunodeficiency Virus 365

Determining the Presence of Antibody Produced When HIV Is Present 366
Requests for HIV Testing 366
Reasons for HIV Testing 366
Immunodiagnostic Techniques for Detecting Antibodies to HIV 367
ELISA HIV Antibody Test 367
Western Blot Assay 377
Other Screening and Confirmatory Tests 378
Rapid HIV Testing 381
Viral Load: Measuring HIV RNA 383
FDA Approves Two Home HIV Antibody Test Kits 383
Who Should Be Tested for HIV Infection? 384
Why Is HIV Test Information Necessary? 386
Immigration into the United States 386
Testing, Competency, and Informed Consent 388
Summary 393
Review Questions 394

13 AIDS and Society: Knowledge, Attitudes, and Behavior 395

The New Millennium and HIV/AIDS 395
HIV/AIDS Is an Unusual Social Disease 396
AIDS Is Here To Stay 397
AIDS 397
Blame Someone, Déjà Vu 398
Fear: Panic and Hysteria over the Spread of HIV/AIDS in the United States 398
Whom Is the General Public To Believe? 400
AIDS Education and Behavior 404
The Character of Society 414

Federal and Private Sector Financing: Creation of an AIDS Industry 425
U. S. Government Believes HIV/AIDS Is a Threat to National Security: The United Nations Believes It is a Weapon of Mass Destruction 429
Global HIV/AIDS Funding for Underdeveloped Nations 429
Forms of U.S. Foreign and Other Assistance for HIV/AIDS 433
Summary 435
Review Questions 436

Answers to Review Questions 438

Glossary 447

References 455

Index 477

Preface

I hope that this 2005 edition of *AIDS Update* will help you gain a clearer perspective on HIV and AIDS and the ways in which the disease both fuels widespread controversy and suffers under a silence that restricts some of the most important information about it. HIV/AIDS is a slow, progressive, and permanent disease. With the disease, there is no loss of infectivity, no development of either individual or group immunity, and no recovery. At present, there is no known biological mechanism that can stop the continuing expansion of the disease. The progressive increase in the pool of people carrying HIV, the virus that causes AIDS, will lead to an exponential increase in the number of newly infected individuals. Until an effective vaccine can be developed or other interventions are at least moderately successful, the infection will continue to spread and will remain a crucial health issue. With the onset of this real human tragedy, we have been forced to learn about our social contradictions and examine our moral judgments. We have had some success in this venture, yet despite great improvements in our understanding of the scientific and social aspects of the disease, the AIDS crisis is not nearly over.

PURPOSE

While this volume is intended for use in college-level courses on HIV/AIDS and as a supplemental HIV/AIDS resource in medical and nursing schools, it is suitable for any situation in which information about the various aspects of HIV and AIDS is desired or required. This text reviews the most important information on all facets of HIV infection, HIV disease, and AIDS. It provides readers with a detailed background in the current biological, medical, social, economic, and legal aspects of this modern-day pandemic. Medical and social anecdotes bring a personal perspective to the worldwide HIV/AIDS tragedy, while other parts of the text detail the history of the pandemic. Special focus has been given to HIV transmission, including risk behaviors and means of prevention, and on the innumerable social issues surrounding this disease. Many of these issues are accompanied by open-ended questions to stimulate class discussions that help students reflect on what they have read.

IMPORTANCE

This book is important for two reasons. First, as educators, it is our job to expose our students to the new concepts in biology that are shaping humanity's future. To do that, we must expose them to new vocabulary, new methodologies, new information, and new ideas. It is my hope that this text will help in this endeavor. Second, because misinformation has increased the danger of the HIV/AIDS pandemic, this text attempts to counter common distortions and myths by presenting the most current, consistent, and scientifically acceptable information possible. The text also provides students with a strong conceptual framework for the issues raised by the HIV/AIDS pandemic so they will be better able to deal with new information as it arises.

During the 24 years since HIV/AIDS was defined as a new disease, more effort and money have been poured into research efforts surrounding HIV and AIDS than into any other disease in history. As a result, information on the virus that causes AIDS has accumulated at an unprecedented pace. The constant stream of new information about the changing social and scientific circumstances surrounding the disease makes it necessary to publish an updated edition of this text every year. The majority of references herein are dated between 1990 and 2004, with outdated information being replaced in each edition. What has been learned and what must still be learned to bring HIV infection, HIV disease, and AIDS under control is valuable information to all of us.

SPECIAL FEATURES

- Each chapter begins with a **Chapter Concepts** section that allows students to preview the material and stay focused on key topics.
- A variety of **boxed essays** provide supplemental information on each chapter's topic. These features illustrate important events and information about HIV/AIDS, including stories from doctors and AIDS survivors, points of scientific note, and discussions of historical context that reveal how knowledge and perceptions of AIDS have evolved over time.
- **Discussion Questions** encourage students to think critically about difficult and controversial issues associated with HIV and provide a starting point for lively class discussions.
- **Further information** sources are listed at the end of relevant chapters. These internet addresses and national hotline telephone numbers give students resources for additional information on prevention, therapy, health-care providers, and specific focuses such as women, parents, children, and teenagers.
- Chapter **Summary** and **Review Questions** provide students with a quick review of the key concepts of the chapter and help students test their understanding of the chapter's material. Answers appear in an appendix.
- A list of **References** allows students to see the sources of all scientific and social data within the book, as well as to engage in further research on their own.

RESEARCH NAVIGATOR

Each copy of *AIDS Update 2005* now includes access to Research Navigator, a powerful research tool combining three exclusive databases of credible and reliable source material, including *AIDS Weekly*, the *American Journal of Public Health*, *New England Journal of Medicine*, *Social Science & Medicine*, and the *New York Times*. The access code for your complimentary subscription can be found in the front of the text.

IN APPRECIATION

The help of the following organizations and people is most deeply appreciated: The Centers for Disease Control and Prevention (CDC) Atlanta, for use of slides and literature produced in their National Surveillance Report and the Morbidity and Mortality Weekly Report; Mara Lavitt and *The New Haven Register* for information on William Bluette; from the *Sun Sentinel* of Ft. Lauderdale, Fla., Director of Photography Jerry Lower and Photo Archivist Britt Head for photographs out of the Caribbean; from the CDC, Richard Salik and Patricia Sweeney (Surveillance Branch); Robin Moseley, Patricia Fleming, Todd Webber (Division of HIV/AIDS Prevention); Barry Bennett, head of Retroviral Testing Services, for his permission and guidance on photographing HIV testing procedures at the Florida Health and Rehabilitation Office of Laboratories Services, Jacksonville, Fla.; Denise Reddington, Department of Health and Human Services, Washington, D.C.; PANOS Institute for their help in obtaining data on the economics of HIV/AIDS worldwide; personnel of the National Institute for Allergy and Infectious Diseases, the National Center for Health Statistics, Brookwood Center for Children with AIDS in New York; the George Washington AIDS Policy Center in Washington, D.C.; the National Institutes for Health; Hoffman-LaRoche Co., Abbott Laboratories, the Pharmaceutical Manufacturing Association; the National Cancer Institute; Pan American Health Organization; the Office of Technological Assessment; the Physicians' Research Network; Teresa M. St. John, University of North Florida for illustrations; the individuals who have contributed photographs; Eric Barnes, for help with tables; the text reviewers whose work has been greatly appreciated; Judy Kaufman, whose reviews and comments over the course of many editions have been invaluable; a special thank you to Gilbertine Yadao who word-processed the updated material; Guy Selander, M.D. and Jack Giddings, M.D., who, over the years, have shared their medical journals with me; James Alderman, Mary Davis, Signe Evans, Cynthia Jordan, Paul Mosley, Sarah Philips, and Barbara Tuck—reference/research librarians at the University of North Florida; and to my wife Delores, who helped with proofreading and demonstrated a great deal of patience and understanding and gave up family weekends so the text could be completed on time.

This book has benefited from the critical evaluation of the following reviewers:

- Dr. Paul R. Elliott, Florida State University
- Dr. Robert Fulton, University of Minnesota
- Dr. Robert M. Kitchin, University of Wyoming

- Dr. Richard J. McCloskey, Boise State University
- Dr. William H. Fenn, Western Michigan University
- Dr. Michael Sulzinski, University of Scranton
- Dr. Judy Kaufman, Monroe Community College
- Dr. Seth Pincus, Montana State University
- Dr. Wayne B. Merkley, Drake University
- Dr. Linda L. Williford Pifer, University of Tennessee
- Dr. Bernard P. Sagik, Drexel University
- Professor James D. Slack, Cleveland State University
- Dr. Carl F. Ware, University of California, Riverside
- Dr. Phyllis K. Williams, Sinclair Community College
- Dr. Charles Wood, University of Kansas

Gerald J. Stine, Ph.D.
904-641-8979

The Face of AIDS

The face AIDS wears is both many and one. The face of AIDS is women and men, children, youth, and adults. It is our sons and daughters, brothers and sisters, husbands and wives, mothers and fathers. Sometimes the face AIDS wears is that of a person without a home or a person in prison. Other times it's the face of a pregnant woman who is fearful she will pass HIV to her unborn child. Sometimes it's a baby or child who has no caregiver and little hope of adoption or being placed in foster care.

The faces of AIDS come from all walks of life. They represent all racial and ethnic groups, religious backgrounds, and countries of the world.

It is two boys, ages 10 and 12, caring for their mother, who is dying from AIDS. The boys get themselves off to school in the morning, make dinner for their mother after school, carry her to the tub to bathe her, softly towel dry her body, which is covered with sores, then help her into bed. How do you go to school and compete academically when your mother is home dying, your family is broke, and you are afraid to tell anyone about the problem?

It is a young gay man in Orange County, Calif., who is having difficulty breathing. At the local hospital, he is diagnosed with *Pneumocystic carinii* pneumonia (one of the characteristic infections afflicting people with AIDS) and is told, "We don't take care of this kind of pneumonia here." Packed into a car for a six-hour drive to San Francisco, he's given an oxygen tank—only the oxygen runs out after four hours. Arriving at the emergency room at 2 A.M., in a city with no friends or family, he can barely breathe. Three months later he is dead.

It is a young woman who is knitting a scarf for her two-year-old daughter. She said, "I want to leave some legacy for her." She did not yet know her daughter was HIV positive and most likely would not live long enough to understand the legacy. Most likely the daughter would die before her mother.

It is an HIV-infected child born to a woman who is an injection-drug user. With parents unable to care for her, hospitals too crowded to board her, and foster agencies unable to place her, no one knows where or how this child will endure what promises to be a brief, sickly life.

It is Maria, 26 and pregnant, who found out that she had AIDS while on a respirator for PCP. She died still muted by the plastic tubing that was her last lifeline. Or John, an artist, who had his life choked out of him by tonsilar lymphoma from one end, while intractable isospora diarrhea drained it away from the other. Or Kathy, who clung desperately to life in order to continue caring for her two children, the children who helplessly watched their mother waste away and die.

It is about a 25-year-old male who looks 40 less than a year after his AIDS diagnosis. When he first tested HIV positive, he decided to kill himself when his T4 cell count dropped to 100. When it did, he prepared his last meal, said goodbye to close friends, and bought a handgun. After dinner he put on his favorite cassette, raised the gun—but could not end his life. "What now?" he asks. "This was my ace in the hole!"

It is a young man who lives in Miami, and he's not ready to come home. He wants to remain independent. Eventually he will come home. His family must be prepared to care for him and to deal with all the issues they will face. His mother is a high school teacher, but she has never told anyone about her son. She says that when he comes home it will be impossible to hide the fact that he has AIDS. "I am not sure how the community and our friends are going to respond."

The Face of AIDS Is Aging. In the United States about 13% of AIDS cases are among people over age 50. As of the beginning of the year 2006, that's about 126,000 people—33,000 are between ages 60 and 69 and about 5500 are age 70 and older.

The Face of AIDS Is Now a Woman's. In 2002, for the first time in the pandemic's history, the number of women living with HIV rose to 50% of the global total. One could say there's been a rapid feminization of AIDS. In some parts of Africa, women make up 70% of the HIV-infected population. Elsewhere around the world, women make up a growing proportion of people living with HIV because sex between men and women has taken over as the primary way the disease spreads in many regions. Of global adult HIV/AIDS deaths in 2003 and 2004, about 50% were women. The face of AIDS globally is now a non-white female, poor, undereducated, and living outside a major urban center. This description is also the description for the face of AIDS in the southern United States.

The Face of AIDS Is Changing. For 1997 through the year 2005 and beyond, the face of AIDS, due to the successful use of AIDS drug cocktail therapy, takes on new and different looks—those of rejuvenation and of guilt. Many who have recovered, even from their deathbeds, and feel "new" again, feel guilty because many are still dying and many do not have access to these drugs. This face reveals the strain of establishing new relationships, giving up disability insurance, finding employment, and experiencing a life they thought they had lost. It shows the fear that all this may be just short-term—that the virus will win in the end!

The Political Face of AIDS. In April 2000, the U.S. government declared HIV/AIDS a threat to American national security, marking the first time ever a disease had been entrusted to the National Security Council. Less than three months later, the United Nations Security Council affirmed Resolution 1308, which delineated the dangers that HIV/AIDS posed to the maintenance of international peace and security. At the Fourteenth International AIDS Conference in July 2002, UNAIDS Executive Director Peter Piot noted the beginning of a "new era: the era of AIDS as a global political issue." Amid all this discourse intertwining a virus with domestic and world affairs, it is important to understand the threat created by HIV/AIDS to a nation's political, economic, and personal security.

The face of AIDS is surely around us. It is a disease that invites commentary, requires research, and demands intervention. Inevitably linking sex and death, passion and politics, it continues to generate controversy. Appearing as a new and lethal infectious disease, at a time when the biomedical establishment in the United States has shifted its focus to chronic disease, HIV/AIDS has called into question many of the premises of the biomedical profession.

TWENTIETH ANNIVERSARY OF AIDS PLUS FOUR: JUNE 6, 2005

The AIDS crisis is not over; it has just begun. The statistics over the last 24 years numb one beyond comprehension. Who can understand the meaning of 41 million people throughout the world carrying the virus? Or that the rate of infection continues to explode throughout much of the world? It is only when people realize that by this anniversary 24 million people have died of AIDS, or that a 15-year old in South Africa has a 50:50 chance of dying of AIDS before the age of 30, that the horror begins to take on a human scale. The American perception of AIDS has gone through a cycle of meanings over the last two decades. In the early days there was no name for the disease; today we continue to ask how to cure it. In the early days there were major protests against the FDA and the medical and pharmaceutical companies. Today, there is quiet acceptance. The early demonstrators influenced basic research, the federal drug approval process, and the massive federal funding for a single disease. This disease as no other before it has changed the way people interact with their physicians, demand medical services, and take control of their own treatments regardless of illness. An entire generation has now been born that has never known an AIDS-free world. Children born with HIV have now themselves become mothers of newborns! The disease has changed the personal as well as the political—how we think and how we love, what we teach our children and what words we say in public. More than anything else, HIV/AIDS has changed the way we view the threat of emerging diseases. Until HIV/AIDS, most of us thought of catastrophic plagues, like the Black Death and the Spanish flu epidemic of 1918, as things of the past. We lost sight of the fact that every once in awhile a new disease emerges. It happened with HIV/AIDS and it can happen again. Witness the current outbreaks of the Hanta, West Nile, SARS, and monkeypox viruses in the United States. AIDS also changed what it means to be a patient. People with AIDS stormed scientific conferences, banded together in ways no other patients ever had, helped revolutionize the process of testing experimental drugs, and inspired others. There is no question that breast cancer activism started because of AIDS activism. Those with cancer saw its success and decided to emulate it, deploying thousands of people to lobby for increased research funding.

HIV/AIDS also changed what it means to be gay in America. The images of gay men dying of this disease rendered them objects of sympathy and opened the doors to compassion. In the eyes of straight America, death gave gay men a humanity they had long been denied. Homophobia, and attacks on gays, became less acceptable. Although some argue that HIV/AIDS divided gays—positives from negatives—it seems more likely that a united gay community was forged in the crucible of HIV/AIDS. People facing mortality responded courageously and seized the chance to proclaim their identity. And it forced society's institutions—from hospitals that barred gay men from seeing their dying lovers to employers that denied them bereavement leave—to recognize gay relationships. On June 25, 2001, the United Nations General Assembly convened a session on HIV/AIDS, the first ever devoted to a public health crisis. Delegates tried to agree on a global action plan, and grappled with how to fund it: $7 billion to $10 billion a year is needed to fight the spread of HIV. If lucky with prevention, treatment, and a vaccine, by 2021 AIDS will be killing 5 million people a year. If not, the toll could be 12 million. HIV/AIDS is 24 years young. The worst plague in modern history is far from finished with us. While searching for a vaccine, people continue to die from HIV/AIDS or the drugs intended to treat it.

Introduction: Histories of Global Pandemics, World AIDS Conferences, Overview of HIV/AIDS, and the AIDS Quilt

THE NEW MILLENNIUM

Now into the third decade of AIDS, the disease HIV/AIDS remains a great challenge to public health, human rights, development, and national security. This is the first plague in the era of globalization! It has become the ultimate terrorist! The figures are truly alarming—over the last six years, 2000 through 2005, about 30 million people became infected with HIV, bringing the total number of infections to about 65 million since the outbreak of this pandemic. While 95% of the cases are concentrated in developing countries, industrialized countries are experiencing about 100,000 new infections each year. By the end of 2005, about 24 million people will have died from AIDS worldwide.

HIV/AIDS is devastating for the individual who may be infected, and its impact on communities and society at large is enormous. This disease disproportionately affects those population groups that are already vulnerable: children, women, the poor, the destitute, and millions of others whose life situations are further degraded by the denial of basic human rights. The majority of those infected are unable to afford the cost of effective health care. Clearly, political and economic solutions must be found if millions of lives are to be saved. Of all the promises the new millennium holds, of all the secrets it will tell, a cure or a vaccine for HIV disease will surely unfold among them.

DÉJÀ VU: A TIME OF AIDS

> It was the best of times, it was the worst of times, it was the age of wisdom, it was the age of foolishness, it was the epoch of belief, it was the epoch of incredulity, it was the season of Light, it was the season of Darkness, it was the spring of hope, it was the winter of despair . . .
>
> Charles Dickens, *A Tale of Two Cities*

Nothing in recent history has so challenged our reliance on modern science nor emphasized our vulnerability before nature. We live with the Acquired Immune Deficiency Syndrome (**AIDS**) pandemic, witnessing its paradoxes every day. People living with the Human Immunodeficiency Virus (**HIV**) live with the fear, pain, and uncertainty of the disease; they also endure prejudice, scorn, rejection, and despair. **This must change!**

HISTORY OF GLOBAL EPIDEMICS

There has never been a time in human history when disease did not exist. The history of epidemics dates at least as far back as 1157 B.C. to the death of the Egyptian pharaoh, Ramses V, from smallpox. Over the centuries, this extraordinarily contagious virus spread around the world, changing the course of history time and again. It killed 2000 Romans a day in the second century A.D., more than 2 million Aztecs during the 1520 conquest by

LOOKING BACK I.1

June 1981—Twenty-four years ago, a report written by UCLA scientist Michael Gottlieb and published by the Centers for Disease Control and Prevention alerted physicians to an unusual constellation of fungal infections and pneumonia in otherwise healthy young gay men. The condition was later named the Acquired Immune Deficiency Syndrome (AIDS). Activists took to the streets and convinced a skeptical public that AIDS was a deadly crisis.

Cortez, and some 600,000 Europeans a year from the sixteenth through the eighteenth centuries. Three out of four people who survived the high fever, convulsions, and painful rash were left deeply scarred and sometimes blind. Because victims' skin looked as if it had been scalded, smallpox was known as the "invisible fire." Even now, malaria in underdeveloped countries afflicts 300 million people, killing between 2 and 3 million each year. The problem is compounded by development of drug-resistant malarial strains of protozoa. Thus epidemics are not new to humankind, but the fear they impose on each generation is.

The major recorded pandemics (global) and epidemics (regional) that have devastated large populations are described in Table I-1.

SILENT TRAVELERS: THEY NEED HUMANS AND THEY CAUSE DISEASE

First Recognition of an Infectious Disease

In 1773, Charles White, an English surgeon and obstetrician, published his "Treatise on the Management of Pregnant and Lying-In Women." In it he appealed for surgical cleanliness to combat childbed or puerperal fever. (**Puerperal fever** is an acute life-threatening condition that may follow childbirth. Its cause, at this time in history, was unknown.) In 1795, Alexander Gordon, a Scottish obstetrician, published his "Treatise on the Epidemic Puerperal Fever of Aberdeen," which demonstrated for the first time that the disease was contagious. In 1843, Oliver Wendell Holmes, a noted American physician, published a paper entitled "On the Contagiousness of Puerperal Fever." In the paper he appealed for surgical cleanliness to stop the spread of this disease.

However, the first person to realize that a pathogen could be transmitted from one person to another was the Hungarian physician Ignaz Phillip Semmelweis. Between 1847 and 1849, Semmelweis observed that women who had their babies at the hospital with the help of the medical staff were four times more likely to contract puerperal fever than those who gave birth with the help of midwives outside the hospital. He concluded that the physicians were infecting women with material remaining on their hands after autopsies and other hospital activities. Semmelweis thus began washing his hands with a calcium chloride solution before examining patients or delivering babies. This simple procedure of hand washing led to a dramatic decrease in the number of cases of puerperal fever and saved the lives of many women. Hand washing remains basic to medical procedures in medical establishments and restaurants in developed nations.

EMERGING DISEASES

The 1970s witnessed the emergence of Lyme disease (1975), Legionnaires disease (1976), and Toxic Shock Syndrome (1978). In the 1980s, HIV and three new human herpes viruses (HHV 6, 7, and 8) were identified. In 1994, a physician in Gallup, N. M., was called to care for a man who had collapsed at a funeral. Soon, several others in New Mexico had also suffered from sudden respiratory failure. As victims contracted the disease, they went from seemingly perfect health to flu-like symptoms, to a horrific state where their lungs began to fill with blood. Most of them died.

Within a few weeks, the Centers for Disease Control and Prevention identified this killer: **hantavirus**—a rodent-borne organism that had infected many GIs during the Korean War and killed hundreds. But the Korean type of hantavirus causes death through kidney failure. Somehow the hantavirus had found its way to the United States and mutated (changed) into a virus that invades the lungs. The mortality rate is between 30% and 40%.

The outbreak of hantavirus in America occurred in May 1993, first in Arizona, Colorado, New Mexico, and Utah. Beginning year 2005, 400 people have been infected in 31 states with the rodent-borne deermouse virus and 135 of these people have died of acute respiratory distress. (There is no vaccine for this virus.)

In mid-1994 there was an outbreak of **necrotizing fasciitis** (fash-e-i-tis)—tissue destruction caused by a tissue-invasive strain of Group A streptococcus, a bacterium. This bacterium is a deadly variant strain of the Group A streptococcus that causes strep throat. The infected may die from bacterial shock, disseminated blood coagulation (clotting), or respiratory or kidney failure.

EMERGING VIRUSES

Although emerging viruses may contain new mutations or represent gradual evolution, more often

Table I-1 Plagues in History[1]

Disease	Dates	Place	Number Killed		Causative Organism	Time to Prevention/Cure (in years)
Measles	from 430 B.C.	Greece/Rome/World	Millions		Paramyxovirus	1712
Plague	542–1894	Europe/Asia/Africa	71 million		Yersinia pestis	580
Cholera	1826–1837	New Jersey	900,000		Vibrio cholerae	75
	1849	United Kingdom	53,293			
	1947	Egypt	11,755			
Tuberculosis	1930–1949	United States	1,000,000		Mycobacterium tuberculosis	85
	1954–1970		150,000			
Malaria	1847–1875	Africa/India	20 million+		Plasmodia	100
Scarlet Fever	1861–1870	United Kingdom	972 per million people		Streptoccous pyogenes	45–44
Polio	1921–1970	North America	37,000		Polio viruses Types I, II, III	30–50
Typhus	1917–1921	Russia	2,500,000		Rickettsias	25
Influenza	1918–1919	United States/Europe	21,640,000		Influenza virus	57
Smallpox	from 1122 B.C.	Europe (middle ages)	Hundreds of millions		Smallpox virus	3050
	1926–1930	India	423,000			
	from 590 B.C.					
Gonorrhea	1921–1992	United States	57,477		Neisseria gonorrhoeae	1832
	from A.D.					
Yellow Fever	1986–1988	Nigeria	10,000		Arbovirus	488
HIV/AIDS[2]			*Deaths*	*HIV Infections*	Human	Treatments
	1981–2004	United States (estimated)[3]	543,000	1,555,000	Immunodeficiency Virus	but no
	1981–2004	Global (estimated)	24,000,000	65,000,000		cure

1. Historical time line on first suspected cases of the above diseases are: Plague, eleventh century B.C.; Cholera, 1781; Tuberculosis, 451 B.C.; Malaria, 1748; Scarlet Fever, 1735; Polio, 1894; Typhus, 1083; Influenza, 1580; Smallpox, 429 B.C.; Gonorrhea, 1768; Yellow Fever, 1647; HIV/AIDS, United States in the 1970s, in Africa, 1959.

2. Total AIDS data estimated through year 2005. Living in some state of HIV disease (9 million living with AIDS + 32 million with HIV disease = 41 million people living with HIV infection + 24 million dead = 65 million total HIV infections worldwide).

3. United States: of the estimated number of people infected with HIV, about 34% or 530,000 will have died through 2004.

POINT OF INFORMATION I.1

INFECTIOUS DISEASE PARADIGM

Plagues and pestilence have evoked fear and awe since time began. Often viewed as divine retribution, these scourges are mentioned in many cultural and religious texts, including the Bible, the Koran, and the Talmud. History itself is punctuated and shaped by epidemics, whose accounts are at the center of such literary works as Boccachio's *Decameron,* Daniel Defoe's *A Journal of the Plague Year,* and Gabriel Garcia Marquez's *Love in the Time of Cholera.* These works provide rare insight into the impact of real epidemics. Accounts of fictional epidemics, such as Albert Camus's *The Plague* or Edgar Allan Poe's "The Masque of the Red Death," are even more fascinating and debatable. For example, Poe's work opens with the description of a mysterious epidemic. "'The Red Death' had long devastated the country. No pestilence had ever been so fatal, or so hideous. Blood was its seal—the redness and the horror of blood. There were sharp pains, and sudden dizziness, and then profuse bleeding at the pores, with dissolution. (Poe's description of this disease fits a form of viral hemorrhagic diseases, especially Ebola.) The scarlet stains upon the body and especially upon the face of the victim, were the pest ban which shut him out from the aid and from the sympathy of his fellow men. And the whole seizure, progress and termination of the disease, were the incidents of half an hour." To escape death, Prince Prospero secludes himself and a thousand noblemen in a castellated abbey. The epidemic rages and kills the poor who were left outside to fend for themselves. Six months into their successful bid to avoid the contagion, the callous prince and his friends celebrate within the sealed confines of the abbey, when quite suddenly the disease invades their sanctuary and kills everyone.

WARS

Although death in battle is the most obvious form of war-related mortality, in many eras the deadliest killer was not the opposing army but epidemics. Throughout history, where armies go, disease has followed. Much of military history is actually the story of how battles and campaigns correlated with diseases occurring in both the military and the civilian populations. Indeed, medieval and early modern history is a continuing chronicle of the cycle of invasions and epidemics. In the late eighteenth and early nineteenth centuries, Napoleon faced his greatest resistance not from the British navy but from microorganisms. French forces were routed in Syria and Egypt by plague. Of Napoleon's wars, his Russian campaign stands out as a classic example of devastation by disease. Of the 600,000 soldiers who left France, 70% succumbed to typhus before they reached Moscow. Russian delaying tactics allowed typhus to ravage the remainder. Thus, Napoleon was left with only 50,000 soldiers to occupy Russia. After the winter killed still more troops, only 3000 returned to France. During the American Civil War (1861–1865), the vast majority of deaths on both sides were due to disease. Prison camps like Andersonville were notorious for the prevalence of disease; incarceration was a virtual death sentence.

SOCIETY IN GENERAL

Most new infections are *not* caused by truly new pathogens. They are caused by microorganisms (bacteria, fungi, protozoa, and helminths) and viruses that have been present for hundreds or thousands of years, but they are newly recognized and named because of recently developed techniques to identify them. Human activities drive the emergence of disease and a variety of social, economic, political, cultural, climatic, technological, and environmental factors can shape the pattern of a disease and influence its emergence into populations. Trade caravans, religious pilgrimages, and military campaigns facilitated the speed of plague, smallpox, cholera, and now AIDS. Global travel is a fact of modern life and, equally so, the continued evolution of microorganisms and viruses; therefore, seemingly new infections will continue to emerge, and known infections will change in distribution, frequency, and severity.

Diseases caused by viruses and microorganisms have had 3 billion years to evolve—we should not expect to conquer them easily.

they emerge because of changes in human behavior or the environment whereby they are introduced to a new host. Viruses that are pathogenic in human and nonhuman hosts include Marburg and Ebola viruses; hantaviruses; human immunodeficiency virus; Lassa virus; dolphin, porpoise, and seal morbilliviruses; feline immunodeficiency virus; and bovine spongiform encephalopathy agent, which from 1996 through 2004 threatened the possible destruction of the entire cow population of Europe and the United States, causing economic loss of billions of dollars and a major unemployment problem.

New and Old Infectious Diseases Exploding Worldwide

Some 35 newly identified disease-causing agents have emerged over the past 30 years, and six old diseases are reemerging: tuberculosis (TB), diphtheria, cholera, dengue (den´ge) fever, yellow fever, and bubonic plague.

Cases of **dengue** or **breakbone** fever have been recognized for over 200 years. Diseases caused by the dengue virus, such as dengue and dengue hemorrhagic fever (DHF), are a greater burden to human health than any other mosquito-borne viral disease. An estimated 80 million persons are infected with dengue virus each year. In contrast, the virus that causes **hepatitis C** (HCV) was identified in 1989. An effective blood screen for it was in use by 1992. It is estimated that 4 million Americans, 9 million Europeans, and about 200 million people worldwide are infected with hepatitis C, marking an epidemic much larger than HIV. About 70% of infected people eventually develop a chronic liver disease. There are an estimated 30,000 new infections annually in the United States, with 10,000 deaths attributed to the virus. It is the leading cause for liver transplant in America.

The fact that so many infectious diseases remain out of control represents a shared failure—of individuals, health-care providers, communities, industry, and governments worldwide.

The world is not becoming a safer place in which to live. More people are crowding into cities, and humans continue to intrude into once remote areas. Clearly some diseases are more lethal than others, and AIDS may be the most lethal of all "new" diseases to strike humans in the twentieth and twenty-first centuries.

UPDATE 2003/2004—SEVERE ACUTE RESPIRATORY SYNDROME (SARS) There was a time when in Asia or in most any country a person would "catch" a cold, spread it to a few neighbors, go home, perhaps infect his/her family and climb into bed until he/she was well. Now, a person catches a cold in his or her hometown, and boards a 747, say, from Hong Kong to Singapore to Bangkok, to Germany, or to the United States. He or she spends hours sharing air and spreading droplets inside the plane with hundreds of others. The person emerges, hacking up phlegm, into an entirely new community of people ripe for infection. Globalization—the twenty-first century reality of humans reaching other continents and disparate communities of millions within hours—is also a global opportunity for disease, a reality dramatically underscored by the swift spread of SARS. In the early 1980s, it took two years to identify HIV as the cause of AIDS. In 2003, the World Health Organization created an extraordinary network of 13 laboratories in 10 countries, including the CDC, which identified a coronavirus associated with SARS in two weeks and had its entire RNA genome of about 30,000 nucleotides sequenced in two more. Those labs shared their knowledge in an unprecedented fashion, to the benefit of everyone.

Through 2004 about 900 people died from SARS (none in the United States) and about 10,000 people have been infected in 29 countries. China has 66% of reported SARS cases. Over 400 cases have been reported in the United States. The numbers of infections and deaths continue to increase as this new disease continues to spread. Globally SARS has a 10% to 15% mortality rate.

Monkeypox Virus

In early May 2003, the first cases of monkeypox in the United States occurred in an outbreak in three states—Illinois, Indiana, and Wisconsin. Monkeypox is an exotic African disease brought here via the Gambian rat, which in turn infected prairie dogs, native to the American Plains. In June 2003, the U.S. government banned the sale of prairie dogs, prohibited the importation of African rodents, and recommended smallpox shots for people exposed to monkeypox. At press time, prairie dogs infected with monkeypox virus have been reported in at least 15 states. Each infected prairie dog was traced back to a single pet shop in Villa Park, Ill.

The Importance of Infectious Diseases: Up Then Down and Up Again

In the early 1900s there were thousands of typhoid fever cases, and many people died of the disease. Most of these cases arose when people drank water contaminated with sewage or ate food handled by or prepared by individuals who were, unknowingly, shedding the typhoid fever bacterium, *Salmonella typhi.* An interesting case of one person spreading typhoid fever occurred between 1896 and 1906 in New York City. Mary Mallon was infected but did not become ill—she just carried and shed the bacterium. She worked as a cook in seven homes in New York City. Twenty-eight cases of typhoid fever occurred in these homes while she worked in them. As a result, the New York City Health Department had Mary arrested and admitted to an isolation hospital on North Brother Island in New York's East River. Examination of Mary's stools showed that she was shedding large numbers of the typhoid bacteria. An article published in 1908 in the *Journal of the American Medical Association* referred to her as "Typhoid Mary," an epithet by which she is still known today. Upon being released, she pledged not to cook for others or serve food to them, but Mary changed her name and began to work as a cook again. For five years she managed to avoid capture while continuing to spread typhoid fever. Eventually the authorities tracked her down. She was held in custody for 23 years until she died in 1938. As a lifetime carrier, Mary was positively linked with 10 outbreaks of typhoid fever, 53 cases, and 3 deaths.

Complacency and Disregard for the Microbial World Leads to a Resurgance of Infectious Diseases

In 1962, Sir MacFarlane Burnet, the 1960 Nobel Prize recipient for his work on immunological tolerance, wrote, ". . . one can think of the middle of the 20th century as the end of one of the most important social events in history, *the virtual elimination of the infectious disease as a significant factor in social life.*" This statement represents the complacency that developed in the United States and some other countries in regard to threats to the human race posed by the microbial world. More recently, Joshua Lederberg, president of the Rockefeller University from 1978 to 1990, wrote, "The ravaging epidemic of AIDS has shocked the world . . . *We will face similar catastrophes again* . . . We have too many illusions that we can . . . govern the remaining vital kingdoms, the microbes, that remain our competitors of last resort for domination of the planet." This statement represents a more realistic assessment of the current situation, by recognizing the fact that improvements in sanitation and introduction of antimicrobial agents and vaccines have failed to eliminate the health risks from exposure to infectious agents.

Infectious Diseases Cause Death

Infectious disease has climbed to the third leading cause of death in the United States. Worldwide, infectious diseases account for 28% of the 54 million global deaths per year and are the leading cause of death among children.

Since the 1980s, we have come to view infectious diseases with a humbler eye. The victories of a quarter-century ago ring hollow as AIDS ravages, disease-causing bacteria become resistant to all standard treatments, and the once easily treated disease-causing bacteria demonstrate antimicrobial drug resistance. Old and new diseases continue to develop with discomforting frequency. As was recognized in 1892 when the first international sanitary convention on cholera was adopted, infectious diseases cannot be observed, battled, or understood street by street or country by country. A global approach is necessary [WINKER, M.A., et al. (1996). Infectious diseases: a global approach to a global problem. *JAMA*, 275:245–246].

Far from being invulnerable, humans are at the mercy of innumerable microscopic agents that can erupt at any time by mutating (changing) into a virulent form, crossing a species barrier, escaping from an environmental niche, or being unleashed in bioterrorism.

There have been some great achievements in the battle against these microbes. Yet the only complete victory over any infectious disease so far has been the eradication of smallpox. However, one of the achievements of this millennium will be the global eradication of poliomyelitis by the end of 2005, the target date set by the World Health Organization.

VIRUS ASSOCIATED WITH CANCERS AND OTHER KNOWN DISEASES

In the mid- to late 1980s, numerous correlations were discovered between viruses and various types of cancers. For example, Epstein-Barr virus was associated with nasopharyngeal carcinoma and B-cell lymphoma, hepatitis B virus with liver cancer, and human papillomavirus with cervical cancer. Now, a decade later, scientists are finding out that viruses may also play a role in an array of other diseases, like atherosclerosis, diabetes, Alzheimer's disease, and even neuropsychiatric disorders such as schizophrenia. Additional connections will be made between viruses and many types of diseases in the future because virologists are turning their attention to diseases to which there is no known cause.

FEAR AND IGNORANCE ABOUT HIV/AIDS AND OTHER PANDEMICS

During the Black Plague, instead of being concerned about providing care to the victims, people spent their time deciding how deep to dig the graves so that none of the horrid fumes would come up and infect others. It was determined that a grave should be six feet deep; and that is exactly how deep it is today. Plague victims were herded together into cathedrals to die or to pray for faith healing to save them. In fourteenth-century Germany and Switzerland, the Christians blamed the Jews for the outbreak of bubonic plague, believing that the Jews were poisoning the water—the very same water the Jews were drinking. As a result, entire communities of Jews were slaughtered. In the 1400s and 1500s, when syphilis was spreading across the world killing thousands, the Italians called it the French disease. Of course, the French called it the Italian disease. In the 1930s, cholera was considered a punishment for people unwilling to change their lives—the poor and the immoral. In New York the Irish were blamed. In the early twentieth century, polio in America was believed to be caused by Italian immigrants.

Placing Blame

The history of epidemics teaches us, again and again, that blame is a central component of these events, whether it is cast upon socially ostracized groups of people, water supplies, politics, or religious or cultural beliefs. The sixteenth-century rise of Protestantism, especially Calvinism, increased public intolerance toward the ill. Victims of syphilis were condemned as targets of God's wrath and were ignored by many medical and charitable institutions. Religious artists often depicted Jesus striking down the unjust, raining murderous arrows down from heaven upon syphilis and those who suffered from bubonic plague. It was during this early modern period that the notion of "guilty" and "innocent" victims of disease arose.

The fear of HIV infection and ignorance about its causes have also created bizarre behavior and at times barbaric practices, strange rituals, and attempts to isolate those afflicted.

Deaths from Disease

Measles, Plague, Smallpox, Polio, Influenza, Tuberculosis, Hepatitis B, and Malaria—These diseases are responsible for many hundreds of millions of human deaths throughout history. The population of the indigenous peoples of Central and South America dropped from an estimated 13 million to about 1.6 million, more as a result of measles and plague than from war. In the nineteenth century measles has been held responsible for the total annihilation of the indigenous population of the island of Tierra del Fuego. Plague swept through Europe in the fourteenth century destroying a quarter of Europe's population. Between 1600 and 1650 the population of Italy actually fell from 13.1 million to 11.4 million. In Venice, an average of 600 bodies were collected daily on barges. More than 50,000 Venetians died in the plague of 1630–1631, leaving a population smaller than at any time during the fifteenth century. Recent outbreaks of bubonic plague occurred in India in August 1994, in Mozambique in June 1997, and in Uganda in November 1998.

The smallpox and polio viruses may well go back to the early beginnings of humans but would have mostly been concealed by the susceptibility of those infected to the more common (or more recognizable) infections.

Worldwide, smallpox killed over 100 million people by 1725. Smallpox is the *only* viral disease ever eradicated. The smallpox death of Egyptian

BOX I.1

OTHER LETHAL VIRUSES IN OUR TIME

Richard Preston, in his 1994 book *The Hot Zone* describes the progression of illness as the Ebola virus kills humans:

> Ebola . . . triggers a creeping, spotty necrosis that spreads through all the internal organs. The liver bulges up and turns yellow, begins to liquefy, and then it cracks apart. . . . The kidneys become jammed with blood clots and dead cells, and cease functioning. As the kidneys fail, the blood becomes toxic with urine. The spleen turns into a single huge, hard blood clot the size of a baseball. The intestines may fill completely with blood. The lining of the gut dies and sloughs off into the bowels and is defecated along with large amounts of blood.

The Ebola virus resists all drug therapy to date and kills up to 80% or more of those infected within one to three weeks after infection. (HIV kills over 90%—it is slower but more lethal.)

HISTORY

There have been at least 12 confirmed outbreaks of Ebola. The first was in 1976 when Belgian doctors discovered the virus at a hospital in northern Zaire; it was named Ebola Zaire after the Zairian river flowing through the region. In December 2001, it struck in Gabon and in the Republic of Congo. In January, October, and December of 2003 it struck again in the Republic of Congo. Each time the virus quickly ran its course of human destruction and disappeared. Beginning in 2005, there were 1650 confirmed cases of Ebola infection, causing about 1000 deaths. In 1989, a strain of Ebola virus appeared in Reston, Va., just a short distance from Washington, D.C. One hundred monkeys had arrived at the Reston Primate Quarantine Unit from the coastal forests of the Philippines on October 4. Two monkeys were dead on arrival and 10 more died during the next few days. Their tests proved positive for Ebola Zaire. The investigation quickly became a potentially volatile political crisis. The quarantine unit was a hot zone containing an organism classified as "biosafety level four": one for which no cure or vaccine existed.

In what has to be considered as an *extraordinary* piece of luck, this strain of Ebola virus, even though it appeared identical to the lethal form of Ebola Zaire, was harmless to humans. Because the viruses appeared to be spread among the monkeys via the air, if that virus was the lethal form of Ebola, the consequences for this heavily populated area of the Eastern Coast would have been unimaginably horrific.

DISCUSSION QUESTION: In what ways would the entire United States and other countries have been affected?

SOME OTHER LETHAL VIRUSES THAT INFECT HUMANS

Ebola is just one of a number of viruses to have emerged from the jungle in the past few decades; others include **Lassa** and **Marburg** viruses in Africa; **Sabía, Junin,** and **Machupo** viruses in South America. In 1995 a virus, yet to be named was found in Brazil and causes lung destruction and death. HIV probably originated in Africa, but unlike Ebola, it is ideally suited to spread around the globe. The HIV-infected can remain symptom-free for years, which provides the opportunity for them to infect others.

In early 2004, a number of African ape hunters were found infected with another retrovirus. The hunters are infected with a simian (monkey) foamy virus (SFV). To date, this virus does not appear to cause a human disease. This is yet another example of cross-species transmission of a retrovirus, animals to humans. Given that the SFV is a retrovirus, as is HIV, it has raised an alarm of possible future epidemics of new diseases.

West Nile Virus

A virus never seen before in America, the **West Nile-like encephalitis (brain inflamination) virus,** was identified after infecting 61 humans and causing the deaths of 7 people and about two-thirds of the crows in and around New York City during the summer of 1999. This virus has now been found in over 40 mosquito species and in some 120 species of birds ranging from robins to broad-winged hawks that migrate long distances. An outbreak of this virus has occurred simultaneously in Russia. Reasons for the spread of this virus out of Africa are unknown. From 1999 through 2002, the virus spread from Canada down the East Coast to the Florida Keys, from the Atlantic Coast going West passing the Rocky Mountains. Through 2002, the virus had spread

BOX I.1 *(continued)*

to 44 states and Washington, D.C. For about 30% of the infected, the virus offers a life-threatening experience. This virus crossed the United States in 2004.

Other New Viruses?

Are there other viruses as dangerous as HIV—or even more dangerous—lurking on the edge of civilization? That question haunts public health officials. Scientists have identified over 200 viruses capable of causing human disease at some level, and they believe there may be 5 to 10 times that many still to be identified!

[HORTON, RICHARD. (1995). *Infection: The Global Threat.* New York: *Reviews,* April 6, 24–28.]

DISCUSSION QUESTION: Should the public be informed about the dangers of research on viruses lethal to humans before it is allowed to begin? To continue?

pharaoh Ramses V in 1157 B.C. is the first known. Invading armies then spread smallpox through Africa and Europe. The Spanish brought it to Mexico. Edward Jenner used a cowpox pustular material from a cow named **Blossom** to create a vaccine against smallpox in the 1870s. Only humans got smallpox. It disappeared after a global vaccination campaign in the 1960s. The last known natural case was in Somalia in 1977. We now live with the fear of smallpox being spread by bioterrorists.

In early 2004, the World Health Organization predicted that polio will be eradicated sometime in 2005. This will be only the second virus to be eradicated from planet earth by human medical intervention.

Influenza—Each year as winter approaches, about 70 million shots of influenza (flu) vaccine are given, mainly to people who are especially susceptible to the virus (the very young, the elderly, asthmatics, and diabetics). Each year in the United States, from 20 million to 50 million people contract the flu, and an average 36,000 die from it! In recent history many thousands have died from influenza infection; in the 1957–1958 epidemic, for instance, 1 in 300 over age 65 died. And according to the CDC, during the year 1998–1999, 64,684 Americans died from influenza. Influenza was first described by Hippocrates in 412 B.C. The first well-described flu pandemic occurred in 1580. Thirty-one flu pandemics have occured since then. In 1890, in China, the duck-borne flu virus crossed over into swine, then into humans, causing thousands of deaths. This epidemic was followed by **severe flu pandemics** in 1900, 1918 (the Spanish flu), 1957 (the Asian flu), 1968 (the Hong Kong flu), and in 1977 (the Russian flu). The flu pandemic of 1918 has recently been estimated

SIDEBAR I.1

AMERICA ATTACKED BY TERRORISTS: BIOTERRORISM

Bioterrorism is the intentional use of microorganisms, viruses, or toxins derived from living organisms to cause death or disease in humans, animals, or plants on which we depend. Perhaps the first case of bioterrorism in America occurred in 1763. To quell Chief Pontiac's rebellion in the Great Lakes and Ohio River Valley, two blankets from Fort Pitt's smallpox ward were purposely given to the Delaware Indians while they were in negotiations to surrender. A short time after the blanket exchange, a large and lethal outbreak of smallpox occurred. The anthrax incident discussed below is believed to be the second act of bioterrorism in America.

Shortly after foreign terrorists destroyed the World Trade Center towers in New York City on September 11, 2001, the United States Postal Service, a newspaper office in Florida, and several government buildings in Washington, D.C., received Bacillus anthracis-(which causes anthrax) laden mail from bioterrorists. Twenty-one infected people were placed on antibiotics, but five died of anthrax, and several buildings were quarantined and had to be disinfected. The loss of life this time was limited to five people. Vaccine production against Bacillus anthracis received a high priority. What infectious agent may be used next time? The overwhelming choice, according to medical bioterrorism experts, is the dissemination of the smallpox virus. Sufficient smallpox virus vaccine is being produced to inoculate every person in the United States should it become necessary. Problem: Telling bioterrorists we now have sufficient vaccine against smallpox only means they will select another infectious agent. And as discussed in this introduction, there are many to choose from.

to have caused about 100 million deaths worldwide. (See *Flu: The Story of the Great Influenza Pandemic of 1918*, Gina Kolata, 1999.) This virus was antigenically similar to the swine virus and it appears to have disappeared from the human population following the 1918–1919 pandemic. This virus is, however, still carried in the swine population with an occasional crossover into humans without person-to-person transmission. The current theory from the Armed Forces Institute of Pathology on where the 1918 flu virus originated is: The virus most likely traveled back and forth between pigs and people until there was a single mutation or a sufficient number of lesser mutations that allowed the virus to cause a serious disease in humans and pigs!

In 1918, America was caught up in the last year of World War I. Deadly influenza, the so-called "Spanish flu," was sweeping the country, spreading terror everywhere. The first documented deaths were in Boston. Explanations were offered, but in fact no one had an answer. Viruses were still largely unknown. That particular strain of flu virus is still one of the mysteries of the 1918 pandemic. It had a mortality rate of about 3%. Once started, the disease moved west in lethal waves that appeared to follow railroad lines. The speed with which it killed was appalling, the loss of life unimaginable. The pandemic took 675,000 lives in just a few months. It would be as if today, with our present population, more than 1,400,000 people were to die in a sudden outbreak for which there was no explanation and no known cure. It was said every family lost someone. Time wise, this is the worst epidemic/pandemic America has ever known. It killed more Americans than all the wars of the 20th century combined. There were so many people dying from the flu that communities nationwide were running out of caskets.

The Beginning of the 1918 Flu Pandemic

Some scientists believe that the 1918 pandemic began in the spring of 1918, when soldiers at Fort Riley, Kan., burned tons of manure. A gale kicked up. A choking dust storm swept out over the land—a stinging, stinking yellow haze. The sun went black in Kansas. Two days later—on March 11, 1918—an Army private reported to the camp hospital before breakfast. He had a fever, sore throat, headache—nothing serious. A minute later, another soldier showed up. By noon, the hospital had over 100 cases; in a week, 500. That spring, 48 soldiers died at Fort Riley. The cause of death was listed as pneumonia. The sickness then seemed to disappear—leaving as quickly as it had come. That same summer and fall, over 1.5 million Americans crossed the Atlantic for war. Some of those soldiers came from Kansas, and it is believed that they carried the virus with them. Almost immediately, the Kansas sickness resurfaced in Europe. American, English, French, and German soldiers became ill. As the sickness spread, it appeared to get worse. By the time the disease came back to America, it was a deadly killer.

On a rainy day in September, Dr. Victor Vaughan, acting surgeon general of the Army, received urgent orders: Proceed to a base near Boston called Camp Devens. On the day that Vaughan arrived, 63 men died at Camp Devens. An autopsy revealed lungs that were swollen, filled with fluid, and strangely blue. Doctors were stunned. When the strange new disease was finally identified, it turned out to be a very old and familiar one: influenza, the flu. But, it was unlike any flu that anyone had ever seen.

During the epidemic of 1918, all San Francisco citizens were required to wear masks. One form of therapy was "cabbage baths." People did not jump into a tub with some cooked cabbage; they ate the cabbage and urinated into the tub. The flu victims then got into the tub. There may have been a positive side to the bath because once you got out, a distance was created between you and other people. In this way, the baths may have helped stop person-to-person transmission.

Tuberculosis, Hepatitis B, and Malaria—In 1937, 112,000 Americans contracted tuberculosis and 70,000 died. About *1 billion* people worldwide have died from Mycobacterium tuberculosis (a bacterium 1/25,000th of an inch long) since 1770. **Over 120 million people will have died of TB from 1990 through the year 2005!** One in three people worldwide carry this organism. Currently, about 10 million people in the United States are infected with TB, but only 5% develop the disease. Each year over 2 million adults worldwide die of hepatitis B. About 8000 people die every day from malaria. Despite over 50 years of global experience in malaria control, more people are dying of malaria now than when malaria control campaigns began. Malaria, like all of the major infectious diseases, is a disease of poverty. Parents and children quietly cope; the ill were and are served, not shunned—that is, until the HIV/AIDS epidemic.

AIDS: ITS PLACE IN HISTORY

As the statistics on AIDS cases mounted, its identity as an inescapable plague seemed confirmed. It appeared to mimic the frightening epidemics of the past: cholera, yellow fever, leprosy, syphilis, and the plague or Black Death. The history of AIDS—the history that seemed relevant to understanding the new pandemic—would be the history of the epidemics of the past. Medical history suddenly gained new social relevance; policy analysts, lawyers, and journalists all wanted to know whether past epidemics could provide some clues to the current crisis. How had societies attempted to deal with epidemics in the past? The contemporary meaning of past plagues is read in the face of AIDS.

The AIDS pandemic is certainly one of the defining events of our time. There are stories to be told from it, stories of the people infected and affected by it—the well, the ill, the dying, and the survivors. There are the stories of scientific discovery, of HIV and viral mechanisms, and of genetic mysteries being understood. Then there are stories of scientific politics, claims and counterclaims, and the manipulating that goes on in the stratosphere of high-level science.

AIDS is a "Weapon of Mass Destruction!"

America's secretary of state, Colin Powell, said in April 2004 that "HIV/AIDS is the greatest threat of mankind today, the greatest weapon of mass destruction on the earth."

The AIDS Pandemic Consists of Two Major Parts

The HIV/AIDS pandemic consists of two parts: one medical, the other social. HIV/AIDS infection has provoked a reassessment of society's approaches to public health strategy, health care resource allocation, medical research, and sexual behavior. Fear and discrimination have affected virtually every aspect of our culture. Both the medical challenge and, in particular, the social challenge will continue in the foreseeable future. **Arthur Ashe,** a world-class tennis player, so feared discrimination against himself and family that he lived with AIDS for 3-1/2 years before he was forced to reveal he had the disease (Figure I-1).

OVERVIEW ON HIV/AIDS

AIDS: A Big Disease With a Little Name, An Earthquake In Slow Motion

The global numbers of people who are HIV-infected along with those who have died from AIDS is almost incomprehensible and very difficult to express. (For the current number of AIDS cases and deaths, see Chapter 10). The numbers are too large to grasp and the implications too terrifying to contemplate. In addition, at this time the future appears grim: First, there is no near-term prospect of a vaccine. Second, existing drug therapies, though relatively effective, are too expensive for the vast majority of those infected, and third, those preventive interventions that can be effective in reducing the numbers of people becoming infected annually are not in place in countries most in need, because the majority of the infected are undereducated and their countries still lack the infrastructure to deal with many infectious diseases including AIDS.

What Is AIDS?

AIDS is defined primarily by severe immune deficiency, and is distinguished from virtually every other disease in history by the fact that it has no constant, specific symptoms. Once the immune system has begun to malfunction, a broad spectrum of health complications can set in. AIDS is an umbrella term for any or all of some 26 known diseases and symptoms. When a person has any of these diseases or has a CD4 or T4 lymphocyte count of less than 200/μL (microliter) of blood and also tests positive for antibodies to HIV, an AIDS diagnosis is given. (See Table 7-2 for a list of diseases and the symptoms of AIDS.)

AIDS: A Human Affair

The history of AIDS is a human affair and is part of a cultural process of attempting to come to terms

FIGURE I-1 Arthur Ashe—Winner of Two U.S. Tennis Championships. On April 8, 1992, Ashe announced that he had AIDS. He died on February 6, 1993, at age 49. He became infected with HIV from a blood transfusion during heart bypass surgery in 1983. In 1988, his right hand suddenly became numb. Brain surgery revealed a brain abscess caused by toxoplasmosis. He asked two questions, "Why do bad things happen to good people?" and "Is the world a friendly place?" In 1975 he became the first black to win a Wimbledon men's title. After learning that he had AIDS, Ashe completed his third book, *A Road to Glory*, a three-volume history of black athletes in the United States. *(Photograph courtesy of AP/Wide World Photos)*

with a new and often terrifying series of events—of young people dying before their time, of the intermingling of sex and death—in a period in which the world itself is changing before our eyes.

The social meaning of the history of AIDS intimately touches upon ideas about sexuality, social responsibility, individual privacy, health, and the prospect of living a normal life span. Understanding how to respond to AIDS and how to think about this pandemic is important not only for what it reveals about the ways in which health policy is created in the United States and elsewhere, but also for what it implies about the human ability to meet the challenge of future emerging diseases and long-standing public health problems. The HIV/AIDS pandemic is a current and long-term public health problem worldwide.

AIDS is reversing decades of public health progress, lowering life expectancy, and significantly affecting international businesses. Lost productivity and profitability, the cost of sickness and death benefits, and the decline in a skilled workforce in the developing world will have economic effects worldwide. AIDS is affecting the military capabilities of some countries as well as international peacekeeping forces.

Lessons Learned

From the lessons of history it is difficult to conceptualize how the AIDS epidemic will be halted, let alone reversed, in the absence of a cheap curative drug or a cheap and effective preventative vaccine. The syphilis epidemic at the early part of the century displayed a similar kind of epidemiology to the present-day AIDS epidemic. The campaigns that were initiated then closely paralleled those in place at present for AIDS. There were vigorous educational programs to reduce sexual high-risk behavior, which were targeted at brothels and prostitutes as well as at military recruits to the U.S. Army. **Scare tactics** were spread through the use of posters, pamphlets, and radio—today it is television. Serological testing became mandatory before marriages could be licensed in certain states. However, these measures had little appreciable effect on the expansion of the syphilis epidemic. It was only the advent of a cheap, safe, and effective drug, penicillin, that eventually brought the epidemic under control.

One lesson learned from HIV/AIDS is that any disease that is occurring in a distant part of the globe may be in the United States, in your state, or in your town, tomorrow. Twenty or even 15 years ago one would not have expected to read that statement.

The advent of miracle drugs and vaccines that conquered the plagues of polio, smallpox, and measles led many people—including scientists—to believe that the age of killer diseases was coming to an end.

AIDS: A UNIQUE DISEASE

The impact of this pandemic is unique. Unlike malaria or polio, previous modern pandemics, it mostly affects young and middle-aged adults. This is not only the sexually most active time for individuals, but also their prime productive and reproductive years. Thus the impact of HIV/AIDS is demographic, economic, political, and social. HIV/AIDS is a disease of human groups and its demographic and social impacts multiply from the infected individual to the group. In the most affected areas, infant, child, and adult mortality is rising and life expectancy at birth is declining rapidly. The cost of medical care for each infected person overwhelms individuals and households.

AIDS: A CAUSE OF DEATH

Peter Piot, (Figure I-2) executive director of the United Nations Program on HIV/AIDS, speaking at the Thirty-seventh Interscience Conference on Antimicrobial Agents and Chemotherapy, said "HIV has transformed the world, joining tuberculosis and malaria as a major cause of death worldwide. This epidemic won't be under control in any country until it is brought under control everywhere."

AIDS is now the seventh leading cause of death among 1- to 4-year olds, sixth among 15- to 24-year olds, and second among 25- to 44-year olds in the United States.

There is the expectation that parents will die before their children. Because of the HIV/AIDS epidemic, it is not working out that way for many thousands of parents. They are watching their children die in the prime of life.

The facts on HIV infection, disease, and AIDS that are presented in the following chapters, when understood, clearly place the responsibility for avoiding HIV infection on *you*. You must assess your lifestyle; if you choose not to be abstinent, you must know about your sexual partner and you must practice safer sex. **Never think that you are immune to HIV infection.**

ANOTHER ANNIVERSARY

June 5, 2005, will mark the beginning of the 24th year of the AIDS pandemic. No cure has been found and, although AIDS is now called a "manageable illness," people who are sick must make

FIGURE I-2 Peter Piot, Executive Director of the Joint United Nations Program on HIV/AIDS (UNAIDS). He is a Belgian physician, microbiologist, and codiscoverer of the Ebola virus in 1976. He is the world's leading advocate for HIV/AIDS control and prevention. *(Photograph courtesy of UNAIDS/Yoshi Shimizu)*

endless compromises to this disease. The AIDS pandemic forces people to face their mortality daily, for months and years.

WORLD INTERNATIONAL AIDS CONFERENCES

The World International AIDS Conference is believed to be the single most important meeting on AIDS.

The International AIDS Conference has been the timepiece of the pandemic. The venue, content, style, and mood of each meeting temporarily freezes in time the state of the worldwide HIV/AIDS pandemic.

In 1983, there were relatively so few groups involved with HIV/AIDS research that they could stay in contact by telephone. As the virus spread, many countries became involved in HIV/AIDS research and clinical care. International, national, and state meetings were formed as a way to exchange new information. The international meetings continue to receive the most press coverage. Since the first international AIDS conference in 1985, the conferences have grown in size to the point that scientists questioned their usefulness.

> **LOOKING BACK I.2**
>
> **TWO INTERESTING DETAILS FROM THE FIRST INTERNATIONAL AIDS CONFERENCE**
>
> First, in the exhibit area there was a huge map of the United States that said on the top: "The Problem: AIDS. The Solution: Banish All Homosexuals." It had little pushpins on all the major cities of the United States indicating how many homosexuals lived in each. Second, then-Vice President George Bush addressed the assembly. As soon as he started to talk, he was shouted down. (Remember, this was seven years into the Reagan administration and it would be another year before Reagan uttered the word AIDS.) When people started to boo and shout, Bush stepped back from the podium a bit, turned around and said to the Secret Service agent standing behind him (thinking he was turned away from the microphone), "It must be a gay group." The words reverberated throughout the auditorium of the Washington Hilton.

As of 1994, the International Conference on AIDS is held every two years (Table I-2). The year 2000 Thirteenth International AIDS Conference was held in an underdeveloped nation for the first time. It was held in Durban, South Africa, a country where a significant proportion of the population has HIV/AIDS. Over 12,000 people attended this conference. The International AIDS Conferences are usually grim, but the Thirteenth International AIDS Conference was the grimmest yet, reporting death and infection rates that carry near-apocalyptic implications for the Third World, and particularly Africa.

The theme for this conference was "Breaking the Silence" but it is likely the conference will be remembered for the "Durban Declaration," a document signed by over 5,000 physicians and scientists testifying to their belief that HIV causes AIDS.

The theme for the Fifteenth International AIDS Conference held in Bangkok, Thailand, was, "Access For All." This is the first conference of its kind to be held in Southeast Asia. About 20,000 delegates attended this conference (read Box 10.1, page 292, for important highlights from this conference).

THE AIDS MEMORIAL QUILT

Conception of the Quilt

The quilt was the idea of Cleve Jones, of San Francisco, who in 1985 feared AIDS would become known for the number of people it killed. He wanted a way of remembering the people, who were, in many cases, his friends.

During the eighth candlelight march in San Francisco, Jones asked fellow marchers to write on placards the names of friends and loved ones who died of AIDS. At the end of the march, Jones and others stood on ladders, above the sea of candlelight, taping these placards to the walls of the San Francisco Federal Building. The walls of names looked to Jones like a patchwork quilt.

Purpose and Dimensions of the Quilt

The purpose of the quilt is to educate. The AIDS Quilt is made up of individual fabric panels, each the size of a grave, measuring three feet by six feet,

Table I-2 International AIDS Conferences

Presented is a list of past International AIDS Conferences by month, year, and location. The number of *reported* AIDS cases and deaths are accumulative by year in the United States.

Number	Month	Year	Location	Deaths
1st	June	1985	Atlanta, Georgia	12,576
2nd	June	1986	Paris, France	24,686
3rd	June	1987	Washington, D.C.	41,088
4th	June	1988	Stockholm, Sweden	62,207
5th	June	1989	Montreal, Canada	90,000
6th	June	1990	San Francisco, California	121,536
7th	June	1991	Florence, Italy	157,000
8th	July	1992	Amsterdam, Netherlands	198,246
9th	June	1993	Berlin, Germany	243,627
10th	August	1994	Yokohama, Japan	293,496
11th	July	1996	Vancouver, Canada	381,896
12th	June	1998	Geneva, Switzerland	421,768
13th	July	2000	Durban, South Africa	454,185
14th	July	2002	Barcelona, Spain	486,185
15th	July	2004	Bangkok, Thailand	530,481
16th	August	2006	Toronto, Canada	550,185

Based on UNAIDS data, during the six days of the Fourteenth conference, about 85,000 new HIV infections occurred and 43,000 people died of AIDS worldwide.

AIDS deaths are minimum estimates for the United States, based on data reported to the CDC and updated data accumulative for 2006 are estimated. See Figure 10-6 for numbers of AIDS cases in each year.

UNAIDS Executive Director Peter Piot reported that between the Eleventh 1996 AIDS Conference and the Twelfth in 1998, 10 million people were infected with HIV! This, he said "represents a collective failure of the world." The rate of over 5 million new infections per year continues.

stitched together into 12 foot × 12 foot sections. In October 1987, the AIDS Quilt was first put on display on the mall in Washington, D.C. At that time it contained 1920 panels and covered an area larger than two football fields. It took less than two hours to read all the names. In 1992 it took 60 hours. As of mid-2005 the quilt weighed about 60 tons with about 60,000 panels containing about 90,000 names representing about 16% of those who have died of AIDS in the United States. Displayed in its entirety, it would cover over 45 football fields. It is the largest piece of folk art in the world and continues to increase in size daily. If the 3-by-6 foot panels were laid end to end, the quilt would stretch over 50 miles. There are about 50 miles of seams and 26 miles of canvas edging. There are panels from each of the 50 states and Washington, D.C. Each day new panels arrive from across the United States and 37 foreign countries to be added to the quilt (Figure I-3). For those left behind, the panels represent an expression of love and a sign of grief—a part of the healing process. A few of the well-known people who have died of AIDS are seen in Figures I-4 and I-5.

Portions of the quilt tour in major cities. About 16 million people have visited the quilt at thousands of displays around the world. Donations made for viewing the quilt are being used to support local Names Project chapters and their staffs.

Each panel has its own story. The stories are told by those who make the panels for their lost friends, lovers, parents, and children. The complete quilt was displayed in Washington, D.C., October 11–13, 1996, for the last time—it is too large to view again as a whole. It took 10 boxcars and a freight train to transport this work of art to the nation's capital.

Ending year 2005, over 540,000 Americans have died from AIDS. If each of their names made up a separate panel, imagine the size of the quilt. If a name is read every 10 seconds, it would take over 62 days of calling names 24 hours a day!

FIGURE I-3 World AIDS Day, December 1, 2000. A member of the New Hope Church AIDS Quilt Committee helps spread out AIDS quilt panels at their church in Delray, Fla. Their program included songs, prayer, and candle lighting. *(Photo by Scott Fisher/* South Florida Sun-Sentinel)

UPDATE 2004 In June 2004, 8000 panels of the quilt were put on display on the Ellipse adjacent to the National Mall in Washington, D.C. The entire quilt will be put on display in D.C. in June 2006. The purpose of bringing the quilt back to Washington is to alert the president and Congress to the continuing needs brought about by the pandemic.

Remembrance

People nationwide celebrated the seventeenth anniversary of the AIDS Quilt on July 21, 2004. In mid-2001, the AIDS Quilt was moved from San Francisco to Atlanta, where it is housed in a climate-controlled warehouse.

For more information about the Names Project's AIDS Memorial Quilt, call (404) 688-5500, ext. 223; www.aidsquilt.org.

CANDLELIGHT MEMORIAL

The first candlelight march occurred in San Francisco in November 1978 as a response to the killing of Harvey Milk, a city superintendent. The fact that he was gay was reported to be a primary motive for his death. The first candlelight march for AIDS occurred in 1981. The memorial march is now international. The International AIDS Candlelight Memorial honors the memory of those lost to AIDS, shows support for those living with HIV and AIDS, raises community awareness of HIV/AIDS, and mobilizes community involvement in the fight against HIV/AIDS. In 2004, the Candlelight Memorial was observed in about 1500 locations in 85 nations on every continent but Antarctica. The Candlelight march takes place on the third Sunday of May. The theme for 2004 was "Turning Rememberence into Action." (For additional information contact Matthew Matassa, email: candlelight@globalhealth.org or visit the website: http://www.globalhealth.org.)

WORLD AIDS DAY

As in years past, rituals of candles and quilts coincide on World AIDS Day with the release of grave new statistics and heartrending personal stories—not only from distant countries but also from our own backyards. While humanitarian concerns demand that developed nations take action in the developing world, a growing apathy about the HIV pandemic in America helps perpetuate a pandemic that has already claimed over a half-million lives and counting.

December 1 is now traditionally called World AIDS Day, a time of remembrance for those individuals who have died from AIDS, a time to assess the current status of the epidemic, and a time to evaluate the effectiveness of our efforts to con-

FIGURE I-4 Rock Hudson, Movie and Television Star. A Hollywood legend and undisclosed homosexual, he was the first major public figure to reveal he had AIDS. The conventional wisdom, expressed by journalist Randy Shilts in 1987 (Figure I-5) is "that there were two clear phases to the disease in the United States: there was AIDS before Rock Hudson and AIDS after." This common observation has been confirmed by others who also note that Hudson's July 1985 disclosure that he was suffering from AIDS led to a permanent increase in media attention to the disease. Over 6000 Americans had already died of AIDS, but now with Hudson's famous face on the front page, the AIDS epidemic suddenly counted. The press developed an instant passion to cover AIDS. Hudson died in 1985 at age 59. *(Photograph courtesy of AP/Wide World Photos)*

FIGURE I-5 Randy Shilts, Author of "And The Band Played On" (Viking Penguin Press, 1987), Died of AIDS February 17, 1994, at the Age of 42. *(Photograph courtesy of AP/Wide World Photos)*

trol it. Now well into the third decade since the recognition of AIDS, it is estimated that about 75 million people (by the end of 2005) have been infected with HIV, clearly making this one of the worst epidemics of our time. Projections for the next 10 years (2005–2015) suggest that the situation will become even more serious as we reach over 125 million infected individuals.

December 1, 1988, was the first acclaimed *World AIDS Day* (WAD). The theme was "Join the Worldwide Effort." World AIDS Day is a day set aside to pay tribute to those who have AIDS and to those who have died of AIDS. Its purpose is to increase our awareness of AIDS. The first WAD did not attract much attention outside the gay community. For 2004 the theme is "Women and HIV and AIDS." For information on World AIDS Day 2005, write 108 Leonard Street, 13th Floor, New York, NY 10013 or call (212) 513-0303, or World AIDS Day Public Information Office, WHO-GPA, 1211 Geneva 27, Switzerland.

POINT OF VIEW I.1

WHY ALL THE FUSS ABOUT HIV/AIDS?

Why has the U.S. federal government spent over $185 billion over the past 17 years (1989–2005) on just this one disease? More money has been spent on HIV/AIDS in its relatively short history than on any other human disease in our history. In addition, the public/private sector has spent a sum equal to or greater than the government over the same time period. (See Chapter 13 for a breakdown of the billions spent on this one disease.)

First, over the past years, AIDS has been killing over 2.5 million people each year worldwide. Through 2005, an estimated 24 million people will have died of AIDS worldwide. Malaria, TB, heart disease, cancer: No other disease is spreading or growing at this rate. In spite of HIV/AIDS campaigns in most countries, there is little evidence so far of any slowdown in the spread of the epidemic. By the end of 2005, about 65 million people are expected to have been infected by HIV.

Second, this virus is mainly transmitted during sexual intercourse. Since few human societies talk openly and honestly about sex, this makes this disease difficult to discuss; and since sex is a very private activity, it makes the transmission of HIV very difficult to control.

Third, it has an extraordinary capacity for change and rapid global spread.

Fourth, there is a long asymptomatic period between infection and illness. On average, it takes about 10 to 12 years for someone infected with HIV to develop AIDS. During this time, the HIV-infected will show few if any recognizable symptoms, but they are able to infect other people.

This long asymptomatic period is *rare* in human infectious diseases. People dying today represent those infected 10 to 20 years ago; the results of anything we do now to reduce transmission will not be apparent for years.

Fifth, HIV/AIDS is more serious than many more common diseases because of the age groups it attacks. Some diseases—measles and diarrhea, for example—affect mainly infants and children; others, such as heart disease and cancer, affect mainly the old. But because HIV is predominantly transmitted sexually, AIDS mainly kills people in their 20s through their 40s. A major increase in deaths among these age groups, society's most productive group, presents a much greater impact socially and economically than deaths that occur among children or old people would have.

Sixth, in the past, plagues were often marked by their lack of discrimination, by the way in which they killed large numbers of people with little regard for race, wealth, sex, or religion. But AIDS was different from the beginning. It immediately presented a political as much as a public health problem. Homosexuals, who until this pandemic had been mostly closeted in the United States, were suddenly at the heart of a health crisis as profound as any in modern American history.

Seventh, with respect to therapy, HIV disease/AIDS requires the use of some of the most expensive and toxic drugs in medical history.

Eighth, it is a disease that has severely stigmatized those who have it.

Ninth, it is a disease that has parents burying their children.

Tenth, and lastly, this virus means there are other viruses in waiting—this has awakened our most primal fear: dying a horrible death by an unknown agent, for example, the **West Nile encephalitis virus** that struck the New York City metropolitan area in the summer and fall of 1999 or the SARS virus that struck in 2003.

So why all the fuss about HIV/AIDS? It is caused by a unique virus that changes itself faster than a rumor making the gossip column of a tabloid. It is lethal, it is transmitted, most often sexually, it affects people in their reproductive and most productive years, it is exceptionally expensive to treat, it defies our best scientists who are working to create a vaccine for preventing infection and transmission, and it is a disease that has severely stigmatized those who have it.

HIV has one requirement to continue its presence on earth: a human host!

THE WORLD HEALTH AND UNAIDS ORGANIZATIONS

In January 2003 Jong Wook Lee, a tuberculosis expert from South Korea, was elected to a five-year term as director-general of the World Health Organization (WHO).

The WHO has established a global program on HIV infection and AIDS. The program has three objectives: to prevent new HIV infections, to provide support and care to those already infected, and to link international efforts in the fight against HIV infection and AIDS.

In 1995, the United Nations secretary-general chose Peter Piot to begin a Joint United Nations Program on HIV/AIDS (UNAIDS). Piot's job is to coordinate actions and reduce duplication among the eight cosponsors of UNAIDS: United Nations Children's Fund (UNICEF), the United Nations Development Programme (UNDP), the United Nations Educational, Scientific and Cultural Organization (UNESCO), the United Nations Population Fund (UNFPA), the United Nations Drug Control Program (UNDCP), the World Health Organization (WHO), the World Bank, and the International Labor Organization.

THE FUTURE

The history of HIV/AIDS is one of remarkable scientific achievement. Never in history has so much been learned about so complex an illness in so short a time. We moved into the new millennium, the third decade of AIDS, with hope and the determination to find better therapies and a vaccine. The task is formidable but it has to be done and it will be accomplished. The evolving story of the HIV/AIDS epidemic has been one of the major medical news events of the past 24 years. It is getting hard to imagine medicine or society without HIV/AIDS.

HIV/AIDS is a truly persistent global pandemic and will require a proportionate response to bring it under control. It is the plague of our lifetimes—and probably that of our children's lives as well. We already have people age 23 and younger who don't know what an HIV-free world is. They were born into this recognized pandemic (1981). To survive this pandemic, society must prevent the face of AIDS from becoming faceless. The chapters on HIV infection, HIV disease, and AIDS in this book will help bring widespread information on the virus into focus. The information within these chapters should also help eliminate many of the myths and irrational fears, or FRAID (**fear of AIDS**), generated by this disease. There is much work to be done by both scientists and society.

Perhaps a borrowed anecdote says it best:

> *As the old man walked the beach at dawn, he noticed a young woman ahead of him picking up starfish and flinging them into the sea. Finally catching up with the youth, he asked her why she was doing this. The answer was that the stranded starfish would die if left to the morning sun.*
>
> *"But the beach goes on for miles and there are millions of starfish," countered the old man. "How can your effort make any difference?"*
>
> *The young woman looked at the starfish in her hand and threw it to safety in the waves. "It makes a difference to this one," she said.*
>
> (Adapted from *The Unexpected Universe* by Loren Eiseley. Copyright 1969, Harcourt Brace, New York)

Too easily we can become overwhelmed by the enormity of the AIDS pandemic. The numbers of patients and their constant needs have caused many to become paralyzed into inactivity and lulled into indifference. Like the old man, many ask "Why bother?"

For the sake of every individual living with HIV, we must focus on what each one of us can do. Each person can make a difference. Believing this, we are empowered to cope with the larger whole.

TOLL-FREE NATIONAL AIDS HOTLINES

- Centers for Disease Control and Prevention/ National AIDS Clearing House, 1-800-458-5231.
- National AIDS, for the English-language service (open 24 hours a day, 7 days a week), call 1-800-342-AIDS (2437).
- The Spanish service (open from 8 A.M. to 2 A.M., 7 days a week) can be reached at 1-800-344-SIDA (7432).
- A TTY service for the hearing impaired is available from 10 A.M. to 10 P.M. Monday through Friday at 1-800-243-7889.
- National Prevention Information Network, 1-800-458-5231.
- National Herpes Hotline, 1-919-361-8488.
- National Native American AIDS Prevention, 1-510-444-2051.
- National Association of People with AIDS, 1-202-898-0414.
- National HIV Telephone Consultation Service for Health Care Professionals, 1-800-933-3413, San

Francisco General Hospital, Bldg. 80, Ward 83, Room 314, San Francisco, CA 94110.

- HIV/AIDS Treatment Information Service, 1-800-HIV-0440 (448-0440), 9 A.M. to 7 P.M. Eastern time, Monday-Friday. 1-800-243-7012 Teletype number for the hearing-impaired, 9 A.M. to 5 P.M. Eastern time, Monday-Friday, Box 6303, Rockville, MD 20849-6303.
- AIDS Clinical Trials Information Service, 1-800-TRIALS-A 874-2572, 9 A.M. to 7 P.M. Eastern time, Monday-Friday. Information on clinical trials of AIDS therapies.
- National Gay and Lesbian Task Force AIDS Information Hotline, 1-888-843-4564.
- Gay Men's Health Crisis AIDS Hot Line, 1-212-807-6655.
- National HIV/AIDS Education & Training Centers Program, 1-301-443-6364, Fax: 1-301-443-9887.
- AIDS National Interfaith Network, 1-202-546-0807, Fax: 1-202-546-5103, 110 Maryland Ave., NE, Room 504, Washington, D.C. 20002.
- National Hemophilia Foundation, 1-800-424-2634.
- Pediatric AIDS Foundation, 1-310-395-9051.
- National Pediatric HIV Resource Center, 1-800-362-0071.
- AIDSLINE via the National Library of Medicine. Free access via Grateful Med, obtained from NLM at 1-888-346-3656.
- National Institute on Drug Abuse Hotline, 1-800-662-HELP 1-4357.
- National Sexually Transmitted Diseases Hotline 1-800-227-8922.
- American Civil Liberties Union Guide to local chapters, 1-202-544-1076.
- AIDS Policy and Law, 1-215-784-0860.
- National Conference of State Legislatures HIV, STD and Adolescent Health Project, 1-303-830-2200.
- United States Conference of Mayors, 1-202-293-2352.
- Centers for Disease Control and Prevention, Public Inquiry, 1-404-639-3534.
- Food and Drug Administration, Office of Public Affairs, 1-301-443-3285.
- American Red Cross, Office of HIV/AIDS Education, 1-800-375-2040.
- World Health Organization, 1-202-861-4354.

USEFUL INTERNET ADDRESSES

- AIDS Treatment Data Network, 1-800-734-7104, 611 Broadway, Suite 613, New York, NY 10012. http://health.nyam.org:8000/public_html/network/index:html, email: AIDS-TreatD@AOL.COM. A home page on the internet for people with AIDS and their caregivers, it provides information on approved and experimental treatments for AIDS-related conditions. It also publishes a quarterly directory of clinical trials on HIV and AIDS in English and Spanish.
- AMA HIV/AIDS Information Center website (http://www.ama-assn.org) offers clinical updates, news, and information on social and policy questions. Cosponsored by Glaxo Wellcome Inc.
- Gay Men's Health Crisis (GMHC), website (http://www.gmhc.org) provides online forums hosted by GMHC representatives.

For additional help you may wish to consult with your college or community library. They may have access to the following AIDS-related data bases:

- AEGIS (AIDS Education Global Information System): http://www.aegis.org
- HIV Info Web: http://www.infoweb.org
- Kaiser Daily Global HIV/AIDS information: www.kaisernetwork.org
- Southern Africa AIDS Information Dissemination Service: www.safaids.org.zw
- Immunet: http://www.immunet.org
- Project Inform: http://www.projinf.org
- The Body: http://www.thebody.com/cgi/treatans.html
- HIVInSite: http://www.hivinsite.ucsf.edu/medical/tx-guidelines
- Search AIDSLINE, MEDLINE: http://www.igm.nim.nih.gov
- Vaccines: http://www.avi.org
- Women, children, health-care workers, hemophiliacs, blind, deaf, and other affected groups: http://beaconclinic.org/website/groups
- AIDS/HIV statistics: http://www.avert.org/statindx.htm
- http://www.healthcg.com/hiv/links.html (provides linkage to nine major U.S. Guidelines for HIV Testing, OIs, Treatment, etc.)
- The Centers for Disease Control and Prevention's (CDC) National Prevention Information Network *(NPIN) Links*: http://www.cdcnac.org/hivlink. html and http://www.cdcnac.org/daynews.html
- National Institute of Allergy and Infectious Diseases (NIAID) online at: http://www.niaid.nih. gov.

- *Critical Path AIDS Project,* a Philadelphia organization for people with HIV disease, provides another online source for the latest news in HIV disease prevention, research, clinical trials, and treatments. The publication's hot link leads to a directory of AIDS-related publications: (http://www.critpath.org)
- UNAIDS Global HIV/AIDS information: www.unaids.org
- WHO HIV/STI Surveillance: http://www.who.int
- European Center for the Epidemiological Monitoring of AIDS: http://www.ceses.org
- AIDS MAP Global HIV/AIDS information: www.aidsmap.com
- *Managing Desire: HIV Prevention Counseling for the 21st Century* targets the HIV test counseling community as well as the general consumer. The site is produced by Nicholas Sheon, the prevention editor of the *HIV Inside* website of the UCSF Center for AIDS Prevention Studies (http://hivsinsite.ucsf.edu) and an HIV test counselor at the Berkeley Free Clinic: (http://www.managingdesire.org)
- *AIDS* offers abstracts from recent issues: http://www.aidsonline.com
- *AIDS Weekly Plus* contains more than 35,000 articles on health-related topics. Full access is available by subscription: http://www.newsfile.com
- *AIDS Treatment News* posts the contents of every issue since the publication began in 1986: http://www.immunet.org/immunet/atn.nsf/homepage
- The *Bulletin of Experimental Treatment on AIDS (BETA)* published by the San Francisco AIDS Foundation, is free online: http://www.sfaf.org/beta.html
- *Treatment Issues,* published by the Gay Men's Health Crisis, provides free access to issues dating back to 1995: http://www.gmhc.org/aidslib/ti/ti.html
- *Project Inform* email (INFO@projinf.org), website, established in 1985 as a national, nonprofit, community-based HIV/AIDS treatment information and advocacy organization, serves HIV-infected individuals, their caregivers, and their health-care and service providers through its national, toll-free treatment hotline: http://www.projinf.org
- The Synergy APDIME ToolKit is a user-oriented, electronic one-stop-shop of HIV/AIDS programming resources. Developed in collaboration with the University of Washington Center for Health Education and Research (CHER), the ToolKit contains five modules of the programming cycle covering Assessment, Planning, Design, Implementation Monitoring, and Evaluation: http://www.synergyaids.com/apdime/index.htm#

The following is a sampling of general internet resources for community research and HIV/AIDS information in Canada:

- Western Canada's largest AIDS group (in Vancouver, BC) has launched its redesigned website featuring online publications, a map of provincial resources, and links to over 100 AIDS websites. The website, published by the British Columbia Persons with AIDS Society, is one of the most popular sites in Canada and has operated for two years. The site carries information about:
  - *Treatments:* http://www.bcpwa.org/treat.htm;
  - *AIDS news:* http://www.bcpwa.org/news.htm;
  - *Organizational activities:* http://www.bcpwa.org/AboutBCPWA/board.htm; and
  - *Links:* http://www.bcpwa.org/Resources/links.ht.
  - For more information: Pierre Beaulne, Developer, Communications and Marketing, British Columbia Persons With AIDS Society, mail to: pierreb@parc.org.
- The Community-Based HIV/AIDS Research Program National Health Research and Development Program, Health Canada: http://www.hc-sc.gr.ca/hppb/nhrdp/cdr.htm
- The HIV/AIDS Aboriginal Research Program National Health Research and Development Program, Health Canada: http://www.hc-sc.gr.ca/hppb/nhrdp/abrfp.htm
- Community-University Research Alliances (CURAs), Social Sciences and Humanities Research Council of Canada: http://www.sshrc.ca/english/programinfo/grantsguide/cura.html
- Canadian Strategy on HIV/AIDS: http://www.hc-sc.gc.ca/hppb/hiv aids/
- Canadian HIV/AIDS Clearinghouse: http://www.cpha.ca/clearinghouse e.htm
- Canadian AIDS Society: http://www.cdnaids.ca/
- Canadian Aboriginal AIDS Network: http://www.caan.ca/
- Community AIDS Treatment Information Exchange: http://www.catie.ca/
- Canadian HIV/AIDS Legal Network: http://www.aidslaw.ca/
- Bureau of HIV/AIDS, STD and TB, Health Canada: http://www.hc-sc.gc.ca/hpb/lcdc/bah/epi/epie.html
- Global Network of People Living With HIV/AIDS (GNP+): email: gnp@gn.apc.org.

Acquired Immune Deficiency Syndrome
AIDS

What do we know about AIDS? The next 13 chapters will present the many faces of the AIDS pandemic in the United States and other countries. Unlike people, the AIDS virus (HIV) does not discriminate; and it appears that most humans are susceptible to HIV infection, its suppression of the human immune system, and the consequences that follow. The viral infection that leads to AIDS is the most lethal, the most feared, and the most socially isolating of all the sexually transmitted diseases. We must, as a people, fight against AIDS, not against each other (Figure I-6).

FIGURE I-6 The Loneliness of AIDS. "Skip" Bluette, diagnosed with AIDS July 1986, died July 1988. He suffered the indignity of having to lie to a dentist to get treatment. He suffered the ignorance of nurses afraid to touch him. He suffered the loss of his greatest pleasures—discos, gourmet meals, movies, and sex with men. Family was vital to Skip. So vital that on July 17, the day he died, their presence was his final wish. Skip Bluette wanted his story told. Photographer Mara Lavitt interviewed and photographed him during the last eight months of his life. Portions of Lavitt's article, which appeared in *The New Haven Register*, are presented in the following chapters. *(By permission of Mara Lavitt and* The New Haven Register*)*

1 Discovering AIDS, Naming the Disease

Chapter Concepts

- The letters A, I, D, S (AIDS) are an acronym for Acquired Immune Deficiency Syndrome.
- AIDS is a syndrome, not a single disease.
- The first cases of AIDS-related *Pneumocystis carinii* pneumonia (PCP) were reported by the Centers for Disease Control and Prevention (CDC) in June 1981; the first case of Kaposi's sarcoma in July 1981.
- Luc Montagnier discovered the AIDS virus in 1983.
- The first CDC definition of AIDS was presented in 1982 and expanded in 1983, 1985, and 1987, and again on January 1, 1993.

A = Acquired = a virus received from someone else
I = Immune = an individual's natural protection against disease-causing microorganisms
D = Deficiency = a deterioration of the immune system
S = Syndrome = a group of signs and symptoms that together define AIDS as a human disease

AIDS: A DISEASE OR A SYNDROME?

AIDS has been presented in journals, nonscience magazines, newspaper articles, and on television as a disease. However, a disease is a pathological condition with a single identifiable cause. As we learned from the days of Louis Pasteur and Robert Koch, there is a single identifiable organism or agent for each infectious disease.

AIDS patients may have many diseases. Most AIDS patients have more than one disease at any given time. Each disease produces its own signs and symptoms. Collectively, the diseases that are expressed in an AIDS patient are referred to as a **syndrome**. The number of different diseases an AIDS patient has and the severity of their expression reflects the functioning of that person's immune system.

FIRST REPORTS ON AIDS CASES IN THE UNITED STATES

In January 1981, while Ronald Reagan was taking his first oath of office as president, doctors around the country were just discovering the pattern of symptoms and infections in patients that was to become a very new disease.

Initially, AIDS appeared among homosexual males; most frequently those who had many sex partners. Further study of the gay population led to the conclusion that the agent responsible for AIDS was being transmitted through sexual activities. In July 1982, cases of AIDS were reported among hemophiliacs, people who had received blood transfusions, and injection drug users. These reports all had one thing in common—**an exchange of body fluids.** In particular, blood or semen was involved. In January 1983, the first cases of AIDS in heterosexuals were documented. Two females, both sexual partners of injection-drug users (IDUs), became AIDS patients. This was clear evidence that the infectious agent could be transmitted from male to female as well as from male to male. Later in 1983, cases of AIDS were reported in Central Africa, but with a difference. The vast majority of African AIDS cases were among heterosexuals who did *not* use injection drugs. These data supported the earlier findings from the American homosexual population: that AIDS is primarily a sexually transmitted disease. Also, the risk for contracting AIDS increased with the number of sex partners one had and the sexual behaviors of those partners. Early empirical observations on which kinds of social behavior placed one at greatest risk of acquiring AIDS were later supported by surveillance surveys, testing, and analysis.

FIRST REACTION TO AIDS: DENIAL

When the disease that would eventually be called AIDS first emerged in 1981, a few officials within agencies like the Centers for Disease Control and Prevention realized that a new infectious agent was at work and that it could well be spreading rapidly. They most cautiously tried to sound the alarm, but the nation was not ready to talk about subjects like **anal sex, needles**, and **condoms.** Among those most heavily in denial were gay men, who were most at risk. They were still enjoying sexual liberation won in the 1970s, and nobody was in a mood to call the party off, even as close friends and sexual partners began dying.

New York playwright Larry Kramer attempted to break through this denial in early 1983 with an article in a widely read gay magazine. Headlined "1,112 and Counting" Kramer warned: "If this article doesn't rouse you to anger, fury, rage and action, gay men have no future on this Earth."

As they turned their fears into political engagement, the activists confronted a Washington that resisted action. Blood banks denied that any extra precautions were needed to prevent transmission. AIDS was buried deep inside newspapers and seldom mentioned on television. The death of movie star Rock Hudson in 1985 finally put AIDS on the front pages. But still, three young hemophiliacs, Ricky, Robert, and Randy Ray, were firebombed out of their Florida home when their neighbors learned they were HIV positive two years later.

LOOKING BACK 1.1

SILENCE = DEATH: SILENCE AND STIGMA

The silence comes because we are talking about sex, we are talking about needles and injection drug use—things we are not good at talking about as a community. But silence is also perpetuating this pandemic, and it must be talked about everywhere, by everyone. By the time former president Ronald Reagan gave his first speech on the AIDS crisis in America, hundreds of thousands of Americans were HIV-infected, 36,000 men, women, and children had been diagnosed with AIDS, and over 25,000 had died of AIDS—the year, 1987.

While there was already convincing evidence that AIDS could not be transmitted by casual contact, people were nevertheless fired from their jobs across the country because of fears that they posed a threat to coworkers.

WHAT CAUSES AIDS?

In the Beginning, 1980, 1981, 1982

The appearance of AIDS in distinctly different populations including young gay men, intravenous drug abusers, hemophiliacs, Haitians, infants, and blood transfusion recipients argued for an infectious agent. But what kind of infectious agent would destroy the immune system of so many different groups of people?

There were very few facts, but many plausible theories about the causes of AIDS. Perhaps the cytomegalovirus (sito-meg-ah-lo-virus) had mutated to cause a more severe illness. Maybe the illness was related to "poppers" (amyl and butyl nitrite) and other drugs popular among gay men for enhancing sexual pleasure. Some researchers thought the other sexually transmitted infections that many gay men contracted somehow overwhelmed the immune system to cause the mysterious disease. Government researchers suggested that sperm in the male bowel caused the disease, a theory that made little sense because homosexuality is probably as old as society. Few doctors immediately considered the possibility of a new infectious agent. Prior to the outbreak of this strange disease, infectious disease scientists arrogantly believed that virtually all diseases were known. It was just that many things were unknown about the individual diseases. In fact, an editorial in *The New England Journal of Medicine* in December 1981 on possible causes of AIDS, omitted the whole idea that AIDS might be caused by an unknown infectious agent! And in 1982, this journal refused to publish Michael Gottlieb's work on the very first cases of this new disease that devastated the immune system. The article was said to be rejected because the disease was considered unusual and not of much importance!

Early in 1983, the agent that destroys an essential portion of the human immune system was identified by French scientists as a virus. From that point on there was a specific infectious agent associated with the cause of AIDS. The symptoms of viral-induced AIDS can begin *only* after one has been infected with a specific virus. This virus is now called the ***H*uman *I*mmunodeficiency *V*irus (HIV).** The specific viral induced disease is referred to specifically as HIV/AIDS because there are other reasons for a suppressed immune system, like congenital inherited immune deficiencies, exposure to radiation, alkylating agents, corticosteroids, certain forms of cancer, and cancer chemotherapy that also produce AIDS-like symptoms (Stadtmauer et al., 1997).

After years of insults and innuendos, U.S. President Ronald Reagan and French Prime Minister Jacques Chirac agreed to name co-discoverers—America's Robert Gallo and France's Luc Montagnier. Naming co-discoverers was a political solution to end a dispute over patent rights covering the blood test for HIV.

HOW DOES HIV CAUSE AIDS?

Over time, HIV depletes a subset of lymphocytes called **T4 helper cells, or CD4+ cells,** that are essential in the proliferation of cells necessary to cell-mediated immunity and in the production of antibodies (Figure 1-1). Thus, cell-mediated immunity and antibodies are critical components of the human immune system. Without the ability to produce a sufficient number of immune-specific cells and immune-specific antibodies, the body is vulnerable to a large variety of infections caused by organisms and viruses that normally do not cause human disease. These infections create the symptoms and progression of illnesses that eventually kill AIDS patients. Thus AIDS begins with HIV infection. Technically it can be called **HIV disease, HIV T4 helper cell or CD4+ cell disease,** but the popular press, scientists, and others still refer to HIV disease as AIDS. AIDS is the end stage of chronic HIV infection. **AIDS IS NOT TRANSMITTED, THE VIRUS IS.** People do not die of AIDS per se. They die of opportunistic infections, cancers, and organ failures brought on by the results of a failed immune system.

It is believed that eventually almost everyone who is *correctly* diagnosed with HIV/ AIDS will die from AIDS. But all who become HIV-infected may not progress to AIDS. Estimates are that some 5% of the HIV-infected population will not progress to AIDS. This implies that there is a percentage of the population that is resistant to HIV-associated immune system suppression. Since

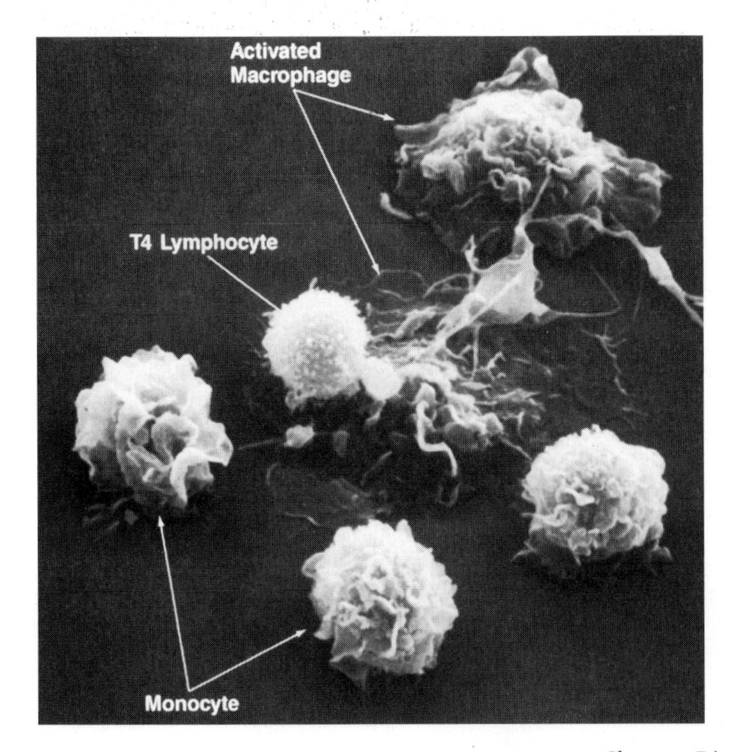

FIGURE 1-1 Normal Human T4 Lymphocytes, Monocytes, and Macrophages. Scanning electron micrograph of monocytes, macrophages, and a T4 or CD4+ lymphocyte, magnified 9000 times. These white blood cells are the targets of HIV infection. Note that the T4 lymphocyte (round cell, at the center) is adhered to a flattened macrophage. *(Photograph courtesy of Dr. M. A. Gonda)*

mid-1996, several genes that offer resistance to HIV infection have been identified. (See Chapter 4.)

Naming the Disease

Early in 1981, practically coincident with the report of the first cases of a new disease in the male homosexual community in the United States, there were reports of 34 cases of a new disease among Haitian immigrants to the United States and 12 cases of a disease previously unrecognized in Haiti—an aggressive form of **Kaposi's sarcoma** (kap-o-seez sar-ko-mah). Michael Gottlieb, who had identified the new disease that seemed to target gay men, found that although each of the cases was different, **all had one thing in common: Whatever was making the men sick had singled out the T lymphocyte cells for destruction.** Eventually the body's battered defenses couldn't shake off even the most harmless microbial intruder. The men were dying from what doctors termed opportunistic infections, such as ***Pneumocystis* pneumonia,** which attacks the lungs, and **toxoplasmosis,** which often ravages the brain.

THE CENTERS FOR DISEASE CONTROL AND PREVENTION REPORTS

In June 1981, the Centers for Disease Control and Prevention (CDC) first reported on diseases occurring in gay men that previously had only been found to occur in people whose immune systems were suppressed by drugs or disease [*Morbidity and Mortality Weekly Report (MMWR)*, 1981].

The report stated that five young men in Los Angeles had been diagnosed with *Pneumocystis carinii* pneumonia (PCP) in three different hospitals. Because cases of PCP occurred almost exclusively in immune-suppressed patients, five new cases in one city at one time were considered as unusual. The report also suggested "an association between some aspects of homosexual lifestyle or

LOOKING BACK 1.2

THE NEW YORK TIMES AND A.I.D.S., AUGUST 8, 1982

An article, "A Disease's Spread Provokes Anxiety," alerted its readers to the growing health crisis in the homosexual community that was baffling medical science. While the *Times* had previously reported on a disease causing opportunistic infections in gay men, this was the first time the term "Acquired Immune Deficiency Syndrome" or A.I.D.S. (the punctuation had not been dropped) appeared in a major newspaper. Later in 1982, the *Washington Post* joined the *Times* in reporting on the death of an infant who had received a blood transfusion from an AIDS-afflicted donor. With that, a second major national newspaper was officially in the business of covering the AIDS story.

The Centers for Disease Control (CDC): 1982/1983

Regardless that this new disease was increasing rapidly and killing all those who displayed the strange set of symptoms, the epidemiologists at the CDC in 1982 had other things to worry about: President Ronald Reagan was moving to reduce the size of the federal government and CDC staff members were facing dismissal. Harold W. Jaffe, who scurried to stay employed, later went on to become chief epidemiologist for the center's AIDS team. Progress was also impeded by scientists who withheld information from health officials because they were competing against one another to be first to publish articles in medical journals that demanded exclusive information. Communication to the public was also poor. People had misplaced fears and misconceptions about the disease even after epidemiologists found, by late 1982, that this disease was transmitted through sex, blood transfusions, injecting drugs with contaminated needles and syringes, and from mother to child. Because health officials and journalists used the phrase "bodily fluids" instead of specifying semen, blood, and vaginal secretions, many people feared that they could contract AIDS from toilet seats or drinking fountains, or by eating at restaurants or going to school.

At a meeting of the World Health Organization (WHO) in Geneva in November 1983, it was clear that AIDS was a global health problem and that cases were occurring in many countries that had officially denied it. But these warning signs were disregarded.

disease acquired through sexual contact and PCP in this population. Based on clinical examination of each of these cases, the possibility of a cellular immune dysfunction related to a common exposure might also be involved."

In July 1981, the CDC (*MMWR*, 1981) reported that an uncommon cancer, Kaposi's sarcoma (KS), had been diagnosed in 26 gay men who lived in New York City and California. This was also an unusual finding because KS, when it occurred, was usually found in older men of Jewish or Italian ancestry. The sudden and dramatic increase in pneumonia cases, all of which were caused by a widespread but generally harmless fungus, *P. carinii*, and KS cases indicated that an infectious form of immune deficiency was on the increase. At first, the new disease was referred to as the "4 H disease" because the first cases of the disease were found among homosexuals, Haitians, heroin users, and hemophiliacs. Later this immunodeficiency disease was called **GRID** for **Gay-Related Immune Deficiency.** This new mysterious and lethal illness appeared to be associated with one's lifestyle. These early cases of immune deficiency heralded the beginning of an epidemic of a previously unknown illness. By 1982 and 1983, the disease was reported in adult heterosexuals and children. Because a ***cellular deficiency of the human immune system*** was found in every AIDS patient, along with an assortment of other signs and symptoms of disease, and because the infection was ***acquired*** from the action of some environmental agent, it was then named **AIDS** for **Acquired Immune Deficiency Syndrome.**

DISCOVERY OF THE AIDS VIRUS

There was no shortage of ideas on what caused AIDS. It was believed by some to be an act of God, a religious curse or penalty against the homosexual for practicing a biblically unacceptable lifestyle that included drugs, alcohol, and sexual promiscuity. The reverend Billy Graham said "AIDS is a judgment of God." Jerry Falwell stated that AIDS is God's punishment, the scripture is clear, "we do reap it in our flesh when we violate the laws of God." Some believed AIDS was due to sperm exposure to amyl nitrate, a stimulant used to heighten sexual pleasure (Gallo, 1987). Others believed there was no specific infectious agent. They believed that certain people who *excessively stressed their immune systems* experienced immune system failure, and before it could recover other infections killed them. But many scientists who had expertise in analyzing the sudden onset of new human diseases thought the cause of this form of human immune deficiency was an infectious agent. They believed that the agent was transmitted through sexual intercourse, blood or blood products, and from mother to fetus. They also believed that this agent, which led to the loss of T4 or CD4+ cells, was smaller than a bacterium or fungus because it passed through a filter normally used to remove those microorganisms. This agent fit the profile of a virus.

In January 1983, **Luc Montagnier** (Mont-tan-ya) and colleagues at the Pasteur Institute in Paris isolated the virus that causes AIDS. (Hobson et al., 1991). In May of that year, they published the first report on a T cell retrovirus found in a patient with **lymphadenopathy** (lim-fad-eh-nop-ah-thee), or swollen lymph glands. Lymphadenopathy is one of the early signs in patients progressing toward AIDS. The French scientist (Figure 1-2) named this virus **lymphadenopathy-associated virus** (LAV) (Barre-Sinoussi et al., 1983).

Naming the Viruses That Cause AIDS: HIV-1, HIV-2

During the search for the AIDS virus, several investigators isolated the virus but gave it different names. For example, Robert Gallo (Figure 1-3) named the virus HTLV III (For the Third Human T Cell Lymphotropic Virus). Because the collection of names given this virus created some confusion, the Human Retrovirus Subcommittee of the Committee on the Taxonomy of Viruses reduced all the names to one: **Human Immunodeficiency Virus** or **HIV.** This term is now used worldwide.

In 1985, a second type of HIV was discovered in West African prostitutes. It was named HIV type 2 or **HIV-2.** The first confirmed case of HIV-2 infection in the United States was reported in late 1987 in a West African woman with AIDS. By December 1990, 16 additional cases of HIV-2 infection were reported to the CDC (*MMWR,* 1990). Beginning in 2005, a total of 94 HIV-2 infections have been reported from 22 states of the United States and the District of Columbia. Of the 94 infected persons, 70 are black and 54 are male. Sixty-six were born in West Africa, 13 in the United States (including 3 infants born to mothers of unspecified nationality), 2 in India and 3 in Europe.

FIGURE 1-2 Luc Montagnier, from the Pasteur Institute in Paris. He and colleagues discovered the Human Immunodeficiency Virus (HIV), the cause of AIDS. In October 1997, he moved to Queens College, New York, to run a new AIDS research institute. In 2002 he became a consultant to Calypte, makers of an HIV urine screening test. *(Photograph courtesy of AP/Wide World Photos)*

FIGURE 1-3 Robert Gallo, Director, Institute of Human Virology. *(Photograph courtesy of AP/Wide World Photos)*

The region of origin was not identified for 10 of the persons. Seventeen have developed AIDS-defining conditions and 11 have died. These case counts represent minimal estimates because completeness of reporting has not been assessed; reporting varies from state to state according to state policy.

First Reported Case of HIV-2 Infection

The earliest evidence of an individual exposed to HIV-2 comes from a case report on an infection most likely occurring in Guinea Bissau in the 1960s. Anne-Mieke Vandamme of the Catholic University of Leuren in Belgium and colleagues believe that HIV-2 first moved into humans near the town of Canchungo in western Guinea-Bissau, since that is where the largest proportion of people carry it. The researchers calculate that HIV-2 jumped into humans in about 1940 for the A subtype and 1945 for the B subtype, a time when there were still Mangabey monkeys around Canchungo. A local fondness for eating the primates could have both wiped them out and exposed people to mangabey viruses. In genetic terms, HIV-2 is much more closely related to the Simian Immunodeficiency Virus (SIV), a group of monkey viruses, than to HIV-1. Both HIV-2 and HIV-1 are said to have been derived from ancestral SIV variants that were from distinct regions and species and do not appear to be direct genetic descendants of each other (Marlink, 1996; Hahn et al., 2000). Clinically, what has been learned about HIV-1 appears to apply to HIV-2, except that HIV-2 appears to be less harmful (cytopathic) to the cells of the immune system and it reproduces more slowly than HIV-1. Because HIV-2 has now been found in

18 African countries with a prevalence of over 1%, the question is why is there no HIV-2 epidemic or pandemic similar to that caused by HIV-1? According to scientists working with HIV-2, most likely an epidemic has not occurred because of different behaviors of HIV-2 in relation to viral load. Surveys in West Africa found that HIV-2 never reaches a high viral load in the blood as found with HIV-1. Thus, it is reasoned that fewer HIV-2 cases means there are less to transmit to other people and fewer viruses to attack the immune system.

UNLESS STATED OTHERWISE, ALL REFERENCES TO HIV IN THIS BOOK REFER TO HIV-1.

DEFINING THE ILLNESS: AIDS SURVEILLANCE

The CDC reported that through 1983 there were 3,068 AIDS cases in the United States and 1,478 of these had died (48%). All demonstrated a loss of CD4+ or T4 lymphocytes, and all died with severe opportunistic infections. Opportunistic infections are caused by organisms and viruses that are normally present but do *not* cause disease unless the immune system is damaged. Clearly there was an immediate need for a name and definition for this disease so that a rational surveillance program could begin.

Definitions of AIDS for Surveillance Purposes

The initial objective of AIDS surveillance was to describe the epidemic in terms of time, place, and individuals and to recognize immediately changes in rate and pattern in the spread of AIDS.

The First Definition of AIDS

In order to establish surveillance, a system for monitoring where and when AIDS cases occurred, a workable definition had to be developed. The definition had to be ***sensitive*** enough to detect every possible AIDS patient, while at the same time ***specific*** enough to exclude those who may have AIDS-like symptoms, but were not infected by HIV.

In 1982, there was no ***single characteristic*** of AIDS that would allow for a useful definition for surveillance purposes. And so, the first AIDS surveillance definition was based on the **clinical description** of symptoms. The first of many criteria for the diagnosis of AIDS were: (**1**) the presence of a reliably diagnosed disease at least moderately predictive of cellular immune deficiency; and (**2**) the absence of an underlying cause for the immune deficiency or for reduced resistance to the disease (*MMWR,* 1982). Because the symptoms varied greatly among individuals, this was a poor first definition.

AIDS Definition Modified

The initial definition of AIDS was thus an arbitrary one, reflecting the partial knowledge of the clinical consequences that prevailed at the time. Various systems for classifying HIV-related illnesses have been devised since 1982 to take into account increasing knowledge about the spectrum of those illnesses. Had the whole picture of HIV infection and its clinical consequences been known in 1982, the term **"AIDS"** would not have been used. Instead, it would have been called **"HIV disease"** (or perhaps, following an older tradition, "Gottlieb's disease," after Mike Gottlieb, who first described it).

The 1982 definition was modified in 1983 to include new diseases then found in AIDS patients. With this modification, AIDS became reportable to the Centers for Disease Control and Prevention (CDC) in every state. In 1985, additional diseases were included in the AIDS case definition.

Broadly speaking, the term **AIDS** may be understood as referring to the onset of life-threatening illnesses as a result of HIV disease that results from an HIV infection. AIDS is the end stage of a disease process that may have been developing for 5, 10, 15, or more years, for most of which time the infected person will have been well and quite possibly unaware that he or she has been infected.

Thus the number of AIDS cases reported from 1987 through 1992 reflects the revisions of the initial surveillance case definitions.

Problems with AIDS Definitions, 1981–1992

One major drawback to all the CDC AIDS definitions is the fact that through 1992, the Social Security Administration (SSA) used the CDC AIDS definition to determine disability. But all the definitions were primarily based on symptoms and opportunistic infections in men. Therefore, about 65% of women with HIV/AIDS symptoms were excluded from Supplemental Security Income (SSI) benefits. They were excluded because of failure to be diagnosed with AIDS by the CDC AIDS definition (Sprecher, 1991).

AIDS Redefined in 1993

On January 1, 1993, the newest definition of AIDS was put into the surveillance network. The reason for the new CDC definition was that epidemiologists felt the 1987 definition failed to reflect the true magnitude of the pandemic. In particular, it failed to address AIDS in women. Thus, the CDC expanded the AIDS surveillance case definition to include all HIV-infected persons who have less than 200 CD4+ or T4 lymphocytes/μL of blood, or a T4 lymphocyte percent less than 14% of total lymphocytes. In addition to retaining the 23 clinical conditions in the 1987 AIDS surveillance case definition, the expanded definition includes (1) pulmonary tuberculosis, (2) invasive cervical cancer, and (3) recurrent pneumonia. The **objectives** of these changes are to simplify the classification of HIV infection and the AIDS case reporting process, to be consistent with standards of medical care for HIV-infected persons, to better categorize HIV related morbidity, and to reflect more accurately the number of persons with severe HIV-related immunosuppression who are at highest risk for severe HIV-related morbidity and most in need of close medical follow-up.

Impact of the 1993 Definition

The expanded AIDS surveillance case definition had a substantial impact on the number of reported cases in 1993 (*MMWR*, 1993).

Of the 106,618 Adult/Adolescent AIDS cases reported in 1993, 57,574 (54%) were reported based on conditions added to the definition in 1993; and 49,044 (46%) were reported based on pre-1993 defined conditions. A substantial increase in the number of reported AIDS cases occurred in all regions of the United States.

When compared to 1992 data the increase in reported cases in 1993 was greater among females (151%) than among males (105%). Proportionate increases were greater among blacks and Hispanics than among whites. The largest increases in case reporting occurred among persons aged 13–19 years and 20–24 years; in these age groups, a greater proportion of cases were reported among women (35% and 29%, respectively) and were attributed to heterosexual transmission (22% and 18%, respectively).

Compared with homosexual/bisexual men, proportionate increases in case reporting were greater among heterosexual injecting-drug users (IDUs) and among persons reportedly infected through heterosexual contact.

The pediatric AIDS surveillance case definition was not changed in 1993.

If all the approximately 1 million persons in the United States with HIV infection were diagnosed and their immune status known, it is estimated that 120,000–190,000 persons who do not have AIDS-indicator diseases would be found to have T4 lymphocyte counts <200/μL. However, since not all of them are aware of their HIV infection status, and of those who are, not all have had an HIV test, the immediate impact on the number of AIDS cases will be considerably less. Under the 1987 AIDS surveillance criteria, approximately 49,106 AIDS cases were reported in 1992, but 106,618 were reported for 1993. In 1994, there were 80,691 new AIDS cases. For 1995, the number of new cases was 74,180; and less than 40,000 in 2004.

Problems Stemming from Changing the AIDS Definition for Surveillance Purposes

Each time the definition of AIDS has been altered by the CDC, it has led to an increase in the number of AIDS cases. In 1985, the change in definition led to a 2% increase over what would have been diagnosed prior to the change. The 1987 change led to a 35% increase in new AIDS cases per year over that expected using the 1985 definition. The 1993 change resulted in a 52% increase in AIDS cases over that expected for 1993. Such rapid changes alter the baseline from which future predictions are made and make the interpretations of trends in incidence and characteristic of cases difficult to process. **For the first time because of the 1993 AIDS definition, one could be diagnosed with AIDS and remain symptom-free for years (become HIV positive and have a T4 or CD4+ cell count of less than 200).**

SUMMARY

Much continues to be written about HIV/AIDS. Some of it, especially in lay articles, has been less than accurate and has led to public confusion and fear. **HIV infection is not AIDS. HIV infection is now referred to as HIV disease. AIDS is a syndrome of many diseases,** each resulting from an opportunistic agent or cancer cell that multiplies in humans who are immunosuppressed. The new 1993 CDC AIDS definition will allow, over the long term, earlier access to federal and state medical and social services for HIV-infected individuals.

REVIEW QUESTIONS

(Answers to the Review Questions are on page 438.)

1. The letters A, I, D, and S are an acronym for?
2. Is AIDS a single disease? Explain.
3. When was the AIDS virus discovered and by whom?
4. In what year did CDC first report on a strange new disease that later was named AIDS?
5. Name one different acronym for HIV.
6. How many times has CDC changed and expanded the definition of AIDS? In what years?
7. Why did the CDC do away with the ARC definition?
8. What is one major advantage of the new CDC AIDS definition for the HIV-infected?

2 What Causes AIDS: Origin of the AIDS Virus

CHAPTER CONCEPTS

- AIDS dissidents say AIDS is not caused by HIV infection.
- HIV/AIDS scientists say AIDS is caused by HIV infection.
- An unbroken chain of HIV transmission has been established between those infected and the newly infected.
- HIV is believed to have crossed into humans from chimpanzees.
- A viral precursor to HIV may have entered humans 300 years ago.
- A third new HIV strain is found.
- The earliest AIDS case to date is reported in 1959.

THE CAUSE OF AIDS: THE HUMAN IMMUNODEFICIENCY VIRUS (HIV)

The unexpected appearance and accelerated spread of an unknown lethal disease soon raised several important questions: **What** is causing the disease? **Where** did it come from? These questions will be answered.

This section has a subtitle that states that AIDS is the result of HIV infection. A relatively small number of scientists and nonscientists claim that HIV does *not* cause AIDS. They are referred to as AIDS dissidents. For a balanced HIV/AIDS presentation, this claim will be presented first.

HIV Does Not Cause AIDS: A Minority Point of View

Fronted locally by ACT UP/San Francisco, a renegade chapter long ago disowned by the rest of the AIDS activist movement (and not to be confused with ACT UP/Golden Gate, which respects most conventional AIDS research), members have repeatedly plastered the Castro district with stickers reading, "AIDS Is Over," and "Don't Buy the HIV Lie." In May and June 1999, the Los Angeles-based dissident group Alive & Well ran a series of full-page ads in several gay/lesbian and alternative papers, including the *Bay Area Reporter, Bay Times*, and *Bay Guardian*, arguing that AIDS is not contagious, HIV is harmless, and that HIV/AIDS drugs are the real danger. The first of the ads states that, "What we have experienced for 20 years is not a sexually transmitted epidemic but a tragic medical mistake. Contrary to popular beliefs, AIDS is not a new disease, AIDS is a new name given by the Centers for Disease Control (CDC) to a collection of 29 old illnesses and conditions. . . . These illnesses and conditions are called AIDS only when they occur in persons who also have certain protective, disease-fighting proteins called antibodies in their blood." In her book, *What If Everything You Thought You Knew About AIDS Was Wrong?*, Christine Maggiore elaborates, "None of these diseases appear exclusively in those who test (HIV) positive. . . . All 29 indicator diseases have established causes and treatments unrelated to HIV." AIDS, in other words, isn't an epidemic at all; it's a phony construct. Maggiore, who tested HIV positive in 1992, has never taken antiretroviral drugs, remains relatively symptom-free and believes that those who have died of AIDS actually died from prescription or recreational drugs, or fear. Her explanations for the global AIDS pandemic: In Africa, people are dying at the same rate as before; in America, people are victims of the prescription drugs.

Disbelieving Scientists

Scientist Peter Duesberg is perhaps the most vocal in his concern that the scientific community is investigating the wrong causative agent. Duesberg is a molecular biologist at the University of California at Berkeley and a member of the National Academy of Sciences. Duesberg has advanced his anti-HIV/AIDS hypothesis at great expense to himself. He states that "I have been excommunicated by the retrovirus-AIDS community with noninvitations to meetings, noncitations in the literature and nonrenewals of my research grants, which is the highest price an experimental scientist can pay for his convictions."

Duesberg's financial and laboratory problems reflect his lonely battle with HIV/AIDS investigations and drug companies who, he believes, have invested so much in the HIV/AIDS theory that they cannot afford to entertain an alternative theory.

In 1971 at age 33, Duesberg codiscovered cancer-causing genes in viruses. In the March 1987 issue of *Cancer Research*, he published "Retro-viruses as Carcinogens and Pathogens: Expectations and Reality." The article provoked wide-based scientific discussion and received a lot of popular press coverage. In the article Duesberg argues that there is *no evidence* that HIV causes AIDS. He has published additional articles in *Science* (1988) and in the *Proceedings of the National Academy of Sciences* (1989) stating that HIV is not the cause of AIDS. In May 1992, Duesberg was the featured speaker at an alternative AIDS conference, "AIDS: A Different View," promoted by homeopathic physician Martien Brands. The week of meetings was spent giving new interpretations to some of the data used to establish HIV as the causative agent of AIDS. In short, Duesberg suggests that there is no single causative agent, that the disease is due to one's "lifestyle." He marshals arguments to support his theory that, in the United States and probably in Europe, AIDS is a collection of noninfectious deficiencies predominantly associated with drug use, malnutrition, parasitic infections, and other specific risks. To read more on Duesberg, see the Wayt Gibbs article, "Dissident or Don Quixote?" in the August 2001 *Scientific American.*

Duesberg believes the tests that detect HIV antibodies are useless. In the June 1988 issue of *Discovery* he said, "If somebody told me today that I was antibody positive, I wouldn't worry one second. I wouldn't take Valium. I wouldn't write my will. All I would say is that my immune system seems to work. I have antibodies to a virus. I am protected."

In June 1990, Robin Weiss and Harold Jaffe wrote a critical refutation of Duesberg's theory that HIV cannot be the cause of AIDS. Duesberg's response suggested that he was unaware of published data that clearly answer the questions he raises concerning HIV involvement in AIDS. For example, one of Duesberg's major points is that no one has yet shown that hemophiliacs infected with HIV progress to AIDS. The data on matched

BOX 2.1

SEX AND HIV DO NOT CAUSE AIDS!

Medical Doctor Puts His Life on the Line to Prove It

On October 28, 1993, Robert Willner held a press conference at a North Carolina hotel, during which he jabbed his finger with a bloody needle he had just stuck into a man who said he was infected with HIV. Willner was a physician who had his medical license revoked in Florida for, among other infractions, claiming to have cured an AIDS patient with ozone infusions. He is also the author of a book, *Deadly Deception: The Proof that SEX and HIV Absolutely DO NOT CAUSE AIDS.* He insists that jabbing himself with the bloody needle, which he described as "an act of intelligence," was not meant to sell books. "I'm interested in proving to people that there isn't one shred of scientific evidence that HIV causes any disease."

On October 3, 1994, *USA Today* carried a full-page advertisement promoting Willner's book. If carried nationally, a full-page ad in the newspaper, regardless of location, then cost $57,500.

DISCUSSION QUESTION: Did *USA Today* management provide socially responsible advertisement or did money talk and responsibility walk? What is the potential medical downside of this advertisement?

The scientific journal *Science* (1994; 226: 1642–1649) presented a series of six articles by Jon Cohen concerning the *question* of whether HIV causes AIDS. The articles are a balanced review of the scientific facts as they relate to the question.

Robert Willner died April 15, 1995, from an apparent heart attack.

groups of homosexual males and hemophiliacs, which show that *only* those infected with HIV develop AIDS, have been available for a number of years (Weiss et al., 1990).

Duesberg's arguments and disagreements with the vast majority of prominent scientists who have researched the causal agent of AIDS are many. But they pale when placed next to the overwhelming evidence that leaves no doubt in the opinion of most scientists that HIV causes AIDS (see Andrews, 1995; Cohen, 1995; Moore, 1996).

Based on an August 1992 report in *Newsweek*, a father discussed his decision, based on Duesberg's claims, to counsel his infected hemophiliac son to avoid Zidovudine (ZDV) treatment. This situation is similar to what happened when desperate cancer patients followed the advice of a credentialed academician who recommended vitamin therapy as the cure for cancer. Based on such advice, some people failed to undergo truly effective therapy.

DISCUSSION QUESTION: Is Duesberg's opinion on this issue inadvertently harmful to humans? To the scientific process? Will the use of his idea, that HIV does not cause AIDS, provide a course of action that will stop the Acquired Immune Deficiency Syndrome?

Duesberg's Belief that HIV Does Not Cause AIDS Continues

In 1996, Duesberg's book, *Inventing the AIDS Virus,* was published and in 1998 he and David Rasnick published a paper, "*The AIDS Dilemma: drug diseases blamed on a passenger virus.*" Through 2004, Duesberg still insisted that HIV does not cause AIDS. He believes that HIV is just another opportunistic agent like those that cause other opportunistic diseases (Duesberg, 1993, 1995a, 1995b; Moore, 1996). With each new scientific report, it becomes more difficult for Duesberg to maintain his position. However, he remains unconvinced that HIV causes AIDS regardless of the reports that newborn infants with HIV got HIV *only* from HIV-infected mothers and progress to AIDS while noninfected newborns from the same mothers do not progress to AIDS and that some 50% of HIV-infected hemophiliacs have developed AIDS yet *no* HIV-negative hemophiliac has ever developed AIDS (Darby et al., 1995; Levy, 1995; Sullivan et al., 1995). In addition, Duesberg claims that the drug AZT (Zidovudine) causes AIDS. What does he make of the AIDS Clinical Trials Group (ACTG) Protocol 076 that demonstrated AZT treatment of women during pregnancy and delivery reduced transmission from mother to infant from 25% in the placebo-treated mothers to 8% in those who received AZT (Connor et al., 1994)?

For those who wish to know more on the rebuttal of Duesberg's arguments, read the study reporting on the death rate among HIV-positive and HIV-negative British hemophilia patients (Baum, 1995; Darby et al., 1995; Editorial, 1995). For more by Duesberg see his Web page, www.duesberg.com.

Impact of HIV/AIDS Dissident Thinkers on the President of South Africa

David Rasnick, an American chemist, is a leader of the HIV/AIDS dissidents in the United States. He is also the person who is thought to be, at least partially, responsible for South African President Mbeki's belief that HIV does not cause AIDS. In February 2002, Rasnick and South African computer science professor Philip Machanick, who believes that HIV causes AIDS and that the antiretroviral drugs are not as toxic as Rasnick says, have agreed, after a heated exchange of letters, to a challenge. The date, at this time, has not been set. But should this bizarre game of chicken occur, the scenario will go something like this. Rasnick will intentionally infect himself with HIV to prove that it does not cause disease, and Machanick will take drugs to prove they are not toxic. Under the agreement, Rasnick will inject himself with HIV on television and the two will meet annually to compare health status. It is doubtful that the action of this challenge will ever occur. Machanick is never going to get a doctor to prescribe him medication for a disease he doesn't have and Rasnick stipulated that he be injected with a highly purified virus, a condition that is impossible to meet. Results, if any, will be offered after they occur.

WHAT DO YOU BELIEVE AND WHY?

A Member of the U.S. Government Enters the Dispute

In March 1995, a letter was sent to eight government scientists and officials who influence AIDS research and policies, in which freshman Representative Gil Gutknecht (R-MN) questions the HIV=AIDS hypothesis. His query echoes the arguments of Peter Duesberg. Gutknecht's initial

BOX 2.2

SCIENCE TAKES BACKSEAT TO POLITICS: SOUTH AFRICAN PRESIDENT THABO MBEKI SAYS HIS GOVERNMENT HAS A RIGHT/OBLIGATION TO DOUBT WHETHER HIV CAUSES AIDS

Beginning in March 2000, two tragedies began unfolding simultaneously in South Africa. The first is epidemiological, with millions of men, women, and children infected with HIV destined to develop AIDS. The second is political, with President Thabo Mbeki (Figure 2-1) seriously entertaining a discredited view that challenges the role of HIV as the cause of AIDS. Together, the tragedies may well increase the AIDS pandemic in South Africa. The outcome can only be measured in untold suffering, death, and orphans.

DENIAL THAT HIV CAUSES AIDS

President Thabo Mbeki appears to be actively flirting with a new form of denial, one that claims that HIV is harmless, AIDS is a phony epidemic, and the only thing endangering the health of Africans is the same set of poverty-related health issues the continent has long faced. To the chagrin of doctors and scientists in his own country and around the world, Mbeki has tried to turn this into an issue of free speech rather than one of science.

Mbeki and other HIV dissidents believe that if HIV exists it does not cause AIDS. The disease AIDS does not exist. There is no epidemic nor are there deaths from AIDS. There is just mass hysteria caused by a conspiracy among pharmaceutical multinationals, aided and abetted by political and medical self-interest. The views of the dissidents—or rethinkers as some like to call themselves—appear to be backed by an impressive range of scientific citations from reputable journals and leading scientists, including at least one Nobel Prize winner. But close examination of the sources dissidents cite often reveals extremely selective quotations from articles or use of outdated research.

Orthodoxy—People who believe HIV causes AIDS have demonstrated a rapidly expanding foundation of scientific and medical understanding that rests upon a detectable virus, and have documented evidence of the impact of this virus on the human immune system. There is an impressive array of leading scientific names—the signatories to the Durban Declaration include many Nobel laureates (see following).

FIGURE 2-1 South African President Thabo Mbeki. In September 2003, Mbeki said he did not know anyone with HIV infection or AIDS. It is believed that members of his staff died of AIDS. *(Photograph courtesy of AP/World Photos)*

Believing in the orthodox position suggests that things can be done to prevent people from becoming HIV-infected and dying of AIDS.

Belief in the dissident view tends to suggest that because HIV does not cause AIDS, there is no epidemic; nothing needs to be done. People are dying of diseases exacerbated by poverty, as they have always done. It is simply that these are being recorded more often. There is no infectious agent at work, in South Africa or in the world, causing a new and different disease!

HIV either causes AIDS or it does not, and the answer must come from science—not politics! The reasons supporting HIV as the cause of AIDS are too numerous to list here but many of those reasons are found in this chapter and in Chapter 10 (see HIV/AIDS statistics for Africa). Nelson Mandela, (Figure 2-2) past president of South Africa said, "Mbeki is a man of great intellect who takes scientific thinking very seriously, while on the other hand the scientific community was committed to freedom of inquiry, unencumbered by undue political interference. Let us get on with what we know works!"

FIGURE 2-2 Former South African President Nelson Mandela holding a news conference at the Thirteenth International AIDS Conference. He believes that HIV causes AIDS. *(Photograph courtesy of AP/Wide World Photos)*

In April 2002, President Mbeki began to distance himself from the AIDS dissidents. Mbeki has decided to cut informal contact with them and communicate with them only when the advisory panel meets. The president has instructed the Health Ministry to write to the dissidents telling them to stop using Mbeki's name when signing their correspondence. Several dissident members of Mbeki's AIDS Advisory Panel, among them Americans David Rasnick and Peter Duesburg, have taken to using this designation when writing documents and signing letters to newspapers. They have also been using Mbeki's name at international platforms when questioning the link between HIV and AIDS.

In July 2003 the South African Cabinet instructed Health Minister Manto Tshabalala-Msimang to develop a plan, by the end of September, to make antiretroviral drugs available to all HIV-infected people, and she did. In developing their plans, they worked on the assumption that treatment should be made available to people with a CD4+ cell count below 200 cells/mm^3, and that clinical condition would also affect whether individuals were recommended for antiretroviral treatment. Even in its most limited form, it will be the largest developing-world treatment program within three to four years, and in its most comprehensive form would treat over a million people by 2008.

In October 2003, President Mbeki announced his withdrawal from discussions on the cause of AIDS. He said his cabinet will lead the fight against AIDS. South Africa, with a population of 45 million people has about 5.3 million HIV infected and about 600 of those infected die everyday while 1,500 a day are newly infected. About a half million South Africans have a CD4+ cell count of less than 200. They require immediate drug therapy. Each year about 100,000 babies are born HIV positive. Politically, the AIDS pandemic and the government's denial and overwhelming number of people infected and dying did little to stop Thabo Mbeki's election to a second term in office as of April 27, 2004. He has retained his controversial health minister, Manto Tshabalala-Msimang, who still does not believe that HIV causes AIDS. In May 2004, Mbeki promised that at least 50,000 people with AIDS would have access to free drugs by March of 2005 and that 1.5 million would have access by 2009.

SUMMARY

The current political and scientific furor in South Africa fueled largely by dissidents' theories on HIV/AIDS and the seeming support of President Mbeki has much broader implications for South Africa and South Africans than some are prepared to admit. The current controversy is undermining the constructive public health messages this government has put in place. It is sending mixed messages to all those who have dedicated themselves to the alleviation and eradication of this epidemic and is having a negative impact on the morale of affected patients and families. The undermining of scientists and the scientific method is especially dangerous in a developing country still in the process of establishing a strong scientific research base. Furthermore, it may erode investor confidence in the country, causing dire economic consequences.

DISCUSSION QUESTION: What if you were president of a country in which 1 in 9 people were infected with a virus that you were told would kill them unless they were treated with exceptionally expensive medications that will always be outside the range of your health care system finances? Add to that a mandate from your country's constitution guaranteeing each citizen the right to health care. What would you do? Might you search for a way to deny that HIV is a major factor in AIDS? Offer pro/con discussion.

inquiry asks whether recreational drug use and anti-HIV drugs might be the true cause of AIDS and questions whether AIDS is contagious.

Gutknecht's staffers think the federal AIDS effort—based on the conclusion that HIV causes AIDS—"will be seen as the greatest scandal in American history and will make Watergate look like a no-fault divorce" (Stone, 1995).

Talk Show Host Claims AIDS Is a Myth—*October 1996, Fort Lauderdale, Fla.*—A talk show host and two medical professionals proclaimed over the airwaves on Miami's WLQY-AM, one of South Florida's most popular Haitian radio stations, that AIDS is a myth, and urged an estimated 155,000 Haitian listeners to discontinue medication and refuse treatment.

The commentators told the Fort Lauderdale *Sun-Sentinel* that they had ample proof that AIDS does not exist and that the health care industry made it up to increase business. People should stop taking AIDS drugs and wearing condoms, they said in interviews for the article.

"We are not spreading lies. What we are saying are facts," said Henri-Claude Saint-Fleur, a North Miami Beach psychologist who was educated in France. "I'm happy to tell people the truth. I know lots of people who stopped taking their AIDS medication and are living well."

HIV/AIDS counselors say Haitian patients from at least four treatment centers stopped seeing doctors because of the show.

EVIDENCE THAT HIV CAUSES AIDS

It has been firmly established that there is a high correlation between HIV infection and the development of AIDS. With respect to establishing HIV as the causative agent of AIDS, look at some of the evidence:

1. The one common denominator is the presence of HIV within the entire range of people with this particular disease: individuals who are HIV positive and have symptoms of HIV disease and individuals who are HIV positive but appear to be healthy.
2. The virus has been identified by electron microscopy inside and on the surface of T4 cells only in HIV-positive and AIDS patients (Figure 2-3 A, B).
3. Recent work of Bruce Patterson and Steven Wolinsky has shown that the genetic material of HIV (HIV-DNA) can be found in as many as 1 in 10 blood lymphocytes of persons with HIV disease (Cohen, 1993).
4. Antibodies against the virus, viral antigens, and HIV RNA have been found in HIV-positive and AIDS patients.
5. There is an absolute chronological association between the emergence of AIDS and the appearance of HIV in humans worldwide.
6. There is a chronological association of HIV-positive individuals who progress to AIDS. People who are truly HIV negative, and without the need for chemotherapy or radiation treatments, do not demonstrate HIV antibodies and never demonstrate AIDS. For example, people with hemophilia, unlike homosexual men, represent a well-defined group with long-term documentable changes in morbidity and mortality, since they had been well studied as a group before the era of AIDS. This research shows that people with hemophilia began to die of dramatically different things starting about 1982. Looking back shows little evidence of a special incidence of opportunistic diseases in people with hemophilia in the United States from the turn of the century up to 1979. Significantly, however, in the years before AIDS, people with hemophilia had never been noted to be particularly susceptible to the more obvious fungal infections, such as candida esophagitis, common to AIDS patients and others with low-lymphocyte type immune deficiency. After 1984, this type of AIDS-associated opportunistic infection and immune failure rapidly became the single most common cause of death in people with hemophilia in America.

   The rise in total mortality in people with hemophilia was sudden: Death in this population, which had been stable in 1982 and 1983, suddenly increased by a factor of approximately 900% in the first quarter of 1984. This increase was consistent with an epidemic, or some new very toxic contamination of the clotting factor supply. Mortality figures in hemophilia patients also showed something else important, that the new deaths of the late 1980s, by virtue of all being diagnosed with AIDS, demonstrated that most or all of them occurred in people with hemophilia who were HIV positive. Since these deaths accounted for almost the entire new increase in mortality, it could be inferred that the mortality rate for HIV-negative people with hemophilia did not increase much in the 1980s, if at all (Harris, 1998).
7. Hemophiliacs from low- and high-risk behavior groups were equally infected from HIV-contaminated blood factor VIII concentrates.

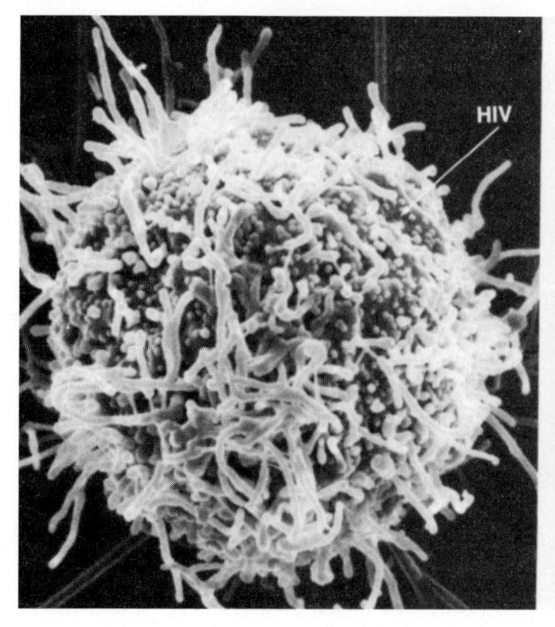

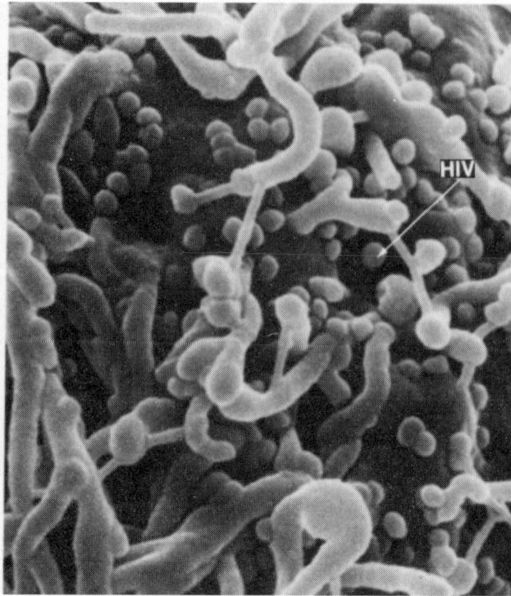

FIGURE 2-3 Viral Replication in Human Lymphocytes. Scanning electron micrograph of HIV-infected human T4 lymphocyte. **A.** A single cell infected with HIV showing virus particles and microvilli on the cell surface (magnified 7000 times). **B.** Enlargement of a portion of the mountain-like cell surface in **(A.)** showing multiple virus particles budding out of the cell surface (magnified 20,000 times). As each HIV exits the cell, it leaves a hole in the cell membrane. *(Photograph courtesy of K. Nagashima, Program Resources, Inc., NCI-Frederick Cancer Research and Development Center)*

8. With the exception of persons who had their immune systems suppressed due to genetic causes or by drug therapy, prior to the appearance of the virus, there were no known AIDS-like cases. The virus has been isolated worldwide—but only where there are HIV-positive people and AIDS patients.
9. An HIV-positive identical twin born to an HIV-positive mother developed AIDS, but the HIV-negative twin did not.
10. Only HIV-positive mothers transmit HIV into their fetuses and only these HIV-positive newborns progress to AIDS. HIV-negative newborns from HIV-positive mothers *do not* get AIDS!
11. Drugs developed specifically to inhibit the replication and/or maturation of HIV, thereby lowering the level of HIV found in HIV-infected people, have delayed the onset of HIV disease and, for HIV-infected pregnant women, have decreased the birth of HIV-infected infants by 66%.
12. If HIV does not cause AIDS, how do HIV dissenters explain the positive effects of drugs used to affect the early and late stages in the life cycle of HIV that have lowered viral load to unmeasurable levels in the blood of those with HIV disease and AIDS and have virtually restored life for those at death's door and who are now back at work? Or, how do they explain the positive effects offered by HIV phenotype- and genotype-resistant drug testing?
13. Finally, there have been numerous reports in the literature on HIV-infected individuals (homosexual, bisexual, and heterosexual) transmitting the virus to their sexual partners and both eventually dying of AIDS. *The unbroken chain of HIV transmission between prostitutes and their customers, between injection drug users sharing the same syringe, from infected mothers to their unborn fetuses, and so on all lead to the inescapable conclusion that HIV does cause AIDS.*

In short, Koch's Postulates (see Box 2.3) have been satisfied: HIV disease meets all four criteria.

1. The causative agent must be found in all cases of the disease. (It is.)
2. It must be isolated from the host and grown in pure culture. (It was.)
3. It must reproduce the original disease when introduced into a susceptible host. (It does.)

BOX 2.3

THE CAUSE AND EFFECT OF DISEASES: THE UNBROKEN CHAIN

For more than a century, four postulates set down by the German microbiologist Robert Koch have guided the hunt for disease-causing microbes. Koch argued in 1890 that to prove an organism causes a disease, microbiologists must (a) show that the organism occurs in every case of the disease; (b) that it is never found as a harmless parasite associated with another disease; (c) the organism must be isolated from the body and grown in laboratory culture; and (d) it must be introduced into a new host and produce the disease again.

Koch's criteria for cause and effect with respect to identifying a disease-causing agent (bacterium) are now being swept aside by new technology. Many of the microbes isolated in recent years by molecular techniques that pull segments of their DNA or RNA directly out of infected tissues *cannot be grown in culture.* That makes it impossible to fulfill all of Koch's postulates. As a result, many biologists are modifying Koch's strict requirements. Researchers can now build a convincing case against a microbe by examining a wide variety of molecular circumstantial evidence, such as how tightly the suspect microbe is associated with infected tissues and how closely the time course of the disease correlates with the amount of microbial genetic material present. But, as with Koch's methodology, the newer methodology for microbe identification with a specific disease also has its pitfalls.

4. It must be found in the experimental host so infected. (It is.)

In summary, HIV is the singular common factor that is shared between AIDS cases in gay men in San Francisco, well-nourished young women in Uganda, hemophiliacs in Japan, and children in Romanian orphanages. The identification of HIV as the causative agent of AIDS is now firmly accepted by scientists worldwide.

The Durban Declaration Admonishes HIV Dissidents—A statement signed by over 5,000 HIV/AIDS scientists and physicians from 50 countries and 5 continents was released to the press on July 1, 2000 (www. nature.com). Their statement that HIV causes AIDS is their answer to those who believe otherwise. They believe the evidence that HIV causes AIDS is clear, concise, exhaustive, and unambiguous. To conquer AIDS, everyone must understand that HIV is the enemy. (The Durban Declaration can be found at http://www.eurakalert.org/releases/hte-uncc063000.html. The evidence that HIV causes AIDS can be found at http://www.thebody.com/niaid/hivcauseaids.html.)

In short, the vast majority of HIV/AIDS scientists and medical personnel may feel as George Orwell stated in his book *1984*, "What can you do against the lunatic . . . who gives your arguments a fair hearing and then simply persists in his lunacy?" *HIV DOES CAUSE AIDS!*

DISCUSSION QUESTION: You have just read some of the evidence for and against HIV being the cause of AIDS. Assuming you agree with the vast majority of HIV/AIDS investigators worldwide that HIV does cause AIDS, do you think there comes a time at which dissenters should forfeit their right to make claims on other people's time and trouble by the poverty of their arguments and by the wasted effort and exasperation they have caused?
NOW discuss the value of the dissenter.
NOW discuss the danger of the dissenter's information or claims.

ORIGIN OF HIV: THE AIDS VIRUS

Clarification of the term "Origin of HIV." Scientists are searching for the source of HIV or HIV-like ancestor. Finding this source will give us the origin of HIV as it pertains to where and in which animal the virus was housed prior to entering humans. But it does not mean the beginning of the virus per se—that will most likely never be known.

Tracking the origins and early history of a newly recognized disease is more than just an academic exercise. Unless we understand where HIV came from we run the risk of new emergencies, and unless we understand the ecology that allowed it to spread, we will be unable to control newly identified diseases. A classic example of tracking a source of a disease and the local epidemic it caused is John Snow's investigation of the cholera epidemic in Golden Square, London, in 1854; his removal of the handle of the Board Street pump contained the outbreak.

More than virtually any other disease, AIDS has generated myths and far-fetched theories about its origin, its causes, and even its very existence. These are probably linked to fear and denial prompted by a virus that is fatal, incurable, and sexually transmitted—and can infect people for years before they show any signs of illness.

Why Do Scientists Want to Know Where HIV Originated?

The object of determining the origin of the AIDS virus is to gain insight into how the virus may have evolved the unique set of characteristics that enable it to destroy the human immune system. Such information will offer valuable clues as to how rapidly the virus is evolving and how to combat it and perhaps help prevent future viral plagues.

Is HIV a New or Old Virus?

The terms "new" and "old" are relative to time and age. Viruses known, say less than 50 years, are generally considered new. Whether a virus is considered new or old is of considerable importance.

If, for example, HIV is a new virus, say less than 30 years old, the many different varieties of HIV now infecting people worldwide probably evolved from a common ancestor sometime after World War II. New varieties can be expected to continue evolving at a frightening pace for several more decades, possibly producing new strains of the virus that are even more dangerous than those now infecting people. This could mean that vaccines now being developed based on current virus strains may not be useful in 10 to 20 years. But, should the known strains of the virus prove to be hundreds or thousands of years old, it might be possible that the current types of HIV are in a state of global balance and most likely they would not offer scientists any shocking evolutionary surprises in the future.

Some Ideas on the Origin of HIV: UFOs, Biological Warfare, Cats and Other Ideas

UFOs—Fear stimulates the imagination. Out of human fear have come some rather strange explanations for the origin of the AIDS virus. Early reports had unidentified flying objects (UFOs) crashing to Earth and releasing a "new organism" that would wipe out humanity.

Biological Warfare—There were frequent reports in the Soviet press linking AIDS with American biological warfare research. The Soviets agreed in August 1987 to stop these reports (Holder, 1988). There are also reports of extremism, as in the case of Illinois State Representative Douglas Huff of Chicago who told the *Los Angeles Times* that he gave over $500 from his office allowance fund to a local official of the Black Hebrew sect to help the group investigate its claim that Israel and South Africa created the AIDS virus in a laboratory in South Africa. Huff said AIDS is "clearly an ethnic weapon, a biological weapon" designed specifically to attack nonwhites (*CDC Weekly*, 1988). Kenyan ecologist and the 2004 Nobel Prize Winner, Wangari Maathal said in 2004 that HIV was created by scientists "for the purpose of mass extermination. We know that developed nations are using biological warfare, leaving guns to primitive people. AIDS is not a curse from God to Africans or the black people. It is a tool to control them designed by some evil-minded scientists." A few days after this statement, in another interview, she said "HIV was deliberately devised to destroy black people."

Domestic Cats—Still another myth to surface is that the AIDS virus came from domestic cats. Because of its similarities to human AIDS, feline immunodeficiency virus has been called "feline AIDS." The cat retrovirus may damage cats' immune systems leaving the animals vulnerable to opportunistic infections *or* it may cause feline leukemia. However, the cat virus has never been shown to cause a disease in humans.

Other Ideas—The origin of HIV has been attributed to HIV-contaminated polio, smallpox, hepatitis, and tetanus vaccines; the African green monkey; African people; their cattle, pigs, and sheep—and the United States CIA. With respect to the use of HIV-contaminated vaccines, a number of articles suggest that early monkey kidney cultures used to produce the polio vaccine carried HIV. Review of the literature offers *no* evidence that this occurred. The argument for the safety of the polio vaccine lies in the absence of any

AIDS-related diseases among the hundreds of millions of persons vaccinated worldwide (Koprowski, 1992).

Some Scientific Ideas on the Origin of HIV

Vanessa Hirsch and colleagues (1995) presented evidence that a virus isolated from a species of West African monkey, the Sooty Mangabey (an ash-colored monkey), may have infected humans 20 to 30 years ago. They believe this virus subsequently evolved into HIV-2. Hirsch et al. studied a virus known as the Simian Immunodeficiency Virus (SIVsm) that infects both wild and captive Sooty Mangabey monkeys. They molecularly cloned and sequenced the DNA of the virus and constructed an evolutionary tree of the several known primate immunodeficiency viruses. This tree showed SIVsm to be more closely related to HIV-2 than to HIV-1.

Gerald Meyers of Los Alamos National Laboratory states that SIVsm and HIV are so closely related that when HIV-2 is found in a human, it may be the Sooty Mangabey virus. However, HIV-1, which causes AIDS, does not sufficiently resemble HIV-2 or SIVsm, thus HIV-1 probably did not evolve from SIVsm/HIV-2. The prevailing theory is that humans were first infected through direct contact with precursor HIV-infected primates. The primate to human scenario is easier to accept than humans infecting primates.

Be a Virus, See the World

Albert Osterhaus of Erasmus University Rotterdam, The Netherlands, believes that *all* human viral infectious diseases ultimately have an animal origin, and natural transfer of these infections is a common event in animal populations.

Humans have hunted, handled, and even eaten primates for thousands of years. Recent laboratory accidents have shown that SIV can infect humans. Even though at the moment, no identifiable disease has been associated with the SIV/human infections, such accidents have demonstrated the potential for cross-species or zoonotic transmission of HIV-related viruses. Why not believe the same for the origin of HIV-1?

The possibility remains that HIV has been present but remained an obscure virus in the human population for a long time before it changed into a lethal agent.

The first recorded AIDS case in America was that of a 15-year-old male prostitute who demonstrated Kaposi's sarcoma and died in 1969. Frozen tissue samples contained HIV antibodies. These findings were reported at the Eleventh International Congress of Virology in August 1999.

The first documented case of AIDS in Europe was seen in a Danish surgeon who had worked in Zaire. She died in 1976.

The first documented case of AIDS in Africa occurred in 1959 (see Point of Information 2.1).

POINT OF INFORMATION 2.1

SCIENTISTS REPORT ON EARLIEST AIDS CASE

Scientists have pinpointed what is believed to be the earliest known case of AIDS—an African man who died in 1959. The scientists looked for signs of HIV in 1213 blood samples that were gathered in Africa between 1959 and 1982. They found clear signs of the virus in one taken from a Bantu man who lived in Leopoldville, Belgian Congo—what is now Kinshasa, Republic of Congo—in 1959. Scientists compared the genes from the 39-year-old sample of HIV with current versions of HIV. They realized that if they had an old sequence of HIV genes it would serve as a yardstick to measure the evolution of the current HIV. HIV has mutated over the years to form 11 distinct subtypes, lettered A through K. One of these, subtype B, is the dominant strain in the United States and Europe, while subtype D is most common in Africa. The family tree of HIV looks like a bush with the various subtypes forming the limbs. Scientists believe the 1959 HIV is near the trunk, around the point where the subtypes B and D branch off and that this virus is an ancestor to B and D. These data suggest that all HIV subtypes evolved from one introduction of HIV into people, rather than from many crossovers from animals to humans, as some have speculated.

However, the question scientists are trying to answer is why the disease took so long to start manifesting itself, ultimately hitting the homosexual and hemophiliac populations in the United States in the late 1970s. Current thinking is that HIV probably first infected rural areas of Africa, slowly moving into the cities and around the world until it hit homosexual communities where conditions were sufficient for rapid transmission of the disease.

The Current Theory on HIV Origin: A Chimpanzee

Beatrice Hahn and colleagues (1999; 2000; 2003) from the University of Alabama at Birmingham say they have gathered sufficient evidence to believe the origin of HIV-1 in humans to be cross-species transmission from a particular subspecies of chimpanzee—meaning that a simian virus closely related to HIV moved from monkeys to humans, and later mutated into its current form, following one of many interspecies transmissions (Figure 2-4).

Hahn presented three lines of evidence in support of their thesis. First, the genes of all four **SIVcpz** isolates cluster on evolutionary trees according to their subspecies or origin, either *Pan troglodytes troglodytes* from West Africa or *Pan troglodytes scheinfurthii* in East Africa. By sequencing parts of the virus's genetic material, the scientists found that more segments of the simian and human virus overlapped than had been identified in three previous simian viruses isolated in recent years from other chimpanzees. The investigation revealed that the other previously studied simian viruses were more closely related to very rare forms of HIV that don't account for most of the infections in the world.

Second, all known HIV-1 strains, including the M group that accounts for about 99% of all HIV-1 infections, as well as the O group and the N group, form a genetic cluster with the West African chimpanzee viruses. This clustering is also geographic and consistent with the likely equatorial Central African origin of HIV-1 (Figure 2-5). Hahn said she had initially been equally ready to accept the idea that chimps had gotten HIV from humans, rather than vice-versa. However, a third line of evidence convinced her that HIV-1 was introduced into the human population from at least three cross-species transmission from chimpanzees—she found evidence of genetic recombination among the SIVcpz strains of the troglodytes' lineage.

Two reports from the Eighth Annual Retrovirus Conference (2001) show that several monkey species carry HIV-like viruses. Eric Delacorte of the University of Montpelier, France, took blood samples from various West African monkey species that were sold as bush meat (meat from the African wilderness), as well as blood from animals kept as pets. Laboratory analyses of the viruses in those animals showed that about 17% of the specimens harbored SIVcpz strains, including four strains not seen before. In addition, scientists in the United States found evidence in chimpanzees that SIVcpz is found in animals in East Africa as well as West Africa, suggesting that not only is the virus widely distributed in Africa, but that continued hunting of the species for food and keeping the animals as pets could result in another outbreak of an AIDS-like disease. These data document, for the first time, that humans are continuously exposed to an unprecedented variety of SIVs through the consumption of bush meat (Santiago et al., 2002; Peeters et al., 2002).

The question that remains is, where did SIVcpz come from?

FIGURE 2-4 Photograph of *Pan troglodytes troglodytes*. This photograph was taken in the Gabon region of Africa where the sole species of Pan troglodytes is found. Their numbers are declining rapidly due to the "bush meat" trade—their slaughter and selling of their body parts. *(Photograph courtesy of Karl Ammann, wildlife photographer)*

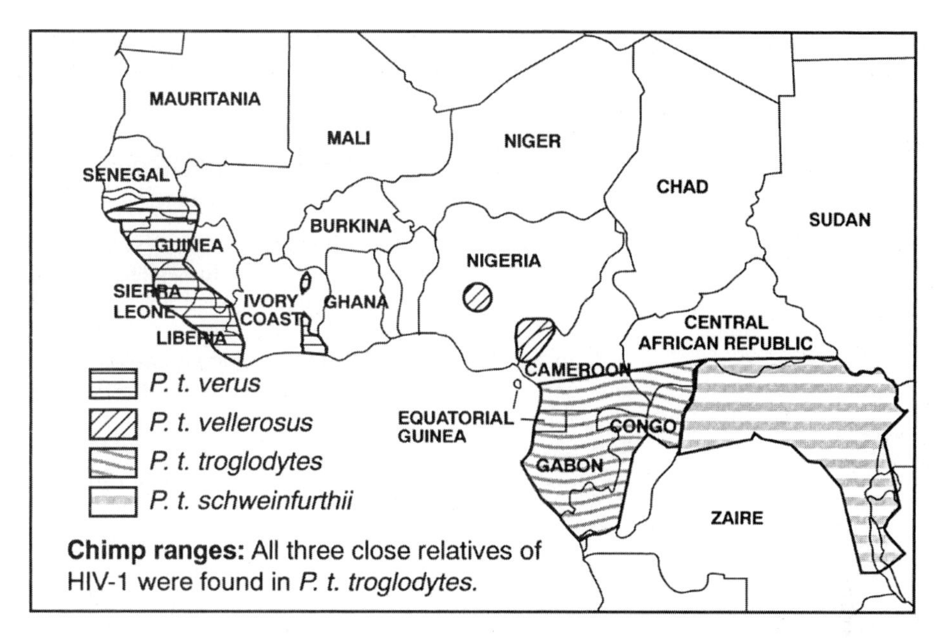

FIGURE 2-5 Geographic Ranges of Four Subspecies of Chimpanzee *Pan troglodytes.* All three close relatives of HIV-1, as determined by genetic sequencing of their genetic material, were found in *Pan troglodytes troglodytes. (Adapted from Hahn, 1999)*

Is the Search for the Origin of HIV Over?

It is nice to think that the issue of HIV origin has been solved. However, time will tell if Hahn's current effort will be the final word on the subject. Some questions still remain. Why, for example, does HIV-1 appear to be so benign in chimpanzees, as has been shown in numerous infection experiments over the years? For scientists, the new chimpanzee finding is just as much a beginning as an end. Although researchers may have the best evidence so far that HIV came from chimpanzees, no one can yet say how the virus became lethal to humans. Chimpanzees share over 98% of the genes that exist in humans, yet they don't get AIDS. So which of the remaining genes protect chimpanzees from an infection that has turned out to be almost universally fatal in people? Also, researchers are still investigating how, after taking root in just a few people, HIV gradually traveled all over the world.

How Often Did SIVcpz Cross Over into Humans?—Researchers now believe that AIDS-like viruses moved from chimp to human more than once, creating different strains of HIV. This means a vaccine against one strain of HIV may not control a new epidemic. As it turned out, *Pan troglodytes troglodytes* live in the region of Africa where HIV-1 was first recognized. Since there are three separate kinds of HIV-1, the scientists believe that HIV-1 crossed into people from chimps at least three times. Hahn said that SIV appears to have dwelled in primates for hundreds of thousands of years before turning into the deadly human virus, HIV. Hahn believes hunting, which exposes people to excessive amounts of blood during slaughter, allowed precursor HIV to infect humans.

Why Do Chimpanzees Resist HIV Infection?—With regard to the question of why chimpanzees resist HIV infection, in late 2002 Dutch investigators presented a theory that an AIDS-like epidemic killed large populations of chimps about 2 million years ago. The survivors of this viral (SIVcpz-like?) eradication were selected for survival because they carried the genes necessary to resist the viral infection. Over the years, this genetic selection gave rise to chimpanzees that are resistant to viruses closely related to HIV, for example SIVcpz and HIV itself.

UPDATE 2004 If humans become infected with SIVcpz from chimpanzees, where or how did the chimpanzees get SIVcpz? Elizabeth Bailes and colleagues (2003) believe that SIV-infected monkeys gave rise to SIVcpz through monkey to monkey cross-species SIV transmission and recombination among slightly genetically different SIV to produce SIVcpz. Because chimpanzees feed on these monkeys, they become SIVcpz-infected, much like humans who consumed the SIVcpz-infected chimpanzees. Thus, chimpanzees and humans acquired their versions of the pre-HIV virus the same way—by killing and eating animals infected with similar viruses.

Where Did HIV Begin to Circulate Among Humans?—According to Nicole Vidal and colleagues (2000), because of the unprecedented degree of HIV-1 group M genetic diversity found in the Democratic Republic of Congo (9 of the 11 different group M subtypes), the HIV-1 pandemic must have originated in Central Africa.

Daniel Vangroenweghe (2001) states that the earliest cases of HIV infection and AIDS in the 1960s and 1970s occurred in Congo-Kinshasa (Zaire), Rwanda, and Burundi. These countries appear to be the source of HIV group M epidemic, which then spread outward to neighboring Tanzania and Uganda in the East, and Congo-Brazzaville in the West. Further spread to Haiti and onward to the United States can be explained by the hundreds of single men from Haiti who participated in the UNESCO educational program in the Congo between 1960 and 1975.

When Did HIV Begin to Circulate Among Humans?—Scientists do have some idea when the virus began to circulate among people. From looking at samples of HIV taken at different times and in different parts of the world, researchers have constructed a type of genetic clock for HIV-1. The speed of the clock is determined by how much the virus changes over time. A key to setting this clock came with the discovery of the oldest known HIV infection, found in the man who lived in what is now Congo in 1959. That year, he was one of a number of Africans who had given a blood sample as part of a study of the immune system. After the AIDS epidemic began, the frozen samples were thawed and screened for the presence of HIV.

Timing the Crossover: From Chimpanzees into Humans

By comparing HIV-1 samples taken over time, Steven Wolinsky and his colleagues determined that HIV began to spread within decades before the 1959 sample was drawn (Balter, 1998; Korber et al. 1998). The estimate is based on the assumption that the rate of change for HIV has remained about the same. This would place the crossover into humans around 1924 to 1946. Studies have also indicated that HIV-2 began to spread about the same time.

From the Seventh Conference on Retrovirus and Opportunistic Infections—February 2000, Bette Korber from Los Alamos National Laboratory presented data attempting an exact estimate as to when the virus SIVcpz from *Pan troglodytes* crossed into humans. Summarizing months of work involving incredibly complicated mathematical modeling of thousands of primate lentivirus gene sequences on a Los Alamos supercomputer that can perform 1 trillion computations/second, Korber said the timing of the origin of the HIV-1 M group (from which all other HIV species have evolved) is near 1930, with the 95% CIs (confidence intervals) spanning from 1910 to 1950. That is, this pandemic got started when the precursor SIVcpz crossed into one or a small group of humans at that time. More recently Korber and colleagues reported that the best estimate for the date when the common ancestor of HIV-1 M group came into existence was between 1915 and 1941. Molecular clock analysis provided the date of 1931 (Korber et al., 2000). This report states that the HIV-1 M group had a *single origin*—from one species of animal. The most common form of HIV in America, M sub-group B is believed to have evolved between 1960 and 1971 with 1967 being the most likely year. Kenneth Robbins and colleagues (2003) estimate 1968 as the date of origin of Americans' HIV subtype B, roughly a decade before the earliest documented U.S. infection in 1977. Because the median clinical latency from HIV infection to AIDS is typically 10 years, the time interval of the introduction of HIV to the first recognition of AIDS in 1981 is consistent with the long latency period for the virus. Ten years is only the median latency period, so AIDS cases in the early 1970s likely existed. AIDS cases are difficult to confirm retrospectively, yet there are a few possible and probable cases from Haiti and the United States in the early 1970s.

Twice This Century!

Preston Marx of the Aaron Diamond AIDS Research Center said, "Never before in thousands of years of exposure to SIV did these viruses cross over to cause (widespread) disease in people, and yet it happened twice this century. . . . Not only did this happen in chimpanzees, it happened in Sooty Mangabeys 1,000 miles away." Many scientists have speculated about why the virus might have become amplified in post-World War II Africa. Among the possibilities: the exploding population of urban centers; social upheaval and commerce that led to migration across the continent; and widespread vaccination campaigns that could have reused needles. In the end, HIV's early accomplices aren't known for sure. Nor is it clear how HIV/AIDS came to the United States.

Continued Evolution of HIV in Humans?— Possible *good* news (?) is that HIV would, after killing millions of humans, become a harmless passenger as most likely its precursor SIV did thousands of years ago in chimpanzees. The bad news would be that HIV is evolving into a virus that is more easily spread and/or that it becomes more lethal. Time will tell.

SUMMARY ON THE ORIGIN OF HIV

In summary, there are at least three ideas on the possible origins of the AIDS virus: (1) It is a human-made virus, perhaps from a germ warfare laboratory; (2) it originated in the animal world and crossed over into humans; and (3) HIV has existed in small isolated human populations for a long time and, given the right set of conditions, it escaped into the larger population. Computer modeling of DNA sequence in HIV and SIVsm and the recent work of Beatrice Hahn and colleagues suggests that HIV evolved within the last 100 to 300 years. So for now, the question remains: Is AIDS a new disease or an old disease that was late being recognized—so late that we will never know its true source, the origin of HIV? Continued investigations may answer these questions.

Has HIV Always Caused a Lethal Disease—AIDS?

An additional question to where HIV came from is whether it has always caused disease. From the study of human history, as it relates to human disease, scientists have numerous examples that show that as human habits change, new diseases emerge. Regardless of whether HIV is old or new, history will show that social changes, however small or sudden, have most likely hastened the spread of HIV. Increased rounds of HIV replication presented humans with new HIV mutations, some of which were to become lethal. In the 1960s, war, tourism, and commercial trucking forced the outside world on Africa's once isolated villages. At the same time, drought and industrialization prompted mass migrations from the countryside into newly teeming cities. Western monogamy had never been common in Africa, but as the French medical historian Mirko Grmek notes in his book, *History of AIDS* (1990), urbanization shattered social structures that had long contained sexual behavior. Prostitution exploded, and venereal diseases flourished. Hypodermic needles came into wide use during the same period, creating yet another mode of infection. Did these trends actually turn a chronic but relatively benign infection into a killer? The evidence is circumstantial, but it's hard to discount.

Can Humans Be Creating Scenarios for the Introduction of New Diseases?

Whatever the forces are that brought us HIV and AIDS, they can surely bring other diseases. By encroaching on rain forests and wilderness areas, humans are placing themselves in ever-closer contact with animal species and their deadly parasites. Activities, from irrigation to the construction of dams and cities, can expose humans to new diseases by expanding the range of the rodents or insects that carry them. Stephen Morse, a Rockefeller virologist, studies the movement of microbes among populations and species. He worries that human activities are speeding the flow of viral traffic. More than a dozen new diseases have shown up in humans since the 1960s, nearly all of them the result of once-exotic parasites exploiting new opportunities. (Certain of these new disease-causing agents, for example, the Ebola virus, are discussed in Box I.1 on page 8.)

SUMMARY

The AIDS virus was discovered and reported by Luc Montagnier of France in 1983. Identifying the virus that caused the immunosuppression that caused AIDS allowed for AIDS surveillance definitions that began in 1982. The recent recognition of non-HIV AIDS cases is not unexpected and can be explained. Presently, there is no new threat of another AIDS-causing biological agent.

The recent work of Beatrice Hahn and colleagues may have pinpointed the reservoir of a precursor HIV-like virus in the chimpanzee, *Pan troglodytes troglodytes.* The work of Bette Korber and colleagues may have set the time frame in which precursor HIV crossed into humans.

REVIEW QUESTIONS

(Answers to the Review Questions are on page 438.)

1. What may be the strongest nonlaboratory evidence for saying that AIDS is caused by an environmental agent?
2. Where might HIV have originated and where did the first HIV infections appear?
3. HIV/AIDS Word Search

C	S	M	O	T	P	M	Y	S	M	O	D	N	O	C
T	I	E	A	F	R	I	C	A	N	C	E	R	M	P
H	M	T	R	A	N	S	M	I	S	S	I	O	N	B
E	I	Y	S	P	B	E	H	A	V	I	O	R	A	L
R	A	C	D	I	O	H	C	R	A	E	S	E	R	O
A	N	O	I	O	N	L	O	I	L	O	P	P	O	O
P	I	H	S	P	B	U	I	Y	H	I	V	L	D	D
Y	A	P	E	N	O	I	T	C	E	F	N	I	E	V
E	R	M	A	L	E	S	T	R	I	S	K	C	N	M
K	T	Y	S	Z	E	G	I	N	O	E	K	A	T	O
N	S	L	E	F	T	C	A	T	A	P	S	T	S	T
O	D	B	I	G	E	N	E	T	I	C	P	I	O	H
M	I	L	S	S	E	N	L	L	I	V	V	O	H	E
L	A	U	X	E	S	P	J	W	L	V	E	N	R	R
T	E	J	E	N	U	M	M	I	N	S	E	C	T	S

AIDS
AZT
AFRICA
ANTIBODY
BEHAVIORAL
BLOOD
CANCER
CELLS
CONDOMS
DISEASE
GENETIC
HIV
HOST
ILLNESS
IMMUNE
INFECTION
INSECTS
LIFESTYLE
LYMPHOCYTE
MONKEY
MOTHERS
NEGATIVE
OPPORTUNISTIC
POLICIES
POLIO
POSITIVE
REPLICATION
RESEARCH
RISK
RODENTS
SEXUAL
SIMIAN
SIV
STRAIN
SYMPTOMS
THERAPY
TRANSMISSION

3 Biological Characteristics of the AIDS Virus

CHAPTER CONCEPTS

- ♦ Retroviruses are grouped into three families: oncoviruses, lentiviruses, and foamy viruses.
- ♦ HIV is a lentivirus.
- ♦ HIV contains nine genes; its three major structural genes are GAG-POL-ENV.
- ♦ HIV contains 9749 nucleotides, its genetic code.
- ♦ Six HIV genes regulate HIV reproduction and at least one gene directly influences infection.
- ♦ HIV RNA produces HIV DNA, which integrates into the host cell to become proviral DNA.
- ♦ HIV undergoes rapid genetic changes in infected people.
- ♦ The reverse transcriptase enzyme is very error prone.
- ♦ HIV causes immunological suppression by destroying T4 helper cells or CD4+ cells.
- ♦ HIV is classified into major (M), outlier (O), and type (N) genetic subtypes. Type M is responsible for 99% of HIV infections worldwide.

VIRUSES NEED HELP

Viruses are microscopic particles of biological material, so small they can be seen only with electron microscopes. A virus consists solely of a strip of genetic material (nucleic acid) within a protein coat.

Viral genomes (DNA and RNA that contain the genetic information to replicate) are very small and contain few genes compared to living cells. With their limited coding capacity, viruses must borrow or hijack host or cell proteins to complete their replication. The interactions between a virus and its host have been shaped by long stretches of coevolution, and the borrowed proteins do not necessarily serve the same function in viral replication as they serve in the host. Identification of these borrowed

SIDEBAR 3.1

HOW HIV ESCAPES THE INFECTED CELL

This is a primary example of how a virus, in this case, HIV, uses living cells to achieve its mission—to infect, replicate, and exit a cell. In October 2001, Jennifer Garrus and colleagues reported on how HIV takes over the human cell's normal processes in order to leave a cell. This discovery could lead to new drugs to control HIV disease in those who are HIV-infected. In a key part of their study, scientists crippled the cells' machinery by silencing a gene that normally makes the protein (Tsg101) necessary for HIV to escape or bud out of the cell. Thus, clusters of connected HIV particles were stuck at the cell membrane and could not get out. The new study showed the Tsg101 protein within cells acts as a key link to the budding process. One part of Tsg101 protein connects to an HIV protein while another part of the Tsg101 protein links to other proteins within a cell—proteins the virus uses to exit the cell. Shutting down the means by which HIV leaves the cell means that other susceptible cells will not become infected. Blocking Tsg101 protein after a person becomes infected could keep the virus from spreading to other cells. These scientists speculate that many other viruses may use the same exit pathway. After 24 years, scientists still do not know just how HIV kills cells bearing a CD4 protein and scientists still do not know why T4 or CD4+ cells fall from about 1000 per microliter of blood to less than 100 per microliter of blood over a 10-year period while viral loads may remain constant or show an enormous increase over a short time interval. And in many people, the relationship of viral load to the rate of their disease progression is questionable at best.

host proteins, which perform essential functions during viral replications, is important for understanding virus-host interactions, which influence the host range and virulence of the virus.

VIRUSES ARE PARASITES

Viruses are parasitic agents; they live inside the cells of their host animal or plant, and can reproduce themselves *only by forcing the host cell to make viral copies.* The new virus leaves the host cell and infects other similar cells. By damaging or killing these cells, some viruses cause diseases in the host animal or plant. Genetically, viruses are the simplest forms of "life-like agents"; the genetic blueprint for the structure of the **Human Immunodeficiency Virus (HIV)** is 100,000 times smaller than that contained in a human cell. The complete sequence of 9749 nucleotides that form the genetic code for HIV have been identified and their arrangement sequenced (mapped).

Scientists have produced a great deal of information about HIV over a relatively short time. In the history of medical science, the immediate involvement of so many scientists followed by the rapid identification of the causative agent of AIDS is equaled only by the rapid identification of the SARS virus in 2003. More is known about HIV than about the viruses that cause such long-standing human diseases as polio, measles, yellow fever, hepatitis, flu, and the common cold. Humankind is fortunate that HIV began spreading through the human population as a pathogen in the mid- to late-1970s. By then scientists had discovered and begun to exploit the molecular aspects of biology. Molecular methodologies necessary to begin the immediate molecular study of HIV were in place to define and refine our knowledge of viruses and, in particular, to learn about HIV.

RETROVIRUSES

Viruses that belong to the Retroviridae family have a Janus-like relationship with modern medicine. Some members of the family have proven to be valuable tools in molecular biology research; their ability to import foreign genes into cells has been explored as a possible vehicle for gene therapy. Retroviruses also are potent disease agents,

---

POINT OF INFORMATION 3.1

VIRAL SPECIFICITY

1. **Viruses are very specific with regard to the types of cells they can enter/reproduce.** Not all viruses can attach to or enter all cells. Humans survive in a world full of viruses that *only* enter or reproduce in a variety of bacteria, protozoa, fungi, and higher forms of plant and animal life. It appears that most of these viruses are harmless to humans—if they enter the body they cannot reproduce in human cells—they do not cause human damage. For those viruses that do enter human bodies from animals such as pigs, chickens, rabbits, mice, cows, monkeys, etc. that cause a human disease, most are very specific as to which cells in a human body they can enter/reproduce and cause damage. For example, the flu virus enters the human respiratory tract cells. Epstein-Barr virus infects cells in the nose and throat. The hepatitis viruses enter liver cells, but each of the hepatitis viruses causes a different degree of human cell damage over time. Some cause a more immediate disease—for example, hepatitis A virus—while hepatitis B or hepatitis C may not cause significant cell damage for years. Polio virus enters cells of the human nervous system that are different from those cells of the nervous system that herpes virus invades. HIV enters and reproduces in cells of the human immune system.
2. **Why do different viruses enter specific cell types?** Each of the viruses that causes human disease does so by finding a cell type that carries a receptor molecule (a protein or a protein attached to a sugar molecule) that *fits* with a projection of a surface molecule of a given virus—much like the key to a lock. That's why specific viruses are known to be associated with certain types of human tissue. For example, there are viruses that attach only to the receptors of heart, gut, eyes, throat, liver, and other specific human cell tissues. Find a way to block either the human cell receptors (CD4, CXCKR-4, and CCKR-5—to which HIV attaches), without harming the cell or block the given viral receptor (in HIV its gp120) and one has a therapy for the given viral disease.

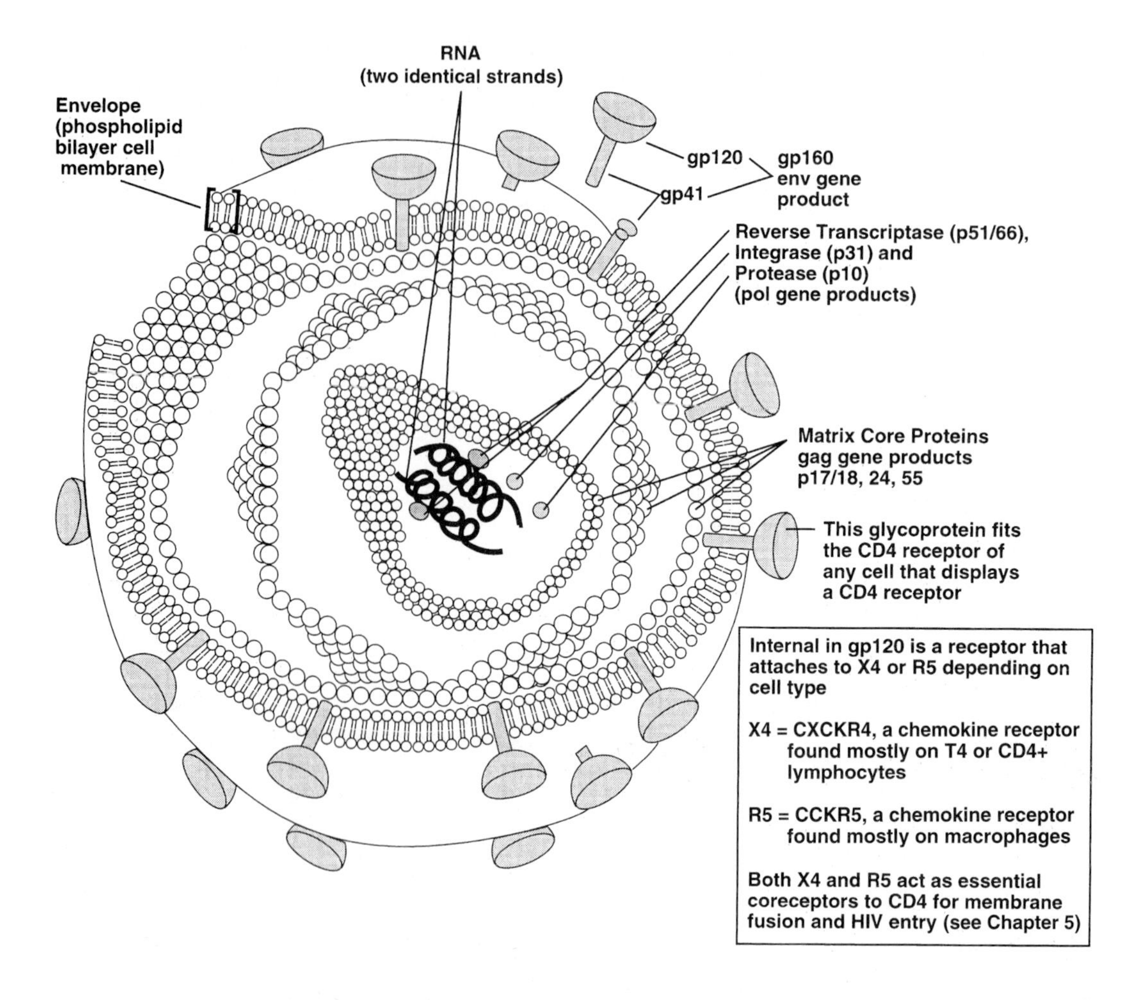

associated with arthritis, systemic lupus erythematosis and a form of leukemia in adult humans. Members of this family also infect chimpanzees and other simian species (Simian Immunodeficiency Viruses or SIVs), and they infect and cause a deadly immunodeficiency syndrome in humans: AIDS.

Data that researchers have gathered during years of work comparing the genetic make-up of SIVs and HIV have shown that while primate species infected with SIVs are the long-time natural hosts to those viruses, humans are not the natural hosts of HIV.

There are three subfamilies of retroviruses, two of which are associated with human disease: **oncoviruses** (cancer-causing) and **lentiviruses** or slow viruses of which HIV-1 and HIV-2 are members. **Spumavirus** is the third group but is not known to be associated with human disease. As a lentivirus, HIV has genetic and morphologic similarities to other animal lentiviruses such as those infecting cats (Feline Immunodeficiency Virus), sheep (visna virus), goats (caprine arthritis-encephalitis virus), and nonhuman primates (Simian Immunodeficiency Virus or SIV).

The retroviruses and in particular the lentiviruses have presumably been present for thousands of years. Modern sociodemographic changes and other human factors have allowed rapid mutating (changing) lentiviruses, such as HIV, to find new niches and host populations in which to reproduce. Now let us review some of the characteristics of HIV, the virus that causes AIDS.

HUMAN IMMUNODEFICIENCY VIRUS (HIV): ITS LIFE CYCLE

HIV is a retrovirus (Figure 3-1). Retroviruses are so named because they *REVERSE* the usual flow of genetic information within the host cell in

FIGURE 3-1 Human Immunodeficiency Virus. It infects cells by a process of membrane fusion that is mediated by its envelope glycoproteins (gp120-gp41, or Env) and is generally triggered by the interaction of gp120 with two cellular components: CD4 and a coreceptor belonging to the chemokine receptor family (CXCKR-4 or X4 and CCKR-5 or R-5). The virus is a sphere measuring 1000 Å or 1/10,000 mm in diameter. The cone-shaped core in a spherical envelope is the dominant feature. In this diagram the virus has been sectioned to better visualize its internal structure. The membrane of HIV is derived from the host cell. HIV gains the membrane while "budding" out or exiting the cell. Each free HIV leaves a hole in the cell membrane. The membrane, acquired from its host cell, consists of two lipid (fat) layers impregnated with some human proteins, for example Class I and Class II human lymphocyte antigen complexes important for controlling the immune response. The external viral membrane also contains molecules of viral glycoproteins **(gp)**—a sugar chain attached to protein. Each glycoprotein appears as a spike in the membrane. Each spike consists of two parts: **gp41,** which contains a coiled up protein and extends through the membrane. On interaction between HIV envelope and T4 cell coreceptors (CD4/chemokine receptors X4 and R-5—depending on cell type) the gp41 coiled protein is unsprung and like a harpoon, pierces the cell membrane initiating the first step in HIV replication. **gp120** extends from the end of gp41 to the outside and beyond the membrane (the numbers 41 and 120 represent the mass of the individual gps in thousands of daltons). As a complete unit, gp41 plus gp120 is called **gp160.** These two membrane or **envelope** proteins play a crucial role in binding HIV to CD4 protein molecules found in the membranes of several types of immune system cells. The gp160 precursor is cleaved into envelope (gp120) and transmembrane (gp41) proteins in the cell's Golgi compartment. The HIV envelope complex is transported via vesicles to patches in the outer cell membrane. Full-length HIV RNA is complexed with capsid proteins and the nucleocapsid is transferred to the cell surface membrane at envelope-containing sites. The binding of gp to CD4 receptors makes such immune system cells vulnerable to infection. Other HIV proteins are located and described in this figure.

Within the cone-shaped core there are two identical strands of viral genomic RNA, each coupled to a molecule of transfer RNA (tRNA) that serves as a primer for reverse transcription of viral RNA into viral DNA. Also present with the RNA are an **integrase,** a **protease,** and a **ribonuclease enzyme.** The released virus is processed internally by HIV protease to form the characteristic dense lentivirus core. Most HIV appear to have initiated DNA synthesis prior to completion of budding and maturation. Actual maturation of HIV takes place after it buds out of the cell (see Figure 4-9).

---

order to reproduce themselves. In all living cells, normal gene expression results from the genetic information of DNA being copied into RNA (Figure 3-2). The RNA is translated into a specific cellular protein. In all living cell types, the directions for protein synthesis come from the species' genetic information contained in its DNA:

DNA $\rightarrow$ RNA $\rightarrow$ Protein Synthesis

THE HIV LIFE CYCLE

Understanding how HIV works inside the human cell gives scientists important clues about how to attack it at its most vulnerable points. Knowing the secrets of how the virus functions and reproduces itself—a process called its life cycle—can help scientists design new drugs that are more effective at suppressing HIV and have fewer side effects. For people with HIV, knowing how HIV works can make it easier to understand the way drugs work in their bodies. Retroviruses have RNA as their genetic material —Not DNA! In brief, retrovirus RNA is copied, using its reverse transcriptase enzyme, into a complementary single strand of DNA (Figures 3-3, 3-4). The single-strand retroviral DNA is then copied into double-stranded retroviral DNA (this replication occurs in the cell's cytoplasm). At this point the viral DNA has been made according to the instructions in the retroviral RNA. This retroviral DNA migrates into the host cell nucleus and becomes integrated (inserted) into the host cell DNA. A recent study indicates that the viral DNA integrates only into transcriptionally active genes. The integrated viral DNA is now a **provirus**. From this point on, the infection is irreversible—*the viral genes are now a part of the cell's genetic information.* In this respect, HIV can be considered as an acquired dominant genetic disease! A provirus, like the "mole" in a John LeCarre spy novel, may hide for years before doing its specific job. But for HIV, there is evidence that in some human cells the provirus begins to produce new copies of HIV RNA immediately after becoming a provirus or shortly thereafter.

Central Dogma

Prior to 1970, cell biologists thought that genetic information flowed only in one direction:

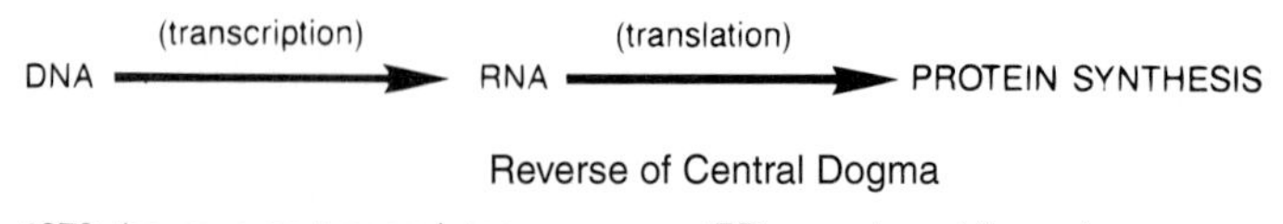

In 1970 the reverse transcriptase enzyme (RT) was found in a virus.
These viruses became known as retroviruses.

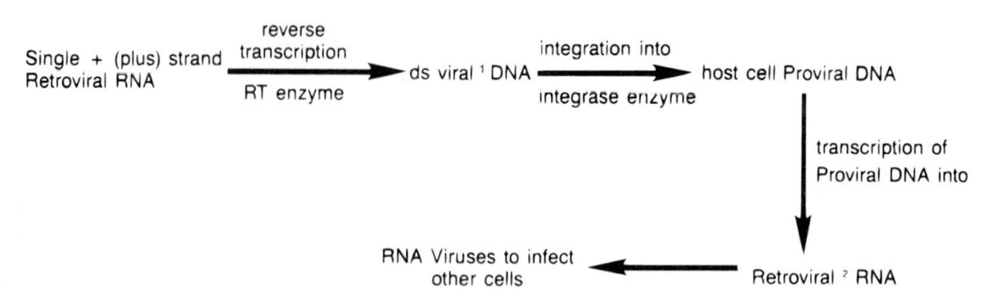

[1] ds = Double stranded DNA

[2] If the RNA transcripted is spliced, RNA base sequences are rearranged. This RNA = messenger RNA and is used to make retroviral protein. If RNA transcripted is not spliced, it becomes the RNA genome of the new virus.

FIGURE 3-2 Retroviral Flow of Genetic Information. The general directional flow of genetic information in all living species is from DNA, where the information is stored, into RNA, which serves as a messenger for the construction of proteins that are the cells' functional molecules. This unidirectional flow of genetic information has been referred to as the **"central dogma"** of molecular biology. In the 1960s, Howard Temin and colleagues discovered an enzyme that copied RNA into DNA, a reverse of what was normally expected, thus the name *reverse transcriptase.*

Before the HIV provirus's genes can be expressed, RNA copies of them that can be read by the host cell's protein-making machinery must be produced. This is done by **transcription.** Transcription is accomplished by the cell's own enzymes. But the process cannot start until the cell's RNA polymerase is activated by various molecular switches located in two DNA regions near the ends of the provirus: the **long terminal repeats.** This requirement is reminiscent of the need of many genes in multicellular organisms to be "turned on" or "off" by proteins that bind specifically to controlling sequences.

Production of Viral RNA Strands or RNA Transcripts

Within the host cell nucleus, proviral DNA, when activated, produces new strands of HIV RNA. Some of the RNA strands behave like messenger RNA (mRNA), producing proteins essential for the production of HIV. Other RNA strands become encased within the viral core proteins to become the new viruses. Whether the transcribed RNA strands become mRNA or RNA strands for new viruses depends on whether or not the newly synthesized RNA strands undergo complex processing. RNA processing means that after the RNA is produced, some of it is cut into segments by cellular enzymes and then reassociated or **spliced** into a length of RNA suitable for protein synthesis. The RNA strands that are spliced become the mRNA used in protein synthesis. The unspliced RNA strands serve as new viral strands that are encased in their protein coats (capsids) to become new viruses that bud out of the cell (Figure 3-4).

Two distinct phases of transcription follow the infection of an individual cell by HIV. In the first or early phase, RNA strands or transcripts produced in the cell's nucleus are snipped into multiple copies of shorter sequences by cellular splicing

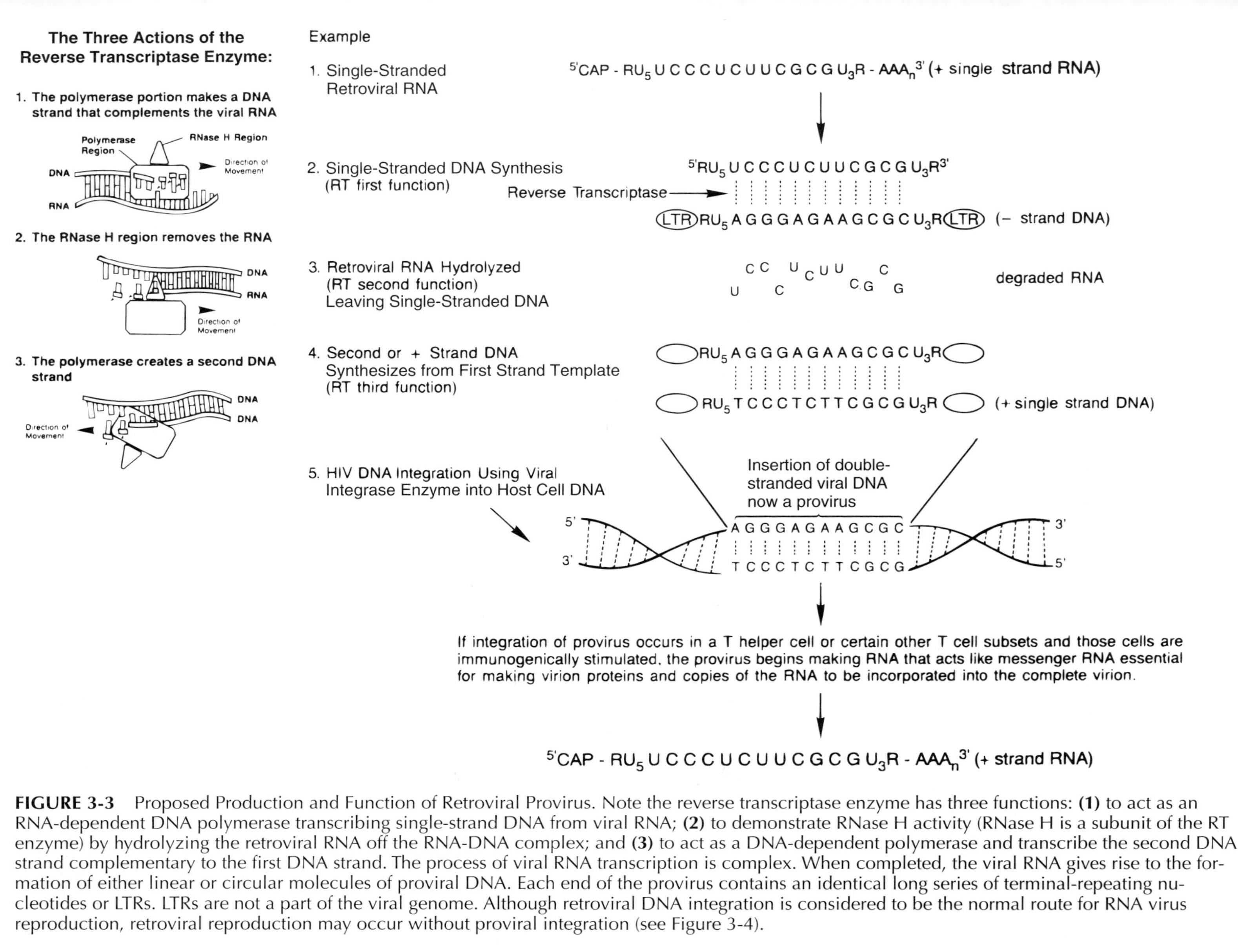

FIGURE 3-3 Proposed Production and Function of Retroviral Provirus. Note the reverse transcriptase enzyme has three functions: **(1)** to act as an RNA-dependent DNA polymerase transcribing single-strand DNA from viral RNA; **(2)** to demonstrate RNase H activity (RNase H is a subunit of the RT enzyme) by hydrolyzing the retroviral RNA off the RNA-DNA complex; and **(3)** to act as a DNA-dependent polymerase and transcribe the second DNA strand complementary to the first DNA strand. The process of viral RNA transcription is complex. When completed, the viral RNA gives rise to the formation of either linear or circular molecules of proviral DNA. Each end of the provirus contains an identical long series of terminal-repeating nucleotides or LTRs. LTRs are not a part of the viral genome. Although retroviral DNA integration is considered to be the normal route for RNA virus reproduction, retroviral reproduction may occur without proviral integration (see Figure 3-4).

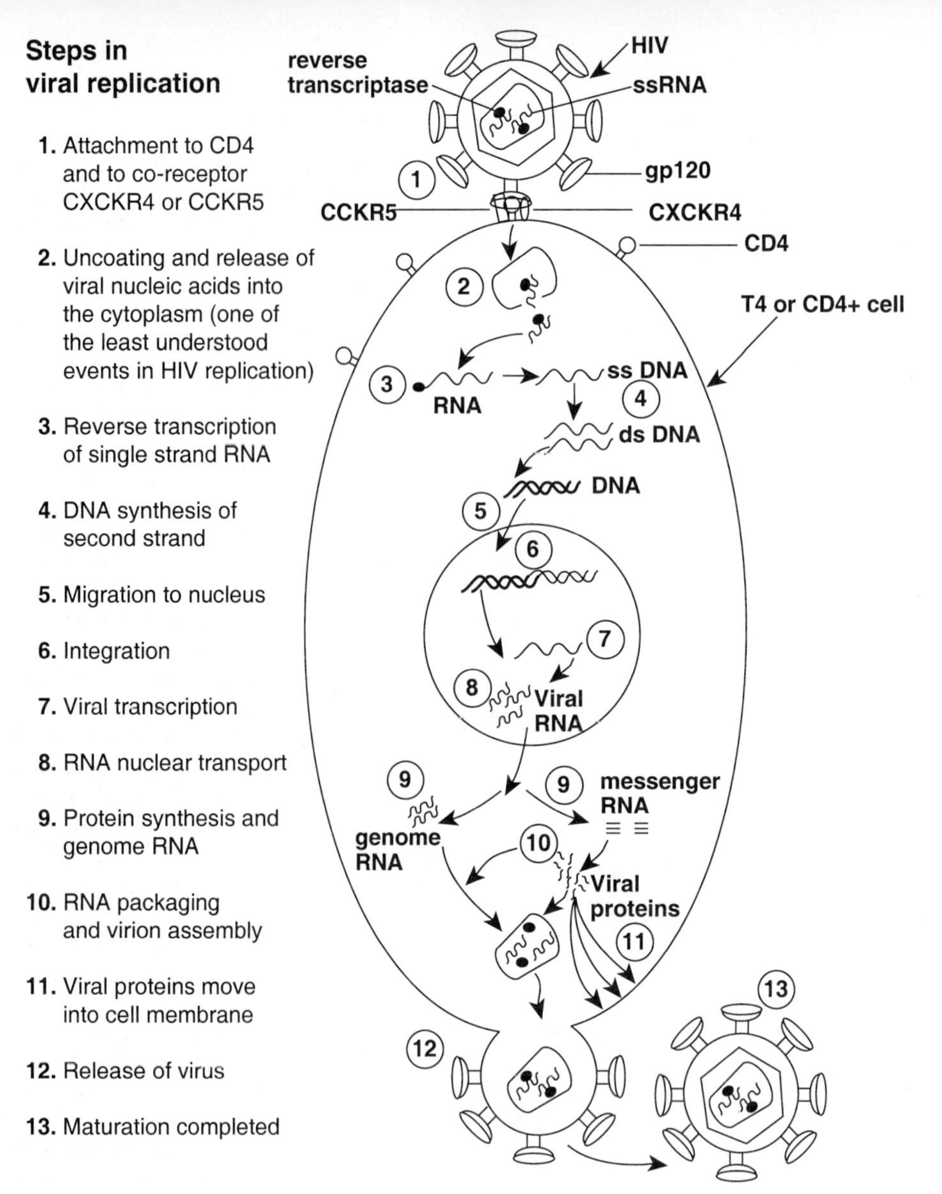

FIGURE 3-4 Life Cycle of HIV in a T4 or CD4+ Lymphocyte. On average, the life cycle of HIV in an infected T4 cell is about two days. This means that one HIV, in one person in one year can produce about 180 generations of HIV. The lifetime of HIV in the blood is about six hours. After fusion of viral and cellular membranes, the inner part of HIV, called the **core,** is delivered into the cell cytoplasm. Uncoating of the core occurs, releasing two identical RNA strands, accompanying structural proteins and enzymes necessary for HIV replication. HIV RNA is then copied into HIV DNA, which is then transported into the cell's nucleus for insertion into the host cell's DNA. This feature of the virus life cycle is essential for the spread of HIV in vivo, because it allows infection of nondividing cells such as monocytes and terminally differentiated macrophages and dendritic cells. There is still some confusion as to whether HIV becomes a latent infection once HIV DNA becomes inserted or integrated into the host DNA. Evidence indicates that whether HIV is latent depends on the tissue that one is investigating. For example, in the T4 lymphocytes within the lymph nodes HIV is constantly being replicated, while some T4 lymphocytes in the blood carry HIV in the latent state. Depending on cell type, T4 or macrophage, chemokine coreceptors CXCR-4 (FUSIN) and/or CCCKR-5 are used by HIV to enter T4 and macrophage cells respectively. (See Chapter 5 for details on chemokine receptors and HIV cell entry.)

enzymes. When they reach the cytoplasm, they are only about 2000 nucleotides in length. These early-phase short transcripts encode only the virus's **regulatory proteins;** the structural genes that constitute the rest of the genome are among the parts that are left behind. In the second or late phase, two new size classes of RNA—long (unspliced) transcripts of 9749 nucleotides making up the new viral genome and medium-length (singly spliced) transcripts of some 4500 nucleotides—move out of the nucleus and into the cytoplasm. The 4500 nucleotide transcripts encode HIV's structural and enzymatic proteins (Greene, 1993).

Experimental results reported by Somasundaran et al. (1988) showed that when lymphoid cell lines or peripheral blood lymphocytes were infected with a laboratory strain of HIV, up to 2.5 million copies of the viral RNA were produced by cells; and within three days of infection, up to 40% of the total protein synthesized by the cells was viral protein. This is an unprecedented takeover for a retrovirus that typically makes only modest amounts of RNA and protein.

Much of what HIV does after entering the host cell or while integrated as a retroprovirus depends on the activity of its genes.

BASIC GENETIC STRUCTURE OF RETROVIRAL GENOMES

The first retrovirus was isolated from a sarcoma (a cancer) in chickens by Peyton Rous in 1911 and named the Rous sarcoma virus. The basic genetic structure of the Rous sarcoma virus and all animal retroviruses is the same. They all contain retroviral RNA sequences that code for the same three genes abbreviated GAG, POL, and ENV (defined here). Flanking each end of the retroviral genome is a sequence of similar nucleotides called long terminal repeats or redundancies (LTRs).

$$5'\underset{\text{LTR}}{=}\overline{\quad\text{GAG POL ENV}\quad}\underset{\text{LTR}}{=}3'$$

Some of the animal retroviruses such as the Rous sarcoma virus contain an additional **oncogene** (ONC) that, along with its LTRs, causes a rapid form of cancer in chickens that kills them in one to two weeks after infection. Without the **onc** gene the virus causes a slow progressive cancer.

Retroviral Genome of HIV

What sets the HIV genome apart from all other known retroviruses is the number of genes in HIV and the apparent complexity of their interactions in regulating the expression of the GAG-POL-ENV genes (Figure 3-5).

The Nine Genes of HIV—The HIV genome contains at least nine recognizable genes that produce at least 15 individual proteins. These proteins are divided into three classes: (1) GAG, POL, and ENV, the three major **structural** proteins; (2) Tat and Rev, the two **regulatory** proteins; and (3) Nef, Vif, Vpu, and Vpr, the four **accessory** proteins. Five of the nine genes are involved in regulating the expression of the GAG-POL-ENV genes.

The letters **GAG** stand for group-specific antigens (proteins) that make up the viral nucleocapsid. The GAG gene codes for internal structural proteins, the production of the dense cylindrical core proteins (**p24,** a nucleoid shell protein with a molecular weight of 24,000), and several internal proteins, which have been visualized by electron microscopy. The GAG gene has the ability to direct the formation of virus-like particles when all other major genes (POL and ENV) are absent. It is only when the GAG gene is nonfunctional that retroviruses (HIV) lose their capacity to bud out of a host cell. Because of these observations, the GAG protein has been designated the virus particle-making machine (Wills et al., 1991).

The **POL** gene codes for HIV enzymes, protease (p10), the virus-associated polymerase (reverse transcriptase) that is active in two forms, and endonuclease (integrase) enzymes. The integrase enzyme cuts the cell's DNA and inserts the HIV DNA. Evidence from retroviral deletion studies shows that the loss of LTRs on the 3′ side of the POL gene stops viral DNA integration into the host genome. However, nonintegrated DNA, without its LTRs and integrase enzyme, can still produce new viruses. This clearly demonstrates that viral DNA integration is not essential for viral multiplication even though integration is the normal course of events (Dimmrock et al., 1987).

The regulation of HIV transcription appears to be intimately related to the onset of HIV disease and AIDS. Thus interruption or inactivation of the POL gene would appear to have therapeutic effects (Kato et al., 1991).

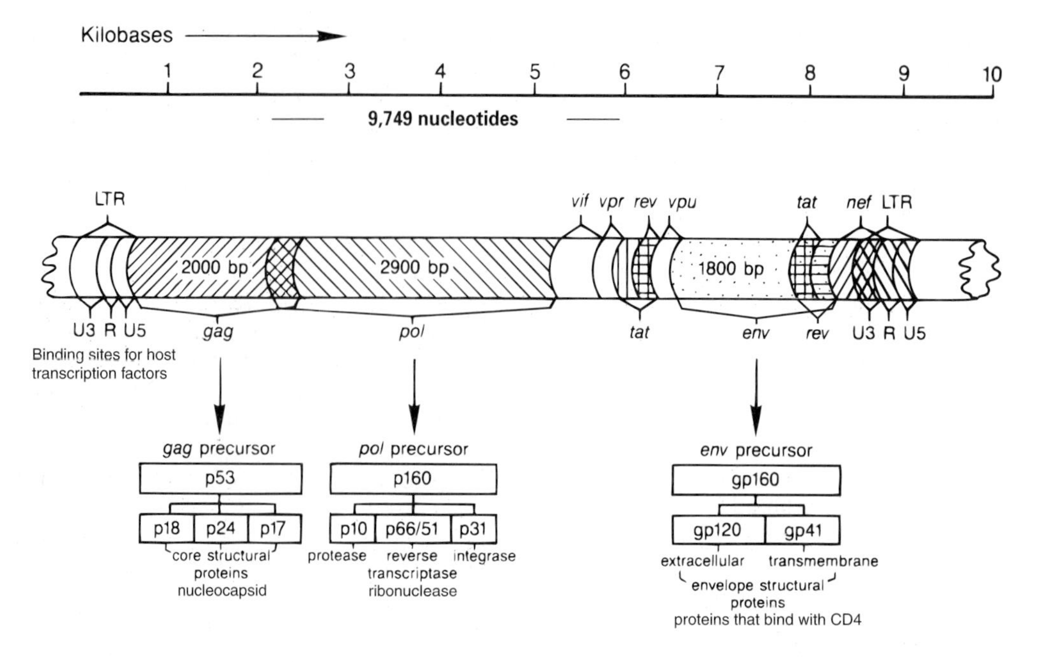

FIGURE 3-5 Genome of HIV. Nine of the genes making up the HIV genome have been identified. They are positioned as shown. Five are essential for HIV replication and six control reproduction (see text for details). The maxtrix protein, p17, forms the outer shell of the core of HIV, lining the inner surface of the viral membrane. Key functions of p17 protein are: orchestrates HIV assembly, directs GAG p55 protein to host cell membrane, interacts with transmembrane protein, gp41, to retain envelope-coded proteins within HIV, and contains a nuclear localization signal that directs HIV RNA integration complex to the nucleus of infected cells. *This feature permits HIV to infect nondividing cells, a distinguishing feature of HIV* (Matthews et al., 1994).

The **ENV** gene codes for HIV surface proteins, two major envelope glycoproteins (gp120, located on the external "spikes" of HIV and gp41, the transmembrane protein that attaches gp120 to the surface of HIV) that become embedded throughout the host cell membrane, which ultimately becomes the **envelope** that surrounds the virus as it "buds" out (Figure 3-6). Studies on how HIV kills cells have revealed at least one way that the envelope glycoproteins enhance T helper cell death. The envelope glycoproteins cause the formation of **syncytia;** that is, healthy T cells fuse to each other forming a group around a single HIV-infected T4 or CD4+ cell. Individual T cells within these syncytia lose their immune function (Figure 3-7). Starting with a single HIV-infected T4 helper cell, as many as 500 *uninfected* T4 helper cells can fuse into a single syncytium. Continued creation of these syncytia could deplete a T4 cell population.

Several studies have demonstrated that the appearance of syncytium-inducing (SI) HIV strains during the chronic phase of HIV disease heralds an abrupt loss of T4 or CD4+ lymphocytes and a clinical progression of the disease. Although the SI phenotype is detected in many patients with AIDS, it has also been isolated from individuals who have not gone on to an early development of AIDS. However, the marked decrease in the T4 lymphocyte counts after shifting from non-SI (NSI) to SI strains as well as the negative effect SI strains have on primary HIV infection suggests that their appearance may not be just a consequence, but the actual cause of immune system alterations (Torres et al., 1996).

The Six Genes of HIV that Control HIV Reproduction—Collectively, the six additional HIV genes tat and rev (regulatory genes), and nef, vif, vpu, vpr (auxiliary genes) working together with the host cell's machinery actually control the reproductive retroviral cycle: adhesion of HIV to a cell, penetration of the cell, uncoating of HIV genome, reverse transcription of the RNA genome producing proviral DNA and immediate production of new viral RNA, or the integration of the provirus and later viral multiplication. The six genes allow for the entire reproductive scenario—from infection to new HIV—to occur in 8 to 16 hours in dividing cells.

Gene Sequence—The HIV proviral genome has been well characterized with regard to gene location and sequence (Figure 3-5), but the function of each gene is not completely understood. The genes for producing regulatory proteins can be grouped into two classes: genes that produce proteins essential for HIV replication (**tat** and **rev**), and genes that produce proteins that perform accessory functions that enhance replication and/or infectivity (**nef, vif, vpu,** and **vpr**) (Rosen, 1991).

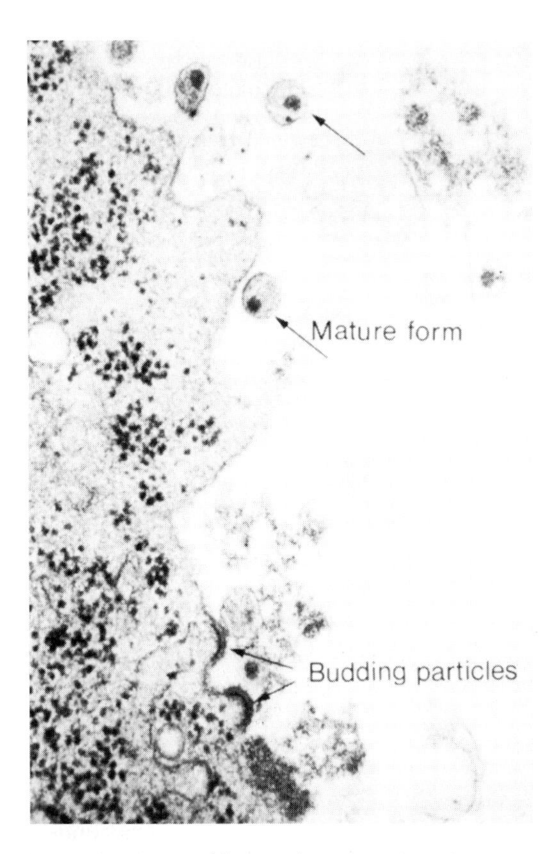

FIGURE 3-6 Budding and Mature Retroviruses. HIV buds from infected cells only at special points on the cellular membrane known as "lipid rafts." The rafts are rich in cholesterol. Without the cholesterol HIV cannot fuse with new CD4+ or T4 cells (Hildreth, 2001). This is a photograph of HIV taken by electron microscopy. Note the difference between the free or mature HIV and those that are just budding out through the membrane of a T4 helper cell. This cell came from an HIV-infected hemophiliac. Closely observing the mature HIV, one can make out the core protein area surrounded by the cell's membrane (virus envelope). *(Courtesy CDC, Atlanta)*

Gene Function—Each end of the proviral genome contains an identical long sequence of nucleotides, the long terminal repeats. Although these LTRs are not considered to be genes of the HIV genome, they do contain regulatory nucleotide sequences that help the six regulatory/auxiliary genes control GAG-POL-ENV gene expression (Figure 3-5). For example, the **vif** gene is associated with the infectious activity of the virus. Currently, the predominant view is that vif acts at the late stages of infection to promote HIV processing or assembly (Potash et al., 1998).

The **tat** gene is essential for HIV infection of T4 cells and HIV replication (Parada et al., 1996; Li, 1997; Stevenson, 1998). It is one of the first viral genes to be transcribed. The tat gene produces a transactivator protein, meaning that the gene produces a protein that exerts its effect on viral replication from a distance rather than interacting with genes adjacent to **tat** or their gene product. **Tat** contains two coding regions or exons—areas that contain genetic information for producing a diffusible protein—which, through the help of the LTR sequences, increases the expression of HIV genes thereby increasing the production of new virus particles. The tat protein interacts with a short nucleotide sequence called TAR located within the 5′ LTR region of HIV messenger RNA (mRNA) transcripts (Matsuya et al., 1990). Once that tat protein binds to the TAR sequence, transcription of the provirus by cellular RNA polymerase II accelerates at least 1000-fold.

The **rev** gene (**r**egulator of **e**xpression of **v**iral protein) selectively increases the synthesis of HIV structural proteins in the latter stages of HIV disease, thereby maximizing the production of new viruses. It does this by regulating splicing of the HIV RNA transcript (cutting out nucleotide sequences that exist between exons and bringing the exons together) and transporting spliced and unspliced RNAs from the nucleus to the cytoplasm (Patrusky, 1992; Fritz et al., 1995).

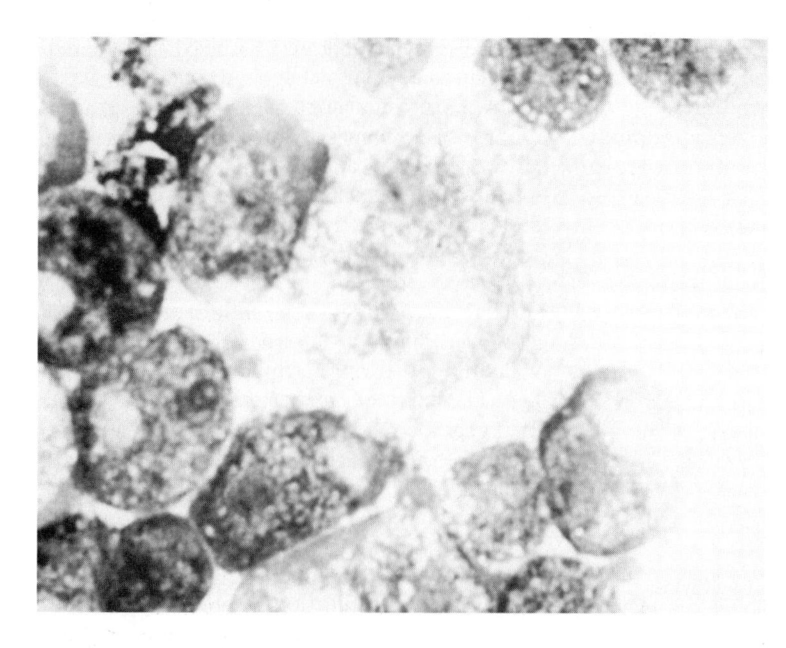

FIGURE 3-7 Formation of T4 Helper Cell Syncytia. A single infected T4 helper cell can fuse with as many as 500 uninfected T4 helper cells. The formation of syncytia leads to a loss or depletion of T4 helper cells from the immune system. *(Courtesy of Tom Folks, National Institute of Allergy and Infectious Diseases)*

The **nef** gene produces a protein that is maintained in the cell cytoplasm next to the nuclear membrane. It is believed **nef** functions by protecting the cell from dying, allowing the cell to continue producing HIV. Several antigenic forms of nef protein have been found, which suggest multiple activities of nef within HIV-infected cells (Kohleisen et al., 1992; Sagg et al., 1995). Olivier Schwartz and coworkers (Cohen, 1997) showed that **nef** can prompt cells to yank down from their surfaces a molecule known as the major histocompatibility complex (MHC), which displays viral peptides to the immune system. The group predicted that this "down-regulation" of MHC would make HIV-infected cells resistant to cytotoxic T cell killing (Collins et al., 1998) (see Chapter 5).

In August 2001, Yuntao Wu and colleagues reported that after HIV RNA manufactures a DNA copy, but prior to its integration into the cell's DNA, the HIV DNA stimulates the production of **tat** and **nef** viral proteins. These proteins awaken T4 cells out of their dormant state. Once the T4 cells are activated, they become vulnerable to nuclear invasion by HIV DNA, allowing the virus to integrate into the T4 cell DNA structure. Not only will the activated T4 cell allow HIV DNA to enter the nucleus and insert itself into human DNA, an activated T4 cell produces a higher rate of HIV RNA replication (Wu et al., 2001).

The functions of the **vpr** gene, which codes for a **v**iral **p**rotein **R,** is associated with the transport of cytoplasmic viral DNA into the nucleus. Vpr is also involved in steroid production that in turn helps produce HIV, is required for the efficient assembly or release of new HIV viruses, and stops T4 cell division. Although vpr protein is not needed for HIV to reproduce in T4 or CD4+ cell cultures, it appears to be very important for HIV to reproduce in macrophages. Vpr can induce CD4+ cell death even when the cells are not HIV-infected and is also poisonous to cell mitochondria, which may be important in killing cells that are HIV-infected. It has recently been shown that HIV-infected people who are long-term nonprogressors to AIDS have a mutant or nonfunctional form of the vpr gene. (Lum et al., 2003)

Vif or the Viral Infectivity Factor—Vif is produced using one of the smallest and least understood of the nine genes that make up HIV. In 2002, scientists probing the secrets of Vif reported two startling developments. They found that human cells contain a powerful enzyme known as APOBEC3G (pronounced APPO-beck) that can stop the production of HIV. Simultaneously, they discovered that HIV itself has overcome this natural defense by using Vif to neutralize that enzyme. It has been shown that as HIV buds out of a cell it carries the APOBEC enzyme, that is, the enzyme stows away inside the new virus. When HIV infects a new cell and begins making genetic copies of itself, APOBEC becomes active, causing HIV to mutate at such a rapid pace that resulting HIV copies are inactive. But, HIV then produces Vif that destroys APOBEC—HIV's solution to survival! Some HIV/AIDS scientists believe that those findings are the most important new information in HIV/AIDS research since the identification of HIV in 1983. Once scientists gain an understanding of how Vif blocks this protective enzyme, they can devise tests to measure how well a potential new drug interferes with that process. Presumably a chemical that disables Vif or prevents it from attacking APPOBEC could become a potent antiretroviral drug. **Vpu** codes for a **v**iral **p**rotein **U** that destroys the CD4 protein within the T4 lymphocyte, thus helping HIV to bud out of the cell. Vif, Vpr, and Vpu appear to be necessary for HIV to cause disease.

Of the HIV proteins, tat, rev, and nef are termed *early* proteins because their production results from the cutting and splicing of full-length HIV mRNA; vif, vpu, and vpr are termed *late* proteins, since their production results from unspliced or single-spliced mRNA.

In summary, HIV/AIDS researchers have picked apart the AIDS virus, decoding its 9 genes and isolating its 15 proteins. Now they are trying, piece by piece, to understand how these components work together to produce one of the most lethal microbes in history. As biologists learn what makes HIV tick, they get new ideas about how to destroy it.

How Genes Store Genetic Information and the Importance of Mutations or Change Within Genes—Genes store the information necessary for creating living organisms and viruses. In sexually reproducing organisms, the information is stored in the form of DNA organized into structures called chromosomes. Apart from sex cells (eggs and sperm) and mature blood cells, every cell in the human body contains 23 pairs of chromosomes. One of each pair is inherited from the mother, the other from the father. Each chromosome is a packet of compressed and twisted DNA. Genes are sections of DNA containing the blueprint for the whole body, including such specific details as what kind of receptors cells will have, for example, CD4, CD8, X4, R-5, and so on.

DNA is made up of a double-stranded helix held together by hydrogen bonds between specific pairs of bases. The four bases A, T, G, and C (adenine, thymine, guanine, and cytosine) bond to each other in fixed and complimentary patterns that give humans and other species their individuality. If a gene is thought of as a sentence, and the nucleotides in DNA as letters, a change or mutation of only one letter can affect the entire sentence or the information the DNA gives the cell. To get an idea of how many mutations can occur in a cell, consider that humans have about 3.2 billion base pairs that make up about 35,000 genes or sentences that contain the information that makes a human. There are about a million differences between your 3.2 billion-letter DNA alphabet and that of another person. The kind, number, and sequence of nucleotides (bases) in human DNA and that of other species is much greater. Another way of saying this is that the closer a species resembles a human or vice-versa, the closer will be the DNA base sequences.

For example, the DNA base sequence in a chimpanzee is about 98% identical to humans. That means the difference between human DNA and chimpanzee DNA is only a base sequence difference for about 60 million base pairs out of 3.2 billion human base pairs.

Importance of Genetic Stability to a Species—The individual or collective characteristics (phenotypes) of any virus, cell, or multicelled organism depend on the expression of their genes and the interaction of gene products within a given environment. From a biological point of view, changes in phenotype (observable characteristics) that are inheritable are by definition genetic changes. Such changes occur due to changes in the kind, number, and sequence of bases in DNA. Base changes may occur by addition, substitution, and deletion. These changes are referred to as **genetic mutations.** Genetic mutations provide biological heterogeneity and genetic diversity (similarity as a species but dissimilarity with regard to certain characteristics). Investigations on the rate at which genetic mutations occur in living species indicate that DNA is a stable molecule with relatively low mutation rates for any given gene. Because of low mutation rates within the DNA of a species's gene pool (all the genes that can be found in the DNA of a species) and selection pressures by a slowly changing environment, species evolution is constant but very slow.

Genetic mutations in the strain of an organism or virus produce genetic and phenotype **variants** (different members) of that strain. Regardless of the rate at which mutations occur, *they are genetic* *mistakes*—they are not intentional, they just happen by chance. Most mutations or genetic mistakes either make no difference to an organism or virus (silent mutations) or they cause a change. Few genetic mistakes within a stable environment improve the species. After all, the species arrived at this point in time via genetic and environmental selection pressures—those with the best constellation of genes survived to reproduce those genes. In species that produce large numbers of offspring, genetic mistakes that are lethal or lead to an early death are of little consequence to the species. A genetic mistake that improves the chance of survival and reproduction is retained.

Genetic Instability of HIV—A virus like HIV can produce hundreds of replicas within a single cell. Genetic mistakes during viral replication produce variant HIVs. In biological economic terms, HIV replicas are inexpensive to make. Even if most of these mutant HIV replicas are inactive or throw-away copies, it makes little if any difference to the HIV per se. However, if a few HIV replicas received environmentally advantageous mutant genes, these HIV mutants would survive as well as or better than the parent HIVs. Both parent HIV and mutant HIV can reproduce in the same cell and exchange genes. Over time, only the most fit mutant or variant HIVs are transmitted among people and undergo still further genetic changes. These variant HIVs could, with sufficient accumulative genetic changes, become a new type of HIV—for example, an HIV-3.

Investigations of some of the RNA viruses revealed relatively high mutation rates. Thus some of the RNA viruses are our best examples of evolution in "real" time. Because of their high error rate during replication they show, as expected, both high genetic diversity and biological heterogeneity in their host, and a rate of evolution about a million times faster than DNA-based organisms (Nowak, 1990). HIV, in particular, fits this category. Heterogeneity of HIV is reflected by: (1) the difference in the kinds of cells variant HIV infects; (2) the way different HIV mutants replicate; and (3) the way different variants of HIV harm infected cells.

It is now known that HIV is capable of enormous genetic flexibility, which allows it to become resistant to drugs, to escape from immune responses, and to avoid potential HIV vaccines. What is not known are all the factors contributing to viral diversity in individual infections. Clearly the high error rate of the reverse transcriptase (because this enzyme lacks a 3′ to 5′ exonuclease proofreading ability, that is, it cannot correct mistakes once made) and the high turnover rate in infected cells generate vast numbers of different virus mutants. The diversity of newly produced variants, however, is shaped by a combination of mutation, recombinations of HIV RNA, and selection forces. The main selective forces that have been proposed to drive HIV diversity are the immune response, cell tropism (cell types most likely to attract HIV), and random activation of infected cells. At the time of seroconversion a person may carry a homogeneous virus population, but then diversification occurs as HIV infects many different cell types and tissues in the body (Bonhoeffer et al., 1995).

For further information on HIV evolution, diversity, and HIV disease progression, see Chapter 7, Box 7.2.

GENETIC REVOLUTION

Genetic revolution is when an organism or virus such as, say, SIV leaves one host (chimpanzee), enters a new host (humans), and rapidly diversifies its genes in an attempt to interpret and adapt to its new environment, as when SIVcpz became HIV in humans. As HIV continued to replicate or reproduce it became more fit to survive in humans. That fitness occurs in two ways, through mutation or change in individual nucleotides making up the genes, and through recombination, a process in which there are large exchanges of sections of HIV diploid RNA.

FORMATION OF GENETIC MUTATIONS IN HIV

One mechanism for producing HIV variants or HIV mutants results from a *highly error-prone* reverse transcriptase (RT) enzyme of HIV. Preston and colleagues (1988) found that HIV transcriptase makes at least one replication error (mutation) in each HIV genome per round of replication in a cell! Other investigators say that reverse transcriptase, on average, introduces a replication mutation once in every 2000 nucleotides or about five mutations per round of replication! This unusually high rate of nucleotide misincorporation (substitution, addition, and deletion of nucleotides) as proviral DNA is being made by RT is responsible for generating the genetic diversity and heterogeneity found among the isolates

BOX 3.1

IMPACT OF HIV ON HUMAN EVOLUTION?

Scientists are looking at the course of infections among wild animals and wondering what the history of their viral infections say about the future of HIV as it evolves in humans. Why, they ask, are wild cats, African monkeys, and, it seems, chimpanzees so impervious to illness from many of their viruses? For example, Mark Feinberg of Emory University asked why HIV-1 and HIV-2 behave so differently in the human population. It is believed that the precursor virus, HIV-1 (SIVcpz), crossed over into humans from chimpanzees and that the precursor virus to HIV-2 (SIVsm) crossed over into humans from the Sooty Mangabey monkeys. Both precursor SIVs evolved in humans to cause AIDS, but HIV-2 has not caused a worldwide pandemic—why? Scientists cannot answer. These primates have, over time, learned to live with their precursor HIV viruses. With respect to HIV-2, it was found that the Sooty Mangabeys infected with SIV-2, and African green monkeys infected with SIV-1, carried billions of SIV per drop of blood when tested, and that these viruses are constantly killing large numbers of their cells, but the monkeys just replace them. It appears that to a lesser extent this is already happening with HIV-2 in humans. The virus is there and kills human cells, but infected humans appear to be far less clinically affected by HIV-2 than by HIV-1. When HIV-1 is injected into chimpanzees—human beings' closest genetic cousins—the animals become infected but usually remain unharmed, showing no sign of disease. Only one research group in the world, the Yerkes Primate Center outside Atlanta, has caused AIDS in chimpanzees using an HIV/SIV hybrid laboratory strain about 15 years ago. In 1996, only one of the chimps developed symptoms similar to human AIDS.

Stephen O'Brien of the National Cancer Institute said that the heavily viral-infected monkeys who remain healthy present a picture similar to that found in wild cats. The animals have enormous numbers of viruses in their blood and tolerate the viruses' constant killing of their white blood cells. O'Brien says this is a predictable adaptation with a clear implication for the human AIDS epidemic. O'Brien believes that when a virus gets into a population, like HIV that crossed into humans, it can kill, be 100% lethal, or just partially lethal, leaving survivors to reproduce. Over time, O'Brien said, either the virus becomes weakened or the species *changes* in order to live with that virus. Many scientists now believe that it is inevitable that HIV will alter humanity, that is, our species will change! For example, so many people are infected in some parts of Africa that future generations will have a disproportionate number of people with genetic predispositions to live with the virus. And that is how evolutionary changes occur. In short, the severity of the epidemic in some parts of the world is so profound that it will clearly impact human evolution. This is occurring in our lifetime. In the past, we've been left to infer what the impact of infection was on human evolution. But now we have the opportunity to observe it (Levin et al., 2001). At the moment, there is no evidence that HIV is weakening.

of HIV. In short, the high error rate in producing proviral DNA means that each HIV-infected cell will carry variant or mutant HIV, most of which will be genetically unique (Vartanian et al., 1992). This high rate of mutation underlies HIV's remarkable ability to become resistant to drug therapies and to hamper the production of an HIV vaccine!

FORMATION OF GENETIC RECOMBINATION IN HIV

Subtypes are a result of the selective accumulation of mutations produced during the process of reverse transcription. Like other retroviruses, HIV produces mutations as a natural part of its replication process. However, HIV produces mutations more frequently than most retroviruses. If you drop the same virus, say in Alaska and Africa and come back 20 years later, the separate viruses will have developed their own different particularities, depending upon the environmental factors or constraints they have confronted. With HIV, there is considerable mixing of the viruses because of the large movement of infected people worldwide. If there are enough people being infected and exposed to viruses, this process will lead to the generation of intersubtype recombinants. Both mutation and recombination by RNA strand switching during transcription may contribute equally.

When HIV replicates, each of the new virus particles being constructed incorporates two strands of viral RNA. Because two RNA molecules are packaged in each virion, reverse transcriptase (RT) may switch from one template to another during reverse transcription. If two RNAs with sequence differences are copackaged in one virion, a mosaic HIV genome containing genetic information from both RNAs could be generated, yielding novel viral genomes. Recombination can also salvage genome damage or detrimental mutations in one RNA molecule by adopting genetic information from the intact template. Thus, recombination may not only introduce genetic diversity, it may also serve as a repair mechanism for the HIV genome (Jetzt et al., 2002).

Mutations create the changes in the genetic material of HIV, but recombination quickly reshuffles these changes.

Forming a recombinant HIV is the same concept as a plant or animal hybrid. The hybrid or new species has picked up genetic information from two distinctly unrelated parents. For example, each of us is a recombinant. Each parent has its own genetic information, and the children are their recombinants. So recombination is a natural process. For us, the process doesn't usually produce large genetic changes. For HIV, the recombination process can produce tremendous change in a single replication cycle. In the case of HIV intersubtype recombinants, each parent is a different subtype, the recombinant picks up part of its genetic information from one parent, part from the other. They are also genetic hybrids.

Completing the Picture on Producing HIV Genetic Variability: HIV Clades or Subtypes

Michael Saag and colleagues (1988) examined the generation of molecular variation of HIV by sequential HIV isolates from two chronically infected people. They found 39 distinguishable but highly related genomes (HIV variants). These results indicate that HIV heterogeneity occurs rapidly in infected individuals and that a large number of genetically distinguishable but related HIVs rapidly evolve in parallel and coexist during chronic infection. That is, whenever a drug or the immune response successfully attacks one variant, another arises in its place. Pools of genetically distinct variants that evolve from the initial HIV that begin the infection are often referred to as **quasispecies** (Delwart et al., 1993; Diaz et al., 1997). However, even though a person possesses diverse quasispecies of HIV, only a very narrow population (perhaps only one) of HIV is transmitted from mother to child or between sexual partners (Derdeyn et al. 2004).

Additional evidence indicates that some HIV genetic variants demonstrate a preference (tropism) as to the cell type they infect. This means that one genetic change may allow the virus to enter cells that were once immune to the virus. Also, a report by Helen Devereux and colleagues (2002) states that a wide variety of quasispecies circulate in each individual and that there may be HIV evolving independently in different body compartments. The rapid genetic change, which results in altered viral products, makes it very difficult to design a vaccine or drug that will be effective against all HIV variants. To date, HIV drug-resistant mutants have been found for all FDA-approved nucleoside and nonnucleoside analogs and protease inhibitors used in the treatment of HIV-infected people and AIDS patients! (Chapter 4 presents a discussion of currently used drug therapies for HIV/AIDS.)

HIV Antigenic Variation—This refers to differences in a protein that occur as a result of changes in the DNA. Each variation of protein produced elicits the production of a different antibody. Such protein changes occur, for example, as a result of DNA changes in the GAG-POL-ENV genes. Antibodies are made against the different GAG-POL-ENV gene proteins, but the antibodies made against the ENV gene protein appear to be the most important. These antibodies neutralize the envelope glycoproteins that seem to be an essential part of HIV's infecting process. The ENV gene is subject to the most frequent mutations, producing HIV with different envelope glycoproteins within a given individual.

It now appears that HIV can constantly change its surface antigenic composition, thereby allowing it to escape antibody neutralization. Thus, HIV mutations allow the virus to persist in the presence of an immune response. This immune selection viral phenomenon is not new. It is what the influenza virus does yearly so that last year's vaccine will not protect people from this year's variation. However, HIV differs from viruses such as influenza. Influenza and many other viruses do not have an RNA to DNA replication step so they are not as mutable as the retroviruses. Because of

SIDEBAR 3.2

CROSSING FROM PRIMATES INTO HUMANS

The three major groups of HIV-1, groups M, N, and O genetically differ from each other as much as they differ genetically from SIVcpz. These data strongly suggest that individual groups M, N, and O entered humans as separate events. HIV-2, which is very closely related to the SIV of Sooty Mangabeys, may have crossed over into humans on at least six occasions (Gao, 1999; Gao, 1992).

Although it may appear odd that distinct strains of HIV should have colonized humans from different animal species on different occasions, natural cross-species transfer has occurred frequently for other retroviruses, not just HIV. There is also a precedent with the flaviviruses, yellow fever and dengue, and with malaria. *Plasmodium falciparum* malaria, arising in Africa, is closely related to the parasite of chimpanzees, whereas *P. vivax* came from Asian monkeys.

Why Has HIV Shown Up Now—Is This a New Virus?

Based on scientific evidence, different SIV strains apparently crossed into humans several times during the twentieth century. There may have been many earlier introductions or precursor HIV into humans but, like Ebola or Lassa fever viral outbreaks, they may only have occurred locally, temporarily, and soon vanished. What helped HIV to become endemic, and group M to become epidemic, might have been the huge expansion of needle and syringe use in the mid-twentieth century during periods of mass vaccination and injecting antibiotic use, as suggested by Preston Marx and colleagues (2001). In other words, it is not the cross-species transfer event that is new or different but the social conditions and medical practices that allowed the evolving HIV eventually to adapt to sexual transmission which allowed for its survival in humans. As described in Chapter 2, strains of SIV most likely crossed into humans around 1931, plus or minus 15 years. So, yes, this is, according to scientists, a new virus!

---

the error-prone reverse transcriptase enzyme used by the retroviruses, the possibility of genetic change far exceeds that for any other known nonretroviral human pathological virus.

DISTINCT GENOTYPES (SUBTYPES/CLADES) OF HIV-1 WORLDWIDE BASED ON ENV AND GAG PROTEINS

HIV is actually an umbrella term for two genetically distinct types of virus: HIV-1 and HIV-2. HIV-1 embraces three genetically distinct groups: M, N, and O. Of these, group M viruses, the ones that are perhaps most studied and are known to be most responsible for infections worldwide, have been subdivided into 11 genetically distinct subtypes, identified by letters A through K. Genetic analyses of HIV-2, the less virulent form of HIV, have yielded six subtypes, identified as A through F.

As stated previously, this book is about HIV-1 or HIV. In order to better understand globally circulating strains of HIV, HIV investigators have placed, based on genetic diversity, the various HIV strains into three major groups, **M,** for the **main** HIV genotypes or **clades** found in different populations, and groups **O** and **N** for HIV genotypes or clades that are significantly different from those in the **M** group.

Group M HIV causes over 99% of the world's HIV/AIDS. **Group O** and the more recently discovered **Group N** (discussed on page 65), cause less than 1%.

Group M Subtypes

There are 11 subtypes or clades of group M: A, B, C, D, E, F, G, H, I, J, and K. The M subtypes have been analyzed with respect to the differences between them based on the variations found in their GAG and ENV proteins (Robertson et al., 1995).

Within a subtype, envelope gene sequences vary from 7% to 12%. Between subtypes the genetic variation is up to 30%. There is the belief that a vaccine against one subtype won't offer protection against the rest. (For discussion of HIV vaccines see Chapter 9 on prevention). Subtypes B and E together make up about 14% of HIV infections worldwide. Subtypes A, C, and D make up about 84% of HIV infections.

The 11 different M subtypes or clades and O have been globally mapped (geographically located) in Figure 3-8 (Brix et al., 1996; Workshop Report 1997). This figure demonstrates the dissemination of the different subtypes from 1990 through 2004. Clearly, subtype B, closely related to subtype D, predominates in the Americas although in Brazil, Argentina, and Uruguay substantial proportions of infections are caused by F subtype or BF recombinant viruses. Apart from these three countries, infections with non-B genetic forms are unusual; however, recently, it was reported that env subtypes A, C, and H had been identified in 5 of 11 samples in Cuba. The identification of diverse non-B subtypes in Cuba is not unexpected, considering that large numbers of Cuban military and civilian personnel had been stationed in the 1970s and 1980s in Angola, a country neighboring the Democratic Republic of Congo (DRC), where the highest group M diversity is found, and that many of the early cases of HIV infection in Cuba were detected among these individuals. The explosive epidemic in Southeast Asia is chiefly attributable to subtype E. There are ample documented introductions of subtypes C, E, and F from endemic areas into the United States and the Western Hemisphere. Subtype C now accounts for about 40% of *new* HIV infections worldwide—particularly in China and India. There is no one subtype that is representative of this global pandemic.

International Genetic Subtypes

Numerous groups have shown an increase in intermixing of HIV genetic subtypes in multiple international locales. HIV subtypes A, B, C, D, F, G, and H have appeared throughout Europe; B, F, C, and D have been found in Brazil; one-third of new infections in Germany are non-B; 15% of infections in a New York City hospital are non-B; increasingly, E is found among injecting drug users in Thailand, from 24% subtype E in 1991 to 44% in 1995, and among heterosexual risk groups, a change from 86% subtype E in 1991 to

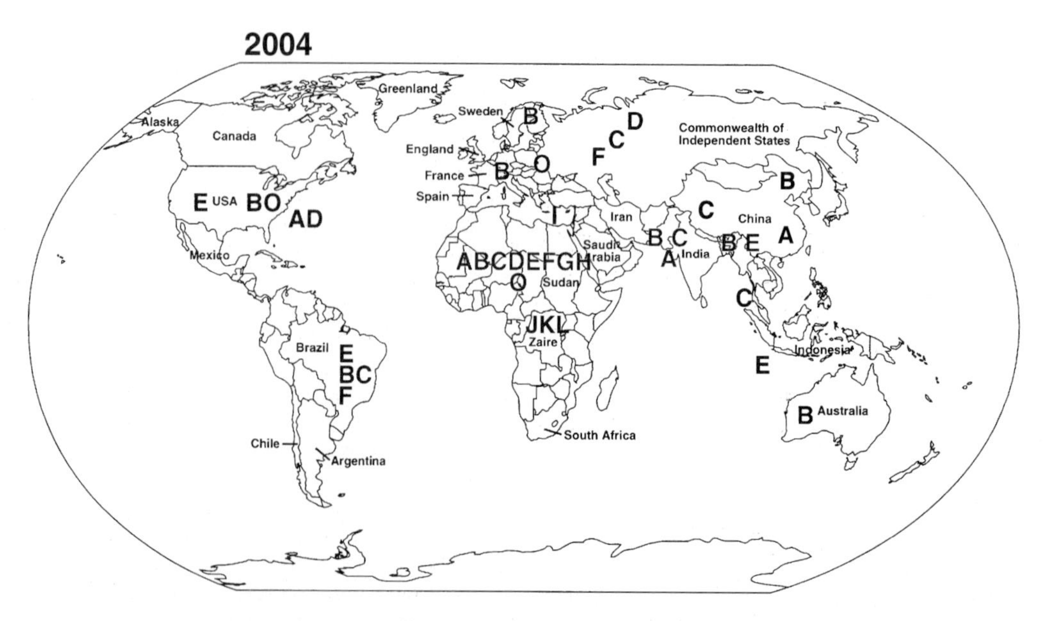

FIGURE 3-8 Global Distribution of the 11 M Subtypes and O in 2004 in Areas of Highest Prevalence. Clearly this global map shows that HIV subtypes are no longer continent based—all subtypes are now present on all continents. No satisfactory explanation exists for the skewed worldwide distribution of HIV-1 subtypes. The worldwide spread of viral subtypes makes it clear that a world vaccine will be required, that is, one that is not subtype specific. None appears to be on the horizon, although large-scale trials with existing envelope vaccines are underway in some developing countries *(Brodine et al., 1997 updated)*.

98% in 1995; and group O has been detected in at least 12 African countries, France, Spain, and in the United States, as well as M/O dual infections. In South Africa, there is spreading of subtypes B and D throughout the homosexual population and C/E throughout the heterosexuals studied. The impact of this is an increasing spread of non-B subtypes into previous B areas, which has significant implications for vaccine development.

Subtypes Related to Transmission Route

While the current geographic distribution of subtypes is generally acknowledged to have arisen from accidents in viral trafficking, the perpetuation of these patterns may be due to the affinity of certain subtypes for certain targets. Max Essex of Harvard has studied Langerhans cells as a target of viral entry for subtypes E and C. He has proposed that Langerhans cells, prevalent in the female genital tract, are less hospitable to subtype B. He concluded that non-B subtypes are dominant in countries where the major mode of HIV transmission is through heterosexual contact and that this is due to the greater accessibility of subtypes E and C through the Langerhans cells. In China, in a large group of injection drug users, a recombinant subtype AB was found to the exclusion of other subtypes. It is estimated that over half of all HIV infections worldwide are from subtype C. It is the most common subtype found in Southern Africa, China, and India.

Global Predominant HIV-1 Subtypes

A. W. Africa, E. Africa, Central Africa, East Europe, Mideast
B. N. America, Europe, Mideast, E. Asia, Latin America
C. S. Africa, S. Asia, Ethiopia
D. E. Africa
E. S.E. Asia

HIV-2 primarily West Africa.

HIV RNA Recombination

Robertson and colleagues (1995) reported that eight of the group M subtypes appear to have been involved in recombination with each other, giving rise to genetic hybrids. These studies raise questions of whether a vaccine that works against one of the subtypes will work against a subtype hybrid.

Group O Subtypes

Group O contains at least 30 genetically different subtypes of HIV. Group O subtypes are referred to as **"outlier"** because their RNA base sequence is only 50% similar to the known genotypes of the M group. The O variants have been known since 1987 but have been found mainly in Cameroon, Gabon, and surrounding West African countries that, to date, have only been marginally affected by AIDS. Group O viruses were of concern primarily because their divergence from group M was sufficient to miss their detection by ELISA HIV testing. (The ELISA test is discussed in Chapter 12.) Seven people in France were identified with group O HIV in 1994. The first documented case of group O HIV infection in the United States was found in April 1996 in Los Angeles. By the beginning of 2003, only three cases of HIV subtype O infection have been reported in the United States. This subtype was found in a Los Angeles county woman and a Maryland woman; both came into the United States from Africa. HIV subtype O is not routinely tested for in the United States.

New Group N Subtype?

In August 1998, Francois Simon and colleagues reported the discovery of a *new* HIV-1 isolate that *cannot* be placed in the M or O subtype category. The authors suggest that the new isolate be classified as group **N** for **"new"** or **"non-M non-O."** The new isolate, designated **YBF30**, was found in a small number of people in the West African nation of Cameroon. The first isolate was from a 40-year-old woman who died of AIDS. This virus is similar to SIV (cpz-gab) and it either branched with the SIV strain or between it and HIV-1 group M. The authors state that future strains of SIV (cpz) could be found and strains that are closely related to YBF30 might circulate among Cameroonian chimpanzees.

SIV retroviruses are endemic in many species of primates but usually don't cause disease, which makes them natural hosts for what are known as lentiviruses or slow-acting viruses. But when they cross the species barrier into a new host like humans, these viruses may become pathogenic, causing illness. Based on evolutionary studies, researchers think SIV has existed in sub-Saharan Africa for thousands of years, adapting to several species including the African green monkey and

POINT OF INFORMATION 3.2

MYSTERIES OF HIV SUBTYPES

Two epidemiologic mysteries have eluded HIV/AIDS researchers for years. The first is why heterosexual intercourse has accounted for approximately 10% of HIV transmission in the United States and Western Europe to date, but more than 90% of HIV transmission in Asia and Africa.

The second mystery has centered around a subtype B HIV epidemic among injection drug users in Thailand. This epidemic began in the mid-1980s, and plateaued with about 100,000 people infected. Several years later, however, a subtype E HIV epidemic took hold in Thailand. This epidemic has exploded among heterosexuals, with over a million people already infected. Most surprising, perhaps, is that subtype B—while present in Thailand, India, and several African countries—has not caused heterosexual epidemics in those places.

Although differences in the rate of heterosexual transmission may also be due to factors such as sexual behavior and the presence of other sexually transmitted diseases, none of these considerations has yet adequately accounted for the widespread heterosexual epidemics in Asia and Africa. Investigators have begun to focus on how biological characteristics of the individual subtypes might also play a role. These findings about subtype cellular affinities do not suggest that subtype B cannot be transmitted heterosexually—it is throughout the United States and elsewhere. The point is that subtype B seems to be transmitted through vaginal intercourse far less efficiently than the subtypes that predominate in Asia and Africa. It has been shown that Langerhans cells are located in specific areas, such as the vagina, cervix, and penile foreskin. They do not appear to localize in the rectal or colon mucus membranes. Experiments with HIV-infected Langerhans cells and uninfected T4 cells showed that HIV-infected Langerhans cells readily formed clusters and syncytia with T4 cells, resulting in efficient cell-to-cell transmission. It was also demonstrated that during the cell-to-cell interaction, Langerhans cells not only effectively infect T4 cells but also activate them, resulting in enhanced virus replication (UNAIDS, 1997). Thus, the enhanced efficiency of subtype E to attach to Langerhans cells may help account for the explosive heterosexual spread of the epidemic in countries such as Thailand, while the heterosexual epidemic in the industrialized world has thus far remained at a comparatively low rate (Essex, 1996).

POINT OF INFORMATION 3.3

FIRST CONFIRMED CASES OF HIV GROUP O-CAUSED AIDS

A Norwegian sailor died of AIDS in 1976 at the age of 29. His wife and youngest daughter, born in 1967, also died of AIDS. The members of this Norwegian family represent the earliest confirmed cases of AIDS and the first case of HIV type O infection. The first symptoms appeared in 1966 in the sailor, in 1967 in his wife, and in 1969 in their daughter. Between 1961 and 1965 he traveled the world's oceans, calling at ports in all six inhabited continents. On his first voyage, which began in August 1961 just after his fifteenth birthday, he worked as a kitchen hand on a Norwegian vessel that sailed down the West African coastline, calling at ports in Senegal, Liberia, Cote d'Ivoire, Ghana, Nigeria, and Cameroon. A gonorrheal infection during this trip shows that he was sexually active. He returned home in May 1962, and never returned to Africa. No known evidence suggests that the sailor was bisexual, which means that sexual contact with a woman is the most straightforward explanation for his infection. This would suggest that HIV group O has been circulating in that part of Africa for at least 35 years.

SECOND CONFIRMED CASES OF HIV GROUP O-CAUSED AIDS?

The second case of group O infection found in the literature is the second child of a French barmaid from Reims, who died in 1981. The child's clinical history is highly suggestive of neonatal AIDS. In 1992 a group O virus was isolated from the mother, who by then had AIDS.

Sooty Mangabey. There are six closely linked strains of SIV. Chimpanzees carry at least three different viral predecessors of HIV-1 that gave rise to groups M, O, and N. Sooty Mangabeys carry an ancestor of HIV-2.

Simon's discovery of group N also raises other critical questions and challenges. Looking ahead, how many other SIV strains may have crossed the species barrier? Should we be mass screening for this new virus N or increasing surveillance of the other SIV strains? What about the human recombinant viruses? How well are we tracking them?

SUMMARY

HIV is a retrovirus. It has RNA for its genetic material and carries reverse transcriptase enzyme for making DNA from its RNA. HIV, using its enzyme, copies its genetic information from RNA into DNA, which becomes integrated into host cell DNA and may remain silent for years, or until such time as it is activated into producing new HIV. HIV contains at least nine genes; three of them, GAG-POL-ENV, are basic to all animal retroviruses. The six additional genes are involved in the infection process and regulate the production of products from the three genes. HIV, because of its error-prone reverse transcriptase enzyme, mutates at an unusually high rate. With time, many mutant HIV variants can be found within a single HIV-infected person. A vaccine against one mutant HIV may not work against a second—like the vaccines made yearly against different mutant influenza viruses.

A new group of HIV has been identified, group N.

REVIEW QUESTIONS

(Answers to the Review Questions are on page 438.)

1. Why is HIV called a retrovirus?
2. What are the three major genes common to all retroviruses? How many additional genes does HIV have?
3. Why are retroviruses, and HIV in particular, believed to be genetically unstable? Give two reasons for your answer.
4. What is believed to be the major reason for the high rate of genetic mutations in HIV production?

4 Anti-HIV Therapy

Chapter Concepts

- There may never be a cure for HIV disease/AIDS.
- From March 1987 through early year 2005, 20 individual anti-HIV drugs and 5 combination drug groups received FDA approval and at least 2 other drugs have received expanded access approval.
- There are no anti-HIV drugs free of clinical side effects.
- All nucleoside analogs and non-nucleosides work by inhibiting the HIV reverse transcriptase enzyme, which functions *early* in the HIV life cycle.
- All protease inhibitors work by inhibiting the HIV protease enzyme from its function *late* in the HIV life cycle.
- Viral load is the number of HIV RNA strands found at any one time in human plasma.
- Viral load is associated with HIV disease progression.
- Viral load is associated with HIV transmission.
- Pediatric anti-HIV therapy is improving.
- Anti-HIV drug combinations are extending lives.
- AIDS deaths are dropping in developed nations due in part to drug combination therapy.
- When to begin anti-HIV drug therapy is the big question.
- Strict adherence to drug regimens is essential but difficult to maintain.
- Salvage therapy, the kitchen sink approach—trying to save the patient.
- Dispensing current anti-HIV drug therapy requires an HIV/AIDS specialist.
- The *best* anti-HIV drug combinations are still unknown.
- Morning-after drug therapy to prevent HIV infection is questionable.
- Scientists do not expect the production of an effective anti-HIV vaccine before 2012.
- HIV/AIDS therapy hotlines and resources are listed.
- See disclaimer on page 105.

We are constantly humbled by the devastation that something so small, HIV, can cast upon something so large, a human (Figure 4-1). This chapter provides no final answers; there is no curative therapy, no truly outstanding therapies (drugs that benefit all HIV-infected without major side effects) against HIV, and, with the number of expensive anti-HIV drugs that are available, debate continues about the details on which drugs offer the best combination therapy and on the standard of care for HIV/AIDS patients. Because some 90% to 95% of HIV-infected people develop AIDS, and because 90% of all *new* HIV infections are occurring in developing nations, an inexpensive, easily taken, nontoxic, effective HIV-directed therapy is essential. Physicians need a drug that works against HIV as antibiotics once worked against a large variety of disease-causing bacteria.

ANTI-HIV THERAPY

The ideal solution would be to prevent HIV from causing an infection. Then anti-HIV therapies would not be necessary. However, there are no means available to stop HIV from entering the body and infecting a limited number of cell types—primarily those cells displaying CD4 antigen receptor sites. Following HIV infection, there is a depletion of cells carrying CD4, especially the **T lymphocyte**

DRUG THERAPY FOR WILLIAM "SKIP" BLUETTE

William "Skip" Bluette (continued from page 22) three weeks before he died. Treatment to maintain Skip was failing—the multiple opportunistic infections were defeating the best medicine had to offer.

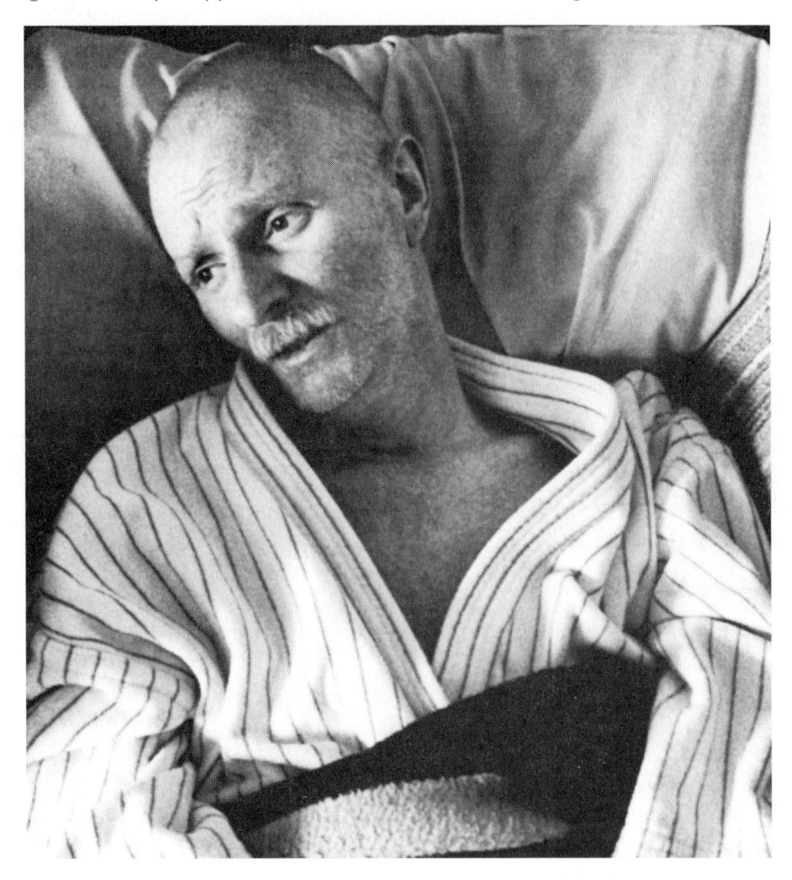

FIGURE 4-1 Drug Therapy for William "Skip" Bluette. Skip's treatment included amphotericin B, an antifungal drug, to combat meningitis. Starting in July 1986, he had to visit one of the hospital's clinics at least once a week to receive his treatments, which were given over several hours. Skip started taking Zidovudine, a drug known to slow the progress of the disease, during the summer of 1987. Unfortunately, Zidovudine also made it difficult for his body to produce blood cells. Skip stopped taking Zidovudine when this occurred. In June 1988, Skip started cleaning out his closets. He said it was part of the dying process to give belongings away "so you know where they're going." Partly, in jest, he said he wanted his ashes scattered on 42nd Street in New York—he thought that's where he "got AIDS."

He was hospitalized July 5, 1988. It was presumed that he had *Pneumocystis carinii* pneumonia, which Skip called "the killer." His breathing and speaking were labored. He had inflammation of the pancreas, for which he received morphine, and his kidneys began to fail. *(By permission of Mara Lavitt and* The New Haven Register*)*

(T4) cells of the human immune system. (See Chapter 5 for an explanation of cell types in the human immune system and their function.) With T4 or CD4+ cell loss, over time immunological response is lost. Loss of immunological response leads to a variety of opportunistic infections (OIs). The suppression of the immune system and increasing susceptibility to OIs and cancers give

VIGIL FOR WILLIAM "SKIP" BLUETTE

The vigil for William "Skip" Bluette (continued from page 69) 48 hours before he died. Time and therapy were running out on Skip. His last wish was to have his sisters Arlene and Nancy near when he died and they were there. On July 15, 1988, the doctors told Nancy and her mother that Skip's kidneys had failed—this time Skip would not recover. He died on July 17. He was 42 years old.

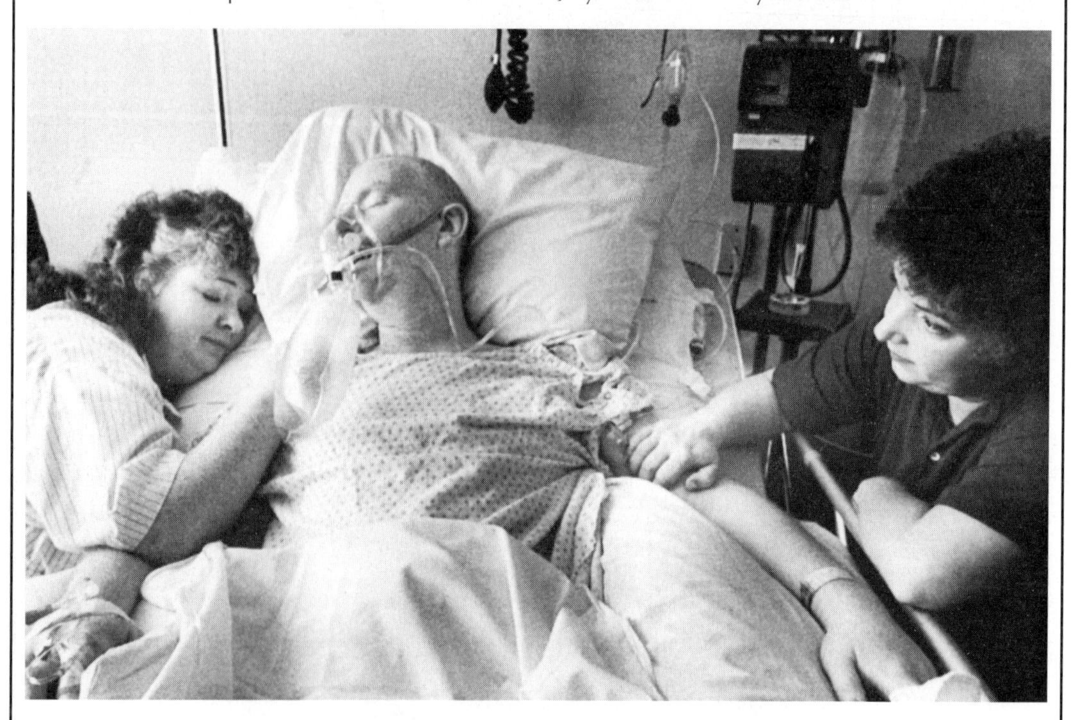

FIGURE 4-2 The Vigil for William "Skip" Bluette. The Bluette family rallied around Skip from the moment they found out he had AIDS. In April 1988, his mother had a heart attack but she survived it. Skip celebrated his 42nd birthday on May 18; it was his second birthday since his diagnosis. *(By permission of Mara Lavitt and* The New Haven Register*)*

HIV/AIDS a multidimensional pathology. (OIs are presented in Chapter 6). Because HIV/AIDS is a multidimensional syndrome, it is unlikely that a single drug will provide adequate treatment or a cure (Figure 4-2).

An ideal goal for HIV drug investigators would be to find a drug that excises all HIV proviruses from the cell's DNA. It is very unlikely that this kind of drug will soon, if ever, be available. Alternatively, the elimination of all HIV-infected cells might be of comparable benefit, as long as irreplaceable cells are not totally lost through the process. In essence, this is the goal that the human immune system sets for itself, yet falls short of reaching, in the vast majority of HIV-infected individuals.

In the absence of a curative weapon (Figure 4-3), therapies must be designed to prevent the spread of the virus in the body. Some of these anti-HIV therapies are presented in this chapter.

Can HIV Infection Be Cured?

An effective cure means cleansing the body of the HIV. To find a cure, HIV incorporation into human DNA and subsequent reproduction of new virus must be understood. Without interference, the **HIV provirus** will remain in host-cell DNA for life. The provirus can, at any time, become activated to mass produce HIV, which leads to cell death and eventually to the individual's death. As of this writ-

WILLIAM "SKIP" BLUETTE: MAY 18, 1946–JULY 17, 1988

FIGURE 4-3 Free of Pain—One with the Universe—William "Skip" Bluette. On July 20, 1988, Skip's body was cremated. There was a memorial service. On July 21, Skip's ashes were buried in Evergreen Cemetery in Connecticut. *(By permission of Mara Lavitt and* The New Haven Register*)*

ing, there is no way to prevent either the proviral state or proviral activation.

President George W. Bush's Goal: An AIDS Cure?

In March 2001, President Bush, making his first public comments on the global AIDS pandemic said his administration's primary goal is to find a cure for AIDS, a feat thought by scientists today to be all but impossible within the boundaries of contemporary biological science. The president's comments were echoed in April by Secretary of State Colin Powell in testimony before Congress in which he said the aim of the administration is to ultimately find a cure for AIDS. Health and Human Services Secretary Tommy Thompson, in his first staff briefing at the agency said, "The best thing that this department can do is find a cure for HIV and that is the best solution to a lot of our health problems." In mid-May at the World Health Assembly annual meeting in Geneva, Thompson reiterated his call for a cure. What is striking about these comments is that few, if any, AIDS researchers speak today of searching for a cure—that word hasn't been part of the epidemic language for the last 12 years.

To cure HIV infection would require one of two options: Kill every single infected cell in the body and destroy all viruses floating outside of those cells in the human bloodstream, or spare the infected cells but destroy the viruses hidden inside human DNA without damaging human genes in the process. So daunting are these tasks, so far beyond the current intellectual level of biological sciences, that the search for a cure effectively ended on the day in 1983 when French scientist Luc Montagnier announced his discovery that AIDS was caused by a retrovirus. Through 2005 an estimated 65 million people will have been infected with HIV. Number of people cured via therapy—zero. Science has yet to discover a genuine cure for any viral disease, though it is possible to restrain some infections through the use of drugs and vaccinations.

> **POINT OF VIEW 4.1**
>
> **MEDICAL BENEFIT: A CURE?**
>
> Many diseases cause death. "We can cure those diseases," people say, "but we can't cure AIDS." Tell me, how many diseases can we cure? We cannot cure very many viral diseases, nor can we cure many of the diseases that kill people, like heart disease and forms of cancer. The concept of curing people is relatively new. The strength of physicians used to come from their capacity to accompany. Accompanying a person through an illness towards health or towards death was central to the practice of medicine, but it disappeared when the issue became one of curing.
>
> —*Jonathan Mann, in Thomas A. Bass,* Reinventing the Future: Conversations with the World's Leading Scientists *(Addison-Wesley, 1994)*

ANTI-HIV DRUGS WITH FDA APPROVAL

Most Americans have no direct contact with the U.S. Food and Drug Administration (FDA), but their lives literally depend on the effectiveness of the agency. Charged with assuring the safety of specific foods and all medicines, the FDA has oversight of 25% of the U.S. economy. One of the FDA's key functions is approving and monitoring prescription drugs—a job that is particularly crucial for people with HIV infection or other life-threatening diseases. The FDA is the American Gold Standard of endorsement! (See Point of Information 4.1.)

Gold Standard of Therapy

The **gold standard** for determining the efficacy (effectiveness) of a new treatment is that it alters the disease in a way that is beneficial to the patient. Therefore, the end points most often used in clinical trials of therapies for a chronic disease such as HIV include prolongation of life or the extension of time to a significant disease complication.

But studies using these end points require large numbers of patients and/or the passage of considerable amounts of time. **Surrogate markers** (that is, physiological measurements that serve as substitutes for these major clinical events) can eliminate this problem if their validity and correlation with clinical outcome in people can be confirmed. The use of surrogates has the potential to shorten the duration of clinical trials and expedite the development of new therapies.

Major Surrogate Markers Used to Evaluate Anti-HIV Drug Therapy

The **T4 or CD4+ immune cell number** and **viral load** are the two most studied and commonly used surrogates (alternates) for clinical efficacy of anti-HIV therapies. They are imperfect measurements, however, because changes in T4 cell number per microliter of blood (a very small drop) and RNA strands per milliliter of blood are only partially explained by the effects of therapy. T4 or CD4+ cell counts exhibit a high degree of day-to-day variation in individuals, and methods used to count these cells are difficult to standardize. Recent investigations have shown that there is a T4 cell level and viral load level **disconnect.** This means that viral suppression by itself does not always predict immunological and clinical benefit. T4 cell counts and viral load can be clearly dissociated during treatment: T4 cells can increase in the presence of a high viral load, remain stable, or drop. **Clinical benefit** is more closely associated with the level of T4 cells rather than viral load. Recent data support the hypothesis that T4 cell depletion during HIV infection occurs largely as a result of the immune system's inability to generate new mature cells, and that the main effect of antiretroviral therapy is to help restore the immune system's ability to produce new T4 cells. (Perrin et al., 1998; Telenti et al., 1998; Deeks et al., 1999; Clark et al., 1999; Hellerstein et al., 1999.)

THE HIV MEDICINE CHEST: HIV/AIDS DRUGS RECEIVING FDA APPROVAL

The number and kind of HIV drugs, their recommended mixtures, and their timing and dosage change almost continuously. This can be a source

POINT OF INFORMATION 4.1

THE DRUG DEVELOPMENT, APPROVAL PROCESS, AND CLINICAL TRIALS

In the United States, drug discovery and FDA approval currently takes an average of 12 to 15 years and costs about $400 million (from the laboratory to the drugstore). In 1996, Merck & Co. claims to have spent $1 billion over 10 years to bring the protease inhibitor **Crixivan** to market! The odds against a new drug making it to the market are about 10,000 to 1. The clinical trial process for a drug usually includes 3½ years of preclinical testing using laboratory and animal studies and 6 years of studies (clinical trials) in humans. **Phase I** includes 20 to 80 healthy volunteers and takes about a year to test the drug's safety. **Phase II** takes about two years and involves 100 to 300 persons with the disease to assess the drug's effectiveness and to look for side effects. **Phase III** of clinical trials lasts about three years and includes 1000 to 3000 patients to verify effectiveness and identify adverse reactions of a drug. Since 1989 the FDA has allowed phases I and II to be combined to shorten the approval process on new medicines for serious and life-threatening diseases. It now takes about 18 months for a drug to go through the review process for approval by the FDA. About one in five medicines that begin a clinical trial is approved for consumer use.

Global annual drug research and development spending is about $70 billion, but only 10% of that money goes to research on diseases that account for 90% of the world's infections.

THE HIV/AIDS DRUG MARKET

The anti-HIV drug therapy market is estimated to be about $8 billion a year. The $350 billion American pharmaceutical industry has been investing over $1 billion a year since the late 1980s to bring new and improved HIV drugs to market. Since 1987, a new FDA-approved antiviral drug has entered the market on average of every 9.4 months.

Beginning in 2005, 84 companies had 85 medicines and vaccines for AIDS and AIDS-related conditions in testing. This is in addition to 84 HIV and HIV-related FDA-approved medicines already on the market.

The most difficult part of presenting anti-HIV therapies to the public is the analysis of the incredible amount of new information continuously coming out in the scientific literature and attempting to make sense out of some very contradictory findings (*The Drug Development and Approval Process*, Pharmaceutical Manufacturers Association, December 1998 updated).

Clinical Trials

A clinical drug trial is a government funded and organized study of an experimental or unproven drug to determine the drug's safety and efficacy (whether or not it works). Drug trials for HIV/AIDS, sponsored by the Food and Drug Administration (FDA), are organized into the AIDS Clinical Trial Group (ACTG). Guidelines are set to determine how the ACTGs may be run and who is eligible to participate. Taking part in a trial may entitle persons to receive medical examinations and checkups and to have their overall health monitored. For many people with AIDS, the ACTGs are the only means of access to certain potentially life-saving drugs, and the only form of health care they may ever receive. Historically, FDA policy on admitting women into clinical trials has been confusing and discriminatory at best. Until recently, the FDA had an outright ban on all women of child-bearing potential from participation in early stages of drug trials. In March 1994, FDA took two important steps to ensure that new drugs are properly evaluated in women. First, it provided formal guidance to drug developers to emphasize its expectations that women would be appropriately represented in clinical studies and that new drug applications would include analyses capable of identifying potential differences in drug actions and value between the sexes. Second, the agency altered its 1977 policy to include most women with child-bearing potential in the earliest phases of clinical trials.

of tremendous confusion for patients and medical practitioners alike.

From March 1987 through 2004, 20 individual anti-HIV drugs and 5 combination drugs have received FDA approval for use in persons infected with HIV. There are at least 25 FDA-approved drugs when combination pills and reformulations are included. Other drugs have been FDA-approved for **expanded access use** where standard regimes have failed (Table 4-1).

Seven of the 20 FDA-approved noncombination anti-HIV drugs are nucleoside (nuk-lee-o-side)

Table 4-1 Choices: Anti-HIV Therapy; Nucleoside Analogs, Non-nucleosides, Protease Inhibitors and Fusion Inhibitors

Name	FDA Approved	Phase Testing	Cost/Year[a]
Nucleoside analog (reverse transcriptase inhibitor)			
Zidovudine, ZDV (Retrovir)/AZT	March 1987		$4,340.18
Didanosine, ddI (Videx)	October 1991		$3,686.85
Zalcitabine, ddC (Hivid)	June 1994		$3,013.89
Stavudine, d4T (Zerit)	June 1994		$4,061.28
Lamivudine, 3TC (Epivir)	November 1995		$3,818.75
Combivir[d]	September 1997		$7,940.61
Abacavir (Ziagen)	December 1998		$4,930.75
Trizivir[d]	November 2000		$12,895.71
Tenofovir (Viread)[b]	October 2001		$5,198.00
Emtricitabine (Emtriva)	July 2003		$3,625.56
Truvada	August 2004		$9,683.45
Nucleoside analog reverse transcriptase inhibitors are potent in combination with other drugs; used alone, they lead to HIV resistance. *ZDV (AZT), d4T, 3TC,* and *abacavir* penetrate the blood-brain barrier. Common side effects: lactic acidosis. *Seven* new nucleoside analogs are in some phase of testing in the United States.[c]			
Non-nucleoside compounds; (non-nucleoside reverse transcriptase inhibitors)			
Nevirapine (Viramune)	June 1996		$4,395.36
Delavirdine (Rescriptor)	April 1997		$3,842.13
Efavirenz (Sustiva, Stocrin)	September 1998		$5,258.69
Non-nucleoside analog reverse transcriptase inhibitors (NNRTIs, or non-nukes) may interact with other *cytochrome p450-processed drugs:* protease inhibitors, oral contraceptives, etc. NNRTIs have a mixed ability to penetrate the blood-brain barrier. Common side effect: mild rash. Some doctors build up drug doses slowly to avoid rash; the other worry is that dose building increases risk of drug resistance. *Thirteen* new non-nucleoside analogs are in some phase of testing in the United States.[c]			
Protease inhibitor drugs[d]			
Saquinavir mesylate (Invirase)	December 1995		$5,233.39
Ritonavir (Norvir)	March 1996		$9,001.32
Indinavir (Crixivan)	March 1996		$6,310.46
Nelfinavir (Viracept)	March 1997		$8,641.23
Saquinavir (Fortovase)	November 1997		$2,954.93
Amprenavir (Agenerase)	April 1999		$4,372.71
ABT-378 (Kaletra)	September 2000		$8,040.05
Atazanavir (Reyataz)	June 2003		$9,959.88
Fosamprenavir (Lexiva)	October 2003		$7,800.00
Protease inhibitors (PIs) are very potent and may interact with other drugs using cytochrome p450 metabolic pathways. Potentially life-threatening if taken with Seldane, Hismanal, Propulsid, Halcion, or Versed. Avoid rifabutin, Nizoral, rifampin. Poor absorption may affect potency. Common side effects: liver toxicity, hypoglycemia, flatulence, bloating, lipodystrophy (fat distribution). Seven new protease inhibitors are now in some phase of testing in the United States. In addition, there are at least 28 other antiretroviral drugs being investigated.			
NEW DRUG CLASSES		**Phase Testing**	
Attachment and Fusion Inhibitors			
-prevents HIV entry			
T-20 (Enfuvirtide, Fuzeon)	March 2003	FDA Approved	$20,000(WAC)[e]
FP21399		I	
PRO 542 and 140		I	
Schering C		I	
TNX-355		II	
UK-427,857		I	
Integrase Inhibitors			
-prevents HIV DNA from entering human DNA			
S-1360		II	
Zinc Finger			
-disrupts polyprotein formation essential for HIV replication			
Azodicarbonamide (ADA)		I/ II	
Antisense Drugs			
-prevents viral function			
HGTV43		II	
Assembly Inhibitors[f]		I	

analogs (each drug resembles one of the four nucleosides that are used in making DNA), one is a nucleotide analog and three are non-nucleosides. All 11 are **reverse transcriptase inhibitors.** Eight drugs are HIV-**protease** (pro-tee-ace) inhibitors. One is a fusion or HIV entry inhibitor and there are three drug combination pills available. Cellular targets of these drugs can be seen in Figure 4-4.

FDA-APPROVED NUCLEOSIDE/NON-NUCLEOSIDE ANALOG REVERSE TRANSCRIPTASE INHIBITORS AND PROTEASE INHIBITORS

All the known mechanisms by which HIV impairs the human immune system depend on HIV reproduction. Therefore, the development of anti-HIV replication drugs would appear to be a positive first step in controlling HIV reproduction.

Each of the eight nucleoside/nucleotide analogs, sometimes referred to as **nukes,** on entering an HIV-infected cell interferes with the virus's ability to replicate itself (Figure 4-5). That is, when any of the eight nucleoside/nucleotide analogs are incorporated into a strand of HIV DNA being newly synthesized, it stops further synthesis of that DNA strand. The nucleoside analog stops the HIV enzyme, reverse transcriptase, from joining the next nucleoside into position.

Focus on the Eight Nucleoside/Nucleotide Analogs

Each of the eight drugs has limited effectiveness as a **monotherapy.** The principal limitations are: **(1)** they are not 100% effective in stopping HIV reverse transcriptase from making HIV DNA; **(2)** positive clinical effects are short-term, they are not sustained; **(3)** each drug has its own set of toxic side effects; **(4)** individually they do not delay the onset of AIDS; and **(5)** HIV rapidly becomes resistant to each of them.

Side Effects: "First Do No Harm"

Each of the eight reverse transcriptase inhibitors (RTIs) is associated with severe to moderate side effects. For example, Zidovudine causes anemia (lowered blood cell count), nausea, headache, and lethargy and inhibits mitochondrial DNA replication in humans (de Martino et al., 1995); didanosine causes peripheral neuropathy (PN) (a burning pain or pins and needles-like sensation in hands and feet) and pancreatitis (a dangerous swelling of the pancreas) and is very hard for people with poor appetites to take; zalcitabine causes PN and mouth ulcers; stavudine causes PN; and lamivudine causes hair loss and PN.

In 2000 both **Zerit** and **Videx** were associated with the lactic acidosis—a large increase in acid buildup in the body damaging the liver and/or

Notes for Table 4-1

[a] Cost is based on average prescription prices as found in Jacksonville, Fl., pharmacies, 2005.

[b] Tenovir is the first nucleotide analog approved for HIV treatment. It blocks HIV replication similar to the nucleoside analogs.

[c] Expanded Access drugs are for patients failing standard regimes. At the beginning of 2005 there are at least 7 new nucleoside and 13 non-nucleoside RTI and 7 PI drugs in early anti-HIV drug trials in the United States.

[d] Kaletra is a combination of protease inhibitors Lopinavir and Norvir. Lopinavir is available only in Kaletra. Combivir is a combination of Zidovudine/Lamivudine; Glaxo Smith Kline charges 90 cents/day in underdeveloped countries.

Truvada is a combination of the nucleosides Emtricitabine and Tenofovir.

Trizivir is a combination of Abacavir/Zidovudine/Lamivudine.

Saquinavir mesylate was the first nucleotide RTI in Phase III trials. Saquinavir mesylate, ritonavir, and indinavir were approved in 97, 72, 42 days, respectively. Abacavir is the first 2-deoxyguanosine analog reverse transcriptase inhibitor.

Atazanavir is the first once-a-day PI for use with other anti-HIV drugs.

[e] Fuzeon, on launch day (USA) 2003, cash and carry price was $2,200/month or $26,400/year. Wholesale acquisition cost (WAC) was $20,000. May 2003, of the 142 largest insurers, 94% have agreed to cover Fuzeon, as have Medicaid programs in 48 states.

[f] Prevent the viral proteins from assembling into the HIV capsid that houses viral RNA, enzymes, etc.

There are now drugs that interfere with at least 9 different mechanisms in the process by which HIV attaches itself to specific cell types, enters them, enters the cells' DNA, makes copies of itself, and exits the cell.

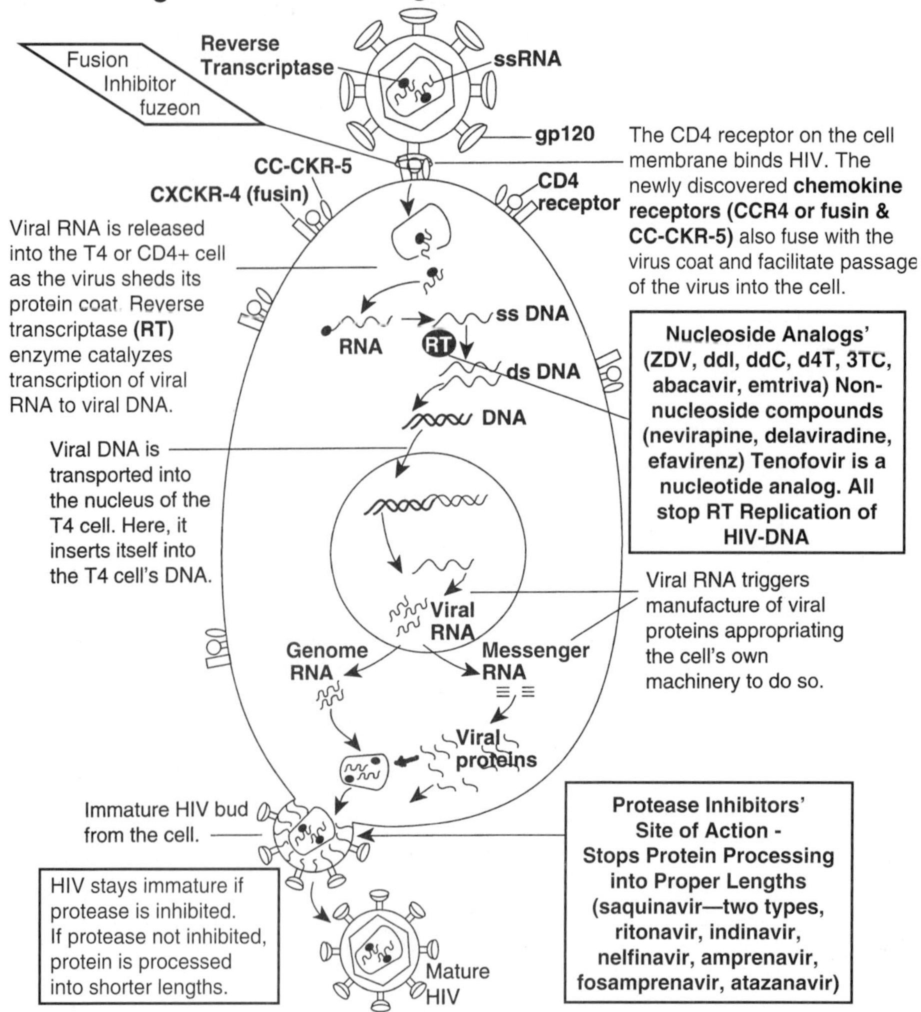

FIGURE 4-4 Diagram of Anti-HIV Therapy. All 20 FDA-approved individual antiretroviral compounds are represented with respect to their anti-HIV activity. The nucleoside analogs act early after infection, while the protease inhibitors act later in the HIV life cycle, after viral proteins have been synthesized into long strands. Those strands of amino acids contain the individual HIV proteins that become functional after they are cut into their appropriate amino acid sequence lengths. NOTE: The enzyme integrase is required for HIV DNA to enter human DNA. Drugs called **INTEGRASE INHIBITORS** are in development.

pancreas—which can be deadly. Pregnant women are at highest risk according to the FDA. These two drugs are to be used in pregnant women *only* when benefits outweigh risks.

With respect to the non-nucleoside analogs ending 2000 and beginning 2001, the FDA has posted warnings on the use of **Nevirapine** (viramune) because of reported serious and sometimes fatal liver toxicity. Most of the cases associated with liver failure occurred in people who took the drug on first exposure to HIV—health-care workers who stuck themselves with an HIV-contaminated needle.

Collectively, some 15% of HIV-infected people cannot tolerate the nucleoside/non-nucleoside an-

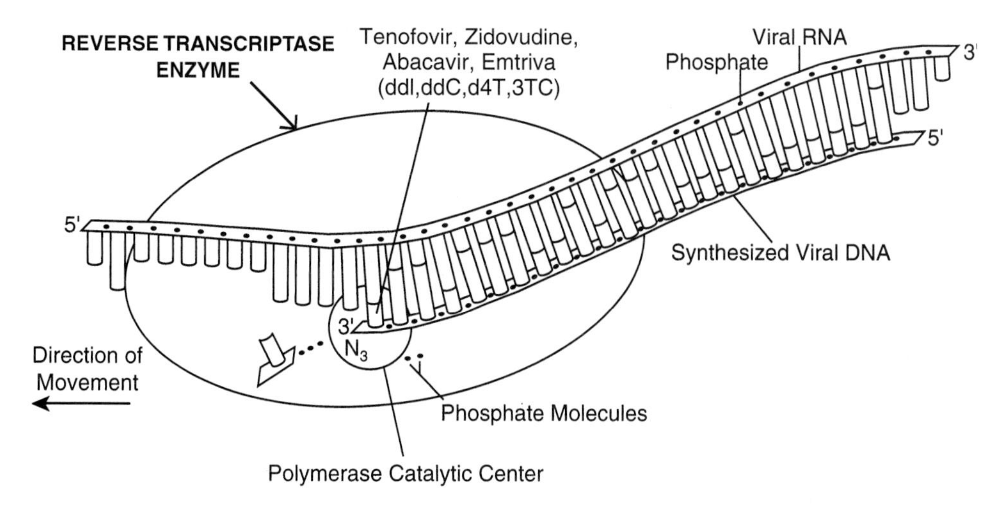

FIGURE 4-5 Incorporation of FDA-Approved Nucleoside/Nucleotide Analogs Preventing HIV Replication. Zidovudine (ZDV) is a synthetic thymidine analog that is widely used in the treatment of HIV disease and AIDS. HIV-RT is 100 times more sensitive to ZDV inhibition than is the **human transcriptase.** The incorporation of Zidovudine-triphosphate into HIV DNA by the action of HIV reverse transcriptase terminates DNA chain extension because polymerization or the incorporation of the next nucleotide adjacent to the azide N_3 group cannot occur. Similar blockage of DNA replication occurs when FDA-approved ddl, ddC, d4T, or 3TC are used. Abacavir is the first 2′deoxyguanosine analog to be used.

tiretroviral drugs. Even those who can tolerate the drugs are time-limited and right now according to Anthony Fauci, director of the National Institute of Allergy and Infectious Diseases, about 80% of those on drug therapy are running out of time.

USE OF NON-NUCLEOSIDE ANALOG REVERSE TRANSCRIPTASE INHIBITORS

Non-nucleoside reverse transcriptase inhibitors (NNRTIs), sometimes referred to as **non-nukes,** are a structurally and chemically dissimilar group of antiretrovirals that can be used effectively in triple-therapy regimes. The mechanism of action of NNRTIs is distinct from that of nucleoside analogs, even though both *prevent* the conversion of HIV RNA into HIV DNA. Nucleoside RT inhibitors constrain, or stop, HIV replication by their incorporation into the elongating strand of viral DNA, causing chain termination. In contrast, NNRTIs are not incorporated into viral DNA but inhibit HIV replication directly by binding noncompetitively to RT. NNRTIs at first appear to offer hope. All four tested did inhibit the RT-HIV enzyme (atevirdine, delvaridine, loviridine, and nevirapine). But this inhibition was short-lived because HIV-resistant mutants for each of the non-nucleoside RTIs were found within weeks of their use. Combination therapies using, for example, Zidovudine and nevirapine, have shown some success but it turned out that various drug combinations do not extend survival time. But they do extend the asymptomatic period! The FDA approved nevirapine (Viramune) for market in June 1996. The primary advantage of using NNRTIs in therapy is to delay use of protease inhibitors.

The key to future success in using the reverse transcriptase inhibitors (RTIs) is finding RTIs that work in cells that are not undergoing division. Most RTIs currently in use work only in cells that are dividing.

FDA-APPROVED PROTEASE INHIBITORS

Recall from Chapter 3 that some HIV genes code for **reverse transcriptase** (RT), **integrase,** and **protease** enzymes. Later on in the reproductive cycle of HIV a specific **protease** is required to process the precursor GAG and POL polyproteins into mature HIV components, GAG proteins, and the enzymes integrase and protease (Erickson et al., 1990). If protease is missing or inactive, noninfectious HIV are produced. Therefore, inhibitors of protease enzyme function represent an **alternative** strategy to the inhibition of reverse transcriptase in the treatment of HIV infection (Figure 4-6).

As research progressed, scientists found that HIV protease is distinctly different from human protease enzymes, so a drug that blocks HIV protease should not affect human normal cell function. This means that HIV protease blockers are specific to HIV protease. Beginning 2005, some 20 HIV protease-inhibiting drugs were under study, six are in clinical trials, and eight have received FDA approval.

A. Why HIV Protease is Essential for HIV Replication

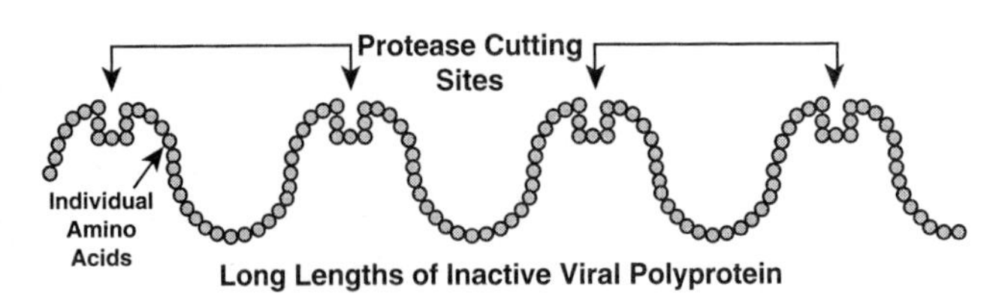

Units of Protease that Cut the Polyprotein into Individual Active Proteins

Protease Protease Protease Protease

A Different Active Protein From Each Separate Length of Polyprotein

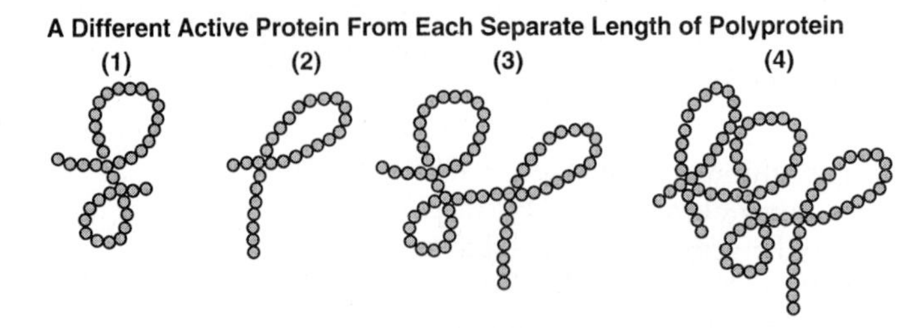

B. How HIV Protease Inhibitors Inhibit Viral Protease Function

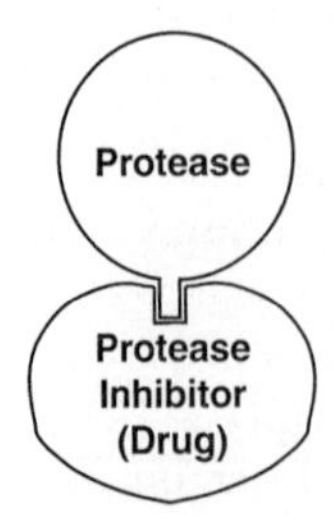

A Protease Inhibitor Binds to the Cutting Site of the Protease. The Inactivited or Blocked Protease Cannot Cut the Polyprotein.

FIGURE 4-6
Representation of Protease and Protease Inhibitors. **A.** The function of HIV protease is to release the individual replication enzymes, core proteins, and envelope proteins so that HIV can develop into mature, infective HIV. **B.** Blocking the production of these essential HIV proteins using protease inhibitors produces immature, noninfective HIV. The action (cutting of protein lengths into active HIV components) of HIV protease occurs during and just after HIV buds out of the cell.

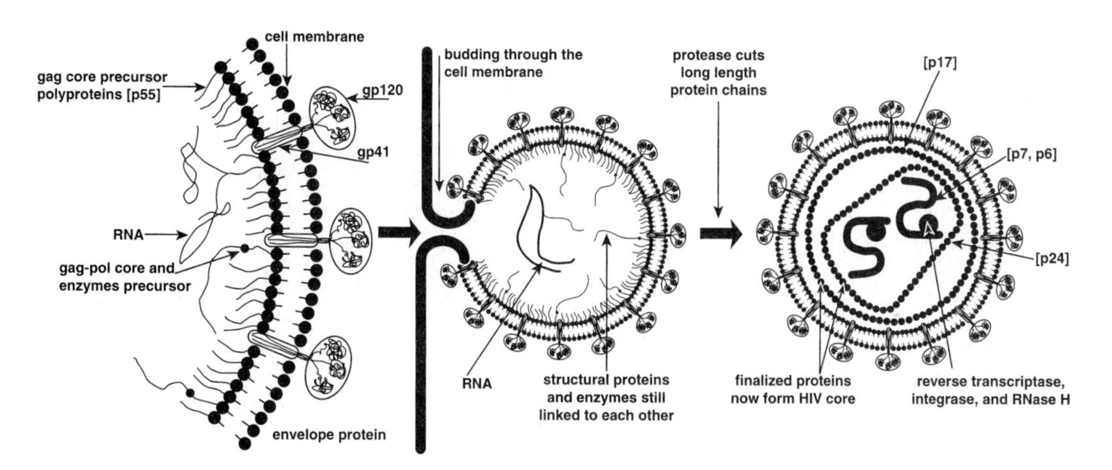

FIGURE 4-7 Representation of HIV Assembly. During the budding process, the viral GAG and GAG-POL polyproteins assemble at the cell membrane together with viral RNA to form immature HIV. These polyproteins are then cleaved by the HIV-coded protease enzyme to provide the structural and functional (enzyme) proteins essential to form the *mature,* infectious viral core. *(Adapted from Vella, 1995)*

Problem: at least 15% of people must give up protease inhibitors because of severe side effects. About 15% of people do *not* respond to protease inhibitors and new reports indicate that protease inhibitors are failing to suppress HIV replication in over half those people in treatment from one to three years (Valdez et al., 1999; Clough et al., 1999). HIV is an impressive enemy. In late November 2003, Megan O'Brien and colleagues reported that over 50% of AIDS patients on first line therapy (taking anti-HIV drugs for the first time, two nukes and a PI) changed or stopped taking their drugs during their first year! Major reason was gastrointestinal problems.

PROTEASE INHIBITORS

In 1988, protease inhibitors with potent and selective antiretroviral activity in cell culture were identified, but the insolubility, poor oral absorption, and rapid liver metabolism of candidate drugs delayed the identification of suitable therapeutic agents until 1992. The **protease inhibitors** made up of a small number of amino acids (up to 15) bind to the protease active site and inhibit the activity of the enzyme. This inhibition prevents cleavage of the long HIV proteins, resulting in the formation of immature *noninfectious* viral particles.

The Eight Protease Inhibitors in Use

1. Saquinavir mesylate (Invirase)—It was the first protease inhibitor (PI) that FDA approved for use in combination with nucleoside analog drugs in persons with advanced HIV disease. At the time of approval, there were no clinical trial data on survival or progression of HIV disease using saquinavir (Nightingale, 1996). *The cost is about $440/month.*

2. Saquinavir (Fortovase)—It is a formulation of saquinavir mesylate. It is sold in a softgel capsule and is more easily absorbed by the body, making it much more effective than S. mesylate. *The cost is about $250/month.*

3. Ritonavir (Norvir)—Ritonavir is the second PI to be FDA-approved. *The cost is about $750/month.*

4. Indinavir (Crixivan)—This PI is perhaps the most often used of the six FDA-approved protease inhibitors. *The cost is about $525/month.*

5. Nelfinavir (Viracept)—It has been evaluated at doses similar to the previous four protease inhibitors. *The cost is about $725/month.*

6. Amprenavir (Agenerase)—It is approved for use in adults and children. It is also the *first* anti-HIV drug to appear as a result of computer imaging—a structure-based drug design. *The cost is about $367/month.*

7. ***Atazanavir (Reyataz)***—First of two FDA-approved PIs in 2003. *The cost is about $830/month.*

8. ***Fosamprenavir (Lexiva)***—Second of two PIs FDA approved in 2003. More easily absorbed than Amprenavir. *The cost is about $650/month.*

FDA-Approved Combination Therapy

Four drug combinations have been FDA-approved. Combivir is a combination of nucleosides Zidovudine and Lamivudine. *The cost is about $662/month.* Kaletra is a protease inhibitor combination containing Lopinavir/Novir. *The cost is about $667/month.* Trizivir is the first to contain three nucleoside analogs—Ziagen (abacavir), Retrovir (Zidovudine or AZT), and Epivir (3TC); *the cost of Trizivir is over $1080 per month.* Truvada is a combination of the nucleosides Emtricitabine and Tenofovir. *The cost is about $796/month.*

The Continuous Debate on the Use of Protease Inhibitors

Protease inhibitors have been hailed as life-saving antiretroviral drugs since their debut in 1996. But there is a major rift in the HIV/AIDS community concerning their use as new information emerges. The new drugs have saved people's lives, that's true. But the new drugs have also killed people. That's also true! They have damaged and deformed some people so badly that although they are alive, they say they wish they were dead. Drug cocktails comprised of protease inhibitors—now considered to be the standard treatment for infected patients who are both sick and healthy—have caused many to suffer liver damage, kidney failure, strokes, heart attacks, and other devastating side effects that are deemed inevitable by advocates of the new retroviral treatments, but called horrific and unnecessary by opponents. According to Joseph Sonnabend, an HIV/AIDS physician at St. Luke's– Roosevelt Hospital Center, "There is absolutely no question whatsoever that protease inhibitors have helped people. But they've probably hurt more people than they've helped. The people for whom benefit has been proven beyond a doubt are really sick people who would have died without them. But the target population for the drug therapy are those that are still healthy, and these people will almost certainly have their lives shortened by these drugs."

DEVELOPMENT AND SELECTION OF HIV DRUG-RESISTANT MUTANTS

Development of Drug-Resistant Nucleoside Analog Mutants

HIV reverse transcriptase, the enzyme that copies (transcribes) HIV RNA into HIV DNA, is *unable* to edit or eliminate transcription errors during nucleic acid replication (Roberts et al., 1988). Because there is no repair or correction of mistakes (as occurs in human cells), there are about one to five mutations in each new replicated HIV DNA/HIV RNA strand (Coffin, 1995). This means that each new virus is different from all other HIV that is being produced because new virus is being produced at a rate of 1 billion to 10 billion per day! There will be 1 billion to 10 billion new HIV mutants produced each day in one person. Thus, virtually all possible mutations, and perhaps many combinations of mutations, are generated in each patient daily (Ho et al., 1995; Wain-Hobson, 1995; Wei et al., 1995; Hu et al., 1996; Mayers, 1996).

DEVELOPMENT OF DRUG-RESISTANT PROTEASE INHIBITORS

HIV resistance to all eight PIs has been found in clinical trials. The development of HIV resistance to PIs is a major concern. For example, recent studies (Schmit et al., 1996) found that long-term (one year) use of **ritonavir** is associated with the production of at least nine different mutations that make HIV resistant to ritonavir. These ritonavir-resistant mutants also demonstrated cross-resistance to both **indinavir** and **saquinavir.** John Mellors (1996) reported that within 12 to 24 weeks after indinavir therapy, at least 10 different HIV-resistant indinavir mutations occurred and five HIV saquinavir-resistant mutations occurred with the use of the individual drugs. Cross-resistance to ritonavir occurred in *at least* 19 isolates of HIV-indinavir resistant mutants.

Rapid Production of HIV Protease Inhibitor Mutations

The high rate of viral replication found throughout the course of HIV infection and the high frequency of virus mutations occurring during each replication cycle, due to the lack of RNA proofreading or correction mechanisms, are the basis for the emergence of drug-resistant variants under the selective pressure of antiretroviral drugs. In fact, a Darwinian model can be applied to HIV dynamics, with the continuous production of genetically different HIV (variants) and the continuous selection of the fittest virus.

With daily production of billions of HIV and a mutation rate that produces billions of new mutants each day, it is likely that any single mutation already exists before any drug is introduced.

Selection of Drug-Resistant Mutants

Based on the large number of mutant HIV produced in any one person, it is not surprising that HIV emerges with resistance to drugs used in antiretroviral therapy. Such strains are referred to as **drug-resistant mutants.** HIV drug-resistant mutants are selected to reproduce most effectively under conditions of selective pressure exerted by the presence of the drugs. Those HIV able to resist the drugs continue to multiply; those that are sensitive to the drugs are destroyed (Figure 4-8).

The Demise of Monotherapy

Prior to 1987 antiretroviral therapy was an unknown. People became infected with HIV and were diagnosed with AIDS, on average, 10 years later. They died of AIDS, on average, three years after their AIDS diagnosis. This, however, is no longer true. In developed countries like the United States, available antiretroviral drugs are extending lives beyond what we can currently predict. Eighteen years ago (1987) the first antiretroviral drug, Zidovudine (AZT as it is often referred to) entered the fight against HIV replication. The infected person took 12 pills per day—two pills every four hours around the clock. Unknown to patients and physicians at the time, the Zidovudine dosage being prescribed was too high and proved toxic for some people. Some died prematurely. Others became resistant to Zidovudine because HIV needs more than one drug to suppress it. The use of Zidovudine and newer antireovirals as single agents had to stop.

The use of *one* anti-HIV RTI drug at a time (monotherapy) leads to HIV drug-resistant

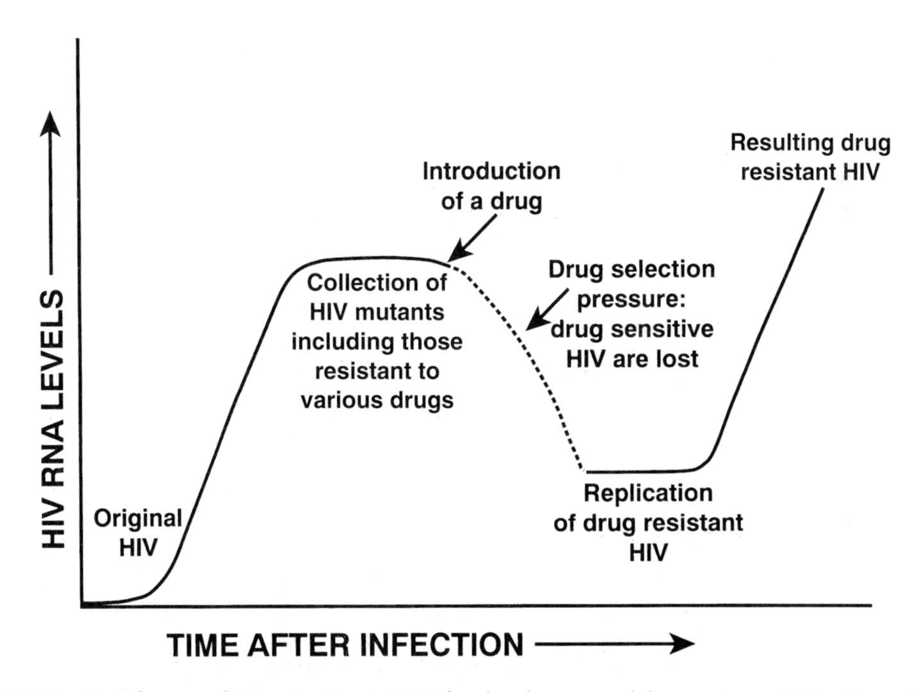

FIGURE 4-8 Enrichment of Drug-Resistant HIV. The development of drug-resistant HIV variants in an individual limits available treatment. This drug-resistant HIV is transmitted such that the recipient becomes infected with one or more drug-resistant HIV variants.

BOX 4.1

PROTEASE INHIBITORS: EXTENDING LIFE; THE DOWNSIDE OF THIS GIFT—RECOVERY?

THE LAZARUS EFFECT: A RETURN TO FUNCTIONAL STATUS

In November 1995, one man in his late forties wrote his obituary—he had been fighting off HIV disease, then AIDS, for over 13 years. In another case, the man's T-cell count was zero. He was on oxygen and morphine. Funeral arrangements were made, and his friends and family were on a death vigil. They are but two examples of several thousand men, beginning in 1999, all under age 50, who had given up hope—they believed they were a short step away from death. Some ran up huge debts—maxed out all their credit cards and gave lavish gifts. Some regretted the way they had lived or not lived to this point, some became very angry, some made peace with themselves and others, but *all* felt death was imminent. *Then it happened*—the first results of combination therapies using nucleoside and non-nucleoside reverse transcriptase inhibitors *dropped viral load counts.* Then, a few months later, the protease inhibitors used alone or in combination with two or three of the nucleoside or non-nucleoside reverse transcriptase drugs dropped viral load counts to unmeasurable or undetectable levels in people with HIV disease and AIDS patients. And, those with significantly lowered viral loads demonstrated surprising recovery—their T4 cells rebounded, in some cases from below 200 back up to normal! These **"AIDS cocktails,"** as the combination therapies were soon called, gave people with AIDS a new chance at a productive life for the first time since the beginning of the pandemic in the United States in 1981.

About 50% of people diagnosed with AIDS in 1995 are still alive, meaning the median survival rate on therapy is now at least 10 years—we don't know how long it will stretch out to, but that is progress.

CAN THERE BE A DOWNSIDE TO SUCH A MIRACLE?

Of course the most tragic downside is the fact that not all who would benefit from the drug cocktail can afford it—but that is not the issue that is relevant to the question per se. No, this question pertains to a downside, if any, for people who have access to these drugs.

WHAT KIND OF PROBLEMS EXIST?

Guilt

Many who now feel "new" again feel *guilty* because these drugs are not available to all who need them—the poor and the uninsured in the United States and those in underdeveloped nations.

Reestablishing Relationships

Depending on how long and how severe the illness, the affected became more or less isolated—even "best" friends stopped calling or dropping by. Some who also had AIDS died, while some were too sick to care or mourn them; but now that they have recovered? **What now? At the beginning of year 2004, about 500,000 people in the United States were on anti-HIV therapy, which includes protease inhibitors. These drugs are not yet deemed necessary to the other 500,000 HIV-infected in the United States. To date they are mostly unavailable in developing nations.**

Ability to Work Again—Loss of Disability Pay

Disability insurance has been a cocoon of safety for many now experiencing "new" life. Traditionally, people with AIDS received disability checks until death. But suddenly it is not so certain that that will be the case. People who were expected to die are going to be coming back. Nobody is prepared for this. Now that drug cocktails have extended life expectancies, it is expected that these people will be reevaluated and lose their disability payments. They will have to go back to work.

Across the United States, AIDS groups are deluged with calls from patients who are excited, confused, or frightened about the prospect of ending disability status and returning to work. In Miami, a psychologist has begun weekly seminars on résumé writing and job interviews for AIDS patients who have not worked in years. At AIDS Project Los Angeles, counselors field 100 calls each week on return-to-work issues.

AIDS patients face a barrage of challenges as they think about rejoining the workforce. Job skills sometimes have atrophied. Old careers do not seem fulfilling anymore after a brush with death. Potential employers may not want to hire someone with a costly disease, even though federal law bars discrimination against people with AIDS. Without clear knowledge of the new drugs' staying power, some patients fear that a return to work could be premature, leaving them in dire shape if protease inhibitors fail and they lose their jobs again.

Dilemmas of Returning to Work: Two Examples

1. A 38-year-old software engineer shudders to think about the myriad new technologies he must master if he is to resume work. He used to be a star computer programmer, deluged with consulting assignments. Then AIDS forced him to go on long-term disability.

   "I'm going to need new software, new tools, new products" to be successful. He estimates that retraining courses will cost $10,000. Where will the money come from?
2. He considered taking a part-time job at a friend's hat-embroidery shop but decided against it, in part for fear of jeopardizing his government benefits, and also because he isn't sure how long his health will hold up. "The biggest downside of this disease is that you stop being a dependable person," he said. "I couldn't be sure that I would be there every day. That's very hard to acknowledge."

TREATMENT DOWNSIDE

From mid-1996 into year 2005, optimism remains high on the medical advances using AIDS cocktail therapy. Though researchers applaud recent drug and immunology findings, many worry about the implications. Treatment successes may create a false sense of security, and lead to complacency in prevention efforts. If we are close to a cure, people may say, why bother with politically sensitive issues like needle exchange programs, distribution of free condoms, sex education programs in our schools, and so on? But the therapy is not available to the majority of HIV-infected, even in the developed countries like the United States. What of the other 44 million living HIV-infected worldwide? If too many glowing promises are made too soon—before treatment is available for all, or most of the world—people may revert back to unsafe sex and other behaviors that will increase their risk of becoming HIV-infected. No, it is not time to stop teaching and providing preventive measures!

POLITICAL DOWNSIDE

As scientists dare ask whether HIV eradication may be possible in some people in the near future (two to five years), the press may be promising more than the scientists can deliver, but the press can and does influence political will. If the political will to fight this disease **weakens** due to the **premature declaration of victory,** this may be the greatest downside of all!

mutants. But, it took seven years before David Cooper, an AIDS-drug therapy researcher, declared 1995 the year of the "demise of monotherapy for HIV and the rise of combination drug therapy" (Simberkoff, 1996; Stephenson, 1996). The **standard of therapy** now is that *all* anti-HIV drugs must be used in combination and that each combination include two reverse transcriptase inhibitors and a protease inhibitor. The use of three or more drugs in combination is referred to as HAART or **Highly Active Anti-Retroviral Therapy** (Merry et al., 1998 updated). The goal of using HAART is to suppress HIV replication or viral load. As a result of a large reduction in HIV fewer T4 or CD4+ cells become infected and die, so the end result is more CD4+ cells remain to divide. Thus, it appears that HAART causes an increase in T4 or CD4+ cells over time, but the increase is not from an increase in new CD4+ cell production. Rather, the increase is due to cell division of existing CD4+ cells saved by HAART drugs.

Jean-Paula Viard and colleagues (2004) have shown that the HAART after the third year does not lead to additional suppression of viral load or increase in CD4+ cell count. In fact, they found that the majority (87%) achieved their maximum CD4+ cell count and viral load reduction at 18 months and after five years showed no improvements.

After fourth line therapy, people, on average, live three years. Total life extension for those going on HAART after symptomatic onset of AIDS with therapy is currently about 6.6 years.

The Function of HAART: Suppression of HIV Replication and Restoration of Immune Competence

Suppression

The function of HAART is to suppress HIV replication regardless of the number of drugs used. The reasoning is to keep the viral load low so that sexual transmission of HIV becomes unlikely, and to prevent HIV mutation by reducing its ability to replicate. A number of clinical trials have demonstrated that only about half the patients placed on HAART, for the first time, are able to achieve maximal suppression—undetectable or less than 50 copies of HIV RNA per milliliter of blood. Where HAART works, most people reduce their viral load to 50 copies within 24 to 32 weeks. Less is known about the long-term durability of this suppression (Table 4.2). Few existing clinical trials are sufficiently long lasting to provide this information. One recent report from the Frankfurt Clinic Cohort found low rates of viral breakthrough or viral rebounds in some patients continuing to take therapy for up to 3.3 years. Of those who achieve this undetectable viral load, the majority will experience a viral rebound during the first year of therapy and most everyone will experience the rebound if they stop taking the drugs (Back, 2001). Currently, about 30% or 330,000 Americans living with HIV infection are untreated. Worldwide, at the beginning of 2005, of the 40 million people living with HIV, about 10% or about 4 million people in developing nations need antiretroviral drugs now. But the World Health Organization (WHO) estimates that less than 3% of these people are receiving the therapy. The WHO's goal, based on a 3 by 5 policy, is having 3 million people in Africa and Asia on therapy by year 2005. It did not happen.

Table 4-2 HAART Increases Longevity

	SURVIVAL AT 10 YEARS	
Age	% Before HAART	% After HAART
15–24	66	84
25–34	57	82
35–44	43	83
45–66	34	67

HOW COMBINATION DRUG THERAPY CAN REDUCE THE CHANCE OF HIV DRUG RESISTANCE

Combination drug therapy works because a single strand of RNA, the genetic material of HIV, must be a multiple mutant, that is, it must carry a genetic change to become resistant to *each* new drug used. So, the greater number of drugs used, each capable of stopping HIV replication by interfering with the function of reverse transcriptase, the greater the number of genetic changes that must occur in a single RNA strand of the virus. For reasons of explanation, say that the chance of change in one nucleotide of HIV for resistance to a single drug occurs in one RNA strand during the production of 10,000 (10 thousand) or 1×10^4 such strands. This is a reasonable figure! Then to have a second nucleotide mutation occur in that same single strand for resistance to a second drug, again at 1×10^4 means that only one RNA strand in 100 million or 1×10^8 would carry both genetic changes. For three separate nucleotide mutations occurring in one RNA strand then, it would be $(1 \times 10^4) \times (1 \times 10^4) \times (1 \times 10^4)$ or only one RNA strand out of 1,000,000,000,000 (1 trillion) would carry a resistance to all three drugs at the same time. But as small as this number is, recall that 1 billion to 10 billion genetically different RNA strands are produced each day in one individual. Also, individual RNA strands can exchange nucleotides in a process called recombination. This increases the chance of multiple resistant RNA strands. This is why it is so important to slow the replication of HIV—fewer rounds of replication mean fewer RNA strands, thus fewer possibilities of producing HIV that are resistant to AIDS drug cocktails. Current AIDS drug cocktails, especially those using one or more protease inhibitors, quickly and significantly reduce HIV replication—this is why people on these combination therapies must take them on schedule, in prescribed doses, and maybe for the rest of their lives.

Problem—Multidrug Cocktails

Multidrug cocktails work well initially because they are attacking the strains of HIV that are least resistant to the drugs. But the result of that success is to encourage the rapid spread of HIV strains that are highly resistant to the drugs, which could give rise to a new and even more dangerous AIDS epidemic among people most at risk for the disease.

Several separate research investigations, using two new types of sensitive diagnostic tests, found evidence suggesting that about 30% of individuals newly infected with HIV are carrying forms of the virus that are already resistant to one of the 20 drugs, and 14% are resistant to two of the drugs used in American combination therapy. In Great Britain, 13% of the HIV infected are resistant to all three main classes of antiretroviral drugs.

Mutations that confer resistance to nucleoside analog reverse transcriptase inhibitors, nonnucleoside reverse transcriptase inhibitors, and protease inhibitors have all been identified in HIV-infected Americans who have never been treated with antiretroviral drugs. The first Zidovudine-resistant HIV was reported in 1989. Thus, with the wider use of antiretrovirals, pretreatment HIV-resistant strains will occur in the population with increasing frequency.

It is clear that the appearance of drug-resistant HIV is a function of the number of HIV replication cycles that take place during infection, and that combination therapy that suppresses HIV replication to undetectable levels can delay or prevent the emergence of resistant strains (Figure 4-9). Abundant evidence has linked the presence of HIV drug resistance to therapy failure (Moyle, 1995).

During 10 to 20 plus years of a person's infection, HIV can undergo as much genetic change (mutations) as humans might experience in the course of millions of years, which makes it extremely difficult to develop treatments that have long-term effectiveness.

In summary, the frequency of genetically variant HIV within an HIV population is influenced by the mutation rate, fitness of the mutant to survive, the size of the available HIV pool for genetic recombination, and the number of HIV replication cycles.

Coinfection and Superinfection

Coinfection—This is an infection with at least two genetically different strains of HIV from the same clade or group, for example clade B. In February 2000, the first documented case of **HIV coinfection** was reported at the Seventh Conference on Retroviruses and Opportunistic Infections. An HIV-infected male became infected with a second strain of HIV, one more aggressive than the HIV he carried. He acquired the second strain from a different sexual partner who harbored the variant HIV. Prior to his second infection he demonstrated good health. His health declined markedly after he acquired the second infection.

FIGURE 4-9 Production of HIV. Each dark hole in the T4 cell membrane represents the emergence of one new HIV that came off the membrane and is loose in the body. Each HIV-infected T4 cell is producing about 3000 to 4000 viruses at any time. Each HIV in this photograph is genetically different from any other HIV being released from this T4 cell and most likely genetically different from any other HIV in the body. T4 or CD4+ cells live for about 80 days in a healthy person and about 25 days in an AIDS patient. *(Source: CDC, Atlanta)*

Superinfection—This is an infection by strains of HIV from different clades or subtypes, for example A and E or B and C.

At the 2004 Eleventh Conference on Retrovirus and Opportunistic Infections, investigators from Los Angeles and San Diego presented evidence that the annual rate for superinfection in their studies was 5%.

HIV TREATMENT REGIMEN FAILURE

Regimen failure occurs when the anti-HIV medications you are taking do not adequately control infection. There are three basic types of regimen failure:

Virologic failure: Regimens should lower the amount of HIV in your blood to undetectable levels. Virologic failure has occurred if HIV can still be detected in the blood 48 weeks after starting treatment or if it is detected again after treatment had previously lowered your viral load to undetectable.

Immunologic failure: An effective regimen should increase the number of CD4 cells in your blood or at least prevent the number from going down. Immunologic failure has occurred if your CD4+ count decreases below the initial count before starting therapy or does not increase above the initial count within your first year of therapy.

Clinical failure: Clinical failure has occurred if you experience an HIV-related infection or a decline in physical health despite at least three months of anti-HIV treatment.

Virologic failure is the most common kind of regimen failure. People with virologic failure who do not change to an effective drug regimen usually progress to immunologic failure within about three years. Immunologic failure may be followed by clinical failure. Clinical failure results in the use of third line regimens.

Third Line Regimens: Salvage, Mega, or Giga HAART Therapy

Under current definition, a treatment-experienced, or so-called salvage HIV-infected patient, is one who has failed at least three HIV regimens that include at least one drug from each approved drug class. Although the HIV salvage population is on the rise as the life expectancy of patients with HIV increases, they are still underrepresented in new clinical trials because sponsors of new drugs do not expect the drugs to be effective in salvage patients. The reasons that HIV drug regimens lose their effectiveness are as varied as the patients who are undergoing treatment with them. Treatment failures, which can occur in up to 60% of patients, can develop in those with a prolonged history of sequential HIV monotherapy and in those who have built up resistance. They also can occur in patients who fail to comply with their prescribed therapy or who are given a poorly suited regimen at the outset.

Drug testing is underway to design fourth line regimens or **salvage therapies** for patients harboring resistant HIV. Salvage therapy is the use of substitute drugs that will continue to suppress viral replication when standard therapy fails.

The Kitchen Sink—By early 2005 some patients had used all 20 FDA-approved, anti-HIV drugs in order to suppress HIV replication! Clearly some of the drugs were being recycled in a variety of new combinations. The try anything approach—the kitchen sink!

The number of HIV-infected persons in America and Europe requiring salvage therapy is increasing rapidly. In America, between 30% and 50% of those in therapy are now in salvage therapy. Because these therapies are not proving effective, deaths due to AIDS are increasing. Fifteen percent who fail therapy die within three years.

Throughout the developed world, 1 in 10 AIDS patients now requires third line or salvage therapy because they carry HIV resistant to the drugs. About 1 in 50 of these patients is resistant to all 20 FDA-approved antiretroviral drugs (Kenyon, 2001). In 2002, French scientists began to use mega HAART or giga HAART therapy for those with limited treatment options. The theory behind using a larger number of drugs is that not all the virus in a person's body is going to be resistant to all the drugs. By using many drugs with different mechanisms of blocking HIV from reproducing, it may still be possible to achieve a potent anti-HIV effect. This salvage or rescue regime consists of the following drugs: three to four nucleoside reverse transcription inhibitors (NRTIs), hydroxyurea, one or two non-NRTI, and three to four protease inhibitors. Continued studies will

SIDEBAR 4.1

CAN AN HIV DRUG PREVENT HIV INFECTION—PRE-EXPOSURE PROPHYLAXIS?

In 1995, scientists from Gilead and the University of California at Davis published studies in the journal *Science* that showed that a preventive shot of Viread (tenofavir) could block HIV from infecting monkeys (monkeys can be infected with HIV but there is no progression to AIDS as in humans). But it wasn't clear that the drug would block sexual transmission of the virus in people. Gilead chose to focus on its core mission of getting Viread FDA-approved for the treatment of AIDS, which it accomplished in 2001. (Table 4.1)

A number of HIV/AIDS physicians in New York, San Francisco, and Florida began giving Viread to people who believed they might have been exposed to HIV Pre-exposure Prophylaxis (PREP). Although none of these people became HIV positive at the time, it is not hard evidence that Viread can prevent HIV infection or HIV transmission—still the possibility is intriguing. And interesting enough for the National Institutes of Health, Centers for Disease and Prevention, and Bill & Melinda Gates Foundation to fund three separate human studies of the drug to determine whether the drug can prevent HIV infection. The Gates Foundation has awarded a $6.5 million grant to fund a randomized, placebo-controlled clinical trial to evaluate whether Viread is effective at reducing the risk of HIV infection. The trial includes 2,000 volunteers in Cambodia, Ghana, Cameroon, Nigeria and Malawi. All of the study participants received "safer" sex counseling and condoms even though their use may make it more difficult to prove whether the drug works to prevent HIV. NIH awarded $2.1 million grant to University of California-San Francisco researchers to test Viread in 960 Cambodian women—most of whom are prostitutes. In addition, the CDC granted $3.5 million to fund a third study examining the drug's safety as a preventive among sexually active men who have sex with men in San Francisco and Atlanta. The race is on; if the prevention trials are a success, there will be a tremendous number of people who would require Viread. Results of these clinical trials should be available in 2006.

---

determine safety and effectiveness of mega/giga HAART therapies. In November 2003, the CDC published its most recent "Guidelines for Using Antiretroviral Agents Among HIV-Infected Adults and Adolescents."

Genotypic and Phenotypic Drug Resistant/Susceptibility Profiles

Determining the specific drug-resistant and drug susceptibility profiles for a strain of HIV can be very useful in order to establish when a change in therapy will best suit the patient and, if necessary, to begin a plausible salvage therapy.

Establishing a list of genetic mutations or changes in HIV or RNA related to drug resistance is called **genotypic analysis** and indicates a given HIV's genetic resistance to a given drug. Each HIV genetic change or mutation to a drug is listed by a license plate-looking name such as "K103N," which means that none of the non-nukes will work in the patient. Note that when HIV becomes resistant to one drug, it can at the same time become resistant to others that function in the same manner (cross-resistance). The suspected degree of drug resistance, as suggested from genotypic analysis, can be measured directly by adding the drug in question to an HIV-cell culture and determining the HIV's ability to reproduce. A virus-culture-drug assay is called **phenotypic analysis.** A check is made to see if HIV's resistance to a given drug correlates to a given genetic mutation and if other mutations offer a cross-resistance to that drug. In this way, drugs that work against the different HIV mutants can be used in salvage therapy.

Phenotypic results are easier to interpret than genotyping results because they do not require the expert interpretation of complex mutation patterns. The drug susceptibility data provide information for the clinician to select a treatment effective against the viral population circulating in the patient's blood. The main use of phenotypic assays at present is to identify those antiretroviral drugs that still retain activity against the patient's virus. They are also useful to detect transmission of drug-resistant virus and to monitor HIV patients during early viral rebound. In essence, phenotypic testing provides information to target antiretroviral therapy against the predominant HIV variant in the patient.

However, phenotype testing is very expensive at \$800 to \$1000 per test. Because of the time element (about two to five weeks) and cost, such testing is not readily available to the patient. Thus, patients rely on the cheaper (about \$400 to \$500) and faster (one week) genotypic testing. Most third-party payers do not cover these tests. Genotype analysis is now available through five commercial laboratories and phenotype testing in two.

In general, resistance tests can't predict which drugs will work—only the ones that don't—which is why they're mostly recommended to help people whose regimens are failing. As with viral load tests, it's critical to be consistent about which tests you use. For now, experts say, the decision to change drug therapy should be based primarily on increases in viral load.

Measuring HIV Replication

In June 2002, ViroLogic, Inc. announced the launch of the first-ever commercial laboratory test to evaluate the **replication capacity** or **viral fitness** of HIV. Their assay measures the ability of a patient's virus to make copies of itself and is designed to provide useful additional information to physicians to select optimal antiretroviral therapy cocktails for their patients.

Replication Capacity Test

Some drug-resistant HIV cannot replicate as well as non-drug-resistant virus (also called "wild type" virus). As a result, the drug-resistant virus is weaker and may be less able to harm the immune system. There is a test to measure the ability of the virus to replicate in the blood. It's call a viral replication capacity test and is available with certain drug-resistance tests. The replication capacity test removes HIV from the patient's blood and compares it with wild type virus, which replicated quickly. Wild type virus has a replication capacity (RC) of 100%. If the patient's virus has an RC of over 100%, it is a stronger strain of HIV. If it has a RC capacity of less that 100%, it is a weakened strain of HIV. Even if a person has a weakened virus, he or she may still have a high viral load. However, the T4 or CD4+ cell count might not drop. This is because the weak virus is reproducing slowly, and cannot cause as much damage to your CD4+ cells.

Research studies suggest that replication capacity data, in combination with drug resistance information, and viral load and CD4+ cell count, may allow for a more informed prediction of HIV disease progression.

VIRAL LOAD: ITS RELATIONSHIP TO HIV DISEASE AND AIDS

Viral load refers to the number of HIV RNA strands in the blood plasma or serum of HIV-infected persons (a discussion of HIV RNA production is provided in Chapter 3; also see referenced material, Jurriaans et al., 1994; Henrard et al., 1995; Goldschmidt et al., 1998). In general, two to six weeks after HIV exposure, infected individuals develop a high level of blood plasma HIV RNA. Methods now exist to quantitate the amount of HIV RNA in the blood plasma or serum of HIV-infected people. (**Plasma** is a transparent yellow fluid that makes up about 55% of blood volume. Removing fibrinogen and blood clotting factors from plasma results in **serum.**) The reduction of viral load with the use of antiretroviral drugs can slow the destruction of the immune system.

The Purpose of HIV Viral Load Testing

If a patient is not taking anti-HIV drugs, his or her viral load will be monitored during regular clinic visits to provide clues about the likely course of HIV infection if left untreated. Among people with the same T4 or CD4+ cell count, those with higher viral loads tend to have more rapid disease progression than those with lower viral loads.

Physicians recommend viral load testing for the following reasons:

- to help you make decisions about starting or changing drug treatment for HIV;
- to find out your risk for disease progression;
- to show how well your drug regimen is working;
- to help determine your HIV disease stage.

Laboratories report HIV viral load test results as the number of copies of HIV per milliliter of blood plasma (copies/ml). Until recently, HIV viral load tests could only accurately measure HIV levels down to 400 copies. New viral load tests can measure HIV levels down to 50 copies. **An "undetectable" viral load, however, is not enough. The goal is for people with HIV to live out their normal life span, free of opportunistic infections and with the fewest possible drug-related side effects.**

Pitfalls of Viral Load (HIV-RNA) Testing

Despite the potential advantage of HIV-RNA testing over that of conventional HIV-antibody testing—particularly in a high-risk population with symptoms of acute HIV—the specificity of all commercially available HIV-RNA technologies is less than 100% and, as a result, likely to yield a significant number of false positives. According to Chiron Diagnostics, the manufacturer of one such assay, the specificity of Qantiplex-branched DNA technology version 3.0 is approximately 97% at a copy number of less than 150 copies/ml. More conservative published data suggest that the specificity of the three commercially available viral load assays (**bDNA, PCR,** and **NASBA**) may be similar, with a false-positive rate of approximately 5% to 10%. The emotional impact of a false-positive screening HIV-RNA test in recently exposed persons can be extraordinarily damaging. (These tests are explained in Chapter 12).

When Is Viral RNA Found in the Plasma?

HIV RNA strands are present during *all stages* of the disease, and the viral load increases with more advanced disease (Piatak et al., 1993; Saksela et al., 1994). Following infection with HIV, there is usually a rapid increase of HIV proteins and RNA followed by a lengthy period of viral RNA replication at lower but measurable amounts (Figure 4-10). In well-characterized groups with known dates of HIV seroconversion, a high viral load immediately

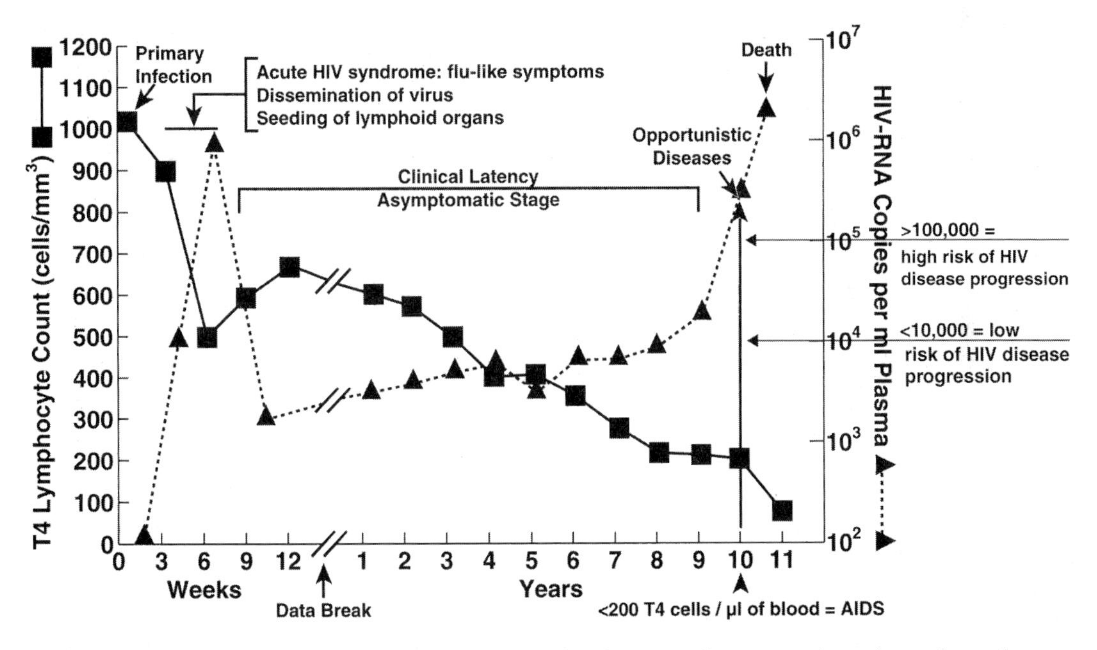

FIGURE 4-10 Clinical Course of HIV Disease as Related to T4 Cell Count and Level (number) of HIV RNA Copies (Viral Load) in Plasma. In an Average Patient Without Antiretroviral Therapy: Primary Infection to Death. During primary infection (the time period between infection and development of HIV antibody), HIV RNA levels spike *(triangles)* and HIV disseminates throughout the body. This is followed by an abrupt drop in measurable HIV RNA in the blood (probably due to the production of HIV antibody), followed by a steady rise in HIV RNA until death. Over these same time periods, there is a continuous loss of T4 cells *(squares).* Note that the asymptomatic stage can be quite long—on average about 10 to 12 years, at which time T4 cell counts drop and HIV RNA copies increase to levels where there is high risk of opportunistic infections.

SIDEBAR 4.2

PERSONAL TESTIMONY ON HIV DRUG-RESISTANT MUTANTS

January 2001—The disparaging news from my doctor was to be expected: My latest phenotype test, a blood test that shows how much the HIV in my body has mutated, was the worst he'd ever seen. In fact, he said, the test had to be run several times because the mutations were off the map and could not be read by the technicians. It wasn't really surprising to me; since 1989 I have been on all but one anti-HIV drug. I knew that I was not responding well to the armament of drugs available today. My CD4+ cell count, after being stable for five years, had begun to drop. Plus, I knew that my virus levels were rising. I feared multiple drug failure and an ultimate virus victory.

Even though I have slowed HIV for 10 years with various combinations of the imperfect drugs, I was forced to once again come up with a new plan. I had been searching for an easier, more tolerable combination of some of the newer drugs because I was tired of the relentless popping of over 20 pills a day. I was weary of the constant bowel changes, nausea, and insomnia from my current regimen. I needed at least two new drugs, preferably three to come up with a new, viable treatment option that would once again buy me more time. Adding just one anti-HIV drug to older drugs I was now resistant to would be like taking a single drug. Fortunately, several new drugs are now available for people with my dilemma. However, getting access to the drugs would prove to be troublesome. I soon realized it is just as difficult to obtain new drugs today as it was in the earlier days of the epidemic.

One new drug, Kaletra, a protease inhibitor, was recently approved and thankfully a prescription from my doctor would ensure I get my hands on it. An easy choice, given current positive data on the effectiveness of the drug, and I knew it was one protease inhibitor I had not burned up. The other drugs I needed are not yet approved and adding them to my new combination has been a painful process of persuasion, deliberation, and stressful anticipation. Even though the desperation to get new drugs is not the same as it was even five years ago, the companies have come up with clinical trial and program designs that are not always in the best interest of the patients most in need. As usual, the companies have their own concerns with drug development.

It's a very different time today in gaining access to experimental anti-HIV drugs. The drug access programs are simply not what they used to be. Activists' negotiations and constant pressure have only forced imperfect programs for people with few options. The definition of "constructing a viable treatment option" has sadly changed as activists become less involved in the fight, and a more seasoned drug industry is able to have its way. *(With permission. Matt Sharp,* Bay Area Reporter, *January 11, 2001, pp. 1–2.)*

after seroconversion (Mellors et al., 1995) and at three years after seroconversion (Jurriaans et al., 1994) appear to be strong predictors of HIV disease progression. However, for most patients with asymptomatic HIV infection, the time of the primary infection (when they were infected) is unknown. The levels of viral load prior to the diagnosis are also unknown.

MEDICAL COMPLICATIONS ASSOCIATED WITH ANTI-HIV DRUG THERAPIES

The term "medical complications" is used herein to describe clinical problems in the management of HIV infection. Most doctors now agree that not everyone infected with HIV needs to take antiretroviral drugs. But everyone with HIV does need medical monitoring and care and access to treatment when and if appropriate. Because those taking antiretroviral treatment have had a fraction of the death rate of those without treatment does not necessarily mean that one's chance of survival will be correspondingly increased by antiretrovirals. The reduced death rates reflect the benefit of treatment for those who have needed and responded to it. (Bozette et al., 2001; Freedberg et al., 2001).

For those who think AIDS is over—it is not.

For those who think the drugs are working well—they do for some, for a while.

Drug Holidays

Over the past several years, those who take or have taken antiretroviral drugs know it is a hard life—for many there is no alternative. But the longer one uses these drugs, the greater the chance of suffering from the toxic side effects. It has been reported that some people are taking eight or more of the antiretroviral drugs in their cocktail. They are running out of options. One cannot say how long an HIV-infected person on antiretroviral drugs can expect to live—or how well. Because each person's lifestyle, healthcare, ability not to become depressed, and genetic immune response varies, there is no way to predict how well or long one can expect to stay alive with current anti-HIV treatments. The 5- or 10-year death sentence has been reprieved for many on treatment. But how long before their drug toxicities kill, or force them off treatments? When People Living With AIDS (PLWAs) started experimenting with going off drugs in order to avoid toxic buildup, healthcare experts coined the phrase **"drug holidays"** to describe such behavior. Pressure to remain compliant was so severe that when patients stopped taking the drugs (because they were often being hurt by them), doctors became angry and patients were accused of laziness or lack of fortitude. Only later did activists and doctors coin a new phrase **"strategic drug interruption"** or **"structured treatment interruptions,"** stopping HIV drugs in order to function or survive. It is now common practice. However, one cannot stay off the drugs too long because HIV will ravage the immune system.

Side Effects of HAART

AIDS really is not yet manageable if it is not survivable. Rounding out what has been learned about anti-HIV drug toxicity is that HIV drugs are implicated in lipodystrophy, heart problems, bone loss (hip replacement), intercellular malfunctions called mitochondriosis (my-toe-con-dree-o-sis) lactic acidosis, and liver and kidney dysfunction and failure. Mitochondriosis is mitochondrial dysfunction—some side effects include fat redistribution syndrome, commonly referred to as lipodystrophy. Lipodystrophy includes increased fat in the neck or stomach; loss of subcutaneous fat from the face, arms, or legs; or enlarged breasts. Additional side effects associated with mitochondrial toxicity that interrupts cellular energy production that leads to fatigue are shortness of breath, weight loss, rapid heart beat, hair loss, numbness and pain in the hands, arms, feet, and legs, muscle disease, heart disease, inflammation of the pancreas, increased blood acidity, and kidney irregularities. Add in various opportunistic infections and the drug toxicities required to treat them, and one finds many surviving AIDS patients have become antiviral and antibiotic multidrug resistant. These AIDS patients require stronger and more toxic remedies month after month and year after year. Ultimately one eventually dies, either from infection or from complications from treatments.

In 2003, N. Friis-Moller of the Copenhagen HIV program and his colleagues presented new data at the Tenth Conference on Retroviruses and Opportunistic Infections demonstrating that the risk of myocardial infarction increases by 27% for each year that HIV-positive patients remain on antiretroviral therapy.

Management of HIV/AIDS: Treat Early, Treat Late?

Building the Ship as It Sails

There are at least three major decisions that have to be made early in the management of HIV/AIDS patients. The **first** is when to begin antiretroviral therapy, the **second** is which drugs to use to initiate treatment, and the **third** is how long to administer the therapy.

Clinicians face a dilemma of choice when advising asymptomatic patients with established HIV infection on when to begin highly active antiretroviral therapy (HAART) and what drugs to use. Begin early, some researchers advise, because later there will be a higher virologic hurdle to overcome. Begin later, others recommend, and save potent drugs until the patient's immune system begins to fail. If therapy is started too early, cumulative side effects of the drugs used and the development of multidrug resistance may outweigh the net benefits of the lengthening of life. If therapy is started too late, increases in disease progression and mortality outweigh the risk of adverse events (Table 4-3).

In order to access the very best time to initiate antiretroviral therapy, one needs to know the goal of antiviral therapy. Most HIV/AIDS physicians would agree that the goal is multidimensional: to prolong life while improving the quality of life; to suppress HIV replication to the limits of detection for as long as possible; to select the best possible therapy for the individual; and to minimize costs

Table 4-3 Risks and Benefits of Early Versus Delayed Initiation of Antiretroviral Therapy

Benefits	Risks
EARLY THERAPY	
Control of viral replication may be easier to achieve and maintain	Drug-related reduction in quality of life
Possible delay or prevention of immune system compromise	Greater cumulative drug-related adverse events
Lower risk of resistance with optimal viral suppression	Earlier emergence of drug resistance if viral suppression is suboptimal
Possible decreased risk of HIV transmission	Limitation of future antiretroviral treatment options
DELAYED THERAPY	
Avoid negative effects on quality of life	Possible risk of irreversible immune system depletion
Avoid drug-related adverse events	Possible greater difficulty in suppressing viral replication
Delay emergence of drug resistance	Possible increased risk of HIV transmission
Preserve maximum number of future drug options when HIV disease risk is highest	

(Adapted from U.S. Department of Health and Human Services, 2002)

The optimal time to initiate antiretroviral therapy in asymptomatic HIV-infected patients with CD4+ cell counts above 200/μL is not known and remains a controversial issue. The rationale for starting therapy early includes the potential for improved virologic suppression, preservation of immune function, and reduction in sexual and perinatal HIV transmission. The rationale for later initiation of therapy includes the potential avoidance of drug resistance and adverse effects and difficulty in adherence to complex drug regimens, as well as reduced cumulative cost of treatment. Commonly considered risks and benefits associated with the approaches of early versus later initiation are shown in the table. At the Tenth Conference on Retroviruses and Opportunistic Diseases, Martin Hirsch provided an excellent summary of the current data. Patients do not do as well on HAART if they start when their CD4+ count is <200 cells/μL or viral load is >100,000 cells/mL, so it is clear that therapy should be started before patients reach these benchmarks. However, the T4 or CD4+ count above 200 cells/μL at which therapy should be started is unclear, and opinions will likely change as therapy becomes more convenient, tolerable, and less toxic.

POINT OF INFORMATION 4.2

DETERMINATION OF VIRAL LOAD AND QUESTIONS PEOPLE ASK

Viral load blood tests have been developed to show how much HIV is in the blood. Through 2003 only two have been licensed by the FDA. Not all doctors order viral load tests because not all insurance companies pay for them. However, as viral load tests become more widely available, insurance coverage will be there.

The amount of HIV in a patient's blood is important, especially since only about 2% of HIV is in circulating blood. The other 98% is in the lymph system and other body tissue. A low level indicates the disease is stable and a high level may mean treatment should begin or be changed.

AVAILABLE VIRAL LOAD TESTS

Four tests have been developed: through 2004 only two have received FDA approval, Amplicors Q-PCR (quantitative polymerase chain reaction, Figure 12-5) and Amplicor's HIV Monitor Ultra Sensitive test that measures viral load down to 50 copies. The other two tests are the bDNA (branch-chain DNA, Figure 12-7) and NASBA (nucleic acid sequence-based amplification). All tests measure the amount of HIV RNA present in the blood of HIV-infected patients. Test results are usually given per milliliter (mL) of blood. Because each HIV carries two copies of RNA, if there are 100,000 copies of HIV RNA it means there are 50,000 copies of the virus present.

RESULTS OF VIRAL LOAD TESTS

Results of viral load tests can range from almost zero to over a million; low numbers mean fewer viruses in the blood and less active disease; high numbers mean more active disease.

The three tests usually give similar results. However, if the patient's viral load is very low, the bDNA may not detect the HIV RNA. Newer versions of all tests may be more accurate and more sensitive. In general, it is best to stay with one type of test for consistency.

RELEVANCE OF VIRAL LOAD TESTING

The test is new and researchers are still trying to decide on the important values for viral load. In general, though, results can be interpreted as follows:

HIV Viral Load Result	**Interpretation** Death within 6 years
5000 or fewer	6%
10,000 or fewer	18%
10,000 to 30,000	35%
Over 30,000 copies	70%

Viral load numbers will depend on the viral load test used. For example, a 25,000 RNA strand number determined using a branched DNA (bDNA) test might be read as 50,000 RNA strands using Amplicors' FDA-approved PCR-RT (polymerase chain reaction-reverse transcriptase) test and higher or lower numbers using other HIV RNA testing procedures. Overall, however, the lower the viral load, the better. But no single test of any type gives a complete picture of health. Trends over time are important, as are nutrition, physical condition, psychological outlook, and other factors. Experimental viral load tests can now detect as few as 5 to 20 RNA strands/mL. Current tests in use measure between 50 and 400 RNA strands/mL.

DETECTABLE VERSUS UNDETECTABLE LEVELS OF HIV RNA

The most immediate measure of the effectiveness of antiretroviral therapy is the viral load test. The lower a person's viral load, the less likely they are to progress to AIDS. The effectiveness of anti-HIV drugs is usually measured in terms of the percentage of people who achieve viral loads below 50 and 400 copies/mL after a set period of time. For example, a trial will find out how many people on a particular combination are below 50 copies after six months of treatment.

What does an **undetectable** level of HIV RNA mean? Many individuals now have an undetectable level of HIV RNA in their blood after taking combination therapy. But undetectable RNA levels do not mean that the person is cured or no longer infectious. An undetectable RNA level means that too few RNA strands are present to measure. The number of RNA strands is below the level of tests of sensitivity.

Achieving undetectable viral loads is now the gold standard for successful HIV treatment. The more rapidly one's viral load becomes undetectable and the longer it remains suppressed, the better the drug and chance for long-term survival. A measurable viral load after it was once undetectable is regarded as a sign of drug failure. Never reaching undetectability is also considered a drug failure for that person.

Jane Simoni and colleagues (2003) reported that, on average, only 50% of AIDS patients on antiretroviral therapy (HAART) achieve HIV RNA levels below detection limits. The major reason for not doing so is patients' poor adherence to medical regimens (see Point of Information 4.3). A few studies have shown that some people have maintained undetectable levels of HIV for the past 10 years.

Resistance Testing

If the viral load rises above 1000 copies/mL then resistance tests can be performed to see which drug or drugs being taken are becoming less effective. HIV that has developed resistance to one drug may also be cross-resistant to other similar drugs not taken. A resistance test should also indicate which drugs will be most effective. In order to keep future treatment options as open as possible, some doctors argue that the aim of treatment should always be to reach an undetectable viral load. However, some doctors argue that in some people, particularly for those who are on their second or later combinations, this might not be possible. This means that people must switch to drugs that are still useful until they eventually run out of treatment options.

TYPICAL QUESTIONS ABOUT VIRAL LOAD

1. **What is the viral set point?** The viral set point represents an equilibrium between the virus, which wants to replicate, and the host, which tries to control or contain the replication (viral reproduction is about equal to viral clearance).

It appears that each patient establishes his or her own set point within two to six months after initial infection with HIV. HIV drugs appear to alter the set point for HIV by reducing its ability to replicate. This helps lower the body viral load and gives the body a chance to restore some of the lost immune function.

2. **What is the cost of a viral load test? Does insurance pay for it?** Beginning 2003, a viral load test cost between $150 and $250. Some insurance policies pay for the test.
3. **If a viral load test result was 12,000 and six months later it is 18,000, is the change important?** Viral load may rise and fall without any change in health. The rule of thumb is that viral load change is not significant unless it more than doubles.
4. **How often should viral load be measured?** Measuring viral load every four months, at the same time T4 cells are measured, has been suggested.
5. **Can viral load be associated with determining anti-HIV therapy?** Measuring viral load about four weeks after changing antiretroviral medications can be useful for finding out whether the new treatment is working.
6. **A person's T4 cells had been between 400 and 600 for several years, and suddenly his viral load rose to 50,000 and his physician recommended drug therapy. Why?** Research suggests that people with viral loads over 10,000 are at higher risk for the progression of HIV disease, regardless of T4 cell counts.

and all side effects to drug therapy. With these goals in mind, and the drugs available to suppress HIV replication and extend life with quality, the first question remains "When should one begin antiretroviral therapy?"

2004 U.S. Department of Health and Human Services (DHHS) Guidelines and International AIDS Society-USA (IAS-USA) Recommendations for the Use of Antiretroviral Therapy

There are no "one size fits all" treatment guidelines. Over the last 10 years it has been learned that antiretroviral therapy is becoming more individualized.

In general, the DHHS guidelines and IAS-USA recommendations are the most aggressive on when to start therapy, followed by French guidelines. The Brazilians are the least aggressive, recommending dual nucleoside therapy for those with earlier disease and lower viral loads. Overall, the various guidelines involve expensive strategies that can only be used in more developed countries, or in the wealthier sectors of less developed countries.

The current guidelines now list T4 or CD4+ count below 350 cells/mm^3 as the most significant threshold to consider in deciding when antiretroviral treatment is recommended, or should at least be offered to the patient. All HIV-positive individuals who are symptomatic or are aymptomatic and have a CD4+ count below 200 cells/mm^3—irrespective of the viral load—should be receiving antiretroviral treatment. Patients with a CD4+ count above 200 cells/mm^3, but below 350 cells/mm^3, should be offered treatment, irrespective of the viral load, although this is still considered to be a controversial recommendation. Recent studies published in 2004 (SAX, 2004) suggest that waiting until the CD4+ count falls below 200 causes an irreversible immunosuppression meaning that certain aspects of the immune system have been damaged beyond repair regardless of which drugs are used in therapy. Important in the Updated Guidelines is the list of antiviral drugs and drug combinations that should not be offered at any time. Theoretically, a physician has 8,000 separate three-drug combinations available to him. But many of the combinations will be harmful. The question is which? (20 FDA-approved drugs used in combinations of three at a time equals 20^3 equals 8000 possibilities.)

Current DHHS policy suggests that the virus should be **Hit Later-Hit Carefully** rather than **Hit Early-Hit Hard.** This suggests that relatively healthy, HIV-infected people should not receive early HAART therapy. The new British guidelines delay the start of therapy even longer. They recommend therapy when the T4 or CD4+ cell count drops below 200, regardless of viral load.

POINT TO PONDER 4.1

RATIO OF HIV/AIDS PHYSICIANS TO THE POPULATION AND THE HIV-INFECTED

In recent opinion polls of physicians experienced in treating HIV-infected people, the question asked was, "What is the most important factor to be considered once one becomes HIV infected?" **Answer:** Overwhelmingly the physicians agreed that the first thing was to select a **physician experienced** in treating HIV-infected people. Given that as the best advice, just how many HIV therapy experiments are available to the infected?

In America there are an estimated **5,000** HIV/AIDS-experienced physicians. **Given that an estimated 1,000,000 HIV-infected need the care of these physicians, the ratio drops to 1 experienced physician for every 200 HIV-infected people!** (Data from the CDC.) In contrast, in Northern India within a land mass of the size and population of the United States, there are an estimated **7** experienced HIV/AIDS physicians (Katz, 2000). **Given that an estimated 2 million HIV-infected need the care of these 7 physicians, the ratio is only 1 experienced HIV physician for every 286,000 HIV-infected people!**

In South Africa the ratio is 1 experienced HIV physician for every 212,000 HIV-infected people! In Vietnam there is 1 HIV specialist for 11,250 HIV infected people; in Thailand the ratio is 1 for 6,750; in China the ratio is 1 for 5,000 HIV; in Indonesia the ratio is 1 for 4,630; in the Philippines the ratio is 1 for 875; in Singapore the ratio is 1 for 375; in Hong Kong the ratio is 1 for 300; in Taiwan the ratio is 1 for 100; and in Japan the ratio is 1 for 24. (Data from the World Health Organization.) These data indicate that even if the antiretroviral drugs were available and affordable, in countries with poor medical and distribution infrastructure, the presence of these drugs may not in itself be of great significance.

In November 2002, the state of California became the first in the nation to enact a law requiring Health Maintenance Organizations (HMO) to make HIV expert physician consultation available to their members.

Traditional Medicine, 2004

Long before medically schooled doctors became available, medical help came from those who, for one reason or another, were believed to have special spiritual powers or knowledge that if followed would lead to healing. Such healing was taken as a matter of faith or faith healing. There were also those who became familiar with use of herbs to heal. They concocted a wide variety of potions that, in many cases, did help in the healing process. Over the centuries, spiritual comfort and herbal potions became known as Traditional Medicine (TM). Traditional healers practice in all countries but are used less in Western nations who have substituted manufactured drugs to help the healing process. In Africa, about 80% of the population use Traditional Medicine to help meet primary health care needs. In Latin America, 71% of the population of Chile and 40% of Colombians use TM. In China, TM accounts for 40% of all health care delivered. In Ghana and Kenya, a course of antimalarial drugs can cost several U.S. dollars, but per capita out of pocket health expenditures are only about $6. Herbal medicines for malaria are much cheaper and often paid for in kind or according to the wealth of the client. In Uganda, the ratio of TM practitioners to population is between 1:200 and 1:400.

DISCUSSION QUESTION: Offer pro/con information of significance that would alter or eliminate the infrastructure problems in getting the antiretroviral drugs to those who need them. How best can we produce HIV-experienced physicians rapidly?

The Future for HAART=SMART?

The National Institute of Allergies and Infectious Disease (NIAID) is funding a study to resolve which is the best HAART strategy—Hit Early-Hit Hard or Hit Later-Hit Carefully or Hit and Run, that is, Interrupted Therapy. The major reasons for the study are the growing concerns about the drugs' toxic side effects and the increase in resistance to them. This critical long-term study to determine which of the common HIV treatment strategies ultimately is better began in January 2002 at 21 national locations in 14 states and the District of Columbia, and several sites in Australia. The study called SMART, or Strategies for Management of Anti-Retroviral Therapies, will

SIDEBAR 4.3

MONETARY REASONS FOR PRODUCING ANTIRETROVIRAL DRUGS

From a drug developer's standpoint, AIDS must be the most attractive disease on Earth. There are four obvious reasons: (1) the drugs are very expensive, (2) many patients are privately insured because they are healthy and working, (3) thanks to advocacy groups the government pays for drugs for many people who can't afford them, and (4) the drugs don't cure anybody—they appear to be necessary for a lifetime! Where else in American medicine is there a population of uninsured patients of whom 79% have access to HIV/AIDS antiretroviral drug therapy that costs between $12,000 and $20,000 a year? Is there anywhere else in American medicine where such expensive health care has been made available to the poor and disenfranchised populations? According to the U.S. Department of Health and Human Services, 60% of those needing HAART therapy receive it.

Economics of Producing Antiretroviral Drugs

In 1998, the markets for HIV/AIDS therapeutics totaled more than $2 billion in U.S. revenues and according to industry consultants Frost & Sullivan (http://www.frost.com), these markets are projected to reach nearly $13 billion by 2005. These figures represent revenues derived from the following HIV drug product classes: protease inhibitors (PIs), nucleoside reverse transcriptase inhibitors (NRTIs), and non-nucleoside reverse transcriptase inhibitors (NNRTIs).

Death Prevention Profits

In 1995 there were 51,000 AIDS-related deaths. Using the new antiretroviral drugs, in 1996 the number of deaths fell to 38,000. By 1998 deaths dropped to 18,000, and by 2002 to 16,000. Using 1995 as a baseline for number of deaths, each year thereafter the fewer that died, the greater the number that must continue to take the drugs for the rest of their lives. Until there are preventative and therapeutic vaccines, or some form of cure available, the number of drug consumers continues to grow. Just in the United States there were an estimated 50,000 new HIV infections in 2001, 2002, and in 2003. Why wouldn't a company commit itself to producing HIV/AIDS drugs? Toby Casper (2000) reported that the profits on antiretroviral drugs are enormous. He gave the following examples: In 1999, Glaxo Wellcome made $589 million on $4.1 billion of Combivir (AZT + 3TC) sales. Profits between 1997 and 2001 for AZT, about $800 million; for 3TC about $1.6 billion; for ddI about $400 million and for d4T $1.4 billion. Yielding to political and foreign competitive drug production and pricing, the world's major drug companies like Boehringer, Glaxo SmithKline, Roche, Merck, and Bristol-Myers Squibb have drastically lowered their drug pricing policies in the underdeveloped nations in Africa and Asia. They say that their prices are near to or at cost.

eventually enroll 6000 people who will be monitored for up to nine years. The study is being conducted by the Community Programs for Clinical Research on AIDS (CPCRA), a network of community-based researchers.

SMART will compare two distinct treatment approaches and will follow participants for an average of seven years. In addition, while most AIDS treatment trials measure indirect indicators of AIDS development, such as the amount of virus or number of CD4+ T cells in the blood, SMART will measure clinical events such as progression to AIDS or to death, which take longer to occur. The study's length is a major reason for great emphasis on patient and physician education both prior to enrollment and the study. SMART's enrollment criteria are broad—teenagers as well as adults are eligible—so the findings will be applicable to as many people as possible.

TWO MOST SOBERING REPORTS ON AIDS-RELATED DEATHS

James Witek and colleagues reported that "among an urban population representative of today's pandemic, mortality continues to be significant even with early access to HIV care and HAART," and that "patients who died in 1999 had lower viral loads on presentation to care, longer time in care, and higher final T4 counts. Those who died in 1999 had taken more antiretroviral regimens (3 versus 2), had better adherence and appeared more likely to have had a virologic response to HAART (59% versus 16%). Eleven

POINT OF INFORMATION 4.3

CURRENT PROBLEMS USING ANTI-HIV THERAPY

First, duration: Many patients ask how long they will have to continue therapy; how long will they have to be harnessed to pills and doctors? But why would they ask when it's keeping them alive? Often individuals take 20 to 70 or more antiretroviral and opportunistic disease pills a day that must be swallowed according to a rigid schedule. The timing centers around an empty stomach, and the drugs often cause severe stomach irritation. Pills rule life (see Figure 4-11).

Second, Adherence or Compliance: You don't have to like them, you just have to take them! Health care providers have been dealing with the issue of client **adherence** or **compliance** for centuries. The medical literature shows that it is difficult for patients to adhere to even the simplest treatment regimens. Factors associated with poor adherence include unstable housing, mental illness, and major life crises. **Adherence** to a drug regimen means taking all the prescribed anti-HIV drugs at the scheduled times and not missing any doses. Any time people are asked to change and/or maintain new behaviors to treat an existing condition or to prevent a threatened one, there is a good chance that they will not comply, consistently and correctly, to the prescribed activities. Combination drug therapy for HIV infection does not cause new compliance issues; it just highlights the known difficulties of an existing problem. **The Achilles heel of anti-HIV therapy:** Skipping only a few pills can trigger the emergence of drug-resistant strains of HIV. Such a development could create a worse problem than the initial infection because the resistant virus could overwhelm the individual taking the drugs and anyone else to whom the individual transmitted the virus.

There is a present danger that the behavior of underdosing (not taking enough) or partial compliance (taking the protease inhibitor when they feel like it) or patients who modify their dosage regimens—to extend their prescription—may create HIV strains resistant to all currently available drugs. This would lead to an even more devastating AIDS pandemic! The major reason given by HIV-positive people on anti-HIV therapy for missing doses was, I FORGOT! (33%). Clearly, the degree of compliance to therapy affects treatment outcome.

Stephen Becker and colleagues (2002) report a mean overall adherence rate of only 53% among a population of over 3700 Medicaid recipients. Barely one-quarter of subjects achieved an adherence rate of 80% or higher, and adherence was worst among those aged 18–24 years. For an HIV class experience on adherence at the University of North Florida, 50 students aged 18–24 were asked to take either three or four different-colored M&M candies representing four different antiretroviral drugs, at three specified times per day between six in the morning and midnight for three days (Friday through Saturday). They were given the colored M&Ms and a dosing schedule sheet to record

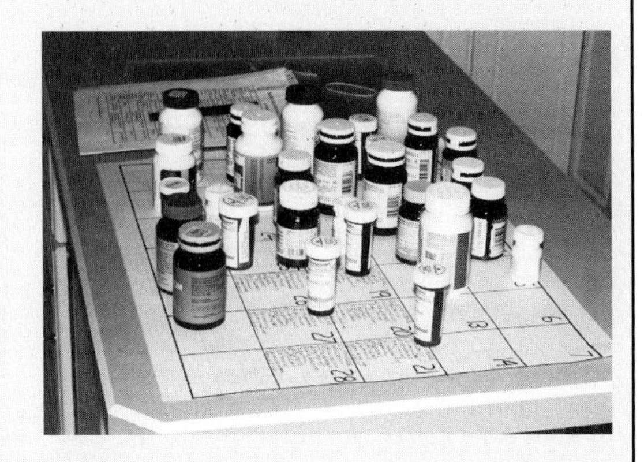

FIGURE 4-11 Patient Compliance. People with HIV disease and AIDS take many different drugs daily, depending on their prescribed medications for the various opportunistic infections and combinations of drugs necessary for anti-HIV therapy. The patient's drug intake, when this photograph was taken, numbered 28 different pills daily. If this patient took 28 pills each day for 10 years, this person would ingest 102,200 pills. Is it any wonder that people fail to take all of their medications? By mid-2003, at least three nucleoside transcriptase inhibitors (didanosine, lamivudine, and Zidovudine), one non-nucleoside (efavirenz), and one protease inhibitor (atazanavir) were approved for once-a-day dosing and other antivirals have also been reduced in daily dosing, making adherence easier. Starting in 2003, patients began taking reformulated and combination drug pills, which has reduced the 10 to 20 pills taken two or three times a day to taking one, two, or three pills twice a day and some pills once a day. *(Photograph courtesy of the author)*

their compliance. After completion of the experiment, class compliance at 100%, 90%, 80%, and 70% was zero! Their overall comment—now we can better relate to what it must be like for those who must be compliant.

DISCUSSION QUESTION: Do you think the results would be much different in your college or university HIV/AIDS class?

Clearly much more needs to be done to improve rates of adherence, or many patients will be embarking on a rapid route to resistance and treatment failure.

Third, costs: AIDS is a disease of poverty. Many of the HIV-infected in the United States and worldwide will *never* get a first dose of a protease inhibitor or an AIDS cocktail. The cost of *just* protease inhibitors in the United States is between $12,000 and $20,000 per year! To treat the estimated one million HIV-infected would cost about $15 billion per year. Over half the costs are for antiretroviral medication—the pills can cost between $5 and $16 each. Worldwide, *if* people received equal treatment, the bill for all people living with HIV/AIDS outside the United States beginning year 2003 would be about $160 billion per year.

Currently (2005) less than 3% of the 40 million people in developing nations who need HAART are receiving the drugs. According to Peter Piot, executive director of UNAIDS, beginning in 2003 in China with about 1.5 million HIV-infected, about 100 were on HAART, in Africa about 40,000 HIV-infected people are receiving HAART and 2.3 million died of AIDS in 2003. In North America, over 500,000 are on HAART with 25,000 deaths. Worldwide over 6 million people should be on HAART. About 8% are. Under new pricing arrangements, HAART drugs can be purchased in Africa for as little as 67 cents a day. But for most, this sum can be too costly. In Uganda, for example, gross national income per capita is about $300 a year, or 83 cents a day.

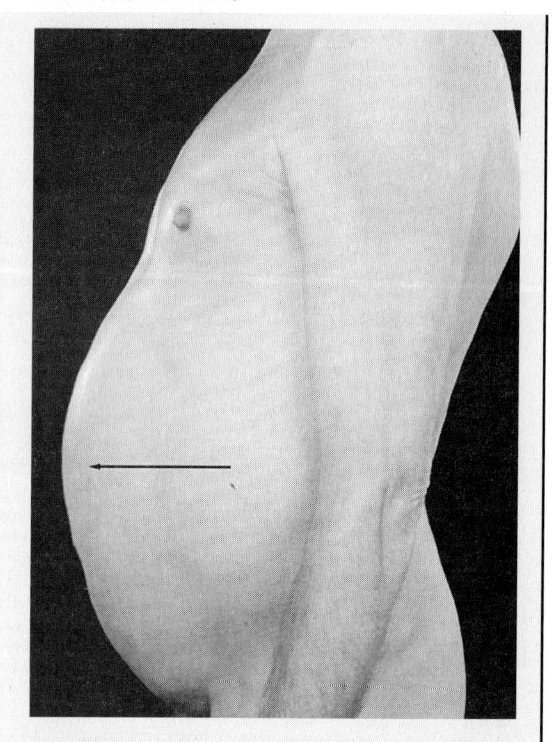

FIGURE 4-12 Protease Paunch—Also Referred to as Crix Belly. This lipid disorder (lipodystrophy) occurs in many patients using protease inhibitors for anti-HIV therapy. In addition to the visual effects of antiretroviral therapy, there are a number of dangerous side effects—see text for details. *(Photograph courtesy of Dr. David Cooper)*

Antiretroviral Drugs as a Viable Option in Africa

In addition to the cost of drugs, another major obstacle that has to be considered is food or the lack of it. Hundreds of thousands get to eat, at most, every other day! Many of the drugs have to be taken with food and clean water, another problem. But what if suddenly the drugs were available to all? Who would monitor those on drugs as to their timely adherence? Most do not own a watch. Who would diagnose the disease and prescribe the drugs? As found in developed nations, drug regimes are based on the individual's needs. Yet, as seen in Point of Information 4.1, HIV/AIDS physicians are in extremely short supply in Africa. Who would monitor the change in drug regimens as drug resistance emerges? Medical problems in addition to AIDS abound in Africa—tuberculosis, malaria, measles, and a large list of other adult and childhood diseases that currently kill millions each year. The problem of health and health care in Africa has been out of control for centuries. HIV/AIDS appears to have awakened the continent but Western-style medicine and care is a long way off.

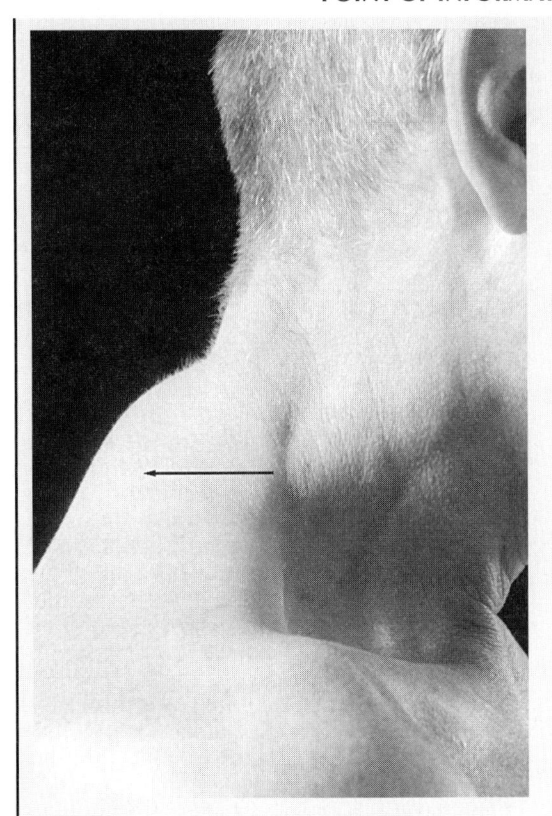

FIGURE 4-13 Buffalo Hump. The enlargement of a cervicodorsal fat pad. Buffalo hump develops after use of protease inhibitors for anti-HIV therapy. In addition to the visual effects of antiretroviral therapy, there are a number of other dangerous side effects—see text for details. *(Photograph courtesy of Dr. David Cooper)*

Drug Costs in America

Larry Kramer, a cofounder of Gay Men's Health Crises, states that the cost of his drugs to combat AIDS, which do not include a protease, amounts to about $19,000 a year; this does not include visits to his doctor or the batteries of blood tests he routinely requires. And he is asymptomatic. A *New York Times* article in 1997 estimated that drugs for someone with symptomatic AIDS cost about $70,000 a year; in response, a New York University adjunct law professor and gay-rights advocate wrote a letter to the editor saying that his drugs cost $84,000 a year using protease inhibitors; the annual drug cost can exceed $150,000. At these prices, how many of the nation's 1,000,000 HIV-infected will be able to afford proper HIV therapy? **AT THESE PRICES PEOPLE HAVE TO CHOOSE WHETHER TO PAY RENT, BUY FOOD, OR PAY FOR THEIR MEDICINE—SOME CHOICE!**

Fourth, side effects: The paradox of HIV treatment is that sometimes the cure feels worse than the disease, especially when treatment begins before symptoms arise. Sometimes the cure also looks worse than the disease. For example, fat redistribution can appear at a time when a person's HIV might otherwise be invisible—both to others and to him or herself. To date, all FDA-approved protease inhibitors (PIs) have important side effects in a large number of PI users (25% to 65% in different surveys). To name a few, PIs have been associated with: dry skin, cracked lips, loss of body hair, sexual dysfunction in men, oral warts, loss of bone mineral density, gastrointestinal problems (diarrhea, etc.), high triglycerides and cholesterol levels associated with coronary artery disease, increased blood sugar levels and diabetes, and abnormal body fat metabolism and distribution or **lipodystrophy.** This occurs particularly on the abdomen, central obesity, referred to as the **protease paunch** (Fig-ure 4-12), and on the back between the shoulder blades, called **"buffalo hump"** (Figure 4-13) and may include a loss of fatty tissue in the arms, legs, and face. Women may experience narrowing of the hips and breast enlargement (Lo et al., 1998; Miller et al., 1998; Lipsky, 1998; Carr et al., 1998; A-E and Gervasoni et al., 1999). Although abnormal fat distribution may not be considered serious in life and death situations, it probably does matter to the asymptomatic person who cares about body image and who cares about cardiovascular problems in the future. HIV-infected children placed on HAART therapy also experienced lipodystrophy (Vigan et al., 2003).

Some of the newer, not yet FDA-approved PIs may cause kidney failure and liver problems. PIs also appear to increase blood levels of Viagra, the new male impotence drug by inhibiting liver enzymes that would normally eliminate Viagra.

Fifth, weighing the unknown: Without taking any drugs, about 50% of the HIV-infected will be well some 10 years after they have become infected!

Help: for those experiencing drug side effects or who want to know more about the drugs, call:

AIDS Treatment News: 1-800-341-2437
Project Inform: 1-800-822-7422
Women Alive: 1-800-554-4876
World: 510-658-6930

TESTIMONY ON SIDE EFFECTS: ONE WOMAN'S EXPERIENCE

(This anecdotal account was sent to the author from Ms. Colleen Perez and is used with her permission and the author's appreciation.)

I am a single parent with three children—they are the medicine that lights my heart afire and keeps me going! The side effects I lived with from taking Saquinavir/Fortovase (not simultaneously), 3TC, and AZT were a nightmare. Firstly, I suffered from a severe peripheral neuropathy in my arms. I'd awaken drenched with sweat, unable to feel my arms, that is until the shooting pains hit! Then came severe edema, I went from 145 lbs. to 180 and still climbing in less than one year. I looked full term 9 months pregnant. In fact, not a day went by that someone didn't ask me when the baby was due? You may initially think me vain, but let me tell you that I labored to breathe, and had great difficulty making the smallest movement. I also suffered from acid reflux disease that was so bad I was nauseous, burping, and farting uncontrollably, and constantly! Having always been fairly fit and athletic in the past, those drugs made my life little more than a living hell. The cost of the drugs is around $1,200 a month. The tab is picked up by our local MIP (Medically Indigenous Person) and somewhat federally funded health insurance plan. The pharmacist here in Guam tells me health care is going to collapse.

It was at the 12th World AIDS Conference in Geneva that I became incensed over the unfairness of North/South, bridging the gap! The gap is just getting wider. Thanks to the giant pharmaceutical companies and their medical marketers. I had never seen such bullshit, it made me ashamed to be human. The drug companies all had their stuff strutting displays, talking to me because I'm kind of white, and ignoring my Asian companions. It made me sick and I flushed all my Fortovase down a Swiss toilet. Why the organizers chose to hold a global conference in one of the world's most expensive cities is beyond me. By the way, I had a breakdown there, right at the Geneva airport, I read my ticket wrong and missed my flight by 12 hours. I spent the night in a Swiss hospital, well at least that was free! I have stopped taking the drugs and started vitamins and mineral supplements, lost the "baby" (fat belly induced from the drugs) and at last I feel OK. My viral load is high at 532,000, my T cells low at 193, my doctor is a bitch and our ONLY HIV/AIDS specialist here. But the worst thing that has happened to me is a whopping coat of white esophagal thrush! I would like to try something that will bring my viral load down. I really don't want to take protease again. Please don't ask me to consult my doctor because she told me to find my own treatment. I'm tough and ornery and you'll continue to come across me and my philosophies like, "no pain, no pain!", far into the next millennium.

out of 40 patients died with viral loads less than 5000 copies, seven of whom had viral loads less than 400 copies." Tanvir Chowdhry and colleagues reported that "though mortality has fallen progressively from 1995 to 1998, there is a recent trend to an increase in death rates in our large HIV clinic," and "deaths are occurring in persons with greater levels of immune capacity as reflected in T4 cell counts and also in persons under good virologic control. These observations underscore the need both to monitor the etiologies of HIV-associated mortality and to better understand the relationship between immune defenses, treatment related toxicities and end organ failure in HIV disease." Such data along with other information presented in this chapter do not bode well for the clinical course of this disease in the year 2005 and beyond.

Structured Treatment Interruptions (STI): Also Called Intermittent Treatment or Drug Holidays

There are two main justifications for STI. The first is simply to provide patients with a break from the rigors of HAART (treatment fatigue and HAART drug toxicities). The second justification is that rebound HIV that occurs during an STI may actually boost the immune response to HIV (autovaccination or autoimmunization) in the same

POINT TO PONDER 4.2

COMMENTS FROM AN ASYMPTOMATIC HIV-INFECTED MALE

He asks about the image the drug companies producing the anti-HIV drugs are presenting to help those who are infected. And he quickly answers his own question: "Open any gay magazine and see the major drug company ads depicting sexy people living, thriving, partying, and having sex with AIDS. They climb mountains (without nearby toilets) and sail across oceans while popping handfuls of happy AIDS pills. Surely AIDS is no longer a bad disease. And check out those young handsome 'come love me' AIDS drug models. If I take the drugs will I have sex, too? Of course you will, unless you become disfigured by lipodystrophy or find yourself on dialysis. You can always date in between IV therapy and emergency room visits. Pardon the sarcasm but the point is this: The image presented in the media depicting AIDS patients is a lie invented to sell AIDS drugs. Living with HIV sucks. It starts out sort of OK but it goes downhill much too quickly."

CONCLUSION

For many thousands of people living in America with HIV, successful treatment has raised the bar of expectations, and this can be unsettling for clinicians. Over the past nine years (1996–2004), people with HIV were relieved not to be dying or hospitalized; medication side effects were a welcome exchange for longer lives. Now, as their lives have been extended over the long term, side effects have become less tolerable. Many people with HIV have unexpectedly sensed the possibility of normal lives and are reacting to the limitations of a progressively distorted figure, diarrhea, peripheral neuropathy, and constant fatigue. Some clinicians have responded to this with impatience, and some disdain; their belief is that the side effects are a small price to pay for being alive. In some cases, this response stems from clinicians' feelings of not being able to be of more help, in others, from a fear that altering a successful regimen might compromise an individual's HIV status and cause their progression toward AIDS.

BOX 4.2

CAN COMBINATION THERAPY USING NUCLEOSIDE ANALOGS WITH PROTEASE INHIBITORS CURE HIV DISEASE? AN ALTERNATIVE—REMISSION?

MISSION IMPOSSIBLE

Beginning in 2002, studies showed that HIV-infected people have compartments or sanctuaries of replication-competent HIV residing in resting (nondividing) T4 and T4 memory cells, macrophage, and other cell types and tissues (for example, brain, nervous system, kidney, and testes). HIV recovered from the pool of cells/tissues, although they did not show resistance to anti-HIV drugs, were still present. That is, after three years on combination drug therapy, HIV were still present in persons that had no measurable HIV RNA in their blood. Such data clearly indicates that the body's reservoir of HIV has a very slow rate of cleansing itself of HIV. The worst case scenario: It may take between 5 and 20 years of uninterrupted aggressive anti-HIV therapy before the pool of latent HIV is cleared from the body (AIDS Clinical Care, 1998; Schrager et al., 1998; Balter, 1997; Wong et al., 1997; Chun et al., 2000). Anthony Fauci (Figure 5-10) and other HIV/AIDS scientists told those at the Twelfth International AIDS Conference (1998) that every patient, even those in successful therapy for three or more years, have a latent HIV pool that is formed early in infection and remains despite the most effective treatment.

People attending the conference were also told that

- HIV/AIDS vaccines are at least 10 years away,
- drug resistance and often unbearable side effects from the most powerful of the new drugs are causing many patients to discard them,
- although prevention campaigns are succeeding in a few countries, those successes are spotty even in the developed nations, and
- cases of HIV resistant to all protease inhibitors and reverse transcriptase inhibitors are being transmitted from one person to another.

BOX 4.2 *(continued)*

In mid-1999, Diana Finzi and colleagues reported that HIV can remain dormant in resting T4 cells. They described the decay rate of the inactive reservoir of HIV in 34 adults who are taking combination therapy and whose plasma levels had fallen to below-detectable level. On average, the half-life of the latent HIV was 43.9 months, which could mean eradication could take as long as 60 years, a lifetime if the source of the latent virus holds only 100,000 cells. As a result, the researchers say, latent infection of dormant T4 cells provides an opportunity for HIV to remain in all patients, including those who consistently take antiretroviral therapy. This work implies that most HIV-infected people will not be able to stop their anti-HIV medications. Yet, current drug therapy most likely cannot be tolerated for a lifetime.

AN ALTERNATIVE—REMISSION

HIV/AIDS investigators remain reluctant to use the word "cure," but they are beginning, in increasing numbers, to believe in *long-term remission* for most HIV-infected people. Cecil Fox, one of the first HIV/AIDS investigators to search for HIV in human lymphoid tissue, believes that the HIV/AIDS scientific community should be relating the treatment and results of treating HIV/AIDS patients much the same as people undergoing cancer therapy with the goal of getting the patient into remission.

To this point HIV/AIDS therapists have been treating HIV/AIDS patients for an infectious disease. They are looking for a cure. Get rid of every last microbe—in this case HIV. But remission, keeping HIV under control, is much more reasonable. Make HIV/AIDS into a long-term, manageable disease with medications that are tolerable—fewer side effects. The model for this alternative outlook on anti-HIV therapy can be seen in long-term nonprogressors. These people make up a small percent of the HIV/AIDS population, but they are infected and remain healthy.

way that a booster immunization would. This enhancement in HIV-specific immunity might allow better control of viral replication in the future. This sounds good in theory, but in reality it does not appear to work for the majority of patients placed on STI. After reviewing many controlled studies on STI the conclusion is that there are no significant lasting beneficial changes in T4 or CD4+ cell counts in patients placed on STI. And there are no consistent data to suggest that STI will lead to durable control of viremia when HAART is discontinued (Blankson et al., 2001; Dybul, 2002). In HIV infection, a dynamic equilibrium exists between viral replication and the HIV-specific immune response. In the vast majority of cases, the immune system is eventually overwhelmed by the extraordinary rates of viral replication. While HAART dramatically inhibits HIV replication, it also dulls or deadens the HIV-specific immune response. So, when HAART is discontinued, HIV replication occurs in the presence of diminished immunity, which may explain the tendency of viral loads to exceed retreatment levels during STIs. The overall dynamics of STI is not understood. Clearly some patients are helped but it appears, at the moment, most are not (Fagard, 2003), and especially those who carry drug resistant HIV (Lawrence et al, 2003; Deeks, 2003). The theoretical advantages of enhancing HIV-specific immunity must therefore be weighed against the real risk of losing T4 cells and the potential risk of drug resistance.

Summary to Structured Treatment Interruptions (STI)

In general, structured treatment interruptions in which patients cycle on and off their drugs for days, weeks, or months, are not yet being recommended by experts for use in the clinical setting, except in patients who have high CD4+ cell counts and low viral loads. That is, a patient may want to stop therapy because of side effects or the fear of developing drug toxicities. These patients need to be followed carefully, just as clinicians would follow untreated patients. For everyone else, which includes patients in the salvage stage of treatment and patients who want to base treatment interruptions on predetermined times or laboratory changes, STIs are not recommended outside of clinical studies. It has been clearly shown that long-cycle intermittent therapy carries substantial risk of drug resistance and, quite possibly, an increased risk of transmission because of the increase in viral load. The num-

BOX 4.3

TESTING OF ANTI-HIV DRUGS FOR POST-EXPOSURE PREVENTION (PEP)

The Forum— A woman has sex with her HIV-positive husband and the condom breaks. A woman is raped by a man who is HIV-positive. A child is sodomized by an HIV-positive male. A prison guard is bitten by an HIV-positive inmate. A couple has unprotected sex—a one-night stand. These were some of the cases brought up by experts as they debated whether doctors should be prescribing AIDS drugs as a morning-after or post exposure prophylaxis or prevention (PEP) treatment for those exposed to HIV.

In mid-October 1997, San Francisco became the first city in the United States to offer new PEP drugs to individuals trying to prevent HIV infection.

72-Hour Window— The standard of practice for about four years at most medical centers is to offer antiviral drugs promptly when health care workers of AIDS patients are stuck with needles or come into contact with body fluids from infected patients. Many researchers believe the drugs must be given within 72 hours of exposure—the time it takes for HIV to integrate into human cells' DNA after infection. The treatments are given for four weeks.

Beginning in 2005 over 1400 American health-care workers have received PEP care—12 have become HIV positive, but they admitted to continuing to practice high-risk behavior. Regardless, over the past 11 years, PEP has become the standard of care for occupational exposure to HIV.

Treatment Guidelines— There are guidelines published by the CDC for prophylaxis following needle stick or occupational exposures to HIV in the health-care setting. Except for the 2003 comprehensive nonoccupational guidelines published by the state of Rhode Island, there are no other published guidelines for the initiation of PEP after sexual exposure to HIV. To date, recommendations for nonoccupational PEP come through inference, judgment, and thoughtful intention. There is a wide spectrum of situations to consider, but there are no absolutes when it comes to the right choices regarding PEP. PEP after sexual exposure is recommended because of the knowledge that in some cases it works. It is well recognized that the use of ZDV in pregnant women dramatically reduced vertical transmission of HIV to newborns. It is therefore conceivable that the use of antiretroviral medication following sexual exposure to HIV will also be effective. However, studies to date remain controversial (Pinkerton, 2004).

For those who think PEP is or will be a magic bullet like some of our antibiotics were—*no way*—better to focus on practicing medical procedures correctly or being safe the night before.

Conclusion—It cannot be overstated that the key to reducing sexual transmission of HIV is not PEP, but safer sex practices. Public health officials, physicians, and patients must continue working toward reducing the need for PEP by advocating and encouraging safer sex practices for everyone.

More information on PEP research is available by calling 1-800-367-2437 (1-800-FOR-AIDS). Information is also available on the internet at hivinsite.ucsf.edu by clicking on "Key Topics." As yet, there is no national list of PEP programs. Your local AIDS service organization may be able to help. The CDC published the "Updated U.S. Public Health Service Guidelines for the Management of Occupational Exposures to HBV, HCV and HIV and Recommendations for Post Exposure Prophylaxis" in June 2001, a must-read for those with questions on PEP.

DISCUSSION QUESTION: With your current knowledge about HIV/AIDS, if you knew that you had just been exposed to HIV would you ask for immediate therapy? Why?

ber of new HIV infections in the United States has increased from approximately 40,000 infections per year since 1988 to about 50,000 infections in years 2001 through 2004. Although antiretroviral treatment has dropped the annual number of AIDS-related deaths from 40,000 to about 16,000, many people have stopped therapy because of the side effects of the drugs or because they are experiencing resistance to treatment.

Novel Drugs in the Pipeline but Some Are Years Away from Clinical Prescription

Entry Inhibitors: Lockouts or Bouncers at the Doors (See Figure 5-5)—Entry inhibitors or the bouncers work at HIV's first point of contact with a human cell. This occurs at an earlier stage in the HIV lifecycle than the other antiretrovirals. To successfully infect a cell, HIV needs to attach to and

SIDEBAR 4.4

CURRENT STATE OF HIV DRUG THERAPY

The Present—Nothing better describes the current state of this epidemic and the treatment of it than to say a job half done. While so much has been accomplished, still there is the lack of ability to truly save lives; they only extend lives. The infected continue to die prematurely. At best, today's treatment and care programs offer a respite in the fight against AIDS—a time in which the virus is not gone but at least subdued into temporary submission. But the price for this, in both dollars and quality of life, is much too high. It is still too early to know how long people will be able to live with the current drugs. For some, it is but a matter of a few years before drug side effects and viral resistance begin to outweigh the benefits. For others it has been years since potent triple-drug therapy became available (1996) and shifted the balance in the battle between virus and the immune system. Some are still doing well on drug therapy and experiencing only minor side effects, others are dying. With time it has become very clear that most people will not be able to stay on treatment for the rest of their lives. Between cumulative drug resistance, long-term side effects and simple weariness with the demands of the various regimens, it is almost naïve to expect people to be able to succeed for periods of 20 to 50 years or more. But that's what it will take to allow people to live a normal life span despite HIV.

The Future— In theory, better drugs are coming, but their reality seldom meets expectations. Even more worrisome is that a variety of economic and social factors are rapidly making HIV/AIDS a less than attractive target for the pharmaceutical industry. AIDS activists may debate the extent of this problem, or its possible causes, but not its reality. Two companies, Pharmacia & Upjohn and Dupont Pharmaceuticals, have already sold off their HIV product lines. Several others quietly ended their HIV research projects after protease inhibitors were first approved. Another major firm has narrowed its HIV research program and will only continue with one or two drug candidates already in development, foregoing any investment in new approaches or viral targets. Still others have shifted their interest to vaccine work. More worrisome, from the companies' point of view, is that few of the recently approved drugs have been successful in the marketplace. Some argue that, despite their improvements over current therapy, new drugs will have a rough time facing off against the 20 better-known drugs already available unless they offer clear-cut advantages. It is more advantageous at the moment to reformulate currently FDA-approved drugs into combination pills lowering the number that have to be taken daily.

NEW DRUGS

The pipeline of new drugs for the next few years according to the number of new HAART drugs in the pipeline presented at the 2004, Eleventh Annual Retroviral and Sexually Transmitted Disease Conference, looks very promising. However, the economic trend is for more of the available drugs to become concentrated within fewer companies. This places the power over production and pricing in the hands of the remaining companies and greater dependence upon them for future drug development.

LONG TERM DILEMMA

With current HIV/AIDS drugs, unfortunately as with most potent drugs, the curse or warning of **Paracelsus** is particularly apt—**"The poison is in the dose."** The idea of hit hard and hit early has forcibly given in to questions of balance—when to start, what drugs to use, how much, when or if to interrupt them—all because the HAART drugs have proven more toxic than originally thought. These side effects have been accepted as the price of preventing death from an otherwise inevitably lethal disease. But as larger and larger populations are expected to use these drugs for the rest of their lives, the need to minimize such short-term discomfort and long-term problems is critical. Such is the case for patients with diabetes today whose susceptibility to other diseases increases the longer they live with their condition. All these problems are also independent of the issues of drug access and drug compliance. Considerable work is being done to create single or less frequent dosage versions of the classic AIDS cocktail compounds.

enter the cell. HIV enters cells in a complex, three-part series. HIV first anchors itself to a receptor on T4 lymphocytes, called CD4, then to a coreceptor, CCR-5. Once hooked, it fuses its membrane to a cell membrane, easing itself inside. Several different compounds under investigation are able to make it more difficult for the virus to enter the cell. One such drug that attempts to block this first step is Bristol-Myers Squibb's experimental medicine, code named BMS805. The drug works by blocking gp120 on the surface of HIV that attaches to the CD4 receptor of the cell. In the test tube, it appears to work against strains of HIV that can resist all the other anti-HIV drugs. However, it has not yet been tried on people. It is an orally ingested drug.

Another drug, SCH-C, developed by Schering Plough, an orally ingested drug, blocks the next step in viral entry, the attachment of HIV to CCR-5. In a pilot study of 12 people, the drug used alone for one week dropped viral levels dramatically. Furthest along in testing is Roche's and Trimeris's T20 or Pentafuside (enfuvirtide or fuzeon), which blocks the third step of viral entry. T20 binds to the gp41 protein on the surface of HIV, stopping HIV from binding to the CD4 receptor and the cell membrane. Fuzeon received FDA approval in March 2003. This is the first FDA-approved drug to block HIV-cell fusion or HIV entry into the cell. T20 or fuzeon must be kept refrigerated and injected twice daily, abdominally, and costs up to $20,000 a year (see Table 4.1). Initially, T20 is being used as a third line or salvage therapy. To date, fuzeon is in short supply. Roche pharmaceuticals say there is only enough of the drug to treat 15,000 people through year 2003 and 40,000 per year by 2005.

In June 2002, Xiping Wei and colleagues (2002) reported that T20 HIV-resistant mutants rapidly emerged during the first 10 days of a clinical trial in which 16 HIV-infected people took part. Other studies are in progress.

Robert Gallo, director of the Institute of Human Virology said at the 2002 Fourteenth International AIDS Conference that "in two or three years, the protease inhibitors will go away, replaced by viral entry inhibitors and more intelligent use of reverse transcriptase inhibitors. The entry inhibitors that are being tried now—five of them—will cause the next wave of excitement in the next conference in Bangkok, Thailand in 2004."

Laurie Garrett, writing in *Esquire* in March 1999, "The Virus at the End of the World," quotes Spencer Cox, New York AIDS activist involved in Treatment Action Group (TAG), who said our worst case scenario now is that it's 1987 all over again for an ever-increasing number of HIV-infected people—for them there are no effective treatments. Clinical studies now suggest that 60% of patients experience HIV suppression *failure* during first line antiviral drug treatment. About 70% to 80% experience second line drug treatment failure.

DISCLAIMER

The author does not accept any responsibility for the accuracy of the information or the consequences arising from the application, use, or misuse of any of the information contained herein, including any injury and/or damage to any person or property as a matter of product liability, negligence, or otherwise. No warranty, expressed or implied, is made in regard to the contents of this material. This material is not intended as a guide to self-medication. The reader is advised to discuss the information provided here with a doctor, pharmacist, nurse, or other authorized health-care practitioner and to check product information (including package inserts) regarding dosage, precautions, warnings, interactions, and contraindications before administering any drug, herb, or supplement discussed herein.

SUMMARY

New results in the field of HIV therapy have given HIV-infected persons new hope in their battle against HIV disease and AIDS. First, combinations of nucleoside analogs were shown to have a clinical benefit (prolonged survival and fewer AIDS-defining events) when given to symptomatic individuals. Next came the demonstration of the extraordinary capability of protease inhibitors. The introduction of the protease inhibitors is the most powerful intervention against HIV to date. Marked reductions in viral load, striking clinical improvement, and reduction in mortality have been observed among patients able to take these medications *properly*. Whether to initiate protease inhibitor therapy in the very early stages of HIV disease is controversial.

Neither protease inhibitors, nucleoside, or non-nucleoside inhibitors work for all HIV-infected persons; the reasons for this lack of efficacy are not completely understood. Long-term adverse effects of protease inhibitors are beginning to show.

Antiviral drugs have certainly shown tremendous success in reversing the trend of HIV infection that leads to disease. But with up to 40% of treated individuals now showing HIV strains that are strongly resistant to anti-HIV drugs, the reality of the dangerous cellular reservoirs of HIV must be better understood.

Looking at the problem of resistance, it's now known that unless the virus is virtually eliminated from the blood—again, an unlikely prospect—it's only a matter of time before a viral mutant emerges that is resistant to therapy. Current estimates are that the virus mutates once each time it copies itself—over 1 billion times a day. It's like going to Las Vegas. HIV just keeps spinning its bases (its genetic building blocks), looking for a jackpot.

Over the past year, progress in HIV has continued to resemble an electrocardiogram. The sharp dips represent difficult roadblocks that have blunted our hopes for a cure. The peaks mark scientific breakthroughs and new approaches used to bring HIV under control. As we closed year 2004 and move within the new millennium, the overall picture remains mixed and varies greatly among individuals battling the virus. But there are many reasons to be optimistic and even excited about the state of research, at least from a scientific point of view, without losing sight of the real battles we face. The brief three-year honeymoon that began with the arrival of protease inhibitors has now experienced significant setbacks. Concerns about serious drug side effects and drug resistance linked to long-term use of HIV combination therapies have replaced optimism, causing a growing number of people on therapy to reconsider the benefits versus the risk of potent antiviral therapies and to seek alternative approaches. In response, pharmaceutical manufacturers have worked to simplify drug regimens and create more potent, less toxic products. It appears now that HIV/AIDS scientists and clinicians may have been a bit too optimistic about what the drugs could do. Today the treatment community is unanimous about one thing: The next generation of drugs must be more powerful and easier to take. They should also improve the problems associated with the present generation of drugs, such as toxic side effects, resistance, dosing, absorption, and bioavailability (the active part of a drug that's available to fight the virus).

At the moment, current treatments are inadequate because they do not lead to a cure; at best, they slow disease progression. They are complicated to administer, require close medical monitoring, and can cause significant side effects. They are also very costly and, as a result, are inaccessible to the majority of people living with HIV/AIDS.

SOME AIDS THERAPY INFORMATION HOTLINES

For HIV/AIDS treatment information, call:

The American Foundation for AIDS Research: 1-800-39AMFAR (392-6237)

AIDS Treatment Data Network: 1-800-734-7104

AIDS Treatment News: 1-800-TREAT 1-2 (873-2812)

National HIV Treatment: 1-800-822-7422

For information about AIDS/HIV clinical trials conducted by National Institutes of Health and Food and Drug Administration-approved efficacy trials, call:

National AIDS Clinical Trials Information Service (ACTIS): 1-800-TRIALS-A (874-2572)

For more information about HIV infection, call:

Drug Abuse Hotline: 1-800-662-HELP (4357)

Pediatric and Pregnancy AIDS Hotline: 1-212-430-3333

National Hemophilia Foundation: 1-212-219-8180

Hemophilia and AIDS/HIV Network for Dissemination of Information (HANDI): 1-800-42-HANDI (424-2634)

National Pediatric HIV Resource Center: 1-800-362-0071

National Association of People with AIDS: 1-202-898-0414

Teens Teaching AIDS Prevention Program (TTAPP) National Hotline: 1-800-234-TEEN (8336)

General information:

English: 1-800-342-AIDS (2437)

Spanish: 1-800-344-7432

TDD Service for the Deaf: 1-800-243-7889

General information for health-care providers:

National Clinician's Post Exposure Treatment: 1-888-448-4911

HIV Telephone Consultation Service: 1-800-933-3413

REVIEW QUESTIONS

(Answers to the Review Questions are on page 438.)

1. Can HIV infection be cured?
2. What is a surrogate marker?
3. From _____ 1987 through _____ 2004 _____ individual and combination anti-HIV drugs were FDA-approved.
4. How many FDA-approved anti-HIV drugs are nucleoside or nucleotide analogs?
5. What is the proper drug name for the following drug acronyms: ZDV, ddI, ddC, d4T, and 3TC?
6. How do nucleoside analogs inhibit HIV replication?
7. What are the two major problems in the use of nucleoside and non-nucleoside analogs in HIV therapy?
8. What is HIV viral load? What can its quantitative measurement reveal?
9. Name the eight protease inhibitors that have FDA approval for use in the United States.
10. Which of the eight FDA-approved protease inhibitors appears to be the most effective and why?
11. How do non-nucleoside drugs inhibit HIV replication?
12. Briefly describe the focus of HIV combination drug therapy.
13. After starting antiretroviral therapy, what is an acceptable target for viral load that indicates the therapy is effective?
    A. 5000 copies/mL
    B. 5000–10,000 copies/mL
    C. 10,000–15,000 copies/mL
    D. No acceptable target level has been set.
14. Which of the following statements is correct regarding HIV pathogenesis?
    A. HIV contains two copies of the viral DNA genome.
    B. Viral polyproteins are cleaved by protease.
    C. The RNA copy is integrated in the host-cell chromosome.
    D. gp120 envelope glycoprotein facilitates entry of HIV core particle into the cell's cytoplasm.
15. Some medical centers are offering drug therapy that may be able to:
    A. cure AIDS
    B. help prevent HIV infection shortly after exposure
    C. effectively treat HIV/AIDS without side effects
    D. prevent HIV infection
    E. prevent AIDS
16. Go to the library/internet and write a report on the most recent advance in antiretroviral therapy.
17. HIV escapes the effects of HIV antiretroviral drugs by __________.
    A. gobbling up the chemicals
    B. disguising itself as an immune cell
    C. pretending to be a harmless virus
    D. changing its genetic makeup
18. When HIV mutates in response to a drug, it can cause __________.
    A. the virus to become less "fit"
    B. the drug to become less effective
    C. other drugs to become less effective
    D. all of the above
19. True or False: Currently, at least two types of tests measure resistance.
20. Single mutations are __________ worse than a group of mutations.
    A. always
    B. never
    C. sometimes
    D. none of the above

21. When your virus becomes resistant to a drug in your drug combination, your __________ may go up.

A. temperature

B. CD4 cells

C. doctor bills

D. viral load

22. True or False: Resistance is an all-or-nothing deal—the drugs work perfectly or not at all.

23. A high genetic barrier to resistance means a drug requires __________ to reduce its effectiveness.

A. multiple viral mutations

B. a single viral mutation

C. experimental gene therapy

D. cross-resistance with another drug in the same class

24. True or False: Mutations take years to develop, so missing one day's dose of meds will not contribute to resistance.

25. Cross-resistance is a problem for which class of drugs? __________

A. protease inhibitors

B. nukes

C. non-nukes

D. all three classes

26. Which of the following is not an important issue when choosing a drug regimen? __________

A. potential side effects

B. easy dosing schedule

C. high genetic barrier

D. all three are important

5 The Immunology of HIV Disease/AIDS

CHAPTER CONCEPTS

- ♦ Immunity, a protection against infection and disease.
- ♦ HIV attaches to CD4 receptor sites on CD4+ or T4 lymphocyte, monocyte, and macrophage cells.
- ♦ HIV uses the CD8 receptor to infect CD8+ cells.
- ♦ CC-CKR-2, CC-CKR-3, CXCKR-4(FUSIN), and CC-CKR-5 are coreceptors that allow HIV to enter cells.
- ♦ Mutant CKR-5 gene confers resistance to HIV infection.
- ♦ Secrets to HIV cell entry revealed.
- ♦ Monocytes, macrophage, follicular, dendritic, and T4 cells serve as HIV reservoirs in the body.
- ♦ Basic immune system terminology is defined.
- ♦ T4 or CD4+, T8 or CD8+, and B cell function is related to HIV disease.
- ♦ Apoptosis is a normal mechanism of cell death.
- ♦ Cofactors may enhance HIV infection.
- ♦ Impact of T4 or CD4+ cell depletion is immune suppression.
- ♦ Latency refers to inactive proviral HIV DNA.
- ♦ Clinical latency refers to infection with low-level HIV production over time.

THE IMMUNE SYSTEM

All living organisms are continually exposed to substances that are capable of causing harm. Most organisms protect themselves against such substances in more than one way—with physical barriers, for example, or with chemicals that repel or kill invaders. Animals with backbones, **vertebrates,** have these types of general protective mechanisms, but they also have a more advanced protective system called an **immune system.** The immune system is a mind-numbing, incredibly complex network of organs containing cells that recognize foreign substances in the body and destroy them. It protects vertebrates against pathogens, or infectious agents, such as viruses, bacteria, fungi, and other parasites.

Although there are many potentially harmful pathogens (agents that cause diseases), no pathogen can invade or attack all organisms because a pathogen's ability to cause harm requires a susceptible victim, and not all organisms are susceptible to the same pathogens. For instance, the virus that causes AIDS is strictly a human pathogen; it does not cause a disease in animals such as dogs, cats, and mice. Similarly, humans are not susceptible to the viruses that cause canine distemper or feline leukemia.

Function of the Immune System

A man dies because his body rejected a heart transplant; a woman is crippled by rheumatoid arthritis; a child goes into a coma that is brought on by cerebral malaria; another child dies of an infection because of an immunodeficiency; an elderly man has advanced hepatic cirrhosis caused by iron overload. These five clinical situations are diverse, yet all have one thing in common: a malfunction of the human immune system.

The immune system filters out foreign substances, removes damaged and dead cells, and acts as a security system to destroy mutant and cancer cells. It is composed of a number of specialized cells, several organs, and a group of biologically active chemicals. The human immune system is like a jigsaw puzzle—many parts come together to form an overall defense against disease-causing agents. If parts of the immune system are missing or damaged, illness may occur due to an immune deficiency.

Separating Friend from Foe: How the Immune System Decides

Skin prevents disease-causing agents from entering the body. But they can enter through body openings, cuts, or wounds. Whether the invader is a life-threatening bacterium or a relatively harmless cold virus, your immune system must control it. A single infectious microorganism or virus that survives and multiplies causes illness. Some infectious agents resemble the body's own cells. How do immune cells know which to attack and which to ignore? The answer is, once an agent enters the body, it triggers an **immune response** if the body does not recognize the substance or agent as a part of itself or **"self."** All body cells have special molecules, called **class I proteins,** on their membranes that are like flags or barcodes with the word "self" on them (Figure 5-1). The cells of the immune system try to destroy anything present in the body that is not carrying the self molecules—anything that is **"nonself."** Nonself is any substance or agent that triggers the creation of antibodies. Such nonself substances are called **antigens.** Antigens may be whole virus or organisms or parts of virus, organisms, or their products.

In general, most cellular organisms that damage cells do so from the outside by producing toxic chemicals or in some way externally interfering with the cell's metabolism. But viruses generally invade or enter different cell types, forcing them to produce viral replicas at the expense of the cells' own essential metabolic functions. Gradually, like a machine wearing out, host cells start to malfunction and die. The best thing a virus can do is find a host cell that does not die and that can produce replicas indefinitely. In a biological time frame, new disease-causing viruses are often very deadly to new hosts. If a new virus strain is too deadly, it kills its host before other hosts are infected, and the ensuing epidemic dies out. For example, the **Ebola virus,** makes its victims very weak shortly after infection and kills them in 7 to 14 days. This virus kills quickly and vanishes. Its origin is still unknown. Over biological time, successful viruses like the human herpes virus and their new hosts learn to accommodate each other. This will most likely happen with human-HIV associations, but how many people over how many years will have to die before human cells learn to accommodate HIV is unknown. Perhaps it will never happen. Smallpox virus has been infecting humans for thousands of years and has never been accommodated by humans.

Cooperation and Coordination Within the Human Immune System

The basic premise of the immune system is simple: to coordinate the activities of various cell types in order to provide extended, if not lifelong, protection against disease-causing pathogens. Usually, this cooperating system works flawlessly, quashing diseases before they can kill their host and sparking an immunity to provide protection against future attacks. Sometimes, however, the system fails and infection and disease prevails—and there is no greater example of this than HIV, a pathogen that almost always succeeds in circumventing and manipulating the body's immune defense to facilitate its own survival.

HUMAN LYMPHOCYTES: T CELLS AND B CELLS

The hallmark of the human immune system is its ability to mount a highly specific response against virtually any foreign entity, even those never seen before in the course of evolution. It is able to do this because of the number of different kinds of cells called lymphocytes. The human immune system contains about 2 trillion (2×10^{12}) lymphocytes, a relatively small number when compared to the 100 trillion cells in the body (see Box 5.1, Figure 5-2). Most mature lymphocytes recirculate continuously, going from blood to tissue and back to blood again as often as one to two times per day. They travel among most other cells and are present in large numbers in the thymus, bone marrow, lymph nodes, spleen, and appendix (Figure 5-3). By 1968, lymphocytes had been divided into two classes: **lymphocytes called B cells** that are derived from and mature in bone marrow, and **lymphocytes called T cells** that are derived from bone marrow but travel to and mature in the thymus gland (Figure 5-4). T cells make up 70% to 80% of the lymphocytes circulating in the body. Circulating T cells are a heterogeneous group of cells with a wide range of different functions. When a T cell encounters another cell, it uses various probes on its surface—known as receptors and CD4+ coreceptors—to examine protein fragments arrayed on an antigen-presenting cell's surface (Figure 5-1). The fragments tell the T cell whether the cell being scanned is normal and to be left unharmed, or infected and to be de-

POINT OF INFORMATION 5.1

CLASS I AND CLASS II PROTEINS

THE HUMAN LEUKOCYTE ANTIGEN (HLA)

This system is, in some respects, the immunological equivalent of a sophisticated alarm system. HLA molecules are produced within human cells, and act as receptacles for fragments of cellular or foreign (for example, viral) proteins. The HLA molecules then display these fragments (known as peptides) on the outside of the cell; a single cell is typically adorned with several hundred thousand different HLA-peptide complexes. This process allows circulating T cells to survey the HLA-peptide complexes for signs of any foreign peptides that might indicate the presence of a pathogen. HLA molecules are divided into two major classes (I and II), which are recognized by different subsets of T cells. The CD8 molecule on CD8+ or T8 T cells interacts with class I HLA molecules. Likewise, the CD4 molecule on CD4+ or T4 cells interacts with class II molecules. In both cases, the peptide associated with the HLA molecule is recognized by a structure on the T cell called a T cell receptor (TCR). The critical aspect of the HLA system for immunity is that both class I and II molecules come in hundreds of different versions, dependent on the HLA genes inherited from our parents. The precise shape and size of an HLA molecule governs its ability to associate with a diverse array of peptides and to present them to T cells. HLA molecules thus exert a profound influence on the body's ability to mount a broad and effective T cell response to any given pathogen.

MORE DETAIL ON HLA

There is a series of some 40 different genes located on human chromosome 6 that is referred to as the *Human Leukocyte Antigen (HLA)* or as the **major histocompatibility** (tissue type) **complex (MHC).** The genes in this region produce a series of proteins that are *almost* unique to the individual. Because these proteins belong to the individual, they would be recognized as foreign or as an *antigen* when placed in another human, for example, as in a transplanted heart (transplantation antigens). Almost all cells in an individual have a sample of their own HLA proteins or transplantation antigens located on the surface of each cell's membrane. Thus, a person's transplantation antigens are recognized as **self-proteins** by his or her immune system. Such HLA proteins are referred to as **class I** (type 1) tissue compatibility proteins. All individuals have their own set of class I proteins on their cell membranes. These proteins cause transplant graft rejections. With minor exceptions, there is not much of your body tissue that would not be rejected if placed in another human body because your class I proteins would be recognized as foreign by the recipient body's immune system. But how does the recipient's immune system recognize these class I proteins as foreign? Answer: By using a second class of HLA proteins, **class II.** Relatively few human cell

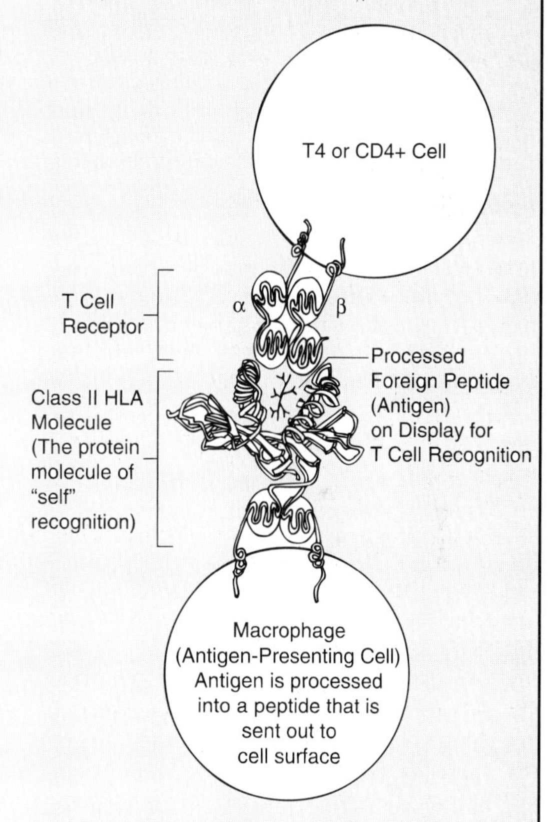

FIGURE 5-1 Interaction Between a Human Leukocyte Antigen "Self" Protein (HLA), a Foreign Peptide (Antigen), and a T Cell Receptor. The diagram represents the trimolecular interaction of a processed antigen into a peptide that is being presented on the surface of the antigen-presenting cell to a T4 cell. The presence of the class II HLA self identity marker and the presence of the foreign peptide stimulate the T4 cell into action. *(Adapted from Sinha, 1990)*

types carry this protein on their cell membranes. All such cells carrying the class II protein are derived from a monocyte-macrophage cell series, which includes macrophages located in the lung, liver and spleen, dendritic cells of the gut, skin, spleen and lymph nodes, Langerhans cells of the skin, and microglial cells of the nervous system.

The class II protein, located on the membrane of these cell types, allows these cell types to act as the body's police force. These cells *digest* foreign protein into smaller products. Some of these products, a small series of amino acids or **peptides,** are then escorted to the cell surface alongside a molecule of the class II protein. In this way, the cells carrying a class II protein serve up or present the foreign substance or antigen alongside or adjacent to the class II protein or "self molecule" so that immune system cells, notably the T4 cells and B cells, can sense the presence of the foreign protein and initiate an immune response against it—the activation of a variety of immune system attack cells (cytotoxic T cells, killer T cells) and the production of very specific antibodies.

In medically important viral infections, **neutralizing antibodies** are generated within 6 to 14 days. In contrast, such protective antibodies generally appear 50 to 150 days after infection with HIV and the hepatitis B virus (HBV) in humans (Pianz et al., 1996).

PROCESSING "FOREIGN" PROTEIN AND LOADING OF THEIR PEPTIDES ONTO CLASS I AND CLASS II MOLECULES

Protein processing and loading of peptides onto class I molecules that are carried out to the cell membrane is taking place all the time in most body cells. There is always plenty of material to feed the processing machinery, because worn-out, damaged, and misfolded internal proteins are continuously being degraded and replaced by new ones.

By contrast, the processing of proteins from outside the cell and the loading of peptides onto class II molecules are normally restricted to B cells, macrophages, and dendritic cells, which are very efficient at taking in material. Although most class I and class II molecules form complexes with peptides derived from proteins inside and outside the cell, this demarcation is by no means absolute.

As a consequence of protein processing, the surfaces of a body cell becomes adorned with peptide-laden HLA molecules, amounting to about 100,000 to 300,000 class I or class II products of each of the HLA genes. Because each HLA molecule has one peptide bound to it, an uninfected cell displays hundreds of thousands of self peptides on its surface. Some of these peptides are present in the thousands, whereas others are represented by a few copies; most peptide species have 100 or so copies on the surface of each cell. Each cell thus displays a heterogeneous collection of peptides, and the surface of a cell resembles rows of well-stocked stalls at a bazaar, with bargain hunters looking over the stock. But if, in this metaphor, the vendors are the HLA molecules and the peptides the goods, who are the potential buyers? They are a group of lymphocytes reared in the thymus and then turned loose to roam the body—**the T cells** (Klein et al., 2000a; Klein et al., 2000b).

stroyed. The protein fragments are cupped inside tiny holders called class II major histocompatibility complexes (class II MHCs). To be more specific, each individual T cell expresses a receptor **(T cell antigen receptor, TCR)**, which recognizes a ligand (a compound that fits a particular receptor) composed of an antigenic peptide, 8–15 amino acids long, bound to a **self-major-histocompatibility-complex (MHC) molecule** (also **referred to as HLA-human leukocyte antigen system**) (see Point of Information 5.1). Thus, a T cell does not directly recognize a soluble antigen, but rather recognizes an antigen displayed on the surface of an **antigen-presenting cell (APC)** like a B cell or macrophage (Table 5-1). There are about 10 billion APCs located in the lymphoid organs. The receptors of T cells are different from those of B cells because they are "trained" to recognize fragments of antigens that have been combined or complexed with HLA class II molecules. As T cells circulate through the body, they scan the surfaces of body cells for the presence of foreign antigens that have become associated with the HLA molecules. The antigen present on the APC signals the T cell on whether the APC should be left unharmed or that the APC is infected and should be destroyed. This function is sometimes called **immune surveillance.**

BOX 5.1

HOW LARGE IS A TRILLION?

The human body is made up of many trillions of cells. For example, the human immune system contains at least a trillion lymphocytes dedicated to destroying foreign substances that endanger health. But, a trillion of anything is a very large number. Can we really appreciate just how large a trillion of something is? Perhaps the following will help.

1. A stack of 1 trillion one-dollar bills would reach a height of 69,000 miles.
2. It would take a person 11.5 days to count to 1 million and 31,688 years to count to 1 trillion—1, 2, 3, 4, 5 . . . !
3. A stack of 1 trillion HIV (Figure 5-2) would be over 62 miles high (diameter of HIV=1,000 angstroms: $(1\times10^{12})(1000)(1\times10^{-10})=1\times10^{5}$meters=62.15 miles).

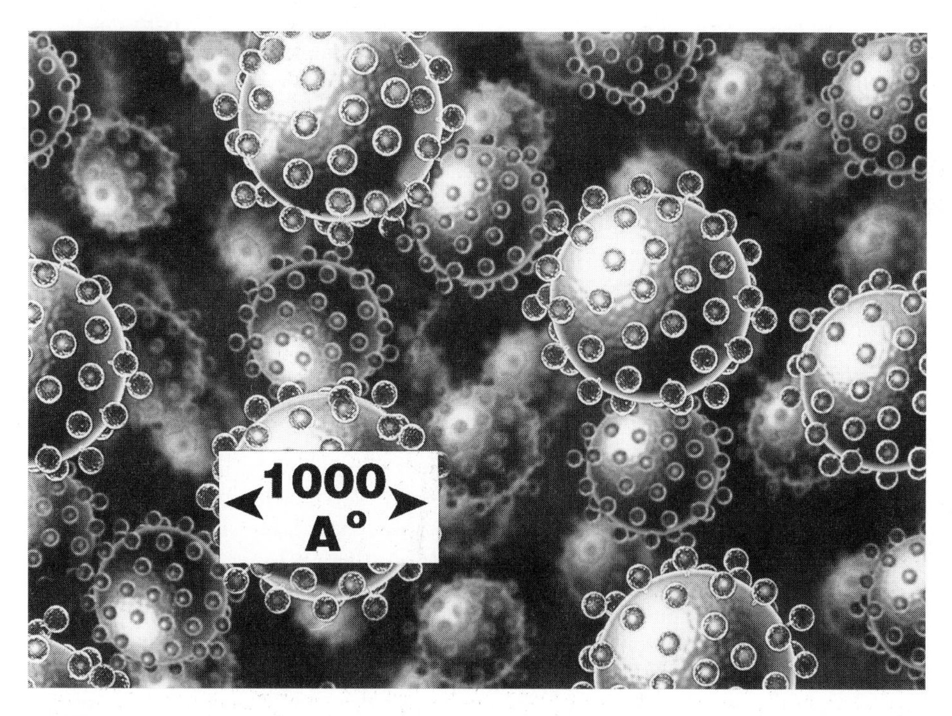

FIGURE 5-2 Dimensions of HIV in Angstroms. HIV has a diameter of 1000 angstroms (Å). It would take 254,000 HIV laid side by side to equal 1 inch in length. *(Illustration courtesy of the author)*

The Cytotoxic and Helper (T4 or CD4+) Lymphocytes

Two of the important kinds of T cells are **cytotoxic** or **killer T** lymphocytes, (CTL) and **helper T** lymphocytes. Killer T cells carry the **CD8 antigen.** Helper T cells carry the **CD4 antigen** and are called **T4** or **CD4+ cells** (for an explanation of CD8 and CD4 antigens, see Box 5.2, T lymphocytes). The CD4 and CD8 proteins (antigens) located on their respective T lymphocytes act as a bar code or number on a football jersey; the proteins or antigens identify the cell type. Killer or CTL T cells bind to cells carrying a foreign antigen and destroy them. T4 cells do not kill cells; they interact with B cells and CTL cells and help them respond to foreign antigens. The

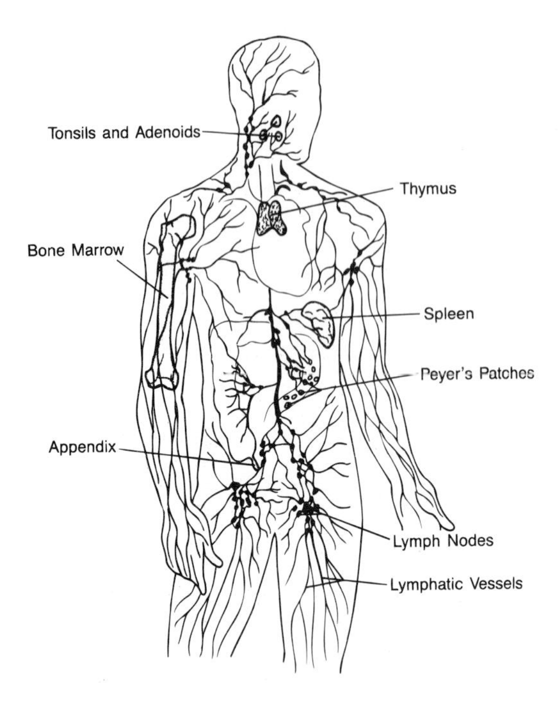

FIGURE 5-3 Organs of the Human Immune System. The organs of the immune system are positioned throughout the body. They are generally referred to as lymphoid organs because they are concerned with the development, growth, and dissemination of lymphocytes, or white cells, that populate the immune system. Lymphoid organs include the bone marrow, thymus, lymph nodes, spleen, tonsils, adenoids, appendix, and the clumps of lymphoid tissue in the small intestine called Peyer's patches. The blood and lymphatic fluids transport lymphocytes to and from all the immune system organs. Gut-associated lymphoid tissue contains 50% to 60% of the body's lymphocytes, and those lymphocytes, according to Peter Anton at UCLA, contain six times more CD4 receptors than do CD4+ or T4 cells circulating in the blood. These data may mean that the gut is the preferred site of HIV replication in untreated patients.

CD4+ or T4 cell has the role of a quarterback in football; it calls the plays for the rest of the lymphocyte team.

The Loss of T4 or CD4+ Lymphocytes and Its Biological Impact

Ashley Haase, director of the University of Minnesota Microbiology Department, estimated that healthy young adults harbor approximately 200 billion mature CD4+ cells. In HIV-positive patients, this total number is halved by the time the CD4+ cell count falls to 200 cells/μL of blood. In the more advanced stage of HIV disease, the destruction of parenchymal lymphoid spaces is so extensive that a total body CD4+ cell count has not even been attempted (Haase, 1999).

It is believed that T4 cells recognize only those antigens of viruses, fungi, and other parasites; and trigger only those parts of the immune system necessary to act against these agents. Indeed, the viruses, fungi, and other parasites produce the majority of opportunistic infections when the T4 cells have been depleted by HIV.

It was unknown if B cells begin to malfunction soon after HIV infection. In late 2001, investigators at the National Institute of Allergy and Infectious Diseases (NIAID) reported that the presence of HIV in infected patients causes B cells (a) to produce excessive amounts of nonessential antibodies, (b) to fail to respond to physiological signals, and (c) to be at risk of becoming cancerous. Some of these changes can be reversed with the use of antiretroviral drugs (Moir et al., 2001).

In Summary

The two types of lymphocytes, B cells and T cells, play different roles in the immune response, though they may act together and influence one another's functions. The part of the immune response that involves B cells is often called **humoral immunity** because it takes place in the body fluids. The part involving T cells is called cellular immunity because it takes place directly between the T cells, other cells, and the antigens. This distinction is misleading, however, because strictly speaking, all adaptive immune responses are cellular—that is, they are all initiated by cells (the lymphocytes) reacting to antigens. **B cells** may initiate an immune response, but the triggering antigens are actually eliminated by soluble products that the B cells release into the blood and other body fluids. These products are called **antibodies** and belong to a special group of blood proteins called **immunoglobins**. When a B cell is stimulated by an antigen that it encounters in the body fluids, it transforms, with the aid of T4 cells, into a larger cell, a blast cell. The blast cell begins to divide rapidly, forming a clone of iden-

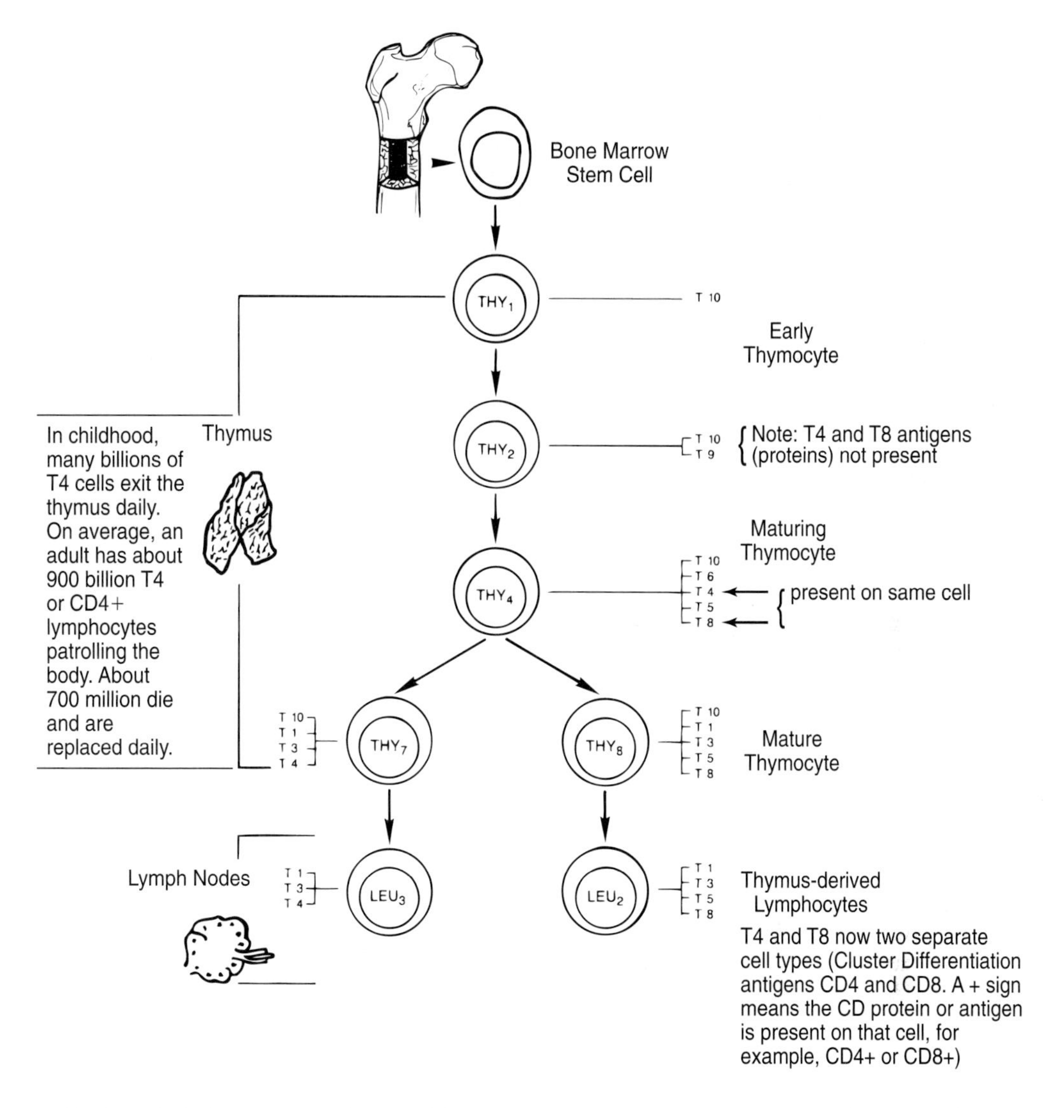

FIGURE 5-4 T Lymphocyte **Cluster Differentiation (CD)** Antigens in Humans. The CD marker or antigen on a T cell tells something about what the cell does. For example the CD4 and CD8 T cells have different jobs to do. Stages of thymic differentiation (that is, the presence of the different antigens or proteins on their membrane surfaces) are defined on the basis of reactivity to monoclonal antibodies. Schematic pictures of cells represent thymocytes within specific stages of a defined phenotype: T1-T10.

tical cells. Some of these transform further into plasma cells—in essence, antibody-producing factories. These plasma cells produce a single type of antigen-specific antibody at a rate of about 2000 antibodies per second. The antibodies then circulate through the body fluids, attacking the triggering antigen.

What Happens to the Immune System after HIV Infection

HIV enters the body via infected body fluids: blood, semen, and vaginal secretions. Once inside, HIV specifically infects the T4 or CD4+ cell. These cells direct the body's immune response against

Table 5-1 Cells of the Human Immune System

Cell Type	Function
Stem cells	Self-perpetuating cells that give rise to lymphocytes, macrophages, and other hematological cells
T helper cells (T_H), also called T4 or CD4+ cells	Cells that interact with an antigen before the B cell interacts with the antigen
Cytotoxic T cells (Tc)	Recruited by T helper cells to destroy antigens
Suppressor T cells (Ts)	Cells that dampen the activity of T and B cells; they somehow inhibit the immune response
B cells	Bone marrow stem cells that differentiate into plasma cells that secrete antibodies
Plasma cells	Cells devoted to the production of antibody directed against a particular foreign antigen
Monocytes	Precursors of macrophages
Macrophages	Differentiated monocytes that serve as antigen-presenting cells to T cells; they can also engulf a variety of antigens and antibody-covered cells
Killer cells (K)	Lymphocytes that recognize and kill any cell that is coated with antibody
Natural killer cells (NK or null cells)	Lymphocyte cells that detect and kill tumor cells and a broad range of foreign cells

infection. As HIV takes over the T4 cells, it alters their growth and reproduction through a complicated process that leads to the T4 cells' destruction. The ratio of T4 cells to T8 cells then changes. In healthy people, the number of T4 cells is greater than the number of T8 cells. In HIV-infected individuals, a decline in the T4 count signals the progress of immune system deterioration. The results are debilitating: The T4 cells are not as responsive to antigen identification, macrophages become less responsive, and B cells produce fewer specific antibodies and lose their normal responsiveness. At this point, the immune system becomes dysfunctional and the host becomes vulnerable to attack from opportunistic infections.

Uses of the T4 or CD4+ Cell Count

Doctors use a test that counts the number of T4 or CD4+ cells in a cubic milimeter or μL (microliter) of blood. A normal count in a healthy, HIV-negative adult can vary but is usually between 600 and 1200 T4 cells/ μL. About 2% of the body's T4 cells are in the blood; the rest are in tissues such as lymph nodes. Changes in your T4 cell count (which looks only at the blood) may reflect the movement of cells into and out of the blood, rather than changes in the total number of T4 cells in your body. In some cases, in order to help understand changes in your absolute T4 count, your doctor may also assess what proportion of all lymphocytes are T4+ cells. This is called the T4 or CD4+ percentage. In HIV-negative people a normal result is around 40%. A T4 percentage, which falls below about 15% is understood to reflect a risk of serious infections. Most people with HIV find that their T4 count falls over time. This often happens at a variable rate, so the count can still be quite stable for long periods. It is useful to have a T4 or CD4+ count measured regularly for two reasons: (1) to monitor your immune system and help you decide whether and when to take anti-HIV drugs and treatments to prevent opportunistic infections (OI); and (2) to help monitor the effectiveness of any anti-HIV drugs you are taking. If your T4 or CD4+ count is persistently below 350, your immune system is weakened and you are at a gradually increasing risk of infections the further it falls. If it drops below 200–250 you are at increased risk for serious OI. At this point your doctor should offer drugs to try to prevent such infections, such as co-trimoxazole for PCP pneumonia. If your CD4+ count starts to drop rapidly or falls below 350, particularly if the viral load is high, anti-HIV treatments may begin. If the T4 count falls below 200–250 it is recommended that treatments with anti-HIV drugs begin. One effect of anti-HIV drugs may be to improve the state of the immune system. This is crudely reflected in an increase in the T4 or

BOX 5.2

TO UNDERSTAND THE DIFFERENCE BETWEEN HEALTH AND DISEASE, PEOPLE NEED TO UNDERSTAND THE IMMUNE SYSTEM

BASIC IMMUNE SYSTEM TERMINOLOGY

Of the many mysteries of modern science, the mechanism of **self** versus **nonself recognition** in the immune system must rank near the top. The immune system is designed to recognize foreign invaders. To do so it generates on the order of 10^{12} or 1 trillion different kinds of immunological receptors so that no matter what the shape or form of the foreign invader there will be some complementary receptor to recognize it and effect its elimination. Understanding how the immune system responds to any foreign substance is puzzling enough, but the added mystery is that the immune system can distinguish foreign carbohydrates, nucleic acids, and proteins from those that exist within us, often in shapes barely distinguishable from the invaders. When the immune system is working well it never gets activated by self substances, but unerringly responds to the nonself substances. When the system is not working well this distinction gets blurred and diseases of autoimmunity occur—our immune cells attack our own tissue!

To understand the human immune system certain basic terminology is reviewed:

Antibody—an immunoglobulin (a protein) that is produced and secreted by B lymphocytes in response to an antigen. Antibodies are able to bind to and destroy specific antigens. Antibodies contribute to the destruction of antigen by their interaction with other components of the immune system.

Antigen—a substance that is recognized as foreign by the immune system when introduced into the body. An antigen can be a whole microorganism (such as a bacterium or whole virus) or a portion or a product of an organism or virus.

Leukocytes—all white blood cells (WBCs) including neutrophils, lymphocytes, and monocytes (phagocytes).

Lymphocytes—mononuclear WBCs that are critical in immune defense because they provide the specificity and memory needed for immune function and long-term or even life-long immunity. The two major classes of lymphocytes are B cells and T cells; both recognize specific antigens. (For an excellent review of lymphocyte life span and memory, see Sprent et al., 1994.)

Lymphocyte Surface Receptors—Proteins are located on cell surfaces to serve a physiologic function. Some are enzymes, some are transport proteins, and many are **receptors**. All cells use specific receptors to communicate with their environment or with other cells. The receptors found on lymphocytes may be loosely classified according to their function. Thus lymphocytes require receptors that recognize antigen. They need receptors for antigen-presenting cells and receptors for the many factors that regulate lymphocyte responses.

Chemokines—also called cytokines, activate and *direct* the migration of leukocytes. There are 28 known chemokines separated into two subfamilies, those whose protein amino acid structure begins with cytosine–amino acid–cytosine, called the CXC chemokines, and those whose protein amino acid structure begins with cytosine–cytosine, called the CC chemokines.

Cytokines—soluble factors secreted by T cells, B cells, and monocytes that mediate complex immune interactions by acting as messengers. Subcategories of cytokines include: (1) lymphokines (secreted by lymphocytes), and (2) monokines (secreted by monocytes and macrophages). Specific cytokines include interleukin-1 (IL-1), interleukin-2 (IL-2), gamma interferons (γINF), B cell stimulating factor (BCSF), and B cell differentiating factor (BCDF).

Two Branches of the Immune System—The two branches of the human immune system are: Cellular Immunity and Humoral Immunity. Basic terms for each are provided, but as the terms will show, the two branches overlap to provide our immunity. (Adapted from *Mountain-Plains Regional HIV/AIDS Curriculum,* 1992.)

I. *Cellular Immunity*—immune protection resulting from direct action of cells of the immune system. Key cell types offering cellular immunity are:

A. *Phagocytes*

1. Phagocytes are leukocytes that are specialized for ingesting particles and molecules.
2. The most active phagocytic cells are monocytes and macrophages. Monocytes circulate in the blood but eventually

move into body tissues (brain, muscle, etc.) where they mature into macrophages.

3. Phagocytes initiate the cellular immune response. When an invader (a virus or bacteria) enters the body, it will be trapped and digested by a phagocyte. This attack against invaders occurs in various places in the body: the lining of the gut, throat, skin, bloodstream, or in organized lymphoid tissue such as the tonsils, other lymph nodes, or the spleen.
4. The phagocyte ingests the foreign invader and partially digests it. Pieces of the invader (antigens) can then be displayed on the phagocyte's surface, making the phagocyte an **antigen-presenting cell.** This process alerts other cells in the immune system that a foreign substance is present.

B. *Lymphocytes*

1. Ninety-eight percent of lymphocytes reside in lymphoid tissue (Pantaleo et al., 1993). Lymphocytes are uniquely specialized. Each lymphocyte has receptors on its surface for one, and only one, of the many millions of possible antigens that can invade the body. When the receptors lock with their matched antigen, the process of neutralizing, inactivating, or destroying the foreign particle begins.
2. This requires that the body have an immense variety of receptors for an immense number of antigens.
3. Humans usually have about 2 trillion lymphocytes. Several types of these lymphocytes play major roles in the immune response.
   a. ***T Lymphocytes*** mature in the thymus and play a central role in the immune response by destroying infected cells, controlling inflammatory responses, and helping B cells make antibodies. There are several subsets of T lymphocytes [Subsets of T cells display different proteins in groups (antigens) on their cell membranes. For this reason they are referred to as **cluster differentiation (CD)** antigens or proteins. At least 180 different cluster-differentiating antigens are known to reside on different human cell types. The different CD antigens located on T lymphocyte subsets allow investigators to distinguish between the different T cell lines. For example, in physical appearance it is nearly impossible to distinguish T cells from B cells. But each carries a different CD antigen. T lymphocytes attached by HIV carry type 4 antigen and are referred to as CD4+ T lymphocytes or T4 cells, likewise for T8 cells. B lymphocytes do not carry the T4 or T8 antigen.]
4. ***Helper T cells*** (T_H) alert the immune system to the presence of an antigen and activate other cells in the system. Helper T cells carry about 10,000 copies of the CD4 protein molecule on the surface of each helper T cell so these **CD4+** cells are often referred to as **T4** cells.
   a. T4 cells have receptors (TCR) on their surface that are specialized for the recognition of antigens found on the surface of antigen-presenting cells. It has been estimated that 10^{11} to 10^{15} different TCRs can be generated in humans. The total of the different TCR specificities of a given individual constitutes their "T-cell repertoire." This is a rough measure of the capacity of a given individual to respond to the myriad of foreign microbes, antigens, and so forth, that he or she will be confronted within his or her lifetime.
   b. When the T4 cell randomly contacts the surface of a phagocyte, its receptor binds to antigen on the phagocyte and the T4 cell becomes activated.
   c. The T4 cell then begins to secrete a variety of stimulatory factors *(lymphokines) into the space around it.*
   d. The T4 cell will eventually divide into two cells of exactly the same specificity. If the foreign substance persists, the two daughter cells will divide, and so on. Thus, the number of cells specific for that foreign antigen is greatly expanded.

e. The T4 cell probably does little to repel the intruding substance on its own, but it is vital for activating the other main classes of lymphocytes (for example, B cells, natural killer cells, and phagocytes). It does this by means of the lymphokines it secretes when activated. The most familiar of these are interleukin-2 (IL-2) and gamma interferon (γIFN).

f. As a result of activation by T4 cells, B lymphocytes multiply and produce specific antibodies. The antibodies attach to the antigens on free organisms and infected cells, leading to inactivation and destruction of organisms and cells.

5. ***Inducer T cells*** (T_i) were previously known as the delayed sensitivity T cells. Inducer T cells also carry the CD4 molecule.

a. When an inducer T cell recognizes antigen on the surface of a phagocyte and is helped by factors from a T4 cell, it also becomes activated and secretes its own family of lymphokines, IL4, IL5, and IL10. These factors attract many more phagocytes from around the body that accumulate in the area where antigen is being recognized.

b. Phagocytes are specialized for ingesting and destroying the foreign invader.

6. ***Cytotoxic T*** lymphocytes (CTL) are also known as killer T cells. CTLs express the CD8 molecule and are most often referred to as **T8** cells. They are crucial for the immune response to viral infections. Although they cannot neutralize free virus, by eliminating virus-infected cells they are largely responsible for recovery from a viral infection.

a. The recognition of invading microbes by CTLs is a complex business. An infected cell must first chemically chop up the microbe's protein into small fragments, or peptides. These peptides are transported to the cell surface, where they become bound to specialized molecules called human leukocyte antigens (HLAs). Special receptors on the CTL recognize the HLA-peptide complex, and the CTL then kills the infected cell by unloading a cocktail of cytotoxic chemicals into it. But even a small mutation can change the microbe's peptide structure enough that it will no longer bind to HLA—or, alternatively, so that the peptide is no longer transported to the cell surface—and the infected cell then becomes "invisible" to the CTL.

b. As the infected cell breaks apart, infectious particles can be released from the cell. The released infectious particles may be phagocytized, neutralized by antibody, or may infect new cells.

c. Cytotoxic T8 cells have antigen specificity; this specificity is determined when the cytotoxic T8 cell is exposed to the antigen.

C. ***B Lymphocytes*** or ***B cells*** arise principally in the bone marrow and are responsible for making antibodies. The B cell response is referred to as humoral immunity.

1. B lymphocytes have receptors for antigen and also require help from T4 cells to become activated.

2. When activated, they release large amounts of immunoglobulins, called *antibodies,* (about 2000 per second) into the bloodstream and mucosal surfaces.

3. Antibodies circulate in body fluids. When they encounter the specific antigen with which they can interact, they bind to it.

4. Antigen-antibody interactions block the antigen's harmful potential in a variety of ways. For example, if the antigen is a toxin it may be neutralized or if it is a virus it may lose its ability to bind to its target tissue.

5. In addition, the complex of antigen and antibody activates a group of proteins in the blood called *complement,* which then facilitates the removal of the antigen by calling in a large number of phagocytes.

BOX 5.2 *(continued)*

D. ***Memory cells*** recall their history quickly.

1. Two groups of T and B lymphocytes become separate T and B memory cells for when the same antigen invades the body again.
2. Memory cells initiate immune response upon reexposure to an antigen.
3. Memory cells remember, recognize, and induce a more rapid immune response against the antigen.
4. Memory cells are responsible for disease resistance after immunization and are also responsible for natural immunity.

E. ***Suppressor T cells***

1. The function of suppressor lymphocytes in the immune system is not well understood, nor has a single kind of lymphocyte been identified that acts as a suppressor cell.
2. Suppressor cells turn off antibody production and other immune responses after an invader has been destroyed or eliminated from the body. This provides a balance in the immune system and allows it to rest when its functions are not required.

F. ***Natural killer (NK) cells***

1. NK cells are antigen nonspecific lymphocytes that recognize foreign cells of many different antigenic types. That is, one NK cell can recognize many different types of invaders. Therefore, NK cells can attack without first having to recognize specific antigens.
2. NK cells are important in fighting viral infections.

II. ***Humoral Immunity***—immune protection by the circulating antibodies that are produced by B cells. (See preceding section on **B Lymphocytes.**)

CD4+ count. Evidence suggests that the cells' ability to fight OI is also improved. For example, people taking anti-HIV drugs who find their T4 count rises and stays above 250 cells may no longer need to take drugs that may have been prescribed to prevent PCP pneumonia or other opportunistic infections. Monitoring the changes in your T4 count while you are taking anti-HIV drugs can help the doctor decide whether the treatment is working or whether it is time to try different options. A fall in T4 count would be a sign the treatment is not working and may need to be switched to a new regimen. However, the T4 count isn't the only consideration when making these decisions; viral load results, how well you feel, which treatments have been used before are also considered.

Understanding the Results

Factors other than HIV can affect your T4 or CD4+ count including infections, time of day, smoking, stress, and which lab tests the blood sample. So it is very important to watch the trend in your CD4+ count over time, rather than to place too much emphasis on a single test that may be misleading. Doctors will normally suggest measuring the T4 or CD4+ count every three to six months if one has a relatively high count, no symptoms, and is not taking anti-HIV drugs. They may suggest more frequent counts if decisions have to be made such as starting treatments, the development of HIV-related symptoms, or if the decline in CD4+ cells appears to be increasing.

The Antibody

Antibodies are Y-shaped molecules that bind to specific foreign proteins, or pieces of protein, called antigens.

When antibodies on the surface of a B cell snag an undesirable or foreign protein, *both disappear inside the cell.* Eventually, a bit of the foreign protein may reemerge, attached to a self recognizable protein molecule called a **class II protein** (Figure 5-1). The pair, the small piece of foreign protein attached to a self class II protein, acts as a red flag to T4 cells, which set off an aggressive immune response.

(For an excellent review of the concept of antigen processing and presentation, see

Unanue, 1995; for class I and class II proteins, Strominger et al., 1995; for the concept of self, Zinkernagel, 1995; for cell-mediated immunity, Doherty, 1995).

B Cells Make Antibodies and Release Them into the Bloodstream

After an antibody and virus join, they are digested by macrophages or cleared from the blood by the liver and spleen. Some B cells and T cells become memory cells, which are stored by the immune system. Memory cells appear to recall their history and remember the antigen they have previously encountered. However, if the antigen, say a virus, has mutated (changed), as the flu virus does yearly, previous antibodies will not affect it. New antibodies must be created to neutralize the new mutant virus. While this antibody production is taking place, the viral invader has time to multiply and infect new cells, and the infected person suffers the symptoms of the flu.

SIDEBAR 5.1

MEASURES AND COUNTERMEASURES: THE HUMAN IMMUNE SYSTEM VS. VIRUSES

A war has existed between humans and viruses that invade human cells and cause disease. Probably from the first encounter, prehuman to first humans, viruses became able to *hide* inside human cells. The immune system in turn adapted by developing a means of surveying and identifying cells containing hidden virus. Virus then adapted by deceiving the immune surveillance system; the immune system then evolved a better means to *mark* cells carrying the virus. This game of hide-and-seek between the virus and the immune system continues. Learning the means by which the virus attempts to trick the immune system carries over into the study of tumor cells. They must also escape the immune system's surveillance system and do so by using several of the same ploys used by the virus. From what microbiologists have learned so far, different viruses have evolved different ways of getting rid of expression of class I proteins, the flagpole of self proteins. But the mechanisms are amazingly different in different viruses. Examples are:

Cytomegalovirus (CMV)—A virus that may cause birth defects in the fetus and the rapid onset of blindness in people with AIDS. On entering a cell, CMV stops self identity proteins from reaching the cell membrane. If the class I proteins can't reach the membrane neither will pieces of CMV. The immune system is blinded to the fact the CMV is inside the cells. The immune system counters with a natural killer cell that destroys all cells that do not contain a self protein on their surface! But CMV over time evolved a means to produce a fake self protein that passes for the real one. The natural killer cells are successfully fooled.

Adenovirus—The virus causing common colds. This virus can also stop self proteins from reaching the membrane carrying identifying pieces of virus. Further, this virus is able to make the cell divide so it can replicate itself. The cell then becomes cancerous (replication out of control), begins to destroy itself—but the virus has evolved a way to stop the cell's self-destruction, so the cell continues to divide. The process is similar to what happens in tumor cells.

HIV—This virus has the most deadly scheme of all. This virus infects cells of the immune system—most often the T4 cells that are essential to the initiation of the immune response. Similar to other viruses, HIV has to disable the self alerting system. One of HIV's genes, call **nef,** makes a protein that attaches to the self proteins, just inside the cell's membrane. The other end of the nef protein carries an address label, readable by the cell's internal sorting system that directs proteins to their proper place. The message carried by the nef protein says, in the cell's sorting code, "Haul to garbage dump and recycle," tricking the infected cell into pulling down its self proteins and destroying them. The nef protein also tags the CD4 proteins for destruction in the same way. Like the self proteins, the CD4's stick up through the cell membrane. What does HIV gain from having the cell destroy CD4 proteins? That's the million-dollar question investigators are working on. The reason may be to prevent other viruses from entering the cell or because in latching onto the CD4 proteins, the nef protein dislodges and energizes another protein that is known to activate the T cells. The cell's activation to make more HIV may be the answer.

Can the Immune System Remember HIV Exposure?

The immune system works because it produces antibody against an antigen and in addition creates immune memory cells. But this does not appear to occur when the immune system is exposed to HIV. Yes, antibody is made, but when HIV disappears (is reduced to below measurable levels using HAART), so does the immune defense against it. There are few if any memory cells to protect us should HIV rebound or break through drug suppression replication. In the mid-1990s, Francis Plummer and colleagues at the University of Nairobi stunned scientists when they announced discovery of a group of Kenyan female prostitutes who appeared resistant to HIV infection, surviving infection for years despite more than six customers a week without condom protection in a society where upward of 20% of their customers were likely to be HIV-positive. Vaccine researchers were ecstatic because the discovery offered evidence that people could successfully become immune to HIV—something cynics had argued might be impossible. But at the Seventh Conference on Retroviral and Opportunistic Infections (2000) Plummer's group had distressing news: 10% of the apparently resistant prostitutes became HIV positive from 1996 to 1999. Their infections coincided with the women's decisions to decrease their exposure to HIV by having fewer customers or insisting that the men wear condoms. Kevin De Crock, an HIV investigator, said, "It suggests that HIV antigen stimulation is required for the maintenance of resistance." De Crock said it may be that the immune system is not capable of remembering without a constant presence of HIV (too few memory cells made?). This would appear to be an HIV catch 22—to control HIV it must be present, but the control will over the long run be insufficient to stop the progression to AIDS.

ANTIBODIES AND HIV DISEASE

Resistance to HIV does not seem to be the same as the more common examples of immunity. The body's protective countermeasures against measles and mumps are absolute. Years after exposure, there is no hint within the body of the foreign agents that cause those diseases. After children become immune to mumps, they can no longer infect other people. Immunity to these diseases occurs because the immune system makes neutralizing antibody that destroys the infecting agent. Memory immune cells are able to produce neutralizing antibody whenever these infecting agents enter the body. In contrast, it has been shown that although the immune system of most people initially produce neutralizing HIV antibodies, the continued evolution of HIV in their bodies results in a series of mutant HIV that are not efficiently neutralized via antibody response. Thus over time, the ratio of effective neutralizing antibody production to new virus production becomes disproportionate, that is, more virus than effective neutralizing antibody and at this time the level of T4 or CD4+ cells begin to drop. T4 cells drop because antibodies to HIV reduce circulating HIV in the plasma (viral load) without affecting

POINT OF INFORMATION 5.2

THE BIG QUESTION

- How does HIV, which at first glance does not appear to be a highly formidable foe, persist in the body for such long periods of time and continue multiplying and progressively causing more and more damage until a fatal outcome is reached?
- Why does the immensely powerful immune system of the body, an organ system that has evolved over millennia of challenges from a wide variety of infectious and noninfectious invaders to become an exceedingly effective defender of the body against agents far more virulent than HIV, now appear to be powerless against it?

The answer: Because HIV, unlike other agents that enter the human body, (virus, bacteria, fungus, and protozoa) infects immune system cells, in particular **the T4 or CD4+ cell that governs the response of the entire immune system.** Without T4 cells the immune system cannot function—like a car without gas, all the parts are there but nothing happens.

HIV replication or cell-to-cell spread of HIV. This means that the production of mutant HIV to existing antibody never ceases, resulting in continued T4 cell infection and loss. Eventually there are too few T4 cells to ward off opportunistic infections (see Chapter 7 for additional information on T4 cell replacement and HIV production).

The rapid production of HIV mutants without the same rapid production of a neutralizing antibody against each mutant means that sometime after infection, much if not most of the antibody in some people may be nothing more than useless antibody copy (antibody to the initial strain or strains of HIV). In these people, HIV disease would most likely progress more swiftly than in persons whose immune system can keep up with the production of new antibody to match the formation of HIV mutants. This may be one important reason why some people progress to AIDS and death so rapidly when compared to other HIV-infected people.

Infection-Enhancing Antibodies

Ramu Subbramanian and colleagues at the University of Montreal (2002) reported that infection-enhancing antibodies (IEAs) make up most of the antibody humoral response to HIV infection. Their results show that the anti-HIV humoral immune response consists of a mixture of antibodies that may inhibit or enhance HIV infection and whose ratios may vary in different stages of the infection. Seventy percent of blood serums from HIV-infected persons contained IEAs. Such antibodies enhance HIV's ability to infect cells.

HIV Protected from Human Antibodies

Humans create antibodies against a number of HIV proteins, namely the envelope proteins (gp120), the transmembrane protein (gp41), and the proteins of HIV's core (gp24). But, *antibodies cannot enter cells.* The antibody can only attack HIV in the plasma. Plasma is the fluid part of the blood and does not include the blood cells. Once inside a host cell, *HIV is protected from antibody.* Such cells, monocytes, and macrophage and dendritic cells carry HIV internally. All these cell types serve as **HIV reservoirs** in the body. In addition, these cells travel to all distant points within the body and deliver HIV. The self antiretroviral chemicals that these cells generate appear to be ineffective against their hidden traveling companion, HIV (Moir et al., 2000; Olinger et al., 2000). Perhaps the most disturbing discovery concerning HIV reservoirs in 2001 is the work of Robert Siliciano at Johns Hopkins University. This discovery is disturbing because little is known about memory cells' lifespan and whether they can be eradicated. Memory cells are programmed to sit and wait

---

POINT OF INFORMATION 5.3

WHAT ARE CELL RECEPTORS?

In the study of cells, the term **receptor** is used to describe any molecule that interacts with and subsequently holds onto some other molecule. The receptor is like the hand and the object held by the hand is commonly called the **ligand.** The interaction between the receptor and ligand implies specificity; a receptor known to bind with substance X would not normally bind with a different substance. For example, a two-slotted electrical wall outlet is a receptacle (receptor) for a two-pronged plug (ligand). A three-pronged plug will not fit into this receptacle. Depending on the type of two-slotted receptacle, even some two-pronged plugs may not fit.

Where Are Cell Receptors Located?

Receptors can be found inside a cell, and especially embedded within all the membranes that a given cell may have. In humans there are organ systems present, and a given receptor may be found associated only with a particular type of cell that comprises a particular type of tissue that makes up a particular organ.

What Do Receptors Do?

Receptors are critical to the life of all cells, whether or not the cells represent an animal, a plant, a fungus, or a bacterium. Every function, response, interaction, pathway, process, and any other term you might think of that concerns the moment-to-moment existence of a cell, is controlled by various receptor/ligand-induced systems. Essentially, you are what your genetically coded receptors allow you to become.

---

for viruses to attack; their job is to keep a record of the germs that the body has previously encountered so that the immune system will be ready the next time it is confronted with the antigens. What HIV has done is tap into the most fundamental aspect of the immune system—a person's immunological memory. It's the perfect mechanism for the virus to ensure its survival. Because the cells are the immune system's memory, they must survive for a long time, creating a latently infected reservoir of HIV in the body. This reservoir is the single biggest obstacle to getting rid of HIV. Anthony Fauci, director of the National Institute of Allergy and Infectious Diseases said, "We are not going to be eliminating this reservoir. Whether you can measure it or not doesn't seem to have a significant impact on the clinically relevant phenomenon of what happens when you stop taking the drug."

Once HIV gets inside these cell reservoirs as a **provirus,** it is likely to remain there for the rest of the cell's life (person's life) unless some other antiviral mechanism within the body or some chemical agent is able to destroy the provirus. To date, no such drug has been found to be effective against the HIV provirus.

IMMUNE SYSTEM DYSFUNCTION

When HIV first arrives in the body, there are no immune memory T cells that know how to deal with it. The body has never experienced HIV. As with any other first exposure to an infection, it is the job of **naïve** or **unexposed T cells** to respond. Returning to the scene in the lymph nodes, naïve T4 or CD4+ cells get recruited to fight HIV and in turn they become infected. As these infected T cells begin to divide, HIV is able to replicate and release new viruses from each infected cell. Viral load counts in the blood will usually rise rapidly during this period called **primary** or **acute infection.** During the acute phase HIV also seeds latent HIV reservoirs in lymphoid tissue. During this period, a small fraction of the T cells that are activated to fight HIV revert back to a resting, or memory, state. As the cells rest, so do the viruses they harbor. In effect, HIV causes two infections: an active infection that spreads via infected T4 cells to other parts of the body, including the brain; and a dormant, or latent infection that persists in lymphoid tissue cell reservoirs.

If this process is allowed to continue unchecked, the large pool of naïve T cells becomes slowly drained. To understand why, think of the lake metaphor. The water flows out of the lake at a rate greater than the amount of water coming in, so if the siphoning isn't stopped, the lake eventually empties. Studies have shown that the naïve T cell pool does indeed dwindle in HIV infection. In fact, a recent study shows that the loss of newly produced naïve T4 cells is strongly linked to disease progression.

HIV also affects the **memory T4** or **CD4+ cell pool.** Studies have shown that people with HIV lose their ability to respond to a variety of infections. When a person progresses to AIDS, he or she often gets sick from old infections that were previously controlled by memory T cells. A critical fact is that the body only has a certain amount of room for memory T cells. Immunologists have proposed that when a new memory T cell matures in response to a new infection, an existing memory cell has to die off to make room. This sounds like a recipe for disaster, but an average person has enough room in their memory T4 cell pool to accommodate all the memory T cell squads they need to stay healthy (for example, the herpes zoster squad, the Pneumocystis carinii pneumonia squad, the cytomegalovirus squad). If a particular memory T cell squad gets low on members when new memory T cells show up, other T cells can always be recruited if the infection reappears.

HIV infection, however, appears to lead to the continual addition of defective HIV-specific memory T cells to the pool. This accumulation causes a reduction in the number of functional memory squads to control other infections. This may explain the occurrence of different infections, including opportunistic microorganisms, during the more advanced stages of HIV infection. It also explains why some people on HIV therapy remain vulnerable to opportunistic infections.

How Invading HIV Gets to the T4 or CD4+ Cell

Based on the recent investigations of American and Dutch scientists, it appears that HIV hijacks immune cells to enter the body's immune system.

Dendritic cells are the watchdogs of the immune system. They are located just below the skin surface and under moist mucosal tissue on

POINT OF INFORMATION 5.4

THE FIRST LOOK AT HOW HIV'S gp120 ENVELOPE GLYCOPROTEIN CHANGES SHAPE TO ATTACH TO CD4 AND CHEMOKINE RECEPTORS

In three landmark studies published in two scientific journals, Carlo Rizzuto and colleagues (1998), Richard Wyatt and colleagues (1998), and Peter Kwong and colleagues (1998) revealed and discussed the crystal structures of gp120, HIV's surface (envelope) glycoprotein. These reports have important implications for the virology, immunology, and vaccine development against HIV and most other viruses that infect humans. These papers came after 10 years of frustrating efforts to crystallize gp120—a prerequisite to determining its structure by x-ray crystallography. David Baltimore, noted laureate and head of a U.S. government advisory panel on AIDS vaccines said, "This is a big deal." Other notable HIV researchers believe these findings are a major advance in the collective knowledge of the virology and immunology of HIV infections. Revealing the structure of gp120 was to reveal the **passkey** on how HIV escapes antibody attack and invades T lymphocytes. First, **gp120** attaches to the T cell receptor, CD4; then gp120 changes shape, unveiling a binding site that allows HIV to attach to a second T cell receptor, one of the chemokines. With both attachments secured, HIV enters the cell. A computer-generated snapshot of the HIV infection process is very informative (Figure 5-5). The picture shows, for example, that some of the virus's most stable—and therefore vulnerable—structures are either located at the bottom of crevices, where the relatively bulky antibodies of the immune system can't reach them, or obscured by great forests of sugar molecules. It appears that the vulnerable structures are covered by a top similar to a moveable roof on a stadium. One possible antibody target comes out of hiding but only in that brief moment after gp120 latches onto the CD4 receptor and before it attaches to the chemokine receptor—much too briefly for the immune system to react. Because HIV keeps its immune-vulnerable structures hidden until the last possible moment, HIV is now called the **"Houdini of viruses."** However, most drugs are very small molecules—much smaller than antibodies. Properly designed drugs might be able to infiltrate some of those crevices in HIV's outer walls.

UPDATE 2003—A New Understanding on How HIV Evades Antibodies Viruses, through mutation, typically vary or change the protein sequence, or epitope, of their viral envelope that acts as a docking station for neutralizing antibodies made against them. This protein variation alters the docking region on the virus and prevents antibodies from grabbing hold and targeting the virus for destruction. It is a kind of cat (the changing epitopes) and mouse (the changing antibodies) game where in most human viral infections the immune system is able to keep up with the viruses' new appearance. HIV, in contrast, continuously changes the arrangement of large sugar molecules studded across its protein coat so that those docking regions, the epitopes, for neutralizing antibody attachment are blocked. In other words, it would appear, contrary to much speculation, that the immune system does make abundant HIV neutralizing antibody but it quickly becomes ineffective because of HIV's ability to mutate or change, creating a new barrier to the neutralizing antibody faster than the immune system can alter or change the antibody to the rapidly changing HIV barriers. In March 2003, Xiping Wei and 14 colleagues (2003) reported that the barrier on HIV to neutralizing antibody was not a change in the protein epitopes, they stayed the same. It was very rapid changes in the sugar molecules (glycan) that coat the viral envelope surface that blocked the neutralizing antibody from docking with the HIV epitopes! The layer of sugar molecules or the newly discovered silent face of HIV is now referred to as the **glycan shield** or sugar-coated HIV. This constantly evolving shield stays ahead of the evolving antibody production—the cat versus the mouse.

Putting It All Together

When HIV initially infects a person it is able to replicate unrestricted until the first set of antibodies develops that recognizes proteins or epitopes within or protruding from holes in the glycan shield. By then, the virus has randomly mutated its shield, as well as other regions of the envelope, conferring a strong survival advantage to viral particles that cannot now be seen by the neutralizing antibodies, which also change their structure in pursuit of the virus. But the cat (the immune system) cannot keep up with the wily mouse (the virus). The result is an unexpected means of HIV escape from the human immune system.

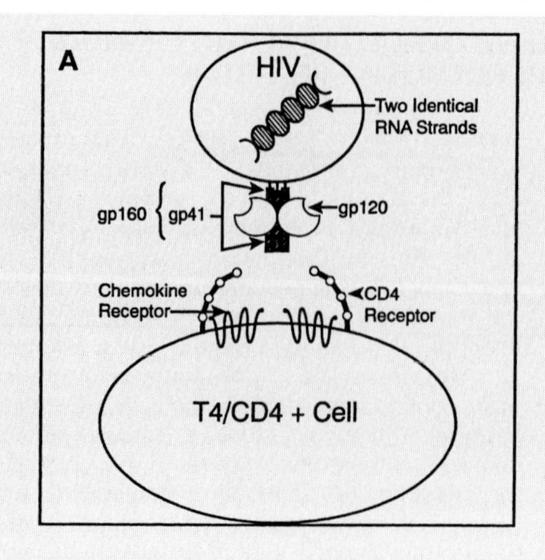

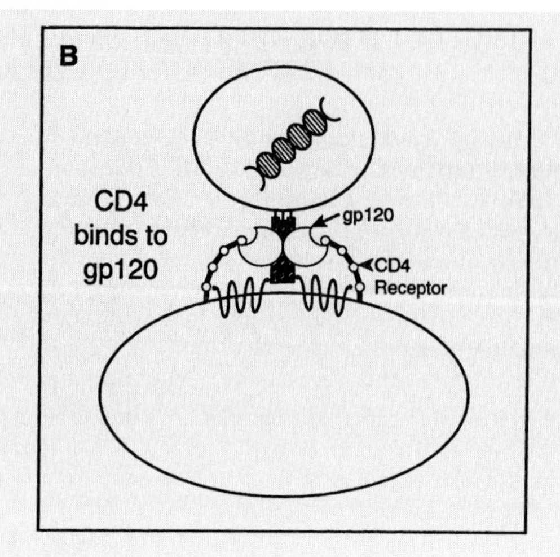

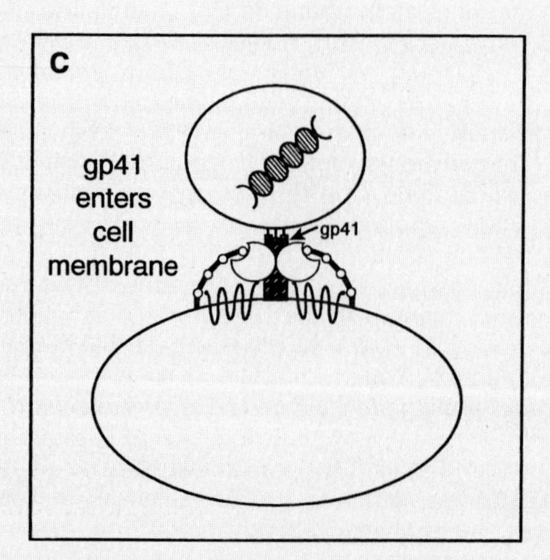

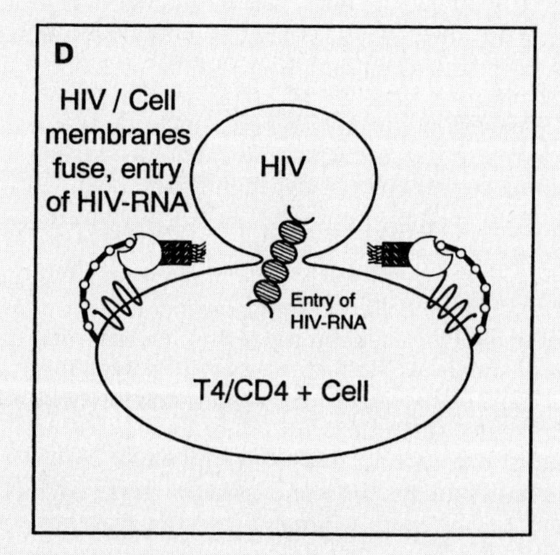

FIGURE 5-5 Unraveling the Secret to HIV Cell Infection: Attachment, Fusion, and Entry. **A.** The process begins with interactions between a cluster of proteins on HIV's outer coat, sometimes referred to as the gp160 spike—and CD4 and one of the chemokine receptors (either CCR-5 or CCR-4) on the cell surface. This interactive complex is made up of three transmembrane glycoproteins (gp41), which anchor the cluster to the virus, and three extracellular glycoproteins (gp120), which contain the binding domains for both CD4 and the chemokine receptors. **B.** The first step in fusion involves the attachment of the CD4 binding domains of gp120 to CD4. CD4 attachment inhibitors (for example, PRO 542) act here. **C.** One gp120 is bound with CD4 protein, the envelope complex undergoes a structural change, bringing gp120 close to the chemokine receptor and allowing for a more stable two-pronged attachment. Antagonists or coreceptor blockades of CCR-5 (SCH-C) and CXCR-4 (AMD-3100) act here. If the virus attaches to both CD4 and the chemokine receptor, additional conformational changes allow gp41 to enter CD4+ cell membrane. **D.** The collapse of the extracellular portion of gp41 to form a hairpin, which is sometimes referred to as a coiled-coil bundle. The fusion inhibitors T-20 and T-1249 act here resulting in a poorly formed hairpin. In the absence of an inhibitor, the hairpin structure brings the virus and cell membrane close together, allowing fusion of the membranes and subsequent entry of viral RNA. *(Illustration adapted from* The PRN Notebook, *Vol. 7, March 2002, www.prn.org)*

SIDEBAR 5.2

THE DENTRITIC CELL AND NEWFOUND FUNCTION

Dendritic cells—among the first cells in the body to encounter infectious organisms—are key players in initiating an immune response. These cells arise in the bone marrow but migrate to tissues throughout the body. Before dendritic cells encounter an infectious agent, they are immature and act as roving sentinels of the immune system. Upon an encounter with an infectious agent, the dendritic cell reaches maturity—capturing the infectious agent and processing it for presentation to the T cell, thus initiating a cascade of immune events that fight infection.

Researchers at the Whitehead Institute for Biomedical Research have discovered that a dendritic cell initiates an immune response that is tailor-made for specific infectious organisms. The researchers found that dendritic cells turn on different sets of genes, or a signature pattern of gene response, depending on whether the organism is a bacteria, virus, or fungus. This study shows that even at the earliest stages of infection, the human body knows the nature of the infectious organism, or pathogen, and responds with a specific type of immune response to eliminate the pathogen.

Dendritic cell maturation—as a result of its recognition of a pathogen—is highly specialized. The dendritic cell fine tunes its response based upon the nature of the pathogen. For every pathogen, there is a specific set of genetic programs that are activated or not activated, which then impacts how the immune system as a whole reacts to the infectious agent. The knowledge that dendritic cells are able to sense and respond specifically to each pathogen could ultimately help clinical scientists detect the presence of particular pathogens and measure the nature of the immune response by looking for the signatures of pathogen-specific genes.

surfaces like those of the vagina, urethra, and penis. When dendritic cells see a foreign invader such as a microorganism, or virus, they capture it, shred it, and display pieces of proteins from the invading pathogen on their surfaces. These displayed proteins serve to alert other immune system cells such as T4 cells that the body is under attack. What recent investigations now show is that HIV attaches to dendritic cells and hitches a ride into the lymphoid tissues where it then infects T4 cells. The dendritic cell has finger-like projections that carry a protein receptor called **DC-SIGN** (dendritic cell-specific ICAM-3-grabbing nonintegrin). HIV adheres to DC-SIGN and is carried from the mucosal lining of the cervix or rectum to the lymph nodes where it transfers HIV to the T4 cells through their coreceptors R-4 and R-5 (Steinman, 2000; Geijtenbeck et al., 2000). Thus, another puzzle piece is in place toward the complete understanding of how HIV successfully attacks the human immune system.

Dendritic Trojan Horses

According to Melissa Pope (2002), as capable as dendritic cells are of kicking off a multipronged immune response to any invading pathogen that crosses their paths, something goes terribly wrong when they are confronted with HIV. Herein lies the paradox. The very cells that should be activating the immune system against this pathogen end up facilitating the virus infection. It is now known that some dendritic cells can actually become infected with HIV and replicate the virus. It also appears that the CCR-5 receptor on immature dendritic cells permits infection and replication to occur.

How HIV Enters T4 or CD4+ Cells and Macrophage

HIV researchers have known since 1984 that human CD4 cell membrane receptors alone are sufficient for binding HIV to the T4 lymphocyte membrane but CD4 receptors are not sufficient for HIV envelope fusion with the T4 cell membrane or for HIV penetration or entry into the cell's interior. This knowledge has enticed many groups of AIDS researchers to search for additional receptors, coreceptors to CD4, that HIV uses to enter a cell after binding to it. Various candidate receptor molecules were put forward, sometimes with more fanfare than fact, but have not stood the test of critical examination.

FUSIN OR CXCKR-4 (R-4)

In May 1996, Ed Berger (Figure 5-6) and colleagues reported finding a receptor that allowed syncytium-inducing (SI) strains of HIV (HIV that causes T4 cells to form clusters—they attach to each other) to enter T4 cells. Strains of HIV that do not induce T4 cell syncytium formation (NSI) could not enter T4 cells. Berger and colleagues named this T4 cell receptor **FUSIN.** In August 1996 Conrad Bleul and colleagues identified a chemokine, **CXC stromal cell-derived factor-1 (SDF-1)** that binds to the FUSIN receptor and blocks HIV entry. They named this chemokine **CXCKR-4.** This coreceptor functions preferentially for T cell line-tropic HIV strains.

FIGURE 5-6 Edward A. Berger, Chief of the Molecular Structure Section, Laboratory of Viral Diseases at the National Institutes of Allergy and Infectious Diseases. In Spring 1996, he and colleagues discovered the first coreceptor **CXCKR-4** or **FUSIN** that HIV needs to complete its attachment and entry into T4 lymphocytes. *(Photograph courtesy of Edward A. Berger)*

What Are β-Chemokine Receptors and β-Chemokines?—β-Chemokine receptors are cell-surface proteins that bind small peptides (a short chain of amino acids) called β-chemokines to their surface. Chemokines are classified into three groups depending on the location of the amino acid cysteine (**C**) in the peptide. These are C_ _, C-C_ _ _ _, and CXC_ _ _chemokines. The X in CXC represents any other amino acid. The chemokine receptors are identified by the individual chemokine(s) that binds to them. In that sense, a reference to a specific chemokine(s) also identifies its receptor.

CC-CKR-5 (R-5) RECEPTOR

Within two months after the FUSIN receptor data were reported, five separate research teams reported on an additional coreceptor called **CC-CKR-5** (R-5). R-5 is the fifth of seven known human chemokine receptors that respond to β-chemokines. The function of β-chemokines is to attract macrophages and other immune cells to sites of inflammation. R-5 is known to be a receptor for three β-chemokine proteins called RANTES, MIP–1α, and MIP–1β. The β-chemokine RANTES (**r**egulated-upon-**a**ctivation **n**ormal **T** **e**xpressed and **s**ecreted), chemokine MIP-1α and 1β (macrophage inflammatory proteins) are all produced by T8 cells and are known to inhibit HIV replication, particularly in **macrophage-tropic HIV strains** (HIV strains that infect macrophage over T4 cells) (Deng et al., 1996; Dragic et al., 1996) and are powerful suppressors of HIV infection, especially HIV infection of macrophages. That is, these three chemokine proteins suppress HIV entry into macrophage by somehow blocking the coreceptor R-5. Members of the research teams who have contributed to the discovery of the R-5 receptor believe that macrophage-tropic or M-tropic HIV strains occur in greatest number early on after HIV infection and then, sometime later during HIV disease the predominant HIV strain shifts to HIV strains that use the FUSIN receptor (R-4) on T4 cells. HIV, by shifting receptors, may be avoiding the suppressive activity of the chemokines that block the R-5 receptor. Figure 5-7 shows a diagram of this suggested **receptor swap** that occurs sometime during HIV disease progression. One of the great unsolved puzzles of HIV disease is *why*, during disease progression, does HIV lose its ability to in-

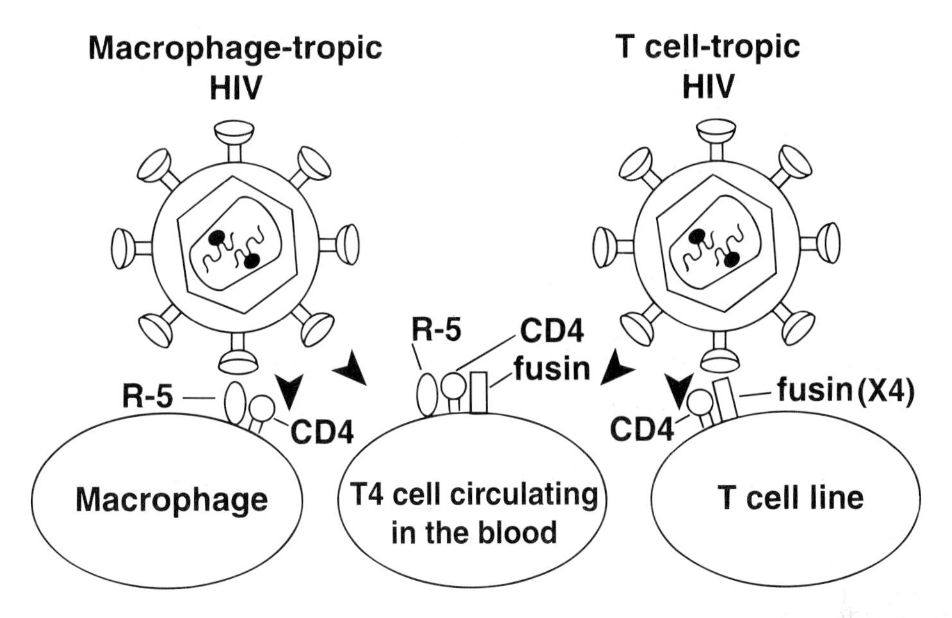

FIGURE 5-7 Coreceptors Required for HIV To Enter Human Cells. HIV can broadly be divided into two classes: those more suitable (tropic) to infecting macrophage, a major reservoir of HIV, and those that infect T4 cells in the lymph nodes or other tissues. Macrophage-tropic nonsyncytium-inducing (NSI) HIV isolates infect macrophages but fail to infect HIV T cells lines, while T4 HIV syncytium-inducing (SI) strains fail to infect macrophage. But HIV of both classes efficiently infect T4 cells isolated from peripheral blood mononuclear cells (PBMC). Macrophage-tropic NSI HIV appear to be preferentially transmitted by sexual contact and constitute the vast majority of HIV present in newly infected individuals (Zhu, 1993). The T-tropic SI viruses generally appear late in the course of infection during the so-called "phenotypic switch" that often precedes the onset of AIDS symptoms (Conner, et al., 1994). The molecular basis of HIV-1 tropism appears to lie in the ability of envelopes of macrophage-tropic and T-tropic viruses to interact with different coreceptors located on macrophage or T4 cells. Macrophage-tropic viruses primarily use R-5 and less often R-3 and R-2, newly described seven-transmembrane domain chemokine receptors, while T-tropic HIV tend to use FUSIN (R-4), a previously identified seven-transmembrane protein (Hill et al., 1996; Moore, 1997).

fect macrophage and become T cell tropic? *Why* does HIV switch to other cell receptors?

CC-KR-2 (R-2) AND CC-KR-3 (R-3) RECEPTORS

Recent studies have shown that at least two additional chemokine receptors participate in allowing HIV cell entry. HIV can use both CC-KR-2 (R-2) and CC-KR-3 (R-3) to help its invasion of cells. For example, dual tropic HIVs use R-2 and R-3 to enter cells and studies with microglial cells **(non-CD4 type cells)** show that both R-5 and R-3 permit HIV to enter these brain cells (McNicholl et al., 1997). Current research involves finding ways to block or fill the chemokine receptors with a harmless molecule, thus blocking HIV's binding site on the cell (Chen et al., 1997).

GENETIC RESISTANCE TO HIV INFECTION

At this point in the HIV/AIDS pandemic it is believed that about 95% of HIV-exposed people are susceptible to HIV infection and HIV disease progression. This statement is made because it has long been known that persons who have deliberately avoided safer sex practice and who have had unprotected sex with HIV-infected persons failed to become HIV-infected! The question that has continued to puzzle HIV investigators is, how can multiple HIV-exposed persons remain uninfected?

Pieces of that puzzle began to fall into place when two gay males who, despite repeated unprotected sex with companions who died from the disease, remained HIV negative. Neither quite understood why he was spared but they pressed scientists to come up with the answer. HIV investigators Rong Liu and coworkers (1996) and Michel Samson and coworkers (1996) reported that repeated HIV-exposed but uninfected people have a 32-nucleotide deletion in the gene that produces the R-5 receptors on macrophage. The protein produced by this gene is severely damaged and does *not* reach the cell surface to act as a chemokine receptor. Without these receptors, the envelope of HIV cannot fuse with the envelope of macrophage to gain entrance into the cell. Thus, people who carry both defective R-5 genes' **homozygotes** (they received one defective gene from each parent) are resistant to HIV infection. If one is **heterozygous** (that is, one carries one defective gene and one normal gene), one will produce the R-5 receptor that HIV needs to penetrate macrophage, but they may be fewer in number. Thus, chemokines (the chemicals that the receptors were made for) would have fewer receptors to fill so there is a greater chance that fewer R-5 receptors are open or available for HIV attachment. As a result, heterozygous people may be *less* resistant to HIV infection than the homozygous mutant (receptorless) people, but may be *more* resistant than people who have two normal genes (homozygous normal) who are the most susceptible to HIV infection. (O'Brien, 1998)

Michael Dean and colleagues (1996) presented evidence that the frequency of R-5 deletion heterozygotes (those with one copy of the gene) was significantly elevated in groups of individuals who had survived HIV infection for more than 10 years, and, in some risk groups, twice as frequent as their occurrence in **rapid progressors to AIDS**. Michael Marmor and colleagues (2001) reported that white gay males who are heterogeneous (have one copy of the mutant gene) were 70% less likely to become HIV-infected than those without the mutation. Survival analysis also shows that disease progression is slower in R-5 deletion heterozygotes than in individuals with the normal R-5 gene. Jesper Eugene-Olsen and colleagues (1997) reported that individuals who are heterozygous for the 32-base-pair deletion in the R-5 gene have a slower decrease in their T4-cell count and longer AIDS-free survival than individuals with the wild-type gene for up to 11 years of follow-up.

Sean Philpott and colleagues reported in 1999 that children who inherit the R-5 mutation are protected from vertical HIV infection (mother to newborn via breast milk).

Will the Mutant R-5 Gene Protect People from All Subtypes of HIV?

This genetic defect, the 32-nucleotide deletion, prevents infection only with the strain of HIV (Subtype B) that is transmitted sexually and is prevalent in the United States and Europe. It does not necessarily protect against other strains of HIV transmitted through intravenous drug use or blood transfusions, or strains prevalent in Africa.

Should Everyone Be Tested for the Presence of the R-5 Gene?

Researchers agree that getting tested for the gene would not be difficult, but it would not be of great value because the tested person could still be infected by other strains of HIV. It must be assumed that most people do not carry a pair of defective R-5 genes because 95% or more of HIV-exposed people become HIV-infected.

Who Carries the Defective R-5 Gene?

Perhaps most surprising, HIV investigators found that the homozygous genetic defect (a person having two defective genes) is common: **It is present in 1% of whites of European descent.** But it appears to be absent in people from Japan and Central Africa; about 20% of whites are heterozygous for this gene (have one copy of this gene).

Recent studies conducted at Stellenbosch University show that this particular mutation, the R-5 gene deletion, is virtually absent in the South African black population. The investigators believe this is but one factor that affects a black person's susceptibility to HIV infection. Most likely there are other genes involved in the complexity of susceptibility to infection. Aside from the issue of susceptibility to infection, there is no consensus on whether blacks or Hispanics, after infection, progress more quickly than whites toward AIDS.

Why Does This Gene Exist?

Scientists speculate that the mutant form of R-5 protected against some disease that afflicted Europeans but not Africans The obvious candidates would be the Black Death of 1346, the plague and/or smallpox. Both lethal diseases are at least 700 years old, enough time for genetic selection to take place. Those without the mutant R-5 gene died, the survivors reproduced and the gene became dispersed within the surviving population. Stephen O'Brien said that the chance of this gene *randomly* reaching its current frequency in the white population is about zero. The idea that a mutant gene can confer protection against a specific infection is not new. The mutation that causes sickle cell anemia provides people carrying one copy of it with resistance (but not immunity) to malaria. There is some evidence that the cystic fibrosis mutation may protect against typhoid fever.

These ideas for why the R-5 genes exists remind us that we carry a genetic record of the diseases of the past and that, at least for some, those genes have once again come to the rescue.

How HIV Kills the T4 or CD4+ Immune Cells

In July 2001, Jacques Corbeil and colleagues at the University of California at San Diego reported for the first time on how HIV assaults the human immune T4 or CD4+ cell. Using sophisticated technology, the team of researchers looked within the body's immune cells and recorded the molecular events triggered by invasion of HIV. They created a detailed account of the devastating progression of cellular injury following HIV infection. For this study, an HIV-infected CD4+ T cell line was monitored following infection at eight time intervals, from 30 minutes to 72 hours after exposure. They monitored some 7000 genes over this time period in an attempt to discover what effect HIV's genes would have on the cell's genes. For each interval, 10 million infected cells were applied to a microarray gene chip. A control sample of healthy cells was analyzed at the same intervals for comparison. According to the researchers, HIV invades and swiftly overpowers the DNA of the immune cells and inserts its own genes. HIV interrupts the cells genes and alters the cellular mitochondrial energy source. In addition, it suppresses the immune cells' DNA repair mechanisms while inducing other genes to initiate their own death. The first observation by the researchers came 30 minutes after infection. At that time HIV had already suppressed more than 500 genes in the infected cells and had activated 200 other genes that included suicide genes, or those activated as part of the normal cycle of cell death, known as apoptosis. Three days after infection of 1400 genes expected to be active, based on healthy cell measurements, half were suppressed—not functioning. The research has moved the study of HIV from observing clinical signs and symptoms of HIV to studying the molecular changes HIV forces on a living cell. Such research may lead to new drug therapies and the next step in finding an effective vaccine.

T4 CELL DEPLETION AND IMMUNE SUPPRESSION

In addition to the new insight as to how HIV kills T4 or CD4+ cells, there are other ways in which these immune cells are depleted.

Recent research using the **polymerase chain reaction** (a method of amplifying present but unmeasurable quantities of HIV DNA in T4 cells into measurable quantities) has revealed that about *1 in every 10 to 100 T4 cells is HIV-infected in an AIDS patient.* Thus T4 cells serve as reservoirs of HIV in the body (Schnittman et al., 1989; Cohen, 1993). Collectively, T4 cell depletion leads to cumulative and devastating effects on the cell-mediated and humoral parts of the immune system.

Means by Which T4 Cells May Be Lost

1. ***Filling CD4 Receptor Sites***—There is evidence from in vitro studies that HIV can attack CD4 receptor sites in at least two ways. First, HIV can attach, via its gp160 "spikes," to CD4 receptor sites. Second, HIV is capable of releasing or freeing its exterior gp120 envelope glycoprotein, thereby generating a molecule that can actively bind to CD4-bearing cells (Gelderblom et al., 1985). As a result of filling the receptor sites on the T4 cells, the T4 cells lose their immune functions; that is, the T4 cell does not have to be infected with HIV to lose immune function. In addition, CD4-bearing cells that attach the free gp120 molecule then become targets for immune attack by antibody-mediated antibody-dependent cell cytotoxicity (ADCC) and non-antibody-mediated cytotoxic T cells. Both events can result in the destruction of uninfected CD4 bearing cells. The extent to which this occurs in vivo depends on the level of gp120 synthesis, secretion, and shedding. Free gp120 has not yet been measured in the circulation, but this is not surprising given its powerful affinity for CD4 (Bolognesi, 1989).

BOX 5.3

COFACTORS EXPEDITE HIV INFECTION

(An excerpt from Michael Callen's article, "Everything Must Be Doubted," *Newsline*, July/August 1988, pp. 44–45.)
(Michael Callen died in 1993.)

By the age of 27 (when I was diagnosed with crypto [cryptococcus infection] and AIDS) I had had over 3,000 different sex partners. Not coincidentally, I'd also had: hepatitis A, hepatitis B, hepatitis non-A, non-B; herpes simplex types I & II; syphilis; gonorrhea; non-specific urethritis; shigella; entamoeba histolitica; chlamydia; fungal infections; venereal warts; cytomegalovirus infections; EBV [Epstein-Barr virus] reactivations; cryptosporidium and therefore, finally, AIDS. For me, the question wasn't why I got AIDS, but how I had been able to remain standing on two feet for so long . . . If you blanked out my name and my age on my pre-AIDS medical chart and showed it to a doctor and asked her to guess who I was, she might reasonably have guessed, based on my disease history, that I was a 65-year-old malnourished equatorial African living in squalor . . . I believe that a small subset of urban gay men unwittingly managed to re-create disease settings equivalent to those of poor Third World nations and junkies.

This excerpt details many of the cofactors that play a role in the development of immune dysfunctions; while any one infection might not cause a problem for a healthy individual, repeated infection and exposure to foreign body fluids takes a cumulative toll. And certain infections are unquestionably more immune-suppressive than others—in particular, the sexually transmitted diseases.

2. *Syncytia Formation*—The formation of syncytia involves fusion of the cell membrane of an infected cell with the cell membranes of uninfected CD4 cells, which results in giant multinucleated cells (Figure 3-7). A direct relation between the presence of syncytia and the degree of the cytopathic effect of the virus in individual cells has been demonstrated in vitro, and HIV isolated during the accelerated phase of infection in vivo has a greater capacity to induce syncytia in vitro. Syncytia have rarely been seen in vivo.

In the asymptomatic phase of HIV infection, predominantly nonsyncytium-inducing (NSI), HIV variants can be detected. In about 50% of the cases SI HIV variants emerge in the course of infection, preceding rapid T4 cell depletion and progression to AIDS (Groenink et al., 1993).

3. *Superantigens*—Superantigens are bacterial or viral antigens that are capable of interacting with a very large number of T4 cells. Unlike a conventional antigen, which usually evokes a response from less than one in a million (0.01%) T4 cells, a superantigen can interact with 5% to 30% of T4 cells. They directly bind to the exposed surfaces of MHC class II molecules on **antigen-presenting cells (APCs)** and to the variable region of the T cell receptors (TCR) β chain ($V_β$) on the responding T cells. Therefore, superantigens can interact with a large fraction of T cells, resulting in cellular activation, proliferation, anergy (absence of reaction to an antigen), or deletion of specific T cell subsets. Bacterial superantigens include staphylococcal enterotoxins and the toxic shock syndrome toxin-1, which are the causative agents for several human diseases such as food poisoning and toxic shock syndrome (Phillips et al., 1995). The superantigen hypothesis regarding HIV infection stems from the observations that endogenous or exogenous retroviral-encoded superantigens stimulate murine T4 cells in vivo, leading to the anergy or deletion of a substantial percentage of T4 cells that have the specific variable β regions (Pantaleo et al., 1993).

4. *Apoptosis*—Programmed cell death, or apoptosis (a-po-toe-sis), is a normal mechanism of cell death that was originally described in the context of the response of immature thymocytes to cellular activation. It is a mechanism whereby the body eliminates autoreactive clones of T cells.

In a typical day, 60 billion to 70 billion cells perish in your body. Much of this normal cellular turnover involves **apoptosis**. When apoptotic pathways are defective, insufficient cell death can lead to cancers and autoimmune diseases, and excessive cell death can result in neurodegeneration or stroke. Accordingly, much work is being done to define key players in apoptosis.

It has recently been suggested that both qualitative and quantitative defects in T4 cells in patients with HIV infection may be the result of activation-induced cell death or apoptosis. Because apoptosis can be induced in mature mouse T4 cells after cross-linking CD4 molecules to one another and triggering the T4 cell antigen receptor, there has been speculation that cross-linking of the CD4 molecule by HIV gp120 or gp120-anti-gp120 immune complexes prepares the human cell for the programmed death that occurs when a MHC class II molecule in complex with an antigen binds to the T4 cell antigen receptor. Thus activation of a prepared cell by a specific antigen or superantigen could lead to the death of the cell, without direct infection by HIV (Pantaleo et al., 1993). For an excellent review of apoptosis and its role in human disease, see Barr et al., 1994; Hengartner, 1995; and Yeh, 1998.

5. *Cellular Transfer of HIV*—Infected macrophages, or antigen-presenting cells, which normally interact with the T4 cells to stimulate the immune function, can transfer HIV into uninfected T4 lymphocytes. In any case, immediate viral or proviral replication kills the T4 cell.

6. *Autoimmune Mechanisms*—One of the older theories, namely that HIV tricks the immune system into attacking itself is back because of the work of Tracy Kion and Geoffrey Hoffmann (1991). Their work showed that mice immunized with lymphocytes from another mouse strain make antibodies to the HIV envelope protein gp120, as do autoimmune strains of mice, even though none of the animals had ever been exposed to the AIDS virus. One implication of these results is that some component on the lymphocytes resembles gp120 closely enough that antibodies directed against it can recognize gp120 as well. The converse is that antibodies to gp120 should also recognize the lymphocyte component so that an immune response directed against HIV might also interfere with normal lymphocyte function. The autoimmune theory is consistent with results that show there is a selective loss of particular subsets of T cells in AIDS patients.

7. *Cofactors May Help Deplete T4 Cells*—HIV-infected people who are asymptomatic show a wide variation in HIV disease time and progression to AIDS. It is believed that cofactors may be responsible for some of the time variation with regard to disease progression (Box 5.3).

Many agents may act as cofactors to activate or increase HIV production. Although, in general, it is not believed that any cofactor is necessary for HIV infection, cofactors such as nutrition, stress, and infectious organisms have been considered as agents that might accelerate HIV expression after infection. Three new human herpes viruses (HHV-6, 7, and 8) may be cofactors and play a role in causing immune deficiency. They have been shown to infect HIV-infected T4 cells and activate the HIV provirus to increase HIV replication (Laurence, 1996). Cytomegalovirus, Epstein-Barr virus, hepatitis B and C viruses, and tuberculosis have also been associated with increased HIV expression. Over time, investigators expect to find other sexually transmitted diseases that behave as cofactors associated with HIV infection and expression.

Drugs may also be cofactors in infection. Used by injection-drug users (IDUs), heroin and other morphine-based derivatives are known to reduce human resistance to infection and produce immunological suppression. *Pneumocystis carinii* pneumonia is about twice as frequent in heroin users as in homosexuals. It is believed that the heroin has an immunosuppressive effect within the lungs (Brown, 1987).

Blood and blood products may also act as cofactors in infection because they are immunosuppressive. Because blood transfusions save lives, their long-range effects are generally overlooked. Transfusions in hemophiliacs, for example, result in lowered resistance to viruses such as cytomegalovirus (CMV), Epstein-Barr (Berkman, 1984; Blumberg et al., 1985; Foster et al., 1985), and perhaps HIV. Seminal fluid (fluid bathing the sperm) may also act as a cofactor in infection because it also causes immunosuppression. One of its physiological functions is to immunosuppress the female genital tract so that the sperm is not immunologically rejected (Witkin et al., 1983; Baxena et al., 1985).

Epidemiologically, homosexuals at greatest risk of AIDS are those who practice passive anal sex, that is, **anal recipients** (Kingsley et al., 1987). It is generally considered that this sexual behavior is a cofactor that enables HIV to enter the bloodstream by means of traumatic lacerations of the rectal mucosa.

Last but not least of the agents that can suppress the immune system and thereby act as a cofactor in HIV infection is **stress**. Stress can be mental or physical, but it is easier to measure the effects of physical stress.

Although moderate exercise appears to stimulate the immune system, there is good evidence that intensive exercise can suppress the immune system. We still do not really know why (Fitzgerald, 1988; LaPierre et al., 1992). The effect of stress on the immune system is one of the reasons why physicians did not want Earvin "Magic" Johnson to play basketball (see Box 8.5 in Chapter 8).

IMPACT OF T4 CELL DEPLETION

The overall impact of T4 cell depletion is multifaceted. HIV-induced T4 cell abnormalities alter the T4 cells' ability to produce a variety of inducer chemical stimulants such as the **interleukins** that are necessary for the proper maturation of B cells into plasma cells and the maturation of a subset of T cells into cytotoxic cells (Figure 5-8). Thus the critical basis for the immunopathogenesis of HIV infection is the depletion of the **T4 lymphocytes,** which results in profound immunosuppression.

Presumably, with time, the number of HIV-infected T4 cells increases to a point where, in terminal AIDS patients, *few normal T4 cells exist.*

Saha Kunal and colleagues reported on the isolation of CD8+ lymphocytes from HIV-infected patients who produced new HIV. The HIV produced by the CD8+ cells were able to infect both the CD4+ (T4) and CD8+ (T8) cells. CD4+ produced HIV-infected CD8+ cells without using the R-4 or R-5 coreceptors that HIV use to infect CD4+ cells. These data are additional evidence of HIV's ability to adapt to existing biology and to continue to replicate itself.

ROLE OF MONOCYTES AND MACROPHAGES IN HIV INFECTION

Some scientists now believe that T4 cell infection alone does not cause AIDS. They believe

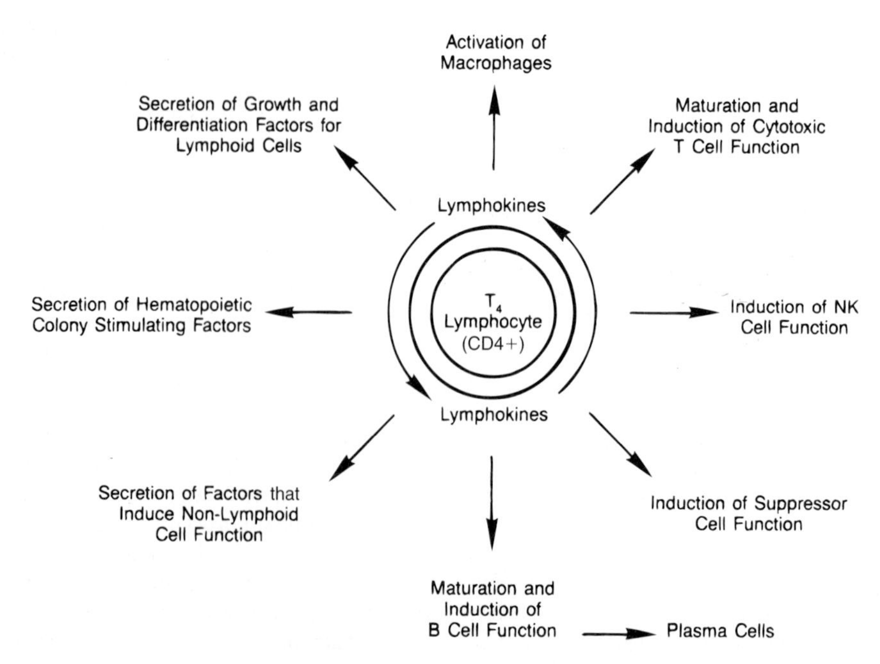

FIGURE 5-8 The T4 Cell Role in the Immune Response. T4 lymphocytes are responsible directly or indirectly for inducing a wide array of functions in cells that produce the immune response. They also induce nonlymphoid cell functions. T cell involvement is effected for the most part by the secretion of soluble factors or **lymphokines** that have trophic or inductive effects on the cells presented in the figure. Lymphokines serve to transfer control of the immune response from the external environmental proteins to the internal regulatory system consisting of ligands and receptors. *(Adapted from Fauci, 1991)*

that equally important to T4 cell infection is **dendritic, monocyte,** and **macrophage** infection (Bakker et. al., 1992). Monocytes change into various types of macrophages (given different names) in order to search and destroy foreign agents within tissues of the lungs, brain, and interstitial tissues, tissues that connect organs. Despite the name changes, all forms of macrophages basically work the same way: They ingest things. Some macrophages travel around within the body, others become attached to one spot, ingesting.

Macrophages, like dendritic cells, are often the first cells of the immune system to encounter invaders, particularly in the area of a cut or wound. After engulfing the invader, the macrophage makes copies of the invader's antigens and displays them on its own cell membrane. These copies of the invader antigens sit next to the self molecules. In effect, the macrophage makes a "wanted poster" of this new invader. The macrophage then travels about showing the wanted poster to T4 or CD4+ cells, which triggers the T4 cells into action. Macrophages also release chemicals that stimulate both T4 cell and macrophage production and draw macrophages and lymphocytes to the site of infection.

Macrophage Trojan Horses

Macrophages may play an important role in spreading HIV infection in the body, both to other cells and to HIV's target organs. First, HIV enters a macrophage and spreads from macrophage to macrophage before the immune system is alerted. Second, macrophages, in their different forms, travel to the brain, the lungs, the bone marrow, and to various immune organs, carrying HIV to these organs.

HIV's ability to infect brain tissue is particularly important. The brain and cerebral spinal fluid (CSF) are specially protected sites. CSF cushions the brain and the spinal cord from sudden and jarring movements. The brainblood barrier, a chemical phenomenon, normally stops foreign substances from entering the brain and the CSF. But HIV-infected monocytes can pass through this barrier. For HIV, monocyte-macrophages are Trojan horses, enabling HIV to enter the immune-protected domain of the central nervous system—the brain, the spine, the rest of the nervous system.

HIV isolates taken from macrophages appear to grow better in macrophages than in lymphocytes. In human cell culture experiments, HIV isolates taken from lymphocytes appear to grow better in lymphocytes than in macrophages. These peculiarities of replication rates may be evidence of separate tissue-oriented HIV strains.

HIV-infected macrophages have proven to be a major problem in efforts to control and stop HIV infection.

Clinical Latency: Where Have All the Viruses Been Hiding?

The immunodeficiency syndrome caused by HIV is characterized by **profound T4 or CD4+ cell depletion.** Prior to the progressive decline of T4 cells and the development of AIDS, there is a symptom-free period that may last for 10 or more years, during which the infection is thought to be clinically latent. Scientists thought little if any viral replication occurred during this period. In recent years, a number of studies have offered an alternative view. Some investigations suggest that there is no real latency period in HIV infection. Instead, starting from the moment of infection, there appears to be a continuous struggle between HIV and the immune system, the balance of which slowly shifts in favor of the virus. Cecil H. Fox of the Yale School of Medicine reported in December 1991 that HIV is not lying dormant at all but is continually infecting immune cells in the **lymph nodes.**

The **lymph nodes,** which are pea-sized capsules that trap foreign invaders and produce immune cells, are all over the body and are connected by vessels much like those that transport blood. Deep within the lymph nodes of 18 HIV-infected patients, researchers found millions of viruses. Based on those data, Fox suggested that HIV uses the lymph nodes as places to meet up with most immune cells. Fox believes that HIV begins replication in T4 cells in the lymph nodes soon after it enters the body. Infected T4 cells leave the lymph nodes and new uninfected cells arrive to become infected. Years later, when enough immune cells have been killed, the patient's defenses become so impaired that he or she is vulnerable to any one of a wide array of opportunistic infections—AIDS has arrived.

Anthony Fauci (Figure 5-9), the head of the National Institute of Allergy and Infectious Diseases (NIAID), and Sonya Heath and colleagues (1995) agree that there are many more HIVs in the lymph nodes than in the blood. Studies in his

FIGURE 5-9 Anthony S. Fauci, M.D. Director, National Institute of Allergy and Infectious Diseases, National Institutes of Health; Associate Director of NIH for AIDS Research. In 2001, he received the Annunzio Humanitarian Award for his impact on the understanding and treatment of infectious diseases/AIDS. *(Photograph courtesy of National Institute of Allergy and Infectious Diseases)*

laboratory have shown there are many millions of HIV particles stuck to what are known as **follicular dendritic cells** in the lymph nodes of an infected individual.

The follicular dendritic cells, which have thousands of feathery processes emanating from them and whose normal function is to filter and trap antigens for presentation to antibody-producing B lymphocytes, serve as highly effective trapping centers for extracellular HIV particles. Virtually every lymphocyte in a lymph node is enmeshed in the processes of these cells. Follicular dendritic cells themselves are susceptible to HIV infection, and appear to place huge numbers of virus particles into intimate contact with other cells that are susceptible to HIV infection (Haase et al., 1996; Knight, 1996).

During the clinically latent phase, the lymph nodes of an infected individual are slowly destroyed. In the final phase of an infection, the follicular dendritic network completely dissolves and the architecture of the lymph node collapses to produce what Fauci called a "burnt-out lymph node." With this collapse, large amounts of HIV are released into the circulatory system and an ever-increasing number of peripheral cells are infected with HIV (Baum, 1992; Edgington, 1993).

In 1995, David Ho and colleagues, Xiping Wei and colleagues, and Martin Nowak and colleagues published separate reports on the rapid production of HIV that begins soon after initial HIV infection.

Rapid viral replication countered by a strong immune response is not unknown; on the contrary, it is characteristic of acute, short-lived viral infections, such as measles or the flu. But in those infections, the immune system most often wins: It rids the body of the virus and builds resistance to future infection. The question is why can't the immune system do the same against HIV? Perhaps it's the length and intensity of the struggle that ultimately does in the immune system by encouraging the emergence of mutant viruses that outstrip the immune system's ability to continue making new types of antibody. During the time when the immune system is functioning well and producing new antibody against each new HIV variant, the struggle between HIV production and immune suppression of the viral load must move repeatedly from the virus to the immune system and back until the immune system is, with time, overwhelmed with the number of different types of antibody it must rapidly produce in large quantity.

Implications of a High HIV Replication Rate

Based on Ho and colleagues' kinetic studies of viral replication, it can be estimated that the population of HIV undergoes between 3000 and 5000 replication cycles (**generations**) over the course of 10 years, producing a minimum of 10^{12} (1 trillion) HIV in an HIV-infected person. This creates a lot of opportunity for HIV RNA evolution because genetic mutations occur most commonly during replication. Some mutations will weaken HIV in such a way as to expose it to attack by the immune system. But other mutations will aid the virus, speeding its replication and increasing the chances it can evade the immune system. Because HIV gets so many chances to replicate, it evolves and mutates, and those strains with the greatest replication efficiency gradually win out. This is Darwinian evolution going on in one patient.

HIV has been shown to have a random (any base has an equal chance of changing) mutation rate of approximately 1 mutation per 10,000 nucleotides and one mutation caused by reverse transcriptase copy error (**transcription error**) per round of replication. The *combination* of events results in too many variants for the immune system to handle. For example, a single HIV-infected person, by the time he or she is diagnosed with AIDS, may have billions of HIV variants in his or her body. With so many variants present, some will be resistant to most if not all drugs currently used in therapy and to drugs that have not yet been created! The constant production of variant HIVs produce the drug-resistant mutants, it's not the action of the drug.

In 1996 the hope was that HAART could wipe out HIV. This hope was based on two ideas, both later proved to be wrong: that the drugs would completely stop the virus from reproducing, and that there were no reservoirs where the HIV could hide. But HIV has the uncanny ability to reestablish itself in a variety of cell types. Even the most rigorous attempts to reduce or eliminate HIV from these reservoirs have failed. Scientists had vastly underestimated the extent of virus activity in an HIV-infected person, particularly during the asymptomatic or clinically latent phase. These and other recent studies should satisfy the major unanswered question concerning HIV disease/AIDS, which was, "Where is the virus?" During the 1980s, it was difficult to find medium to high levels of HIV in persons with HIV disease or even in those persons in the later stages of AIDS. We now know that there are large amounts of HIV present early on in the lymph nodes of HIV-diseased people and in viral reservoirs. Perhaps the most troubled group of researchers now may be those currently developing AIDS vaccines. (See Chapter 9 for a discussion about possible HIV vaccines.)

SUMMARY

After a healthy person is HIV-infected, he or she makes antibodies against those viruses that are in the bloodstream, but not against those that have become integrated as HIV proviruses in the host immune cell DNA. Over time the immune system cells that are involved in antibody production are destroyed. Evidence is accumulating that cofactors such as nutrition, stress, and previous exposure to other sexually transmitted diseases that increase HIV expression are associated with HIV infection and HIV disease. Agents that suppress the immune system may also play a significant role in establishing HIV infection. In short, HIV infection is permanent. HIV attaches to CD4-bearing cells and enters those cells using one or more of at least four chemokine receptors—**CC-CKR-2, CC-CKR-3, CXCKR-4 (FUSIN),** and **CC-CKR-5.** These coreceptors allow HIV to fuse with the cell membrane, and enter the cell after HIV attachment to the CD4 receptor. As seen in Chapter 3, HIV undergoes rapid genetic change, and, as far as is known, attacks only human cells—mostly of the human immune system. It has recently become quite clear that large amounts of the virus are produced within weeks after HIV infection. The virus remains, for the most part, in the lymph nodes until very late in the disease process.

REVIEW QUESTIONS

(Answers to the Review Questions are on page 438.)

1. Which cell type is believed to be the main target for HIV infection? Explain the biological impact of this particular infection.
2. What is CD4, where is it found, and what is its role in the HIV infection process?
3. Is there a period of latency after HIV infection—a time when few or no new HIV are being produced?
4. True or False: HIV is the cause of AIDS.
5. True or False: HIV primarily affects red blood cells.
6. True or False: Lymphocytes have a major role in the immune response to antigens.
7. True or False: All T and B lymphocytes inhibit or destroy foreign antigens.
8. True or False: CD4 and CD8 molecules are antibodies.

9. True or False: HIV belongs to the family of retroviruses.
10. True or False: Cytotoxic and suppressor T lymphocytes are the main targets of HIV.
11. True or False: The latent period is that time between initial infection with HIV and the onset of AIDS.
12. True or False: HIV can spread to infect new cells after it buds out of infected cells.
13. True or False: HIV causes the gradual destruction of cells bearing the CD4 molecule.
14. The most effective use of T4 cell counts in the clinical management of patients with HIV infection is for:
    A. determining when to initiate therapy.
    B. assessing risk of disease progression.
    C. deciding on prophylaxis for opportunistic infections.
    D. measuring the antiretroviral effect of initial therapy.

6 Opportunistic Infections and Cancers Associated With HIV Disease/AIDS

Chapter Concepts

- Suppression of the immune system allows harmless agents to become harmful opportunistic infections (OIs).
- OIs respond well to Highly Active Antiretroviral Therapy (HAART).
- OIs in AIDS patients are caused by viruses, bacteria, fungi, and protozoa.
- In the United States, about 95% of the HIV-infected are coinfected with herpes 1 and/or 2, and hepatitis B. About 40% are coinfected with hepatitis C, and 35% are coinfected with tuberculosis.
- Human Herpes Virus-8 may be the cause of Kaposi's sarcoma.
- In the United States, 14% of AIDS patients are coinfected with tuberculosis.
- The cost of treating OIs can be very high.
- There are two types of Kaposi's sarcoma (KS): classic and AIDS-associated.
- KS is rarely found in hemophiliacs, injection drug users, and women with AIDS.
- The clinical presentation of OIs has been impacted by the use of protease inhibitors.
- Self-evaluation/education quiz is offered.

WHAT IS AN OPPORTUNISTIC DISEASE?

Humans evolved in the presence of a wide range of parasites—viruses, bacteria, fungi, and protozoa that do not cause disease in people with an intact immune system. But these organisms can cause a disease in someone with a weakened immune system, such as an individual with HIV disease. The infections they cause are known as **opportunistic infections** (OIs). Thus, OIs occur after a disease-causing virus or microorganism, normally held in check by a functioning immune system, gets the opportunity to multiply and invade host tissue after the immune system has been compromised. For most of medical history, OIs were rare and almost always appeared in patients whose immunity was impaired by either cancer or genetic disease.

With improved medical technology, a steadily growing number of patients are severely immunosuppressed because of medications and radiation used in bone marrow or organ transplantation and cancer chemotherapy. HIV disease also suppresses the immune system. Perhaps as a corollary to their increased prevalence, or because of heightened physician awareness, OIs seem to be occurring more frequently in the elderly, who may be rendered vulnerable by age-related declines in immunity. New OIs are now being diagnosed because the pool of people who can get them is so much larger, and, in addition, new techniques for identifying the causative organisms have been developed. However, most of the infections considered opportunistic are not reportable, which interferes with a clear-cut count of their growing numbers.

Although OIs are still not commonplace, they are no longer considered rare—they occur in tens of thousands of HIV/AIDS patients. But despite this increase, physicians and their patients have reasons to be optimistic about their ability to contain these infections. The reasons are: (1) In a massive federal effort, driven by the HIV/AIDS epidemic, researchers are finding drugs that can prevent or treat many of the OIs; and (2) various anti-HIV drug therapies have shown promise for warding off OIs by boosting patients' immune systems.

THE PREVALENCE OF OPPORTUNISTIC DISEASES

The prevalence of OIs in the United States is very high. There are some 300,000 HIV-seropositive individuals with T4 or CD4+ cell counts below 200/μL of blood. Worldwide, there are over 6 million HIV infected with a T4 or CD4+ cell count of 200 or less. More than 100 microorganisms—bacteria, viruses, fungi, and protozoa—can cause disease in such individuals, even though only a fraction of these (17) are included in the current surveillance definition for clinical AIDS. In a large survey from the Centers for Disease Control and Prevention (CDC), such OIs were diagnosed in 33% of individuals at one year and in 58% at two years after documentation of a T4 cell count below 200/μL. In 1995, 1997, 1999, and 2002 the CDC presented guidelines for prevention and treatment of OIs in HIV-infected persons (*MMWR*, 2001). The new treatments for OIs have extended the survival of AIDS patients, but they have also opened new issues. With the growing proportion of longer-term AIDS survivors, new OIs have become prominent, together with concerns about cost, compliance, drug interactions, and quality of life (Laurence, 1995; *MMWR*, 1995). The six most common AIDS-related OIs are bacterial pneumonia, candidal esophagitis, pulmonary/disseminated TB, mycobacterium avium complex disease, herpes simplex reinfection, and *Pneumocystis carinii* pneumonia or PCP. The year 1997 marked the first time, since the AIDS pandemic began in the United States, that the incidence of AIDS-defining OIs among HIV-infected persons fell in number from the previous year's total.

PROPHYLAXIS AGAINST OPPORTUNISTIC INFECTIONS

In the United States

Drug prophylaxis against OIs has become a cornerstone of treatment for AIDS patients. For example, the prevalence of PCP dropped from about 80% in 1987 to about 20% by mid-1994 because of excellent drug therapy. The mortality of PCP without treatment is almost 100% (Dobkin, 1995). Researchers at the University of California, San Francisco, found that on average it costs $215,000 to extend by one year, the life of an HIV-infected patient with PCP who is treated in an intensive care unit (Laurence, 1996). That is more than twice the comparable care cost for 1988. Part of the reason given was that as people with AIDS survived longer, they were presenting with third, fourth, and fifth episodes of PCP superimposed on other chronic infections. The downside to OI prophylaxis is that it is difficult to find drugs that work without harmful side effects. In addition, viruses and organisms that cause OIs become resistant to the drugs over time. This is one of the primary reasons researchers are looking for ways in which to boost an immunosuppressed patient's immune system.

In Underdeveloped Nations

Treatment recommendations that are taken for granted in the developed world are not always applicable in Africa or other resource-limited regions and must, at times, be explored in different clinical settings. For example, cofactors such as prevalence of malaria and other parasitic infections or the geographical distribution of various microbial organisms and viruses may alter the pattern or order of clinical events that will be observed as people progress to late stage disease.

OPPORTUNISTIC INFECTIONS IN HIV-INFECTED PEOPLE

AIDS is a devastating human tragedy. It appears to be killing about 95% of those who demonstrate the symptoms. One well-known American surgeon said, "I would rather die of any form of cancer rather than die of AIDS." This statement was not made because of the social stigma attached to AIDS or because it is lethal. It was made in recognition of the slow, demoralizing, debilitating, painful, disfiguring, helpless, and unending struggle to stay alive.

Because of a suppressed and weakened immune system, viruses, bacteria, fungi, and protozoa that commonly and harmlessly inhabit the body become pathogenic (Figure 6-1). Prior to 1998, about 90% of deaths related to HIV infection and AIDS were caused by OIs, compared with 7% due to cancer and 3% due to other causes. Now, with the use of antiretroviral drugs, OIs cause about 50% of deaths. Liver and kidney organ failure, heart disease, and various cancers are on the increase as the cause of death in AIDS patients.

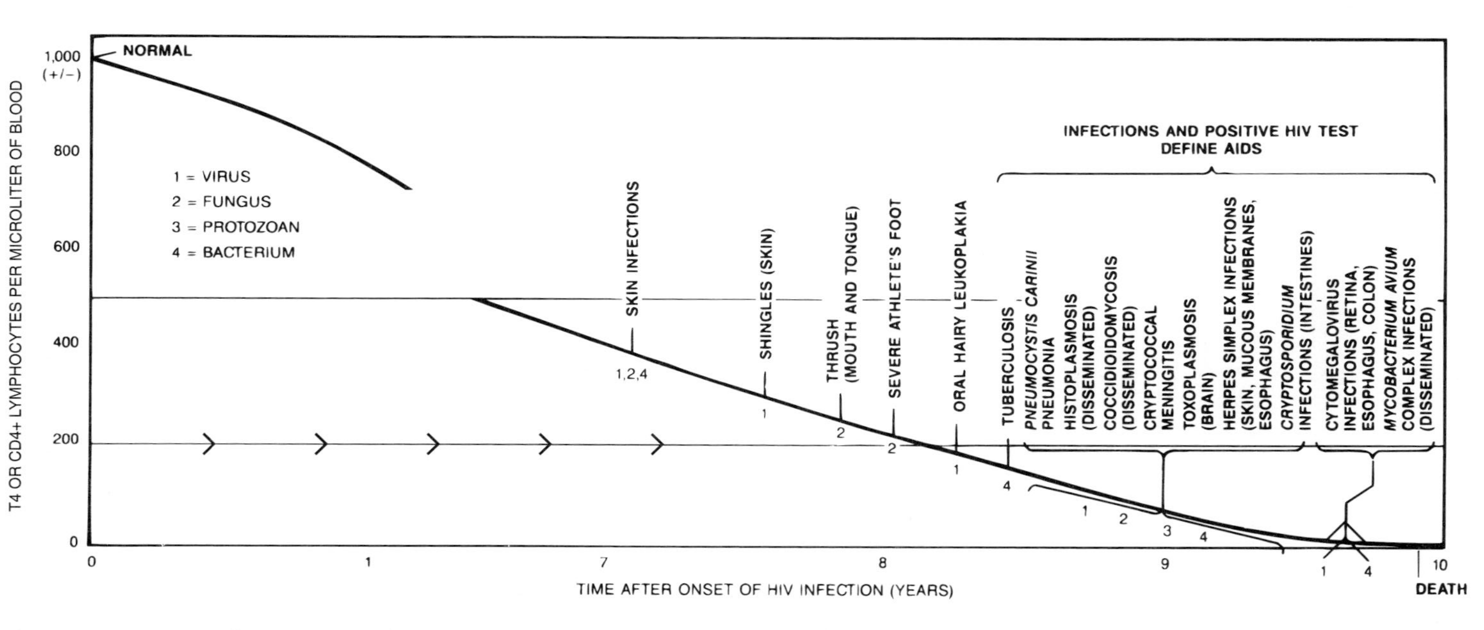

FIGURE 6-1 General Progression of Opportunistic Infections in Untreated Adolescent/Adults After HIV Infection. Normal T4 or CD4+ cell count in adolescent/adults is, on average, about 1000/μL of blood. There is a relationship between the drop in T4 lymphocytes and the onset of opportunistic infections (OIs). The first sign of an OI begins under 500 T4 cells/μL. As the T4 cell count continues to drop, the chance of OI infection increases. Note the variety of OIs found in AIDS patients with 200 or less T4 cells/μL.

POINT OF INFORMATION 6.1

THE CHANGING SPECTRUM OF OPPORTUNISTIC INFECTIONS

Twelve years ago there was hardly any standardized use of protective agents to block the infectious complications or opportunistic infections (OIs) associated with HIV-induced immunodeficiency. Now there is an array of drugs that can be used in strategies to prevent or delay nearly all the major OIs. With this advancement has come the need to weigh the pros and cons of various strategies. Cost, antimicrobial resistance, drug interactions, and pill overload are all important considerations.

EFFECT OF ANTI-HIV THERAPY ON OPPORTUNISTIC INFECTIONS

As presented in Chapter 4, the use of anti-HIV combination drug therapy (AIDS drug cocktails) has produced a number of unexpected results in patient response to those drugs. Soon after combination therapy began, physicians witnessed a rather confusing or unusual presentation of OIs. In some cases certain OIs improved, in others the situation deteriorated. Such changes in OIs expression are occurring now, at a time when hundreds of thousands of HIV-infected Americans are on **Highly Active Antiretroviral Therapy (HAART).**

TREATMENT IN THE HAART ERA

HAART therapy appears to scramble the human immune system. When HIV patients take retroviral medicines that control HIV replication, their immune systems begin to recuperate in ways that are puzzling and controversial. For example, patients recover immunity to some deadly opportunistic infections but appear unable to fight diseases for which they were vaccinated as children, for example, tetanus, or to target HIV itself. Collectively, such observations indicate HAART patients can only raise successful immune responses against pathogens they see regularly. For example, cytomegalovirus is an organism found in almost everybody's blood, so the immune systems of HAART patients see the pathogen constantly and generate cells and antibodies that attack it. But tetanus is something people rarely encounter, so HAART patients, unlike their HIV-negative counterparts, fail to raise immune responses against it. The ultimate irony is that HAART, when successful, destroys all but a few million HIVs that are forced into hiding.

Recent studies suggest that the incidence of esophageal candidiasis and, by inference, other forms of *Candida* infection has fallen by 60% to 70% on patients treated with HAART. The use of HAART dramatically changed the epidemiology of opportunistic infections and is clearly associated with gradual recovery of the immune system (Powderly et al., 1998; Ledergerber et al., 1999).

This information aside, when HIV is controlled with the antiretroviral drugs, immunity to infections—other than HIV—usually starts to return. As a result, some opportunistic infections go away without specific treatment; and sometimes patients can stop prophylactic treatment for certain opportunistic diseases. However, entering year 2004, it is still unclear who can stop prophylaxis safely, and who cannot.

Studies from 1998 through 2004, using protease inhibitor drugs in HAART have shown that virtually no patient whose T4 level rose to and stayed over 200 per microliter of blood (μL) developed an OI. This is the strongest evidence to date that immune reconstruction is occurring with protease inhibitor therapy, and suggests that it occurs early and with quite modest improvements in T4 cell levels. The implication is that the search for immunorestorative therapy other than with the current antiretroviral is somewhat less urgent than previously believed, though still a clear priority.

OMINOUS SHADOWS

It is now clear that sustained viral suppression is not feasible in all patients receiving HAART. A post-HAART era is projected, during which resurgence of opportunistic infections will be seen. This era may be marked by a higher rate of infection by antimicrobial-resistant microorganisms in patients with HIV infection than in the pre-HAART era of the late 1980s to the late 1990s.

VIRAL LOAD RELATED TO OPPORTUNISTIC INFECTIONS

HIV clinicians have recently looked at the predictive value of plasma HIV RNA for the development of three OIs: PCP, CMV, and MAC. Using a database of patients participating in AIDS Clinical Trial Groups (ACTG), for every 1-log increase in plasma HIV RNA level, the risk of developing one of these OIs was increased 2- to 3-fold. Plasma HIV RNA level was predictive of an increased risk of an OI independent of T4 cell count, which also predicted OI risk. This information confirms that maintaining control of viral replication may be a critical component of preventing OIs in HIV-infected patients.

HIV-Related Opportunistic Infections Vary Worldwide

The course of HIV infection tends to be similar for most patients: Infection with the virus is followed by seroconversion and progressive destruction of T4 or CD4+ cells. Yet the opportunistic infections and malignancies that largely define the symptomatic or clinical history of HIV disease vary geographically. People with HIV and their physicians in different regions confront distinct problems, mainly because of differences in exposure, in access to diagnosis and care, and in general health.

Comparisons between the data about opportunistic infections in different countries must be made with care. But most developing nations lack the facilities and trained personnel to identify opportunistic infections correctly; consequently, their prevalence may be underreported. Clinicians in developed countries can order sophisticated laboratory analyses to identify pathogens. Those in developing countries must rely on signs and symptoms to make their diagnoses. Oral candidiasis and herpes zoster are easy to diagnose without laboratory backup because the lesions are visible. While some pneumonias and types of diarrhea can be specified, others, such as extrapulmonary tuberculosis, cytomegalovirus infections, cryptococcal meningitis, and systemic infections such as histoplasmosis, toxoplasmosis, microsporidiosis, and nocardiosis, go underreported due to the lack of laboratory facilities.

Socioeconomic Factors

Geography explains much about the varying patterns of opportunistic infections, but a decisive factor is often financial capacity. On the most fundamental level, money is needed to create an infrastructure that limits exposure to pathogens. Thus, while few people with HIV in wealthy countries develop certain bacterial or protozoal infections, they are a major cause of death in poor areas that cannot provide clean water and adequate food storage facilities.

Financial resources also affect clinicians' abilities to diagnose AIDS and, when appropriate, to provide the proper medicine. AIDS patients in Africa often die of severe bacterial infections because they don't have the antibiotics or the clinical care they need. They don't survive long enough to develop diseases such as PCP.

United States, Europe, and Africa—The United States and Europe on one end, and Africa on the other, represent the global extremes of financial resources for health care. The most common opportunistic infections each region faces reflect the overall quality of health care, sanitation, and diet. For example, Thailand and Mexico belong to the large group of nations that have intermediate incomes and correspondingly intermediate patterns of HIV complications (Harvard AIDS Institute, 1994).

AIDS patients rarely have just one infection (Table 6-1). The mix of OIs may depend on lifestyle and where the HIV/AIDS patient lives or has lived. Thus, a knowledge of the person's origins and travels may be diagnostically helpful. (Note: a number of the symptoms listed in the CDC definition of HIV/AIDS can be found associated with certain of the OIs presented.)

Fungal Diseases

In general, healthy people have a high degree of innate resistance to fungi. But a different situation prevails with opportunistic fungal infections, which often present themselves as acute, life-threatening diseases in a compromised host (Medoff et al., 1991).

Because treatment seldom results in the eradication of fungal infections in AIDS patients, there is a high probability of recurrence after treatment (DeWit et al., 1991).

Fungal diseases are among the more devastating of the OIs and are most often regional in association. AIDS patients from the Ohio River basin, the Midwest, or Puerto Rico have a higher-than-normal risk of histoplasmosis (his-to-plaz-mo-sis) infection. In the Southwest, there is increased risk for coccidioidomycosis (kok-sid-e-o-do-mi-ko-sis).

Table 6-1 Some Common Opportunistic Diseases Associated With HIV Infection

Organism/Virus	Clinical Manifestation
Protozoa	
Cryptosporidium muris	Gastroenteritis (inflammation of stomach-intestine membranes)
Isospora belli	Gastroenteritis
Toxoplasma gondii	Encephalitis (brain abscess), retinitis, disseminated
Fungi	
Candida sp.	Stomatitis (thrush), proctitis, vaginitis, esophagitis
Coccidioides immitis	Meningitis, dissemination
Cryptococcus neoformans	Meningitis (membrane inflammation of spinal cord and brain), pneumonia, encephalitis, dissemination (widespread)
Histoplasma capsulatum	Pneumonia, dissemination
Pneumocystis carinii	Pneumonia
Bacteria	
Mycobacterium avium complex (MAC)	Dissemination, pneumonia, diarrhea, weight loss, lymphadenopathy, severe gastrointestinal disease
Mycobacterium tuberculosis (TB)	Pneumonia (tuberculosis), meningitis, dissemination
Viruses	
Cytomegalovirus (CMV)	Fever, hepatitis, encephalitis, retinitis, pneumonia, colitis, esophagitis
Epstein-Barr	Oral hairy leukoplakia, B cell lymphoma
Hepatitis C (HCV)	Liver cirrhosis or cancer (major reason for liver transplants)
Herpes simplex	Mucocutaneous (mouth, genital, rectal) blisters and/or ulcers, pneumonia, esophagitis, encephalitis
Papovavirus J-C	Progressive multifocal leukoencephalopathy
Varicella-zoster	Dermatomal skin lesions (shingles), encephalitis
Cancers	
Kaposi's sarcoma	Disseminated mucocutaneous lesions often involving skin, lymph nodes, visceral organs (especially lungs and GI tract)
Primary lymphoma of the brain	Headache, palsies, seizures, hemiparesis, mental status, or personality changes
Systemic lymphomas	Fever, night sweats, weight loss, enlarged lymph nodes

Patients with compromised immune systems are at increased risk for all known cancers and infections (including bacterial, viral, and protozoal). Most infectious diseases in HIV-infected patients are the result of proliferation of organisms already present in the patient's body. Most of these opportunistic infections are not contagious to others. The notable exception to this is tuberculosis.

Disclaimer: This table was developed to provide general information only. It is not meant to be diagnostic nor to direct treatment.

(Adapted from Mountain-Plains Regional Education and Training Center HIV/AIDS Curriculum, *4th ed., 1992 updated, 1997; and from* MMWR *2002.)*

In the southern Gulf states, the risk is for blastomycosis. Other important OI fungi such as ***Pneumocystis carinii*** (nu-mo-sis-tis car-in-e-i), ***Candida albicans*** (kan-di-dah al-be-cans), and ***Cryptococcus neoformans*** (krip-to-kok-us knee-o-for-mans) are found everywhere in equal numbers. Because of their importance as OIs in AIDS patients, a brief description of histoplasmosis, candidiasis, *Pneumocystis carinii* pneumonia, and cryptococcosis are presented.

Histoplasmosis (Histoplasma capsulatum)—Spores are inhaled and germinate in or on the body (Figure 6-2). This fungal pathogen is endemic in the Mississippi and Ohio River Valleys. Signs of histoplasmosis include prolonged influenza-like symptoms, shortness of breath, and possible complaints of night sweats and shaking chills. Histoplasmosis in an HIV-positive person is considered diagnostic of AIDS. In about two-thirds of AIDS patients with histoplasmosis, it is the initial OI. Over 90% of cases have occurred in patients with T4 cell counts below 100/μL (Wheat, 1992).

Candidiasis or Thrush (Candida albicans)—This fungus is usually associated with vaginal yeast infections. It is a fungus quite common to the body and in particular inhabits the alimentary tract. It is normally kept in check by the presence of bacteria that live on the linings of the alimentary tract. However, in immunocompromised patients, especially those who have received broad spectrum antibiotics, candida multiplies rapidly. Because of its location in the upper reaches of the alimentary tract, if unchecked, it may cause mucocutaneous candidiasis or **thrush,** an overgrowth of candida in the esophagus and in the oral cavity (Figure 6-3). Mucosal candidiasis was associated with AIDS patients from the very beginning of the AIDS pandemic (Powderly et al., 1992). In women, overgrowth of candidiasis also occur in the vaginal area.

Oral or esophageal candidiasis causes thick white patches on the mucosal surface and may be the first manifestation of AIDS. Because other diseases can cause similar symptoms, candidiasis by itself is not sufficient for a diagnosis of AIDS.

Pneumocystis carinii—The life cycle and reproductive characteristics of *P. carinii* are not completely understood because the organism is difficult to culture for laboratory study. However, the molecular biology and molecular taxonomy are now being rapidly constructed (Walzer, 1993; Stringer et al., 1996). Virtually everyone in the United States by age 30 to 40 has been exposed to *P. carinii.* It lies dormant in the lungs, held in check by the immune system. Prior to the AIDS epidemic, *P. carinii* pneumonia was seen in children and adults who had a suppressed immune system as in leukemia or Hodgkin's disease and were receiving chemotherapy. In the AIDS patient, the onset of *P. carinii* pneumonia is insidious—patients may notice some shortness of breath and they cannot run as far. It causes extensive damage within the alveoli of the lungs (Figure 6-4).

Prior to 1981, fewer than 100 cases of *P. carinii* infection were reported annually in the United States; yet 80% of AIDS patients develop *P. carinii* pneumonia at some time during their illness. This is one of the few AIDS-related conditions for which there is a choice of relatively effective drugs. The first of these to be made available was the intravenous and aerosolized versions of pentamidine. There are frequent recurrences of this infection. (Montgomery, 1992). PCP accounted for a diagnosis of AIDS in over 65% of AIDS cases in 1990. In 1994 it had fallen to 20% due to available therapy (Ernst, 1990; Murphy, 1994). The triad of symptoms that almost always indicates the onset of PCP during HIV disease is fever, dry cough, and shortness of breath (Grossman et al., 1989). *P. carinii* pneumonia is unlikely to develop in people with HIV disease unless their T4 cell count drops below 200 (Phair et al., 1990).

Cryptococcosis (Cryptococcus neoformans)—Since its discovery in 1894, *C. neoformans* has been recognized as a major cause of deep-seated fungal infection in the human host. The infection can affect many sites, including skin, lung, kidney, prostate, and bone. However, symptomatic disease most often represents infection of the central nervous system. Cryptococcal meningitis is the most common form of fungal meningitis in the United States (Ennis et al., 1993). This fungus is shed in pigeon feces, the spores of which enter the lungs. If the lung does not eliminate it, it gets into the bloodstream, travels to the brain, and can cause cryptococcal meningitis. It does not appear to be spread from person to person.

C. neoformans is a fatal OI that occurs in about 13% of AIDS patients (Brooke et al., 1990). It is

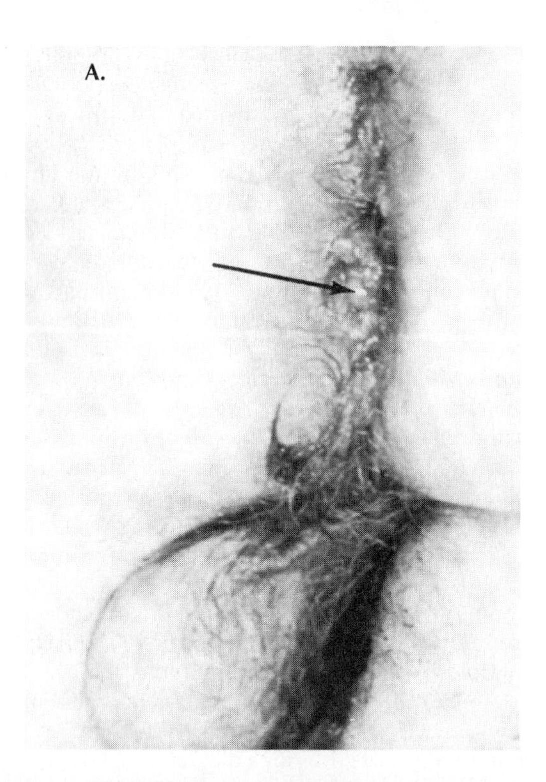

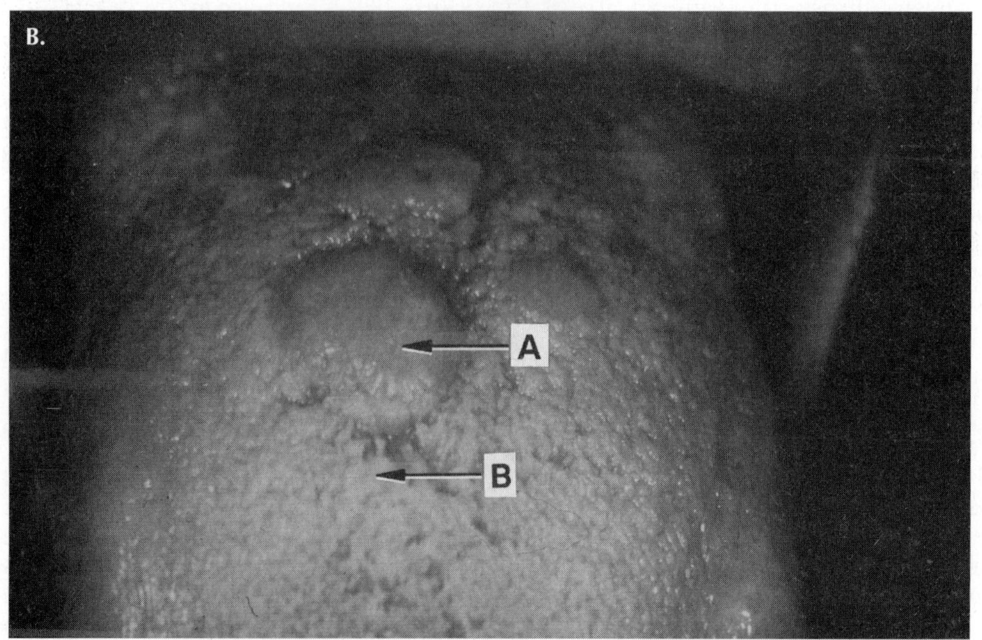

FIGURE 6-2 **A.** Anal Histoplasmosis. Histoplasmosis is caused by *Histoplasmosis capsulatum* and causes infection in immunocompromised patients. *(Courtesy CDC, Atlanta)* **B.** AIDS patient's tongue showing multiple shiny, firm Histoplasma erythematous nodules (see arrow A) and thrush (see arrow B). *(Photograph courtesy of Marc E. Grossman, New York)*

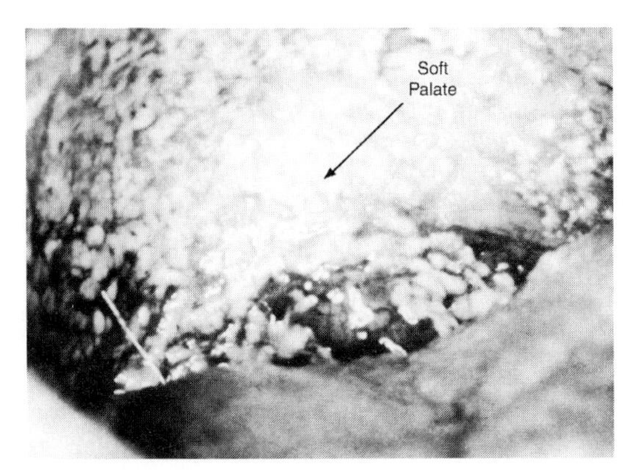

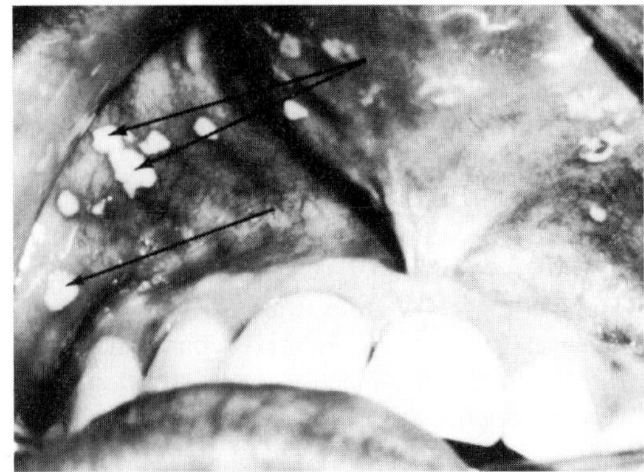

FIGURE 6-3 Thrush. **A.** An overgrowth of *Candida albicans* on the soft palate in the oral cavity of an AIDS patient. **B.** Creamy patches of candida that can be scraped off, leaving a red and sometimes bleeding mucosa. *(**A,** Courtesy CDC, Atlanta; **B,** Schiodt, Greenspan, and Greenspan 1989.* American Review of Respiratory Disease, 1989, *10:91–109. Official Journal of the American Thoracic Society. © American Lung Association)*

acquired through the respiratory tract and most commonly causes cryptococcal meningitis. In AIDS patients, *C. neoformans* causes infection of the skin, lymph nodes, and kidneys. *Cryptococcus* cannot be cured and it does recur. Which anti-fungal drugs should be used is a subject of controversy.

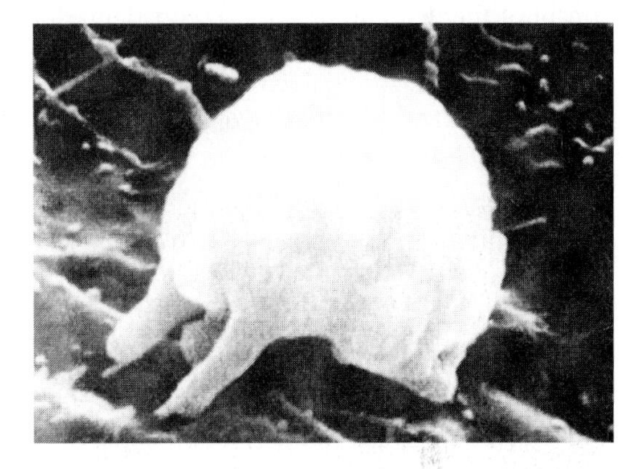

FIGURE 6-4 Scanning Electron Microscope Image of *P. carinii* Attached to Embryonic Chick Lung Cells in Culture. Note the tubular extensions through which it extracts nutrients from host lung tissue. *(Propagation of Pneumocystis carinii in Vitro, © Pediatric Research, 11:305–316)*

Viral Diseases

Hepatitis C—Hepatitis means an inflammation or swelling of the liver. Viruses can cause hepatitis. Alcohol, drugs (including prescription medications) or poisons can also cause hepatitis. In late 1999, **hepatitis C** caused by the hepatitic C virus (**HCV**) was classified as an OI because the relative risk for liver-associated death is increased 7-fold in HIV-positive patients compared to non-HIV-infected individuals. It is estimated that 200 million people worldwide are infected with the hepatitis C virus. Over 4 million of the infected live in America.

HIV/HCV Means Double Trouble

The hepatitis C virus (HCV) is now at least four times as widespread in America than HIV. It kills between 10,000 and 15,000 each year and is predicted to kill 30,000 a year by 2010. Its transmission is similar to that of HIV. About 40% of HIV-infected in America, about 400,000 people, are believed to be coinfected with the HCV. Among some groups, primarily HIV-positive current and former intravenous drug users, the coinfection rate is thought to be about 90%. According to the National Hemophilia Foundation, the coinfection rate among HIV-positive hemophiliacs is equally high. For those who become infected with HCV, the virus produces no symptoms for 10–30 years. Then symptoms like fatigue, joint and abdominal pain, nausea, and lapses in concentration begin to set in. Because doctors have not traditionally screened patients for the virus, many people do not even know they are infected until their livers show signs of serious damage.

Studies now indicate that HIV can greatly speed the progression of hepatitis C. That means that many with HIV may suffer advanced liver disease (cirrhosis of the liver) after just 5 or 10 years, even as the antiretroviral drugs boost their life expectancies. Studies on HCV's effects on AIDS patients have been contradictory, but the coinfection has been associated with a higher risk of progression to HIV disease and AIDS. The antiretroviral drugs used to fight HIV, particularly the protease inhibitors, place a great strain on the liver, the organ whose function is to metabolize them. Some patients infected with both viruses find that their bodies have great difficulty tolerating many HIV/AIDS medications. As yet, there is no national policy for dealing with this virus.

Herpes Viruses

Because of a depleted T4 or CD4+ cellular component of the immune system, AIDS patients are at particularly high risk for the herpes family of viral infections: cytomegalovirus, herpes simplex virus types 1 and 2, varicella-zoster virus, and Epstein-Barr virus.

Cytomegalovirus (CMV)—This virus is a member of the human herpes virus group of viruses. CMV is the perfect parasite. It infects most people asymptomatically. When illness does occur, it is mild and nonspecific. There have been no epidemics to call attention to the virus. Yet CMV is now considered the most common infectious cause of mental retardation and congenital deafness in the United States. It is also the most common viral pathogen found in immunocompromised people (Balfour, 1995).

The virus is very unstable and survives only a few hours outside a human host. It can be found in saliva, tears, blood, stool, cervical secretions, and in especially high levels in urine and semen. Transmission occurs primarily by intimate or close contact with infected secretions. The incidence of CMV infection prior to HAART therapy beginning in 1996, varied from between 30% and 80% depending on the geographical community tested. In the 1980s, over 90% of homosexual males tested positive for CMV (Jacobson et al., 1988).

CMV infection of AIDS patients usually results in prolonged fever, anemia (too few red blood cells), leukopenia (too few white blood cells), and abnormal liver function. CMV also causes severe diarrhea and HIV-associated retinitis resulting in eventual blindness (*Emergency Medicine,* 1989; Lynch, 1989; Dobkin, 1995).

Prior to HAART therapy, 75% of AIDS patients had an eye disease, with the retina the most common site (Russell, 1990). The retina, which is a light-sensitive membrane lining the inside of the back of the eye, is also part of the brain and is nourished by blood vessels. AIDS-related damage to these vessels produces tiny retinal hemorrhages and small cotton wool spots—early indicators of disease that are often detected during a routine eye examination (Figure 6-5). Since the beginning of HAART therapy, CMV retinitis has fallen to about 5% in AIDS patients.

In unusual circumstances, the virus can produce dramatic symptoms, such as loss of vision within 72 hours. Without treatment, CMV can destroy the entire retina in three to six months after infection.

Herpes Viruses Types 1 & 2 (HSV 1 & 2)—Herpes infections are one of the most commonly diagnosed infections among the HIV/ AIDS population. Almost all HIV-infected individuals (95%) are coinfected with HSV 1 and/or HSV 2. Both viruses cause *severe* and progressive eruptions of the mucous membranes. HSV 1 affects the membranes of the nose and mouth. Also, when herpetic lesions involve the lips or throat, 80% to 90% of the time they either precede or occur simultaneously with **herpes-caused pneumonia** (Gottlieb et al., 1987). Bacterial or fungal superinfections occur in more than 50% of herpes-caused pneumonia cases and are a major contributory cause of death in AIDS patients.

Mortality from HSV pneumonia exceeds 80% (Lynch, 1989). Herpes may also cause blindness in AIDS patients. The following is from Paul Monette's *Borrowed Time:*

> I woke up shortly thereafter, and Roger told me—without a sense of panic, almost puzzled—that his vision seemed to be losing light and detail. I called Dell Steadman and made an emergency appointment, and I remember driving down the freeway, grilling Roger about what he could see. It seemed to be less and less by the minute. He could barely see the cars going by in the adjacent lanes. Twenty minutes later we were in Dell's office, and with all the urgent haste to get there we didn't really

reconnoiter till we were sitting in the examining room. I asked the same question—what could he see?—and now Roger was getting more and more upset the more his vision darkened. I picked up the phone to call Jamiee, and by the time she answered the phone in Chicago he was blind. ***Total blackness, in just two hours!***

The retina had detached. (An operation on retinal attachment was successful and sight was restored. The cause of the retinal detachment was a herpes infection of the eyes.)

HSV 2 also affects the membranes of the anus, causing severe perianal and rectal ulcers primarily in homosexual men with AIDS. Herpes of the skin can generally be managed with oral **acyclovir** (Zovirax), Foscavir, or Famvir (Table 6-1).

Herpes Zoster Virus (HZV)—Like herpes simplex, this virus has the potential to cause a rapid onset of pneumonia in AIDS patients. Untreated HZV pneumonia has a mortality rate of 15% to 35%. HZV is now monitored as an early indicator that HIV-positive people are progressing toward AIDS.

Protozoal Diseases

An increasing number of infections that have not been observed in immunocompromised patients are being found in AIDS patients. Two such infections are caused by the protozoans *Toxoplasma gondii*, and *Cryptosporidium muris.*

Toxoplasma gondii—*T. gondii* is a small intracellular protozoan parasite that lives in vacuoles inside host macrophages and other nucleated cells. It appears that during and after entry, *T. gondii* produces secretory products that modify vacuole membranes so that the normal *fusion* of cell vacuoles with lysosomes containing digestive enzymes is blocked. Having blocked vacuole-lysosome fusion, *T. gondii* can successfully reproduce and cause a disease called **toxoplasmosis** (Joiner et al., 1990). It can infect any warm-blooded animal, invading and multiplying within the cytoplasm of host cells. As host immunity develops, multiplication slows and tissue cysts are formed. Sexual multiplication occurs in the intestinal cells of cats (and apparently only cats); oocysts form and are shed in the stool (Sibley, 1992). Transmission may occur transplacentally, by ingestion of raw or undercooked meat and eggs containing tissue cysts or by exposure to oocysts in cat feces (Wallace et al., 1993).

In the United States, 10% to 40% of adults are chronically infected but most are asymptomatic. *T. gondii* can enter and infect the human brain causing **encephalitis** (inflammation of the brain). Toxoplasmic encephalitis develops in over 30% of AIDS patients at some point in their illness

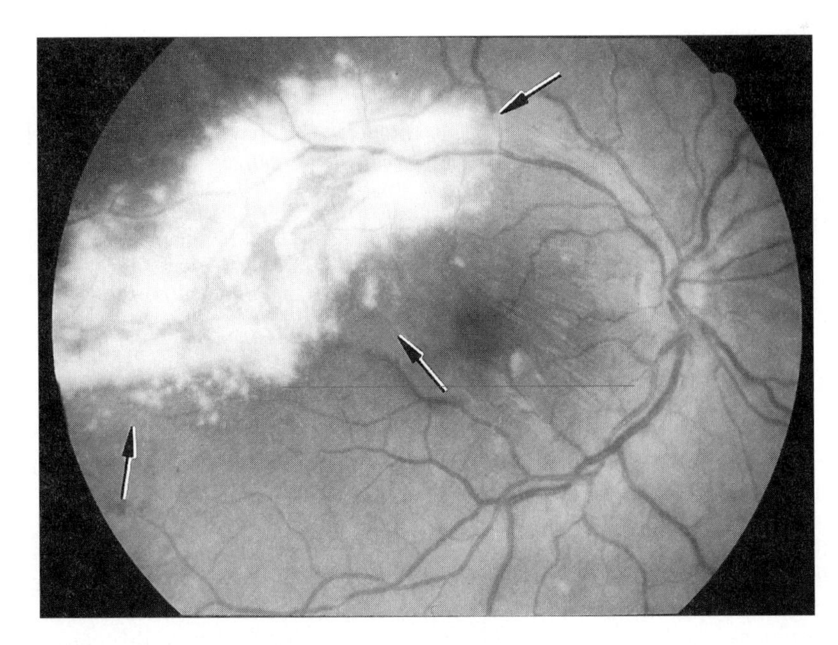

FIGURE 6-5
Cytomegalovirus Retinitis. The disease, as seen in this photograph, involves the posterior pole of the right eye. Fluffy white infiltrate **(cotton wool spots)** with a small amount of retinal hemorrhage can be seen in the distribution of the superior vascular arcade. *(Courtesy of Scott M. Whitcup, M.D., National Eye Institute, National Institutes of Health, Bethesda, Md.)*

BOX 6.1

A PHYSICIAN'S AGONIZING DILEMMA

Opportunistic infections are the primary threat to patients with AIDS; they are the main causes of illness and death. The cruel irony is that although there are 30 or more FDA-approved drugs to treat these infections, most cannot be used in patients receiving Zidovudine (ZDV) *because the combination drug therapy is devastatingly toxic to bone marrow.*

EXAMPLE:

Cytomegalovirus (CMV) Retinitis

CMV retinitis is one of the *worst* of the opportunistic infections (Gottlieb et al., 1987). Both the patient who contracts CMV retinitis and the treating physician confront an almost impossible dilemma: whether to continue the Zidovudine and risk blindness or to treat the retinitis and risk death from another infection. A physician recently said, "In my experience, I have never yet been able to combine Zidovudine with the experimental drug ganciclovir (DHPG), which is the only recognized treatment for CMV retinitis, because the combination destroys the bone marrow. I therefore face the virtually impossible task of asking a 25-year-old person: 'Which of these options do you prefer: to take Zidovudine and keep on living but go blind, or to preserve your sight by having retinitis treatment but risk dying?' That is a truly terrible question to ask of any patient, especially a young one." He continued, "Recently, I treated just such a young patient. He had been taking Zidovudine for AIDS when he contracted CMV retinitis. When I presented him with this agonizing choice, he told me, 'I definitely don't want to go blind. I want to be treated for this retinitis.' So he stopped taking Zidovudine and was started on DHPG treatment. He started to have seizures, which were a consequence of the toxoplasmosis brain abscess that eventually ended his life. Just before he died, he told me that the worst decision he ever made was to stop taking Zidovudine. 'I should have stayed on it,' he said, 'and gone blind."' (Robert J. Awe, M.D., 1988)

Cryptococcal Meningitis

The situation for AIDS patients with cryptococcal meningitis is similarly depressing. Because of the severe bone marrow toxicity, it is virtually impossible to use amphotericin B, the effective treatment for this meningitis, at the same time as Zidovudine. Besides, the use of amphotericin B has its own side effects as can be noted from this excerpt from Paul Monette's *Borrowed Time: An AIDS Memoir:*

> Amphotericin B is administered with Benadryl in order to avoid convulsions, the most serious possible side effect. It was about nine or ten when they started the drug in his veins, and I sat by the bed as nurses streamed in and out. A half hour into the slow drip, the nurse monitoring the IV walked out, saying she'd be right back, and a couple of minutes later Roger began to shake. I gripped him by the shoulders as he was jolted by what felt like waves of electric shock, staring at me horror-struck. Though Cope [the physician] would tell me later, trying to ease the torture of my memory, that "mentation" [mental activity] is all blurred during convulsions, I saw that Roger knew the horror (page 336).
>
> When the nurse returned she looked at him in dismay: "How long has this been going on?" Then she ordered an emergency shot of morphine to counteract the horror. When at last he fell into a deep sleep they all told me to go home, saying they would try another dose of the ampho in a few hours. I was so ragged I could barely walk. So I left him there with no way of knowing how near it was [Roger's death], or maybe not brave enough to know.

Whichever option is chosen, the patient is bound to suffer, and perhaps die, either of cryptococcal meningitis or of some other infection.

(Figure 6-6). The signs and symptoms of cerebral toxoplasmosis in AIDS patients may include fever, headache, confusion, sleepiness, weakness or numbness in one part of the body, seizure activity, and changes in vision. These symptoms can get worse and progress to coma and death unless toxoplasmosis is promptly diagnosed and treated. Thus, for most AIDS patients, it is believed that *T. gondii* is latent within their bodies and is reactivated by the loss of immune competence.

Cryptosporidium—*Cryptosporidium* is the cause of cryptosporidiosis, and is a member of the family of organisms that includes *Toxoplasma*

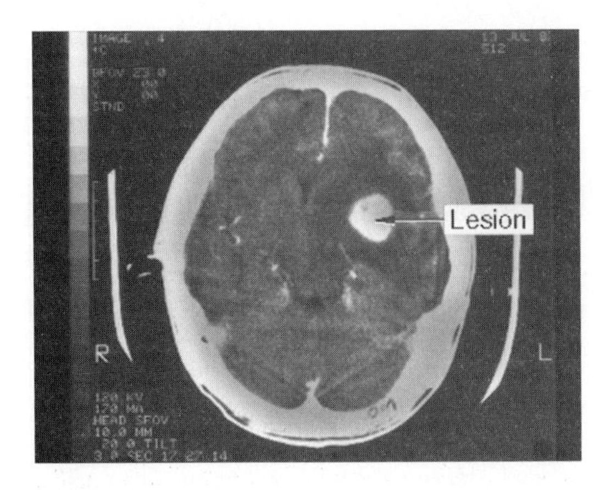

FIGURE 6-6 *Toxoplasma gondii* Lesions in the Brain. Radiographic imaging shows a deep ring-enhancing lesion located in the basal ganglia. *(By permission of Carmelita U. Tuazon, George Washington University)*

gondii and *Isospora.* Its life cycle is similar to that of other organisms in the class Sporozoa. Oocysts are shed in the feces of infected animals and are immediately infectious to others. In humans, the organisms can be found throughout the GI tract. *Cryptosporidium* causes profuse watery diarrhea of 6 to 26 bowel movements per day with a loss of 1 to 17 liters of fluid (a liter is about 1 quart). It is an infrequent infection in AIDS patients, usually occurring late in the course of disease as immunological deterioration progresses.

Studies of transmission patterns have shown infection within families, nursery schools, and from person to person, probably by the fecal-oral route. The infection is particularly common in homosexual men, perhaps as a consequence of anilingus (oral-anal sex).

Bacterial Diseases

There is a long list of bacteria that cause infections in AIDS patients. These are the bacteria that normally cause infection or illness after the ingestion of contaminated food, such as species of *Salmonella.* Others, such as *Streptococci, Haemophilus,* and *Staphylococci* are common in advanced HIV disease. A number of other bacteria-caused sexually transmitted diseases such as syphilis, chancroid, gonorrhea, and chlamydial diseases are also associated with HIV disease.

One difference between AIDS and non-AIDS individuals is that bacterial diseases in AIDS patients are of greater severity and more difficult to treat. Two bacterial species, *Mycobacterium avium intracellulare* and *Mycobacterium tuberculosis* are of particular importance as agents of infection in AIDS patients (Table 6-2).

Mycobacterium avium intracellulare (MAI)—Over the past 40 years, MAI has gone from a rare, reportable infection to something that is common in most large American communities. Unlike tuberculosis, which is almost exclusively spread person to person, MAI is, in most instances, environmentally acquired. MAI exists in food, animals, water supplies, and soil, and enters people's lungs as an aerosol when they take showers.

The fact that MAI produced disseminated disease in AIDS cases was recognized in 1982. The epidemiology of MAI continues to evolve. MAI occurs in 18% to 43% of people with HIV disease and has been implicated as the cause of a nonspecific **wasting syndrome.** AIDS patients demonstrate **anorexia** (inability to eat), weight loss, weakness, night sweats, diarrhea, and fever. Some patients also experience abdominal pain, enlarged liver or spleen, and malabsorption. In contrast to viral infections, this bacterium rarely causes pulmonary or lung problems in AIDS patients. Among persons with AIDS, the risk of developing disseminated MAI increases progressively with time. AIDS patients surviving for 30 months had a 50% risk of developing disseminated MAI. It appears most HIV-infected persons will develop disseminated MAI if they do not first die from other OIs (Chin, 1992).

Mycobacterium tuberculosis—Tuberculosis (TB) infects one-third of the world's population, over 2 billion people, and is now the leading cause of illness and death in people infected with HIV. Tuberculosis is an infectious disease caused by the bacterium ***Mycobacterium tuberculosis,*** which is spread almost exclusively by airborne transmission. TB has been observed in elephants, cattle, mice, and other animal species. In 1993, TB was transmitted from an infected seal to its trainer in Australia. In the United States, monkeys are the primary source of animal-to-human transmission.

The disease can affect any site in the body, but most often affects the lungs. When persons with pulmonary TB cough, they produce tiny droplet nuclei that contain TB bacteria, which can remain

BOX 6.2

LIFE GOES ON!

by Wendi Alexis Modeste

What this epidemic has cost me is the complete faith I had in the medical profession. I was raised believing that doctors were second only to priests and God. They were never to be questioned. Whatever the doctor said was law. If a person didn't get well after seeing the doctor, somehow they (the patient) had done something wrong. This was pretty much standard thinking for middle-class African-Americans. For a variety of reasons (mainly no self-esteem) I became a drug addict, prostitute, convict, battered, homeless woman, in that order! With the exception of emergency room admissions (which are a joke and a whole 'nother story) I had no access to health care.

Now, as a PWA (person with AIDS) fortunately/unfortunately on SSI, Medicaid pays for my nine different AIDS medications, clinic visits, treatment, tests, etc. When I received the "exciting news" that I was eligible for "all" Medicaid benefits, I was still under the impression doctors were those super-intelligent, gifted, Christian, saint-like people. Girlfriend, I am here to tell you, AIDS has totally shot that Marcus Welby theory straight to hell!

Early in my diagnosis, I went to my physician because my tongue was almost completely covered with what looked like cottage cheese. There was also a horrible pink lesion dead center. The first time I showed it to my doctor he said, "Ugh," and made a face. He told me to wait a month. If nothing changed or got worse when I returned he'd have someone look at it. Being ignorant about the disease at that time, and still blindly believing in the medical profession, I waited a month, then returned. Again I showed the doctor my tongue. He asked me if I wanted him to write me a prescription for codeine and Valium. I totally freaked! By this time, I'd done some reading and realized I probably had thrush and some sort of herpes. This physician was aware of my serostatus. He also knew that I was a person with a 20-year history of drug abuse. I'd been in recovery less than a year and this jerk wanted to prescribe for me two of the most addictive and abused prescription drugs. I'd never mentioned being in pain or that I was experiencing any type of anxiety. I contacted the Executive Director of this health facility and asked to have a different doctor assigned to me. In an attempt to make me feel guilty about requesting a change, I was told about the problems doctors have in getting Medicaid reimbursement. I was neither intimidated nor impressed. A new doctor was assigned. My new physician was very nice. After *I* told *him* what I thought my diagnosis was, he prescribed the appropriate medications. He was not trained in AIDS/HIV. I could have dealt with his ignorance because I knew he was trying. His nurse, however, was a different story. Every time she came to do my vitals she'd say the same thing: "I always get nervous when taking the temperature of you guys." She'd then force a little chuckle and go on to say, "Oh, well, I figure we all have to die of something." (I assume this was to show me what a courageous Florence Nightingale she was.)

Let me tell you, when you are burning up with a 103 degree temp. and your bowels haven't stopped running for a week, causing your butt to feel like it's on fire, it's real hard to be the patient, understanding AIDS educator. I really get crazy when the person I'm forced to educate is someone whose been privileged to more information than myself.

But life goes on!

One day I had a toothache. I go to the dentist. After waiting half the day, I'm brought into the treatment room for an X-ray. At first I thought the dental assistant had made a mistake. Surely this room had been prepared for a paint job. Everything was draped in white towels. The entire dental unit, including where my head, arms and feet went, was completely covered. The seat of the unit was securely wrapped up, as was the metal extension arm that holds the overhead dental lamp. All surfaces of the walls were also draped. When I asked the reason for this "painter preparation," I was informed it was done because I have AIDS, and they had to protect their other patients from coming into contact with my contaminated blood. Needless to say, I saw red! I knew I had to protect myself. I mean, what kind of dentistry were they practicing if they were concerned about my blood splattering that far and wide? What was even more frightening was that they'd done all this unnecessary draping and I was only having an X-ray. I filed a complaint with the Human Rights Dept.

I became a patient at the AIDS Care Center in Syracuse. My physician, a woman (need I say more?), is a caring person and well-educated about AIDS/HIV. My nurse/social worker is excellent, but as all of us living with AIDS know, shit happens. One day I awaken with enough yeast in my body to make all the baked goods in Central New York rise. My doctor isn't in. I wait a couple of days but can no longer stand the discomfort. I beg to see a doctor in the AIDS Care Unit. I'm assigned the doctor who sees the HIV-infected prisoners. (My heart and soul truly go out to those guys.) First he talks with me over the phone to find out if I can possibly wait another week when my doctor is due to return. I tell him my tongue is unrecognizable, the yeast in my esophagus is burning like a heart attack, and the Roto-Rooter service couldn't satisfy the itch caused by the yeast in my vagina. HELL NO, I can't wait another week! I go in to see him. I show him my tongue. I can't believe it, but like the first doctor, he uses that medical term, "Ugh," then says, "That does look nasty." At this point I'm ready to French kiss this idiot. But it gets worse. He won't touch me, let alone examine anything. He asks me what I think will work. I feel too badly to curse, so I tell him Mycelex, Myastatin suspension, Monistat 7. He writes the prescriptions and for good measure increases my acyclovir. For this he gets paid? He did nothing!

I'm now as educated as a lay person can be about HIV disease/AIDS. In addition to my doctor at the AIDS Care Center, I have a private primary care physician. Sometimes it's easier to get in to see this doctor. I call his office one day because there is swelling and burning on the sides of my tongue. I cannot eat. My regular doctor isn't in, but one of his associates assures me if I come in to the office he'll see me right away and give me something to ease the pain so I can at least eat. As a fat person who proudly admits a genuine fondness for food, I can tell you not being able to eat registers serious panic in my soul. Nothing stops me from eating. I was probably the only overweight homeless dope addict living on the streets of NYC. Being scared is putting it mildly. I scrape up the carfare and go to the office. After waiting an unreasonably long time, Doogie Howser's twin comes in to see me and announces he's Dr. Jones, whom I spoke to on the phone. OK, I know not to judge a book by its cover. I mean, Doogie is pretty good on TV. Dr. Jones looks at my tongue and proceeds to ask a zillion questions, all of which are answered in my chart. Finally he says he's never seen an HIV-infected person or a person with AIDS, and frankly he doesn't know what to do. He then suggests I eat popsicles for a few days because the cold will soothe the pain and the sugar should help give me energy. I kid you not, this actually happened. This jerk prescribed me popsicles. I truly wished I could transmit the virus by biting at this point. He used me so he could write in his journal or resume (or somewhere) that he'd treated a person with AIDS. He could also charge Medicaid for nothing.

But this type of quackery must stop.

The last gripe I'm going to list is this patient statement I hear all the time when I get an unexplained fever or infection and no one can determine its origin. I'm told, "there hasn't been enough studies done on the paths this disease takes in women. Even less has been done on the effects of different AIDS medications on people of color." This is said to me as if it's my fault they don't know. This disease has been documented for ten years in both men and women. I know African-Americans were dying of AIDS long before the gay white community mobilized and, thank God, refused to lay down and die quietly. There is no excuse for the fact that there's no studies done on these populations.

I am thankful to Dr. Sallie Klemmens and nurse/clinician/social worker Judith Swartout at University Hospital's designated AIDS Care Center here in Syracuse. I am thankful for Dr. Barbara J. Justice who with God's help kept me alive when I lived on the streets of NYC. They are all examples of what the medical profession should be about. Dr. Justice made me feel that I counted and should be assertive about my health care. Though no longer my surgeon, she continues to be a source of inspiration and a fountain of information for me. These three are gems in a field I think is greatly overrated, overpaid, and run by capitalist male chauvinist pigs.

Though medical people wear white, that absence of color symbolizing purity and goodness, I beg you all, "Don't believe the hype!" We need a national health care system for everyone. As PWAs we must be assertive about our health care. Good health care is a right not a privilege.

As a child I cried when I learned there was no Santa Claus. When my illusion about the medical profession was shattered I got angry. I decided to fight with the only ammunition I had, education! **Knowledge gives one power.** A close friend of mine told me I shouldn't submit this

BOX 6.2 *(continued)*

article because I might offend some members of the medical profession. He also felt because I'm on Medicaid I'm not supposed to complain, I should be grateful. To his comments I respond, raised with two college-educated parents, never wanting or needing anything, I wasn't supposed to be a drug addict. Unfortunately I was. After 20 years of addiction, I wasn't supposed to be able to stop. I've been in recovery almost two years now. I'm not supposed to be living with AIDS, but with God's help I'm happy and living large (as the kids say now).

The best things I do have always been what I'm not supposed to do!

Power to all PWAs!

Wendi Alexis Modeste, a PWA who was diagnosed in 1990, died August 25, 1994.

Source: Modeste, W.A., August 1991, Issue 68, PWA Coalition Newsline. *Reprinted with permission.*

suspended in the air for prolonged periods of time. (With respect to transmission, the cough to TB is like sex to HIV.) Anyone who breathes air that contains these droplet nuclei can become infected with TB. It has been suggested that there is a minimal chance of inhaling HIV in blood-tinged TB sputum (Harris, 1993).

About 10% of otherwise healthy persons who have latent tuberculosis infection will become ill with active TB at some time during their lives (*MMWR,* 1992). With HIV disease, the risk is 10% per year (Daar et al., 1993). It has been estimated that through year 2005, 16 million HIV-infected people worldwide will be coinfected with TB. HIV activates latent TB which erodes the lungs. TB in turn agitates HIV, speeding its destruction of the immune system. The result: In many countries, it is TB that most often triggers death in AIDS patients.

Tuberculosis is not generally considered to be an OI because people with healthy immune systems contract TB. After infection with M. tuberculosis about 5% of immunocompetent individuals will develop TB (Daley, 1992). But, people with a depressed immune system are much more likely to develop the disease.

TB AND HIV

HIV infection is now considered to be the single most important risk factor in the expression of TB. HIV disease is associated with the reactivation of a dormant or inactive TB infection (Stanford et al., 1993).

M. tuberculosis infection occurs in about 35% of HIV-infected individuals. The CDC defines extrapulmonary TB combined with an HIV-positive test as diagnostic of AIDS.

According to the World Health Organization, TB worldwide is the leading cause of death in HIV-infected people and among adults from a single, infectious organism. TB has killed at least 200 million people since 1882, the date of the discovery of the bacterium that causes TB. Millions more die from TB each year. Globally, TB was estimated to account for 30% of AIDS-related deaths in 1999, and greater than 30% each year from 2000 through 2004. March 24 of each year is recognized as World TB Day.

In the United States, 14% of AIDS patients are also coinfected with TB. Health officials state that between 40% and 60% of those developing multidrug-resistant TB will die (Ezzell, 1993).

Other Opportunistic Infections

Other opportunistic infectious organisms and viruses and the diseases they cause and possible therapies are listed in Table 6-1. Table 6-2 separates OIs into the body parts most affected by a particular organism or virus.

From diagnosis until death, the AIDS battle is *not* just against its cause, HIV, but against those organisms and viruses that cause OIs. Opportunistic infections are severe, tend to be disseminated (spread throughout the body), and are characterized by multiplicity. Fungal, viral, protozoal, and bacterial infections may be controlled for some time but are rarely curable.

CANCER OR MALIGNANCY IN AIDS PATIENTS

The word "malignancy" means a cancer. Specifically, cancer is an abnormal growth of cells that divide uncontrollably and may spread to other parts of the body. There are many kinds of can-

Table 6-2 Categories of Organism and Viral Involvement in Opportunistic Diseases

Symptoms	Causative Agent
Generally Present	
Fever, weight loss, fatigue, malaise	*Pneumocystis carinii* Cytomegalovirus Epstein-Barr virus *Mycobacterium avium intracellulare* *Candida albicans*
Diffuse Pneumonia	
Dyspnea, chest pain, hypoxemia, abnormal chest X-ray	*Pneumocystis carinii* Cytomegalovirus *Mycobacterium tuberculosis* *Mycobacterium avium intracellulare* *Candida albicans* *Cryptococcus neoformans* *Toxoplasma gondii*
Gastrointestinal Involvement	
Esophagitis (sore throat, dysphagia)	*Candida albicans* Herpes simplex Cytomegalovirus (suspected)
Enteritis (diarrhea, abdominal pain, weight loss)	*Giardia lamblia* *Entamoeba histolytica* *Isospora belli* Cryptosporidium *Strongyloides stercoralis* *Mycobacterium avium intracellulare*
Proctocolitis[a] (diarrhea, abdominal pain, rectal pain)	*Entamoeba histolytica* Campylobacter Shigella Salmonella *Chlamydia trachomatis* Cytomegalovirus
Proctitis[a] (pain during defecation, diarrhea, itching, and perianal ulcerations)	*Neisseria gonorrhoeae* Herpes simplex *Chlamydia trachomatis* *Treponema pallidum*
Neurological Involvement	
Meningitis, encephalitis, headaches, seizures, dementia	Cytomegalovirus Herpes simplex *Toxoplasma gondii* *Cryptococcus neoformans* Papovavirus *Mycobacterium tuberculosis*
Retinitis (diminished vision)	Cytomegalovirus *Toxoplasma gondii* *Candida albicans*

[a]Especially in those persons practicing anal sex.

Adapted from Amin, 1987.

For Sexual Exposures—People should use male latex condoms during every act of sexual intercourse to reduce the risk of exposure to cytomegalovirus, herpes simplex virus, and human papillomavirus, as well as to all other sexually transmitted pathogens. Use of latex condoms will help prevent the transmission of HIV to others. Avoid sexual practices that may result in oral exposure to feces (oral-anal contact) to reduce the risk of intestinal infections such as cryptosporidiosis, shigellosis, campylobacteriosis, amebiasis, giardiasis, and hepatitis A and B (*MMWR*, 2002).

cer, which can involve just about any part of the body.

HIV infection carries with it a high susceptibility to certain cancers. Because of the severe and progressive impairment of the immune system, host defense mechanisms that normally protect against certain types of cancer are lost. Cancers develop in approximately 40% of AIDS patients. Four kinds of cancer that occur with increased frequency are: **progressive multifocal leukoencephalopathy, squamous cell carcinoma** (oral and anal), **non-Hodgkin's lymphoma,** and **HIV/AIDS-associated Kaposi's sarcoma (KS)**. None of these cancers, except for KS, is considered to be an opportunistic infection because the rest are not infections. They are cancers arising from cells that have lost control of their division processes. Of the nine types of AIDS-associated cancers, KS occurs with the greatest frequency and is discussed in some detail. Lymphomas are briefly described (Table 6-3). (For a review of HIV/AIDS-related cancers read Hessol, 1998; Grulich 2000.)

Kaposi's Sarcoma (cap-o-seas sar-com-a)

No other HIV/AIDS-related opportunistic disease attacks and singles out one segment of the population as KS does with HIV-positive gay men. Men with KS outnumber women approximately 95% to 5%; HIV-positive homosexual men with KS outnumber heterosexual men almost as significantly. KS is extremely rare in hemophiliacs with HIV. In fact, in one major study of hemophiliacs with HIV, only 1 in 93 developed KS, and he happened to be a gay man.

Table 6-3 Malignancies Associated With HIV/AIDS

Kaposi's sarcoma (epidemic form)
Burkitt's lymphoma
Non-Hodgkin's lymphomas
Hodgkin's disease
Chronic lymphocytic leukemia
Carcinoma of the oropharaynx
Hepatocellular carcinoma
Adenosquamous carcinoma of the lung
Cervical cancer
Anal cancer
Squamous cell carcinoma
Progressive multifocal leukoencephalopathy
Vulva cancer

HIV infection represents an overwhelming risk factor for the development of KS, which was rare in the United States (incidence less than 1/100,000/year) before the HIV epidemic. Today, KS still remains one of the most frequent diseases affecting HIV-infected individuals. It is an aggressive disease, with involvement of the gut, lung, pleura, lymph nodes and the hard and soft palates.

In the United States, Kaposi's sarcoma is at least 20,000 times more common in people with HIV/AIDS than in the general population, and 300 times more common than in other immunosuppressed groups (Beral et al., 1990).

KS was first described by Moritz Kaposi in 1877 as a cancer of the muscle and skin. Characteristic signs of early KS were bruises and birthmark-like lesions on the skin, especially on the lower extremities. KS was described as a slow-growing tumor found primarily in elderly Mediterranean men, Ashkenazi Jews, and equatorial Africans.

Kaposi's sarcoma as described by Moritz Kaposi is called classic KS and it differs markedly from the KS that occurs in AIDS patients (Figure 6-7). Classic KS has a variable prognosis (forecast), is usually slow to develop, and causes little pain (**indolent**). Patient survival in the United States ranges from 8 to 13 years with some reported cases of survival for up to 50 years (Gross et al., 1989). Symptoms of classic KS are ulcerative skin lesions, swelling (**edema**) of the legs, and secondary infection of the skin lesions.

Kaposi's Sarcoma and AIDS

The HIV/AIDS epidemic has brought a more virulent and progressive form of KS marked by painless, flat to raised, pink to purplish plaques on the skin and mucosal surfaces that may spread to the lungs, liver, spleen, lymph nodes, digestive tract, and other internal organs. In its advanced stages it may affect any area from the skull to the feet (Figure 6-8). In the mouth, the hard palate is the most common site of KS (Figure 6-9) but it may also occur on the gum line, tongue, or tonsils.

KS in AIDS patients comes on swiftly and spreads aggressively. However, there have been *no reported* AIDS deaths due to just KS. But, KS can have enormous psychological impact particularly if the lesions occur on exposed areas.

Some of the most inconvenient and uncomfortable KS targets include the soles of the feet, the nose, and the oral cavity. Lesions on the lower extremities or on the feet are often associated with edema and swelling, causing not only severe pain but difficulty putting on shoes and walking. Swelling can be complicated by bacterial cellulitis, ulceration and skin breakdown, often with gram-negative bacterial infections. In addition to the obvious cosmetic damage, lesions on the face may be accompanied by swelling around the eyes that can sometimes progress to the point where patients literally cannot open their eyes. Finally, oral lesions can be painful and make eating and speaking problematic.

The prevalence of KS among gay men in 1981 was 77%; by 1987, it had fallen to 26% and by 2004 to less than 5% in gay men on HAART. This drop in KS among gay men is paralleled by a fall in CMV cases.

Human Herpes Virus Is the Kaposi's Virus—Most HIV/AIDS researchers now believe that HIV is not the primary pathological agent for Kaposi's sarcoma. Beral and colleagues (1990) at the Centers for Disease Control and Prevention concluded that the epidemiological data on Kaposi's distribution suggest that it is caused by a sexually transmitted pathogen other than HIV.

In December 1994, Yuan Chang and colleagues reported that they found DNA sequences that appear to represent a **new human herpes virus (HHV-8)** in KS tissue. Gianluca Gaidano and colleagues (1996) report that their data confirm that HHV-8 DNA sequences are found, at high frequency, with selected types of AIDS-related KS. Evidence continues to accumulate indicating that HHV-8 is the infectious agent responsible for KS (Kledal et al., 1997; Said et al., 1997).

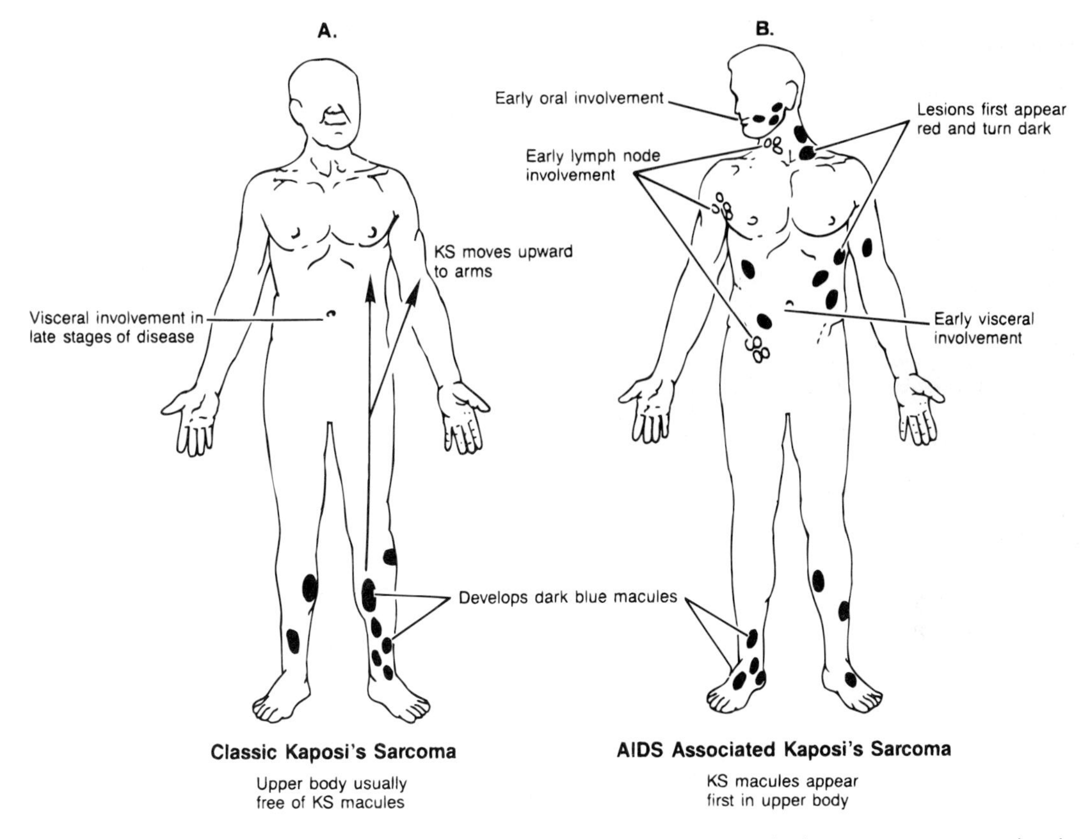

FIGURE 6-7 Classic and AIDS-Associated Kaposi's Sarcoma. **A.** Patients with classic KS (non-AIDS-related) demonstrate violet to dark blue bruises, spots, or macules on their lower legs. Gradually, the lesions enlarge into tumors and begin to form ulcers. KS lesions may, with time, spread upward to the trunk and arms. The movement of KS appears to follow the veins and involves the lymph system. In the late stages of the disease, visceral organs may become involved. **B.** For AIDS patients, initial lesions appear in greater number and are smaller than in classic KS. They first appear on the upper body (head and neck) and arms. The lesions first appear as pink or red oval bruises or macules that, with time, become dark blue and spread to the oral cavity and lower body, the legs, and feet. Visceral organs may be involved early on and the disease is aggressive. However, death is not caused by KS.

Many research papers on whether herpes virus 8 causes KS have been published in recognized scientific/medical journals. It appears there is a cause and effect relationship; HHV-8/KS.

The work of Charles Rinaldo and colleagues (2001) reveals that healthy non-HIV-infected people who carry HHV8 have a healthy immune response and control the virus. HIV-infected persons who carry HHV8 have a poor immune response to the virus, which then becomes a precursor for the expression of KS. In 2002, Michael Cannon and colleagues reported that the risk of an HIV-positive man developing the AIDS-defining cancer Kaposi's sarcoma is linked to the amount of human herpes virus 8 (HHV-8) in peripheral blood monoculear cells (PBMC) and oral fluids, rather than the CD4+ cell count or viral load. The study found that HHV-8 is likely to be orally transmitted.

The Kaposi's virus may have entered the same population in which HIV is endemic, which would explain why the two are often transmitted together (O'Brien et al., 1999). HIV may produce the right conditions for Kaposi's development by causing growth factor production, and possibly by suppressing the body's immune defenses against cancer.

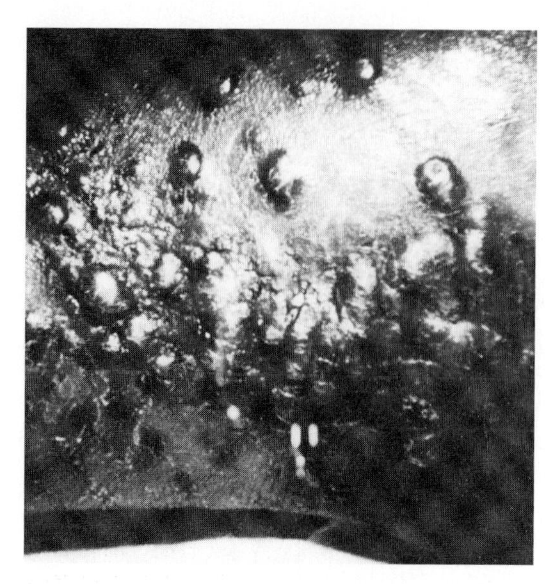

FIGURE 6-8 Kaposi's Sarcoma on Lower Leg of an AIDS Patient. *(Courtesy of Nicholas J. Fiumara, M.D., Boston)*

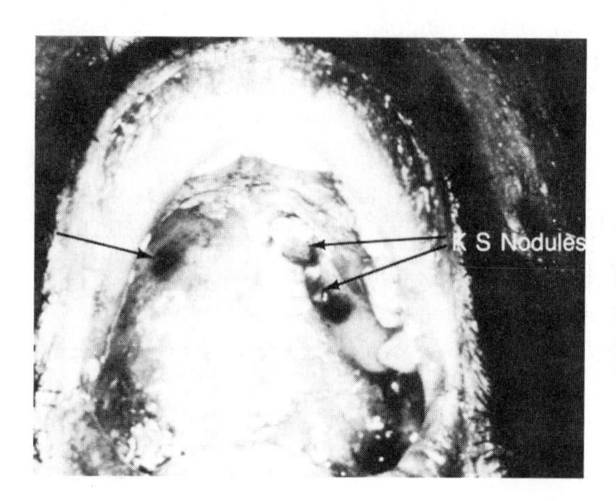

FIGURE 6-9 Oral Kaposi's Sarcoma. KS can be seen on the hard palate and down the sides of the oral cavity. *(Courtesy of Nicholas J. Fiumara, M.D., Boston)*

The work of Dennis Osmond and colleagues states that a KS-associated herpes virus was present in about 25% of gay males in San Francisco in 1978, several years before the CDC reported on the immune deficiency disease later called AIDS. As evidence mounts that genital herpes plays an important role in facilitating HIV acquisition, researchers are debating whether herpes screening and treatment should be considered for an HIV prevention strategy.

Lymphoma (lim-fo-mah): Cancer of the Lymph Glands

Lymphomas are the second most common cancer in HIV and are now the seventh most common cause of death for people with AIDS. A lymphoma is a neoplastic disorder (cancer) of the lymphoid tissue (Figure 6-10). B cell lymphoma occurs in about 1% of HIV-infected people, but makes up about 90% to 95% of all lymphomas found in people with HIV disease (Herndier et al., 1994; Scadden, 2002). Although it occurs most often in those demonstrating persistent generalized lymphadenopathy (swollen lymph glands), the usual site of lymphoma growth is in the brain, the heart, or the anorectal area (Brooke et al., 1990). The most common signs and symptoms are confusion, lethargy, and memory loss. Lymphomas are increasing in incidence primarily due to the extension of the life span of AIDS patients because of medical therapy (Table 6-1).

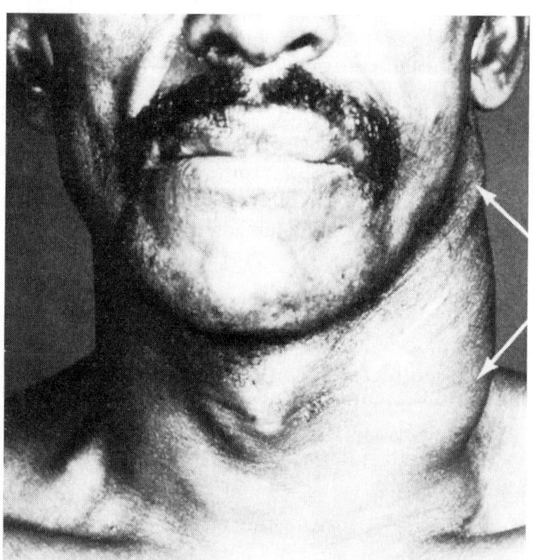

FIGURE 6-10 HIV/AIDS Patient Demonstrating a Lymphoma of the Neck. *(Courtesy CDC, Atlanta)*

Progressive Multifocal Leucoencephalopathy (leuco means white; encephalo means brain; pathy means disease)

Progressive multifocal leucoencephalopathy (PML) is an opportunistic infection caused by a papovavirus [Jamestown Canyon virus (JCV)] affecting 4% of AIDS patients. It is usually fatal within an average of 3.5 months and there is no treatment. In a few patients spontaneous improvement and prolonged survival have been reported. Some observations have indicated that cytosine arabinoside, a potent antiviral, may reverse the symptoms of PML. Symptoms of PML include altered mental status, speech and visual disturbances, gait difficulty, and limb incoordination (Guarino et al., 1995).

HIV Provirus: A Cancer Connection

In early 1994 AIDS investigators reported that HIV, on entering lymph cell DNA, activated nearby cancer-causing genes (oncogenes). The evidence suggests that HIV itself may trigger cancer in an otherwise normal cell.

These findings may mean that a variety of retroviruses that infect humans may also cause cancer (McGrath et al., 1994). Such findings raise concerns for developing an HIV vaccine. Using a weakened strain of HIV to make the vaccine may, when used, increase the incidence of lymphoma and other cancers.

DISCLAIMER

This chapter is designed to present information on opportunistic infections in HIV/AIDS patients. The author does not accept any responsibility for the accuracy of the information or the consequences arising from the application, use, or misuse of any of the information contained herein, including any injury and/or damage to any person or property as a matter of product liability, negligence, or otherwise. No warranty, expressed or implied, is made in regard to the contents of this material. This material is not intended as a guide to self-medication. The reader is advised to discuss the information provided here with a doctor, pharmacist, nurse, or other authorized health-care practitioner and to check product information (including package inserts) regarding dosage, precautions, warnings, interactions, and contra-indications before administering any drug, herb, or supplement discussed herein.

SUMMARY

One of the gravest consequences of HIV infection is the immunosuppression caused by the depletion of the T4 or CD4+ helper cell population; suppressed immune systems allow for the expression of opportunistic diseases and cancers. The OI, end organ failures, and cancers kill AIDS patients, not HIV per se. It is the cumulative effect of several OIs that creates the chills, night sweats, fever, weight loss, anorexia, pain, and neurological problems.

One tragic disease that does not result from an OI is Kaposi's sarcoma (KS), characterized as a cancer that can spread to all parts of an AIDS patient's body. About 20% of AIDS patients, mostly gay men, have KS. It is not usually found in hemophiliacs, injection-drug users, or female AIDS patients.

REVIEW QUESTIONS

(Answers to the Review Questions are on page 438.)

1. Define opportunistic infection (OI).
2. Which OI organism expresses itself in 80% of AIDS patients? Where is it located and what does it cause?
3. Which of the protozoal OI organisms causes weight loss, watery diarrhea, and severe abdominal pain?
4. Which of the bacterial OIs causes "wasting syndrome," night sweats, anorexia, and fever?
5. True or False: Kaposi's sarcoma (KS) is caused by HIV. Explain.
6. Name the two kinds of KS.
7. True or False: KS affects all AIDS patients equally. Explain.
8. True or False: Candidiasis and ulceration may be present in patients with HIV infection.
9. True or False: Oral candidiasis occurs frequently with HIV infection.
10. True or False: The use of combination anti-HIV drug therapy, especially those combinations containing a protease inhibitor, have substantially decreased the severity and number of OIs in AIDS patients.
11. Opportunistic Infections Crossword Puzzle

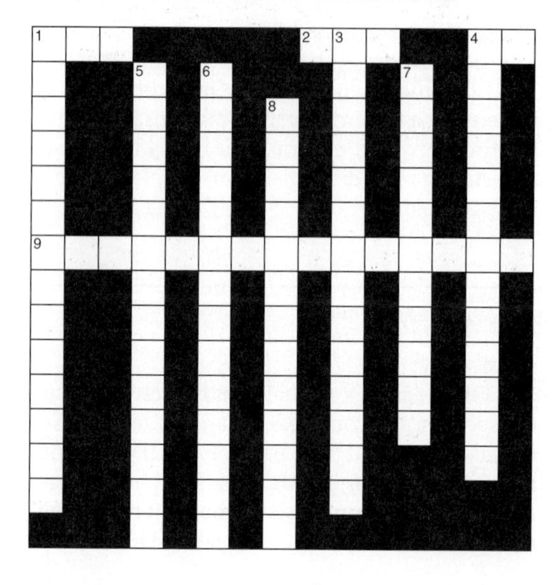

ACROSS

1. Centers for Disease Control and Prevention
2. Cytomegalovirus
4. Opportunistic infection
9. Protozoal infection

DOWN

1. Infection of the central nervous system
3. Bacteria, fungi, viruses, and protozoa
4. Pathogenic when immune system impaired
5. Fungal disease
6. To become serologically positive
7. Associated with yeast infection
8. Protozoan infection

HIV/AIDS SELF-EVALUATION/EDUCATION QUIZ

Before you begin reading Chapters 7 through 13, take this quiz to help determine your current knowledge about HIV infection and AIDS.

1.	HIV infects only homosexual persons.	T	F	?
2.	AIDS affects only adults.	T	F	?
3.	The agent that causes AIDS is HIV.	T	F	?
4.	HIV is a bacterial agent.	T	F	?
5.	The risk of an HIV infection from a blood transfusion is low, but it can happen.	T	F	?
6.	Patients with AIDS have reduction in T4 or CD4+ cells.	T	F	?
7.	People who have antibodies to HIV always develop AIDS.	T	F	?
8.	HIV has been found in vaginal secretions, saliva, blood, and other body fluids.	T	F	?
9.	AIDS is not a disease, but a syndrome.	T	F	?
10.	HIV is easily transmitted to people.	T	F	?
11.	The period from exposure to HIV to the expression of AIDS may be at least a year.	T	F	?
12.	There is no known cure for HIV disease.	T	F	?
13.	There is no known cure for AIDS.	T	F	?
14.	HIV carriers without signs of active disease can transmit the virus.	T	F	?
15.	All HIV-infected persons progress to AIDS.	T	F	?
16.	All persons with AIDS die prematurely.	T	F	?
17.	An effective HIV vaccine will be developed by 2007.	T	F	?
18.	A person can be diagnosed with AIDS but have no outward symptoms.	T	F	?
19.	One can have a zero T4 cell count and still be alive.	T	F	?
20.	AIDS is considered a new disease of the twentieth century.	T	F	?

The correct answers to these questions are: 1.F; 2.F; 3.T; 4.F; 5.T; 6.T; 7.F; 8.T; 9.T; 10.F; 11.T; 12.T; 13.T; 14.T; 15.F; 16.T; 17.?; 18.T; 19.T; 20.T.

7 A Profile of Biological Indicators for HIV Disease and Progression to AIDS

CHAPTER CONCEPTS

- Clinical signs and symptoms of HIV infection and AIDS are presented.
- Stages of HIV disease vary substantially.
- HIV replication is rapid and continuous in HIV-infected lymphoid cells.
- AIDS Dementia Complex presents as mental impairment.
- Viral load indicates current viral activity.
- T4 or CD4+ cell counts indicate degree of immunologic destruction.
- Information on long-term survival is presented.
- Serological changes after HIV infection are presented.
- The rate of clinical HIV disease progression is variable among individuals infected with HIV.
- The development of AIDS over time is discussed.
- Classification of HIV/AIDS progression is presented.
- Clinical indicators to track HIV disease progression are listed.
- Diarrhea is the most common gastrointestinal sign and symptom of HIV/AIDS infection.
- Clues to pediatric AIDS diagnosis are presented.
- HIV/AIDS word search is presented.

HIV DISEASE DEFINED

The diagnosis of AIDS in the 1980s was most often associated with a quick death. Mortality came quickly and was inevitable within a few months of the diagnosis.

By the mid-1980s the CDC had learned enough about HIV infection to call it a disease. That made sense, as the vast majority of those who became infected became ill. HIV infection leads to the loss of T4 or CD4+ cells, which in turn produces a variety of signs and symptoms of a **nonspecific disease** with initial acute fever-associated illness or mononucleosis-like symptoms that may last up to four weeks or longer. After the initial symptoms, most individuals enter a clinically asymptomatic phase (see Case in Point 7.1). This means the infected person feels well while his or her immune system is slowly compromised. It has been shown that long-lasting symptomatic *primary* HIV infection predicts an increased risk of rapid development of HIV-related symptoms and AIDS, but it is not known whether the different responses to HIV infection are caused by viral, host factors, or both. Virulent strains of HIV have been characterized by their rapid replication, **syncytium-inducing (SI)** capacity, and tropism (attraction) for various types of T cells. It is known that the biological properties of HIV strains in asymptomatic HIV-infected individuals with normal T4 or CD4+ cell counts may predict the subsequent development of HIV-related disease, and that patients who harbor SI isolates develop immune deficiency more rapidly. It is not clear whether the appearance of more virulent strains during the symptomatic phase of the infection is a cause or an effect of progressive immune deficiency (Nielson et al., 1993). Several studies have demonstrated that a long period of fever around the time of **seroconversion** (the presence of detectable HIV antibody in the serum) is associated with more rapid development of immune deficiency (Pedersen et al., 1989).

Spectrum of HIV Disease

Because the immune system slowly falters, HIV disease is really a spectrum of disease (Figure 7-1). At one end of the spectrum are those infected with

CASE IN POINT 7.1

VARIATION OF INITIAL SYMPTOMS AFTER HIV INFECTION

Case I: Male, Age 35, Los Angeles, California

One evening, for no apparent reason, John began sweating profusely. Soon after, a red rash began on his arms, face, and legs and then covered his body. Simultaneously, breathing became difficult and he was rushed to an emergency room. By then he was shaking violently. After medication and a battery of tests his problem could not be defined. This brief illness passed, but some years later, during a blood screen for insurance purposes he came up HIV positive. He immediately reflected back on his earlier illness and its cause.

Case II: Male, Age 29, Los Angeles, California

This case is in marked contrast to Case I. Feeling the pinch of a sore throat, this male went to his physician for an antibiotic. On examination, he had a yeast infection that appeared far back in his throat. This raised suspicion and he agreed to an HIV test. It came back positive. He had no other illness. He was treated, the sore throat vanished, and he is thriving in a long asymptomatic period.

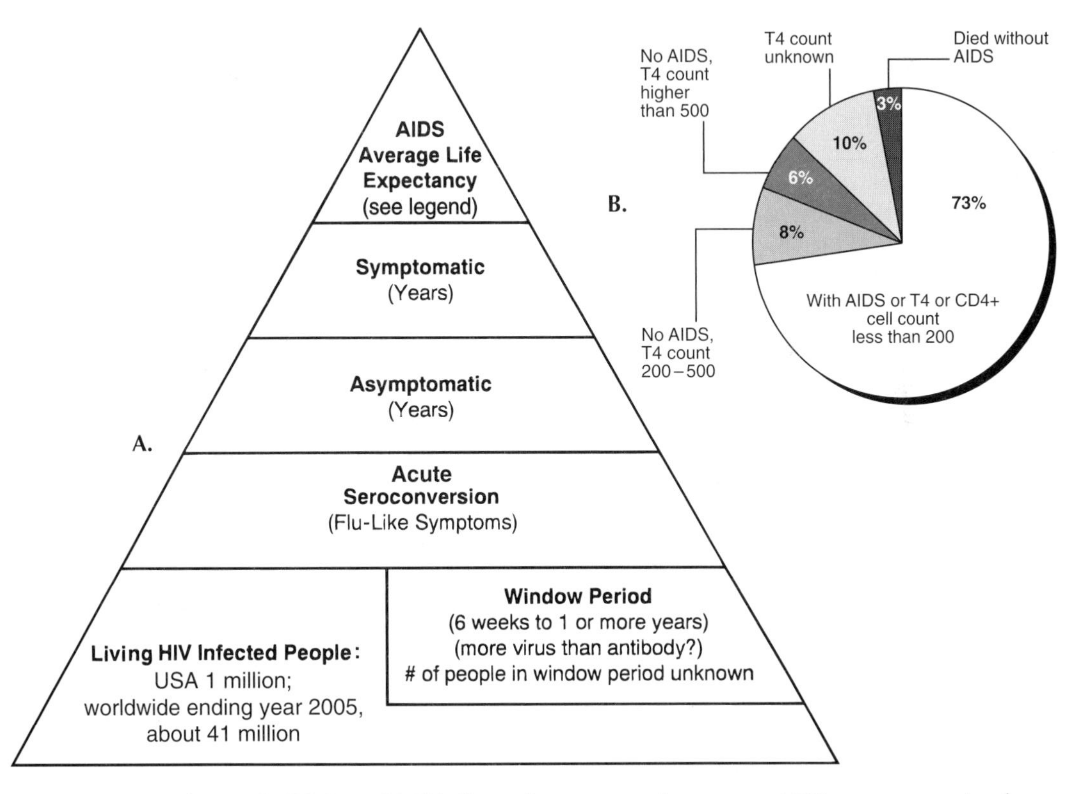

FIGURE 7-1 **A.** The HIV/AIDS Pyramid. This figure demonstrates that current AIDS cases are coming from an existing pool of HIV-infected persons. In the United States, of those infected, about 30% do not yet know they are HIV positive. Although both the asymptomatic and symptomatic periods may last for years, once diagnosed with AIDS the average life expectancy without AIDS drug cocktails is two to three years. Life expectancy for those now on combination drug cocktails depends on the state of the individual's immune system and response to antiretroviral therapy. **B.** Clinical outcomes 10 to 16 years after HIV infection in a population tracked by the San Francisco City Clinic from the beginning of the AIDS outbreak.

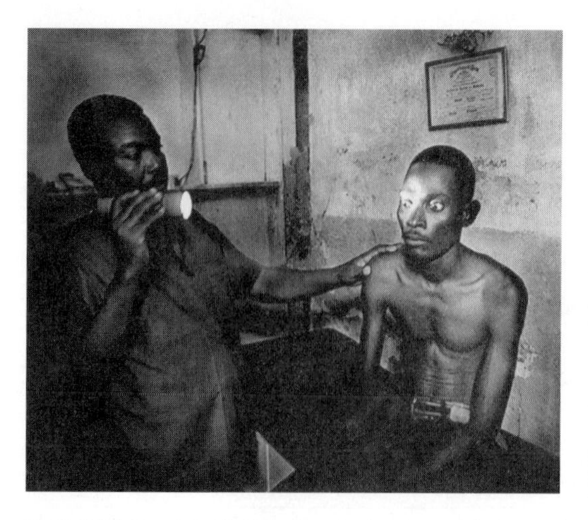

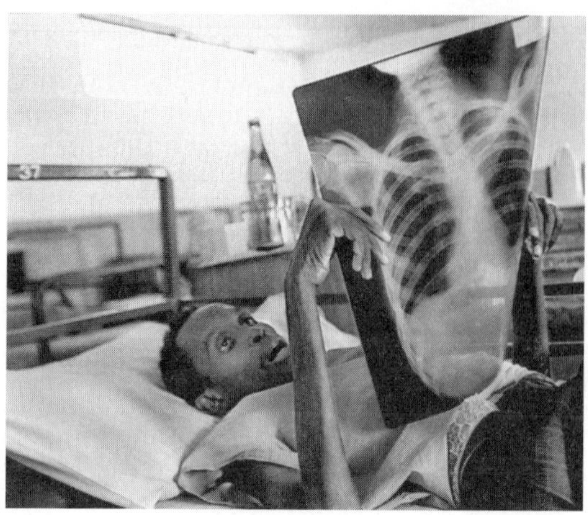

FIGURE 7-2 HIV/AIDS: The Beginning and the End. **A.** Because he has no electricity in his office in Cite Soleil, Haiti, this physician uses a flashlight to examine an HIV-positive patient who is in the first stages of the infection. This physician sees, on average, one new HIV-infected person each day. **B.** At age 28, Jean David looks at his X-ray, which, along with his medical history, according to his physician, indicates his impending death from AIDS. Jean David died soon after this picture was taken. He left an HIV-infected wife and HIV-infected 14-month-old daughter. *(Photos by Mike Stocker/South Florida* Sun-Sentinel)

HIV who look, feel, and are perfectly healthy. At the opposite end are those with advanced HIV disease (**symptomatic AIDS**) who are visibly sick and require significant medical and psychosocial support (Figure 7-2). Between these two extremes, HIV-infected people may develop illnesses that range from mild to serious. Symptoms can include persistent fevers, chronic fatigue, diarrhea, swollen lymph nodes, night sweats, skin rashes, significant weight loss, visual problems, chest pain, and fungal infections of the mouth, throat, and vagina. Illness from these conditions can be severe and disabling, and some people may die without ever being diagnosed with AIDS. Also, people with HIV disease may develop neurologic disorders, which can cause forgetfulness, memory loss, loss of coordination and balance, partial paralysis, leg weakness, mood changes, and dementia. These symptoms may occur in the absence of any other symptoms. The interval between initial HIV infection and the presence of signs and symptoms that characterize AIDS is variable and may range, in those who have not used antiretroviral drugs, from several months to a median duration of about 11 years (Figure 7-3).

Defining Incubation and Latency

Because of the long delay in determining what happened after HIV infection and progression to AIDS, the terms **incubation** and **latency** are used, in many cases interchangeably, causing some confusion. In this chapter, the two terms are used with respect to **clinical** observations as follows: **Clinical incubation** is that period after infection through the window period or when anti-HIV antibody production is measurable. **Clinical latency** is the time period from detectable anti-HIV antibody production (seroconversion—the person now tests HIV-antibody positive) through the asymptomatic period—a time prior to the expression of opportunistic diseases. This time period, being asymptomatic, may last from 1 to 25 years—the average being 11 years. The beginning and end of these periods will vary from person to person and their susceptibility and expression of HIV disease and whether they are on drug therapy.

STAGES OF HIV DISEASE (WITHOUT DRUG THERAPY)

The course of the disease in the infected individual varies substantially. At the extremes are individuals who show either little evidence of progression (loss in T4 cells) 10 to 25 years following infection (about 3%) or extremely rapid progression and death within less than two to three years. In general, HIV-infected

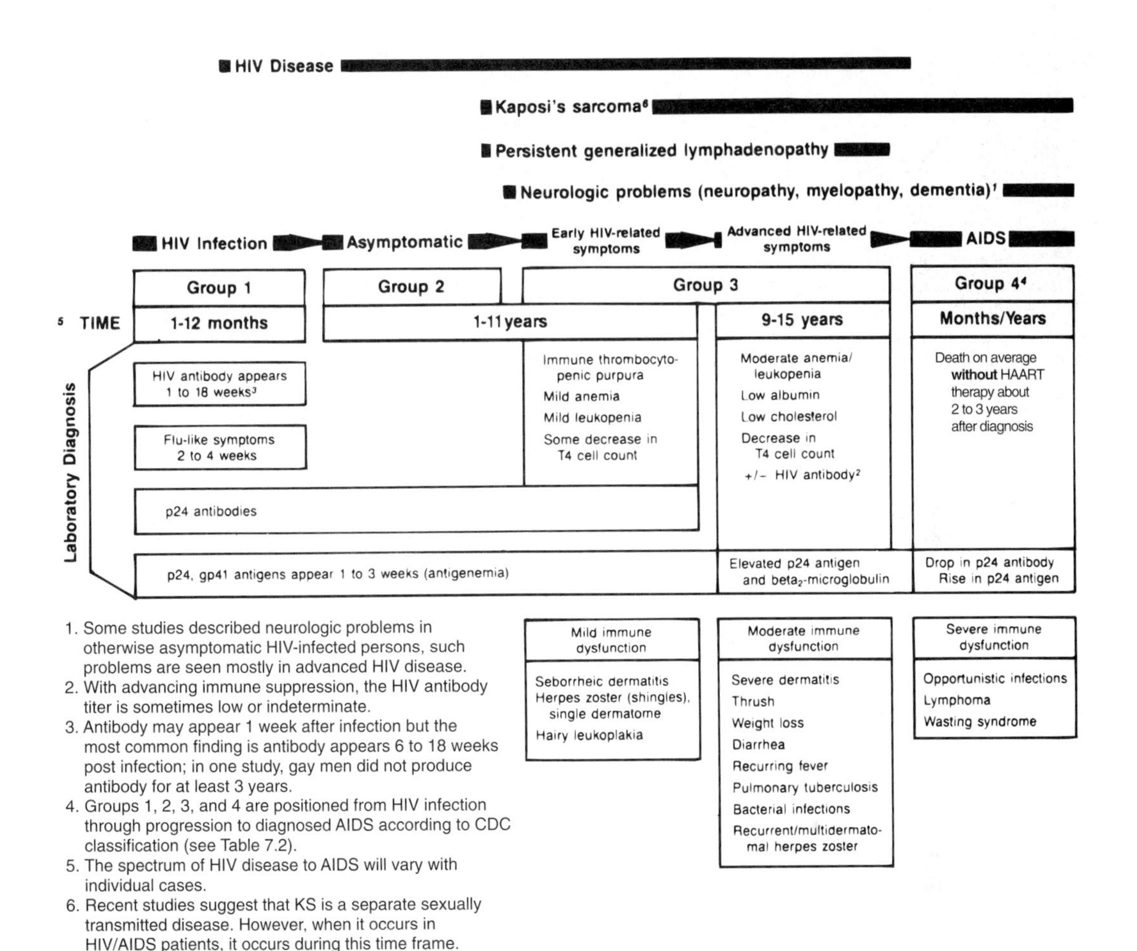

FIGURE 7-3 Spectrum of HIV Infection, Disease, and the Expression of AIDS. Seroconversion means that HIV antibodies are measurably present in the person's serum. With continued depletion of T4 or CD4+ cells, signs and symptoms appear, announcing the progression of HIV disease to AIDS. Although HIV antibodies have been found as early as 1 week after exposure, **most often seroconversion occurs between weeks 6 and 18; 95% within 3 months, 99% within 6 months** (see Figure 7-5). The early stage of HIV infection can be separated from the symptomatic stage by years of **clinical latency.** Early infection is characterized by a high number of infected cells and a high level of viral expression. AIDS is characterized by increased levels of viremia and p24 antigenemia, activation of HIV expression in infected cells, an increased number of infected cells, and progressive immune dysfunction. Stages of HIV disease blend into a continuum ranging from the asymptomatic with apparent good health, to increasingly impaired health, to the diagnosis of AIDS. Thus the spectrum of HIV disease ranges from the silent infection to unequivocal AIDS. Clinical expression moves from one condition to another, often without a clear-cut distinction. The level of an individual's infectiousness is believed to be greatest within the first months after infection and again when the T4 cell count drops below 200. **However, people who are HIV-infected can transmit HIV at any time.**

SIDE ISSUES 7.1

HLABISA, AN AIDS-RAVAGED TOWN IN KWAZULU NATAL PROVINCE

In 1994, 10% of adults in Hlabisa (Sha-BEE-sa) were HIV-infected. Beginning in 2004, over 30% or 75,000 people are HIV positive.

In the desperately crowded clinics and hospital wards, the scale of the epidemic is clear enough. This district of 250,000 people sits amid the hills of KwaZulu Natal, which has the highest adult rate of HIV infection of any province in South Africa. It is one of the few communities in South Africa in which government researchers have kept statistics on HIV infection rates for nearly a decade. Its story offers a rare and intimate look at one community ravaged by the plague. It marks the faces of young widows who trudge the road in somber capes and skirts, traditional mourning garb. It inspires the medicine makers who brew slivers of tree bark and bundles of dried leaves into elixirs sold in used Coca-Cola bottles. "Two spoons in the morning, two spoons in the afternoon," advises a herbalist who charges $1.25 a bottle. Chilled bodies that are often stacked one on top of another swamp the morgue.

Fear, Denial, and Guilt

Professional men in Hlabisa boast over beers about extramarital affairs and the pleasures of unprotected sex. Some church leaders burn condoms and assail people with the virus as sinners. Prominent community members die in silence because the disease is considered so shameful. The disease is so deadly and so frightening that many hospital employees are reluctant to call it by name. They say a patient is immune, compromised, or that he suffers from "that disease." Others simply say, "You know what he's got." It is as if uttering the word might infect the tongue.

Healthcare Choice: Hospital or Healer?

Hospitals—On any given day there are more patients than available beds. In the male ward, 80% of men are HIV positive and about half the nursing positions are vacant. The first event in morning rounds is to count those who died of AIDS or a combination of HIV infection and tuberculosis. Of those who die, about half are under age 30! In the first six months of 2001, of 500 people who took HIV tests, 63% tested positive. One of the staff nurses has six children, four daughters and two sons. She leaves boxes of condoms on their bedroom dressers. The two things that give her the greatest sense of peace are taking her pills and watching those boxes empty. The children do not comment—she watches and replenishes the condom supply as needed.

Healers or Sangomas—In Southern Africa, a sangoma undergoes a long apprenticeship studying plant lore and making diagnoses that can include playing a guessing game with the patient, dancing into a trance, reading cast bones or waiting for the answer to come in a dream. In Hlabisa, physicians believe that virtually all their patients first visit a sangoma—a traditional healer. Donald McNeil describes the residence of one of the preeminent sangomas in Hlabisa. She is 57 years old. The walls of her home are covered with dried birds, the python skins she stabbed herself, and a crocodile her husband speared—from which, she says, she can make a lotion that renders a man's skin bulletproof. The floor is covered with hundreds of medicine containers, as humble as a Yum Yum peanut butter jar and as ornate as a set of gourds covered with beadwork. But there is also a box of latex gloves, which she wears when inspecting sores. When she injects a patient with a porcupine quill tipped with brown gunk from the Yum Yum jar, she makes a point of wiping the quill afterward with alcohol. Asked about those precautions, she proudly displayed the framed certificate from a 1998 workshop on "The Traditional Healer's Role in AIDS: Sexually Transmitted Diseases and Primary Health Care" offered at Hlabisa Hospital. One of the sangoma's patients said that her illness was sent to her from her ancestors as a test. She also has a goat's bladder tied into her hair to remind her ancestors that she had sacrificed a goat to them.

Traditional Therapy—Because traditional Zulu medicine focuses chiefly on digestion, bile, and mucus, sangomas often give emetics or enemas. Some are simply dishwashing liquid or toothpaste. But others contain powerful herbs that can cause serious drug interactions. Enemas are used for anything from constipation to hysterical crying. However, enemas can rapidly dehydrate patients and send them into kidney or liver failure. For those who receive herbs causing them to throw up (emetics), it defeats the use of antiretroviral drugs.

SIDEBAR 7.1

DISGRUNTLED HIV PATIENT COMMENTS ON HIS HEALTH CARE PROVIDER

Patient: What kind of virus do I have? Doctor: Who knows? Is it already drug resistant and to which drugs? Don't know but your insurance will only pay for this kind of information once in a blue moon. Does my immune system have a natural ability to fight my virus? Who knows? One cannot readily have tests run to answer that unless they are being studied by the National Institutes of Health. Are my drugs working? Maybe, but we can't always tell for sure, until you lose a big chunk of your immune system. Are my drugs trying to kill me? That depends; how are you feeling today? Can you tell me if my pain is from my liver, gas, or mitochondriosis? We need more information but it is not available. Sorry your 10 minutes are up. And so it goes. Eventually you learn the best emergency rooms to visit in your area and start shopping for a viatical company to buy your life insurance.

adults experience a variety of conditions, categorized into four stages: **acute infection, asymptomatic, chronic symptomatic,** and **AIDS**.

Primary HIV Infection or the Acute Disease Stage

The clinical syndrome of primary HIV infection was recognized and documented in 1985, about two years after the initial identification of the causative agent of AIDS. By 1991 it was known that this symptomatic period is associated with an explosive replication of the virus, which is then partially controlled as the illness resolves spontaneously. Reports in 1993 further showed the population of HIV during this early period of infection to be quite homogeneous, in distinct contrast to the diverse quasispecies that are typically found in chronically infected persons. The course and time frame of the infection is illustrated graphically in Figure 7-3. The course of Primary HIV Infection (PHI) is limited to a few weeks or months, whereas the entire course of HIV infection can span many years. Specifically, PHI is the period after infection with HIV but before the development of detectable antibodies or seroconversion (see Figure 7-4).

The acute stage usually develops in two to eight weeks during PHI. Up to 70% of infected individuals develop a self-limited (brief) illness similar to influenza or mononucleosis: high spiking fever, sore throat, headaches, and swollen lymph nodes. Some may develop a rash, vomiting, diarrhea, and thrush (yeast infection in the mouth).This is referred to as the **acute retroviral syndrome.** The symptoms generally last about one to four weeks and resolve spontaneously (Table 7-1). The acute stage can be over quickly and easily missed. The acute phase is marked by high levels of HIV production, in excess of 1 million copies per milliliter of blood. During this phase, large numbers of HIV spread throughout the body, seeding themselves in various organs, particularly lymphoid tissues such as the lymph nodes, spleen, tonsils, and adenoids (Figure 7-4). During this time frame, HIV infects monocytes, macrophage, T4 or CD4+ cells, and follicular dendritic cells. Within these safe havens, HIV can persist for years despite Highly Active Antiretroviral Therapy (HAART). This pool of latently infected cells is established very early after HIV enters the body—even if the person takes drug therapy immediately after exposure to HIV.

Table 7-1 Primary HIV Infection: Signs and Symptoms (Department of Health and Human Services [DHHS] Guidelines [*Ann Intern Med* 2002;137:381])

Fever — 96%	Myalgias — 54%	Hepatosplenomegaly — 14%
Adenopathy — 74%	Diarrhea — 32%	Weight loss — 13%
Pharyngitis — 70%	Headache — 32%	Thrush — 12%
Rash* — 70%	Nausea & vomiting — 27%	Neurologic symptoms** — 12%

*Rash – erythematous maculopapular rash on face and trunk, sometimes extremities, including palms and soles. Some have mucocutaneous ulceration involving mouth, esophagus, or genitals.

**Aseptic meningitis, meningoencephalitis, peripheral neuropathy, facial palsy. Guillain-Barre syndrome, brachial neuritis, cognitive impairment, or psychosis.

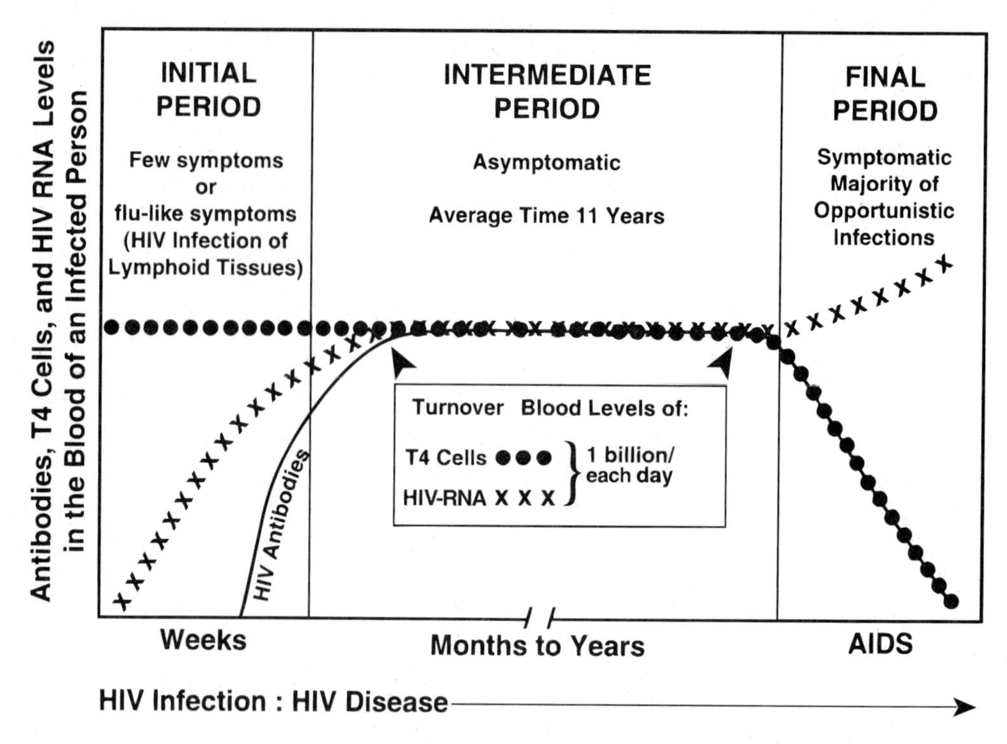

FIGURE 7-4 Relationship of T4 or CD4+ Cells, HIV Antibodies, and HIV RNA Levels (Viral Load) Beginning With HIV Infection Through AIDS. Within a week to weeks after HIV infection, HIV becomes seeded throughout the body's blood and lymph system. HIV reproduction (infection of T4 cells) begins almost immediately in the lymph system. The T4 cell population also begins rapid reproduction to replace T4 cell loss. This is why the T4 graph line stays at the same level through the **asymptomatic** period. The immune system begins to turn out HIV antibodies, in general 6 to 18 weeks later (window period, see Figure 7-5). Note that during the asymptomatic period T4 cell and HIV replication and antibody production keep pace. With time, however, T4 cells fail to replace losses, HIV continues to replicate, antibody levels drop due to loss of T4 signals to B cells to produce antibodies, and opportunistic infections begin—the **symptomatic** period. Without therapy this period lasts on average about two to three years. For people using HAART, the average number of years one can survive in the symptomatic period is now between 5 and 10 years.

During acute primary infection, patients have extremely high levels of viral replication but a variable antibody immune response. The level of cytotoxic T lymphocytes (CTLs) targeted against HIV appears to increase significantly, an attempt by the cellular part of the immune system to contain the high rate of HIV replication. Plasma HIV RNA levels may reach 10^5 (100,000) to 10^6 (1 million) copies/mL. It is now believed that the increase in the presence of CTLs and chemical factors they produce bring about a milder illness and a reduction in HIV to a **lower set point**—a steady state of viral load. The higher this set point the more rapidly HIV disease progresses to AIDS. (The presence of CTLs and long-term survival is presented in Box 7.3.) At this time in HIV infection and in some cases for weeks or months, neutralizing antibodies to HIV are not measurable—thus antibody testing, whether at clinical labs or using a home testing kit, is negative. This period of high viral replication in the absence of detectable antibody is called the **window of infectivity before seroconversion** or window period. During this period, patients may be highly infectious, about ten times more infectious than in the asymptomatic stage. The viral burden in genital secretion is particularly high during this time. Mathematical models suggest that 56% to 92% of all HIV infections may be transmitted during this period of acute infection (Quinn, 1997).

A true state of **biological latency,** according to the work of Xiping Wei and coworkers (1995) and David Ho and coworkers (1995), does not exist in the lymph nodes at any time during the course of HIV infection. The Wei and Ho investigations show that from the time of infection HIV replication is rapid and continuous, and within two to four weeks the infecting HIV strain is replaced by drug-resistant mutants. Each day over 1 billion HIV are produced and mostly destroyed and millions of T4 cells are infected, dying, and replaced. Over time the immune system fails to destroy HIV and replace its T4 cell losses and HIV disease progresses. Also, over time, many T4 cells in the lymphoid organs probably are activated by the increased secretion of certain cytokines such as tumor necrosis factor-alpha and interleukin-6. T4 or CD4+ cell activation allows uninfected cells to be more easily infected and causes increased replication of HIV in infected cells. Other components of the immune system are also chronically activated, with negative consequences that may include the suicide of cells by a process known as programmed cell death or apoptosis and an inability of the immune system to respond to other invaders.

Asymptomatic HIV Disease Stage

Following acute illness, an infected adult can remain free of symptoms from 6 months to a median time of about 11 years. During the asymptomatic period, measurable HIV in the blood drops to a lower level but it continues to replicate and to destroy T4 cells within the lymph nodes while the body continues to produce new T4 cells and antibodies to the virus (Figure 7-5). An asymptomatic individual appears to be healthy and performs normal activities of daily living.

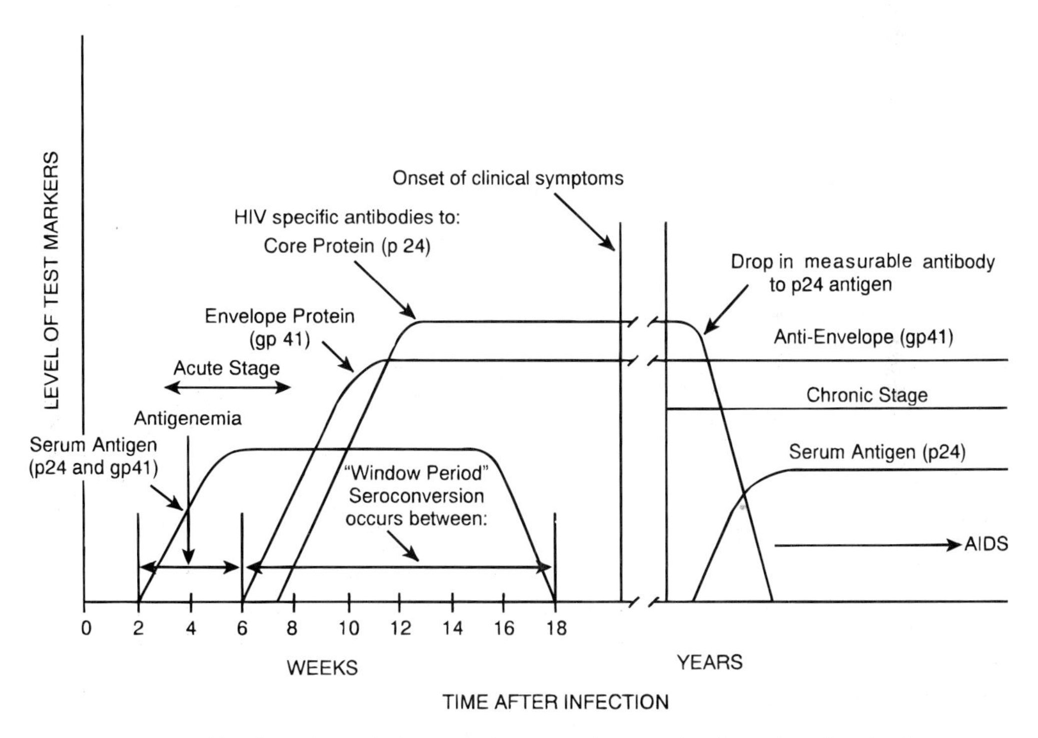

FIGURE 7-5 Profile of Serological Changes After HIV Infection. The dynamics of antibody response to HIV infection were determined by enzyme immunoassays (EIA). Note that during antigenemia, specific HIV proteins (antigens) can be detected before seroconversion occurs. Perhaps other HIV proteins will allow even earlier detection of HIV infection. Once antibodies appear, some antigens like p24 and gp41 disappear only to show up again later on. Note also that although antibody production is a sign that the immune system is working, in HIV-infected people it is not working well enough. Although envelope and core protein antibodies are being produced as clinical illness begins, as the p24 antibody drops, the illness becomes more serious. *(Adapted from Coulis et al., 1987)*

POINT OF INFORMATION 7.1

ONE MISTAKE COST HIM HIS LIFE

I held my son today while he died from AIDS. There is no pain like the pain in a mother's heart. He was 28 years old, and now, he is dead.

This wonderful young man will never have a family. He will never again have a chance to do the things he enjoyed so much—water ski, snow ski, travel. He loved *Star Trek* and music. He loved working for the airlines and traveling all over the world. He was delightful and smart—a computer whiz—could take one apart and put it back together.

He wasted away from a handsome young man to a skeleton—nothing more than skin and bones. His weight dropped from 160 pounds to 80 pounds. His hair fell out. His beautiful teeth fell out. Sores broke out all over his body. He couldn't hold down any food, and eventually, he starved to death.

No, young people, he was not gay, nor was he a drug user. He just went to bed with a girl he didn't know.

His Mom

(*Source: Ann Landers, syndicated columnist, 1994*)

Table 7-2 Correlation of Complications With CD4+ Cell Counts (see *Arch Intern Med* 1995; 155:1537)

CD4+ Cell Count*	Infectious Complications
>500/mm^3	Acute retroviral syndrome Candidal vaginitis
200–500/mm^3	Pneumococcal and other bacterial pneumonia Pulmonary tuberculosis Herpes zoster Oropharyngeal candidiasis (thrush) Cryptosporidiosis, self-limited Kaposi's sarcoma Oral hairy leukoplakia
<200/mm^3	Pneumocystis carinii pneumonia Disseminated histoplasmosis and coccidioidomycosis Military/extrapulmonary TB Progressive multifocal leuko-encephalopathy (PML)
<100/mm^3	Disseminated herpes simplex Toxoplasmosis Cryptococcosis Cryptosporidiosis, chronic Microporidiosis Candidal esophagitis
<50/mm^3	Disseminated cytomegalo-virus (CMV) Disseminated (Myco-bacteri-um avium complex

*Most complications occur with increasing frequency at lower CD4+ cell counts.

Symptomatic HIV Disease Stage

The symptomatic phase can last for months or years before a diagnosis of AIDS occurs. During this phase, as viral replication continues, T4 or CD4+ cells drop significantly. As the number of immune system cells decline, the individual develops a variety of symptoms such as fever, weight loss, malaise, pain, fatigue, loss of appetite, abdominal discomfort, diarrhea, night sweats, headaches, and swollen lymph glands. Ultimately, HIV overwhelms the lymphoid organs. The follicular dendritic cell networks break down in late-chronic-stage disease and virus trapping is impaired, allowing spillover of large quantities of virus into the bloodstream. The destruction of the lymph node structure seen late in HIV disease may stop a successful immune response against HIV and other pathogens as well. Individuals at this stage, with a T4 cell count of 500 or less/μL of blood, often develop thrush, oral lesions, and other fungal, bacterial, and/or viral infections (Table 7.2). The duration of these symptoms varies, but it is common for HIV-infected individuals to have them for months at a time. Of those persons in the symptomatic stage and not using HIV drug therapy, about 30% developed AIDS-associated infections within five years.

Response To HAART

The CD4+ count typically increases ≥50 cells/mm^3 at four to eight weeks after viral suppression with HAART and then increases an additional 50–100 cells/mm^3/year in some patients. Normal CD4+ cell counts range between 500 to 1400 cell/mm^3 (Ann. Internal Med. 1993; 119:55).

AIDS: Advanced HIV Disease Stage

The diagnosis of AIDS is a marker, not an end in itself. Currently, most people recover from their first, second, and third AIDS-defining illnesses. People with AIDS are a very heterogeneous group—some feel well and continue working for several years, others are chronically ill, and some die rather quickly.

Patients with AIDS became an even more diverse group after the 1993 expansion of the Centers for Disease Control and Prevention's definition of AIDS. People in excellent health are diagnosed with AIDS if they test HIV positive and their T4 or CD4+ cell count is less than 200/μL of blood.

The 26 clinical conditions used in the diagnosis of AIDS can be found in Table 7-3.

The final stage of HIV infection is called AIDS. During this time there is continued rapid viral replication that finally upsets the delicate balance of HIV production/T4 cell infection to T4 cell replacement. The virus largely depletes the cells of the immune system. It has been suggested that during the AIDS stage, serious immunodeficiency occurs when HIV diversity exceeds some threshold beyond which the immune system is unable to control HIV replication (Nowak et al., 1990; Wei et al., 1995; Cohen, 1995).

SIDEBAR 7.2

WHAT IS MEANT BY THE TERM WASTING DURING HIV DISEASE

Body wasting during HIV disease is sometimes referred to as AIDS WASTING. Wasting is not well understood, but it involves the involuntary loss of more than 10% of body weight, plus more than 30 days of either diarrhea or weakness and fever. Wasting is linked to disease progression and death. Part of the weight lost during wasting is fat. More important is the loss of muscle mass. This is also called lean body mass, or body cell mass. Lean body mass can be measured by bioelectrical impedance analysis (BIA). This is a simple, painless office procedure. AIDS wasting together with lipodystrophy can cause serious body shape changes. Wasting is the loss of muscle. Lipodystrophy is a loss of fat from one area and the accumulation of fat in another area of the body (see Figures 4-12 and 4-13).

Cause of Wasting—Several factors contribute to wasting—(a) poor appetite, (b) oral opportunistic infections, (c) poor nutrient absorption, (d) an altered food processing metabolism, (e) changes in hormone levels due to the infection. There are no standard treatments.

Table 7-3 List of 26 Conditions in the AIDS Surveillance Case Definition

- Candidiasis of bronchi, trachea, or lungs
- Candidiasis, esophageal
- Cervical cancer, invasive[a]
- Coccidioidomycosis, Disseminated or extrapulmonary
- Cryptococcosis, extrapulmonary
- Cryptosporidiosis, chronic intestinal (>1 month duration)
- Cytomegalovirus disease (other than liver, spleen, or nodes)
- Cytomegalovirus retinitis (with loss of vision)
- HIV encephalopathy
- Herpes simplex: chronic ulcer(s) (>1 month duration); or bronchitis, pneumonitis, or esophagitis
- Histoplasmosis, disseminated or extrapulmonary
- Isosporiasis, chronic intestinal (>1 month duration)
- Kaposi's sarcoma
- Lymphoma, Burkitt's (or equivalent term)
- Lymphoma, immunoblastic (or equivalent term)
- Lymphoma, primary in brain
- *Mycobacterium avium complex* or *M. kansasii*, disseminated or extrapulmonary
- *Mycobacterium tuberculosis*, disseminated or extrapulmonary
- *Mycobacterium tuberculosis*, any site (pulmonary[a] or extrapulmonary)
- *Mycobacterium*, other species or unidentified species, disseminated or extrapulmonary
- *Pneumocystis carinii* pneumonia
- Pneumonia, recurrent[a]
- Progressive multifocal leukoencephalopathy
- Salmonella septicemia, recurrent
- Toxoplasmosis of brain
- Wasting syndrome due to HIV

[a]Added in the 1993 expansion of the AIDS surveillance case definition.

(Adapted from the CDC, Atlanta.)

Finally the Question—How Long Can an HIV-Infected Person Live?

The average time someone survives from the moment of HIV infection until death continues to increase. At the beginning of the epidemic, the average time was about 10 years. Many people confuse the date of diagnosis with the date of actual

SIDEBAR 7.3

HIV: CASE PRESENTATIONS

Case 1—A 28-year-old man visited his family physician complaining of flulike illness. The physician obtained a drug-use and sexual history, through which it was discovered that in the past year the patient had engaged in unprotected sex with several partners. HIV testing was ordered. It showed a T4 count of 550 μL and a plasma HIV RNA level of 735,000 copies/ml.

Diagnosis: Acute HIV Infection—The illness may present dramatically, with features such as painful mouth ulceration, lymphadenopathy, or a rash indistinguishable from that of mononucleosis. It may even have neurologic manifestations, such as Bell's palsy. Typically, however, the illness is hard to distinguish from influenza and an array of nonspecific viral illnesses. It may even be essentially asymptomatic. Except perhaps in retrospect, the patient usually fails to recognize the symptoms as marking acquisition of HIV infection.

Case 2—A 22-year-old woman in the first trimester of her first pregnancy reported being in fair health. She underwent routine HIV screening as part of the antepartum evaluation. The T4 count was 275/μL, and the plasma HIV RNA level was 127,000 copies/mL.

Diagnosis: Asymptomatic Chronic HIV Infection—HIV infection was detected by routine screening. Here the infection has moved from its acute phase to a latent phase during which the virus is residing in sites such as the lymph nodes, its replication still counterbalanced by host immune responses. Today detection by routine screening occurs most often in pregnant women. Indeed, as many as two-thirds of HIV cases in women are diagnosed during pregnancy.

Case 3—A 34-year-old man presented at an emergency department seeking treatment for a chest rash typical of herpes zoster. Thinking that the patient was surprisingly young to have shingles, the physician identified a history of recreational cocaine use. The patient insisted that his drug use was purely intranasal, not intravenous. He said he had never had thrush or pneumonia. The HIV antibody test was positive.

Diagnosis: Symptomatic Chronic HIV Infection Complicated by Herpes Zoster—Although zoster can occur at any time of life, it most often occurs in patients of advanced age or in those with serious conditions such as malignancy or long-term steroid therapy. In chronic HIV infection, zoster is now recognized to have an increased chance of preceding, often by a long period, other opportunistic infections. In this instance, the patient had not had thrush or *Pneumocystis carinii pneumonia.* Measurement of the HIV viral RNA level was deferred because of the presence of inflammation, which may cause a misleading spike in plasma HIV RNA, regardless of the underlying condition causing the inflammation. *(Reproduced with permission. John Bartlett et al., Primary Care of HIV Infection.* Hospital Practice, *1998; 33 (12:53–56, 61–64, 67–69) © 1998. The McGraw-Hill Companies, Inc.)*

infection. The latter date is most often not known. Many years may separate the dates of infection and diagnosis. It currently is estimated at least 25% of persons infected with HIV today, utilizing drugs and treatments available, will survive on average for about 25 years. Average survival of persons infected 5–10 years ago is about 16 years from moment of infection until death. But averages are exactly that—averages. More precise estimates for individuals depend on current and past HIV RNA levels, current and past T4 or CD4+ cell counts, number of antiretroviral regimens used, adherence to therapy, response to therapy, and current health status. At least 5% of HIV-infected persons are estimated to be long-term nonprogressors. That is, in the absence of therapy, these individuals maintain a T4 or CD4+ cell count of about 450 cells per microliter of blood and typically have HIV RNA levels of less than 5000 copies/mL of blood. It is not clear what immunologic features distinguish these individuals from the other 95% of HIV-infected persons (see Box 7.2 and 7.3).

First Federal Clinical Guide In America—2003

In early 2003, Human Health and Services Secretary Tommy Thompson announced the release of "A Clinical Guide to Supportive and Palliative Care for

POINT OF INFORMATION 7.2

AIDS: A MANAGEABLE AND CHRONIC DISEASE?

It would appear, from reviewing the research and applied drug therapy literature that the holy grail of HIV therapy is to contain, control, and force HIV infection into a manageable and chronic disease. Jose Catalan and colleagues (2001) state that HIV disease has always been a chronic disease. But while it is now better managed than ever, HIV infection/disease has never been a truly manageable condition. Prior to the beginning of antiretroviral therapy (HAART) in 1996, the HIV-infected died of a variety of opportunistic infections (OI) and at varying times from their initial infection. Treatment and management of the virus and for many of the OIs was poor at best. Beginning in 1996, many patients, about half, placed on HAART therapy recovered from conditions in which they were close to death. Others, using the drugs, delayed the progression of their HIV disease. But again, who would respond to the therapy, for how long, and why they responded while others did not, still leaves this disease as remarkably unmanageable for at least half the infected population. And as HIV resistance to the drugs continues to increase, those who once responded to therapy are once again becoming desperate for improved management of their disease. No, HIV disease has never really been manageable, rather for most, drug therapy offers a temporary reprieve for a failing immune system.

Jose Catalan and colleagues write that "chronic" is an umbrella term used in opposition to "acute," a time-limited, self-contained disease. The term hides a wide range of diseases with few things in common: They do not go away easily, they last for a long time, and on the whole, they are never successfully eradicated. It is useful to realize that HIV infection has always been a chronic illness, in the sense that it develops over a number of years. In the early days of the epidemic, it was calculated that only about 50% of HIV-infected individuals would develop HIV-related opportunistic conditions within 10 years of infection, although once these problems appeared, decline tended to be rapid—survival after an AIDS diagnosis sometimes lasting less than two years. HIV infection is still a chronic disease, but for those with access to HAART, the time span has extended beyond these original expectations. So, is HIV infection a chronic illness? To the extent that it lasts years and, to date, cannot be cured, it is a chronic illness and has always been one. Longer survival, to date, has meant more time for more adverse consequences to develop because of HAART. And because HAART is allowing some people to live longer, they express some bizarre forms of disease never encountered before in HIV-infected patients. At least 70% of those living with HIV/ AIDS experience mental health problems such as depression, anxiety, insomnia, lethargy, impaired concentration, and mood swings. The majority of physicians attribute their patients' psychiatric problems to the side effects of antiretroviral drug therapy. Management for one AIDS patient, after 10 years, added up to having 700 vials of blood drawn, taking over 51,000 pills, and spending 27 hours with his HIV/AIDS specialist. Cost: about $188,000. His psychiatrist's bill was over $57,000.

HIV/AIDS," which provides practical, experience-based advice and authoritative guidelines for clinicians in providing palliative and supportive services to their patients living with HIV/AIDS. The clinical guide is organized into five parts, focusing on specific aspects of palliative care. Copies of the guide are available online at http://hab.hrsa.gov or may be ordered from the HRSA Information Center at 1-888-ASK-HRSA (1-888-275-4772).

Symptoms and Impairment

In the symptomatic stages of HIV disease, an individual's ability to carry on the activities of daily living is impaired. The degree of impairment varies considerably from day to day and week to week. Many individuals are debilitated to the point that it becomes difficult to hold steady employment, shop for food, or do household chores. It is also quite common for people with AIDS to experience phases of intense life-threatening illness, followed by phases of seemingly normal functioning, all in a matter of weeks. For a good review on the mechanisms of HIV disease, read *The Immunopathogenesis of HIV Infection* by Giuseppe Pantaleo et al. (1993), Bucy (1999), and Yu (2000).

BOX 7.1

EVOLUTION OF HIV DURING HIV DISEASE PROGRESSION

HIV is a unique retrovirus. For example, mitosis, a form of cell division, is a requirement for the nuclear entry of most retroviral nucleic acids. In contrast, mitosis does not appear to be required for nuclear entry of HIV nucleic acids, particularly in terminally differentiated cells (for example, macrophages and dendritic cells) (Freed et al., 1994). Second, HIV lacks any mechanism to correct errors that occur as its genetic material is being duplicated, so a few days or weeks after initial infection, there may be a large population of closely related, but not identical, viruses replicating in an infected individual. In the **quasi**-steady-state condition, there are successive generations of viral progeny, with each generation following the next by about 2.6 days. Approximately 140 generations of virus are produced over the course of a year and 1400 generations over the course of 10 years, allowing production of an extraordinary number of genetic variants. Some variants can provide preexisting drug-resistant forms or enable rapid development of resistance under drug pressure, and some enable the viral population to escape immune activity.

VIRAL POOL

The viral pool in an HIV-infected person is estimated to be about 10 billion viruses and each is genetically different from all other HIV in the pool. It is known from experimental data that about 1 in 1000 particles is infectious, so the infectious viral pool may be on the order of 10 million viruses. With a genome of approximately 10^4 nucleotides, and from 1 to 10 billion HIV variants made daily, mutations most likely occur at every nucleotide position on a daily basis (Ho, 1996). This creates an enormous potential for viral evolution. On average, the HIV that is transmitted to another individual will be over 1000 generations removed from the initial HIV infection. This extent of replication per transmitted infection (transmission cycle) is probably without equal among viral and perhaps bacterial infections (Coffin, 1995). Regardless of the underlying mechanism of immunodeficiency, it is becoming apparent that the force that is driving the disease is the constant repeated cycles of HIV replication.

EVOLVING RESISTANCE TO ANTIRETROVIRAL DRUGS

Simon Wain-Hobson (1995) suggests that because of 24-hour-a-day HIV replication, as shown by Wei and coworkers (1995), an HIV-infected asymptomatic person can harbor at least 1 billion distinct HIV variants and an AIDS patient 10 billion HIV variants. With so many genetic variants, some are going to be resistant to any given drug. (This would include drugs not yet used in therapy). During the investigations of Ho and coworkers (1995) and Wei and coworkers (1995), variants of HIV resistant to the drugs ritonavir (a protease inhibitor), and nevirapine (a non-nucleoside agent that inhibits reverse transcriptase function), occurred within days or weeks. Alan Perelson and colleagues (1996) reported data collected from five HIV-infected people after administering ritonavir through 7 days. Each person responded with a similar pattern of decline in plasma HIV RNA. Their results: Infected T4 cells had an average life span of 2.2 days. Plasma HIV RNA had an average life span of 0.3 days. The results also suggest that the minimum duration of the HIV life cycle in human T4 cells is 1.2 days on average and that the average HIV generation time—defined as the time from release of a virus until it infects another cell and causes the release of a new generation of HIV—is 2.6 days. The lifetime of HIV in resting or latent T4 cells may range from 6 months to perhaps an infinite amount of time. Such cells can produce HIV when activated.

Number	1 to 10 billion HIV produced each day.	Many million of T4 cells infected each day.
After two days	Half of HIV are destroyed and are replaced by about an equal amount of new HIV.	About half the infected T4 cells die and are replaced by new T4 cells.
After 14 days	After mutations, most new HIV are resistant to one or more drugs.	T4 cell count decreases and new T4 cells are infected with the drug-resistant viruses.

BOX 7.1 *(continued)*

HIV CLEARANCE

Douglas Richman (1995) states that the HIV clearance rate can be calculated based on the rate of reduction in viral RNA load. In cases in which viral RNA load attains a constant level, it can be assumed that there is a dynamic equilibrium resulting in a steady-state level, with production rates matching clearance rates. When viral resistance emerges, as in cases of drug resistance, the production of resistant virus may double every two days. The proportion of virus with resistance mutations can also be calculated. According to Richman, in the case of the non-nucleoside reverse transcriptase inhibitors, this has been found to be approximately 1 to 2 per 1000 RNA copies circulating in the plasma; that is, a patient with 60,000 copies/mL plasma has approximately 100 copies/mL of resistant mutants prior to the beginning of treatment. A similar scenario probably holds for all drug-resistant mutations. It is reasonable to expect the HIV-resistant mutants will emerge to any antiretroviral compound.

TREATMENT FAILURE

Virologic Failure

Failure to reduce the viral load to an undetectable level *or*

Failure to reduce the viral load by at least 2 to 2.5 $\log_{10}$ *or*

A persistent increase in viral load following a period of adequate suppression

Immunologic Failure

Failure to restore the T4 or CD4+ cell count to more than 200 cells/mm^3 *or*

Failure to significantly increase T4 cell count *or*

A persistent decline in T4 cell count after a period of immune reconstitution

Clinical Failure

Development of new opportunistic infections (OIs) *or*

Failure to resolve pretreatment OI, wasting, or dementia

(Adopted from Soloway et al., 2000)

With new techniques for quantitating plasma HIV RNA, infected individuals can be evaluated for response to antiretroviral therapy.

MUTANT HIV OVERWHELM THE IMMUNE SYSTEM

As HIV multiplies and mutant forms are produced, the immune system responds to these new forms. But ultimately the sheer number of *different* viruses to which the immune system must respond becomes overwhelming. It's a bit like the juggler who tries to keep too many balls in the air: The result is disastrous. Once the immune system is overwhelmed, the latest escape mutant—which may not necessarily be the most pathogenic one to come along—will predominate and immune deficiency will progress (Japour, 1995).

HIV Can Be Transmitted During All Four Stages

A person who is HIV-infected, even while feeling healthy, may unknowingly infect others. *The greatest risk of HIV transmission occurs within the acute period, the first several months after infection, and again when the T4 cell count drops below 200.*

HIV DISEASE WITHOUT SYMPTOMS, WITH SYMPTOMS, AND AIDS

Michael is a 31-year-old Hispanic male who complains of fatigue, headache, muscle aches, sore throat, and nausea. Physical assessment demonstrates a skin rash on his trunk and swollen lymph glands. His temperature is 98°F (37°C); other vital signs are within normal limits. Laboratory findings including white blood cell count, platelet count, and blood chemistry are normal. Michael states that his symptoms began one week ago. Subsequent laboratory testing will confirm that Michael has acute retroviral syndrome that accompanies primary HIV infection (PHI). But the odds are high that in almost every emergency room or physician's office in the country, Michael will be misdiagnosed. He is likely to be told that he has a viral infection, probably the flu, and sent home with instructions for supportive care.

A person may have no symptoms (**asymptomatic**) but test HIV positive. This means that the virus is present in the body. Although he or she has not developed any of the illnesses associated with HIV disease, it is possible to transmit the virus.

In time, most people with HIV disease progress to AIDS. A person has AIDS when the defect in his or her immune system caused by HIV disease has progressed to such a degree that an unusual infection or tumor is present or when the T4 or CD4+ cell count has fallen below 200/μL of blood. In AIDS patients, a number of diseases are known to take advantage of the damaged immune system. These include opportunistic infections and tumors such as Kaposi's sarcoma or lymphoma, a malignancy of the lymph glands. The presence of one of the opportunistic diseases or a T4 cell count of less than 200, along with a positive HIV test establishes the medical diagnosis of AIDS. Thus the disease we call AIDS is actually the end stage of HIV disease. It is important to remember that AIDS itself is not transmitted—the virus is. AIDS is the most severe clinical form of HIV disease.

The term "full-blown AIDS" is often used on TV and in the press but not in this textbook. Terms such as advanced HIV disease or advanced AIDS are not only more accurate but also more meaningful.

ASPECTS OF HIV INFECTION

HIV infection depends on a variety of events, for example, the amount and strain of HIV that enters the body (some strains of HIV are known to be more pathogenic than others), perhaps where or how it enters the body, the number of exposures, the time interval between exposures, the immunological status of the exposed person, and the presence of other active infections. These are referred to as cofactors that contribute to successful HIV infection. (For an explanation of cofactors, see Chapter 5.)

PRODUCTION OF HIV-SPECIFIC ANTIBODIES

During the 22 years (1983–2005) since the discovery of HIV, scientists have constructed a serological or antibody graph of HIV infection and HIV disease. The graph reveals how soon the body produces HIV-specific antibodies after infection and about when the virus begins its reproduction. Different parts of the graph (Figure 7-5) have been filled in by Paul Coulis and colleagues (1987), Dani Bolognesi (1989), and Susan Stramer and colleagues (1989). The history of the HIV antibody is not yet complete, but the order of appearance and disappearance of antibodies specific for the serologically important antigens over the course of HIV disease has been described.

The time period from seroconversion to the presentation of clinical symptoms is quite variable and may last for 11 or more years in adults and adolescents, but occurs earlier in children and older persons with HIV disease. The time period for moving from clinical symptoms of HIV disease to AIDS is also quite variable (Figure 7-3). Much depends on the individual's genetic susceptibility and his or her response to medical intervention. Note that Figure 7-5 shows that HIV plasma viremia (the presence of virus in blood plasma) and antigenemia (an-ti-je-ne-mi-ah—the persistence of antigen in the blood) can be detected as early as two weeks after infection. This demonstrates that viremia and antigenemia occur prior to seroconversion. Using HIV proteins produced by recombinant DNA methods (making synthetic copies of the viral proteins), antibodies specific for gp41 (a subunit of glycoprotein 160) are detectable prior to those specific for p24 (a core protein) and persist throughout the course of infection. Levels of antibody specific for p24 rise to detectable levels between six and eight weeks after HIV infection but may disappear abruptly. The drop in p24 antibody has been shown to occur at the same time as a rise in p24 antigen in the serum. This strange phenomenon is thought to be due to the loss of available p24 antibody in immune complexes—too little p24 antibody is being made to handle the new virus being produced. It is believed that this imbalance is one of the important factors that moves the patient toward AIDS. A sudden decrease in anti-p24 is considered by many scientists to be a prognostic indicator that people with HIV disease are moving toward AIDS.

Some AIDS researchers and health care professionals believe that 90% to 95% of those persons infected with HIV will eventually develop AIDS. Without antiretroviral drugs, approximately 50% of people with HIV disease will progress to AIDS within 8 years after infection. At 10 years 70% will have developed AIDS. After that, an additional 25% to 45% of the remainder will develop AIDS.

Classification of HIV/AIDS Progression

There are several classifications that spell out the progression of signs and symptoms from HIV infection to the diagnosis of AIDS. The classifications were developed to provide a framework for the medical management of patients from the time of infection through the expression of AIDS. All classification systems are fundamentally the same—they group patients according to their stage of infection, based on signs that indicate a failing immune system (Royce et al., 1991).

The most widely accepted classification because of its greater clinical applicability, comes from the CDC. The CDC classification uses four mutually exclusive groupings (Figure 7-6). The groupings are based on the presence or absence of signs and symptoms of disease, and clinical and/or laboratory findings and the chronology of their occurrence. Group 1, acute infection, means the person is **viremic** (that is, many virus particles are present in his or her blood or serum). There are *no measurable antibodies;* one is HIV positive but lacks HIV antibodies and signs and symptoms of HIV disease.

The majority of people in Group 1 remain asymptomatic. Some may experience flu- or mononucleosis-like symptoms that generally disappear in a few weeks. In relatively few cases the patient moves rapidly from mild symptoms into severe opportunistic infections and is diagnosed with AIDS.

In Group 2, antibodies are present but most patients remain free of HIV disease symptoms. Regardless of the lack of outward clinical symptoms, 90% of those who are asymptomatic experience some form of immunological deterioration within five years (Fauci, 1988).

In Group 3, asymptomatic people from Groups 1 and 2 become symptomatic and demonstrate lymphadenopathy in the neck, armpit, and groin areas. Although a number of other diseases may cause the lymph nodes to swell, most swelling declines as the other symptoms of illness fade. However, with HIV infection, the lymph nodes remain swollen for months, with no other signs of a related infectious disease. Consequently, lymphadenopathy is sometimes called **persistent generalized lymphadenopathy** (PGL). People with PGL may experience night sweats, weight loss, fever, on-and-off diarrhea, fatigue, and the onset of oral candidiasis or thrush (see Figure 6-3). Such signs and symptoms are prodromal (symptoms leading to) for AIDS. Studies have shown that people in Group 3 appear to become more infectious as the disease progresses.

People in Group 4 have been diagnosed with AIDS. They fit the 1987 CDC criteria for AIDS diagnosis. (The CDC AIDS diagnostic criteria are listed in

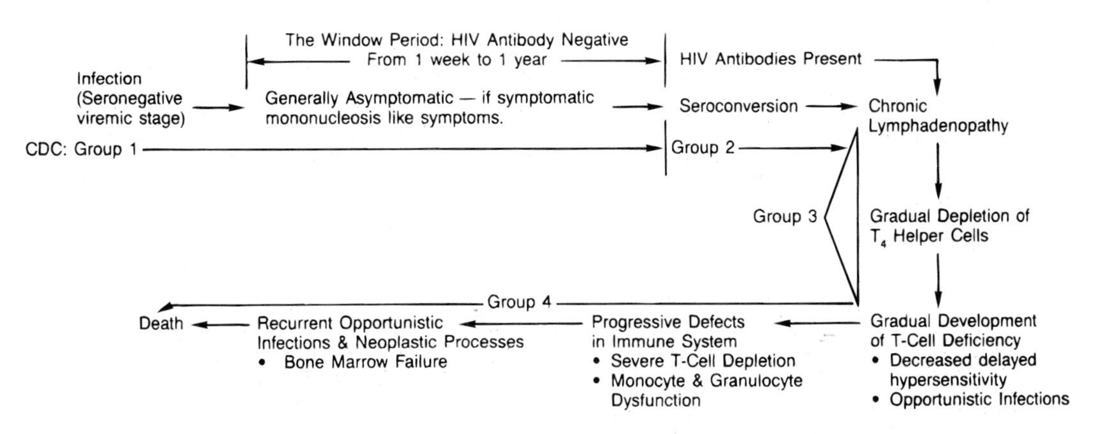

FIGURE 7-6 Clinical History of HIV Infection According to Centers for Disease Control and Prevention Groupings. Seroconversion means that HIV antibodies are measurably present in the person's serum. With continued depletion of T4 cells, various signs and symptoms appear announcing the progression of HIV disease into AIDS. Although HIV antibodies have been found as early as 1 week after exposure, most often seroconversion occurs between weeks 6 and 18 but may not occur for up to 1 year or more.

BOX 7.2

DEVELOPMENT OF AIDS OVER TIME

HIV-INFECTED AND THEIR PROGRESSION TO AIDS WITHOUT ANTIRETROVIRAL THERAPY

A spectrum of clinical expression of disease can occur after HIV infection. Approximately 10% of HIV-infected subjects progress to AIDS within the first 2 to 3 years of HIV infection **(rapid progressors).** People with greater than 50,000 HIV RNA copies/mL at 6 months after infection are the most likely to be rapid progressors. About 60% of adults/adolescents will progress to AIDS within 12 to 13 years after HIV infection **(slow progressors).** Approximately 5% to 10% of HIV-infected subjects are clinically asymptomatic after 20 to 25 years. They have stable peripheral blood T4 cell levels of greater than 500 **(long term non-progressors or LTNPs)** (Haynes et al., 1996; Schanning et al., 1998).

It has been estimated that about 3% or 1 in 25000 infected, termed nonprogressors, will *never* progress to AIDS. Nonprogressors all have one thing in common, a functional squad of **memory T4** or **CD4+ cells** that respond to HIV infection. Such people have an immune system that remembers HIV and controls it over time.

LTNPs are a very important group to study as they control infection without intervention and thus provide valuable information on disease progression and how it is influenced by immunological, virological, and genetic factors.

An important chemical signal (or cytokine) used by cells is interlukin 10 (IL-10). When produced in large amounts by cells, IL-10 can weaken the immune response against viruses, bacteria, and fungi. Some researchers think HIV may trick the immune system into producing large amounts of IL-10, weakening the body's ability to fight the virus. Researchers at Mt. Sinai Hospital in Toronto reported at the Eighth Conference on Retroviruses and Opportunistic Infections that T lymphocyte cells taken from nonprogressors and healthy HIV-negative people produced relatively low levels of IL-10. But such cells taken from people with AIDS produced between two to five times more IL-10. The researchers found that those who took highly active antiretroviral therapy (HAART) were able to significantly reduce their levels of IL-10.

One thing is clear from the studies of nonprogressors thus far: Not all are the same. Some show distinct features not shared by others, yet somehow all experience nonprogression. The reasons behind this phenomenon may be virologic, immunologic, or both. However, no common genetic correlate has yet been identified to explain long-term nonprogression.

AIDS: LONG-TERM SURVIVORS WITHOUT THERAPY

People with AIDS having low T4 cell counts who have survived for 5 years or more without therapy are usually described as long-term survivors. Some 6% of persons diagnosed with clinical AIDS are long-term survivors (Laurence, 1996).

AIDS: LONG-TERM SURVIVORS WITH DRUG THERAPY

Current and contained use of AIDS drug cocktails containing at least one protease inhibitor has increased the number of long-term survivors. Survival after the onset of AIDS, without HAART, has been increasing in industrialized countries from an average of less than 1 year to over 5 years at present. With therapy, on average, survival time has been increased by 5 to 15 years, depending on the case and therapy. Survival time with AIDS in developing countries remains short and is estimated to be, on average, less than 2 years. Longer survival appears to be directly related to routine treatment with antiretroviral drugs, the use of drugs for opportunistic infections, and a better overall quality of health care.

The majority of AIDS cases occurs before age 35, and over 90% of all AIDS deaths occur in people under the age of 50 worldwide.

Philip Rosenberg and colleagues (1994) reported that the length of incubation and progression from HIV infection to AIDS varied according to the age at the time of infection. Younger ages were associated with a slower progression to AIDS. The estimated median treatment-free clinical incubation period was 12 years for those infected at age 20, 9.9 years for infection at age 30, and 8.1 years for infection at age 40.

LONG-TERM AIDS SURVIVORS

Parade Magazine, January 31, 1993, carried a review of 16 long-term survivors who date back to 1982. On April 16, 1995, the same magazine reported that 12 of the 16 had since died of AIDS. On April 6, 1997, the magazine continued its followup on the four remaining survivors. As of July 2004, there are three remaining survivors, seen in Figure 7-7. They are Michael Leonard Marshal, 55, who is studying psychology at Santa Monica College in California; George Melton, 48, who now works as a colorist in a hair salon in San Francisco; and Niro Asistent Markoff, 56, who now works in a school of healing in East Hampton, New York.

Marshal began taking an AIDS cocktail of anti-HIV drugs in 1996. Melton began taking the drugs in 1997. Both men have undetectable levels of HIV in their blood and "feel great." Niro Markoff said that it has been many years since she has had any trace of HIV in her blood—she has never taken HIV/AIDS drugs.

FIGURE 7-7 Long-Term AIDS Survivors from the *Parade Magazine* series. *(© Blake Little)*

Tables 7-3 and 1-1.) Hairy leukoplakia (Figure 7-8, Table 7-3, Category C-2) is virtually diagnostic of AIDS in Group 4 patients. Statistics show that about 30% of all the newly HIV-infected will progress to Group 4 (AIDS) every 5 years, so that about 90% will have been diagnosed with AIDS within 15 years. Not all the opportunistic infections (OIs) or cancers will appear in any one AIDS patient. But some OIs, like *Pneumocystis carinii* pneumonia, occur in some 80% of AIDS patients prior to their deaths.

PROGNOSTIC BIOLOGICAL MARKERS RELATED TO AIDS PROGRESSION

The ideal marker would be able to predict HIV disease progression, be responsive to antiretroviral therapy, and explain the variance in clinical outcome due to therapy. It is, however, unlikely that any one marker will be able to fulfill all these criteria in HIV infection. Therefore, individual markers used to track HIV infection to AIDS are presented.

p24 Antigen Levels

p24 is a specific protein located in the core or inner layer of HIV. Because the immune system produces antibody against foreign protein, antibody is made against p24. A positive test for p24 antigen in the blood means that HIV production is so rapid that it overcomes the available antibody (Figure 7-5). This raised p24 antigen level condition occurs at least twice: once shortly after infection and again during the AIDS period when the immune system is rapidly deteriorating and unable to produce sufficient antibody to deal with newly produced HIV.

Those who, during the early stage of HIV infection, test p24 antigen positive are likely to progress to AIDS earlier than those who test p24 negative. Thus, a positive p24 test is an early and serious warning sign for HIV-infected people (Escaich et al., 1991; Phillips et al., 1991a).

p24 Antibody Levels

High levels of p24 antibody indicate that the immune system is functioning well and clearing the body of free HIVs. High antibody levels appear to slow the progression toward AIDS. Typically, p24 antibody levels are high during a person's asymptomatic or latent stage (Figure 7-4). However, antibody levels begin to decrease and p24 antigen levels rise, indicating a loss of immune function.

Other Markers

Beta-2 microglobulin (B-2M) is a low molecular weight protein that is present on the surface

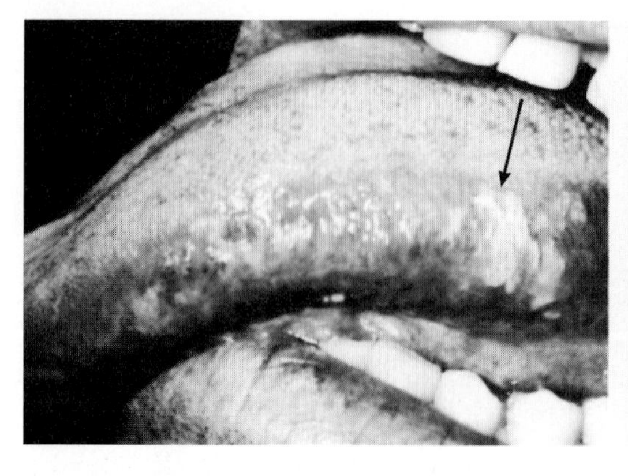
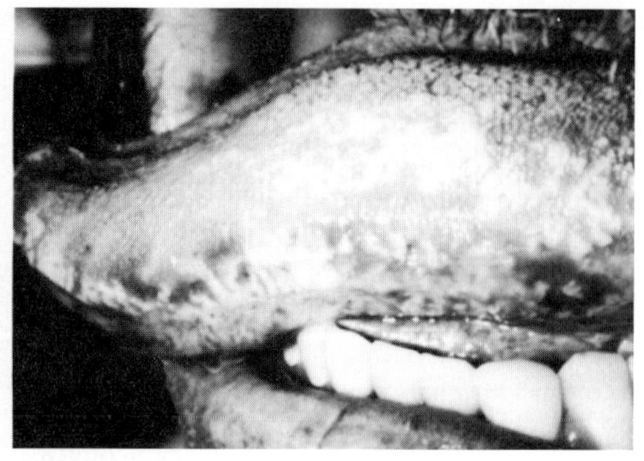

FIGURE 7-8 Oral Hairy Leukoplakia of the Tongue. An early manifestation, it is virtually diagnostic of AIDS. The white shaggy-appearing or corrugated "hairy" patches typically occur on the lateral borders of the tongue and are caused by the Epstein-Barr virus (EBV).These white plaques cannot be removed. **A.** A milder form of the disease. **B.** A more severe manifestation of the disease. Note that the white plaques may cover the entire tongue. *(Schiodt, Greenspan, and Greenspan 1989.* American Review of Respiratory Disease, 1989, 10:91–109. *Official Journal of the American Thoracic Society. © American Lung Association)*

of almost all nucleated cells. As cells die, this compound is released into body fluids. Thus there is always some B-2M in the blood because of normal cell degeneration and replacement. However, in a chronic illness with increased cell destruction, as in HIV infection, B-2M increases beyond normal levels.

To date, a B-2M level of 5 mg/L or higher is the best available indicator of progression to AIDS within three years. Levels below 2.6 are considered normal. B-2M protein increases dramatically shortly after infection occurs, then declines, and finally rises again with AIDS. B-2M can be used with T4 counts to foretell which HIV-positive individuals face the greatest immediate risk of progressing to AIDS.

T4 and T8 Lymphocyte Levels

The most extensive use of data for AIDS progression risk identification involves the number of T4 or CD4+ and T8 or CD8+ cells circulating in the blood (Anderson et al., 1991; Burcham et al., 1991; Phillips et al., 1991b).

T4 cells and the ratio of T4 cells to T8 cells found in the blood have, since 1997, been widely used as prognostic indicators for AIDS progression. T4 cell counts, however, are not ironclad predictors of HIV disease progression. In some cases persons with HIV disease and very low (less than 50 or 100) T4 counts remain healthy; conversely, some HIV-diseased persons have relatively high counts (over 400) and are quite ill. *T4 counts are notoriously fickle*—their counts can vary widely between labs or because of a person's age, the time of day a measurement is taken, and even whether the person smokes (Sax et al., 1995).

Lymph Nodes, the T Cell Zones, T4 or CD4+ Cell Production, and HIV Infection

The T cell zone in the lymph nodes is involved with 98% of the body's T4 or CD4+ cells. It is also where 99% of HIV production occurs. As HIV replication continues it causes inflammation, scarring, and eventual destruction of the T cell zone. Newly produced T4 cells have little room in the scarred T cell zone to divide. If they divide there is a high probability they will become infected. With this new knowledge reported in 2003, researchers now believe they understand why some people respond well to antiretroviral drug therapy and have their T4 cell counts increase while others on drug therapy do not experience T4 cell increase. They believe the increase of T4 cells or lack of it is associated with the amount of T cell zone destruction.

Performing T4 or CD4+ Cell Counts in HIV-Infected People Continues To Be Essential

A major player in the body's immune system is the T4 or CD4+ cell. By binding with CD4+ cells,

BOX 7.3

ARE THERE LONG-TERM ADULT NONPROGRESSORS OF HIV DISEASE—YES!—WHY?

Long-term nonprogressors (LTNP) of HIV disease in the absence of HIV drug therapy are defined as those persons who are still alive 10 or more years after they tested HIV positive, with seroconversion documented by history or stored serum samples, absence of symptoms, and normal and stable T4 cell counts (at least 500 T4 cells/microliter of blood). Studies through 2004 suggest that 12% to 15% of HIV-infected people remain asymptomatic with about normal T4 cell counts for at least 8 to 12 years (Conant, 1995; Levy, 1995; updated). At least two people are known to be asymptomatic now, 27 years after HIV infection! (Shernoff, 1997; updated). The question is: How does their immune system differ from those who do not live as long?

Some Male, Long-Term Nonprogressors

Perhaps the longest long-term nonprogressor is Robert Massie. He was infected in 1978. By the end of 2005, he will have lived for 27 years with his HIV infection. Massie is the current executive director of Ceres, a coalition of environmental, investor, and advocacy groups. In 1994, he went to the Massachusetts General Hospital, urging doctors to study him. He told them that he was infected with HIV, had never been treated, and that his immune system was fine. Massie first learned he was infected in 1984, when he had an HIV test as part of a research study. He was not surprised by the result, he explained, because he is a hemophiliac, and, like virtually every other American with this genetic disease who was alive in the early 1980s, he was repeatedly exposed to HIV through the blood clotting factors made from pooled blood of donors (see Chapter 8). Massie kept waiting to become ill. "I would read a report saying it could be as long as five years before you come down with symptoms of AIDS and die, then it could be as long as seven years, then as long as ten years, then twelve years. I always thought I was at the limit." When Massie's doctors looked back at his blood samples, which they had stored in hemophilia research projects, they found antibodies to HIV as early as 1979. They suspect that he was infected in November 1978 when he became gravely ill. Massie said, "One day I was sitting at my desk and I suddenly felt overwhelmingly tired, incapable of talking, I was so tired." He had a fever and symptoms that led doctors, at that time, to suggest pneumonia, tuberculosis, a cerebral hemorrhage, and a seizure disorder. So, in 1994, when Massie arrived at Massachusetts General Hospital, he had been infected with HIV for 16 years. Now age 50, he has been infected for more than half his life. Since then, Massie has been meeting with Bruce Walker, an infectious disease specialist in HIV infection. The first question for Walker was what was Massie's T4 or CD4+ cell count? He found Massie had near-normal levels of T4 cells. Walker said, "This was the first evidence that HIV followed the rules that other viruses follow." It meant that his body's immune system could control HIV. Massie, in fact, had the AIDS virus under such exquisite control that Walker and Eric Rosenberg, who joined him in the research, could not find HIV in his bloodstream. There were, however, antibodies to HIV in his blood. This left no doubt that he was HIV-infected. For Walker and Rosenberg, Massie is a very unusual case.

The second-longest documented case of a long-term nonprogressor to date is that of a gay male who was infected 25 years ago. His T4 count remains normal.

In 1983, at age 71, a man became HIV positive. He received a contaminated blood transfusion while undergoing colon surgery. Unlike most long-term HIV survivors, he has suffered no symptoms and no loss of immune function. He celebrated his 81st birthday. He is one of five patients who came to the attention of an AIDS researcher as he was preparing a routine update on transfusion-related HIV infections. He has since died of natural causes. All five people were infected by the same donor. And 10 to 15 years later, none of the remaining four has suffered any effects.

The blood donor was a gay male who had contracted the virus during the late 1970s or early 1980s and gave blood at least 26 times before learning he was infected. After locating the donor it was found that the man was just as healthy as the people who got his blood.

A 39-year-old San Francisco artist has beaten the odds by living with the virus that causes AIDS for 20 years. He has only routine medical complaints: the stuffiness of an occasional head cold or the aches and pains of a flu. He has never taken an anti-HIV drug. His own immune system seems to have held the virus at bay.

Susan Buchbinder and colleagues (1992, 1994) reviewed 588 HIV-infected gay men. Thirty-one percent were still AIDS-free 15 years after infection. They attempted to determine why these men lived while others died of HIV/AIDS. Some long-term survivors have low T4 cell counts, some have never taken antiviral therapy, and some have high T4 cell counts. If it can be determined why or how their bodies have delayed the progression of HIV disease, then perhaps new approaches to treating all HIV-infected persons will follow.

Buchbinder and colleagues have found that three aspects of the healthy survivors' immune system appear to delay HIV disease progression: Survivors have strong cytotoxic lymphocyte activity, or T8 cell activity, and have higher levels of antibodies against certain HIV proteins.

Stephen Migueles and colleagues (2002) examined a group of 15 nonprogressors who have controlled HIV for up to 20 years without antiretroviral therapy. They found no significant difference in the number of T8 or CD8+ cells between nonprogressors and the others. Instead, they found that the nonprogressors' CD8+ cells were better able to divide and proliferate when called into action and they also produced higher levels of a molecule called perforin, which helps kill off cells infected with HIV. This study represents the first time scientists have observed a difference in the HIV-specific CD8+ T cell response of nonprogressors and suggests a mechanism whereby the CD8+ T cells of nonprogressors control HIV while those of most HIV-infected individuals do not.

It is also possible that long-term nonprogressors carry a less pathogenic virus or that these men have not been reexposed to HIV through unprotected sexual activities. Nicholas Deacon and colleagues (1995 updated) have sequenced HIV DNA from a blood donor and a group of six recipients who have not shown HIV disease symptoms despite being infected for 14 to 18 years. Deletions were found in the *nef* gene. Because the lack of disease progression appears to depend on the virus instead of the host immune system, these results suggest a possible use of such HIV strains in live vaccines. However, in July of 1998, it was reported that the HIV donor and two of the six blood recipients experienced a drop in their T4 cell counts. Physicians who treat HIV/AIDS persons do not notice any trends that would lead one to recognize the type of patient or factors that would lead to long-term survival. It was concluded that, at the moment, there is a lack of advice for ensuring longer life for persons who have become HIV-infected. No one can tell them how to live longer.

Some Female, Long-Term Nonprogressors

For years, some 18 women have intrigued AIDS investigators because they have remained HIV-free despite having frequent, unprotected sex with an infected partner. Researchers at the University of Medicine and Dentistry of New Jersey found that in most of the women, key immune cells worked in various ways to block HIV from multiplying in their bodies. During the study, only one became infected with HIV, and that took nine years of frequent, unprotected sex. The 17 uninfected women had unprotected sex with their HIV-positive partners for periods ranging from 1 year to 11 years dating as far back as the mid-1980s. Tests on their blood focused on two types of immune cells, cytotoxic or CD8+ and T4 or CD4+ cells, which kill invading organisms and rev up the rest of the immune system. The CD8+ cells and the CD4+ cells separately were mixed with HIV proteins in laboratory dishes mimicking how they would interact in a person's body and researchers watched for reactions. In many samples, the CD4+ cells rapidly reproduced, as if stimulating the rest of the immune system and the CD8+ cells. Also, they produced two different substances that stopped the virus from reproducing. The research team expected to find one immune response protecting the women from HIV infection, but instead found different types of immune responses spread among the 17 women (Skurnick et al., 2002).

In the Pumwani district of Nairobi, a group of 100 women have become well known to HIV researchers around the world by offering evidence that the immune system can, in rare cases, fight off HIV. The evidence derives from a group of women sex workers (prostitutes) established in 1984 by Elizabeth Ngugi and colleagues from the University of Nairobi and the University of Manitoba. Despite an estimated 60 or more unprotected exposures to HIV every year, one of the highest documented exposure rates in the world, 100 of the 2000 women enrolled in the study have tested negative for HIV infection for at least three years, and for some women up to 15 years. Studies of these highly exposed persistently seronegative (HEPS), also referred to as exposed seronegative (ESN), women convinced many skeptics that immunological resistance to HIV and by extension that an HIV vaccine is possible.

More recently, 98 prostitutes of Maniper, India who routinely have had unprotected sex with their clients for years have remained HIV negative, and 54 of these women during the interim have had HIV negative babies. Again, why they remain HIV negative despite repeated exposure to HIV is unknown.

Heterosexual Men Exposed To HIV Remain HIV Negative

In a study by Mario Clevici and colleagues (2003), 14 HIV negative men remained HIV negative after repeated exposure to HIV from HIV-infected female partners after four years of unprotected sex. On average, the 14 couples said they had sex 14 times a year. Why they remain HIV negative is unknown.

Current evidence suggests that between 5% and 10% of HIV-infected people will live up to 20 or more years. It should be mentioned that if the average time from infection to AIDS is 11 years, statistically speaking, survivors at 20+ years are expected to occur. Time will tell. For other accounts of long-term survivors, see the articles by Cayo (1995), Pantaleo (1995), Kirchhoff (1995), Baltimore (1995), Gegeny (2000), and Migueles (2000).

IMMUNE CELL RECEPTORS, MOLECULES THAT BIND TO THESE RECEPTORS (LIGANDS), AND T4 OR CD4+ AND CTL CELL INFLUENCE ON SURVIVAL

Chemokine Receptors: An Association of Coreceptors to Progression to AIDS

As discussed in Chapter 4, the **chemokine receptors** [the four most familiar are CCKR-5 (R-5), CXCKR-4 (X-4), CCKR-3 (R-3), and CCKR-2 (R-2)] reside on immune system cell membranes. The receptors act as host sites for a variety of chemokines that need entry into such cells. But scientists recently learned that these chemokine receptors also act as **coreceptors** to the CD4 receptor, the receptor to which HIV first attaches. That is, HIV anchors to the cell's CD4 receptor, but also needs to bind with a coreceptor, one of the chemokine receptors, in order to complete its entry into the cell. **Genetic mutations** that inhibit given chemokine receptor formation and/or reduce the numbers of such receptors on the immune cell membrane offer such cells a complete or partial resistance to HIV infection. A double or homozygous mutation at R-5 makes a person completely resistant to HIV infection, but the R-2 mutation only slows the progression of HIV disease to AIDS. Such mutations contribute to some people's long-term survival. Both CCKR-5 and CCKR-2 receptor mutations offer this protection (Balter, 1998; Collman, 1997; Cocchi et al., 1995; Feng et al., 1996). Exciting research published in 1998 reports that the genetic mutation that interferes with the X-4 receptor actually interferes with the *production* of the specific chemokine that attaches to the R-4 receptor. It appears that this mutation causes the overproduction of a chemokine called **S**tromal **D**erived **F**actor-1 (SDF-1). With an excess in the environment, SDF-1 fills available R-4 receptor sites, blocking HIV-attachment. So, those persons carrying an SDF-1 mutation, although HIV-infected, progress through HIV disease at a much slower pace, delaying the onset of AIDS as much as 7 to 10 years later than average (Winkler et al., 1998; Balter, 1998). Thus it appears that the progression rate to AIDS is linked to the types of coreceptors people carry on their T4 cells and macrophages (Reynes et al., 2001).

CHILDREN

Micheline Misrashi and colleagues (1998) reported that children carrying one copy (heterozygous) of the mutant R-5 gene demonstrate a substantial reduction in progression of HIV disease to AIDS.

T4 OR CD4+ AND CYTOTOXIC LYMPHOCYTE (CTL) CELLS

One notable immunologic problem that occurs after HIV infection is that most people show no evidence of altered T4 cell activity. But an abnormality in immune function is present, long before a significant decline in T4 cell number is seen. In order to be able to kill off HIV-infected cells, cytotoxic T lymphocytes (CTLs) require the assistance of T4 cells—the very cells that HIV infects. These T4 cells must also be able to recognize HIV antigens in order to direct the cytokine signals that will activate CTLs to kill. During the initial (acute) stage of HIV infection, even before seroconversion when antibodies are generated, the T4 cell population is devastated by HIV.

Without sufficient T4 cells capable of recognizing HIV, the killer CTLs cannot in turn learn to recognize HIV antigens and kill off infected cells. Recently, it has been shown that some long-term nonprogressors with low viral loads have high amounts of HIV-specific T4 cell activity, while some patients with rapidly progressive disease have no detectable HIV-specific T4 cell activity. This suggests again that T4 cell responses play an extremely important role in containing infection.

Studies in long-term nonprogressors show that they have a persistent, vigorous, virus-inhibiting CTL response, and that this response is broad and adaptable. Rapid progressors, on the other hand, appear to have a narrowly directed CTL response that is unable to adapt to changes in the virus (Hay, 1998).

HIV can kill them and stop associated production of antibodies, leaving the immune system weakened and vulnerable to opportunistic disease. Doctors gauge the health of the immune system by counting T4 or CD4+ cells. A healthy, HIV-negative adult has between 600 and 1200 CD4+ cells per cubic millimeter of blood. If the count falls below 350, the immune system has become weakened. Further loss leads to immune suppression and the onset of opportunistic diseases.

Thus, accurate and reliable measures of T4 or CD4+ lymphocytes are essential to the assessment of the immune system of HIV-infected persons. The progression to AIDS is largely attributable to the decrease in T4 lymphocytes. Consequently, the Public Health Service (PHS) has recommended that T4 lymphocyte levels be monitored every three to six months in all HIV-infected persons. The measurement of T4 cell levels has been used to establish decision points for initiating prophylaxis for a variety of opportunistic infections. Moreover, T4 lymphocyte levels are a **criterion** for categorizing HIV-related clinical conditions by CDC's classification system for HIV infection and surveillance case definition for AIDS among adults and adolescents (*MMWR*, 1997).

FIGURE 7-9 David Baltimore, 1975 Nobel Prize Molecular Biologist and President of the California Institutes of Technology. *(Photograph courtesy of AP/Wide World Photos)*

Levels of HIV RNA in the Blood: Viral Load

David Baltimore (Figure 7-9), the Nobel Prize-winning retrovirologist, and coworkers have found a useful clinical predictor of HIV disease progressors, *levels of HIV RNA in the blood. More RNA means more HIV,* and that makes patients get sicker sooner. HIV RNA is a more sensitive measure than other assays and may detect the virus earlier than it would be seen otherwise.

Since the reported work of Baltimore and others on the levels of HIV RNA in the blood, Dennis Henrard and coworkers (1995) have concluded that the stability of HIV RNA levels suggests that an equilibrium between HIV replication rate and efficacy of immunologic response is established shortly after infection and persists throughout the asymptomatic period of the disease (Figure 7-5). Thus, a defect in immunologic control of HIV infection may be as important as the viral replication rate for determining AIDS-free survival. Because individual steady-state levels of HIV RNA are established soon after infection, HIV RNA levels can, as Baltimore suggests, be useful markers for predicting clinical outcome. Viral

POINT OF INFORMATION 7.3

VIRAL LOAD AND T4 CELL COUNTS: THEIR USE IN CLINICAL PRACTICE

VIRAL LOAD

Viral load assays measure only viral RNA present in the plasma. They do not account for **intact virus** in the lymph system or other tissues. Individual measurements are a good surrogate for virus replication. Although the relationships among viral replication, T4 cell count, and disease progression are not entirely clear, viral load correlates fairly well with all of them. Viral load measurements can be effectively used to monitor therapy or the effectiveness of a given drug.

HIT EARLY, HIT HARD?

The level of plasma HIV RNA at six months after primary infection is called the **set point,** but treatment should be considered regardless of set point RNA values. The idea is to stop HIV replication early rather than later. (In February 2000 the previously NIH-recommended hit early, hit hard was changed to hit hard, hit later, see referenced material and Chapter 4 for reasoning.)

T4 OR CD4+ CELL COUNTS

T4 cell counts remain helpful in determining the stage of the disease. But, T4 counts are highly variable, even from hour to hour in some HIV-infected people. They are also significantly influenced by other current viral infections. Although T4 cell counts function well as reliable markers of disease progression and antiretroviral drug efficacy in large populations, they are less reliable for prescribing therapy for the individual. Viral load determinations present a better basis on which to present therapy. Until viral load determination costs fall and until physicians become more familiar with them and insurance companies pay for them, the T4 tests will remain the more practical determinant of disease stage and when to begin therapy. (Medicaid, Medicare, and some HMOs do cover HIV RNA determinations.)

It is suggested that HIV RNA levels be used as a means to prescribe medication. RNA levels should be determined after exposure, at 6 weeks, 12 weeks, and at 6 months.

SUMMARY

The level of circulating RNA in plasma is a direct reflection of ongoing HIV replication in the host. The T4 lymphocyte, a target of HIV replication, is an indirect marker of antiretroviral therapy. The T4 count itself is a measurement of the relative production and destruction of the T4 cell population. For example, if there is a production problem, either due to nutritional, toxicity, or other factors, the T4 count may not rise proportionately to the decrease in viral load. Thus, the change in viral load is more precisely measuring the effect of antiretroviral therapy and, therefore, is the best marker of how well drugs are working.

load measurements indicate the amount of current HIV activity. T4 cell counts indicate the degree of immunologic destruction. Thus, the best monitor of disease progression is the use of both the T4 cell count and viral load (Merigan et al., 1996; Katzenstein et al., 1996; Voelker, 1995; Goldschmidt et al., 1997).

No current viral or immunological markers adequately reflect drug toxicities caused by therapy.

HIV INFECTION OF THE CENTRAL NERVOUS SYSTEM

The nervous system has two parts. The brain and the spinal cord are the central nervous system (CNS). The nerves and muscles are the peripheral nervous system. Peripheral means around the outside. A wide variety of CNS abnormalities occur during the course of HIV infection. They result not only from the opportunistic infections and malignancies in the immunodeficient individual, but also from direct HIV infection of the CNS. It was believed that because some brain cells contain CD4-like receptors they were receptive to HIV infection. In addition, other brain cells contain a glycolipid that allows for HIV infection (Ranki et al., 1995). However, more recent research shows that after HIV-infected monocytes migrate to the brain, become tissue macrophage, and release HIV, the virus *does not* infect brain cells; rather HIV resides in brain spaces, in the cerebral spinal fluid. In 2004, Yan

Table 7-4 Neurological Manifestations Associated With Direct HIV Infection of the Nervous System

AIDS dementia complex
Asymptomatic infection
Acute encephalitis
Aseptic meningitis
Vacuolar myelopathy
Inflammatory demyelinating polyneuropathy
Radiculopathy
Mononeuropathies
Distal sensory neuropathy

Xu and colleagues reported that it is the molecular products of HIV and not the virus itself that cause nerve cell death which leads to neurodegeneration and associated cognitive and motor dysfunctions. HIV investigators believe that HIV may invade the brain within a few weeks to months after HIV infection. Although the precise mechanisms by which HIV produces nerve cell dysfunction are still undetermined, infection of the brain is an integral component of the biology and natural history of HIV infection; the resulting clinical manifestations are summarized in Table 7-4.

The CNS and Antiretroviral Therapy

HIV infection of the CNS has to be treated with antiviral drugs. Yet, the "blood brain barrier" keeps many drugs out of the central nervous system. The barrier is a tight network of blood vessels that protects the brain and spinal cord from most infectious agents or poisons in the blood stream. Several anti-HIV drugs that do get through the blood-brain barriers are: Zidovudine, Stavudine, Ziagen, and Viramune. A special concern is that people with CNS problems may need extra help remembering to take their medications.

Genetic analysis and clinical studies have revealed that HIV in the cerebrospinal fluid (CSF) of some people with AIDS-related dementia evolves *independently* of the HIV in their blood, leading to two genetically distinct forms of the virus. This finding poses a new challenge for treatment of these patients, suggesting that drugs effective against HIV in their blood may not do the job in the central nervous system, and vice-versa. Why HIV evolves independently in the CSF of the demented patients is not clear. It may reflect a separate viral subpopulation thriving in their central nervous system, or maybe those with dementia have progressed further in immunodeficiency that may allow HIV to become more virulent. Scientists at Gladstone Institute of Virology and Immunology also found preliminary evidence that in some AIDS dementia patients HIV in the CSF compartment is more resistant to antiretroviral drugs than is HIV in the blood. This can cause a rebound of virus levels in the CSF while the viral load in the blood is suppressed. Ongoing studies of resistant mutations should clarify this pattern (Haas et al., 2000).

AIDS Dementia Complex

In addition to cancers and opportunistic diseases, a progressive dementia (mental deterioration) due to HIV infection of the central nervous system develops in 55% to 65% of AIDS patients. Pathological changes in the CNS are observed in up to 80% of those autopsied (McGuire, 1993). But the mystery remains as to why some HIV-infected people develop dementia while others do not. Some people with high viral loads were not demented, while others with low viral loads had dementia. A study by Johnson and colleagues (1996) suggests that factors other than viral load lead to dementia. Investigators have found that this dementia is solely associated with HIV infection and progression to AIDS. HIV-caused dementia has a unique set of clinical and pathological features. Some authorities estimate that 90% of AIDS patients in the terminal stages of the disease have AIDS Dementia Complex (ADC) (Hanley et al., 1988). Information presented by the National Institute of Allergy and Infectious Diseases in June 1989 indicated that asymptomatic HIV-infected persons do not demonstrate mental impairment (Figure 7-10). Therefore, the onset of mental impairment must begin sometime after the HIV-infected person becomes symptomatic (1989, updated). Initially, investigators thought that OIs caused ADC, but it was later discovered that HIV is carried into the brain by HIV-infected macrophages.

Early Symptoms—Early symptoms of cognitive dysfunction include forgetfulness, recent memory loss, loss of concentration and slowness of thought, social withdrawal, slurring of speech, loss of balance, deterioration of handwriting, and impaired motor function. An early diagnosis of ADC in AIDS patients is difficult because the early signs of neurological disease are very similar to

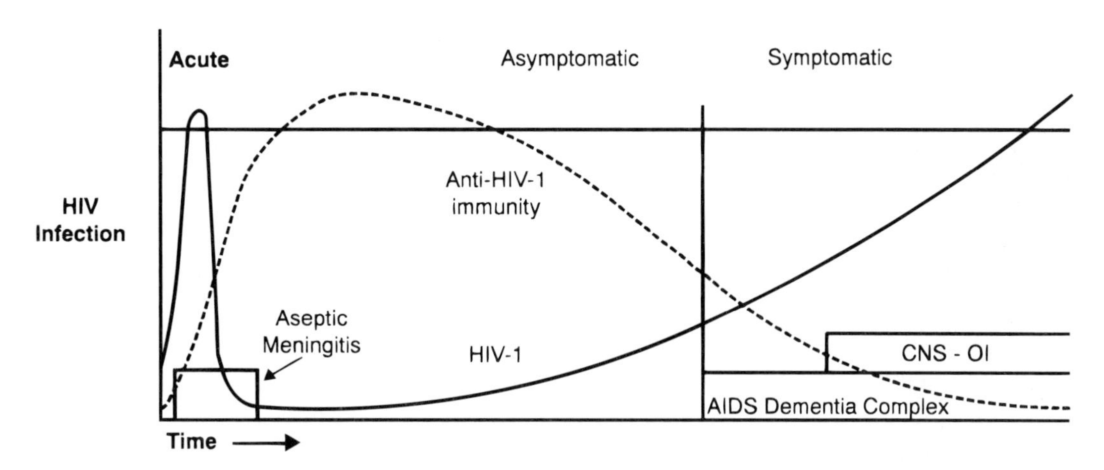

FIGURE 7-10 Central Nervous System Events After HIV Infection. Note that aseptic meningitis (an inflammation of the membrane of the brain and spinal cord in the absence of viral or bacterial infection), when it occurs, occurs early after HIV infection. Although usually apparent later, AIDS dementia complex may begin during the early-late phase. The late phase represents the period during which major AIDS-defining opportunistic infections occur. The headings acute, latent, early-late, and late refer to periods after HIV infection. It appears that opportunistic infection of the brain occurs after the onset of ADC. *(Adapted from Price et al., 1988)*

BOX 7.4

SCIENTISTS NEED A SMALL ANIMAL MODEL IN WHICH TO STUDY HIV DISEASE—WHY ISN'T THERE ONE?

WHY THE NEED FOR A SMALL ANIMAL MODEL

HIV/AIDS researchers typically describe HIV as a wily, stealthy, and clever killer. But researchers who have been struggling for over two decades to get HIV to infect small animals have another adjective for the virus: impotent. HIV causes disease only in humans and chimpanzees. If it could be coaxed to infect mice and rats—and, better yet, make them sick—the payoff could be enormous. Robert Gallo, head of the Institute of Human Virology in Baltimore, said, "Instead of five animals in an experiment, we'd have 500. Instead of waiting two years to get results, you'd wait two months. It would greatly catapult the field forward." But until recently, attempts to develop a rodent model for AIDS have been frustrating. HIV, it seems, is just too picky.

To date, scientists do not really know how HIV kills cells that carry the CD4 protein on their outer membranes—mostly those cells of the human immune system, the T lymphocytes. But, that is not all that is unknown about how this unique virus attacks, inserts itself into human DNA, replicates, and exits the cell. To learn more about the devastation HIV brings to humans after cellular infection, replication, and release into the blood stream, scientists must use chimpanzees and rhesus macaque monkeys. Yet HIV does not cause AIDS in these animals. HIV evolved in humans and to date, only causes AIDS in humans. But, there are no other animals that are closer to the human system in which to study the infection and disease process. So scientists have routinely infected these primates with an HIV/SIV (human/simian or monkey virus/hybrid virus called SHIV). This hybrid virus will cause AIDS-like symptoms in these primates. But, these primates are scarce and expensive. Chimpanzees cost up to $50,000 each. Rhesus macaque monkeys, more abundant than chimpanzees, but still short in supply, cost up to $5000 each. Breeding of either species takes years.

Scientists in a number of laboratories have, for years, been trying to develop specific strains of mice and rats to use as animal models to study HIV. To date, all attempts to develop rodent models have failed. HIV just won't adapt to the biology of species other than humans. Intense research shows that (a) HIV will not infect cells other than human or (b) if HIV enters a cell it will not replicate or (c) if it replicates it will not have the proper proteins produced to create new HIV or (d) if the products are made they cannot be assembled into a new virus. Many steps in HIV's life cycle require the presence of human—and only human—molecules. It is known that viruses are most often species specific. At the moment, however, there are no promising alternatives to using the scarce and expensive primates.

UPDATE 2004 Researchers led by virologist Joseph Sodroski of the Dana-Farber Cancer Institute in Boston reported that a protein from monkey called TRIM5-α (alpha) powerfully restricts HIV's ability to establish an infection. Although HIV can easily enter monkey cells, the virus must convert its RNA into DNA before it can weave itself into the host's chromosomes and copy itself. Experiments demonstrated that TRIM5-α targets HIV's capsid, a sheath of proteins that protects the viral genetic material. The capsid uncoats after the virus enters a cell releasing HIV RNA into the cells cytoplasm, a key step in the conversion of its RNA into DNA. By some unknown mechanism, TRIM5-α appears to disrupt the proper uncoating of the capsid. This protein has the potential to be a natural preventative to HIV infection. Stay tuned.

those symptoms used to identify emotional depression. However, neurological symptoms may be the first sign of illness in about 10% of adult AIDS patients.

Late Symptoms—These symptoms are characterized by loss of speech, great fatigue, muscle weakness, bladder and bowel incontinence, headache, seizures, coma, and finally death. About 95% of patients with ADC have HIV antibodies in their cerebral spinal fluid.

NEUROPATHIES IN HIV DISEASE/AIDS PATIENTS

Soon after it enters the body HIV colonizes in the brain and other nerve tissues. Within the compartment of the nervous system, the virus remains, often at high concentrations and with time can lead to a spectrum of problems spanning from localized disease of the brain to distal peripheral neuropathy.

Neuropathies are functional changes in the peripheral nervous system; therefore, any part of the body may be affected. Although neuropathies are not OIs, they may result from the presence of certain OIs. **Peripheral neuropathy** is caused by nerve damage and is usually characterized by a sensation of pins and needles, burning, stiffness, or numbness in the hands, feet, legs, and toes. It is a common, sometimes painful, condition in HIV-positive patients, affecting up to 30% of people with AIDS. At autopsy, two-thirds of AIDS patients have neuropathies (Newton, 1995). Neuropathy has been a continuous problem for patients throughout the HIV/ AIDS epidemic. It is most common in people with a history of multiple opportunistic infections and low T4 or CD4+ cell counts. There is a wide range of expression among patients with neuropathy, from a minor nuisance to a disabling weakness. The kinds of neuropathies occurring in people with HIV/AIDS are numerous and must be identified before appropriate treatment can be prescribed. The underlying cause of the most common type of peripheral neuropathy remains elusive. What was a common complaint early in HIV infection of severe neuropathy—usually, burning feet, causing patients to walk on their heels—has diminished. The decrease in such complaints may be attributable to the antiviral effects of the drug ZDV. On the other hand, new varieties of drug-induced nerve damage (neuropathies) have been recognized with the use of nucleoside, non-nucleoside reverse transcriptase inhibitors, and protease inhibitor antivirals.

PEDIATRIC CLINICAL SIGNS AND SYMPTOMS

Currently over 90% of pediatric AIDS cases are newborns and infants who received HIV from mothers who were injection-drug users or from the sexual partners of IDUs. For those who become

HIV-infected during gestation, clinical symptoms usually develop within six months after birth. Few children infected as fetuses live beyond two years and survival past three years used to be rare; but with the use of anti-HIV drugs and therapy for opportunistic diseases, some children born with HIV are still alive at 10 and 22 years of age.

The clinical course of rapid HIV disease progression in infants diagnosed with AIDS is marked by failure to thrive, persistent lymphadenopathy, chronic or recurrent oral candidiasis, persistent diarrhea, enlarged liver and spleen (hepatosplenomegaly), and chronic pneumonia (interstitial pneumonitis). Bacterial infections are common and can be life-threatening. Bacterial infection and **septicemia** (the presence of a variety of bacterial species in the bloodstream) is a leading cause of death. An excess of gamma globulin and depressed cell-mediated immunity and T cell function are frequently encountered.

Less than 25% of AIDS children express the kinds of OIs found in adult AIDS patients. Kaposi's sarcoma occurs in about 4% of them. Young AIDS children experience delayed development and poor motor function. Older AIDS children experience speech and perception problems.

SUMMARY

The clinical signs and symptoms of HIV infection and AIDS have been addressed in the previous chapters. However, there are two major classification systems used to diagnose patients as they progress from HIV infection to AIDS. The first is the Walter Reed System. It recognizes six stages of signs and symptoms that a person passes through to AIDS. The CDC uses four groupings to identify the stage of illness from infection to AIDS. Both systems revolve around the recognition of a failing immune system, persistent swollen lymph nodes, and opportunistic infections. Mysteries still to be resolved are exactly why and how HIV kills cells, and why some people stabilize after the initial symptoms of HIV infection while others move directly on to AIDS.

One disorder that was not immediately recognized in AIDS patients is AIDS Dementia Complex (ADC), a progressive mental deterioration due to HIV infection of the central nervous system. ADC develops in over 50% of adult AIDS patients prior to death. Research has shown that some of the symptoms of this dementia can be reversed with the use of the drug Zidovudine.

REVIEW QUESTIONS

(Answers to the Review Questions are on page 439.)

1. Name the two major AIDS classification systems used in the United States.
2. What percent of HIV-infected individuals will progress to AIDS in 5 years; in 15 years?
3. What is the neurological set of behavioral changes in AIDS patients called?
4. Name three body organs and their associated AIDS-related diseases.
5. True or False: Currently the single most important laboratory parameter that is followed to monitor the progress of HIV infection is the T4 cell count.
6. True or False: The average time from infection to seroconversion is two weeks. Explain.
7. True or False: Being infected with HIV and being diagnosed with AIDS are the same thing. Explain.
8. True or False: The average length of time from infection with HIV to an AIDS diagnosis is approximately two years.
9. Write a brief essay on: (a) In general, about how long HIV can reside in the body before one shows signs of HIV infection. (b) In general, about how long it takes the body to generate antibodies to HIV after infection.
10. The general signs and symptoms associated with HIV/AIDS include:

    A. recurrent fever

    B. weight loss for no apparent reason

    C. white spots in the mouth

    D. night sweats

    E. all of the above

11. HIV/AIDS Word Search

C	A	L	C	A	D	V	A	N	C	E	D	H	I	I
I	P	H	H	S	E	P	T	I	C	E	M	I	A	O
T	J	C	R	P	T	D	L	N	X	T	D	V	M	N
A	E	I	O	S	A	I	U	R	C	U	E	U	U	E
M	L	N	N	U	U	S	D	E	N	C	T	S	R	G
O	B	E	I	B	T	E	A	P	O	A	E	N	T	I
T	A	G	C	G	N	A	A	L	T	R	C	K	C	T
P	R	O	G	R	E	S	S	I	O	N	T	L	E	N
M	U	H	S	O	T	E	O	C	D	N	A	L	P	A
Y	S	T	T	U	T	N	O	A	U	S	B	F	S	T
S	A	A	A	P	A	N	K	T	S	G	L	S	N	S
A	E	P	G	A	V	N	O	I	T	C	E	F	N	I
U	M	H	E	E	Y	E	F	O	M	E	T	S	Y	S
T	H	E	R	A	P	Y	C	N	E	T	A	L	M	E
B	N	T	N	O	I	T	A	B	U	C	N	I	G	R

ACUTE
ADULT
ADVANCED
AIDS
ANTIGEN
ASYMPTOMATIC
ATTENUATED
CHRONIC
CLASSIFY
DETECTABLE
DISEASE
HIV
INCUBATION
INFANT
INFECTION
LATENCY
MEASURABLE
MUTATION
PATHOGENIC
PROGRESSION
REPLICATION
RESISTANT
SEPTICEMIA
SEROCONVERT
SPECTRUM
STAGE
SUBGROUP
SYSTEM
THERAPY

8 Epidemiology and Transmission of the Human Immunodeficiency Virus

Chapter Concepts

- ♦ First evidence of HIV-1, 1959, Central Africa.
- ♦ In 1985, HIV-2 was isolated in West Africa.
- ♦ Transmission of HIV into the United States may have been via Haiti.
- ♦ Behavior is associated with HIV transmission.
- ♦ HIV is not casually transmitted.
- ♦ HIV is not transmitted to humans by insects.
- ♦ HIV transmission is being reported from 194 countries and among all ages and ethnic groups.
- ♦ The four basic mechanisms of HIV transmission are: sexual contact, needles and syringes, mother to child, and blood transfusions. All involve an exchange of body fluids.
- ♦ HIV enters the body through the lining of the vagina, vulva, penis, rectum, or mouth during sex.
- ♦ No new routes of HIV transmission have been discovered in the last 24 years.
- ♦ First documented case of HIV transmission via deep kissing is presented.
- ♦ Highest frequency of HIV transmission in the United States is among homosexual and bisexual males and among injection-drug users.
- ♦ It has been found that most HIV-infected people do not infect others.
- ♦ Other countries, HIV infection, and injection-drug use.
- ♦ HIV/AIDS in the Caribbean.
- ♦ Worldwide, highest frequency of HIV transmission is among heterosexuals.
- ♦ Bloodbanks in several countries knowingly allowed the distribution of HIV-contaminated blood.
- ♦ HIV-infected athletes want to compete.
- ♦ Death due to HIV infection is placed in perspective.
- ♦ Interactions between HIV and sexually transmitted diseases are discussed.
- ♦ Prenatal HIV transmission generally occurs after the 12th to 16th week of gestation, most often during childbirth and breast-feeding.
- ♦ Zidovudine and Nevirapine decrease perinatal HIV transmission.
- ♦ National HIV/AIDS Resources phone numbers are listed.
- ♦ An HIV/AIDS word scramble is presented.

When a population becomes infected with a contagious disease, an epidemic results. **Epidemic** is derived from Greek and means "in one place among the people." To understand how an infectious disease can spread or remain established in a population, investigators must consider the relationship between an infectious disease agent and its host population. The study of diseases in populations is an area of medicine known as **epidemiology.**

The danger of complacency has been learned from earlier epidemics. Complacency about HIV infection is especially dangerous because the infection can remain hidden for years. Because many infected people remain symptom-free for years, it is hard to be sure just who is infected with the virus. The more sexual partners, the greater the chances of encountering one who is infected and subsequently becoming infected.

With regard to HIV infection, it is your behavior that counts. The transmission of HIV can be prevented. HIV is relatively hard to contract and with exception, can be avoided.

The presence of HIV/AIDS is not isolated—the transmissibility of HIV between individuals and across borders and populations is what drives this global pandemic and makes it imperative that nations work together to prevent the continued transmission of HIV. On an individual level, IT IS NOT IMPORTANT HOW YOU GOT HIV, WHAT IS IMPORTANT IS HOW YOU LIVE YOUR LIFE WITH THE VIRUS.

How HIV Enters the Body

In all cases, HIV is transmitted in one of two ways. First, it can be transmitted as the virus itself, or second, within an HIV-infected cell. The virus is held within the white blood cells of the immune system and is carried Trojan horse-style within the fluid of one person into the body of another.

HIV and Sexual Transmission

Epidemiological data suggest that sexual transmission, in general, is relatively inefficient, in that exposure often does not produce infection. HIV is transmitted more efficiently intravenously than through sexual routes. However, worldwide the predominant mode of transmission of HIV is through exposure of mucosal surfaces, of the vagina, vulva, penis, rectum, or mouth, to infected sexual fluids (semen, cervical/vaginal, rectal) and during birth. Sexual transmission of HIV now accounts for about 90% of infections worldwide.

Factors Driving Sexual Transmission

There is evidence from around the world that many factors play a role in initiating a sexually transmitted HIV epidemic or driving it to higher levels. Among the **behavioral** and **social factors** are (a) condom use, (b) proportion of the adult population with multiple partners, (c) overlapping (as opposed to serial) sexual partnerships—individuals are highly infectious when they first acquire HIV and thus more likely to infect any concurrent partners, (d) sexual networks (often seen in individuals who move back and forth between home and a far-off workplace), (e) age mixing, typically between older men and young women or girls, and (f) poverty and in particular women's economic dependence on marriage or prostitution, robbing them of control over the circumstances or safety of sex.

Biological factors include (a) high rates of sexually transmitted infections, especially those causing genital ulcers, (b) low rates of male circumcision, and (c) high viral load HIV levels in the bloodstream that are typically highest when a person is first infected and again in the late stages of illness.

While all these factors help spread the virus, it is not known exactly how much each of them contributes and to what extent they need to be combined in order to spread the epidemic. The issue of male circumcision is a good example. Many countries in which all boys are circumcised before puberty have very limited epidemics, and even in some countries with wider epidemics, circumcised men have lower HIV rates than uncircumcised men.

HIV: Other Routes of Transmission?

HIV is not communicable through contact with fomites (inanimate objects) or through vectors. Thus people do not "catch" HIV in the same way that they "catch" the cold or a flu virus. Unlike colds and flu viruses, HIV *is* **not,** according to the CDC, spread by tears, sweat, coughing, or sneezing. The virus *is* **not** transmitted via an infected person's clothes, phone, or toilet seat. HIV *is* **not** passed on by eating utensils, drinking glasses, or other objects that HIV-infected people have used that are free of their blood.

HIV *is* **not** transmitted through daily contact with infected people, whether at work, home, or school. Insects **do not** transmit the virus. Kissing is also considered very low risk: There is only one documented case to prove that HIV is transmitted by kissing. Paul Holmstrom and colleagues (1992) report that salivary HIV antibodies are detected regularly in HIV seropositive subjects. The route of HIV into saliva is not fully understood. Both salivary glands and salivary leukocytes have been shown to harbor HIV. Gingival fluid (fluid seeping out of the gums) has been regarded as the main source of salivary HIV antibodies and infectious HIV.

In its 1990 supplemental guidelines for cardiopulmonary resuscitation (CPR) training and rescue, the Emergency Cardiac Care Committee of the American Heart Association (AHA) noted that there is an extremely small theoretical risk of HIV or hepatitis B virus (HBV) transmission via cardiopulmonary resuscitation (CPR). To date no known case of seroconversion for HIV or HBV has occurred in these circumstances.

LOOKING BACK 8.1

PUBLIC OPINION ON HOW THE DISEASE—AIDS—WAS TRANSMITTED JUNE 1983

A *Newsweek* poll found that 9 in 10 (90%) Americans over 18 had heard of AIDS, but also found that 4 in 10 (40%) believed it was possible to contract the disease through casual contact. From 1983 to 1985, AIDS stories made regular appearances in newspapers and on television news broadcasts. Over 150 AIDS stories were broadcast by the three major TV network evening news programs in 1985—more than double the combined totals of AIDS stories broadcast in 1983 and 1984 (*Source: the Vanderbilt Television Archives*). These initial years of AIDS press coverage were marked by stories that often focused on the dramatic aspects of the disease (for example, its progressive nature), the death toll among higher-risk groups, and its potential to spread to the public at large.

WE MUST STOP HIV TRANSMISSION NOW!

Education and advocacy for risk reduction remain important tools for preventing HIV infection. Used alone however, they will never accomplish the objective of slowing the speed and extent of the virus's spread. This can only be achieved by combining educational and early intervention efforts with specific measures guiding those resources to the people who need them most: those who are already infected and to their sex and drug partners.

Beginning year 2005, there were about 930,000 AIDS cases in the United States. Figure 8-1 breaks this number down according to means of HIV infection—sexual behavior, drug use, medical exigencies, and undetermined causes. Figure 8-2 gives a breakdown by transmission category for 38,000 AIDS estimated cases for 2004.

EPIDEMIOLOGY OF HIV INFECTION

The first scientific evidence of human HIV infection came from the detection of HIV antibodies in preserved serum samples collected in Central Africa in 1959. The first AIDS cases appeared there in the 1960s. By the mid-1970s HIV was being spread throughout the rest of the world. The earliest places to experience the arrival of HIV were Central Europe and Haiti. Transmission into the United States may have been by tourists who had vacationed in the area of Port-au-Prince, Haiti (Swenson, 1988).

On entry into the United States the virus first spread among the homosexual populations of large cities such as New York and San Francisco. The first **recorded** AIDS cases in the United States occurred in 1979 in New York. The first CDC-**reported** AIDS cases were in New York, Los Angeles, and San Francisco in 1981. In all cases, the diagnosis of AIDS was based on clinical descriptions.

According to the CDC's first clinical AIDS definition, at least one case of AIDS occurred in New York City in 1952 and another in 1959. Both males demonstrated opportunistic infections and *Pneumocystis carinii* pneumonia, a hallmark of HIV infection. This early evidence of AIDS suggests that the virus might have been in the United States, Europe, and Africa at about the same time (Katner et al., 1987). If HIV has been present for decades as suggested, its failure to spread may reflect a recent HIV mutation, a major change in social behaviors conducive to HIV transmission, or both. For example, the sexual revolution and the widespread use of birth control pills, which began in the 1960s, and the subsequent decrease in the use of condoms may have contributed to the transmission of HIV.

TRANSMISSION OF TWO STRAINS OF HIV (HIV-1/HIV-2)

The spread of HIV-1 is global. The clinical presentation of AIDS caused by HIV-1 is similar regardless of geographical area.

HIV-2 is a genetically distinct strain. HIV-2 was first discovered in 1985 in West Africa. It is believed to have been present in West Africa as early as the 1960s. Clinical data demonstrate that HIV-2 has a reduced virulence compared to HIV-1 (Marlink et al., 1994).

It appears that HIV-2, like HIV-1, may spread worldwide. HIV-2 has already spread from West Africa to other parts of Africa, Europe, and the Americas. Both HIV-1 and HIV-2 are transmitted or acquired through the same kinds of exposure.

Because over 99% of global AIDS cases are caused by the transmission of HIV-1, only data that pertain to HIV-1 (HIV) will be presented unless otherwise stated.

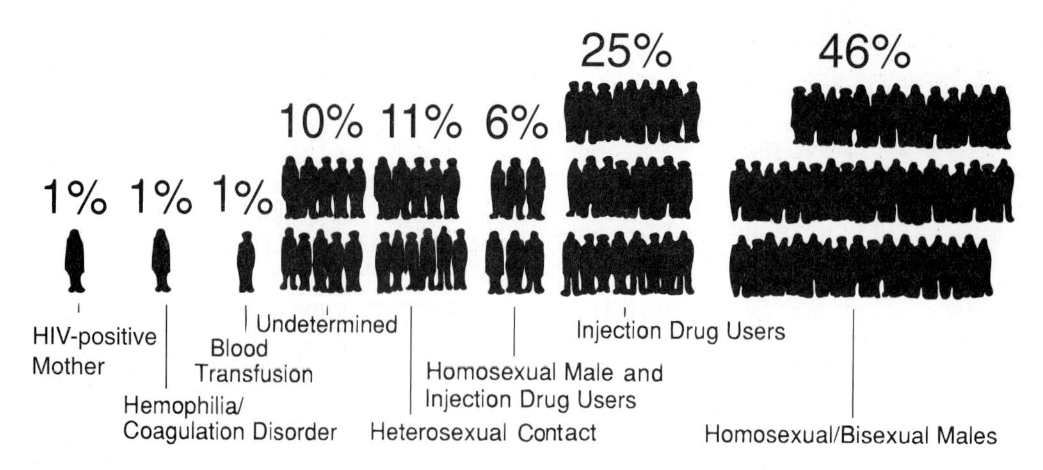

FIGURE 8-1 Cumulative AIDS Cases by Route of Transmission. At the beginning of 2005, there were about 930,000 AIDS cases in the United States. This diagram gives the percentage of adults and adolescents in each group through 2004. Groupings are according to sexual preference, drug use, medical conditions, and others not associated with any of these. *(Courtesy of CDC, Atlanta—updated)*

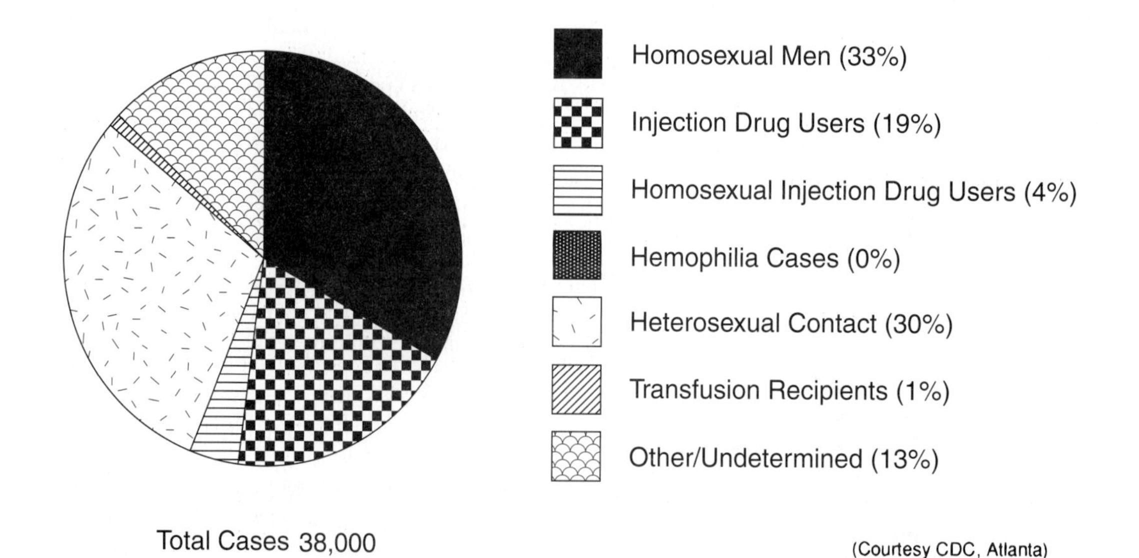

FIGURE 8-2 Adult/Adolescent AIDS Cases by Transmission Category Estimated for 2004, United States.

IS HIV TRANSMITTED BY INSECTS?

In spite of convincing evidence of the ways in which HIV can be transmitted, it remains difficult for the general public to believe that a virus that appears to spread as rapidly as HIV is not either highly contagious or transmitted by an environmental agent. After all, there are many viral and bacterial diseases that are highly contagious and transmitted by insects. The question was asked: Is this virus being transmitted by insects?

Necessary data to resolve the question were available. Epidemiological data from Africa and the United States suggested that HIV was not transmitted by insect bites. If it were, many more cases would be expected among school-age children and elderly people, groups that are proportionally underrepresented among AIDS patients. In one study

POINT OF INFORMATION 8.1

CONSEQUENCES OF IGNORING HIV

You may be offended by the description of how HIV is transmitted. You may feel invulnerable to HIV infection. You may think that given time, HIV infection and AIDS will go away. You may choose to ignore that it exists. *But ignoring HIV may kill you.* The cost of ignorance about any of the STDs is always high, but with AIDS the cost is a disfiguring, painful death.

of the household contacts of AIDS patients in Kinshasa, Zaire, where insect bites are common, not a single child over the age of 1 year had been infected with the HIV, while more than 60% of spouses had become infected.

In 1987, the Office of Technology Assessment (OTA) published a detailed paper on the question of whether blood-sucking insects such as biting flies, mosquitoes, and bedbugs transmit HIV (Miike, 1987). The conclusion was that the conditions necessary for successful transmission of HIV through insect bites and the probability of their occurring rule out the possibility of insect transmission as a significant factor in the spread of AIDS. Jerome Goddard reported (1997) that blood-sucking arthropods (for example, mosquitoes and bedbugs) for good biological reasons, **cannot** transmit HIV. Insects inject their own saliva as a lubricant to assist feeding. They do not inject their own or a previously bitten person's blood. Malaria and yellow fever are transmitted through insect saliva—HIV isn't. The virus has to overcome many obstacles. It must avoid digestion in the gut of the insect, recognize receptors on the external surface of the gut, penetrate the gut, replicate in insect tissue, recognize and penetrate the insect salivary glands, and subsequently escape into the lumen of the salivary duct. Webb and colleagues (1989) inoculated bedbugs intraabdominally (belly) and mosquitoes intrathoracically (chest) with HIV to enable the virus to bypass gut barriers. HIV failed to replicate in either.

HIV TRANSMISSION

If most HIV-exposed people can become HIV-infected, can most infected people *transmit* HIV to others? This is a difficult question to answer. Infection with HIV appears to depend on a large number of variables that involve the donor, recipient, and portal of entry. The most important variables are mode of transmission, viral load, which subtype and variant of HIV is present, and the recipient's genetic resistance.

Global Patterns of HIV Transmission

Worldwide there are now three types or patterns of HIV epidemics unfolding. The first pattern is occurring in wealthy countries, such as the United States, where the epidemics are heterogeneous but dominantly involve **male-to-male** sexual transmission. After a long period of decline, those epidemics are now showing troubling signs of resurgence, largely due to unsafe sexual practices among gay men.

The second pattern is seen in sub-Saharan Africa and Latin America, driven by **heterosexual** transmission. Africa continues to have the largest numbers of people living with HIV and dying of AIDS.

The third pattern and labeled as explosive by UNAIDS has almost nothing to do with sex. It is driven by **needles shared among people who inject narcotics.** All over the world the narcotics-driven HIV pandemic seems to begin, unnoticed by government officials, in isolated communities of injection drug users, spreads like wildfire, and then suddenly takes on national significance. The most disturbing examples of this phenomenon are the epidemics of Eastern Europe and Asia, which regionally are in the midst of an HIV explosion that was predicted some years ago. Injection-drug use associated HIV infections are out of control in Russia, China, and Indonesia.

Across the globe, depending on the developing nation, IDU is associated with the highest incidence of HIV transmission; in other countries it may be via men having sex with men, or heterosexual sex among males and females.

HIV TRANSMISSION IN FAMILY/HOUSEHOLD SETTINGS

Several studies of the family members of AIDS patients have failed to demonstrate the spread of HIV through household contact. The only cases in which family members have become infected involved the sexual partners of AIDS patients or children born to mothers who were already infected with the virus. Even individuals who bathed, diapered, or slept in the same bed with AIDS patients have not become infected. In one study, 7% of the family members shared toothbrushes with the infected person and no one became infected.

Perhaps the best evidence *against* casual HIV transmission comes from studies of household members living with blood-transfused AIDS patients (Peterman et al., 1988). Transfusion infection cases are unique because their dates of infection are known retrospectively. Prior to the onset of AIDS symptoms, the families were unaware that they were living with HIV-infected individuals. Family life was not altered in any way, yet family members remained uninfected. In some cases, the transfusion patients were hemophiliacs who received weekly or monthly injections of blood products and became HIV-infected. From the combined studies of these households, only the sexual partners of infected hemophiliacs became infected. In some cases, the sexual partners of hemophiliacs remained HIV-free after three to five years of unprotected sexual intercourse.

Although contact with blood and other body substances can occur in households, transmission of HIV is rare in this setting. Through 2004, at least 11 reports have described household transmission of HIV not associated with sexual

BOX 8.1

A NEIGHBOR'S STORY: PUBLIC IMAGE, PUBLIC FEAR

We have over the last 23 years witnessed educated people offering misrepresentation and fantasy about HIV infection and AIDS. We have listened to them distort the truth by presenting false perceptions rather than facts. If education is to become a major player in prevention of the spread of HIV, the mix of myth and fantasy must be replaced by reality—and this can be done by giving the proper respect to a new disease in our lifetime.

The Neighbor

Some neighbors were having a garage sale. Their friend came over and asked them a question about AIDS. They mentioned that he should talk to the professor next door and he came over to the house. He said, "I'm looking for the truth about who is, and who isn't HIV-infected and how I can tell people that are HIV-infected from those who are not." I asked if he cared to share the reason for the questions and he immediately told me that he divorced three years ago and has been so frightened by the information he has seen on TV, heard on the radio, and read in the paper that he has remained celibate for three years and that it has truly affected his quality of life. He said, "I'm afraid to have sex with any woman I date because I believe almost every woman in Jacksonville and elsewhere carried HIV. It's driving me insane. My male friends are still having sex. Are they crazy or am I the fool?" I took the second question first. I assured him that his worst fears were correct. You can't tell the HIV positive from the HIV negative just by their appearance. Infected and noninfected all look alike—at least in the early and middle stages of the disease. For the second question of who is and who is not HIV-infected, I gave him some details relating to people's behavior, the particular behavioral risk groups and I said that if he did not belong to such groups or did not have a blood transfusion before 1985–1986, and that if he had some information about his sexual partner and did not change sexual partners too frequently and always used a condom, he could feel as safe as one can. The expression on his face went from concern to relief. He shook my hand repeatedly. He thanked me profusely and he said, "I feel like the media, the government, the gays, and everyone else has cheated me out of three years of my life. I can never get that back." I said, "Yes, to some degree that is true, but here you stand after three years still HIV negative. Would you like to exchange three years of free-wheeling sexual encounters for an HIV infection?" He agreed that he would not, but we both understand his point. We all need to understand his point. Misinformation—data skewed by ignorance or lack of education causes fear. And failure to give an accurate representation of a disease, from both the medical and social vantage points, can cause a great deal of harm—in ways that most of us would not begin to contemplate. Did I give him the right advice? No, because I did not give him advice at all. I simply attempted to put things in their proper perspective. He did not leave me less afraid of dying of AIDS. He left me with the idea that he could have his sex life restored without dying from AIDS, provided he maintains his low-risk behavior.

—The Author

contact, injection-drug use, or breast-feeding. Of these 11 reports, 7 were associated with documented or probable blood contact. In one report, HIV infection was diagnosed in a boy after his younger brother had died as the result of AIDS; however, a specific mechanism of transmission was not determined.

NONCASUAL TRANSMISSION

The routes of HIV transmission were established *before* the virus was identified. The appearance of AIDS in the United States occurred first in specific groups of people: homosexual men and injection-drug users. The transmission of the disease within the two groups appeared to be closely associated with sexual behavior and the sharing of IV needles. By 1982, hemophiliacs receiving blood products, as well as the newborns of injection-drug users, began demonstrating AIDS. By 1983, heterosexual female partners of AIDS patients demonstrated AIDS. **Twenty-four years** of continued surveillance of the general population has failed to reveal other categories of people contracting HIV/AIDS (Table 8-1). It became apparent that the infectious agent was being transmitted within specific groups of people, who by their behavior were at increased risk for acquiring and transmitting it (Table 8-2). Clearly, an exchange of body fluids was involved in the transmission of the disease.

With the announcement that a new virus had been discovered, further research showed that this virus was present in a number of body fluids. Thus, even before there was a test to detect this virus, the public was told that it was transmitted through body fluids exchanged during intimate sexual contact, contaminated hypodermic needles, contaminated blood or blood products, and from mother to fetus. In addition it was concluded that the widespread dissemination of the virus was most likely the result of multiple or repeated viral exposure because the data from transfusion-infected individuals indicated that they did not necessarily infect their sexual partners. In other words, it was concluded early on and later confirmed that this virus was not transmitted as easily as other blood-borne viral diseases such as hepatitis B, or viral and bacterial sexually transmitted diseases. Table 8-3 lists the means of HIV transmission worldwide. Table 8-4 lists the means of HIV transmission in the United States.

Table 8-1 HIV Transmission and Infection[1]

CHAIN OF HIV INFECTION

Agent causing the disease	HIV
Major reservoirs (source of HIV in the body)	Lymph nodes, blood, genitals
Replication site of HIV	Mostly inside T4+ or CD4-bearing lymphocytes
Portal of exit (how does HIV leave the body)	Mucosal openings, skin breaks, bleeding, or expulsion of body fluids
Transfer or transmission of HIV (from one human to another)	Via body fluids
Portal of entry (how does HIV enter the body)	Mucosal openings, skin breaks, areas of bleeding, injection

TRANSMISSION ROUTES

Blood Inoculation

Transfusion of HIV-infected blood and blood products
Needle sharing among injection-drug users
Needle sticks, open cuts, and mucous membrane exposure in health-care workers
Use of HIV-contaminated skin-piercing instruments (ears, acupuncture, tattoos)
Injection with unsterilized syringe and needle (mostly in undeveloped countries)

Sexual Contact: Exchange of semen, vaginal fluids, or blood

Homosexual, between men
Lesbian, between women
Heterosexual, from men to women and women to men
Bisexual men and women

Perinatal

Intrauterine
Peripartum (during birth)
Breast-feeding

[1]To reduce the risk of spreading HIV, use condoms during sexual activity. Do not share drug injection equipment. If you are HIV infected and pregnant, talk with your doctor about taking anti-HIV drugs. If you are an HIV-infected woman, don't breast-feed. Protect cuts, open sores and your eyes and mouth from contact with blood and other bodily fluids. If you think you've been exposed to HIV, get tested and ask your doctor about taking anti-HIV medications.

Table 8-2 Adult/Adolescent AIDS Cases by Sex and Exposure Categories, Estimated Through 2004, United States[a]

Male Exposure Category (80%)	Total No.
1. Men who have sex with men	309,254
2. Injecting drug use	257,712
3. Men who have sex with men and inject drugs	73,632
4. Hemophilia/coagulation disorder	7,363
5. Heterosexual contact:	29,452
a. Sex with injecting drug user	10,930
b. Sex with person with hemophilia	59
c. Sex with transfusion recipient with HIV infection	236
d. Sex with HIV-infected person, risk not specified	18,260
6. Receipt of blood transfusion, blood components, or tissue	7,363
7. Other/undetermined	51,542
Total male AIDS cases	736,320
Female Exposure Category (20%)	
1. Injecting drug use	84,677
2. Hemophilia/coagulation disorder	
3. Heterosexual contact:	90,199
a. Sex with injecting drug user	36,080
b. Sex with bisexual male	4,870
c. Sex with person with hemophilia	631
d. Sex with transfusion recipient with HIV infection	812
e. Sex with HIV-infected person, risk not specified	47,806
4. Receipt of blood transfusion blood components, or tissue	5,522
5. Other/undetermined	3,682
Total female AIDS cases	184,080
Total male/female cases	920,400

[a]Pediatric cases, 9600. These calculations only give a general idea of risk. They can tell you which activities carry a higher or lower risk. They cannot tell you if you have been infected. If, for example, the risk is 1 in 100, it doesn't mean that you can engage in that activity 99 times without any risk of becoming infected. You might become infected with HIV after a single exposure. That can happen the first time you engage in a risky activity.

Table 8-3 How HIV Is Transmitted Worldwide, 2004

Exposure	Efficiency, %	% of Total
Blood transfusion/ blood products	>90	3
Perinatal	20–40	9
Sexual intercourse[a]	0.1–1.0	80
Injection drug use	0.5–1.0	8

[a]Heterosexual intercourse 70%; See Sidebar 8.1

(Source: WHO/Global Programme on AIDS and UNAIDS, 1999, updated)

Table 8-4 An Approximation of How an Estimated 1,527,000 Americans Became Infected with HIV, through 2004

152	health care workers got infected from the blood or body fluids of patients;
20,615	children infected through their mothers;
32,831	people got HIV from infected blood or blood products;
106,890	people did not know how they were infected, did not report their risk, or died before anyone could find out;
91,620	people were infected who had both unprotected sex and shared needles;
381,750	people were infected who shared needles;
893,295	people were infected through unprotected sex.

(Based on data from HIV/AIDS Surv. Rpt, CDC, Year-End Edition, 2003)

Mobility and the Spread of HIV/AIDS

Mobility is an important epidemiological factor in the spread of communicable diseases. This becomes particularly obvious when a new disease enters the scene. In the early stage of the HIV/AIDS epidemic, for example, the route of the virus could be associated with mobility.

The first HIV-infected people in some Latin American and European countries reported a history of foreign travel. In some African countries, spread of the virus could be traced along international roads. Today, increasing numbers of HIV infection have been observed to be associated with the relaxation of travel restrictions in Central and Eastern Europe.

Few countries are unaffected by HIV/AIDS. This has made it clear that restrictive measures such as refusal of entry to people living with

HIV/AIDS and compulsory testing of mobile populations are inappropriate and ineffective measures to stop spread of the virus. In times of increasing international interdependency, it is an illusion to think that the disease can be stopped at any border.

Number of HIVs Required for Infection

Robert Coombs and colleagues (1989) calculated the dose of HIV necessary to cause an HIV infection. They reported that "one infective dose" or 1000 HIV particles is necessary to establish HIV infection in human tissue culture cells. To establish HIV infection in the body, it was reasoned that it would take 10 to 15 infective doses. Their study indicated that a pint of blood from an AIDS patient contains 1.8 million infective doses, or about 2 billion HIV particles per pint of blood, or about 4.2 million HIV particles per milliliter.

Body Fluid Transmission

High levels of HIV have been isolated from blood, semen, vaginal fluids, saliva, serum, urine, tears, and breast milk. Lower levels of HIV are found in lung fluid and cerebrospinal fluid (Friedland et al., 1987). HIV has not been found in sweat.

In cases of low levels of HIV in cell-free body fluids and within the cells of these fluids does not mean that HIV cannot be transmitted via these fluids or cells—it can, but the dose (number of viruses) is so small that the risk of infection is minimal; thus the reason for the low number of health-care workers contracting HIV infection after touching, being splashed by, or needle sticking themselves with blood containing HIV.

HIV is found in greatest numbers within T4 or CD4+ cells, macrophage, monocytes and dendriditic cells, and in vaginal fluids and semen. Laboratory findings, along with overwhelming empirical observations, support the scientific conclusion that the major route of HIV transmission is through human blood and sexual activities involving exchange of semen and vaginal fluids. Semen carries significantly larger numbers of HIV than vaginal fluid. It appears that of all body fluids, these three contain the largest number of infected lymphocytes (Figure 8-3), which provide the largest HIV concentration in a given area at a given time.

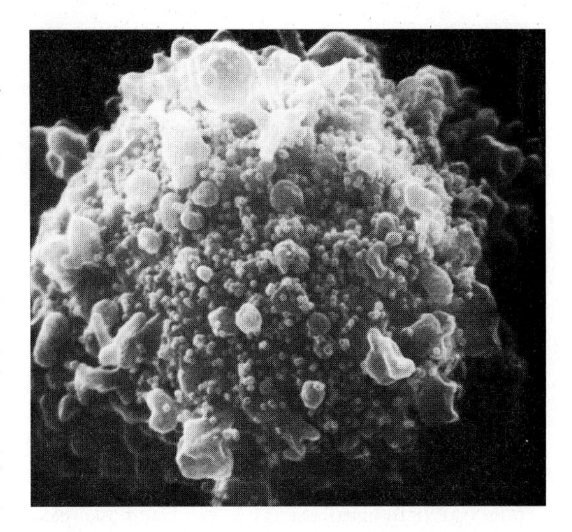

FIGURE 8-3 Electron Micrograph of an HIV-Infected T4 Lymphocyte. The T4 or CD4+ cell has produced a large number of HIVs that are located over the entire lymphocyte. Each HIV leaves a hole in the cell membrane. The photograph shows part of the convoluted surface of the lymphocyte magnified 20,000 times. *(Courtesy of The National Biological Standards Board, South Mimms, U.K. and David Hockley)*

Presence of HIV-Infected Cells in Body Fluids

Blood—Recent observations indicate that both latent and HIV-producing cells are present throughout the course of infection. Considerable numbers of HIV and HIV-infected cells can be detected in the blood. Thus it is probable that HIV-infected people harbor substantial numbers of HIV-infected cells in their blood from soon after the initial infection through the terminal stage of the disease.

Semen—A single specimen of ejaculate contains between 1 million and 10 million nonspermatozoal cells, and many leukocytes including T4 cells. But the number and type of all mononuclear cells in the semen of a healthy man differs considerably from day to day (Kiessling, 1999).

Reducing the numbers of lymphocytes and macrophages in semen could decrease the chances of infection. In one study on 94 semen samples from HIV-infected men, HIV DNA was found in 35% of the mononuclear cells (Bagasra, 1994). It has been suggested that **vasectomy** could reduce the infectivity of HIV-infected men because it would eliminate mononuclear cells or cell-free virus in semen originating from the testes and epididymis. The work of John Krieger and colleagues (1998) shows that sperm can exist in semen for up to three months after vasectomy (see **sperm** below) and that there was no significant difference in pre- and post-vasectomy HIV RNA levels in semen. Collectively, it would appear that a vasectomy stops HIV transmission. Lee Harrison and colleagues (2000) reported on 93 HIV-positive men on antiretroviral drugs. Before drug treatment 74% had detectable levels of HIV in their semen. After six months on therapy 33% had detectable HIV in their semen.

With regard to the presence of HIV in semen and actual transmission, investigators at the University of North Carolina at Chapel Hill have used HIV RNA in semen samples from several hundred men to estimate the effects of semen in viral burden on heterosexual transmission. They concluded that when HIV RNA in semen was low (<5000 copies/mL) transmission was unlikely to occur, at 1 per 10,000 episodes of intercourse. Conversely, when the concentration of HIV in semen was high (for example, 1 million copies/mL) the probability of transmission rose to 3 per 100 episodes of intercourse.

Sperm—Omar Bagasra (1994) reported that 33% of sperm samples from 94 HIV-positive men contained HIV RNA. Many investigators have now reported that the **mitochondria** of various tissue types are infected with HIV. The sperm midpiece contains large numbers of mitochondria.

Saliva—There continues to be some concern over the presence of HIV in saliva because of the exchange of saliva during deep kissing, the saliva residue left on eating utensils, and saliva on instruments handled by health-care workers, especially in dentistry. Results of studies on hundreds of dental workers, many of whom have cared for AIDS patients, have shown no evidence of HIV infection (Friedland et al., 1987 updated). Also, in a CDC study, none of 48 health-care workers became infected after parenteral (IV) or mucous membrane exposure to the saliva of HIV-infected patients (Curran et al., 1988). Studies by Fox (1991), Archibald (1990), and Pourtois (1991) showed that human saliva contains factors that inhibit HIV infectivity.

First Documented Kissing Transmission Case

Although about 65 million people around the world will have been HIV-infected, by the end of 2005, there are remarkably few reports of its spread by kissing, dental treatment, biting, or coughing. Even among those people whose bodies

LOOKING BACK 8.2

FEDERAL STUDY COMMITTEE ADVISES THE FOOD AND DRUG ADMINISTRATION ON THE ACCEPTANCE OF BLOOD DONATIONS IN THE TIME OF AIDS, JULY 1983

The Seattle Times reported that a federal study committee faced a tough task in advising the Food and Drug Administration regarding blood donations and transfusions in relation to the mysterious disease called Acquired Immune Deficiency Syndrome, or AIDS. The committee recommended that the government not have an automatic policy of destroying a large reserve of blood just because one donor is suspected of having AIDS. We're not sure how helpful that recommendation will be to the FDA. Not having an automatic policy could mean anything from destroying 99% of suspect blood reserves to destroying none. But the committee did base its recommendation on three seemingly sound factors. **First,** although a few people have contracted AIDS after receiving blood transfusions, it is unclear what the cause-and-effect relationship is. **Second,** any contaminated blood from a potential AIDS victim would be diluted in a large pool. **Third,** widespread destruction of blood pools could put people who suffer bleeding diseases at a great risk. Two other advisory-committee recommendations to the FDA are of unquestioned validity. The committee called for a public education campaign and the screening of blood donors who are in the high-risk categories. The latter point can scarcely be overemphasized. The government should take every possible step to ensure the screening is thorough and that it is so perceived by a concerned public.

are actively shedding HIV, their saliva usually contains only noninfectious components of the virus.

In July 1997, the first documented case of HIV transmission via deep kissing was reported on by the CDC. In this *one* case, the man was HIV positive, via IDU in 1988. Both he and his female partner had serious gum (periodontal) disease. His gums routinely bled with brushing or flossing. Investigators at the CDC believe that the HIV was transmitted via blood within the man's oral cavity due to oral lesions onto the mucus membrane of the woman (*MMWR*, 1997).

Spitzer and colleagues (1989) reported that a male became HIV-infected during unprotected fellatio with an HIV-infected prostitute. A second similar case has also been reported. The mechanism of transmission, saliva, in these cases may be suspected, but it is not conclusive.

The precise risk of oral sex is difficult to assess because most couples engage in other sexual practices.

Mother's Milk—Several studies have shown that HIV can be transmitted by breast-feeding. Van de Perre and coworkers (1993 updated) found that infection of babies via breast milk was most strongly correlated with the presence of HIV-infected cells in the milk, suggesting that infection might be cell-mediated. However, infection was also correlated with low levels of antibodies to HIV, suggesting that infection was initiated by cell-free virus.

Dentist With AIDS Infects Patient During Tooth Extraction?

In July 1990, the CDC reported on the possible transmission of HIV from a dentist with AIDS to a female patient (*MMWR*, 1990a). This case, like no other before it, sent chills through many. But why this case? Because the vast majority of people go to dentists! They don't inject drugs and are not gay. This case, however, is difficult to re-

POINT OF INFORMATION 8.2

DANGER OF HIV INFECTION VIA ARTIFICIAL INSEMINATION

By early year 2005, 15 women were reported to have been HIV-infected through the use of anonymous donor sperm to initiate pregnancy: one in Germany, two in Italy, four in Australia, two in Canada, and six in the United States. Thirty recipients of semen from HIV-infected donors refused to be HIV tested (Guinan, 1995 updated). All cases except Germany occurred *before* the availability of HIV antibody testing. About 80,000 women each year are artificially inseminated with donor sperm. But 23 years into the AIDS epidemic, the increasingly popular fertility business remains largely unregulated and unmonitored, even though it traffics in semen, long known to be one of the two main HIV transmission routes.

Only a few states (New York, California, Ohio, Illinois, and Michigan) require HIV testing of semen donors. There are no federal regulations.

Medical and public health experts agree that artificial insemination is an HIV risk that somehow fell through the cracks of public education and health regulations. They insist, however, that the risk is low.

In January 1996 Washington State's top educator announced that she had AIDS. She received the virus trying to become pregnant through artificial insemination with donor sperm. This announcement once again focused national attention on the problem, the lack of federal regulations for this industry.

Omar Bagasra and colleagues (1994, 1996) state unequivocally that there is a risk of HIV transmission during artificial insemination. To quote Bagasra, "As we have shown, the sperm easily penetrates other cells—and when this occurs, the midpiece also penetrates. A single ejaculation contains [millions of] sperm. We have data to show that the mitochondria in the midpieces of HIV-infected sperm are loaded with HIV. If 1 in 1000 of these mitochondria contains HIV, that is a significant viral load. The viral load transmitted from a man to a woman in this manner is quite likely to be high."

Lack of federal and state regulations means you must protect yourself! (1) Stay away from private physicians unless you know they are using only certified sperm banks to get their products. (2) Review a doctor's or clinic's testing and record-keeping procedures and demand to see donor medical records, which can be shared even if the donor's identity is protected. (3) Accept only frozen donor semen that has been HIV tested at least twice at three- to six-month intervals.

solve. For example, two years had elapsed from the time of the dental work to when the patient, Kimberly Bergalis (Figure 8-4), was diagnosed with AIDS. Both patient and dentist, David J. Acer of Stuart, Fla., were uncertain of exactly what happened. Some of the pertinent factors in this case are (1) review of dental records and radiographs suggest that the two tooth extractions were uncomplicated; (2) interviews with Bergalis and the dentist did not identify other risk factors for HIV infection. Bergalis, age 22, stated over national TV in 1990 that she was still a virgin. This point was disputed on June 19, 1994, during the TV program *60 Minutes* (with Mike Wallace). Yes, she did engage in sexual foreplay, but not intercourse. Yes, she was infected with the human papilloma virus, which can be sexually transmitted, but this is not at all uncommon in immune-suppressed AIDS patients with no history of sexual intercourse; (3) nucleotide sequence data indicated a high degree of similarity between the HIV strains infecting her and the dentist; and (4) the time between the dental procedure and the development of AIDS was short (24 months), and Bergalis developed oral candidiasis 17 months after infection. At this time, only 1% of infected homosexual/bisexual men and 5% of infected transfusion recipients develop AIDS within two years of infection.

David Acer, a bisexual, was diagnosed with symptomatic HIV infection in 1986 and with AIDS in 1987. He died on September 3, 1990. Since then, 1100 of his 2500 patients were contacted for HIV testing. In January 1991, the test results of 591 of these patients revealed that five were HIV positive: a 68-year-old retired school

FIGURE 8-4 Kimberly Bergalis, Age 23. Bergalis is being comforted by her mother after their train trip from their home in Florida to Washington, D.C. She made the trip to testify before a congressional committee on September 26, 1991. Bergalis favored mandatory HIV testing for health-care personnel. She died December 8, 1991. *(Photograph courtesy of AP/Wide World Photos)*

teacher, a middle-aged father of two, a 37-year-old carnival worker, an unemployed drifter, and a 19-year-old student. As with Bergalis, infection in these patients may have come from some other source. All six patients denied having sexual contact with the dentist or with one another (*MMWR*, 1991a). If an absolute case can be made that Acer transmitted the virus to Bergalis, this will be the first documented case of a health-care professional infecting a patient.

Bergalis died of AIDS on December 8, 1991. She was 23 years old and weighed 48 pounds.

There is no shortage of ideas as to how Acer might have infected his patients. For example, he could have used the same dental instruments on himself or his sexual partners that he used on his patients without sterilizing them.

The actual route of HIV transmission in the Acer-Bergalis case will most likely never be known. There have been suggestions that the dentist did not wish to die alone and chose certain people to infect. It was suggested that he may have attempted to infect still others, but was unsuccessful. A friend of the dentist said that he believed that Acer intentionally infected his patients to call attention to the HIV/AIDS problem in the United States. Acer felt that mainstream America was ignoring the problem.

A CDC estimate put the theoretical risk of HIV transmission from an HIV-infected dentist to a patient during a procedure with potential blood exposure at 1 chance in 260,000 to 1 chance in 2.6 million (Friedland, 1991).

(Read Denis Breo (1993). The dental AIDS cases—murder or an unsolved mystery? *JAMA*, 270:2732–2734.)

Conclusion—Beginning in 2005, two of the six, Lisa Shoemaker and Sherry Johnson, believed to have been infected by Acer have progressed to AIDS but are still alive; Kimberly Bergalis, Richard Driskill, John Yecs, and Barbara Webb have died. To date, the six people have received $10 million from Acer's insurance company.

Sexual Transmission

The predominant mode of HIV transmission is through sexual contact. Not all sexual practices are equally likely to result in HIV transmission. HIV usually gains access to the immune system at mucosal surfaces. Such surfaces include the oropharynx (throat area), rectum, and genital mucosa. Mucosal surfaces are rich in Langerhans cells, dendritic cells that trap antigens and virus particles. In addition, lymphoid aggregates are found throughout the tissue immediately below the mucosal surface.

Sexual transmission of HIV occurs when infected blood, semen, or vaginal secretions from an infected person enter the bloodstream of a partner. This can happen during anal, vaginal, or oral penetration, in descending order of risk. Unprotected anal sex by a male or female appears to be the most dangerous, since the rectal wall is very thin. Masturbation or self sex is the safest. In general, a person's risk of acquiring HIV infection through sexual contact depends on (1) the number of different partners, (2) the likelihood (prevalence) of HIV infection in these partners, and (3) the probability of virus transmission during sexual contact with an infected partner. Virus transmission, in turn, may be affected by biological factors, such as concurrent sexually transmitted disease (STD) infections in either partner. Behavioral factors, such as type of sex practice and use of condoms, or varying levels of infectivity in the source partner (for example viral load) related to clinical stage of disease also increase the risk of HIV transmission/infection. Based on these factors, the risk for HIV infection is highest for an uninfected partner of an HIV-infected person practicing unsafe sex. In February 2001 researchers at the University of North Carolina, Chapel Hill reported that HIV could be transmitted via unprotected sex between 5 and 13 days after infection. Persons who have sex partners with risk factors for HIV infection or who themselves have multiple partners with high rates of injection-drug, "crack" cocaine, and methamphetamine use, prostitution, and other STDs are also at increased risk.

Circumcision and HIV Transmission

As summarized in an editorial authored by Daniel Halperin of the University of California, San Francisco, and Robert Bailey of the University of Illinois at Chicago in a recent issue of *The Lancet*, the highly vascularized prepuce contains a higher density of Langerhans cells—the primary target cells for sexual transmission of HIV—than cervical, vaginal, or rectal mucosa (Halperin, 1999). They also note the foreskin is more susceptible to traumatic epithelial disruptions during intercourse, which allows additional vulnerability to ulcerative STDs and HIV.

A summary on the Technical Meeting on Male Circumcision–Global held in Washington, D.C., in September 2002 states that a synthesis of 28 studies shows that circumcised men are 50% less likely to be infected by HIV than noncircumcised men. Among African studies, there was 69% reduction in HIV infection among higher-risk but circumcised men. Based on all data to date, male circumcision is an emerging issue of importance to HIV prevention programs worldwide.

UPDATE 2004 Currently, there are three randomized controlled trials being conducted in Kenya, South Africa and Uganda to determine whether circumcision does protect males from HIV infection. These results should be available in three to five years (2007 to 2009).

Personal Choice—Personal Risks

Ray Bradbury wrote that "living at risk is jumping off a cliff, and building your wings on the way down." About 90% of the HIV infections that occur within the heterosexual noninjection drug use population occur through one or more sexual activities. Some 90% of the HIV infections that occur among gay males occur through anal intercourse (Kingsley et al., 1990). In any sexual activity, HIV is transmitted via a body fluid.

Eighty Percent of New HIV Infections Come from Risky Behavior by Just 20 Percent of Infected People. Stop Them and We Stop HIV Transmission.

The 80/20 Rule—The fact that specific high risk behavioral groups sustain the current HIV infection rate in the United States has led scientists to name this phenomenon the 80/20 Rule: 80% of the new infections are caused by 20% of the HIV infected population.

David Holtgrave (2004) of Emory University's Rollins School of Public Health analyzed HIV transmission rates between 1978 through 2000. Holtgrave found that transmission rates dropped during the 1980s from essentially 100% to about 5.49%. The rate fell again slightly at the beginning of the 1990s, then remained relatively stable at 4.0–4.34%. Holtgrave believes this rate is surprisingly low. The drop indicates a real success of HIV prevention programs and may help to explain why the number of new HIV infections has been rather stable at 40,000 infections per year for over a decade. But in what he called a potentially important insight, Holtgrave noted that the closer the transmission rate gets to zero percent, the harder it will be to keep making continual reductions.

Following Holtgrave's findings, it means that in the mid-1980s virtually everyone who became HIV positive infected in turn someone else during any given year. With greater education and counseling, by the late 1990s the annual odds of someone passing his or her virus fell to 4.0 to 4.34%. And this would mean that at least 95% of persons living with HIV didn't transmit the disease to another person during any given year in the 90s.

Problem: Transmission calculations are based on **risk** per year. Thus, multiplying the average life expectancy of someone with HIV on therapy, say 18 years, by Holtgrave's transmission rate of 4.17 suggests that the average HIV positive person has a 75% chance of infecting someone else during his or her lifetime. But that conclusion would be misleading because smaller groups of persons living with HIV, for example, gay men and injection-drug users actually infect many people, sometimes dozens per year. These drive up the average transmission rates and contribute to the negative image many people have of HIV positive people.

Below is a presentation of the different behavioral groups and how they put themselves at risk for HIV infection and transmission.

Heterosexual HIV Transmission

Heterosexual HIV transmission means that the virus was transmitted during heterosexual sexual activities. As such, the proportion of HIV infection and AIDS cases among the heterosexual population in the United States is now increasing at a greater rate than the proportion of HIV infections and AIDS cases among homosexuals or IDUs. Persons at highest risk for heterosexually transmitted HIV infection include adolescents and adults with multiple sex partners, those with sexually transmitted diseases (STDs), and heterosexually active persons residing in areas with a high prevalence of HIV infection among IDUs. In 1985, fewer than 2% of AIDS cases were from the heterosexual population; by year 2004 the incidence of heterosexual AIDS cases is over 30% of the total number of reported AIDS cases in the United States. Of the number of the heterosexual AIDS cases reported in the United States, about two-thirds of cases occur among women and one-third among men.

Key Factors in Heterosexual HIV Transmission

The key factors involved in the transmission of HIV heterosexually are: frequent change of sexual partners, having unprotected sexual intercourse, the presence of sexually transmitted infections, the lack of male circumcision, the social vulnerability of women and young adults or adolescents, and the amount of poverty, economic instability, and the prevalence of HIV in a given community.

Heterosexual HIV Transmission Outside the United States—Heterosexual intercourse is the most common mode of transmission of HIV in poor countries. In Africa, over 80% of infections are acquired heterosexually, while mother to child transmission (5–15%) and transfusion of contaminated blood account for the remaining infections (see Sidebar 8.1 on page 229). More females than males are HIV infected in Africa. Homosexuality and injection drug use occurs, but at a very low incidence. Heterosexual contact and injection drug use are the main modes of HIV transmission in South and Southeast Asia. Studies in Haiti and other Caribbean and Third World countries indicate that HIV transmission is most prevalent among the heterosexual population. In late 1991, the World Health Organization stated that 75% of worldwide HIV transmission occurred heterosexually. By the year 2005, at least 90% occurred heterosexually. The high frequency of AIDS cases in Third World countries is thought to be due to poor hygiene, lack of medicine and medical facilities, a population that demonstrates a large variety of sexually transmitted diseases and other chronic infections, unsanitary disposal of contaminated materials, lack of refrigeration, and the reuse of hypodermic syringes and needles due to supply shortages.

Vaginal and Anal Intercourse—Among routes of HIV transmission, there is overwhelming evidence that HIV can be transmitted via anal intercourse. In vaginal intercourse, male-to-female transmission is much more efficient than the reverse. This is believed to be due to (1) a consistently higher concentration of HIV in semen than in vaginal secretions, and (2) abrasions in the vaginal mucosa (lining or membrane). Such abrasions in the tissue allow HIV to enter the vascular system in larger numbers than would occur otherwise, and perhaps at a single entry point.

BOX 8.2

HIV/AIDS ROULETTE

Case I: The Woman Executive

I am a successful executive woman. A year ago I applied for life insurance. I was required to take an HIV antibody test. It came back positive.

I am not a prostitute or promiscuous. I am not and have never been an injection-drug user. I am not a member of a minority group, indigent nor homeless, and I have not slept with a bisexual male (?).

I don't fit any of the stereotypes that people have designated for those infected with HIV. I got HIV from a man I love and have been seeing for five years. He is not homosexual or bisexual. He has never used injection drugs. He had no idea he was carrying the virus. He believes he may have been infected about six years ago by a woman with whom he had a brief, meaningless relationship. For that one indiscretion we will both pay the ultimate price.

Case II: The HIV-Infected Male

A 21-year-old man walked into a sexually transmitted disease clinic and told the doctor he had "the clap," or gonorrhea, but he carried HIV.

When a counselor inquired about his sex partners, he told them about several, including a 12-year-old girl. The girl had gone elsewhere to be treated for gonorrhea and tested positive for HIV.

The man admitted to sleeping with 27 women, including 13 teenagers. Ten of the partners couldn't be found. Of the 17 others, 12 tested positive for HIV.

The man has since died, and the clinic has not been able to track all of his sexual partners.

There is a message to be found in these two cases: People with HIV are much more dangerous to a community than someone with AIDS. On average, the HIV-infected are asymptomatic for 11 years. Those with AIDS and not treated, are symptomatic. They are losing weight. They are sick. They have little or no sexual appetite. Those with HIV are healthy and vigorous. That's where the sexual roulette begins.

The same reasoning explains why the **receptive** rather than the **insertive** homosexual partner is more likely to become HIV-infected during anal intercourse. It appears that the membranous linings of the rectum, rich in blood vessels, are more easily torn than are those of the vagina. In addition, recent studies indicate the presence of receptors for HIV in rectal mucosal tissue. A recent report by Richard Naftalin (1992) states that human semen contains at least two components, collagenase and spermine, that cause the breakdown of the membrane that supports the colonic epithelial cell layer of the rectal and colon mucosa. This leads to the loss of mucosal barrier function allowing substances to penetrate the rectal and colon mucosa.

Homosexual Anal Intercourse—In 2004, over 27% of homosexual men in San Francisco are HIV-infected, probably the highest density of infection anywhere in the developed world. "It colors everything we do out here," said a gay activist. "The gay community, to a large extent, is about addressing AIDS. It has to be, because it's literally a war: Your entire community is under siege."

In a single year, 1982, 21% of the uninfected gay male population became infected, and for some reason not yet known, many of those infected early, died early. "Soon everyone, and I mean everyone, had a friend who was dying" (Science in California, 1993). During 1995 through mid-1997, three gay men died of AIDS each day. From mid-1997 through 2004, using aggressive anti-HIV drug therapy, daily deaths from AIDS dropped to less than two per week (Conant, 1995 updated).

It appears that of all sexual activities, anal intercourse is the most efficient way to transmit HIV (DeVincenzi et al., 1989). Information collected from cross-sectional and longitudinal (cohort or group) studies has clearly implicated receptive anal intercourse as the major mode of acquiring HIV infection. The proportion of new HIV infections among gay males attributable to this single sexual practice is about 90%.

Major risk factors identified with regard to HIV transmission among gay males include anal intercourse (both receptive and insertive), active oral-anal contact, number of partners, and length of homosexual lifestyle (Kingsley et al., 1990).

Heterosexual Anal Intercourse—A number of sexuality-oriented surveys of the heterosexual population indicate that between one in five and one in ten heterosexual couples have tried or regularly practice anal intercourse. Bolling (1989) reported that 70% to 80% of women may have tried anal intercourse and that 10% to 25% of these women enjoyed anal sex on a regular basis. He also reported that 58% of women with multiple sex partners participated in anal sex. James Segars (1989) reported that the highest rates of anal sex occur among teenagers who use drugs and older married couples who are broadening their sexual experiences.

Although it may increase the risk of HIV infection, it must be emphasized that anal intercourse is not necessary for HIV transmission among heterosexuals. In fact, most HIV-infected heterosexuals say they have never practiced anal intercourse.

Risk of HIV Infection; Number of Sexual Encounters—The risk of HIV infection to a susceptible person after one or more sexual encounters is very difficult to determine. In some cases, people claim to have become infected after a single sexual encounter.

In some reported transfusion-associated HIV infections, the female partners of infected males remained HIV negative after five or more years of unprotected sexual intercourse. Television star **Paul Michael Glaser** said that he had unprotected sexual relations with his wife Elizabeth for five years prior to her being diagnosed as HIV-infected (Figure 8-5). He was not HIV-infected. She received

FIGURE 8-5 Paul Michael and Elizabeth Glaser. Elizabeth died of AIDS on Dec. 4, 1994. *(Photograph courtesy of AP/Wide World Photos)*

HIV during a blood transfusion, but was not diagnosed until after their first-born child was diagnosed with HIV. In other studies of heterosexual HIV transmission, many couples had unprotected sexual intercourse over prolonged periods of time with no more than 50% of the partners becoming HIV-infected. There are many instances of heterosexuals and homosexuals who remained HIV negative after having repeated sexual intercourse with HIV-infected partners.

The fact that not all who are repeatedly exposed become infected suggests that biological factors may play as large a role in HIV infection as behavioral factors. For biological reasons, some people may be more efficient transmitters of HIV; while others are more susceptible to HIV infection, that is, require a smaller HIV infective dose. Some people carry genes that may make them resistant to HIV infection (see "Mutant CKR-2 and CKR-5 Genes" in Chapter 5).

Number of Sexual Partners and Types of Activity—One relatively large risk factor for HIV infection in both homosexuals and heterosexuals is believed to be the number of sexual partners. The greater the number of sexual partners, the greater the probability that one of the partners is HIV positive.

The scale of multiple-partnering during the late twentieth century and continuing into the twenty-first century is unprecedented. With over 6 billion people on Earth, an ever-increasing percentage of whom are urban residents; with air travel and mass transit available to allow people from all over the world to go to the cities of their choice; with mass youth movements advocating, among other things, sexual freedom; with a feminist spirit alive in much of the industrialized world, promoting female sexual freedom; and with 47% of the world's population made up of people between the ages of 15 and 44—there can be no doubt that the amount of worldwide urban sexual energy is unparalleled.

The amount of protection one actually obtains from limiting one's number of partners depends mainly on who those partners are. Having one partner who is in a high-risk group may be more dangerous than having many partners who are not. An example of this is seen in prostitutes, who may be more likely to be infected by their regular injection-drug-using partners than by customers who are not in a high-risk group. The risk status of a person who remains faithful to a single sexual partner depends on that partner's behavior: If the partner becomes infected, often without knowing it, the monogamous individual is likely to become infected (Cohen et al., 1989).

The Effects of Sexual Partner Reduction in Uganda—In the mid-1980s Uganda had a 30% rate of HIV infection. Today it is less than 6%. A miracle? No, the bottom line was sexual partner reduction.

In 1986, the Uganda Ministry of Health started a vigorous HIV prevention campaign in which the slogans "Love Carefully," "Love Faithfully," and "Zero Grazing" (Uganda slang for don't have sexual partners outside the home) were posted on public buildings, broadcast on radio and in speeches by government officials, teachers and AIDS prevention workers across the country. Religious leaders scoured the Bible and the Koran for quotations about infidelity. Newspapers, theaters, singing groups and ordinary people spread the message of abstinence. In short, Uganda promoted their A,B,C approach to lowering the HIV infection rates—A for abstinence, B for faithful (stay with one sexual partner) and C if you fail A and B, use a condom. As a result the frequency of casual sex fell by 60% between 1989 and 1995. HIV associated pregnancy rates fell by 50%. There was a drop in the rate of HIV infection across Uganda. All of this was occurring at a time when condoms were not widely available. Some researchers have attributed Uganda's HIV prevention success to increased sexual abstinence among teenage girls, but statistics suggest that partner reduction, especially on the part of men, was far more important. Partner reduction has been an important factor in HIV prevention elsewhere as well. In Thailand, HIV infection rates declined steeply during the 1990s. While condom use increased significantly, visits by men to sex workers also fell by 60%. Among gays in America, HIV rates fell steeply during the 1980s. Part of this decline was attributable to the increased use of condoms, but partner reduction was also very important. In Zambia and northern Tanzania, where churches promoted faithfulness, HIV rates also declined. Meanwhile, in such countries as Botswana, South Africa and Zimbabwe, condoms have been emphasized as the main method of prevention, and HIV rates have remained high.

In 2002 the U.S. and other international organizations began promoting the A,B,C approach to reducing HIV infections.

Sexual Activities—In addition to a high-risk partner or a number of sexual partners, the types of sexual activities that occur are also significant (Table 8-5 on page 215). Any sexual activity that produces skin, anal, or vaginal membrane abrasions (tear) prior to or during intercourse increases the risk of infection. Recent studies show that men who have been circumcised are at less risk of infection. Foreskin of the penis tears as it is forced back over the glans, providing a direct route for HIV to enter the blood stream.

Orogenital Sex—Historically, it has been very difficult to establish the contribution oral genital sex makes to overall HIV transmission since few people engage solely in oral sex. Instead, many people also have vaginal or anal sex, which are recognized routes of HIV transmission. Though there have been a number of cases of apparent HIV transmission via oral genital sex, health professionals have tended to prioritize HIV prevention efforts in the areas of greatest risk. This strategy may have inadvertently downplayed the risks attached to oral genital sex, and left some individuals confused over risk reduction options.

How Risky Is Orogenital Sex?—Orogenital sex may be a greater risk factor for becoming HIV-infected than previously thought. The risk of transmission is related to the presence of HIV at the sexual sites (oral, vaginal, penile, and anal), the amount of HIV present, and whether there are physical openings such as tissue tears or open sores (Rothenberg et al., 1998; Kahn et al., 1998).

After reviewing over 100 research reports Sara Edwards and colleagues (1998) concluded that HIV can be transmitted through oral sex. In a very rigorous study, Beth Dillon and colleagues (2000) appear to have clearly pinpointed oral transmission in approximately 7% of primary infection cases. By mid-2000, strong evidence for at least 30 cases of oral HIV infection were reported in the United States (Dillon, 2000). In Britain, Barry Evans, director of Public Health Laboratory Service, reported that oral sex may account for up to 8% of HIV infections among gay men. He said that oral sex leads to 30 to 50 new HIV infections annually in the United Kingdom. Evans states that the risk of HIV infection via oral sex is greater than previously thought.

Susan Buchbinder and colleagues in 2001 calculated that the odds of acquiring HIV from any single act of oral sex with an infected partner are roughly 4 in 10,000 (1 in 2500) compared with odds of 4 in 1000 (1 in 250) for anal sex with a condom. Jorge del Romero and colleagues (2002) reported on 135 uninfected or HIV-negative men and women whose only risk of HIV exposure was through unprotected orogenital sex with their HIV-infected partners. Collectively, the 135 people had over 19,000 unprotected oral genital contacts with their HIV-infected partners. None of the 135 HIV-negative partners became infected!

It should also be recognized that one can get other sexually transmitted diseases from oral sex, such as gonorrhea, syphilis, chlamydia, and herpes.

Prostitution (Sex Worker): United States

There is little if any evidence that prostitutes in the United States and most other developed nations play a large role in heterosexual HIV transmission (Cohen et al., 1989 updated).

A consequence of the attention prostitutes or sex workers have attracted in relation to AIDS in the United States and other developed countries is that they tend to be seen as responsible for the spread of HIV—an attitude reflected in descriptions of prostitutes as reservoirs of infection or high-frequency transmitters. But, the sex worker is only the most visible side of a transaction that involves two people: For every sex worker who is HIV positive there is, somewhere, the partner from whom she or he contracted HIV. Given the fact that the chance of contracting HIV during a single act of unprotected sex is not high, infection in a sex worker is likely to mean that she or he has been repeatedly exposed to HIV by clients who did not or would not wear condoms. The more accurate way of reading the statistics of HIV infection in prostitutes or sex workers is to view them as an indication of how strong a foothold the epidemic has gotten within a community.

Prostitution—Developing Countries

In a garbage-strewn alley in the city's red-light district, a 24-year-old prostitute slumps on a string bed, weak from tuberculosis and diarrhea. She knows she has AIDS but she has never heard

POINT OF INFORMATION 8.3

DISTRIBUTION OF HIV-CONTAMINATED BLOOD: FRANCE, GERMANY, THE UNITED STATES, AND OTHER COUNTRIES

Near the end he could not bring himself to visit his youngest brother, to see him dying of AIDS. He too was dying of AIDS. Their deaths would close out a family of four HIV-infected hemophiliac brothers, all diagnosed with AIDS. The first died at age 24, the second committed suicide at age 33, the last two brothers died in 1993. The four brothers became HIV-infected about 1985 from using HIV-contaminated blood-clotting factor.

Blood banks, centers for blood transfusions, and companies involved in the creation and distribution of blood clotting factor VIII used to save the lives of hemophiliacs have been sued for selling blood and blood products that they should have known were contaminated with HIV. The use of these HIV-contaminated products by hemophiliacs became a worldwide tragedy. Many thousands of hemophiliacs have died prematurely because of these products. In France, over 4000 have been infected and over 1000 have died. French ministers were charged with collusion in poisoning people who used the contaminated blood products. In Germany, authorities closed blood plasma centers with charges of fraud, greed, and negligent bodily harm. The owners of the different plasma centers were charged with 5837 counts of attempted murder—corresponding to the number of blood products distributed.

In the United States, over 9000 hemophiliacs became HIV-infected after using blood products that either were not tested for HIV or if tested were distributed regardless. France and Germany did the same thing. The evidence showed that the heads of the companies did not wish to discard large batches of blood products because it meant large financial losses! Other countries involved in similar HIV-contaminated blood distribution scandals are: Canada, Switzerland, Japan, India, Iran, Ireland, Pakistan, Russia, and China. Although each country's distribution of HIV-contaminiated blood and blood products is scandalous, China may be the most scandalous of all.

China

In October 1996, China's Ministry of Public Health ordered thousands of state-run medical institutions to stop using a certain brand of serum albumin, a common blood product produced by a military-run factory.

Two years after Chinese journalists in Hong Kong and the United States first published reports that the blood product was contaminated with HIV, China's Foreign Ministry acknowledged that tests of some samples of the blood product indicated the presence of HIV. But, at the Department of Medicine Administration and Control, a Mr. Liu said: "No, no, it has never happened. Some people outside China and some foreign journalists just made their reports on the basis of rumors."

In October 1998, in an effort to reduce the spread of HIV, the Chinese government created a law requiring that all blood products come from volunteers. Many countries banned the practice of payment for blood donation long ago, since there were concerns that the people who would be most likely to be strapped for money—such as drug users—would also be at high risk for blood-borne diseases. However, in China it can be hard to get donors because there is a cultural aversion to blood donation. Traditionally, the donation of blood is considered disrespectful to ancestors and parents; the culture also equates blood levels with health. Because China has a chronic blood shortage, many experts believe it will take several years for the illegal blood trade to die down. The government is trying to confront this problem by actively encouraging officials, students, and soldiers to donate blood.

Chinese health officials now believe that 50% of about 2 million HIV-positive Chinese contracted HIV through the blood trade.

The Secret of China's Henan Province, 2001

Hints of a secret AIDS epidemic in China's countryside first began to surface in 2000. The depth of the tragedy and the unusually high incidence and impact of HIV/AIDS on villages in the Henan Province are now emerging, as disparate, dying farmers have begun to tell their stories of illness and death. All their stories begin with their willingness to sell their blood for needed money. The blood donors earned between $12 and $15 per donation. Some farmers reported donating blood 50 times in two months. Because of the plasma collection methods routinely used at the time throughout China, even those who donated only a few times ran a high risk of becoming ill. Blood from dozens of sellers was pooled and put into a huge centrifuge where it was spun at a speed sufficient to separate out the desired plasma. The remaining fraction, mainly red cells, was divided up and transfused back into the sellers, who felt the process to be healthful because it limited the blood loss. The process was highly unsanitary. The result was once one blood seller in a village was infected with HIV or hepatitis, the rest were quick to become ill because the viruses from other people's bodies rode along with the red cells back into their veins. Because the sellers did not lose enough red cells with any donation to result in anemia, using this method meant that people could sell their blood frequently—raising their chance of infection.

In the village of Donghu, population 4500, every family has one or more members who suffer from fevers, chronic diarrhea, mouth sores, unbearable headaches, weight loss, racking coughs, and boils that do not heal. Residents estimate that more than 80% of adults carry HIV and more than 60% are already suffering debilitating symptoms. That would give this village, and the others like it in the province, localized rates that are the highest in the world.

Physicians who have worked in the Henan province said more than a million people had probably contracted the virus from selling their blood. They add that while the sale of blood has stopped in the most severely affected villages, it continues elsewhere to a lesser extent, both in Henan and other provinces.

Repeats of the Henan tragedy also occurred in the provinces of Qinghai and Shaanxi.

Beginning 2005, China continues to downplay the effect of returning blood cells to paid donors in the Henan province. Officials have stopped villagers from traveling outside of Henan province or from speaking to the media. The epidemic in Henan province is still shrouded in silence because local officials were often involved in the profitable blood business. So while the authorities in Beijing held a big international conference in November 2001 and aired television dramas on the subject, repression and some concealment continues into 2005.

SUMMARY

The issues in these blood scandals were delays in institution of donor screening, delays in screening donated blood with HIV antibody testing, and failure to withdraw contaminated factor from use and replace it with heat-treated factor for economic reasons—after it was known that these actions would protect individuals from infection.

According to the World Health Organization (WHO), the majority of nations are using unsafe blood with regard to HIV and hepatitis. The use of unsafe blood is most common in developing countries, home to 80% of the world's population, or some 4.8 billion people. Because of the high cost of blood testing, between $40 to $50 per unit, more than 13 million pints of blood annually are not checked for transmissible infections including HIV, malaria, and syphilis. Globally, transfusions of contaminated blood are thought to have caused up to 10% of AIDS patients. Moreover, unsafe transfusions and injection practices are estimated to cause up to 160,000 new HIV infections, 16 million new cases of hepatitis B, and 4.7 million new hepatitis C infections each year. In response to these findings, the WHO, in cooperation with the International Federation of Red Cross and Red Crescent Societies, launched a global campaign to promote blood safety.

For the United States, the CDC states that over the past 20 years of blood screening for HIV (1985–2004) 46 adults and three children have developed AIDS after receiving an HIV-negative blood transfusion. This blood was taken in the window period before antibodies were made against HIV (Klein, 2000 updated). The majority of over 9000 HIV-contaminated blood transfusions occurred prior to 1985, the year when HIV blood screening began. It is estimated that about 50 new HIV-contaminated blood transfusions continue to occur each year in America. For additional information on blood scandals, read *Blood Feuds: AIDS, Blood and the Politics of Medical Disaster*, edited by Eric Feldman and Ronald Bayer, Oxford University Press, 1999.

BOX 8.3

GAY MEN PUTTING THEMSELVES AT HIGH RISK FOR HIV INFECTION

Gordon Mansergh reported at the Twelfth International AIDS Conference (June-July 1998) that of gay men in Denver, Chicago, and San Francisco, almost two-thirds engaged in unprotected anal sex in the previous 18 months and that 56% of gay men under age 25 had unprotected receptive anal sex in the same time frame.

Experts believe one reason risky behavior continues among the young is that they have not yet seen their friends die of the disease.

Another is simply the kind of risk-taking common among the young—the same impulse that prompts teenagers to drive fast or take up smoking. Indicators of increased risky sex among gay men included: gay men having **unprotected sex** increased from 35% in 1994 to 55% in early 1999. Thirty percent of 3000 gay men surveyed in four large American cities had **unprotected sex**—7% of them were HIV positive. In three large American cities, 66% of gay men surveyed had **unprotected sex** in the last 18 months. **One in five** gay men in San Francisco and New York City said they had **unprotected sex** with an HIV-negative partner or with a partner whose HIV status was unknown. In 1998 the CDC surveyed about 22,000 gay men in San Francisco. Over 39% said they have **unprotected sex**—this is a 30% increase since 1994. Linda Valleroy and colleagues at the CDC reported at the Eighth Conference on Retroviruses, February 2001, on testing and interview results of 2400 gay men ages 23 through 29. The men were from Baltimore, Dallas, New York, Los Angeles, Miami, and Seattle. Over 12% or 293 of the men were new HIV positives. **Twenty-nine percent** knew they were HIV positive before testing! Forty-six percent said they had unprotected oral sex during the previous six months. These data shocked the researchers.

On January 29, 1999, the *San Francisco Chronicle* carried a story about an $8 admission for a night of communal sex—the rules: no clothes, no condoms, no discussion of HIV. The article also covers an internet link offering gay men the **extreme sex party** where becoming infected or HIV-infecting another is the erotic allure of the party. Still another twist is the **Russian Roulette Party** where three noninfected men have sex with five others, one of the five being HIV positive!

Exit the Condom

Over the last three years, the number of new HIV infections in San Francisco held at about 500 per year. In 2002 that number doubled and continues to rise. The number of new syphilis cases also increased.

During a health inspection of a gay sex club, the inspector felt something crunch underfoot. It was an empty blister pack of **Viagra.** The inspector, San Francisco's director of sexually transmitted disease prevention, began to wonder whether Pfizer's impotence drug was contributing to unsafe sexual behavior and fueling an increase in HIV and other diseases. As a result of follow-up studies, he believes he has his answer. In the first three months of 2003, 43 new cases of syphilis and 14 new HIV infections have been diagnosed in Viagra users in San Francisco. Together these data imply that gay men are reverting to unsafe sex. And all this comes after the state and federal government has spent millions of dollars on AIDS education and prevention.

Effect of Highly Active Antiretroviral Therapy (HAART) on Sexual Behavior and HIV Transmission

Mitchell Katz and colleagues (2002) and William Holmes and colleagues (2002) examined the data from interviews from over 26,000 men having sex with men (MSM) conducted between 1995 and 2002. They found that although the use of HAART reduced the risk of HIV transmission, the success of that therapy might have caused MSM to be less cautious and more likely to engage in unsafe sex. The interviews indicated that, as a result of the availability of HAART, HIV-negative men who have sex with men (MSM) are less concerned about contracting HIV; HIV-positive MSM are less concerned about transmitting HIV; and both groups are more likely to engage in unsafe sex. Also because HAART decreases mortality and improves the quality of life of persons with AIDS, it has increased the number of persons living with HIV/AIDS who are engaging in sexual relations.

Recent studies on HIV infected gay males in San Francisco revealed that the use of HAART reduced HIV transmission by 60%. However, the increase in unprotected anal sex rose from about 8% in 2000 to 25% in 2003 during the same time period. This behavioral change, going from protected to unprotected anal sex has offset the beneficial effects of HAART. In addition to these data, John McGowan and colleagues (2004) reported that of those diagnosed with HIV in New York, 41% of men and women continued to have unprotected sex. Their reasons were trading sex for money or drugs and the availability of HAART. For those who need it, being on HAART has taken away much of their fear, and the uninfected at-risk population no longer sees the pallid face of death in public places as in the past—they never developed the fears of yesterday, the days before HAART. For many with HIV/AIDS the time of disbelief and terror evolved into burnout and despair, which in turn has now become a time of recuperation and salvation. The newly infected mostly see salvation through HAART. They see magazine ads that show hot muscular men living life to the fullest thanks to HAART. Other ads show couples holding hands, sending messages that the road to true love and happiness is being HIV positive. Unlike the photos of buff men in the ads, most who are on drug cocktails are not having the time of their lives. They spend mornings in the bathroom throwing up or suffering from diarrhea. They spend afternoons at doctor's appointments, clinics and pharmacies. And they spend endless evenings planning their estates and trying to make ends meet because they are not well enough to support themselves and HAART. The reality is, AIDS is not fun. It's not sexy or manageable. AIDS is a debilitating, deforming, terminal and incurable disease. And the drugs used can bring on heart, kidney and liver disease, as well as a host of daily discomforts.

The Use of Methamphetamine (Crystal or Tina)

"Whenever I want sex, I want meth and whenever I am high on meth, I want sex." The use of crystal meth by gay and heterosexual men is not new, it is the sudden accelerated use of the drug among gay men that is of major concern.

Meth use and attendant HIV transmission has become such a concern across the nation that in New York City, Gay Men's Health Crisis, one of the nation's largest gay AIDS/HIV groups, has launched a major education campaign. The organization is putting up billboards, sending out mailings, sponsoring workshops and dispatching counselors into the community to talk about meth abuse and HIV. Meth, a psychostimulant that excites pleasure centers in the brain, makes users feel euphoric for hours. The drug impairs judgment, lowers inhibitions, keeps people awake for days and can increase sexual arousal. They go from feeling like wallflowers to feeling like supermen, and safer sex messages are just forgotten. This drug has become almost normalized in the gay community. One gay male said, "It gives people a way to have sex for hours and hours and hours. It's the greatest euphoria you can ever feel."

Party and Play

Meth is so linked with this subculture of gay men engaging in anonymous sex with strangers that men advertise either that they have the drug or want it during sex in personal ads and on the internet. Their notices carry the phrase "PnP" for "party and play," a euphemism for crystal methamphetamine and sex. For one gay male, his friends introduced him to crystal meth on a Thursday evening and he stopped using it on Monday morning. He lost count over the weekend when he hit having sex with 12 men. About a month later he developed a flu-like syndrome, got tested and he was HIV positive. In his mind he was thinking this was just one weekend. One weekend and it will impact him for the rest of his life.

Results of Syphilis Study Among Men Who Have Sex with Men (MSM) in New York City, 2003—A survey of 88 gay or bisexual men with syphilis and 176 gay/bisexual men without syphilis (controls), all aged 18 to 55 who reported at least one sex partner over the last year produced the following summation of their sexual behavior: Many MSMs, both cases and controls, were engaging in high-risk sexual behaviors. Specifically, that meant sex with multiple, anonymous partners, unprotected anal intercourse (barebacking), recreational drug use before sex, and not discussing HIV status prior to sex with a partner. Among men with syphilis, the average number of sex partners in the prior six months was 16 compared to 11 among controls. When barebacking first became popular in the late 1990s, its most visible proponents were HIV-positive men who were seeking other HIV-positive men for unprotected sex. Those early barebackers would often reject men who did not know their HIV status or who were HIV negative. That was not the case in the city study. People aren't talking about HIV before having sex. Forty-three of the 88 men with syphilis and 29 of the 176 men without syphilis were HIV positive, with 24 cases and 12 controls taking HIV drugs, respectively. Nearly all HIV-positive men knew their serostatus. Two cases and four controls were newly diagnosed during the study. On average, the HIV-positive men had been infected seven years earlier. The study suggests that these men have largely abandoned safe sex practices, and notes that most of the men were not exposed to any recent safe sex messages.

Circuit Parties—Circuit parties got their name from those who travel to various cities—the circuit—to attend several parties each year. The parties, which began in 1986, essentially are the gay version of raves, parties that are popular among some young people. There are two or three circuit party weekends a month. Among the cities where they are held: Montreal, San Francisco, Atlanta, Palm Springs, Miami, and Washington. The parties are not universally popular among gay men, although many say they have attended one. The parties generally attract professionals ages 21 to 35. The parties began with the intention of fund raising and community building. But somewhere along the way, the original intent of the party has become diluted. Now, circuit parties have become weekend bashes. They attract thousands of mostly young gay men who dance until dawn and whose admission fees raise millions of dollars for AIDS prevention groups and gay charities. But, according to Ronald Johnson, an official of the circuit party organization, "It (circuit parties) became a social phenomenon above and beyond what (we) intended and beyond what (we) could control." Health officials say the parties have become a reflection of the risky behavior that is contributing to rising rates of HIV infection among gay men. Drugs are so prevalent at the parties that organizers often hire medical teams to treat overdoses. Troy Masters, publisher of a gay newspaper in New York, who now opposes these parties said, "You wouldn't find the American Cancer Society throwing a smoking party." Gay Men's Health Crisis, which was founded in 1981 and serves 11,000 clients annually in the New York area, stopped holding its party in 1998 after it became known for drug use and sex. Two Philadelphia couples said they attend up to eight parties a year and usually take Ecstasy pills with a liquid shot of GHB or some ketamine, which can be liquid or powder. They agreed that "if you don't do drugs, you're not going to enjoy it as much."

BAREBACKING: CHASING THE BUG—INTENTIONALLY SEEKING HIV INFECTION

Bug chasing sounds like a group of children running around chasing crickets, butterflies, or grasshoppers. Enter Robert, age 30. He routinely hooks up with three to four men each week for unprotected sex—he hopes they are all HIV positive! For Robert, bug chasing is mostly about the excitement of doing something that everyone sees as crazy and wrong. Keeping this part of his life secret is part of the turn-on. That forbidden aspect makes HIV infection incredibly exciting for him, so much so that he seeks out sex exclusively with HIV-positive men. "This is something that no one knows about me, it's mine. It's my dirty little secret." He compares bug chasing to the thrill you get by "screwing your boyfriend in your parents' house," or having sex on your boss's desk. You are not supposed to do it, and that's exactly what makes it so much fun. When asked why he wants to become infected, his eyes light up as he says that the actual moment of transmission, the instant he gets HIV will be the "most erotic thing I can imagine." When asked whether he is prepared to live with HIV after that "erotic" moment, he dismisses living with HIV as a minor annoyance. Like most bug chasers, he has the impression that the virus isn't such a big deal anymore. "It's like living with diabetes. You take a few pills and get on with your life."

It has recently been observed that a minority of both HIV-positive and negative men have begun to consciously, willfully, and proudly engage in unprotected anal sex. The new phenomenon is referred to as raw, skin-to-skin, or bareback sex. Such behavior in this group leads to the intentional transmission and reception of HIV! This form of barebacking is referred to as **"bug chasing" or "chasing the bug."** These HIV-negative gay men seek to become HIV-infected.

Brotherhood of the Infected

According to Yvonne Abraham of the *Boston Globe Online,* and DeAnn Gauthier (1999), health professionals are seeing something unexpected in the gay community: a small group of men who want to contract HIV in the belief that it offers community and kinship. Increasing numbers of chat rooms and websites approach the subject of men who want to convert from HIV negative to positive. Marshall Forstein, medical director of mental health at Fenway Community Health Center in Boston, says some of the men feel lonely or shunned and believe that HIV will bring them attention from friends and caregivers. One California activist who stopped using condoms and later contracted HIV also noted, "When I was entering the gay community at the height [of the epidemic], I felt like I'd joined the war. And that somehow transformed HIV into this rite of passage as opposed to something to be reviled and avoided." Some websites even discuss conversion parties, where uninfected men can have unprotected sex with HIV-positive men to try to contract the virus. The groups refer to HIV infection as a rebirth and a way to bond with a new family. According to Gautheir, bug chasing as a form of bareback sex actually involves two categories of participants. There are the HIV-negative men who seek to become infected with the virus (bug chasers), and there are the gift givers, HIV-positive men who seek to share their gift of HIV to others. Chasers typically advertise for partners with statements such as: "Will let you _____ me raw only if you promise to give me all your diseases like AIDS/herpes, etc. Let's do it." Gift givers, on the other hand, typically make comments such as: "Attention neg men! Why stay locked in a boring world of sterile sex when you can join the ranks of the AIDS Freedom Fighters? Let me give you my gift and set you free." Becoming "bug brothers" may occur one-on-one but is just as likely to result at special marathon group sex parties that are held for the purpose of seroconverting as many HIV-negative participants as possible. That individuals would knowingly participate in such events is shocking not only to some members of the gay community, but also the community at large (Gauthier, 1999).

The story of chasing HIV until infected is presented in a film, **"The Gift."** It became available in 2003 and was shown nationally on the Sundance TV network on February 2, 2004. It is a factually powerful presentation of gay men in search of **The Gift,** HIV.

THE HIGH RATE OF HIV INFECTION AMONG GAY MEN ON SOUTH BEACH, MIAMI

Miami's South Beach—hot music, hot spots, hot sun, hot bodies. The once sagging tip of Miami Beach has been turned into the playground of the beautiful, the rich, and the chic. South Beach is also a growing mecca for young gay men. In a world where homosexuality is often confined to the closet, it is a place to be out and to be open. But behind the perfect tans, there is dark reality. According to a new study one in every six gay men between the ages of 18 and 29 in South Beach has tested HIV positive.

William Darrow, AIDS researcher who surveyed 87 gay men ages 18 to 29 and 70 gay men age 30 and older, said, "People seem to think they're on a holiday there from everything—including AIDS." Darrow's survey showed that about 75% of gay men in both age groups had unprotected anal sex in 1996.

Testimonial—One gay male living at South Beach said he fell victim to love. "I was in a relationship, and we both got tested together and we tested negative, so we started having unprotected sex. But during the relationship, my partner was cheating on me and he was infected and in turn infected me."

A majority of young gay men have grown up thinking they won't reach middle age; for those who aren't positive, the specter of AIDS haunts them. For many of those who have grown up with the safer sex message, the words of warning are often lost. In addition, the media covering anti-AIDS drugs, protease inhibitors, are generating hype and new hope, but there is a downside: Young gay men may be letting their guards down because of the hype. Another contributing high-risk behavioral factor is the growing number of gay bathhouses and sex clubs cropping up across the country. People come to South Beach to have anonymous sex; they pay to use the facility and condoms are free. Safer sex posters are everywhere. But, some argue that bathhouses are centers for HIV infection; others argue that shutting down bathhouses will drive men in search of sex to public parks or other unmonitored settings.

Table 8-5 Sexual Activity According to Degree of Risk for Transmitting HIV

Risk	No.	Activity
Lowest Risk	1.	Abstinence
	2.	Masturbating alone
	3.	Hugging/massage/dry kissing
	4.	Masturbating with another person but not touching one another
	5.	Deep wet kissing
	6.	Mutual masturbation with only external touching
	7.	Mutual masturbation with internal touching using finger cots or condoms
	8.	Frottage (rubbing a person for sexual pleasure)
	9.	Intercourse between the thighs
	10.	Mutual masturbation with orgasm *on,* not *in* partner
	11.	Use of sex toys (dildos) with condoms, or that are not shared by partners and that have been properly sterilized between uses
	12.	Cunnilingus
	13.	Fellatio without a condom, but never putting the head of the penis inside mouth
	14.	Fellatio to orgasm with a condom
	15.	Fellatio without a condom, putting the head of the penis inside the mouth and withdrawing prior to ejaculation
	16.	Fellatio without a condom with ejaculation in mouth
	17.	Vaginal intercourse with a condom correctly used and spermicidal foam that kills HIV and withdrawing prior to orgasm
	18.	Anal intercourse with a condom correctly used with a lubricant that contains spermicide that kills HIV and withdrawing prior to ejaculation
	19.	Vaginal intercourse with internal ejaculation with a condom correctly used and with spermicidal foam that kills HIV
	20.	Vaginal intercourse with internal ejaculation with a condom correctly used but no spermicidal foam
	21.	Anal intercourse with internal ejaculation with a condom correctly used with spermicide that kills HIV
	22.	Brachiovaginal activities (fisting)
	23.	Brachioproctic activities (anal fisting)
	24.	Use of sex toys by more than one partner without a condom and that have not been sterilized between uses
	25.	Vaginal intercourse using spermicidal foam but without a condom and withdrawing prior to ejaculation
	26.	Vaginal intercourse without spermicidal foam and without a condom and withdrawing prior to ejaculation
	27.	Anal intercourse with a condom and withdrawing prior to ejaculation
	28.	Vaginal intercourse with internal ejaculation without a condom but with spermicidal foam
	29.	Vaginal intercourse with internal ejaculation without a condom and without any other form of barrier contraception
Highest Risk	30.	Anal intercourse with internal ejaculation without a condom

(Source: Shernoff, 1988)

of the multidrug cocktail that curbs the progress of the disease. Even if she had, she could never afford to buy it. Nobody wants the girls once they get sick, explains a health counselor who visits the alley each week. She says prostitutes with AIDS are shunned by many hospital doctors, brothel operators, and their own families. "We feel helpless; all we can do is comfort them and find them a place where they can go to die." This true vignette is happening in small towns, the suburbs, and in every large city in every country in developing nations. Many of the prostitutes are housewives out to make the difference between starvation and existence for their families. The prostitutes sell their bodies for money or drugs to anyone who can afford their services. In some towns in Africa, Vietnam, and India between 50% and 90% of the prostitutes are HIV positive. Prostitution in developing countries is a large contributing factor to the spread of HIV/AIDS.

In Chennai, India, mobile brothels manned by cell-phone toting operators are proving profitable. For those in the trade, running a mobile unit makes sound business sense as it cuts down on operational costs of renting a building and paying bribes. As for the client, it promises "pick and drop" sex workers at a place of his choice. More than 17 mobile units operate in the city, according to Chennai-based NGO Indian Community Welfare Organization (ICWO), which works with Commercial Sex Workers (CSWs). According to estimates, there are 6300 women sex workers operating throughout the city. Thirty percent said they couldn't use condoms with regular clients. Another 30% said, at times, they would forget to have their clients wear condoms and 10% felt that condom usage would prolong sexual activity. Another 10% said some clients would object to it. About 20% said they knew better techniques to avoid infection! As for reasons why they chose to be sex workers, 31% respondents said they entered the profession due to family debts and 29% said their husbands deserted them. About 29% said their lovers had ditched them.

Risk Estimates for HIV Infection During Sexual Intercourse in the Heterosexual Population—Norman Hearst and Stephen Hulley of the University of California, San Francisco calculated the odds of heterosexual HIV transmission at 1 in 500 for a single act of sexual intercourse with an HIV-infected partner when no condom is used.

Estimates of the risk of HIV infection from a single heterosexual encounter and after five years of frequent heterosexual contact for various types of partners depends on the following: (1) the probability that the sexual partner carries the virus; (2) if one of the sexual partners is HIV positive, the size of their viral load (suppression of viral load reduces the risk of HIV transmission); and (3) the reduction in risk by using condoms and spermicides.

Choosing a partner who is not in a high-risk group provides almost 5000-fold protection compared with choosing a partner in the highest risk category (Figure 8-6).

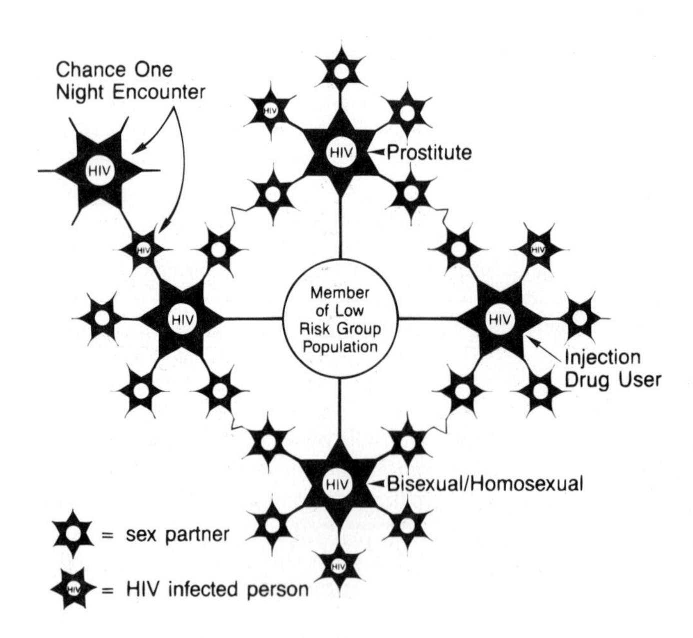

FIGURE 8-6 Risk Transmission of HIV. Sexual transmission can occur among homosexuals or heterosexuals. Prostitutes can be either male or female. The diagram shows possible bridges for transmission of HIV from high-risk groups into low-risk groups. To be safe, *you* must not become part of the chain.

BOTTOM LINE—Researchers have developed estimates of the risk of transmission of HIV. These estimates can give you a general idea of which activities are more or less risky. They cannot tell you that any activity is safe, or how many times you can do them without getting infected.

BOX 8.4

SPORTS AND HIV/AIDS: EARVIN "MAGIC" JOHNSON AND OTHER ATHLETES

HIV-INFECTED ATHLETES AND COMPETITION

The question regarding whether HIV-infected athletes should be allowed to compete has two facets:

1. Should these athletes be banned from competition to avoid the risk of spreading HIV infection?
2. Does the exercise that is demanded in competition accelerate progression of HIV disease?

There is no hard, fast, scientifically supported answer to either question. However, as of the beginning of 2005, there has not been a single reported case of HIV transmission in any sporting event worldwide (Drotman, 1996 updated).

Magic Johnson: Professional Basketball Player, Los Angeles Lakers, HIV Positive

On November 7, 1991, Magic Johnson, age 32, appeared at a nationally televised press conference and said, "Because of the HIV virus I have obtained, I will have to announce my retirement from the Lakers today." He admitted having been "naïve" about AIDS and added, "Here I am saying *it can happen to anybody,* even me, Magic Johnson." He also assured the world that his wife, Cookie Kelly, two months pregnant, had tested negative for the virus.

Some Events Since Magic's Announced Retirement

June 4, 1992—Earvin III is born *without* antibody to HIV. As of August 14, 2004, Magic is age 46. His wife, age 46, and son, age 14, are HIV negative (Figure 8-6).

Magic announced in April 1997 that his AIDS drug combination, which included a protease inhibitor, has reduced his viral load to unmeasurable blood levels.

February 16, 2000—Magic Johnson joined the AIDS Healthcare Foundation (AHF), the largest direct provider of HIV/AIDS care in the United States, in unveiling ***Thrive Magazine,*** a new national magazine targeting minorities living with and impacted by HIV and AIDS. *Thrive Magazine* seeks to address the severity of the impact of AIDS among people of color by providing minority communities a unique, multicultural perspective on resources and information concerning HIV education and treatment. In August 2001, *Sports Illustrated* reported that Magic takes Combivir (AZT+3 TC) and a protease inhibitor. Magic said "I never thought I would die from AIDS." He believes he owes his good health to "keeping stress out of my life."

In July 2001, Magic played in a basketball game in a Summer Pro League in Los Angeles, nearly 10 years after he announced that he was HIV positive. In January 2003, Magic Johnson reached an agreement with Glaxo-Smith Kline (GKS), a leading antiretroviral drug manufacturer to use his image in promoting Combivir for HIV drug therapy among the urban black population. Newspaper, billboard, and subway posters include photos of a robust-looking Johnson and feature messages such as "Staying healthy is about a few basic things: A positive attitude, partnering with my doctor, taking my medicine everyday."

Why Magic Johnson? Peter Hare, vice president of GKS, noted that AIDS is the leading cause of death for blacks between the ages of 24 and 44. One in 50 black men and 1 in 160 black women are believed to be HIV positive. About one in three does not know he or she is infected. And, the traditional methods of marketing HIV drugs do not always reach blacks. Hare said, "This group doesn't particularly trust the health-care system. Research shows that they want someone they believe, and they believe Magic Johnson." His September 2004 checkup confirms he's still asymptomatic after 13 years.

OTHER SPORTS, OTHER ATHLETES

Basketball players are not the only athletes whose behavior may place them at risk for HIV infection. In 1996 there were 6114 professional athletes in the United States involved in boxing (3500), National Football League (1590), National Hockey League (676), and the National Basketball Association (348). Of these professionals, the CDC estimated 30 to be HIV positive.

John Elson (1991) wrote a revealing article for *Time* magazine just after Magic Johnson revealed his HIV status. Elson tells of groupies that follow athletes in all sports. They are usually

FIGURE 8-6 Earvin "Magic" Johnson, wife Cookie, and son Andre arrive for a special ceremony of *Hoodlum* at a Magic Johnson-owned theater. Magic has become a spokesman for those struggling against HIV/AIDS. *(Photograph courtesy of AP/Wide World Photos/Michael Caulfied)*

college-age or older. Mainly they seek money, attention, and the glamour of associating with celebrated and highly visible "hard bodies." According to a 31-year old who has had affairs with athletes in two sports, "For women, many of whom don't have meaningful work, the only way to identify themselves is to say whom they have slept with. A woman who sleeps around is called a whore. But a woman who has slept with Magic Johnson is a woman who has slept with Magic Johnson. It's almost as if it gives her legitimacy."

The Girls

Baseball players call them "Annies." To riders on the rodeo circuit, they are "buckle bunnies." To most other athletes, they are "wannabes" or just "the girls." They can be found hanging out anywhere they might catch an off-duty sports hero's eye and fancy, or in the lobbies of hotels where teams on the road check in. To the athletes who care to indulge them—and many do—these readily available groupies offer pro sport's ultimate perk: free and easy recreational sex, no questions asked. Recently, an HIV-infected female stated publicly that she had had sex with at least 50 Canadian ice hockey players. She could not recall their names. The sex may be free, but now there is a price for the lifestyle—HIV/AIDS.

Sports/Injuries/Blood

Concerns over the transmission of HIV are shared throughout sports, particularly those sports that cause blood-letting injuries—football, hockey, and boxing. In football Jerry Smith, a former Washington Redskin, died of AIDS in 1986; in December 2003, Roy Simmons, an offensive lineman for the New York Giants and Washington Redskins, revealed that he is HIV positive. He tested HIV positive in 1997 and has a homosexual lifestyle. In 1996, the National Football League officials estimated that there was the *possibility* of one HIV transmission from body fluid exchange in 85 million football games played. In boxing, Esteban DeJesus, WBC lightweight champion, died of AIDS in 1989. Four other boxers are known to be HIV positive.

Tommy Morrison

In February 1996 Tommy Morrison, age 27, a heavyweight boxing title contender said, on announcing that he was HIV positive, "I honestly believed I had a better chance of winning the lottery than contracting this disease. I've never been so wrong in my life. I'm here to tell you I thought that I was bulletproof, and I'm not." Morrison went on to describe his promiscuous sexual lifestyle and ignorance about AIDS. Former heavyweight champ Floyd Paterson, former chairman of the New York State Athletic Commission, who was asked why his organization waited until the Tommy Morrison case to institute HIV testing for boxers, said "AIDS just came out, I go back to the '50s. I fought for 23 years. There was no AIDS. I just heard of AIDS a few weeks ago." Just heard of AIDS? Since Morrison's announcement, 12 states have banned HIV-positive professional boxers from the ring. In November 1996, Morrison fought in Japan. He knocked his opponent out in 1:38 (1 minute, 38 seconds). The fight would have been stopped at the first sign of Morrison's blood.

Morrison refers to himself as the most educated person he knows regarding HIV (he read Peter Duesberg's book, *Inventing the AIDS Virus* and Richard Wilner's *Deadly Deception*). "I don't know how I know the things I know, I just know I'm right—I have not been sick in 5 years—if it ain't broke don't fix it." Fighting weight about 240 pounds—early 1998 he weighed about 185.

By early 1998, nine states, (Oregon, Indiana, Washington, Pennsylvania, Nevada, Arizona, Maryland, New Jersey, and New York) and Puerto Rico required all professional boxers to be HIV tested. There are no known cases of HIV transmission via a boxing match. In late 1998 Morrison began taking antiretroviral medications.

There is a certain irony in all of this because as part of his probationary terms of a suspended sentence for weapons violation and assault charges, he gives speeches to high school and college students on HIV/ AIDS!

Morrison had his sperm washed free of HIV and fathered a boy in late 2003. He recently contacted Sylvester Stallone about playing a role in the upcoming Rocky VI. He is also writing his autobiography.

Ice Skating

In professional ice skating, the *Calgary Herald* reported that by 1992, at least 40 top U.S. and Canadian male skaters and coaches have died from AIDS (among them Rob McCall, Brian Pockar, Dennis Coi, and Shawn McGill).

Swimming

In February 1995, Greg Louganis, the greatest diver in Olympic history, announced that he had AIDS.

Race Car Drivers

It was reported in 1996 that Tim Richmond, race car driver, age 34, died of AIDS. He won 13 Winston Cup races on the NASCAR racing circuit. One report states that Richmond may have infected up to 30 women (Knight-Tribune Service, March 27, 1996, A-1). His physician estimated that he was HIV positive for at least eight years. During this time, according to accounts of friends he was sexually promiscuous. (Richmond actually died in 1989 but his story was kept silent until 1996.)

Ice Hockey

Bill Goldsworthy, five-time NHL All-Star, age 51, died of AIDS. He played 14 seasons in the NHL. He was diagnosed with AIDS in 1994. Goldworthy said his health problem was caused by drinking and sexual promiscuity.

Condoms are estimated to provide about 10-fold protection. A negative HIV antibody test provides about 2500-fold protection against false negatives.

The implication of this analysis is clear: ***Choose sex partners carefully and use condoms.***

Other Studies on Heterosexual HIV Transmission

Nancy Padian and colleagues (1991) reported that women are 17.5 times more likely to become HIV-infected from an infected male than men are to contract the disease from an infected female.

Viral Load and Heterosexual Transmission

Recent data from two separate HIV investigations shed light on the issue of gender transmission. Thomas Quinn and colleagues (2000) reported that the lower the level of HIV in the blood, the less likely HIV-infected persons will transmit the virus to their heterosexual sexual partner. This research team studied 400 heterosexual couples over 2.5 years. In each couple, only one person was HIV positive. No one with less than 1500 copies of HIV/milliliter of blood infected his or her partner. In those cases where one partner became infected, 80% of the time the viral load of the other partner was over 10,000. Such data confirm the benefit of lowering HIV levels in the blood via antiretroviral therapy.

Injection-Drug Users and HIV Transmission

While a number of drugs are ancient, their injection is relatively recent. Injecting oneself is mainly a twentieth century innovation because that is when (after 1960) cheap and disposable syringes became available. Experience from IDU transmission of HIV indicates that once HIV enters an injection-drug use population, that country can expect a large and substantial HIV epidemic.

HIV entered injection users during the mid-1970s and spread rapidly through 1983 largely unrecognized and unidentified. HIV transmission via IDU is the second most frequent risk behavior among adults/adolescents for becoming HIV-infected in the developed world (Table 8-6).

Table 8-6 Population[a] Risk of HIV Transmission Based on Type of Activity[b]

Activity	Risk
Vaginal sexual intercourse	1 infection per 1000 acts with HIV-positive partner (0.1%)
Receptive anal intercourse	5 to 10 infections per 1000 acts with HIV-positive partner (0.5%–3%)
Intravenous drug injection with infected needle	10 to 20 infections per 1000 needle uses (1%–2%)
Accidental stick in medical setting with infected needle	3 infections per 1000 sticks (0.3%)
Transfusion of screened blood[c]	1 infection per 450,000 to 650,000 donations (0.0002%–0.00015%)

[a]Rough estimates of the relative risks in the United States and Western Europe of the more risky activities that can transmit HIV. **THESE CALCULATIONS CANNOT BE USED AS A GUIDE TO INDIVIDUAL BEHAVIOR.** Risk to any one person depends on many factors that cannot be reduced to a single number. Recent research, for example, suggests that the infectiousness of HIV can vary greatly over the life of an infected person; infectiousness is likely to be high both at the very outset of the infection, before symptoms have appeared, and several years later. Also, women may be several times more likely than men to be infected during vaginal intercourse, a distinction that the overall risk figure obscures. *(Adapted from Bennett et al., 1996)*

[b]Risks vary because of differences in genital, anal, and oral mucous membranes. Sex is HIV risk-free, with or without a condom if the partner is known to be uninfected. Nonpenetrative sex such as mutual masturbation is nearly risk-free. Sex involving penetration with fingers is low risk. Sex involving unshared sex toys is risk free. Receiving oral sex is less risky than other penetrative behaviors. Penile-receptive sexual behavior is more risky than its penile-insertive counterpart. Anal or vaginal sex without a condom is far more risky than performing oral sex without a condom. *(Source: Voelker, 1996)*

[c]With the implementation of nucleic acid testing (NAT) in late 2002, the risk in transfusion of screened blood dropped to 1 in about 1 million donations.

Illicit drug injection occurs in at least 121 countries and HIV infection has been reported in IDUs in over 100 of these countries. Beginning year 2005, of 930,000 cases of AIDS reported to CDC, 335,000 (36%) were directly or indirectly associated with injecting-drug use. Injecting-drug-user-associated AIDS cases include persons who are IDUs n=288,000, their heterosexual sex partners n=42,000, and children n=6700 whose mothers were IDUs or sex partners of IDUs. About half the females and about one-third of the heterosexual males who were diagnosed with AIDS had a sex partner who was an IDU (*HIV/AIDS Surveillance Report,* 1998 updated). It is estimated that half the 45,000 new HIV infections in the United States, or 64 people per day through the year 2005 will be associated with IDUs. The epidemic among injection-drug users in New York City is an example of a very large, high-seroprevalence HIV epidemic. More than 100,000 injection-drug users have been infected with HIV, and over 50,000 cases of AIDS are reported among injection-drug users, their sexual partners, and their children in New York City (DesJarlais et al., 2000).

UNITED STATES: HETEROSEXUAL IDU

Among injection-drug users in America, about 40% to 45% are HIV-infected, 30% have a positive test result with tuberculin skin testing, 80% to 90% are infected with hepatitis C virus (HCV), 40% are infected with hepatitis B virus (HBV), and 60% abuse alcohol. The frequency of other sexually transmitted diseases ranges from 0% to 80%, and STDs are more common in women than in men. Such data are known only for injection-drug users, who account for approximately 10% of the estimated 6 million active drug users in the United States.

Women and Injection Drug Use

Over 90% of injection-drug users in the United States are heterosexuals. Thirty percent are women, of whom 90% are in their childbearing years. From 1988 to the beginning of 2005, female IDUs made up about 40% of all AIDS cases in women. Of the 41% of women infected by heterosexual contact, 38% were infected by having sex with a male IDU. During this same period about 8000 new cases of AIDS in children occurred—37% were from IDU mothers and 18% were from mothers whose sex partners were IDUs (*MMWR,* 1992; *HIV/AIDS Surveillance Report,* 2001 updated). Women IDUs and sexual partners of male IDUs represent the largest part (61%) of the estimated 100,000 HIV-infected women of childbearing age. Thus, there is a direct correlation between HIV perinatal transmission and pediatric AIDS cases and injection-drug use.

Adult Heterosexual Injection-Drug Use

Over 80% of reported adult heterosexual AIDS cases in America are associated with people who have a history of injection-drug use or have sex with IDUs. Twenty-five percent of all AIDS cases in the United States occur in IDUs; 21% of these cases occur where IDU is the only risk factor (*HIV/AIDS Surveillance Report,* 2002 updated). In New Jersey, about 45% of IDUs are now HIV-infected. Because 70% to 80% of IDUs have sex with non-drug users, IDUs are a major source of heterosexual and perinatal HIV infection in the United States and Europe. HIV is also spreading rapidly among IDUs in developing countries such as Brazil, Argentina, and Thailand (Des Jarlais et al., 1989 updated).

Injection-Drug Use and HIV Infections in Other Countries

Worldwide there are about 6 to 10 million injection-drug users. Eighty percent of IDUs are men. An estimated 3 to 4 million past and current injecting-drug users are living with HIV/AIDS. Accordingly, in many countries, IDUs represent a significant proportion of those individuals who are in need of antiretroviral drugs. The dual epidemic of injection-drug use and HIV particularly affect resource poor countries where there is limited access to HIV prevention measures such as needle and syringe exchange programs.

It appears that IDU is the major mode of HIV transmission in Kazakhstan, Malaysia, Vietnam, China, North America, Eastern Europe, The Newly Independent States, and the Middle East. Additionally, it is becoming more of an issue in West Africa and Latin America. The greatest problem has been seen in the Newly Independent States, in Eastern Europe, and in China.

Eastern Europe and the Former Soviet Union

The countries of Eastern Europe, countries of the former Soviet Union, and Russia are currently experiencing the most rapid spread of HIV anywhere in the world. Previously characterized by very low prevalence rates, the region now faces an extremely steep increase in the number of new HIV infections, up from 420,000 ending 1999 to about 2 million beginning 2005. Russian health officials estimate there may be as many as 4 million IDUs. About half are HIV-infected. They say up to 60% of IDUs are between ages 18 and 30 with precollege teenagers accounting for another 20%. Vadim Pokrovsky, head of the official AIDS prevention center, told a news conference broadcast on internet site www.presscentr.ru. that, "We are currently going through the peak of an epidemic among drug users. Currently about 90% of HIV infections are associated with IDU. In two or three years there will be another upsurge from sexual transmission of the disease."

The hotbed of the disease had moved from the tiny Baltic enclave of Kaliningrad to Moscow and its suburbs. In some parts of Moscow, up to 5% of the young people are already HIV positive. If counted as a percentage of the population, Russia will reach U.S. levels. About 80% of the infected are aged 15 to 25. They will be lost in 10 years, just after they have finished their education. It means one gets infected at 20, graduates at 30, starts working, and dies. Murray Feshbach, a specialist in Russian demographic trends, estimates that 5 to 10 million people will die of AIDS in Russia after year 2015.

At the 2002 Fourteenth International AIDS Conference, Pokrovsky said, "The spread of HIV in Russia is the fastest in the world, there is an average 2000 to 5000 new cases each week." He also said that "his government allocated less than $5 million in 2003 to fight AIDS, when $65 million is needed immediately, plus a further sum of up to $15 million per year." In contrast, he said, "While the United States in 2003 is spending over $500 million on AIDS-related research, Russia had put aside just $190,000 for that purpose."

The Ukraine's AIDS epidemic, the worst of all the former Soviet Union nations is causing alarm among Ukrainian physicians and other neighboring countries. In a nation of 50 million people, about 250,000 (5%) are HIV positive, and the level of infection continues to rise.

China

The lifting of a ban on AIDS-related reporting in 2000 in the state-run media appears to have fed a public panic through alarmist reports that China is on the verge of an epidemic. The government recently projected that the number of people infected with HIV could rise to 20 million by 2015 from an estimated 1.5 million now if preventive measures are not enforced. The current estimate provided by the Chinese health officials is believed, by the WHO and UNAIDS, to be several times that high. By 2001 HIV infections were found in every province with prevalence rates of 77% among injection-drug users, ages 20 to 29, and 11% among prostitutes.

Caribbean HIV/AIDS: Big Problems Among Small Islands

In the Caribbean, an estimated 60,000 adults and children became infected annually in 2000 through 2004, bringing the total number of infected to over a half million. The Caribbean has the highest rate of HIV infection in the world after sub-Saharan Africa and AIDS is already the single greatest cause of death among young men and women in this region. The rate of HIV infection in the Caribbean is about four times that of North America, Latin America, and South and Southeast Asia. The rate is 35 times that found in East Asia and the Pacific and almost 10 times the rate in Western Europe (Figure 8-8). According to UNAIDS, if left unchecked HIV/AIDS will, by year 2020, cause 75% of deaths in the Caribbean. Such a scenario mimics what is happening in some parts of Africa. In some countries of Africa as much as 35% to 40% of the population is infected. Once an epidemic reaches its tipping point—that threshold from which it grows exponentially—it is almost impossible to stop. For example, bodies are routinely stacked up in morgues in Nigeria, where only a few years ago those dying from the virus were largely invisible. The disease has spread over the African continent through mobility. Long-haul truck drivers having sex with prostitutes along their routes have brought it home to wives and girlfriends. Likewise, throughout the Caribbean, tourists, truck drivers, shipmates, and soldiers have been important male agents of the spread of HIV as is the use of injection-drugs. In Puerto Rico, for example, over half of 30,000 plus AIDS cases were/are associated with IDU. Between 35% and 50% of injection-drug users are HIV-infected. According to the CDC, Puerto Rico

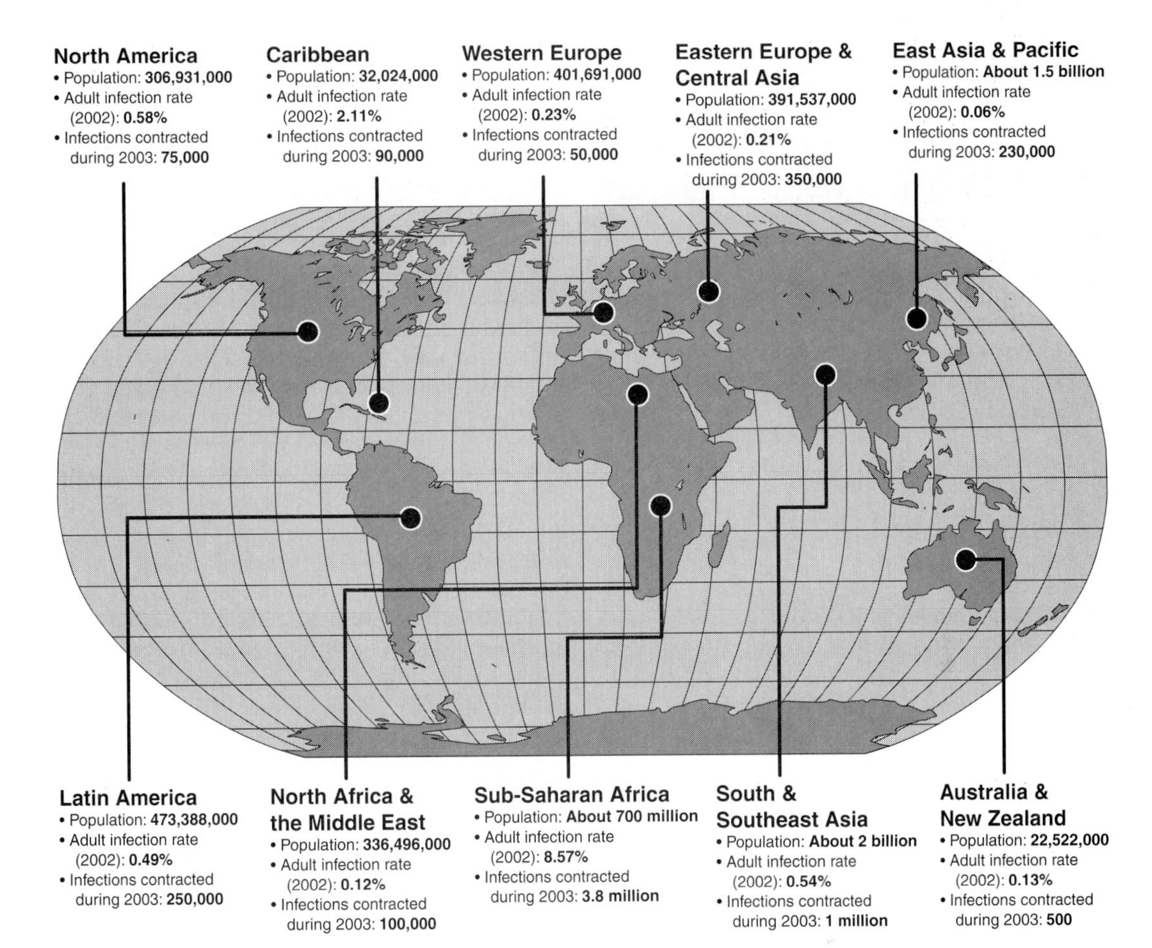

FIGURE 8-8 Global Impact of HIV/AIDS. About 41 million adults and children worldwide will be living with HIV/AIDS by the end of 2005, according to UNAIDS estimates. *(UNAIDS; World Health Organization; U.S. Centers for Disease Control and Prevention; updated)*

has the fourth-highest HIV infection rate of American states and territories, behind Washington D.C., New York City, and the U.S. Virgin Islands. Puerto Rico plays a special role as a global border town in the transmission of AIDS because it is a central transportation and commerce hub for the Caribbean. It is a favorite location for both trafficking drugs and smuggling people, especially from the Dominican Republic.

The Sex Trade, Women and Children: Two Examples, the Caribbean and Southeast Asia

Overall, in the Caribbean, the majority of HIV infection is sexually transmitted (Figure 8-9). According to a 2000 UNAIDS report, the spread of HIV is partly driven by older men who have numerous sex partners, who seek out young women for sex, and infect them. They in turn pass the virus on. If Caribbean women in general refuse unsafe sex and insist on the use of a condom—they often risk spousal abuse because it creates a suspicion of infidelity. Ten years ago there were seven times more infected men than women in the Caribbean. By 2003 that ratio changed to 2 to 1.

Southeast Asia, The Golden Triangle

There are no reliable statistics on the number of children working in the sex industry worldwide, but the lowest figure cited is 1 million. The United

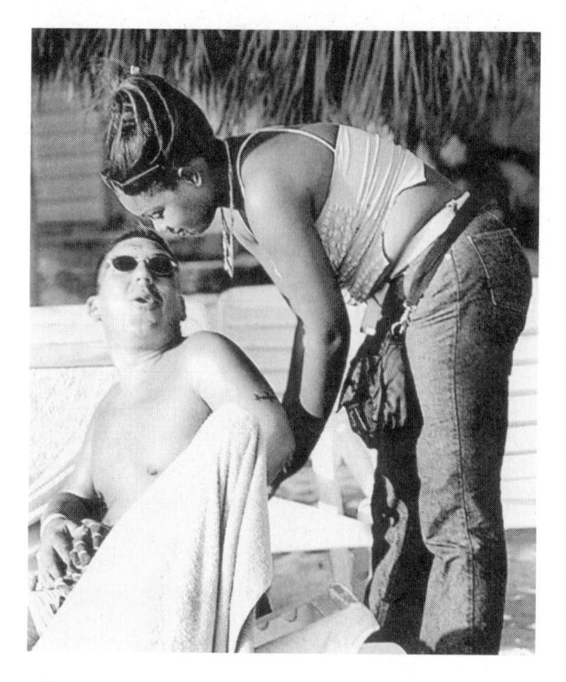

FIGURE 8-9 A Sex Worker Solicits a Tourist In Boca Chica, a Well-known Beach for Prostitution About 20 Minutes From Santo Domingo. About 100,000 women in the Dominican Republic are engaged in sex work. Most sex workers will tell you they will not have sex without a condom, but sometimes clients think that insistence on using a condom is a bargaining tactic based on economics. Since its first appearance in the Caribbean, AIDS has been a mobile epidemic closely linked to the region's underground sex trade and has altered the economy in many regions of the Caribbean and Latin America. *(Photo by Hilda M. Perez/South Florida* Sun-Sentinel*)*

Nations Children's Fund estimates that one-third of sex workers in Southeast Asia are 12 to 17 years old. The smuggling of vast quantities of heroin and amphetamines from Burma and China through Thailand has given the region its infamous tag, "The Golden Triangle," but it's the explosion in the recruitment of girls into the lucrative Thai sex industry that has put this border town on the map. Every year, hundreds of young girls from Mae Sai, a town of 80,000 inhabitants on Thailand's northern-most border with Burma, are spirited away to brothels in Bangkok where they feed the insatiable appetite of the $20 billion commercial sex industry. Mae Sai has two main trades, drugs and daughters. A nongovernmental organization in Mae Sai that works with local girls who are at risk of being sold, estimates that of the village of Pa Tek's 800 families, 7 in every 10 have sold at least one daughter into the trade.

When a Burmese migrant in Pa Tek sold his 13-year-old daughter into prostitution for $114, his wife had one regret—they didn't get a good price for her. She said, "I should have asked for $228. He robbed us." The mother earns about $100 a year selling bamboo bowls in the local market and lives in a thatched hut in Pa Tek village on the outskirts of Mae Sai. With prices varying from $114 to $913 per daughter, the latter figure is equal to almost six years' wages for most families. Parental bonds in impoverished households are easily broken. In fact, child prostitution is so established that many brothel agents live in the village, and are often friends or relatives of the family from whom they buy the children. The director of the Child Protection and Rights Center in Mae Sai said, "Agents will come to the village with orders to fill so people in Bangkok, Thai men, and foreigners, mostly Europeans, can order girls like they order pizza. If they want a girl with thin hips and big breasts, the agents will come here and find her. They always deliver."

The cross-border movement of women who are trafficked from country to country for prostitution is furthering the dispersion of HIV. For instance, outbreaks in Europe of traditionally African strains of the disease have been traced to the import of sex workers from Africa. Due to the nature of the virus, there is the potential for these new geography-jumping strains to recombine with the other versions of the virus found in other parts of the world. This could create new subtypes of the virus that would not only be untreatable by current successful antiviral treatment regimes, but could also be unaffected by vaccines now under development.

Additional Reasons for the Spread of HIV: Politics and Poverty

Political disarray and poverty are also important factors. The poorest country in the region—Haiti—has the highest infection rate, and the country most isolated from tourism in the epidemic's early years—Cuba—has the lowest. In Haiti, the epidemic was concentrated mainly in urban areas until the early 1990s, but attacks on the urban poor by the country's military juntas drove refugees back into the countryside and the epidemic traveled with them.

Caribbean Connection to America: South Florida

Five nations in the Caribbean with the highest infection rates have HIV infections in all segments of their populations. During the last 24 years over 300,000 Haitians have died of AIDS. UNAIDS estimates that Haiti, with 12% of the urban and 5% of the rural population infected (about 400,000: population 8.5 million), has the highest rate of HIV infection and 90% of all AIDS cases in the Caribbean. Next is the Bahamas at over 4% and 3% for Barbados. Guyana and the Dominican Republic are each at 2% with about 8% of their pregnant women HIV-infected. (For data on Cuba, see Chapter 9.)

The Connection

The cruise ships and planes that leave and return to Fort Lauderdale and Miami weekly depend on the allure of tropical beaches, exotic ports, and lusty streets in the islands. Discount department stores depend on shirts sewn in Haitian textile plants, and increasingly, much of the low- and high-tech workforce essential to South Florida comes from Haiti, the Dominican Republic, and other Caribbean countries. In short, there is no way that South Florida can be spared the effects of the AIDS calamity in the Caribbean. The region is so tightly linked through culture, race, ethnicity, and economics that any public health crisis in the islands is felt here. In a highly mobile, global age, there are few boundaries that cannot be crossed. The disease, carried unknowingly among tourists and workers everyday throughout the region, respects no borders.

Other Means of HIV Transmission

Other means of HIV infection have been documented. There has been a reported case of HIV transmission via **acupuncture.** It is believed that HIV-infected body fluids contaminated the acupuncture needles (Vittecoq et al., 1989). Also, **artificial insemination** can be a means of HIV transmission (Point of Information 8.2). Unlicensed and unregulated **tattoo** establishments may also present an unrecognized risk for HIV infection to patrons. If the operator does not use new needles or needles that have been autoclaved (steam sterilized), the possibility exists that infection with HIV or a number of other blood-borne pathogens may take place. In addition, single-service or individual containers of dye or ink should be used for each client.

Human Bites—Brazilian investigators reported at the 2002 Ninth Annual Retrovirus Conference that a woman was infected with HIV when her HIV-positive son, who was suffering from an AIDS-related brain seizure, accidentally bit her on the hand. This case is one of only a relatively few worldwide to show that HIV can be transmitted through human bites. The infection from son to mother was confirmed through sophisticated DNA tests showing the virus in both patients to be virtually identical. In the United States, according to police reports, a Florida woman with a history of arrests for prostitution and who has tested HIV positive, bit a 93-year-old man on his arm, head, and leg while robbing him. The bites required stitches. The man initially tested HIV negative but a test several months later was HIV positive. A complete investigation into his personal life ruled out previous HIV infection.

There have been nine other reports, according to police files, in which HIV appears to have been transmitted by a bite. Severe trauma with extensive tissue tearing and damage and presence of blood were reported in each of these instances. Biting is not a common way of transmitting HIV. In fact, there are numerous reports of bites that did not result in HIV infection (see Box 8.5).

Sexual Assault, United States—The subject of sexual assault, in all its forms, of children by adults or adult on adult is beyond the scope of this text. However, each time there is a rape there is the chance that the rapist may be HIV positive. One in five adult American women has been the victim of a completed rape at some time in her life (Koss et al., 1991). A conservative estimate of the risk of transmission of HIV from sexual assault (that involved anal or vaginal penetration and exposure to ejaculate from an HIV-infected assailant) is greater than 2 infections per 1000 contacts, given a variety of factors, such as clinical stage of the assailant's HIV infection, strength of the strain of HIV, and repeated exposure. The per contact risk is

ASSAULT WITH HIV

Reflecting a growing frustration and fear about AIDS, legislators around the country are passing an increasing number of laws intended to protect the public. This latest wave of legislation shifts the focus from earlier laws that protected the civil liberties of HIV-infected people to laws that seek to identify, notify, and in some cases punish people who intentionally place others at risk of contracting the virus. At least 42 states now make it a crime to knowingly transmit or expose others to HIV, with a third of those states enacting laws within the last three years.

According to Richard Lacayo (1997), Darnell McGee, age 28, through 1995 and 1996, had sex with at least 61 women ranging in age from 12 to 29. According to a Missouri public health report in February 1998, McGee had sex with at least 101 females, including four whose ages were 13 or 14. It is reported that McGee infected 18 women but Missouri officials believe he infected 30 women. Some are pregnant. At least one has given birth to an infected infant. State public health officials, who are trying to track down and notify his sex partners, expect the tally to climb as more women come forward to be tested.

McGee knew what he was doing. According to the *St. Louis Post-Dispatch,* which broke the story, state records show that he tested positive in 1992 and was told the results.

On January 15, 1997, he was shot and killed, assailant unknown.

Darnell "Bossman" McGee is just one of a number of men who recklessly and in some cases, even willfully transmit HIV to their sex partners. Twenty years ago it was Gatean Dugas, or patient zero, a gay male who, over three years knowingly infected an untold number (probably 50) of gays across the United States. Fourteen years ago it was Fabian Bridges, a gay prostitute in Texas who was followed from city to city in the last months of his life by cameras from PBS's *Frontline* as he continued to have unsafe sex.

In 1997, there was Nushawn Williams (Figure 8-10), age 21, who in mid-1997 admitted to having unprotected sex with 50 to 75 women after he was told he was HIV positive. Most of them were teenagers ages 13 and up living in New York's Chautauqua County and in New York City. To date, 13 in Chautauqua are infected, the youngest was age 13; others were ages 15, 16, 18, and 21. Ten are positive in New York City but it is not known if Williams was the source of their infection. After the newspaper and TV carried this story, 625 people showed up at the county health department for HIV testing. Williams told authorities that he did not believe health officials when they told him he was HIV positive in September 1996. Early in 1998, two of the women who had unprotected sex with Williams gave birth to HIV-positive babies. There are now 16 HIV-positive people who are linked to Williams. In April 1999, Williams was sentenced to 4 to 12 years in prison. Only two women agreed to testify against him.

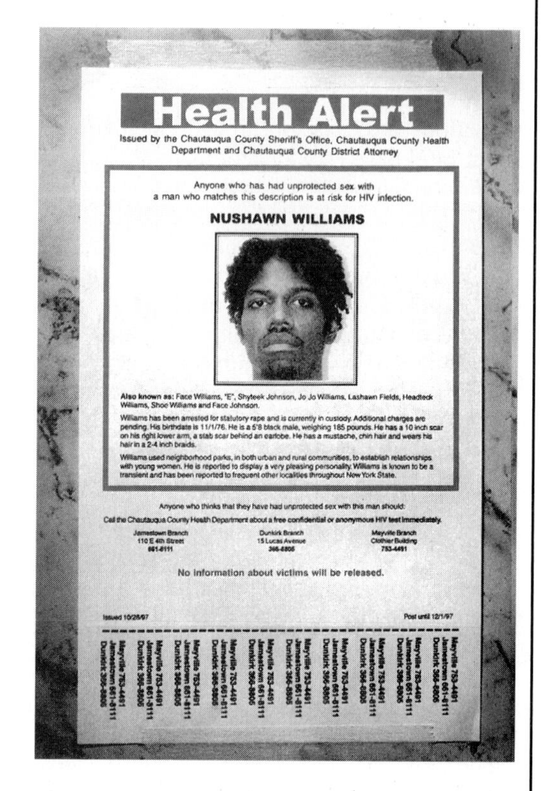

FIGURE 8-10 Nushawn Williams. Chautauqua County officials issued posters like this one in October 1997, for hanging on restroom walls in Jamestown, New York. The poster was to alert the public to his sexual activities throughout the county that may have exposed many to HIV infection. *(Photograph courtesy of AP/Wide World Photos/Bill Sikes)*

In February 2002, the San Francisco Superior Court Commissioner ordered a former San Francisco health commissioner to pay his ex-lover $5 million in damages for knowingly exposing him to HIV and lying about his HIV status. This civil suit judgment is one of the largest awards to date for this kind of damage. The suit also alleged that the commissioner, who was appointed by Mayor Willie Brown to the health commission in 1997, in part to represent people with HIV, was suffering from symptoms of AIDS but told his lover that he had cancer.

In April 2002, an HIV-positive rapist in Florida received three consecutive life sentences, one for each of his victims, who so far have tested negative for HIV. Also, an HIV-positive rapist's life sentence in Louisiana will begin after he finishes a 78-year sentence for raping a teenage girl in Tennessee.

In December 2003, an HIV-infected man in London, England was sentenced to 10 years in prison and deported to the Congo after raping a woman.

In May 2002, a child molester with HIV in Kansas City, Mo., was sentenced to three consecutive life terms, plus 52 years. He pleaded guilty in December to 13 counts, including statutory rape, sodomy, child molestation, and exposing others to HIV. His victims were ages 9, 11, and 14. Also, in May 2002, an inmate was sentenced to 8 years in prison for an assault that prompted an Anne Arundel County, Md., corrections officer to undergo treatment to prevent a possible HIV infection. The inmate threw a bloody gauze pad that hit a guard in the face.

In November 2002, a 35-year-old married Australian man was jailed for seven years for knowingly infecting a teenage girl he met via the internet with HIV. He was convicted of grievous bodily harm after passing the virus to the 18-year-old during a four-month relationship, which began after meeting her through a chat room. The woman learned she was HIV positive after donating blood.

In January 2003, a 39-year-old Bronx, N.Y. second-grade teacher accused of sexually abusing students while HIV positive was sentenced to 10 years in prison.

In April 2003, a 33-year-old Los Angeles male prostitute who had known for 10 years that he was HIV positive continued to have unprotected sex with his clients. He was taken into custody and received a 10-year prison sentence.

In March 2004, in what may be the first verdict of its kind, a Cook County, Ill., jury awarded $2 million to a woman who sued her fiancé's parents for allegedly covering up that he was dying of AIDS. The woman known only as "Jane Doe," was infected through unprotected sex with her fiancé in August 1996, according to the woman's attorney. The woman's lawsuit contended that her fiancé, who died of AIDS in November 1999, did not tell her he was infected with HIV. The suit alleged his parents knew of his infection and lied to her when she asked about his deteriorating health. The suit contended that the parents' misrepresentations prevented the woman from learning that she may have been infected for almost three years, during which time she could have received antiretroviral medicine to treat HIV infection.

In May 2004, a Libyan court sentenced to death by firing squad six Bulgarian health care workers and a Palestinian physician for deliberately infecting over 400 children with HIV using contaminated blood products. This sentence was handed down regardless of HIV/AIDS expert testimonies from people like Luc Montagnier, the scientist who discovered HIV, and stated the infections had to have occurred before these people became associated with these children. The sentence is under appeal. The world community is in shock. Stay tuned.

In May 2004, a 23-year-old Iowa male received a 50-year sentence on four counts of criminally transmitting HIV.

Between 1987 and 2005 at least 18 HIV-infected men and 3 HIV-infected women were incarcerated in 10 states because they **bit another person.** Two of the 18 bitten people became HIV positive. The charges in these cases varied from assault with a dangerous weapon, assault with a deadly weapon, attempted murder, aggravated assault with intent to murder, felony, reckless endangerment, and assault and reckless endangerment. Prison terms varied from 18 months for reckless endangerment to 27 years for attempted murder. In nine cases within seven states, one HIV-infected woman and seven HIV-infected men went to jail for periods of one to five years for spitting on other people. In a seventh case of spitting, in Texas, at trial, a court-recognized AIDS expert testified that HIV could be transmitted through the air! The man got a life sentence and an appeals court upheld the sentence. This man died in prison.

In July 2003, in Oklahoma, a convicted HIV-infected rapist received a life sentence for spitting on a police officer. None of the people who were spit upon contracted HIV.

Sex and HIV: If you have the first without disclosing the second, you can go to prison. In some states sex crimes and sex work can be elevated to attempted murder if the perpetrator (criminal) is HIV-infected. Disclosing one's HIV status to a sexual partner may exempt one from prosecution. But not telling—even if you do protect and don't infect—is still a crime in most states in America.

These cases, which represent just a few of the 500 recorded convictions, through 2004, of people having unprotected sex after being told they were HIV positive and not informing their sexual partners, raise several important issues.

DISCUSSION QUESTION: What is your response to these issues?

1. **Is knowingly transmitting HIV an act of violence?**
2. **Should the reckless or intentional transmission of HIV be a crime? If yes, how severe the penalty?**
3. **Do the cases bolster arguments for more aggressive partner notification and contact tracing? Why?**
4. **Do HIV confidentiality protections help or hinder efforts to alter the course of the epidemic? Why?**
5. **Would more ready access to condoms have helped avert these tragedies? How?**
6. **Who is responsible when an HIV-infected person knowingly continues to have unprotected sexual relations with others? Should the infected person be warned another time, assuming that the educational message was not heard? If so, how many times should warnings go forth? Are public health officials responsible for protecting susceptible spouses or long-term lovers of those who are infected and knowingly refuse to use condoms? Should the police become involved if protective advice is not followed, or should confidentiality remain in effect while educational messages go out that untold persons in the community are infected and all should use condoms?**
7. **Do such incidences support calls for more sex education, or less? Or perhaps different approaches to sexuality education? What approach might work? Why?**

higher if there was violence producing trauma and blood exposure or the presence of inflammatory or ulcerative sexually transmitted diseases (Gostin, 1994; Zierler et al., 2001).

At the beginning of 2005, there were over 500 reported cases of purposeful HIV infection of males by HIV-infected females or vice-versa. This too should be looked on as a form of **sexual assault**—one partner is being sexually deceived by the other. In some of these cases the jury found the HIV-positive persons who kept this knowledge from their sexual partner guilty of attempted murder.

Fear. Many rape victims who know their attacker has HIV are left living in fear and are taking all necessary precautions to protect family and friends from possibly contracting the virus. In other cases the raped woman has no idea whether her assailant is HIV-infected and will continue to worry even though her first or second test is negative. In some cases, people who have been raped cannot be told their attacker's HIV status because laws prevent states from releasing HIV test results; however, some states—including Arizona and Hawaii—have changed their laws in recent years to give victims access to records.

Sexual Assault, Africa—In at least 10 countries in Africa, 20% to 25% of the adults/adolescents are HIV positive. These countries have been reported to have an unusually large number of rape cases. In South Africa, it is estimated that one in three women are raped during their lifetime. In Johannesburg, South Africa, about 40% of men between the ages of 20 and 29 are HIV positive. *Time* magazine, in 1999, estimated that there are 50,000 women raped each year in South Africa. Peter Hawthorne, who wrote the *Time* article, reported that only 1 in 35 rapes is reported; suggesting that there are over 1.6 million rapes annually. Are such numbers credible? If so it will be most difficult to slow the transmission of HIV in South Africa. The high rate of HIV infection in these countries means there is a significant possibility that the assailant is HIV positive. Because of culture norms, the rapes are unlikely to be reported. In addition, these women do not have access to counseling or necessary antiretroviral therapy. They suffer and die in silence from AIDS due to sexual assault.

Transplants—On any given day, about 20,000 Americans are waiting for a transplant. There is a small but present risk of receiving HIV along with the transplant tissue. A CDC report revealed that a bone transplant recipient became HIV-infected from an HIV-infected donor. HIV transmission has also occurred in the transplantation of kidneys, liver, heart, pancreas, and skin (*MMWR*, 1988a). In May 1991, the CDC reported on 56 transplant patients who received organs and tissues from an HIV-infected donor in 1985. A transplantation service company supplied tissues to 30 hospitals in 16 states. All tissues came from a single young male who was shot to death during a robbery. He twice tested HIV negative before his heart, kidneys, liver, pancreas, cornea, and other tissues were removed for transplant. By mid-1991, three recipients of these tissues had died of AIDS and six others were HIV positive. As of mid-1991, 32 other recipients had been located, 11 of whom tested HIV negative. It is unknown whether the others have ever been tested.

In May 1994, the CDC published guidelines for preventing HIV transmission through transplantation of human tissue (*MMWR*, 1994b).

Since then, at least 88 liver and kidney transplant cases have involved the use of human organs that were HIV positive. Of the 88 transplant cases, 66 of the patients were HIV negative prior to transplant and received organs that contained HIV. All became HIV positive. The average time for progression to AIDS was 32 months. The 22 people who were HIV positive prior to receiving an HIV-positive organ progressed to AIDS, on average, within 17 months (Horn, 2001).

SIDEBAR 8.1

HIV TERRORISM, SOUTH AFRICA AND ISRAEL

Johannesberg

In November 1999, two apartheid-era security officers asked South Africa's truth commission for amnesty for their part in the use of turned freedom fighters to spread the deadly AIDS disease in 1990. They implicated four of their former colleagues and four fighters—members of the anti-apartheid African National Congress and smaller Pan Africanist Congress who crossed sides. The *askaris* (fighters) who had AIDS were used to spread the disease by infecting prostitutes in two Hillbrow hotels, the Chelsea and Little Rose, in May 1990.

Jerusalem

In April 2004, an attempt by a Palestinian to carry out a suicide attack using a bomb laced with HIV-infected blood during the Passover holiday was foiled by Israel's Shin Beth security service.

AIDS Patients and Organ Transplants

According to published reports, Larry Kramer, age 67, who cofounded the Gay Men's Health Crisis in 1981 and the AIDS Coalition to Unleash Power (ACT UP), suffered from end stage liver disease. Kramer said his situation was similar to that of a growing number of patients who live long enough with HIV to suffer from a second infection (coinfected). After being rejected by other health centers, Kramer received a new liver at the Thomas E. Starlz Transplantation Institute in Pittsburgh on December 21, 2001. As of December 2002, Kramer's liver transplant has cost Medicare over $500,000 and Empire Blue Cross over $100,000 for the medications he must take, including $10,000 a month for Hepatitis B Immune Globulin, which he will receive for the rest of his life. The University of Pittsburgh Medical Center transplant facility has performed 12 transplants on patients with HIV/AIDS since 1997. Art Kaplan, director of the Center of Bioethics said that the medical community has yet to debate the ethics of transplanting organs into people with HIV/AIDS. This is largely because centers like Starlz are just beginning to create possibilities. Beginning 2005, about 50 HIV-positive people have received liver transplants, one heart, and one kidney transplant in the United States.

DISCUSSION QUESTION: Debate the ethics of giving an HIV-positive person an organ transplant in light of the fact that thousands of young and old noninfected people die each year while waiting for a transplant.

SIDEBAR 8.2

MEXICAN DOCTORS TRANSPLANT HIV-INFECTED KIDNEYS

In February 1999, Mexican health officials fired five physicians and warned two others for transplanting HIV-infected kidneys into two patients. One of the two patients has since tested HIV positive. According to the regional director of the state-run hospital, the physicians did not wait for the results of the HIV test on the kidney donor before making the transplant.

In April 2002, the unofficial estimate was that 150,000 Mexicans required access to HIV/AIDS care and treatment. The epidemic in Mexico is centered among gay men ages 15 to 44. The number of HIV-infected women is rapidly rising.

Influence of Sexually Transmitted Diseases on HIV Transmission and Vice-Versa

Sexual intercourse occurs more than 100 million times daily around the world. Results: 910,000 conceptions and over 600,000 cases of sexually transmitted disease. In the United States about 19 million new cases of sexually transmitted diseases occur each year. Nine million of these cases occur in 15 to 24 year olds. By age 21, about one in five people has received treatment for an STD. At current rates at least one American in four will contract an STD at some point in his or her life. Over 50 organisms that can cause an STD are transmitted through sexual activity (Hooker, 1996). Regardless of these facts, data from the CDC in 2001 revealed that one in four physicians do not screen their patients for STDs. In addition, a GayHealth.com online survey conducted in May 2001 found 4 in 10 gay and lesbian patients have physicians that don't ask about their sexual practices. Over 40% of the men surveyed said they have not been vaccinated against either hepatitis A or B, potentially fatal viral liver infections spread through sexual activity. Men who have sex with men are at higher risk than the general population for these diseases. The annual cost of STDs in the United States is in excess of $10 billion. The latest U.S. government information on STDs and their treatment can be found in the updated STD Treatment Guidelines published by the CDC.

Annually, worldwide, there are over 300 million new cases of STDs. The majority of infections occur in those ages 20–24, followed by those ages 25–29, then ages 15–19.

Association of STDs of HIV Infection

STD researchers have long recognized that the behaviors that place individuals at risk for other STDs also increase their risk of becoming infected with HIV.

STDs are associated in several ways with HIV. Because STDs and HIV are spread by similar types of sexual activity, people who engage in behaviors that transmit HIV are also more likely to contract STDs, and vice-versa. Epidemiological evidence shows that populations and geographical regions in the United States with the highest STD rates also tend to have the highest rates of HIV. According to the CDC, "The geographic distribution of heterosexual HIV transmission closely parallels that of other STDs." In the past decade, high HIV incidence rates have shifted toward women infected through heterosexual activity, young adults, African Americans, and people living in the southeastern United States, all populations with disproportionately high rates of STDs.

Infection by sexually transmitted diseases usually occurs through the mucosal surfaces of the male and female genital tracts and rectum. The mucosal route also accounts for a large percentage of heterosexual and homosexual transmission of HIV. It is known that STDs increase the number of T4 or CD4+ cells (HIV target cells) in cervical secretions, thereby increasing the chance of HIV infection in women.

Most AIDS researchers agree that treating STDs, which cures genital sores and reduces inflammation, can raise the body's barriers against HIV infection. According to a study by Grosskurth and coworkers (1995), researchers working in rural Tanzania saw the number of new HIV infections plummet by 42% after they improved STD health care.

Because HIV can be sexually transmitted, the association between HIV and other sexually transmitted diseases can be in part attributed to the shared risk of exposure and shared modes of transmission.

For the purpose of understanding which STDs best promote HIV transmission, the sexually transmitted diseases can be divided into **genital ulcer** and **genital nonulcerative diseases.**

Genital Ulcer Disease (GUD)—Signs of genital ulcer disease appear as open sores on the penis, vagina, other genital areas, and at times elsewhere on the body. The most widespread genital ulcer STDs are syphilis, genital herpes, and chancroid. About 60 million people in America over age 12 have chronic genital herpes. There are about 1 million new herpes infections and 70,000 syphilis cases annually.

In early 1997, researchers from the University of Washington showed for the first time that herpes sores contain high levels of HIV, which they believe makes the virus especially easy to spread during sexual contact. Additional research by Timothy Schacker and colleagues (1998) also showed that HIV can be consistently found in herpes genital lesions of HIV-infected people. Such data suggest that genital herpes infection likely increases the sexual transmission of HIV.

HIV and Virulence

Virulence, the capacity of a pathogen to produce disease, is a consequence of both pathogen and host factors. Thus, HIV infection and associated immune deficiency disease may account for the increased prevalence of genital ulcer disease; and this in turn may further amplify HIV transmission in a network of social contacts.

Worldwide the estimated number of new STD infections for 2004 includes 12 million new cases of syphilis, 62 million new cases of gonorrhea, 89 million new cases of chlamydia, and 170 million new cases of trichomoniasis. Although trichomoniasis does not increase the risk of acquiring HIV as much as does syphilis, the huge number of trichomoniasis infections makes it at least equally important in the context of HIV transmission. All four of these STDs are curable—and examples from high- and low-income countries show that it is feasible and affordable to achieve a significant reduction of the STD burden everywhere.

Nonulcerative Disease—The nonulcerative STDs include gonorrhea, about 650,000 new cases annually, chlamydia, about 3 million new cases annually, and trichomonal infections, about 5 million new cases annually (also called discharge diseases), and genital warts. There are over 30 million people in the United States infected with genital wart virus, with about 5 million new cases annually. There are about 100 types of genital wart viruses, the human papilloma virus or HPV. HPV is one of the most common sexually transmitted agents. Current estimates are that approximately 75% of the sexually active general population ages 15 to 49 years acquires at least one genital HPV type during their lifetime. Most individuals remain asymptomatic after acquiring the infection, and only about 1% will develop clinically or histologically recognizable lesions.

In most populations, the nonulcerative STDs are much more common than genital ulcer diseases. None causes the noticeable open sores that occur in the ulcer diseases but they do cause microscopic breaks in affected tissue, and are associated with HIV transmission (Laga et al., 1993). The most common symptoms are warty growths on the genitals, discharge from the penis or vagina, and painful urination.

For an example of how nonulcerative STDs may enhance HIV infection, an uninfected woman has about a 0.2% chance of being infected with HIV during vaginal intercourse with an HIV-positive partner. If her partner had gonorrhea instead, she would have a 50% to 70% chance of becoming infected.

Collectively, worldwide there are over 300 million cases a year of just seven major STDs: syphilis, herpes, and chancroid, which cause ulcers; and trichomoniasis, chlamydia, warts, and gonorrhea, which do not (Figure 8-11).

HIV infection and other sexually transmitted diseases share the same risk factors. The major difference between HIV/AIDS and the other STDs is the degree of cell and tissue destruction and the mortality of HIV/AIDS.

HIV is transmitted most often during sexual contact with an infected partner. There is abundant evidence that if a sexual partner has an active STD, especially one that causes an ulcer, he or she is at greater risk of becoming HIV-infected (Laga, 1991).

The types of blood cells, lymphocytes, or macrophages most likely to become infected if exposed to HIV tend to collect in the genital tract of people with STDs. This makes an STD-infected person both more likely to transmit HIV and more vulnerable to it (Laga, 1991).

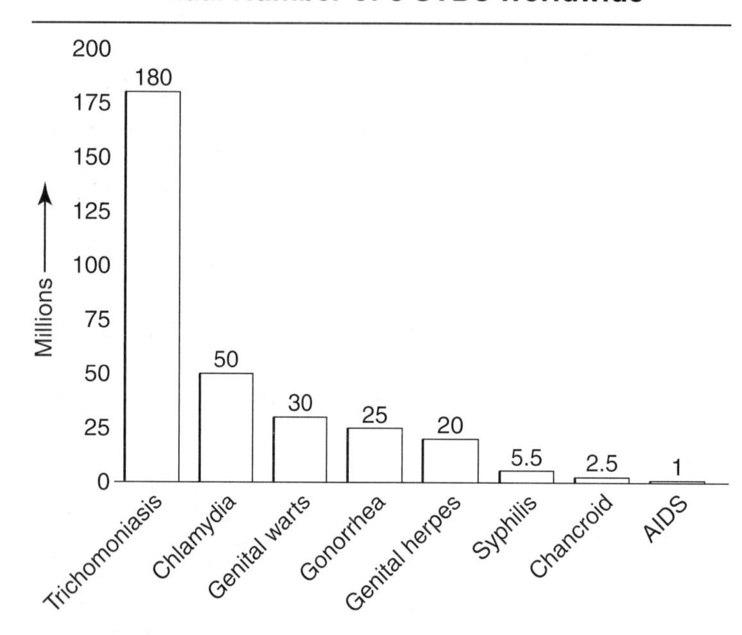

FIGURE 8-11 Global Incidence of Eight Sexually Transmitted Diseases. For 2004, ages 15 to 49. According to the CDC, 50% of STDs in the United States are unreported and 50% to 90% of STDs worldwide are unreported.

Pediatric Transmission

Children can acquire HIV from their mothers in several ways. A pregnant HIV-infected woman can transmit the virus to her fetus in utero (during gestation) as the virus crosses over from the mother into the fetal bloodstream (Jovaisas et al., 1985; St. Louis et al., 1993). At least 50% of newborn infections occur during delivery by ingesting blood or other infected maternal fluids (Scott et al., 1985; Boyer et al., 1994; Kuhn, et al., 1994). If breast-fed, the newborn may also become infected from breast milk (Zigler et al., 1985; deMartino et al., 1992; Van DePerre et al., 1993). In case reports, three women who contracted HIV by blood transfusions immediately after birth subsequently infected their newborns via breast-feeding (Curran et al., 1988). Other studies suggest that the risk of HIV transmission through breast-feeding is increased if the mother becomes HIV-infected during lactation (Hu et al., 1992).

The relative efficiency of these three routes of infection is unknown. However, the data on mothers' milk add to the urgency of learning more about mucosal transmission. The most likely explanation for HIV transmission through breast-feeding is that the virus penetrates the mucosal lining of the mouth or gastrointestinal tract of infants. If this occurs in newborns, then what of older children, adolescents, and adults? Does the mucosal lining change with development and become HIV-resistant?

HIV-Infected Babies—One major problem in perinatal transmission is how to determine which babies are truly HIV-infected as opposed to just carrying the mother's HIV antibodies (which would produce a false-positive test). HIV transmission can occur during pregnancy (in utero), as well as at the time of delivery **(intrapartum)** and through breast milk. HIV transmission is more likely if virus can be cultured from the mother's blood, or if she has later-stage HIV disease, or if her T4 or CD4+ counts are low; and is more likely to occur in the firstborn than in the secondborn of twins. A baby automatically acquires the mother's antibodies and may carry them for two or more years. Usually by 18 months of age, most of the mother's antibodies will be gone. The babies may then begin to show signs of clinical AIDS-related illness. But, even at 18 months, a child cannot be unequivocally diagnosed. The most commonly used HIV antibody test is not sufficiently accurate until the child is at least two years old. A new antibody test in development shows promise in recognizing newborn infection by examining the type of HIV antibodies the infected mother is producing.

Although the rate of perinatal and breast milk HIV transmission is unknown, evidence from 1986 into year 2005 indicates that over 90% of pediatric AIDS cases acquired the virus in utero from an HIV-infected mother after the first trimester or during the birthing process. In 1990, researchers concluded that a fetus can become infected as early as the eighth week of gestation (Lewis et al., 1990). HIV has been isolated from a 20-week-old fetus after elective abortion by an HIV-positive female and from a 28-week-old newborn delivered by caesarean section from a female who was diagnosed with AIDS (Selwyn, 1986).

Reports on the probability of a fetus becoming HIV-infected when the **untreated** mother carries the virus vary widely. The most often quoted estimate in the United States is from 30% to 50%.

Other than viral load, there is little documented information on maternal factors that influence vertical transmission. As with other congenital infections, only one of a pair of twins may be HIV-infected (Newell et al., 1990; Ometto, 1995). A mother's clinical status during pregnancy and the duration of her infection (stage of disease) may be important, but evidence remains circumstantial (see Chapter 11 for update information). Studies to determine mother-to-fetus transmission relative to stage of disease are in progress.

According to the CDC classification, children under 13 years of age are considered pediatric AIDS cases. They make up about 1.3% of all AIDS cases in the United States. Through 2004, about 5% of reported pediatric male AIDS cases occurred due to blood transfusions, 3% received HIV-contaminated blood factor VIII used in treating hemophiliacs, and in 2%, the cause was undetermined.

The largest numbers of pediatric AIDS cases through 2004 were in New York, Florida, California, and New Jersey, in that order. The highest incidence of all pediatric cases occurs in minority populations. By early 2005, there were over 9400 pediatric AIDS cases in the United States. Blacks and Hispanics make up 12% and 6% of the United States population, respectively, yet make up 55% and 20%, respectively, of all pediatric AIDS cases. Thus 75% of pediatric AIDS cases occur within two minority populations.

Vertical Infection: HIV-Infected Childbearing-Age Women

Over 100,000 women of childbearing age are estimated to be infected with HIV in the United States. The majority of these women may not know they are infected; they are identified as infected only after their children are diagnosed as having an HIV infection or AIDS. It is not uncommon for HIV-infected women to go through several pregnancies before expressing HIV disease. Also, there are women who become pregnant knowing they are HIV positive. They want to have a baby regardless (see Chapter 11).

Mother-to-fetus infection or **vertical infection** could be avoided by avoiding pregnancy, but this is possible only in cases where the female is aware of her infection and takes measures to prevent pregnancy (birth control or tubal ligation). In many cases, pregnancy occurred before the mother knew she was carrying the virus. In other cases, the mother has become infected after she has become pregnant.

Antiviral Therapy Decreases Perinatal HIV Transmission

Probably the most important step forward in the use of antiviral agents has been the discovery that antiretroviral drugs can decrease the rate of perinatal transmission of HIV. In a landmark study by Edward Connor and colleagues (1994) and the interim results of the AIDS Clinical Trials Group (ACTG) Protocol 076 (*MMWR,* 1994d; Goldschmidt et al., 1995; *MMWR,* 1995), the intensive use of Zidovudine beginning in the second trimester of pregnancy, including intravenous Zidovudine during delivery, and six weeks of oral therapy in the neonate, *decreased the rate of transmission* from 25% to 8%. (See Chapter 11 for updated information.)

CAESAREAN SECTION

The time-honored surgical maxim, "A chance to cut is a chance to cure," takes on new meaning in the debate now raging about **caesarean section** as a means of reducing perinatal HIV transmission. From 1998 through 2004, clinicians have shown that the use of caesarean section to deliver a newborn from an HIV-infected mother is a significant help in reducing HIV infection of the newborn (Riley et al., 1999; Read et al., 1999).

Possible Antiretroviral Drugs to Prevent Maternal HIV Transmission

Current concepts on the prevention of vertical transmission now focus on the appropriate antiretroviral therapy of infected pregnant women. Despite the data from ACTG 076, Zidovudine monotherapy alone can no longer be considered adequate for pregnant women because some newborns are still HIV positive. Some HIV/AIDS experts recommend the use of combination therapy that includes a protease inhibitor. In March 1999, five important studies were published regarding the use of combination antiretrovirals to reduce mother-to-infant transmission. The conclusion to be drawn from the studies is that antiretroviral prophylaxis of mother and her infant is very effective. With many women on HAART therapy during pregnancy, there were less than 200 HIV-infected infants born in the United States in 2003 and 2004. However, in developing countries, despite recent progress, there is still a great need for shorter, inexpensive, effective regimens. But the debate continues on when to begin therapy—first, second, or third trimester?

(For an excellent review of mother-to-child transmission of HIV, see the article by Peckham and College, 1995.)

HIV Transmission in the Workplace

The idea of contracting HIV from a fellow employee generates fear in many employees regardless of their jobs. Most people in the United States have been exposed to information on the routes of HIV transmission and on how to practice safer sex, but remote possibilities remain worrisome to many in the job force. In 1999, about one in six companies offered employee HIV/AIDS education. Many people still believe that HIV can be transmitted casually via handshakes, coffee cups, and food handling.

It is the anxiety of *uncertainty* that engenders suspicion about the possibility of HIV infection in the workplace—an anxiety that HIV/AIDS scientists could be wrong about the routes of HIV transmission. As Judith Wilson Ross states, "We have spent our lives in a culture in which infectious disease does not represent a significant threat. And thus we had consigned living in fear of life-threatening contagious diseases to the pages of history books." But today, HIV/AIDS forces us to reexamine our faith in the certainty of science. We want to believe, we want to accept—but the fear of death prevents complete surrender to education.

PUBLIC CONFIDENCE: ACCEPTANCE OF CURRENT DOGMA ON ROUTES OF HIV TRANSMISSION

One living legacy is the distrust arising from the Tuskegee syphilis experiments. Black prisoners were purposely given syphilis and not given treatment so that investigators could observe medical signs and symptoms as the infected progressed toward death. Over 30 years after the study was publicized, its impact on black Americans' view on the medical establishment remains strong, experts say. For example, a recent survey of black Americans tested for HIV in a street outreach program found that 80% reported conspiratorial beliefs about the virus. The Tuskegee study has never died and that distrust is continuously there and needs to be dealt with.

Since the start of the epidemic in 1981, lawmakers have championed proposals ranging from quarantines to criminal prosecution for infecting others with HIV. At the beginning of 2005, at least 43 states have some type of law stating that it is unlawful for anyone who knows they are infected with HIV to transmit or expose others to the virus without their knowledge and consent. Public support for such measures is high. Results of one 2002 poll show that over 90% of Americans believe that an individual who knowingly infects another person with HIV should face criminal charges, and half of those surveyed said that people who knowingly transmit the virus should be charged with murder. Critics of criminal penalties say such laws will not help to control the HIV/AIDS epidemic and will likely do more harm than good.

Public Confidence Surveys

The December 2000 issue of *Morbidity/Mortality Weekly Report* carried a public survey by Research

Triangle Institute on beliefs about HIV. The survey was run using the internet. The research team analyzed the responses from 5641 people. The findings:

> **18.7%** strongly agreed that "people who got AIDS through sex or drug use got what they deserve."
>
> **40.2%** said that HIV infection (transmission) could occur through sharing a glass.
>
> **41.1%** said one could get HIV by being coughed on or sneezed on by an HIV-infected person.

Compare these data with the 1997 data presented. In short, after over 23 years into this pandemic, a significant number of people still believe one can contract HIV from casual contact.

Telephone Survey Conducted In Canada, 2003

A market research firm, Ipsos-Reid, administered a phone survey to 2,004 Canadians over age 15. The title: **"HIV/AIDS–An Attitudinal Survey."** The findings were: 25% believed HIV could be transmitted by kissing and mosquito bites; 11% by coughing and sneezing. About 50% cited sharing needles and 33% said blood transfusions could transmit HIV. Eighty-four percent thought that unsafe sex (no condom) was a route of HIV transfer and 20% thought there was a cure for HIV/AIDS.

Telephone HIV/AIDS Survey Conducted in Russia

Having reviewed what Americans feel and believe about HIV/AIDS transmission, a Russian telephone poll revealed an equal lack of effective education. Uri Amirkhanian and colleagues at the Medical College of Wisconsin presented the results of their study, which they believe to be the first population-based, random-digit telephone AIDS survey conducted in Russia. They found that significant proportions of respondents believed that HIV could be spread through mosquito bites (56%), kissing (48%), or cigarette sharing (29%).

Education Survey in the United Kingdom

Forty percent of boys in the United Kingdom have never heard of HIV or AIDS, according to the "Young People in 2000" survey conducted by the Schools Health Education Unit. The survey polled 42,000 people between the ages of 10 and 15 and found that in addition to not knowing about HIV/AIDS, more boys than girls said they would "not take care" to prevent contracting HIV. Lisa Power, head of policy for the Terrence Higgins Trust, said, "Teenagers today were babies when the "Don't die of Ignorance" advertisements and leaflets were used in the mid-1980s, so HIV awareness among today's younger generation is low. We don't want young people to think that HIV is around every corner—but they should be aware of the genuine risks of HIV and other sexually transmitted diseases." The survey also found that a third of 18- to 24-year olds think there is a cure for HIV. Nearly a quarter said they could be infected with the deadly virus through kissing and 1 in 10 people are convinced they can be infected through toilet seats. When asked about using condoms, 10% said they are not worried about unsafe sex, but 50% said they are concerned about HIV infection.

United Nations Survey on HIV Education in Poland

The 2003 survey found that 40% of Poland respondents thought HIV/AIDS could be transmitted through insect bites; 20% believed that someone could become infected by using public toilets and baths; and 26% said that they could be infected by cutlery used by an HIV-positive individual. The low level of public knowledge could increase complacency about HIV and increase chances for a fresh outbreak of HIV in Eastern Europe. Although the survey was limited in numbers of people surveyed, there is clearly a big problem. More than 50% of survey respondents said that they would prefer to hire people with cancer or cardiac problems than HIV-positive individuals. The authors of the survey said that public knowledge of HIV/AIDS in Poland has deteriorated over recent years. The director of Poland's National AIDS Center said that education and prevention methods account for 8% of Poland's HIV/AIDS budget. Most of this budget goes toward care and support. According to the World Health Organization, Eastern Europe and Central Asia had the world's fastest growing HIV/AIDS epidemic in

BOX 8.6

HIV AND SENIOR CITIZENS

Looking at the majority of safer sex workshops and street outreach programs, one would get the impression that only the young are at risk of contracting HIV. It's true that most people with AIDS are under 49. But, according to the CDC, now about 13% of Americans who test positive for the virus are over the age of 50. It's not just with regard to prevention that over-50s are left behind. Older people with HIV are often misdiagnosed and typically learn they have the virus only later in the disease process. Medical treatment is more difficult, both because of the later diagnosis and factors related to age. Few practitioners are expert both in HIV and the health problems associated with aging. When it comes to social support services aimed at their particular needs, older HIV-infected people are all but invisible. Attitudes about HIV/AIDS and aging reflect misconceptions about how people behave in the second half-century of their lives.

1. Old people are no longer interested in sex;
2. If they are interested, no one's interested in them;
3. If they do have sex, it's within a monogamous, heterosexual relationship;
4. They don't do drugs;
5. If they ever did, it's so long ago it doesn't matter.

It isn't hard to see how these misconceptions help erect barriers to effective HIV/AIDS prevention efforts, medical care, and social services for the late middle-aged and elderly. After all, if they're not doing anything risky, there's nothing to worry about, right? **WRONG,** older adults refuse to conform to the stereotypes.

Gender and Age

Older women are becoming HIV-infected at a higher rate than older men. No longer afraid of becoming pregnant, the postmenopausal woman who is uninformed of the danger of HIV transmission may become more sexually active, with more partners, and may give up a decades-old habit of using condoms. Even her biology increases her risk: After menopause, the vaginal walls thin and vaginal lubrication decreases. Thus, the vaginal membranes are more likely to tear during intercourse, providing easier access to HIV.

2002 through 2004, with over one-fifth of the region's 1.6 million HIV patients becoming infected in these years.

HIV/AIDS Survey of Chinese Health-Care Workers from Three Hospitals in Southern China

Allen Anderson of Indiana University, along with a group of Chinese researchers, reported in 2003 that despite the great HIV/AIDS devastation in China, Chinese health-care workers have insufficient knowledge of HIV/AIDS and are apprehensive about treating the HIV/AIDS patients. In the study performed, a questionnaire was administered to the health-care workers in three different hospitals in Southern China. This questionnaire contained questions on the knowledge of HIV transmission, their facility's HIV/AIDS control policies, and their personal attitude to those infected. Though blood, semen, and vaginal fluids were well known as potential transmission vehicles for HIV, the results were: 92% were concerned about becoming infected in their health-care position; 27% did not believe that accidental needle sticks could transmit HIV; 34% considered saliva as a source of transmission; 4% believed transmission occurred by breathing air in the patient's room; 22% believed in transmission via toilet seats; and 33% believed in transmission via mosquito bites.

CONCLUSION

Although no new routes of HIV transmission have surfaced over the last 24 years of this pandemic, many people still do not believe that's all there is. People still make the arguments that: (1) Scientists do not yet know enough about this disease to be certain there are no other routes of transmission; and (2) scientists know other routes exist but are either too frightened to tell the truth, or are under political pressure not to do so for fear of creating a public panic. Many thousands of people in the United States firmly believe that in a few years they will look back and say "I told you so": You can get HIV from HIV-infected people if they breathe on you or if you touch their sweat and so on.

DISCUSSION QUESTION: How do you get everyone to believe what medical and research scientists say? Should we get everyone to believe scientific dogma?

NATIONAL AIDS RESOURCES

AIDS Action Council	1-202-547-3101
Coalition for Leadership on AIDS	1-202-628-4160
Gay Men's Health Crisis	1-212-807-6655
Mothers of AIDS Patients	1-619-234-3432
National AIDS Information Clearinghouse	1-301-762-5111
National AIDS Network	1-202-546-2424
National Association of Persons with AIDS	1-202-483-7979
Project Inform (Alternative AIDS Info)	1-800-822-7422
Public Health Service Hotline	1-800-342-2437
Centers for Disease Control and Prevention Technical Information	1-404-639-2070
American Red Cross, National AIDS Education	1-202-639-3223
Guide to Social Security and SSI Disability Benefits for People with HIV Infection	1-800-772-1213

(You can write or call for this Social Security brochure: Social Security Administration, Public Information Distribution Center, P.O. Box 17743, Baltimore, Md., 21235.)

SUMMARY

The World Health Organization began keeping records of AIDS-like cases in 1980. Beginning 2005, there were an estimated 34 million AIDS cases in 194 reporting countries and territories. About 24 million of these have died. About 2% of AIDS cases have occurred in the United States. At the start of 2005, of the 40 million living with HIV infection worldwide, about 2.3% or over 1,000,000 live in the United States. It has been reported that the first cases of AIDS entered the United States via homosexual men who had vacationed in Haiti in the late 1970s. However, there is evidence of AIDS cases in the United States as early as 1952. While testing West Africans for HIV infection, a second strain of HIV was discovered: HIV-2. Both are transmitted in the same manner and both cause AIDS. However, HIV-2 appears to be less pathogenic than HIV-1.

Nearly all Americans are aware that HIV can be transmitted through unprotected intercourse (99%), the sharing of intravenous (IV) needles (99%), and unprotected oral sex (91%). Fewer than half (42%), however, know that having another sexually transmitted disease (STD) increases a person's risk for HIV. In addition, even after years of public education, unwarranted fears of infection through casual contact persist. For example, one in five (22%) Americans incorrectly believes that sharing a drinking glass can transmit HIV, or are unsure about the risk of this activity. Sixteen percent believe that touching a toilet seat can transmit HIV or are unsure about the risk. Such views contribute to discrimination and stigma, which can interfere with public health efforts to encourage early testing and care.

There are two major variables involved in successful HIV transmission and infection. First is the individual's genetic resistance or susceptibility, and second is the route of transmission. Not all modes of HIV exposure are equally apt to cause infection, even in the most susceptible individual. There have been a number of studies and empirical observations that demonstrate that HIV *is not* casually acquired. HIV is difficult to acquire even by means of the recognized routes of transmission.

HIV is transmitted mainly via sexual activities involving the exchange of semen and vaginal fluids, through the exchange of blood and blood products, and from mother to child both prenatally and postnatally (breast milk). Besides a few cases of breast milk transmission, no other body fluids have as yet been implicated in HIV infection.

The current belief is that anal receptive homosexuals have a higher risk than heterosexuals of acquiring HIV because the membrane or mucosal lining of the rectum is more easily torn during anal intercourse. This allows a more direct route for larger numbers of HIVs to enter the vascular system.

Others at high risk for acquiring and transmitting HIV are injection-drug users. They infect each other when they share drug paraphernalia. Changes in sexual and injection drug use behavior can virtually stop HIV transmission among these people.

REVIEW QUESTIONS

(Answers to the Review Questions are on page 438.)

1. True or False: Africa makes up the largest percentage of *reported* AIDS cases worldwide. Explain.
2. What evidence is there that HIV may have evolved in the United States and Africa at the same time?
3. Are HIV-1 and HIV-2 related? Explain.
4. True or False: HIV-1 and HIV-2 are transmitted differently and therefore are located in geographically distinct regions of the world. Explain.
5. True or False: HIV is *not* believed to be casually transmitted. Explain.
6. Name the routes of HIV transmission.
7. True or False: Deep kissing wherein saliva is exchanged is a direct route for *efficient* HIV transmission. Explain.
8. True or False: Insects that bite or suck have been claimed to be associated with HIV transmission. Explain.
9. True or False: Among heterosexuals, HIV transmission from male to female and from female to male are equally efficient. Explain.
10. True or False: If a person has unprotected intercourse with an HIV-infected partner he or she will become HIV-infected. Explain.
11. What is the percentage of risk that a developing fetus with an HIV-positive mother in America will be born HIV positive, with and without Zidovudine therapy? With Zidovudine and "C" section?
12. Despite the warnings, groups that continue to engage in high-risk sexual activity include:
    A. high school students
    B. black women
    C. injection-drug users
    D. prostitutes
    E. all of the above
13. True or False: Prior to 1985, use of blood component therapy put hemophiliacs at risk for contracting HIV.
14. True or False: Relapse to risky sexual behavior can be an important source of new HIV infection in the homosexual community.
15. True or False: The body fluids shown most likely to transmit HIV are blood, semen, vaginal secretions, and breast milk.
16. True or False: Participation in risky behaviors and not identification with particular groups puts an individual at risk of acquiring HIV infection.
17. True or False: Unprotected receptive anal intercourse is the sexual activity with the greatest risk of HIV transmission.
18. True or False: Women who are HIV-infected always transmit the virus to their fetus during pregnancy or delivery.
19. True or False: A person infected with HIV can transmit the virus from the first occurrence of antigenemia throughout the rest of his/her life.
20. True or False: Women constitute the fastest growing segment of the population with HIV infection.
21. True or False: The majority of HIV-infected women whose source of infection is known became infected through vaginal intercourse.
22. True or False: HIV infection in children is now a leading cause of death in children between the ages of 1 and 4.
23. True or False: Sexual contact is the major route of HIV transmission among black Americans.
24. True or False: Urine is one body fluid that remains an unproven route of HIV transmission.
25. Which of the following is *not* a recognized mode of HIV transmission?
    A. unprotected sex with an infected partner
    B. mosquito bite
    C. contact with infected blood or blood products
    D. perinatal transmission

26. HIV/AIDS Word Scramble

ABMETARUTS _ _ _ _ _ _ _ _ _ _
ACIPMDEN _ _ _ _ _ _ _ _
AERIBRR _ _ _ _ _ _ _
AGSEASM _ _ _ _ _ _ _
ANLA _ _ _ _
ATOUNITM _ _ _ _ _ _ _ _
BENSTCENIA _ _ _ _ _ _ _ _ _ _
BLAUEXSI _ _ _ _ _ _ _ _
BLMSKRITAE _ _ _ _ _ _ _ _ _ _
BYUDILSFOD _ _ _ _ _ _ _ _ _ _
CDRLSAIIMEP _ _ _ _ _ _ _ _ _ _ _
DBOOL _ _ _ _ _
DEIPCIME _ _ _ _ _ _ _ _
EGEIVTAN _ _ _ _ _ _ _ _
FOCTNEINI _ _ _ _ _ _ _ _ _
HEHALIIMPO _ _ _ _ _ _ _ _ _ _
HREUSALXEOET _ _ _ _ _ _ _ _ _ _ _ _
IAVLNAG _ _ _ _ _ _ _
IDSA _ _ _ _
IVH _ _ _
LSNCMIAOOO _ _ _ _ _ _ _ _ _ _
MEENS _ _ _ _ _
MNTSNAOISISR _ _ _ _ _ _ _ _ _ _ _ _
MREPS _ _ _ _ _
NCEIALBCUMMO _ _ _ _ _ _ _ _ _ _ _ _
NITSARS _ _ _ _ _ _ _
NOEUTESRIRC _ _ _ _ _ _ _ _ _ _ _
OEULXMASOH _ _ _ _ _ _ _ _ _ _
OMODNC _ _ _ _ _ _
PESREH _ _ _ _ _ _
PRLIATANE _ _ _ _ _ _ _ _ _
RSNIAUVONET _ _ _ _ _ _ _ _ _ _ _
SAXELU _ _ _ _ _ _
SDT _ _ _
SILAAV _ _ _ _ _ _
SIPEN _ _ _ _ _
SNRISEGY _ _ _ _ _ _ _ _
TFYELSEIL _ _ _ _ _ _ _ _ _
TLACER _ _ _ _ _ _
UNOASNISFTR _ _ _ _ _ _ _ _ _ _ _
VPIETISO _ _ _ _ _ _ _ _
VSURI _ _ _ _ _

9 Preventing the Transmission of HIV

The best time to plant a tree is 20 years ago. The second best time is now.

African Proverb

CHAPTER CONCEPTS

- Prevention is today's vaccine.
- HIV transmission can be prevented; the responsibility rests with the individual.
- No new routes of HIV transmission have been found after 24 years.
- Safer sex essentially means using condoms and not knowingly having intercourse with an HIV-infected person.
- The female condom (vaginal pouch) was FDA-approved in 1993.
- Oil-based lubricants must not be used with latex condoms.
- Plastic condoms are now available.
- Polymer gel condoms are being developed.
- Free syringe and needle exchange programs claim to help lower the incidence of HIV transmission.
- Blood bank screening to detect HIV antibodies began in 1985.
- Universal precautions and blood and body substance isolation are techniques to help health-care workers prevent infection.
- Universal precautions require certain body fluids from all patients to be considered potentially infectious.
- Blood and Body Substance Isolation (BBSI)
- Partner notification is a means of notifying at-risk partners of HIV-infected individuals.
- Vaccines are made from whole or parts of dead microorganisms, inactivated viruses, or attenuated viruses, microorganisms, and naked DNA.
- Experimental subunit vaccines are prepared using recombinant DNA techniques.
- There is no effective vaccine for prevention of HIV infection.
- A moral problem exists in attempting to get human volunteers for HIV vaccine testing.
- Word search on HIV/AIDS prevention.

THE AIDS GENERATION: "I KNEW EVERYTHING ABOUT IT, AND I STILL GOT IT!"

The "magic bullet" to cure or prevent HIV infection has not been found and too many people with or affected by HIV/AIDS are isolated by cultural, geographic, and economic barriers. HIV is a preventable disease and the first step in preventing disease is the transformation of information into knowledge, and getting people to use that knowledge. For example, the slogan " Practice Safer Sex" is now as common as "Buckle Up For Safety" and "Just Say No To Drugs," but HIV infections among age groups over 12 continues at an alarming pace.

PREVENTION, NOT TREATMENT, IS THE LEAST EXPENSIVE AND MOST EFFECTIVE WAY TO STOP THE SPREAD OF HIV/AIDS

Twenty-four years into one of the worst health disasters in human history, the HIV/AIDS pandemic continues to grow exponentially, outstripping prevention efforts and treatment programs; everyday it kills over 8000 people and infects 14,000 more. The global effort is inadequate to check its spread or stop the deaths. **The HIV/AIDS pandemic is almost, if not actually, incomprehensible in size and scope. And, in the developing world this pandemic is in its early stages! Two things are now**

clear: the first is that treatment will have a relatively small effect on the transmission or spread of HIV, and second, treatment/prevention is not an either/or choice—both are vital. But, what can be done now that will change the course of this pandemic? Something has to be done! What can be done and by whom? The world must engage in the largest possible dissemination of HIV prevention strategies.

Primary Goal of HIV Prevention

The primary goal of HIV prevention is to prevent as many infections as possible. This requires allocating HIV prevention resources according to cost-effectiveness principles: Those activities that prevent more infections per dollar are favored over those that prevent fewer. This is not current practice in the United States, where prevention resources from the federal government to the states flow in proportion to reported AIDS cases. Although such allocations might be considered equitable, more infections could be prevented for the same expenditures were cost-effectiveness principles invoked. The downside of pure cost-effective allocations is that they violate common norms of equity. In 2004 the CD4 allocated $788 million for HIV prevention programs in the United States.

GLOBAL PREVENTION

Because we live in a global village, the public health of Africa, Asia, and elsewhere affects the public health of the United States. As there is one global economy, there is one global public health. Prevention of infectious diseases in any country is prevention for all. In each of the years from 1998 through 2004, on average, an estimated 2.5 million people died from AIDS. Worldwide by the end of 2005, about 24 million people will have died of AIDS. Over 80% of these deaths will be in Africa. While waiting for an effective vaccine, how can the out of control spread of HIV be slowed? How can people everywhere be saved from HIV infection? In a word, **prevention** is the only hope short of a vaccine.

Stop, listen and learn all you can about HIV/AIDS. Prevention and life—it's your choice!

Types of HIV/AIDS Prevention

Worldwide prevention efforts focus on two types of prevention: primary prevention—preventing HIV exposure among uninfected individuals, and secondary prevention—preventing HIV transmission by those who are infected.

Through 2002 and many billions of dollars allocated for prevention there remained an imbalance of primary and secondary prevention efforts. Yet, it was the public health strategy of secondary prevention that freed the world from smallpox and greatly reduced the incidence of syphilis and a number of other diseases worldwide. So, in April 2003, the CDC shifted its prevention efforts to mainly funding secondary prevention—preventing those infected from transmitting HIV to those uninfected. That is to shift most of the money being spent on those at higher risk for HIV infection, like HIV-negative gay men to those who are HIV positive.

Investing in Prevention

Spending money on prevention is a smart investment. For example, in America, the CDC's goal is to lower the HIV infection rate from the current 45,000 each year to 20,000. A 2003 study by HIV economists at Emory University estimated that preventing 50% of new HIV infections yearly would save about $22 billion in medical costs by 2010. Current prevention efforts have already averted at least 200,000 HIV infections in America over the last 23 years. That would translate into about $30 billion to $60 billion already saved in medical costs. The potential for HIV prevention interventions to save lives and dollars emphasizes the need to spend money now rather than later, and to maintain consistent, if not increasing, funding to protect those at high risk. At the 2002 Fourteenth International AIDS Conference, Michael Saag reported that in the United States, health care for each patient in the advanced stages of AIDS costs an average of $34,000 a year. Antiretroviral medicines make up the largest part of the cost. The cost of treating the average patient with HIV is about $14,000. Overall, medication costs were about $11,000–$24,000. Hospital costs were second, ranging from an average of $1700 for early stage patients to $7800 for those with advanced AIDS. Taking the average of $14,000 and $34,000 or $24,000 times 1 million people living with HIV/AIDS in America comes to $24 billion! Estimates are that about 450,000 people in 2004 received some measure of care for their HIV infection. Their total health care cost is estimated to be $11 billion a year (see Figure 13-7 for year 2004 federal dollars allocated to HIV/AIDS). America's prevention budget for 2004 was over $1 billion. The bottom line is prevention, regardless of the means of assessment.

The Challenge

Prevention is a hard sell. It is easier to get thousands of dollars to rescue a baby down a well than it is to get a few hundred dollars to cover old wells.

The fact that there is no cure for HIV/AIDS, and that the drugs are costly and cause severe side effects, makes prevention crucial. CDC researchers reviewed 83 studies from 1978 through 1998. They found that as soon as prevention education began in the early 1980s, the rate of new HIV infections plummeted and that it has remained relatively stable, at about 40,000 to 50,000 new infections a year (see Figure 10-7).

The means of preventing HIV infection exist. They must be used effectively to make an impact on this escalating pandemic. **Unlike treatment, HIV prevention does not have to be perfect to be effective.** The existing methods of HIV prevention are presented in this chapter.

Prevention, Is Anyone Listening?

Why do people knowingly engage in sexual behavior that can lead to a slow and painful premature death? Why do the best-intentioned HIV prevention programs often have so little impact?

Robert Smith of HIV Edmonton, Canada, and Michael Yoder, chairman of the Canadian AIDS society reported at the Fourteenth International AIDS Conference that North American prevention programs are failing. They believe it's back to the drawing board for the AIDS community that has discovered, to its horror, that no one seems to care about safer sex anymore. (Safer sex means any sexual activity that prevents HIV or other STDs within semen, vaginal fluid, or blood from entering the bloodstream of another person.) People are not listening. Too few are getting tested. It's a marketing nightmare. Canada spends $42 million a year on HIV/AIDS awareness and prevention programs. The United States spends about 25 times that. After 24 years of being bombarded with safer sex messages, you'd think everyone in North America would know how to protect themselves from HIV and other sexually transmitted diseases. As health professionals are becoming increasingly aware, getting the facts out is one thing—doing it in a way that changes people's behavior is another.

Among the depressing reports out of the 2002 International AIDS Conference in Spain is a study showing that most of the young, gay, HIV-positive men in major U.S. cities are unaware that they're infected. More than half the HIV-positive men who didn't know they had the virus considered themselves at low risk of HIV infection and nearly half of them reported they didn't use condoms. They concluded that in North America, young people seem to be fed up with hearing about AIDS.

Internationally, it seems, there is similar skepticism. At the same conference, the Joint United Nations Program on HIV/AIDS (UNAIDS) quoted grim statistics on a pandemic still in its early stages, with no stabilization of the epidemic in Africa and with exploding epidemics in Eastern Europe and Central Asia. Although the conference was full of stories about how prevention programs across the world are making a difference, the overall message focused on the staggering numbers of people living with HIV and the need for prevention and improved care.

A Virtual Vaccine to Prevent HIV/AIDS: Education

With regard to HIV infection, there is no available vaccine against the virus, but there is a **virtual vaccine** (meaning a procedure as effective as a preventative vaccine): **education.** Thus, the leading primary preventative is education: teaching people how to adjust their behavior to reduce or eliminate HIV exposure. Because the vast majority of HIV infections are transmitted through consensual acts between adolescents or adults, the individual has a choice as to whether to risk infection or not.

Despite widely supported educational efforts at both institutional and street levels, a large number of gay males, drug abusers, and heterosexuals continue to participate in **unsafe sexual practices.** Unsafe sex is defined as having sex without using a condom. This allows the exchange of potentially infectious body fluids such as blood, semen, and vaginal secretions. Unsafe sex most often occurs among gay men and with injection-drug users, by bartering sex for drugs, and by having sex with multiple partners. The sharp increase in the use of crack cocaine and its connection to trading sex for drugs has led to a dramatic rise in almost all sexually transmitted diseases. The idea of ***safer* sexual practices** began in the early 1980s and now refers almost exclusively to the use of a latex or plastic condom with or without a spermicide.

Among the severely drug addicted, concerns about personal safety and survival are secondary to drug procurement and use. Thus, their range of unsafe behaviors leads to random sex and sex without condoms. These behaviors are in part responsible for the increased incidence of HIV and other sexually transmitted diseases (Weinstein et al., 1990).

Comparing HIV/AIDS Prevention to Cancer

AIDS prevention is, in a sense, more essential than, say, cancer prevention. Preventing one HIV infection now will not simply prevent one death from AIDS, as preventing one incurable cancer would prevent one cancer death. Preventing an HIV infection now will help break the chain of transmission, averting the risk that the infected person will knowingly or unknowingly pass the virus on to others who in turn might infect a still wider circle of people.

ADVANCING HIV PREVENTION: NEW STRATEGIES FOR A CHANGING EPIDEMIC

In April 2003 the Centers for Disease Control and Prevention (CDC), in partnership with other U.S. Department of Health and Human Services agencies, other government agencies, and nongovernment agencies decided to change their **primary** prevention strategy, in use for the past 22 years—preventing HIV infection among the at-risk uninfected, to the **secondary** prevention strategy of preventing HIV transmission by those who are infected and their sexual partners. The secondary prevention mission will involve a large federal monetary investment in initiatives that offer HIV testing and counseling to the HIV-infected. This will mark a substantial shift in priorities. At stake is some $90 million that the federal government provides to community groups for HIV prevention each year.

The new strategy is aimed particularly at the estimated 300,000 people who have HIV but do not know it and may be passing it to others unwittingly. The major reasons for this shift in prevention strategy are (1) efforts to reduce the number of annual HIV infections in America have either, depending on one's point of view, stalled or failed. New infections dropped to about 40,000 cases each year in 1988 and remained at that level through 1998 and then **increased** to about 45,000 new infections per year for 1999 through 2004, and (2) the advent of antiretroviral drug therapy in 1996 and its continued success at prolonging the lives of the HIV-infected has made a significant increase in numbers of healthy HIV-infected who, along with those who do not know they are infected, continue the transmission of HIV. The new prevention strategy will begin in July 2004.

Global Prevention Concerns—It is not certain that other countries, even if they have the ability, will follow the prevention strategy shift occurring in the United States. Most countries, at least through 2004 and 2005, will stay with the primary prevention strategy. It is estimated that in order for this strategy to be effective, it will require some $15 billion a year for at least the next 10 years. Where will that money come from? This money does not include the additional monies needed for medical care and living facilities, etc.

DISCUSSION QUESTION: Do you agree with the strategic shift in prevention the CDC initiated in July 2004? Support your decision with known data/facts that can be found within this book or from other sources (see Box 13.8, AIDS Programs: An Epidemic of Waste?), and ask yourself if the CDC is using expanded testing/ counseling to dodge the criticism it has received from conservative politicians about funding "safe sex" programs. Do you hold much hope for a successful prevention program in developing countries? Why?

PREVENTING THE TRANSMISSION OF HIV

The News Is Mostly Bad

We are now into the twenty-fourth year of an epidemic that has touched—directly or indirectly—virtually every person on the planet. We know so much about the virus, yet despite our knowledge, our only option is to *prevent* the initial

BOX 9.1

HIV/STD PREVENTION EDUCATION—UNITED STATES

Eliminating all unsafe sex is not a reasonable goal. Preventing all future HIV and STD infections is impossible, but striving for anything less is unacceptable.

Since 1988, CDC has provided fiscal and technical assistance to state and local education agencies and national health and education organizations to assist schools in implementing effective HIV and STD prevention education for youth. These agencies and organizations develop, implement, and evaluate HIV/STD prevention policies and programs and train teachers to initiate effective prevention efforts and implement curricula in classrooms. As a result of these and other efforts, school-based HIV/STD education is widely implemented in the United States. From 1987 through 2004, the number of states requiring HIV/STD prevention education in schools increased from 13 states to 38 states plus the District of Columbia. This high level of policy support is consistent with public support; 93% of U.S. residents in a 2004 survey reported that information about AIDS and STDs should be provided in school.

The findings in this survey indicate that, despite wide implementation of HIV/STD prevention education in U.S. schools, improvements in prevention programs are still needed. In particular, efforts are needed to increase the percentage of teachers who teach HIV/STD prevention in a health education setting and who receive inservice training on HIV/STD prevention.

HIV/AIDS Prevention and College Students

Ending 2002, CollegeClub.com, a student interest website that provides information about student health, their relationships, academics, and lifestyle issues, ran a poll on American college students and unprotected sex during 2002. The results: Over half had unprotected sex! Julia Davis, from the Kaiser Family Foundation, oversees the Foundation's sex education partnerships with MTV and *Seventeen* magazine. She said, "Young people are more likely to say they'll use a condom in a hook-up than in a longer term relationship. In that sense, casual sex may be safer sex for a lot of young people." From the National Marriage Project at Rutgers University comes a study that reveals that today's singles culture is not oriented to marriage. The current culture is best described as low-commitment, sex without strings—a relationship without rings.

DISCUSSION QUESTION: What is right and who is wrong? What is your position on abstinence-only education? Support your position with factual information—not just opinion.

infection. Prevention is foremost because there is no vaccine, no cure, and, even using the best AIDS drug cocktails available, long-term survival remains questionable, even for those who can afford the drugs. As we face this realization, alarming statistics continue to emerge about the spread of HIV infection.

In San Francisco, estimates suggest that about 50% of homosexual African-American men are infected with HIV. Unsafe sexual practices that could lead to HIV transmission are common among adolescents and young adults, as evidenced by the epidemic of other sexually transmitted diseases in this population. Tens of millions of persons in developing countries will become HIV-infected and most likely die. Entire generations are threatened with extinction in these countries.

The Hard Questions

How can reputable HIV/AIDS scientists explain to the public that the world is being consumed by a disease that is preventable and have it make sense?

The political, social, cultural, economic, and biological factors that have led to the HIV pandemic seem overwhelming. How can a drug user be convinced to use clean needles to prevent an infection that may kill him in 10 years, when he faces an immediate struggle in a violent environment every day? How can condom use be promoted in countries with inadequate supplies of condoms or resources to provide even basic immunizations? Why should young women on the streets of New York, San Francisco, New Delhi, or Bangkok who depend on the sex industry for daily survival care about safer sex when it might lead to rejection by their customers and an end to their livelihood?

DISCUSSION QUESTION: How would you answer the hard questions?

Gender Power

In many societies, there is a large power differential between men and women. Socially and culturally determined gender roles bestow control and authority on males. The subordinate status of women is reinforced by the fact that men are the main or only wage earners in the majority of families. This is compounded further by age differences: In most heterosexual relationships, the man is the older partner.

Wives in many cultures are expected to tolerate infidelity by their husbands, while remaining totally faithful themselves. But AIDS has raised the price of such tolerance, as it puts women at great risk of infection by their husbands. Many women feel powerless to ask their husbands to use condoms at home. Even when they can do this, their need to protect themselves may conflict with a social or personal imperative to have children.

Is There Hope?

Hopelessness threatens reason, but there is reason to believe that education may reduce the number of new HIV infections. In San Francisco, gay men organized grassroots efforts to educate themselves about HIV transmission, and the results are impressive: Less than 1% of the gay male population was infected with HIV after 1985, compared to 10% to 20% in the preceding years. People can change their behavior when educated about the risks of transmission (Clement, 1993).

Educators Given the Job of Prevention

HIV has just as much potential to kill someone infected in 2005 as it did in 1981, but are today's sixth graders and older hearing about HIV/AIDS half as much as did students from the late 1980s through the late 1990s? The great irony is that our fear of HIV now stands in inverse proportion to the damage it does. AIDS is killing millions of people a year—"but those people are in Africa and Asia, so they don't count." To dismiss AIDS as someone else's or some other country's problem is to deny the fundamental reality: Sex is one of the few things that can link you to anyone else on this planet. Remember the 1990s bromide, *If you have sex with someone, you're having sex with everyone they ever had sex with?* It's still true.

Misinformation and Prevention

Millions of people globally have been exposed to a lot of HIV/AIDS misinformation. Because just about everybody thinks they know how HIV/AIDS is spread, they may think they are not at risk because they are not in some specific category, or they may be denying their risk. For example, some may think, "Yes, I sleep with men, which is risky behavior, but I only sleep with men who are married, and they are clean." There are those who will only have sex with very young women because it is more likely they are virgins. Supposedly, having sex with a virgin will cure HIV/AIDS. There is a whole category of people who don't think they are at risk for one reason or another. It is not understood why. Misinformation, denial, and carelessness seriously hamper prevention efforts.

The importance of prevention is especially clear as one comes to understand the limitations in HIV/AIDS therapy. Regardless of what can be medically done for patients with HIV disease, there is no cure. Thus officials from the Centers for Disease Control and Prevention (CDC), the World Health Organization (WHO), the United Nations Joint HIV/AIDS Program (UNAIDS), and the American Health Organization (AHO) have placed the responsibility of prevention in the hands of educators, which include health-care professionals, parents, and teachers.

Current Success of Education Prevention Programs

A growing number of countries have documented the success of their prevention efforts through careful program evaluations and well-designed surveys. There should be no doubt in anyone's mind that prevention programs can reverse a major epidemic, as has been seen in Uganda and Zambia; can contain an emerging epidemic as has occurred in Thailand and Brazil; and can avoid an epidemic all together, as has been well documented in Senegal.

Complacency: The Success of Highly Active Antiretroviral Therapy—HAART

The success of HAART is good news for the people living longer, better lives because of it, but the availability of treatment may lull people into believing that preventing HIV infection is no longer important. This complacency about the need for prevention adds a new dimension of complexity for both program planners and individuals at risk. **First,** while the number of AIDS cases is declining, the number of people living with HIV infection is growing. This increased prevalence of HIV in the population means that even more prevention efforts are needed, not fewer. For individuals at risk, increased prevalence means that each risk behavior carries an increased risk for infection. This makes the danger of relaxing preventive behaviors greater than ever. **Second,** past prevention efforts have resulted in behavior change for many individuals and have helped slow the epidemic overall. However, many studies find that high-risk behaviors, especially unprotected sex, are continuing at far too high a rate. This is true even for some people who have been counseled and tested for HIV, including those found to be infected. **Third,** the long term effectiveness of HAART is unknown. HIV develops resistance to these drugs. If the development of drug resistance is coupled with a relaxation in preventive behaviors, resistant strains can and are being transmitted to others and spread widely.

Prevention Works—Pay Attention!

Sustained, comprehensive prevention efforts begun in the 1980s have had a substantial impact on slowing the HIV/AIDS epidemic in developed countries. While it is difficult to measure prevention— or how many thousands of infections did not occur as a result of efforts to date—in the United States the epidemic was growing at a rate of over 80% each year in the mid-1980s and has now stabilized. While the occurrence of approximately 40,000 to 50,000 new infections annually is deeply troubling, tremendous progress has been made. Also, there is now scientific evidence on which prevention programs are most effective. There is no question that prevention works and remains the best and most cost-effective approach for bringing the HIV/AIDS pandemic under control and saving lives.

Grim Reality

Steven Findlay (1991) wrote that burying those who have died from AIDS has become almost routine. With a 23-year death toll estimated at about 528,000 beginning year 2005 and over 550,000 by 2007, most Americans are indeed becoming accustomed to HIV/AIDS-related deaths. But how many will have died, say, ending in the year 2010 or 2020 in America or worldwide? Will a cure or preventive and therapeutic vaccines be produced? Will our health-care system become swamped and ineffective? The best guess by scientists is that neither an effective vaccine nor cure will be found in the near future. Through the year 2006 it is projected that worldwide about 27 million people will have died of AIDS, over a half million of them in the United States. San Francisco may lose 4% of its population; New York 3%; Central Africa 15% to 20%. To avoid the realization of these projections, people of the world must work on HIV/AIDS prevention.

What We Know

Based on over 24 years of intensive epidemiological surveys, scientific research, and empirical observations, it is reasonable to conclude that HIV is not a highly contagious disease. HIV transmission occurs mainly through an exchange of body fluids via various sexual activities, HIV-contaminated blood or blood products, prenatal events, and in some cases postnatally through breast milk. Since 1981, no new route of HIV transmission has been discovered.

HIV Is a Fragile Virus: Life Span of HIV in Different Environments

The virus is fragile and, with time, self-destructs outside the human body.

The most recent data show that HIV remains active for up to five days in dried blood, although the number of virus particles (titer) drops dramatically. But it is dangerous to assume that there are no infectious viruses remaining in the dried blood or stored body fluids from an HIV/AIDS patient. In cell-free tissue culture medium, the

LOOKING BACK 9.1

THE "MAGIC" EVENT: A FORCE FOR PREVENTION, NOVEMBER 1991

While scientists, the public, and vocal AIDS groups argued that more needed to be done to encourage people to protect themselves from HIV infection, it was not simply a matter of the media providing more information—something had to happen to capture the public's attention. That event occurred in 1991 when Earvin "Magic" Johnson made the stunning announcement that he was HIV positive and was immediately retiring from the National Basketball Association. (See Box 8.4) It resulted in an unprecedented volume of AIDS coverage by the media—259 stories focused on AIDS that week, compared with less than 100 within any other week analyzed.

Johnson's decision to become a spokesperson to promote AIDS awareness and safe sex provided a means to directly link background stories on prevention with the week's major news story. The 1995 Kaiser Family Foundation/Princeton Survey Research Associates survey identified Johnson and Elizabeth Taylor as the two individuals most recognized by the public as national leaders in this area (5% each); and among black Americans, Johnson is particularly likely to be cited as a leader (14%). After this major news event, AIDS coverage was never the same. Celebrity activities, including fundraisers, became an integral part of the AIDS story. AIDS became an issue for reporters on the sports beat as well as those in national affairs or health and science.

virus retains activity for up to 14 days at room temperature (Sattar et al., 1991). According to a recent study, HIV was found to survive between two and four days in glutaraldehyde, a lubricant used to clean surgical instruments. This finding has serious implications for instruments too delicate to be autoclaved (high-pressure steam sterilization), such as endoscopes (Lewis, 1995).

Joseph Burnett (1995) reported that HIV can survive 7 days storage at room temperature, 11 days at 37° C in tissue culture extracellular fluid, and can still be infectious in refrigerated postmortem cadaver tissue for 6 to 14 days. Nadia Abdala and colleagues (1999 updated) reported that HIV recovered in the blood from used syringes can remain active up to at least six weeks. The bottom line is HIV is more resistant to the environment than originally believed.

BEHAVIORAL CHANGE

Perhaps the most difficult area of HIV/AIDS prevention lies in the area of behavior change. Behavior change is certainly difficult to inspire and extremely hard to measure. Furthermore, the theoretical basis for behavior change has been difficult to characterize.

It appears that HIV transmission can be prevented by individual action but it will require change in social behaviors. The best way to protect against all sexually transmitted diseases is **sexual abstinence.** The next best way is a mutually monogamous sexual relationship. Following these two options is the use of a barrier method during sexual activities—male and female condoms or rubber dental dams.

Over 90% of new HIV infections now occur in the developing world. For the foreseeable future, prevention through behavioral change is the only way to slow this epidemic. In fact, history shows that prophylaxis and immunization for most infectious diseases are only partially effective in the absence of behavioral change.

Sexual transmission accounts for the majority of HIV infection in the developing world, but this is the most difficult type of transmission to prevent. The use of condoms, reducing numbers of partners, and abstinence remain the mainstays of preventing sexual transmission of HIV, but they will not be enthusiastically adopted just because health authorities tell people to do so.

Behavior has changed in many populations: among gay men in San Francisco, among injecting-drug users in Amsterdam and New Haven, Conn., and among sex workers and their clients in Nairobi, to name a few. In none of these examples is it clear how the behavioral change took place. Even so, success stories in HIV/AIDS prevention seem to have some elements in common, including consistent and persistent intervention measures over a period of time, a clear understanding of the realities of the target population, and involvement of members of that population in prevention efforts. Successful interventions do far more than provide information: They teach communication and behavioral skills, change perceptions of what is **preventive behavior,** and ensure that the means of prevention, such as condoms or clean needles, are readily available (Hearst et al., 1995).

Table 9-1 provides a number of recommendations for preventing the spread of HIV. These recommendations place the responsibility for avoiding HIV infection on both adults and adolescents. **Lifestyles must be reviewed, choices made, and risky behavior stopped.** The public health service and the CDC have established guidelines that, if followed, will prevent HIV transmission while still allowing individuals to be somewhat flexible in their personal behaviors (*MMWR*, 1989).

Quarantine

With few exceptions, proposals to quarantine all individuals with HIV infection have virtually no public support in the United States. Given the civil liberties implications of quarantine, its potential cost, and the realization that alternative, less re-

Table 9-1 Guidelines for Prevention of HIV Infection

I. For the General Public:

1. Sexual abstinence
2. Have a mutual monogamous relationship with an HIV-negative partner (the greater the number of sexual partners, the greater the risk of meeting someone who is HIV-infected).
3. If the sex partner is other than a monogamous partner, use a condom.
4. Do not frequent prostitutes—too many have been found to be HIV-infected and are still "working" the streets.
5. Do not have sex with people who you know are HIV-infected or are from a high-risk group. If you do, prevent contact with their body fluids. (Use a condom and a spermicide from start to finish.)
6. Avoid sexual practices that may result in the tearing of body tissues (for example, penile-anal intercourse).
7. Avoid oral-penile sex unless a condom[a] is used to cover the penis.
8. If you use injection drugs, use sterile or bleach-cleaned needles and syringes and *never* share them.
9. Exercise caution regarding procedures such as acupuncture, tattooing, ear piercing, and so on in which needles or other unsterile instruments may be used repeatedly to pierce the skin and/or mucous membranes. Such procedures are safe if proper sterilization methods are employed or disposable needles are used. Ask what precautions are being taken before undergoing such procedures.
10. If you are planning to undergo artificial insemination, insist on frozen sperm obtained from a laboratory that tests all donors for infection with the AIDS virus. Donors should be tested twice before the sperm is used—once at the time of donation and again six months later.
11. If you know you will be having surgery in the near future and you are able to do so, consider donating blood for your own use. This will eliminate the small but real risk of HIV infection through a blood transfusion. It will also eliminate the more substantial risk of contracting other transfusion blood-borne diseases, such as hepatitis B.
12. Don't share toothbrushes, razors, or other implements that could become contaminated with blood with anyone who is HIV-infected, demonstrates HIV disease, or has AIDS.

II. For Health-Care Workers:

1. *All* sharp instruments should be considered as potentially infective and be handled with extraordinary care to prevent accidental injuries.
2. Sharp items should be placed into puncture-resistant containers located as close as practical to the area in which they are used. To prevent needle stick injuries, needles should not be recapped, purposefully bent, broken, removed from disposable syringes, or otherwise manipulated.
3. Gloves, gowns, masks, and eye coverings should be worn when performing procedures involving extensive contact with blood or potentially infective body fluids. Hands should be washed thoroughly and immediately if they accidentally become contaminated with blood. When a patient requires a vaginal or rectal examination, gloves must always be worn. If a specimen is obtained during an examination, the nurse or individual who assists and processes the specimen must always wear gloves. Blood should be drawn from all patients—regardless of HIV status—only while wearing gloves.
4. To minimize the need for emergency mouth-to-mouth resuscitation, mouthpieces, resuscitation bags, or other ventilation devices should be strategically located and available for use where the need for resuscitation is predictable.

Table 9-1 *(continued)*

III. For People at Risk of HIV Infection:

1. See the recommendations for the general public.
2. Consider taking the HIV antibody screening test.
3. Protect your partner from body fluids during sexual intercourse.
4. Do not donate any body tissues.
5. If female, have an HIV test before becoming pregnant.
6. If you are an injection-drug user, seek professional help in terminating the drug habit.
7. If you cannot get off drugs, do not share drug equipment.

IV. For People Who Are HIV Positive:

The prevention of transmission of HIV by an HIV-infected person is probably lifelong, and patients must avoid infecting others. HIV seropositive persons must understand that the virus can be transmitted by intimate sexual contact, transfusion of infected blood, and sharing needles among injection-drug users. They should refrain from donating blood, plasma, sperm, body organs, or other tissues. HIV-infected people should:

1. Seek continued counseling and medical examinations.
2. Do not exchange body fluids with your sex partner.
3. Notify your former and current sex partners, and encourage them to be tested.
4. If an injection-drug user, enroll in a drug treatment program and do not share drug equipment.
5. Do not share razors, toothbrushes, and other items that may contain traces of blood.
6. Do not donate any body tissues.
7. Clean any body fluids spilled with undiluted household bleach.
8. If female, avoid pregnancy.
9. Inform health-care workers on a need-to-know basis.

V. Practice of Safer Sex:

Safer sex is body massage, hugging, mutual masturbation, and closed-mouth kissing. HIV-seropositive patients must protect their sexual partners from coming into contact with infected blood or bodily secretions. Although consistent use of latex condoms with a spermicide can decrease the chance of HIV transmission, condoms do break. (Also see 1 through 6 under "For the General Public" in this table.)

[a]Tests show that HIV can sometimes pass through a latex condom. Experts believe that natural-skin condoms are more porous than latex and therefore offer less effective protection. Never use oil-based products such as Vaseline, Crisco, or baby oil with a latex condom because they make the latex porous, causing latex deterioration and breakage, thus nullifying the protection the condom provides against the virus. The use of condoms containing a spermicide is recommended.

pressive strategies can be effective in limiting the spread of HIV infection, quarantine proposals in most countries have been dismissed. Despite claims that HIV/AIDS is similar to other diseases for which quarantine has been used, public health officials have insisted on distinguishing between behaviorally transmitted infections and those that are airborne. HIV/AIDS is not airborne.

Through year 2005, HIV prevention methods in the United States and other developed nations are education, counseling, voluntary testing and partner notification, drug abuse treatment, and syringe exchange programs. To date, the power to quarantine, for any disease, has rarely been used in the United States. In fact, only one country, Cuba, officially used the power of quarantine in 1986 to stem the spread of HIV. Data to date indicate that the use of quarantine of HIV-infected and AIDS persons in Cuba had been very effective. Cuba had 17 sanatoriums holding some 900 persons of which about 200 have AIDS. Cuba stopped the quarantine of HIV-infected persons in mid-1993. Controversial as the sanatoriums were because of serious civil liberties violations, the measures worked.

At the beginning of 2005, about 20% of Cuba's 3900 HIV/AIDS population, by choice, live in the remaining 14 sanatoriums. The rest live outside

POINT OF VIEW 9.1

SAN FRANCISCO PREVENTION PROGRAM FAILING

Despite $20 million spent each year on prevention in San Francisco, many in the AIDS community say HIV prevention strategies are not working, and the growing number of infections seems to support their claim that more men are having unsafe sex. The latest figures from the Department of Public Health show that estimated rates of new HIV infections in the city's gay community doubled in 2001 to 800 from the reported 400 per year in the late 1990s. A recent CDC study found that 14% of gay men intentionally engaged in sex without a condom in the past two years. The CDC reports have put AIDS prevention workers in a quandary. While they struggle to find a solution, the majority of prevention workers agree that the current policy is a failure. In the early days of the epidemic, many in the gay community were confronted daily with the prospect of death. Now, with the aid of medications, many HIV-positive people are living relatively healthy lives, but there is a downside. Marcus Conant, head of the Conant Foundation, and one of the first doctors to work with AIDS patients said, "The messages didn't make it work, only fear made HIV prevention work in the first place." Michael Siever, director of the Stonewall project and adviser to the Castroguys prevention program said, "AIDS prevention has gotten so professional, so slick, and there are so many posters and ads that people don't listen to them. The disease isn't in your face anymore, so people aren't as careful. They are tired of getting tested and they're tired of hearing the same old campaigns to scare them about having sex." AIDS prevention specialists say the urgency to stop infections was removed after 1996 when HIV drugs began to combat the disease. But that's not the only shift. Steven Gibson, Stop AIDS Project spokesman, said beyond the evaporation of urgency is the plain fact that sex without a condom feels better. Tom Coates, director for the Center of AIDS Prevention Studies at the University of California at San Francisco said, "We have all these programs and HIV prevention people are scratching their heads saying, 'How can we help people with HIV?' When all is said and done it's up to the community to say we just don't want HIV to spread." At its peak in the Castro district a decade ago, over 1800 people with AIDS died annually. In 2001, that figure was 218. The sharp decline is a profound relief, but San Francisco is still far from winning its long battle with the disease. In fact, some of the troubles the city thought were over are returning. Cases of syphilis, a disease that significantly increases the risk of being infected with HIV, have quadrupled in the past three years. Cases of HIV infection have doubled in the past five years. The same trends are apparent in other cities with large numbers of gay men. In 2001, a federal study concluded that young, gay men across the country were contracting HIV at a rate not seen in more than a decade. In May 2002, the CDC recommended that all gay and bisexual men be tested for HIV exposure at least once a year. Previously, the CDC had recommended screening only for patients with risky lifestyles. In 2002, gay men in the Castro district were saying, "It is easier than it has been in decades to hide a sexually transmitted disease. Or to forget to ask sexual partners whether they have any illnesses. Or to stop wearing condoms. Everybody around here looks so good now that nobody thinks they're going to catch anything bad."

and receive care at a few specialty centers. A key criterion for living outside the sanatoriums is disclosure of one's sexual partners and providing evidence to health authorities that one is sexually responsible. The authorities actively pursue contact tracing and HIV testing of sexual partners, strategies borrowed from the TB program. There is also mandatory HIV testing of pregnant women, soldiers, and blood donors, but anonymous testing is available for the general public. Last year, over 1 million HIV tests were done in Cuba, out of a population of 11 million.

Cuba has screened all blood donors for HIV since 1986, well ahead of other Caribbean countries. Jorge Perez, who diagnosed the country's first AIDS patients and helped shape the policies of the Santiago sanatorium as its director for 11 years, said, "The Cuban point of view is that you have the right to be sick, but not to transmit it to anyone else." Currently, Cuba has the distinction of having one of the smallest HIV infection rates in the world in a region with one of the highest. It also is the most populous country in the region and according to surveys, has some of the most liberal attitudes toward sex.

While the authoritative response helped contain the spread of HIV, Cuba's political isolation initially prevented access to life-saving antiretroviral drugs. Unable to afford the expensive antiretroviral medicines produced in developed countries, Cuban chemists analyzed the drugs' chemical components and set about creating their own. Cuba now produces sufficient quantities of seven antiviral medications for all its patients. Under Cuba's socialized health-care system, all HIV/AIDS patients receive medical care and drugs free of charge. Cuba hopes to soon begin selling these drugs at cost to other countries.

NEW RULES TO AN OLD GAME: PROMOTING SAFER SEX

The use of barrier methods is one of the few behavioral strategies that individuals can adopt to protect themselves against sexually transmitted diseases. Male, and in some countries female, condoms are currently the only barrier methods widely available. However, there are many cultural, gender, economic, and service-delivery barriers that impede the wide-scale and consistent use of barrier methods for preventing STDs.

Barriers to HIV Infection

The two most effective barriers to HIV infection and other sexually transmitted diseases are **(1) abstinence,** which can be achieved by saying *no* emphatically and consistently; and **(2) forming a no-cheating relationship with one individual, preferably for life.** These solutions to the danger of HIV infection may not be "cool," but they do work. These two apparently safe approaches are endorsed by the surgeon general as the preferred methods. For those who do not practice abstinence, barrier methods are necessary to prevent HIV infection/transmission.

The barrier methods used to prevent HIV infection are the same methods used to prevent other sexually transmitted diseases and often pregnancy. They include latex condoms, plastic condoms (new in 1995), and latex dental dams and diaphragms used in conjunction with a spermicide. Barrier dams or dental dams are thin sheets of latex or similar material placed over the vagina, clitoris, and anus during oral sex. (Ask your dentist to show you a dental dam.) **Spermicides** are chemicals that kill sperm. These same chemicals have also been shown to kill some bacteria and inactivate certain viruses that cause STDs. Spermicides are commercially available in foams, creams, jellies, suppositories, and sponges. Use of these products may provide protection against the transmission of STDs, but the only recommended barrier protection against HIV infection is a condom with a spermicide. National Condom Week is February 14–21. National Condom Day is always on Valentine's Day.

Condom—A Medical Device?

Condoms are classified as medical devices. Every condom made in the United States is tested for defects and must meet quality control guidelines enforced by the federal Food and Drug Administration (FDA).

Choosing the Condom: Manufacturers, Colors, and Shapes

Condoms are intended to provide a physical barrier that prevents contact between vaginal, anal, penile, and oral lesions and secretions and ejaculate.

At least 50 brands of condoms are manufactured in the United States. There are colored condoms—pink, yellow, and gold; flavored condoms; and condoms that are perfumed, ribbed, stippled, and glow in the dark. This assortment of condoms exposes the user and his partner not only to rubber but also to a variety of different chemicals—some that can cause allergic skin reactions (**contact dermatitis**). One to two percent of people are sensitive to latex rubber and demonstrate contact dermatitis.

Condoms are also called rubbers, prophylactics, bags, skins, raincoats, sheaths, and French letters. They can be lubricated or not, have reservoir tips or not, and can contain spermicide.

Condom Size

Most brand name condoms are made in four different lengths and widths (sizes). There is no standard length for condoms, though those made from natural rubber will stretch if necessary to fit the length of the man's erect penis. The width of a condom can also vary. Some condoms have a slightly smaller width to give a closer fit, while others will be slightly larger. Condom makers have realized that different lengths and widths are needed

BOX 9.2

DOES HIV/AIDS PREVENTION WORK?

"**Prevention works"** was the most frequently used phrase at the Thirteenth International AIDS Conference. But are the prevention interventions widely referred to as successful really understood? Twenty-four years into the pandemic, do we actually know what works and what doesn't in controlling the spread of HIV? **If HIV prevention is working, why are infection rates still rising around the world?** Although prevention strategies have been known for over two decades, very few countries or their international partners have taken sufficient action to slow the spread of HIV. **Prevention methods work, yet prevention isn't working.** This pandemic is washing the globe in human suffering and death. Experience in a number of countries has taught that HIV infection can be prevented through investing in information and life skills development for young people. Promoting abstinence, the use of condoms, and ensuring early treatment of sexually transmitted diseases are some of the steps needed. Also needed is an expanded political response and the individual and global acceptance that HIV does cause AIDS.

Undifferentiated Prevention Strategies

All HIV prevention guidelines function as if the same **"safer sex"** mantra applied to every individual in all populations. The basic message is: Protect yourself; use a condom every time. But this has not worked for many reasons including economics, religion, education, poverty, denial, and legal and political will among the global population. In virtually every country people are afraid to take an HIV test or reveal their HIV status because they fear stigma, ostracism, and violent retribution.

In the 1960s there was the peace symbol; it promoted love and loosened sexual attitudes in America. In the 1990s and 2000s, the symbol is a square or round foil-wrapped condom. AIDS revived and catalyzed the use of condoms in developed countries. The dilemma, however, is that the longer the sexual relationship between couples, the less likely they will continue to use the condom—they assume they know each other. But no one knows where their sexual partner is 24 hours a day. In many underdeveloped countries, condoms are either not available, not affordable, or simply shunned by cultural standards. Another problem is that in many poor countries that allow the use of condoms, their pool of many millions of donated condoms is dropping due to donor fatigue.

Prevention Worldwide

Prevention, despite numerous programs around the world, is in its infancy stages. The use of condoms is rare in most regions in the developing world. In nearly all African countries the condom use rate is less than 5%. Entering 2005, in sub-Saharan Africa and China, the estimated total condom supply was sufficient to supply three condoms per man, per year. However, in South Africa, with the largest number of HIV-infected worldwide, condoms are in good supply and relatively easy to get. Current estimates suggest that more than 99% of the HIV infections in Africa in 2004 are attributable to unsafe sex. In the rest of the world, the 2004 estimates for the proportion of HIV/AIDS deaths attributable to unsafe sex range from 13% in East Asia and the Pacific to 94% in Central America. Globally, about 2.9 million deaths a year are attributed to unsafe sex, most of these deaths occurring in Africa. In Russia, from the results of a telephone survey in 2001, 8 out of 10 respondents said they rarely or never used a condom. Only 6% reported regular condom use. One in three believed that condoms provided protection against HIV transmission. In Vietnam in 2003, condom sales of 40 million amounted to a third of their estimated needs. In countries like Iran, Iraq, and other countries in the Middle East, religious teachings do not allow the use of condoms. Only four countries in Asia and four in Latin America and the Caribbean have condom use rates of 10% or more. The highest rate of condom use is in Japan with 46%, and Singapore 24%. In Europe, Denmark, Finland, Sweden, Spain, and Slovakia all show a rate over 20%. The U.S. condom usage rate is 13% and Britain is 18 %.

PREVENTION IN THE UNITED STATES

Over the last 15 years, the number of new HIV infections has remained stable at about 40,000 to 50,000 a year. For the last five years the number of AIDS-associated deaths has stabilized at about 16,000 to 18,000 a year. In February 2001, the CDC, recognizing that such levels of new HIV infections and AIDS deaths are unacceptable, announced a new program to half the number of new HIV infections by year 2005. The CDC's primary strategy has now shifted to identifying people already infected and encouraging them not to spread HIV. The CDC's **Serostatus Approach to Fighting the HIV Epidemic,** or **SAFE** program, is based on the assumption that "Most HIV infections are spread by outwardly healthy people who do not realize they have HIV." The CDC, which already spends over $700 million a year on AIDS prevention, estimated that it will need to increase its budget to $900 million a year to cover the campaign. Within the SAFE program, the CDC also wants to increase the proportion of persons who know they are HIV-infected from the currently estimated 70% up to 95%, and to increase the proportion of HIV-infected people who are linked to appropriate treatment services from 50% up to 80% by 2005. Currently about 300,000 Americans are infected but do not know it. For years 2002 through 2005, about 80,000 HIV infections were discovered in the United States. This number comes from the 45,000 new infections annually and 35,000 of previously infected but newly discovered infections.

and are increasingly broadening their range of sizes. An internet retailer now advertises 55 sizes of "They Fit" condoms.

History of Condoms

Condom use can be traced back to 1000 B.C. when Egyptian men used linen sheaths or animal membranes as a sheath to cover their penises (Barber, 1990). Animal intestines were flushed clean with water, sewn shut at one end and cut to the length of the erect penis. In 1504, Gabrielle Fallopius designed a medicated linen sheath that was pulled on over the penis to prevent syphilis infection. A Japanese novel written in the tenth century refers to the uncomfortable use of a tortoise shell or horn to cover the penis.

It is interesting to note that condoms were used far more often throughout history as protection against STDs than as contraceptives. For example, an eighteenth-century writer recommended that men protect themselves against disease by placing a linen sheath over the penis during intercourse.

The term "condom" came into common usage in the 1700s. According to accounts in the early 1700s, condoms were sold and even exported from a London shop whose proprietress laundered and recycled them in a back room (Barber, 1990). Condoms became more widely available after 1844. The latex condom was first manufactured in the 1930s.

Condoms have been available in the United States for over 130 years, but have never been as openly accepted as they are now. Their sale for contraceptive use was outlawed by many state legislatures beginning in 1868 and by Congress in 1873. Although most of these laws were eventually repealed, condom packages and dispensers until only a few years ago continued to bear the label "Sold only for the prevention of disease," even though they were being used mainly for the prevention of pregnancy.

After the advent of nonbarrier methods of contraception during the 1960s (mainly the use of the birth control pill) there was an ensuing epidemic increase in most sexually transmitted infections. Condoms once again are being marketed for the prevention of disease (Judson, 1989) (Figure 9-1).

Safer Sex, the Choice of Condom

Although a variety of preventative behaviors have been recommended (Table 9-1), the responsibility of safer sex, with a condom, is a personal choice. If one decides to use a condom, then the choice is what kind, and whether or not to use a spermicide.

The Male Condom—The American-made condom most often sold is made of latex, is about 8 inches long, and in general, one size fits all. About 500 million condoms are sold annually in the United States. Seven to ten billion are sold annually worldwide but 24 billion more are needed and most of them in Asia. (Grimes, 1992 updated).

Intact latex condoms provide a continuous mechanical barrier to HIV, herpes virus (HSV), hepatitis B virus (HBV), *Chlamydia trachomatis*, and *Neisseria gonorrhoeae*. A recent laboratory study indicated that latex and **polyurethane condoms** are the most effective mechanical barrier to fluid containing HIV-sized particles (0.1 μm in diameter) available. The male polyurethane condom is thinner than the latex condom, which makes them more agreeable in feel and appearance to some users. However, they also break more easily during use.

FIGURE 9-1 Advertisement for Condoms. A common theme centered around the use of a condom to protect against sexually transmitted diseases. *(Photograph courtesy of the Minnesota AIDS Project, Minneapolis)*

Three prospective studies in developed countries indicated that condoms are unlikely to break or slip during proper use. Reported breakage rates in the studies with latex condoms were 2% or less for vaginal or anal intercourse (*MMWR*, 1993a; Spruyt et al., 1998).

Choice—The best choice for preventing STDs and pregnancy is condoms that are made of **latex** or **polyurethane (plastic)** and contain a **spermicide.** The spermicide is added protection in case the condom ruptures or spills as it is taken off. Although some laboratory evidence shows that some spermicides can inactivate HIV, researchers have found that these products cannot prevent a person from becoming HIV-infected.

During the Thirteenth International AIDS Conference held in Durban, South Africa, July 9–14, 2000, researchers from the Joint United Nations Program on AIDS (UNAIDS) presented the results of a study of a product, Col-1492, which contains nonoxynol-9 (N-9). The study examined the use of candidate microbicide, or topical compound to prevent the transmission of human immunodeficiency virus (HIV) and sexually transmitted diseases (STDs). The study found that the spermicide N-9 did not protect against HIV infection and may have increased the risk of transmission. Women using N-9 gel became infected with HIV about 50% more often than women who used the placebo gel.

CDC has released a "Dear Colleague" letter that summarizes the findings and implications of the UNAIDS study. The letter is available on the Web, http://www.cdc.gov/hiv; a hard copy is available from the National Prevention Information Network, telephone 1-800-458-5231.

Condom Advertising on American Television—On January 4, 1994, condoms danced into America's living rooms as part of the most explicit HIV prevention campaign the nation has ever seen. Previously, condoms were rarely seen or even mentioned on American television. But the Centers for Disease Control and Prevention launched a series of radio and television public service announcements targeting sexually active young adults, a group at high risk for HIV infection.

One of the television ads, entitled ***Automatic,*** features a condom making its way from the top drawer of a dresser across the room and into bed with a couple about to make love. The voice-over says, "It would be nice if latex condoms were automatic. But since they're not, using them should be. Simply because a latex condom, used consistently and correctly, will prevent the spread of HIV." Another, entitled ***Turned Down,*** features a man and woman kissing, when the woman asks the man, "Did you bring it?" When he says he forgot it, she replies, "Then forget it," and turns on the light. There is also a pair of abstinence ads, in which condom use is not mentioned. The ads feature a man and woman talking. She says, "There is a time for us to be lovers. We will wait until that time comes."

Language that was once forbidden on TV has now become routine. Will it prevent HIV infection? Will it save lives?

DISCUSSION QUESTION: What is your opinion on condom advertising? When? Where? Why?

SIDEBAR 9.1

TWO ANECDOTES ON CONDOM USE IN SOUTH AFRICA

Women in South Africa are brought up to be subservient to men. Especially in matters of sex, the man is always in charge. Women feel powerless to change sexual behavior. Even when a woman wants to protect herself, she usually can't; it is not uncommon for men to beat partners who refuse intercourse or request a condom. "Real men" don't use them, so women who want their partners to use a condom must fight deeply ingrained taboos.

Anecdote One

A nurse in Durban, South Africa, coming home from an AIDS training class, suggested that her mate should put on a condom as a kind of homework exercise. He grabbed a pot and banged loudly on it with a knife, calling all the neighbors into his house. He pointed the knife at his wife and demanded: "Where was she between 4 p.m. and now? Why is she suddenly suggesting this? What has changed after 20 years that she wants a condom?"

Anecdote Two

This schoolteacher is an educated man, fully cognizant of the AIDS threat. Yet even he bristles when asked if he uses a condom. "Humph," he says with a fine snort. "That question is non-negotiable." So despite extensive distribution of free condoms, they often go unused. Astonishing myths have sprung up. If you don one, your erection can't grow. Free condoms cannot be safe: They have been stored too long, kept too hot, kept too cold. Condoms fill up with germs, so they spread AIDS. Condoms from overseas bring the disease with them. Foreign governments that donate condoms put holes in them so that Africans will die.

Buying Condoms—Women are taking a more active role in buying condoms. In 1985, women bought about 10% of the condoms sold. Now they purchase 40% to 50%. According to surveys, most women buying condoms are single, and their concern is about HIV infection rather than pregnancy. The fact that more women are willing to buy condoms is evidence that HIV education is working to some degree.

Many condoms are purchased from vending machines. The FDA recommends the following guidelines when purchasing condoms from a vending machine:

1. Is the condom made of latex or polyurethane?
2. Is the condom labeled for disease prevention?
3. Is the spermicide (if any) outdated?
4. Is the machine exposed to extreme temperatures or direct sunlight?

In mid-1992, the first drive-up "Condom Hut" opened in Cranston, R.I. With each purchase the customer receives a brochure on safer sex.

It is generally recommended that condoms be stored below 25° C (77° Fahrenheit; room temperature is 72° Fahrenheit or 22° Centigrade). The packaging should be impermeable to both sunlight and gas. If air, which includes ozone, enters the package, it will affect the condom very quickly—ozone is like rust to a condom. Latex is a natural product—it will go bad if you don't treat or store it properly.

Condoms in Public Schools?

Condoms are now being dispensed without charge in most college and university and public health clinics, and in at least 400 high school health offices in the United States. Some cities in Canada have been providing access to free condoms in high schools since 1984.

In December 1999, the American Medical Association (AMA) adopted a policy that advocates handing out condoms in schools and minimizes the value of abstinence-only sex education. While some doctors and groups have called the policy medically irresponsible, it is supported by the U.S. surgeon general. The policy, based on studies, concluded that safer sex programs are effective in delaying sex in teenagers, and that abstinence-only programs have limited value.

Are Policies or Studies Reality?

Regardless of educational programs on safer sex and condom usage, recent studies indicate that adults and teenagers still refuse to use condoms. What they know is not equal to what they do! Based on their findings, the researchers said information-oriented school- and community-based AIDS prevention programs will not succeed in getting some adults and adolescents to use condoms because there is no association between knowledge and preventive behavior.

POINT OF INFORMATION 9.1

CLARIFYING THE ISSUES OVER CONDOM USE

Two major issues surface in the debate over advocating condom use in the prevention of HIV infection: One concerns the concept of efficacy, the condom's ability to stop the virus from passing through, and the other, the fear that making condoms available will encourage early sexual activity among adolescents and extramarital sex among adults.

EFFICACY (DO THEY WORK?)

No public health strategy can guarantee perfect protection. For instance, the influenza vaccine is only 60–80% effective in preventing influenza, but thousands of deaths could be prevented annually through the wider use of this less than perfect vaccine. The real public health question is not whether condoms are 100% effective, but rather how can we more effectively use condoms to help prevent the spread of disease?

All condoms are not 100% impermeable; they are not all of the same quality. Investigators using different testing methods have reported that latex condoms are effective physical barriers to high concentrations of *Chlamydia trachomatis, Neisseria gonorrhoeae,* the herpes and hepatitis viruses, cytomegalovirus, and HIV (Judson, 1989). But for maximum effectiveness condoms must be properly and consistently used from start to finish (Table 9-2).

Table 9-2 Proper Placement of a Condom on the Penis[a]

1. Open the packaged condom with care; avoid making small fingernail tears or breaks in the condom.
2. Place a drop of a water-based lubricant inside the condom tip before placing it on the head of the penis. Be sure none of the lubricant rolls down the penis shaft as it may cause the condom to slide off during intercourse.
3. Hold about half an inch of the condom tip between your thumb and finger—this is to allow space for semen after ejaculation. Then place the condom against the glans penis (if uncircumcised, pull the foreskin back).
4. Unroll the condom down the penis shaft to the base of the penis. Squeeze out any air as you roll the condom toward the base.
5. After ejaculation, hold the condom at the base and withdraw the penis while it is still firm.
6. Carefully take the condom off by gently rolling and pulling so as not to leak semen.
7. Discard the condom into the trash.
8. Wash your hands.
9. Never use the same condom twice.
10. Condoms should not be stored in extremely hot or cold environments.

[a]Males should practice putting on and removing a condom prior to engaging in sexual intercourse.

To be effective, a condom must be worn on the penis during the entire time that the sex organ is in contact with the partner's genital area, anus, or mouth. Care must be taken that the condom is on before vaginal, anal, or oral penetration, and that it does not slip off. If properly used, the condom provides protection against most of the STDs that occur within the vagina, on the glans penis, within the urethra, or along the penile shaft.

Because the condom covers only the head and shaft of the penis, it does not provide protection for the pubic or thigh areas, which may come in contact with body secretions during sexual activity.

Margret Fischel reported in 1987 that 17% of women whose husbands were HIV positive became HIV-infected while using condoms properly and consistently.

In 2003, the European Commission, after supporting 15 years of intense investigation in Europe, Asia, and Africa on whether condoms prevent HIV transmission reported that condoms are close to 100% efficient at blocking HIV transmission if used properly. However, Norman Hearst and colleague (2004) used computerized searches of peer-reviewed scientific literature, conference presentations, publications of national and international organizations, and lay media to determine the most likely probability that condoms will prevent HIV transmission. He elected to use the most reliable and relevant data available. Hearst determined that if a condom was used properly, it was 90% effective in stopping the transmission of HIV. These data are sure to fuel the ongoing debate. In spite of argument, pro and con on the use of condoms, next to abstinence, condoms are about the only mechanical device available for safer sex. The bottom line is that condoms reduce some of the risks in preventing the transmission of HIV and other STDs, thus condoms are used for safer sex but they cannot guarantee safe sex.

DO CONDOMS ENCOURAGE SEXUAL ACTIVITY?

Many persons assert that those who promote condom use to prevent HIV infection appear to be condoning sexual intercourse outside of marriage among adolescents as well as among adults. Recent data from Switzerland suggest that a public education campaign promoting condom use can be effective without increasing the proportion of adolescents who are sexually active. A 3-year, 10-month study showed condom use among persons aged 17 to 30 years increased from 8% to 52%. By contrast, the proportion of adolescents (aged 16 to 19 years) who had sexual intercourse did not increase over that same period. A report from Deborah Sellers and colleagues (1994) and Sally Guttmacher et al. (1997) also concluded that the promotion and distribution of condoms did not increase sexual activity among adolescents. The study involved 586 adolescents who were 14 to 20 years of age. Douglas Kirby and colleagues (1998) studied 10 Seattle high schools that made condoms available to students through vending machines and school clinics. The study measured the number of condoms students took and the subsequent changes in sexual behavior. The investigators concluded that making condoms available in Seattle schools enabled students to obtain relatively large numbers of condoms but did not lead to increases in either sexual activity or condom use. **Administrators of these studies feel that condoms no more cause sexual activity than umbrellas cause rain.**

Adolescents, Condoms, and Mixed Messages

The AIDS epidemic has brought new dimensions of complexity and urgency to the debate over adolescent sexual activity. Some have urged abstinence as the only solution while others champion condom use as the most practical public health approach. Thus a clear message about condoms may have been obscured by controversy over providing condoms for adolescents in schools while at the same time trying to discourage these same young people from initiating sexual activity.

There must be a common ground: People should be able to agree that premature initiation of sexual activity carries health risks. Therefore, young people must be encouraged to postpone sexual activity. Parents, clergy, and educators must strive for a climate supportive of young people who are not having sex. Let them know it is a very positive and intelligent decision, and so help to create a new health-oriented social norm for adolescents and teenagers about sexuality.

The message that those who initiate or continue sexual activity must reduce their risk through correct and consistent condom use needs to be delivered as strongly and persuasively as the message "Don't do it." Protection of the individual and the public health will depend on our ability to combine these messages effectively (Roper et al., 1993).

ANECDOTE

Presenting facts without understanding won't work. Here is a simple story to emphasize the point: "A minister, following his custom, paid a monthly call on two spinster sisters. While he was standing in their parlor, holding his cup of tea, engaged in their usual chit chat, he was startled by something that caught his eye. There on the piano was a condom! 'Ladies, in all the years we've known each other I have never intruded into your private lives, and never felt the need to. But now I am forced to ask what is that thing doing there?' One of the ladies replied, 'Oh, that's a wonderful thing, pastor, and they really work!' The minister was agitated: 'I'm not talking about their value or effectiveness. I just want to know what that thing is doing on your piano.'

"'Well, my sister and I were watching television. We heard this lovely man, the surgeon general of the whole United States. He said that if you put one of those on your organ, you'll never get sick. Well, as you know we don't have an organ, but we bought one and put it on the piano, and we haven't had a day's sickness since!'"

Equally discouraging is a recent study in the United States. A 2004 online survey by the American Social Health Association of 1,155 people ages 18 to 35 indicated about 84% believed they adequately protected themselves against HIV and other STDs, but nearly half engage in unprotected sex. Approximately 47% of the respondents never used protection for vaginal sex, 82% never used protection for oral sex, and 64% never used protection for anal sex. The survey showed that 93% believed their current or most recent partner did not have an STD, yet one of three people have never discussed HIV or

STDs with their partner, while 68% did not think they would contract HIV or an STD.

Studies in several other countries show a return to high-risk practices in populations, particularly men who have sex with men, in which risk reduction had been previously observed. Surveys conducted in London gay bars, gyms, and health clinics showed that at least one in five gay or bisexual men was practicing unprotected anal intercourse, and many of these men knew that they were HIV-infected. In France, unprotected anal intercourse among gay men increased from 17% in 1997 to 23% in 2000. In the Netherlands, between 1996 and 2000, rates of unprotected anal intercourse with casual partners increased from 25% to 34%, with parallel increases in sexually transmitted infections. In Cape Town, South Africa, one-third of gay men surveyed had recently had unprotected intercourse with partners of unknown HIV status. In Vancouver, a study of gay men found increases in risk behaviors and HIV infection rates. In Australia, rates of unprotected anal intercourse are steadily increasing. Together, these studies also illustrate that levels of risk are ever changing within the populations in which HIV prevention efforts have been most concentrated (Kalichman et al., 2001). Perhaps the following quote will help explain the current rise in HIV infections. "It's not just the fact that young people don't know any older people with HIV. The image of HIV itself isn't bad anymore. We don't walk through the streets like we used to and see people so sick they can hardly walk. Today, people with HIV are buffed, they have boyfriends and they are doing things. It's just not such a terrible thing anymore."

Condom Availability in Developing Nations

Access to condoms in developing nations, although crucial, is poor even though over *1 billion* condoms are distributed in developing nations each year. The World Health Organization spends some $70 million a year for male condoms distributed in 14 countries. But, it would cost $460 million to provide just Africa with an adequate condom program. In 1992, the United Kingdom purchased 66 million condoms for Zimbabwe, but the country needs at least 120 million condoms per year.

In May 1999, researchers on sexually transmitted diseases started to close the gap between the number of condoms now produced and the numbers needed. It has been estimated that it will take an additional 24 billion condoms a year used mostly in Asia to slow the transmission of HIV and other STDs.

Religion Prohibits Use of Condoms—The Catholic church teaches that the conjugal act must always leave open the possibility of conception. It is the purpose of devices such as condoms to rule out impregnation. Consequently, the church holds that the use of contraceptives in order to prevent the origination of new life is immoral. Peter Piot, executive director of UNAIDS, said, "Catholicism has had a major role in virtually all aspects of the global response to AIDS since the disease was identified. With its hospices and hospitals, orphanages and parish outreach, the Catholic church provides more direct care for people with AIDS and their families and communities, particularly in Africa and Latin America, than any other institution. Yet, while the UN, most governments and governmental agencies, and almost all of the international organizations working in AIDS agree that condoms are the most effective means of slowing the spread of HIV, the Vatican has remained steadfast in its opposition to their use." In 1999, 2000, and again in November 2002, the Vatican reiterated the Catholic Church's opposition to condom use as a way to prevent the spread of HIV. The church promotes a moral education to protect its members. Given that 1 billion, 30 million people worldwide are Catholics and that not all are low-risk individuals, the sanction against condom use presents a difficult problem in slowing the transmission of HIV.

In October of 2003, Cardinal Alfonso Lopez Trujillo, President of the Vatican's Pontifical Council for the Family, said on the British Broadcasting Corporation's *Panorama* program "Sex and the Holy City" that "HIV can easily pass through the net that is formed by the condom. Sperm pass through the net and HIV is 450 times smaller than sperm. Condoms do not reduce the risk of HIV infection." The cardinal also said, speaking of condoms, that "safe sex is a form of Russian roulette." Other Catholic leaders on the program claimed that because of condom permeability, condoms can spread HIV.

First Global Campaign To End Catholic Bishops' Ban on Condoms

The first global campaign to end the Catholic bishops' ban on condoms was being launched by Catholics for a Free Choice (CFFC) on the eve of World AIDS Day 2003. Billboards and ads in subways and newspapers carrying the message "Banning Condoms Kills" began appearing around the world on November 30. This unprecedented worldwide public education effort is aimed at Catholics and non-Catholics alike to raise public awareness of the devastating effect of the bishops' ban on condoms. It invites the public to join a global campaign to end the ban—Condoms4life at www.condoms4life.org. People who join the campaign are being asked to contact local policy makers and express their support for the availability of condoms and their concern that the bishops should not undermine responsible public health policy on HIV/AIDS. Advertising in the United States and in countries with a significant Catholic population or AIDS crisis, such as Mexico, the Philippines, Kenya, South Africa, Chile, and Zimbabwe, is the first phase of a sustained mobilizing effort to change the Vatican's policy and its aggressive lobbying against availability and access to condoms, especially in areas of the world where HIV transmission and AIDS deaths are rising dramatically.

A Message to All Catholics

Using the core message that "Good Catholics Use Condoms," the campaign presents a positive message to sexually active Catholics about responsibility and caring for others. The ads appeal to people of faith noting that "We believe in God. We believe that sex is sacred. We believe in caring for each other. We believe in using condoms." The campaign aims to counter the message sent by Catholic bishops worldwide that condoms are immoral and unsafe.

The World Health Organization believes these incorrect statements about condoms and HIV are dangerous when the world is facing a global pandemic which will kill an estimated 24 million people, ending year 2005 and affects about 41 million living with HIV/AIDS. The World Health Organization maintains consistent and correct use of condoms reduces transmission by 90%.

In October 2004, Cardinal Cormac Murphy-O'Conner, the head of Roman Catholics in England and Wales, and Cardinal Godfreid Daneels, his Belgian counterpart, have suggested that the use of condoms may be morally obligatory in certain circumstances. For many in Africa and Asia, sex is often the only commodity people have to exchange for food, school fees, exam results, employment, or survival itself in situations of violence.

Regardless of argument, one thing is certain: without the use of a condom, there is nothing to block HIV transmission.

Condom Quality

Some have touted condoms as a bulletproof vest for preventing HIV infection. This is simplistic and inaccurate because condoms may break or be used improperly. Condoms have been shown to reduce significantly but not eliminate the transmission of HIV. Condom use is a form of safer sex, but not absolutely safe sex. The March 1989 issue of *Consumer Reports* did a rather extensive study on the quality of different brands of condoms. The reader survey revealed that one in eight people who used condoms had two condoms break in one year of sexual activity; one in four said one condom had broken. Calculated from its data, about 1 in 140 condoms broke, with condom breakage occurring more often during some sexual activities than others. For example, the breakage rate during anal sex was calculated at 1 in 105 compared to 1 in 165 for vaginal sex. One in 10 heterosexual men admitted to engaging in anal sex while using condoms (*Consumer Reports*, 1989).

In a separate U.S. Public Health Service study (1994), if a couple used male condoms consistently and correctly, there was a 3% chance of unintentional pregnancy. Other studies have found that condoms reduce the risk of gonorrhea by as much as 66% in men, but by no more than 34% in women. It doesn't appear as though they prevent papilloma virus infection—which causes genital warts—in either sex (Stratton et al., 1993).

In 1987, FDA inspectors began making spot checks for condom quality. They fill the condom with 10 ounces (about 300 mL) of water to check for leaks. If leaks are found in more than 4 per 1000 condoms, that entire batch is destroyed.

Over the first 15 months of spot inspections, 1 lot in 10 was rejected. Import condoms failed twice as often—one lot in five.

Polyurethane (Plastic) Condoms—For the 1% of the general population that is sensitive to latex and for those who have a variety of other reasons not to use a latex condom, there is now a clear, thin, FDA-approved polyurethane (plastic) condom for sale in the United States. The condoms are colorless, odorless, and can be used with any lubricant. The current cost is about $1.80 each. Manufacturers of at least five other plastic condoms are waiting FDA approval. A report by Ron Frezieres and colleagues (1999) reported that although polyurethane and latex condoms provide equivalent levels of contraceptive protection, the polyurethane condom's higher frequency of breakage and slippage suggests that this condom may confer less protection from sexually transmitted infections than do the latex condoms.

A Polymer Gel Condom?—In November 1997 Michael Bergeron of the Laval Infectious Disease Research Center in Canada created a liquid condom that is nontoxic, tasteless, and transparent. The condom, made out of a polymer gel, is spread over the vagina or anus using an applicator. It gelifies in response to body heat and surrounds the penis. According to the inventor: "It's a physical barrier, but it moves—that's the advantage." The polymer gel can be flushed out with water. Bergeron developed the gel due to patient complaints about tight condoms. A major benefit is that the gel condom is somewhat invisible—either sexual partner can use it as a protective measure. Trials to determine the contraceptive's efficacy in preventing HIV, herpes, and pregnancy are ongoing.

Why Some Men Don't Use Condoms: Male Attitudes About Condoms and Other Contraceptives—This is the name of a study published by the Henry J. Kaiser Family Foundation (1997). The study shows that men, especially teenage males, don't use condoms mostly because of embarrassment: buying the condom, talking about the condom with their sexual partner, putting the condom on in front of their sexual partner, offending or scaring their sexual partner, losing an erection while putting the condom on, and the reduction of sexual pleasure. (This free report can be obtained at 1-800-656-4533 monograph #1319.)

The Need for Education on Condom Use Among College-Age Men

In a study on errors in condom use among college-age men, researchers found that condom use errors were commonplace. The study, conducted from November 2000 through January 2001, explored condom use errors and problems among college men at Indiana University. Of 362 men, 158 met the study inclusion criteria (never married, and reporting putting a condom on for sex at least once in the past three months). Forty-two percent of young men participating in the study reported that they wanted to use a condom but did not have one available. Some of the other basic problems highlighted by the study included not checking the condom for visible damage (74%), not checking the expiration date (61%), and not discussing condom use with their partner before sex (60%). In addition, various technical errors were found, including putting on the condom after starting sex (43%), taking off the condom before sex was over (15%), not leaving a space at the tip of the condom (40%), and placing the condom upside down on the penis and then having to flip it over (30%). In addition, 29% of the study participants reported condom breakage and 13% reported that the condom slipped off during sex. The problems highlight the need for better condom education and instruction.

The Female Condom (Vaginal Pouch)—The female condom was FDA-approved in May 1993, and has become available to the general public. Before giving the Reality condom final approval, the FDA asked that two caveats be put into the labeling. First, the agency required a statement on the package label that male condoms are still the best protection against disease, and second, that the label compare the effectiveness of female condoms with that of other barrier methods of birth control. According to the FDA, in a study of 150 women who used the female condom for six months, 26% became pregnant. The manufacturer contends that the pregnancy rate was 21%—and only because many women did not use the condom every time they had sex. With "perfect use," company officials say, the rate is 5%, in contrast to 2% for male condoms. Both conditions were met to the FDA's satisfaction. Gaston Farr and coworkers (1994) concluded that the female condom provided contraceptive efficacy in the same range as other barrier methods, particularly when used consistently and correctly, and has the added advantage of helping protect against sexually transmitted diseases.

Design of the Female Condom

The female condom is now called the **vaginal pouch.** However, the female condom is being used by gay men for anal sex. A report from gay men using this condom says that they are having problems with the condom's design and experience usage difficulties. In short, of the gay men interviewed, none believe the female condom will replace their use of the male condom for anal sex. It is 17 cm (about 6-3/4 inches) long and it consists of a 15-cm polyurethane sheath with rings at each end (Figure 9-2). The closed end fits into the vagina like a diaphragm. The outer portion is designed to cover the base of the penis and a large portion of the female perineum (the area of tissue between the anus and the beginning of the vaginal opening) to provide a greater surface barrier against microorganisms. Studies of acceptability, contraceptive effectiveness, and STD prevention are currently underway. Potential advantages of this product are: (1) it provides women with the opportunity to protect themselves from pregnancy and STDs; (2) it provides a broader coverage of the labia and base of the penis than a male condom; (3) its polyurethane membrane is 40% stronger than latex; and (4) it is more convenient; it can be inserted hours before sexual intercourse. Disadvantages are (1) not aesthetically pleasing and (2) can be difficult to insert and remove.

Because of the polyurethane used to make it, the female condom is both strong and durable. No special storage arrangements have to be made because polyurethane is not affected by changes in temperature and dampness. The expiration date on the female condom is 60 months (5 years) from the date of manufacture.

Global Use of the Vaginal Pouch (Female Condom)—Female Health Company (FHC) of Chicago is the sole manufacturer of the female pouch. Under agreement between UNAIDS and FHC the female pouch is sold for between 50 and 90 cents (U.S.) in the developing world to encourage its use and provide greater access to women. It is also marketed in the Americas and Europe for about $2.50 (U.S.). In March 2002, France unveiled its first female pouch (condom) machines, blue for men and pink for women. At the beginning of 2005, over 40 million female condoms were distributed in 88 countries in Africa, Asia, and Latin America. Distribution began in India in early 2004. They cost Third World governments 12 cents each. Male condoms cost 3 cents each! UNAIDS hopes to get the pouch into all developing nations.

Redressing the Balance of Power

The symbolic importance of the female condom should not be understated: It is the first woman-controlled barrier method officially recognized as a means for the prevention of sexually transmitted disease. The female condom allows women to be able to deal with the twin anxieties—AIDS and unwanted pregnancy—with a method that is under their own control.

Many women become HIV-infected not because of their own behavior, but because of their partner's. Because of the nature of gender relations, women may have little influence over their partner's sexual behavior.

Commentary: Positive messages for women have been spread worldwide on the advantages of using the female condom. But convincing women, in large numbers, to use this relatively new device may not be so easily accomplished. For example, it took 17 years before women in developed countries accepted and began to routinely use the tampon.

Condom Lubricants

It has been demonstrated that petroleum or vegetable oil-based lubricants should not be used with latex condoms. Nick White and colleagues (1988) have reported that these lubricants weaken latex condoms. Latex condoms exposed to mineral oil for 60 seconds demonstrated a 90% decrease in strength (Anderson, 1993). There are a number of water-based lubricants that do not

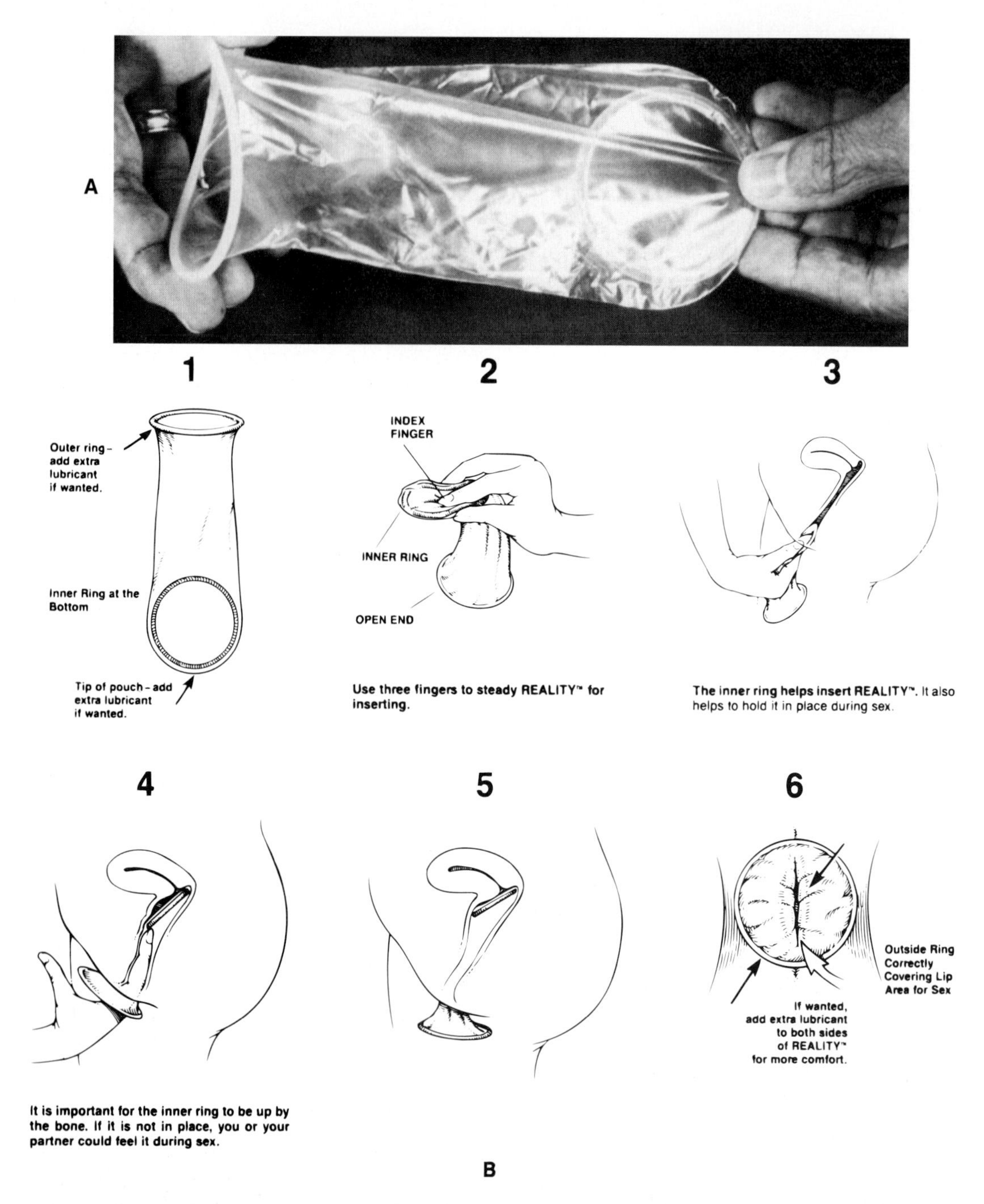

FIGURE 9-2 **A.** The Vaginal Pouch. **B.** The 7-inch female condom or vaginal pouch is made of lightweight, lubricated polyurethane and has two flexible rings **(1)**, one at either end. It is twice the thickness of the male latex condom. The inner ring **(2)** is used to help insert the device and fits behind the pubic bone. The outer ring remains outside the body. Unlike the diaphragm, the vaginal condom protects against the transmission of HIV, which can penetrate the vaginal tissues. The pouch can be inserted anytime from several hours to minutes prior to intercourse. The vaginal pouch is inserted like a diaphragm and removed after sex. FDA-approved in May 1993. *(Courtesy of Female Health Co., Chicago)*

Table 9-3 Water-Based[a] and Oil-Based[b] Lubricants Often Used With Latex Condoms

Lubricants Recommended	Lubricants Not Recommended
Aqualube	Petroleum jellies
Astroglide	Mineral oils
Cornhuskers Lotion	Vegetable oils
Forplay	Baby oil
H-R Jelly	Massage oil
K-Y Brand Jelly	Lard
RePair	Cold creams
Probe	Hair oils
Today Personal Lube	Hand lotions containing vegetable oils
	Shaft
	Elbow Grease
	Natural Lube

[a]Water-based lubricants can be used with latex condoms.

[b]Oil-based lubricants will chemically weaken latex, causing it to break.

These lists are not exhaustive of all available lubricants used by consumers.

(Source: The STD Education Unit of the San Francisco Department of Public Health)

adversely affect latex condoms; these should be the lubricants of choice (Table 9-3).

DISCUSSION QUESTION: Reaching this point in the text and having just read the section on male and female condoms, do you think the danger for life, with respect to HIV and other life-threatening STDs, is high enough that condoms should be as familiar to everyone as are toothpaste and toilet paper and also as available? Defend your view with credible information.

An Alternative to Condoms: Vaginal Microbicides

While condoms, when used consistently and correctly, are effective in preventing the sexual spread of HIV, there is an urgent need for chemical barrier methods that women can use for HIV prophylaxis, such as vaginal microbicides. Vaginal microbicides are products for vaginal administration that can be used to prevent HIV infection and/or other sexually transmitted diseases. An ideal vaginal microbicide would be safe and effective, and also tasteless, colorless, odorless, nontoxic, stable in most climates, and affordable. It must be pointed out that like condoms, microbicides will not protect injection-drug users.

It will take generations to change male sexual behaviors, and women, especially in the underdeveloped nations, do not have generations—they are dying now in very large numbers 24 hours a day, seven days a week, 365 days a year! The idea that women will have a way of reasserting control over their own sexuality, the idea that they will be able to defend their bodily health, the idea that women will have a course of prevention to follow which results in saving their lives, the idea that women may have a microbicide which prevents infection but allows for conception, the idea that women can use microbicides without bowing to male dictates, the idea that men will not even know the microbicide is in use—these are ideas whose time has come.

If used by only 30% of women, microbicides could save 6 million lives over five years. These data are based on microbicide that is 60% effective. But, the first generation of vaginal microbicide is not expected to receive FDA approval until 2010.

What Are Microbicides?

Microbicides are chemical substances that, when applied to the skin, help prevent the spread of disease. When applied in the vagina or the rectum, hopefully, they would substantially reduce transmission of HIV or other STDs. They could be produced in many forms, including gels, creams, suppositories, films, or as a sponge or vaginal ring.

In developing a vaginal microbicide, scientists must be sure that the substance is safe, does not kill microbes naturally present in the vagina that benefit female hygiene, and does not impair a woman's ability to conceive. Any microbicide will have to be tested to determine whether it damages spermatozoa, which could result in birth defects.

Application

A woman can apply a microbicide vaginally or rectally to protect herself and her partner. Current techniques to prevent HIV exposure—condom use, monogamy, reducing the number of partners, and treatment of other sexually transmitted diseases—often are not feasible or available for women. Many women face cultural barriers that prevent them from requiring their partners to use condoms, as well as cultural and

POINT OF INFORMATION 9.2

QUESTIONS AND ANSWERS ABOUT THE FEMALE CONDOM

1. **Does one have to be fitted for use of a female condom?** The female condom is offered in one size and is available without prescription. Unlike using a diaphragm, the female condom covers not only the cervix but also the vagina, thereby containing the man's ejaculate.
2. **Should a lubricant be used with the female condom?** A lubricant is recommended for use with the female condom to increase comfort and ease the entry and withdrawal of the penis. The female condom is prelubricated on the inside with a silicone-based, nonspermicidal lubricant. Additional water-based lubrication is included. The lubricant can be placed either inside the female condom or on the penis.
3. **Can oil-based lubricants be used with the female condom?** The female condom is made of polyurethane, which is not reported to be damaged by oil-based lubricants.
4. **Can a spermicide be used with the female condom?** Use of a spermicide has not been reported to damage the female condom.
5. **How far in advance of sexual intercourse can the female condom be inserted?** The female condom may be inserted up to 8 hours before sexual intercourse. Most women insert it 2 to 20 minutes before engaging in vaginal intercourse.
6. **Can the female condom be reused?** Ideally, no. A new female condom must be used for each act of vaginal intercourse. After intercourse, the condom must be removed before the woman stands, to ensure that semen remains inside the pouch. Despite the fact that the World Health Organization (WHO) does not recommend reuse of the female condom (FC), it has developed a protocol. This protocol recommends that the FC not be reused more than five times, and that it be sterilized in weak bleach solution (1:20 parts water), rinsed in water, and patted dry. The reason the WHO has developed a protocol for reuse is that in a number of developing African/Asian countries, the high cost of the female condom is forcing women, particularly commercial sex workers, to reuse the device to save money, despite the risks associated with reuse. Because of cost, commercial sex workers have revealed that many of the women were reusing the condom after cleaning it with substances such as beer, urine, water, and detergent.
7. **Should a female condom and a male condom be used at the same time?** The female condom and male condom can be used at the same time but it is not recommended, because the condoms may not stay in place due to friction between the latex in the male condom and the polyurethane in the female condom.
8. **Does Medicaid cover the female condom?** Currently Medicaid covers this device in 40 states, as it does the male condom, spermicide, and other barriers. *(Information provided by the New York State Department of Health AIDS Institute Division of HIV Prevention,* Info Bulletin, *Jan. 1994, Number Five Updated)*
9. **What is the main problem for women in using the female condom?** Ninety percent of women who found their first experience using the condom difficult had trouble with insertion. As an aside, young girls in Lesotho, Africa strip the rings out of the condoms and use them as bangles.
10. **Gay Men and Female Condoms?** Cristina Renzi (2003) reported that of men trying the female condom for anal sex, one in five (20%) of active or passive partners promote their use. Problems reported were: spillage and slippage during removal was reported more frequently with female condoms than with male latex condoms. Receptive partners more frequently reported pain or discomfort and rectal bleeding using female condoms than with male condoms.
11. **Materials on the Web** The following materials are available from UNAIDS at http://www.unaids.org/publications/documents/care/index.html#female:

The Female Condom: A Guide for Planning and Programming Information Update on Re-use of the Female Condom

Launching and Promoting the Female Condom in Eastern and Southern Africa

Use of the Female Condom: Gender Relations and Sexual Negotiations

The Female Condom and AIDS

economic barriers to securing other types of protection and treatment. Meanwhile, nearly 6 of every 10 new HIV infections occur in women worldwide. It is estimated that in the United States, about 22 million women would use an effective microbicide. To be an effective prevention tool in poor countries, microbicides would have to be cheap, easily manufactured, universally available, and culturally acceptable. Because of this the fast-track research is concentrated on very simple chemicals that disrupt or block HIV in crude ways.

The Hi-Tech Microbicides

In 2003 and 2004, at the Tenth and Eleventh Retrovirus Conferences, a number of investigations in progress toward more chemically sophisticated HIV-specific microbicides were presented. The new hi-tech microbicides in contrast incorporate already-developed or very new and experimental anti-HIV drugs. Some even have a systemic effect, meaning that they block HIV-infection sometime after application by proofing cells against HIV rather than by acting as a simple barrier. One trial already incorporates a currently available HIV drug into a microbicide. Tenofovir is under investigation in gel form as a possible vaginal microbicide. Tenofovir gel has undergone a Phase I human safety study among U.S. women and is about to start a Phase II African trial. Two candidate microbicides incorporate non-nucleoside HIV drugs (NNRTIs). The highest tech end of microbicide research is starting to look at what could be topical vaccines that actually incorporate anti-HIV antibodies. The broadly effective neutralizing antibody b12 was incorporated into a hydroxymethyl cellulose gel and was used to protect half of a group of 25 Rhesus monkeys from infection. It lowered the infection rate by about 67%. Further research is in progress.

Male Tolerance to Microbicides

If men are to be exposed to unknown products used to prevent HIV infection, the product should not cause open genital lesions or major discomfort. To this end, studies on the products mentioned above are in progress. The results to date are that those products, at most, cause minor itching, small lesions, or minor pain upon urination. Studies continue.

INJECTION–DRUG USE AND HIV TRANSMISSION: THE TWIN EPIDEMICS

About 25% of the 930,000 (at the start of 2005) U.S. AIDS cases recorded since 1981 were transmitted through injection-drug use. Now about half of all new HIV infections occur from IDU. About 75% of all people with IDU-related AIDS are either black (50%) or Latino (24%).

Syringes and needles used to inject drugs or steroids, or to tattoo the body or to pierce the ears, should *never be shared.* If an individual is going to assume the risk of HIV transmission through needle sharing, the risk can be marginally reduced by sterilizing the needle and the syringe, if one is used, in undiluted chlorine bleach. The needle and syringe should be flushed through twice with bleach and rinsed thoroughly with water. Although using bleach to clean syringes after use may offer some protection, it has now been shown that using bleach and water is not nearly as effective as believed through the mid-1990s. Since then several studies have questioned the value of using bleach for disinfection of syringes. Collectively, these studies report that bleach has little or no protective effect (Abdala et al., 2001). **(NOTE: FROM THIS POINT FORWARD, NEP STANDS FOR NEEDLE EXCHANGE PROGRAMS AND INVOLVES A NEEDLE AND A SYRINGE.)**

Both injection-drug use and HIV infection are on the increase. They are twin epidemics in the United States and Europe because the virus is readily transmitted by injection-drug users and then from infected drug users to their non-infected sexual partners. Stopping injection-drug-associated HIV transmission in theory is easy—just avoid injection-drug use. But, that is a difficult proposition for most of the estimated 8 million IDUs worldwide (2.5 million in the United States). The number of countries reporting IDU in 1989 was 80, in 2003, 140. IDUs will remain a major HIV connection to the homosexual, heterosexual, and pediatric populations (Anderson et al., 1998 updated).

HIV PREVENTION FOR INJECTION-DRUG USERS

What can be done and what is being done to prevent HIV transmission by this population?

Available drug rehabilitation programs are far too few. It is estimated that only 15% of injection-drug users in the United States are receiving treatment at any given time. Many addicts want to quit their habit but may become discouraged because of having to wait so long before getting treatment because of lack of money. Even if there were a sufficient number of treatment centers, there will always be the hard-core IDUs who will not enter a program.

IDUs have an economic motive to share equipment (Mandell et al., 1994). At the beginning of 2005, studies continue to show that over 50% of IDUs share the equipment. Perhaps the most important drawback may be that IDUs have little interest in health care or changing their behaviors. In addition, there is always the problem of legality. IDU is illegal throughout the United States (see Box 9.2).

IDUs know this and fear incarceration without the possibility of a "fix." A catch-22 situation also exists for those who want to help make injection-drug use safer: Many governmental agencies and law enforcement officers interpret the intention of making drug use safer as advocating drug use. As a result, many proponents of safer drug use have avoided becoming involved in the issue. The 2000 Kaiser Family Foundation National Survey of Americans showed that 58% of those polled were in favor of syringe exchange programs. At the beginning of 2005, federal money could still not be used to fund NEPs.

Peter Lurie and colleagues (1998) state that each year, over 1 billion syringes would be required for IDUs to have a sterile syringe for each injection.

Sexual Behavior: A Greater Risk Than Needle Sharing Among Injection-Drug Users?

After reviewing the many studies that support NEPs worldwide, the report by Steffanie Strathdee of the Bloomberg School of Public Health, Johns Hopkins University comes as a surprise. The results of a 10-year study, 1988 through 1998, revealed that the biggest predictor of HIV infection for both male and female injection-drug users is sexual behavior, not sharing needles used to inject drugs. High-risk homosexual activity was the most important factor in HIV transmission for men; high-risk heterosexual activity was the most significant for women. Risky drug use behaviors also were strong predictors of HIV transmission for men but were less significant for women. Roland Foster, aid for the U.S. Congress subcommittee on Criminal Justice, Drug Policy, and Human Resources said, "While needle exchange advocates have been arguing that needle distribution was the answer to ending HIV infection among drug users, it turns out their assumptions were wrong: sexual behaviors rather than needle sharing are the most significant HIV risk behaviors for drug abusers."

The Needle Exchange Program (NEP) Strategy

The idea of syringe-needle exchange program is based on the established public health policy of eliminating from any system potentially infectious agents or, where possible, carriers of infectious agents. The rationale of NEPs is similar, wherein active injection-drug users exchange used, potentially contaminated syringes for new, sterile syringes (Figure 9-3). In general, these exchanges are done on a one used needle and syringe for one new needle and syringe basis, though some programs will add an additional number of needles and syringes on top of those already exchanged. NEPs also provide other paraphernalia and supplies including cotton, cookers, water, and sterile alcohol prep pads. In addition, NEPs offer a variety of other services to IDUs including education, HIV testing and counseling, referrals to primary medical care, substance abuse treatment, and case management.

The world's first NEP on record began in 1984 in Amsterdam, The Netherlands. It was started by an IDU advocacy group called the Junkie Union.

Jon Parker is believed to be the first person in the United States to distribute free drug injection equipment publicly. He did so in North Haven, Conn., and in Boston in November 1986.

Evaluation of Needle/Syringe Exchange Programs

Entering year 2005, NEPs operating in American cities were exchanging about 30 million syringes annually. An IDU makes about 1000 drug injections each year (*MMWR*, 1997). The San Francisco AIDS Foundation operates the largest NEP in the United States. There are too many NEPs to list, but a few are presented here.

FIGURE 9-3 Getting a Fix. An injection-drug user shoots up in a shooting gallery in the La Perla neighborhood of Old San Juan, Puerto Rico. Another man sorts used syringes to be exchanged for new. *(Photo by Enrique Valentin/South Florida* Sun-Sentinel*)*

Tacoma, Washington—Their NEP began in August 1988. It began as a one-man program by Dave Purchase, a 20-year drug counselor.

The NEP in Tacoma held the HIV infection rate to under 5% over a five-year study period (1988–1992). During that same five-year study period, the prevalence of HIV infection among IDUs in New York City, with few syringe exchange programs, increased from 10% to 50%! About 1.5 million syringes are exchanged annually.

New York City—In November 1988, after many delays, New York City began its NEP. The program was canceled in early 1990—the reason: because over 50% of NYC's 240,000 IDUs were HIV-infected, the program offered too little too late to have an impact. IDUs make up about 38% to 40% of NYC AIDS cases. The NEP was resumed in 1992. In 1998 there were at least five NEPs operating in New York City. In May 2000, New York State passed a law making it legal to buy needles without a prescription, the forty-third state to do so. About 3 million syringes are exchanged annually.

New Haven, Conn.—Their 13-year-old program has demonstrated that NEPs dramatically slow the rate of infection without encouraging new injection-drug use. Some indicators even suggest that the program has been responsible for a decrease in both crime and the amount of drugs used illegally. These results have enabled policymakers elsewhere to call for NEPs. Edward Kaplan (1993) of Yale University reported at the Eighth International AIDS Conference that the New Haven syringe program cut the rate of new HIV infections by 33% to 50%. His data came from comparing the number of HIV-contaminated syringes found on the streets versus those turned in at syringe exchange points. (There is some controversy over these data.)

After passing a 1992 law permitting pharmacies to sell syringes without a prescription, syringe sharing has dropped 40% in the state of Connecticut. Seventy-five percent of AIDS cases in Connecticut occur among IDUs, their sex partners, and their children.

California—Each year about 8000 Californians are infected with HIV, and injection-drug use is the second leading cause of those infections. In October 1999 a law was passed that would allow cities and counties to establish NEPs. By early 2005 at least eight cities and counties had funded NEPs. About 4 million syringes are exchanged annually.

Hawaii—In 1990, Hawaii became the first state to legalize a statewide NEP. The state legislature felt it was necessary to stem the rate of HIV infection in women and newborns. About 1 million syringes are exchanged annually.

Needle Exchange Programs in Other Countries

There are an estimated 7 to 8 million IDUs worldwide. Many countries are now involved in NEPs to lower the spread of HIV.

Needle exchange program results from England, Austria, The Netherlands, Sweden, and Scotland, presented at the Fourth International AIDS Conference (1988) suggest that the European programs attracted IDUs who had no previous contact with drug treatment programs; and that IDUs were drawn from NEPs into treatment programs, thus the decrease in drug use. There was no indication in these studies of an increase in injection-drug use in cities with exchange programs. Where HIV testing had been done, the rate of HIV infection showed a marked decline after the introduction of NEPs (Raymond, 1988; Hagen, 1991).

Some of the countries with active NEPs are: Canada, England, France, Ireland, The Netherlands, Australia, New Zealand, and Italy.

Summary

Through 2004, 84% of IDUs in Glasgow, 82% in Lund, 84% in Sydney, 73% in Tacoma, and 87% in Toronto reported they had changed their behavior in order to avoid HIV/AIDS. The most commonly mentioned specific behavior change: reduced sharing of injection equipment.

PREVENTION OF BLOOD AND BLOOD PRODUCT HIV TRANSMISSION

A combined fear of disease and lawsuits have led most wealthy developed nations to adopt a zero tolerance policy regarding HIV contamination of the blood supply. However, 10% of all new HIV infections in developing countries are now due to transfusions of tainted blood. In the early 1980s, 1 of every 50 bags of blood collected in San Francisco contained HIV. The chance of acquiring either HIV or hepatitis C from a blood transfusion, in America, is now about 1 in 1 million.

Blood Donors

In the United States, there are at least 52 medically related restrictions for donating your blood. Thirteen of these reasons place a person on permanent restriction from donating blood—for example, being HIV positive, having multiple sclerosis, being a hemophiliac, men having sex with other men since 1977 (even once), having used injection drugs (even once), or having had a stroke.

There should be no risk in the United States or in other developed nations of contracting HIV by donating blood if blood centers use a new, sterile needle for each donation. Yet a 1998 survey revealed that 25% of those polled believed that they could become HIV-infected by donating blood.

Over 8 million people are donating some 18 million units of blood that are transfused into about 5 million people annually in the United States. Blood can carry HIV in a cell-free state or HIV may be carried within cells of the blood. A large dose of HIV received via blood transfusion almost universally results in HIV infection. A small dose of HIV-contaminated blood, such as the blood received by a needle stick, seldom results in infection (Francis et al., 1987). The mean volume of blood injected by a needle stick has been calculated to be 1.4 μL (1.4 millionths of a liter). It is difficult to determine exactly how large a viral dose is necessary to cause infection. However, it is known that infection is more likely to occur if blood is donated close to the time the HIV-infected donor becomes symptomatic. Entering year 2005, 6% of HIV-infected adults and 10% of HIV-infected children in the United States are believed to have become infected via blood transfusions or by the use of contaminated blood products. The majority of HIV infections occurred prior to the initiation of the blood screening program in 1985. To date only whole blood, blood cellular components, plasma, and blood clotting factors have been involved in transmitting HIV.

Blood Collection and Screening Blood for HIV

Testing blood for infectious diseases began with syphilis screening in the 1940s. Hepatitis B antibody screening was added in the 1970s, HIV antibody screening in 1985, the hepatitis B-core antigen in 1986, HTL V-1 and II-antibodies in 1988, and hepatitis C antibodies in 1990. Inclusion of the HIV antigen (p24) test in 1996 provided detection of HIV infection sooner than antibody testing. At least eight tests for infectious diseases are now routinely performed on each unit of blood collected.

No Blood Purchases for Transfusion—All blood transfused in the United States comes from volunteer donors. Blood from paid donors is used for pharmaceuticals such as Rh Ig, albumin, and intravenous immunoglobulins. Under the current standards for blood banks and transfusion services of the American Association of Blood Banks, all units must be clearly labeled volunteer, paid, or autologous (donated for self-use).

BOX 9.3

THE CONTROVERSY: PROVIDING STERILE SYRINGES AND NEEDLES TO INJECTION-DRUG USERS

As the HIV pandemic enters the new century in the United States, the profile of those affected is beginning to change. HIV is slowly changing from a disease of men who have sex with men to a disease of injection–drug users, their sexual partners, and children. From 1997 through 2004, 50% of all new HIV infections were associated with IDUs. The syringe is a Typhoid Mary! Accordingly, should IDUs be provided with free syringes and needles at public expense to help stop the spread of HIV?

There has been no other HIV prevention activity, with the possible exception of condom distribution in public schools, that has generated as much controversy as needle exchange programs **(NEPs.) Pro forces** believe that it is impossible to eliminate injection–drug use. Providing IDUs with free sterile equipment is an attempt at slowing the spread of HIV among drug users and their sexual partners, subsequently reducing the number of people in the general population that will become infected. In support of the pro forces, the CDC and the National Institutes of Health reviewed many of the NEPs operating in the United States (at the start of 2005, there were over 600 sites in 85 cities in 35 states) and others in Canada and Europe. The most important conclusion is that NEPs are preventing HIV infections in drug users, their sex partners, and their children. That finding was based, in part, on evidence collected in Tacoma and Seattle, Wash., and in New Haven, Conn., where there was a 33% reduction in the rate of new infections among NEP clients. In another key finding, they found no evidence that distributing syringes leads to an increase in drug use.

Gibson and colleagues (2001) analyzed 42 publications published between 1989 and 1999 that evaluated the efficacy of syringe exchange programs. They found significant evidence for the conclusion that syringe exchange is effective in preventing both risky behavior and HIV seroconversion. Twenty-eight studies reported positive effects regarding the use of needle or syringe exchange, two found negative associations, and 14 found either no association or a combination of positive and negative effects.

Besides saving lives, these NEPs deliver a huge financial payoff. Consider the case of an HIV-positive addict who infects eight others in a one-year period (a reasonable estimate). If each turns to Medicaid to pay his or her lifetime medical costs (at an average $119,000 plus), that's about a $1 million burden for taxpayers—money that could have been saved if the one addict had been in a NEP. In addition, arguing for the availability of NEPs is *not* an argument against using police and criminal justice mechanisms to deal with drug-use related robbery, assault, or other crimes.

Con forces argue that any move to make injection drug use safe is a move to make it attractive or socially acceptable and thus represents a step backward in the long-declared war on drugs. Those opposing the needle exchange programs say that to provide NEPs is to condone drug addiction and perhaps promote injection drug use among those who would not otherwise participate. Some feel that free syringe programs are just a Band-Aid that attempts to prevent HIV infection but does nothing to cope with the underlying drug addiction problem. (These people may feel it's too little, too late.)

A side issue is the fact that it is illegal to sell needles and syringes over the counter in 5 of the 50 states and the District of Columbia through 2004 (*MMWR,* 1993b; 1997). Laws restricting these sales were intended to discourage drug use (Fisher, 1990; Gostin et al., 1997). Law enforcement and community leaders (police, churches, businesses, parents, teachers, and residents) are concerned that allowing access to needles and syringes sends the wrong message, encourages initiation into drug use, and accelerates the disintegration of families. Residents and business owners fear increased street crime, lower property values, and health risks from discarded needles and syringes.

Middle Ground

There is a **third force** of people who take the middle ground. They advocate providing free bleach to IDUs to clean their drug equipment. People from many AIDS-action groups in metropolitan areas are involved in the free bleach dispensing program. With each bottle of bleach is a set of directions. But since many IDUs cannot read, they are given verbal instructions, though some are too "spaced out" to listen or too uncomfortable to care. Third force people have to understand that NEPs and distribution of bleach kits have to be accessible privately, off the street and out of sight, as well as publicly. Use of a public, streetcorner NEP is a statement to the entire community that the person exchanging a syringe is an IDU. This has potentially very different consequences for women than for men. Women who have children have good reason to try to conceal their addiction as they risk losing custody or contact with their children if the state finds out they are actively using drugs—and if they are using a public NEP the child welfare bureaucracy is more likely to find out than if they are not getting needles and syringes in public. Perhaps NEPs should provide some type of informal child care so that a mother doesn't have to literally exchange needle and syringes in front of her children. Addiction among women is often taken as a statement of sexual availability.

Legalize Drugs

A **fourth force** suggests that drugs be legalized so that they can be used openly, thereby reducing the threat of HIV transmission through illegal "shooting galleries." Discussion of the fourth force's position is beyond the scope of this text. However, a quasi-fourth force situation did exist in Zurich, Switzerland, until 1992.

Zurich gained notoriety in 1987 for an open Needle Park. The city of 250,000 people experimented in setting aside Platz Pitz park where cocaine and heroin addicts could openly buy, sell, and use injection drugs. The park attracted up to 4000 drug users/day. In early 1992, the park was closed because the park became a magnet for professional dealers, especially Lebanese, Yugoslav, and Turkish gangs that overran small dealers in a violent price war. Some of the park's inhabitants clustered around the central train station, others headed off in search of methadone. Others went back to the alleys and shelters from which they came. With sales suddenly back underground, addicts complained that the price of heroin had doubled overnight to $214 a gram (454 grams/pound=$97,156). Health workers said efforts to prevent the spread of HIV/AIDS will now be much more difficult.

UPDATE Switzerland is currently providing IDUs with **heroin** three times a day in 18 treatment centers across the country. So far Swiss health authorities say the program has reduced criminal activity among the participants about 60%. The program also reduced their homelessness from 12% to zero and death rate by 50%. Through late 2004, following Switzerland's lead, Germany, The Netherlands, Spain, Portugal, Australia, Luxembourg, and Vancouver, British Columbia, (North America's first consumption room) have implemented consumption or injection rooms for the legal use of heroin. There are about 50 legal heroin consumption rooms or clinics worldwide. Requests for heroin consumption rooms in Austria and the United States have been denied.

DISCUSSION QUESTION: The United States has zero tolerance for such activities and will not become involved in what the Swiss term "an innovative program." Do you think the United States should, based on the Swiss data, offer heroin to the addicted in a controlled environment similar to the Swiss? Present credible reasons/data to support your stand.

LIFTING THE BAN ON THE USE OF FEDERAL FUNDING TO SUPPORT NEEDLE EXCHANGE PROGRAMS

Since 1990, seven national reports have reviewed the scientific evidence and recommended that the federal ban for NEP funding be lifted. Needle exchange programs are supported by the President's Advisory Council on HIV/AIDS, the American Medical Association, the National Academy of Sciences, the Centers for Disease Control and Prevention, and the American Public Health Association, as well as other prestigious medical and public health organizations. In addition, the American Bar Association and the U.S. Conference of Mayors have urged the federal government to allow states and localities to use federal HIV-prevention funds to implement NEPs. Polls have shown that the American public supports lifting the ban on federal funding of NEPs. A recent Harris poll found that 71% of Americans believe that cities and states—and *not* the federal government—should decide whether federal HIV-prevention funds can be spent on NEPs.

On April 20, 1998, Health and Human Services Secretary Donna Shalala informed President Clinton that scientific research has proven that NEPs effectively prevent the transmission of HIV and hepatitis, and do not lead to increased drug use. Her long-awaited action cleared the way for the president to lift the 10-year ban on federal funding for NEPs. Clinton accepted the findings, but stated that in spite of them he would continue to block the use of federal funds for NEPs. President George Bush continued the ban. As of the start of 2005, the ban remains in effect.

DISCUSSION QUESTION: Do you agree with current federal policy on money for NEPs? List any scientific facts or studies that could be used to support the president's action. List any that disagree with his decision.

COMMENTARY

It is hard to avoid the suspicion that the concerns about free syringe programs have less to do with science or public health than with politics: specifically, a reluctance to **detract** or **distract** from the "Just say no" message. But there is another message that both sides would agree on—that no one should die needlessly! A leading University of California researcher estimates that nearly 10,000 lives could have been saved over the past few years by an aggressive expansion of syringe exchange programs. Isn't the war on drugs supposed to be about saving lives? Isn't a part of government about saving lives? Your response is?

Blood Screening for HIV—If the test indicates that the blood carries antibodies to HIV, it is treated to destroy the HIV and then discarded. In March 1983, the major U.S. blood bank organizations instituted procedures to reduce the likelihood of HIV transmission through blood transfusions. People with signs or symptoms suggestive of HIV/AIDS, sexually active homosexual and bisexual men with multiple partners, recent Haitian immigrants to the United States, past or present IDUs, men and women who have engaged in prostitution since 1977 or have patronized a prostitute within the past six months, and sexual partners of individuals at increased risk of HIV infection are asked to refrain from donating blood.

U.S. FDA Approves Blood Screening Test for HIV—Blood screening for HIV and HIV-testing procedures are presented in Chapter 12. The risk of becoming HIV-infected from a blood transfusion has dropped by more than 99% from 1983 to 2003. Regardless, a male living in Durango, Tex., received contaminated blood during heart bypass surgery in August 2000. In March 2002, he filed a $100 million lawsuit against those thought to be responsible for his HIV infection. This is reported to be the first case of HIV-positive blood transfusion since restrictive blood-screening tests for HIV were put in place in April 1999. In 2002, two people became HIV-infected from blood transfusions in Florida's Tampa Bay area. Tracing the infectious blood back to its donor source revealed that, in both incidents, the donors were in the **window period;** they were infected but the blood test failed to detect the virus.

Blood Transfusions Worldwide—Fifteen years after the industrialized world began to screen all blood used in transfusions for HIV, about 1 in 10 people in developing countries are still being infected through this route.

A combination of the lack of screening with high levels of infected donors turns transfusion into a form of roulette. As 2005 began, blood transfusions accounted for 5% to 10% of HIV infections worldwide.

Blood Safety—From 1985 into 2005, about 500 million units of blood or plasma have been screened for HIV antibody in the United States. By excluding those who test HIV positive and by asking people from high-risk behavior groups not to donate blood, the incidence of HIV transmission from the current blood supply is relatively low. With faster and more accurate testing procedures now in use, the risk is becoming even lower. However, the probability or risk of receiving HIV-contaminated blood will never be zero. The reason a small risk still exists is because some people infected with HIV may donate blood during what is known as the window period. During that period, a person may be infected with the HIV, but the test cannot yet detect the infection. And, the test is not 100% accurate.

Eve Lackritz and colleagues (1995) reported that of 4.1 million blood donations from 19 regions served by the American Red Cross, 1 donation in every 360,000 was made during the window period. They further estimated that 1 in 2,600,000 donations contained HIV but was not identified (false negative) because of laboratory

BOX 9.4

AMERICANS SUPPORT SEX EDUCATION AND CONDOMS IN SCHOOLS AND IN TV ADS, AND NEEDLE EXCHANGE PROGRAMS

In 1966, and again in 2004, the Kaiser Foundation reported the following facts after reviewing the results of its **"Survey of Americans and HIV/AIDS":** Virtually all Americans (95%) think AIDS information should be provided in the schools, including 69% who think children should start receiving AIDS education *at the latest* by age 12; 7 out of 10 Americans think the major television networks should accept condom advertisements; two-thirds favor providing clean needles to IDUs; and 63% say there should be more references to condom use in movies and television. However, when asked whether they thought high schools should provide condoms or AIDS information to teens, Americans were almost evenly divided: 49% say just HIV/AIDS information and 46% say making condoms available should be the priority. Support for AIDS prevention measures persists across demographic groups.

Eight out of ten parents with children under 21 years of age say they are concerned about their child getting AIDS, including 53% who say they are very concerned. When asked about the information they want most about AIDS, 5 out of 10 parents say they need more information about "what to discuss with children about prevention." Getting information about where to go for help if exposed to HIV was the number two information needed, picked by 12% of parents.

Americans Generally Informed About HIV/AIDS, Although Some Gaps in Knowledge Persist

Minorities and those with low incomes and less education are most likely to be personally concerned about AIDS. More than a third (35%) of people with annual incomes under $10,000 say they are "very concerned" as compared with only 14% of those with incomes over $50,000 per year. Similarly, while 34% of those without a high school degree say they are "very concerned" about getting AIDS, only 14% of college graduates say they are.

Most people (54%), however, think 100,000 or fewer Americans have died from AIDS, when in fact it's over five times that number. While underestimating the number of lives lost to AIDS, many Americans (51%) overstate the impact here relative to the rest of the world, believing that half or more of all AIDS cases occur in America. In reality it's less that 2%. Half of the population (51%) incorrectly thinks a person can get AIDS while "giving or donating blood for use by others." (There are no known cases of HIV transmission as a result of *donating* blood. Some of the confusion may be related to perceived risk of receiving blood transfusions.

Significant Percentages of Americans Distrust Information About HIV/AIDS

Although most Americans believe they are receiving accurate information about the AIDS epidemic, significant percentages doubt what the government and media are telling:

- 34% do not believe the government is telling the whole truth about AIDS; and
- 25% do not believe the media is telling the whole truth about AIDS.

In addition, fewer, but still some Americans question the origins of the epidemic:

- 18% believe there is some truth to reports that the AIDS virus was produced in a germ-warfare laboratory; and
- 12% believe that AIDS came from God to punish homosexual behavior.

error. Their report states that collectively, of 42 Red Cross regions collecting 9 million units of blood, the risk of HIV transmission was one transmission for every 450,000 to 660,000 available units of screened blood. Based on their data they conclude that there are about 18 to 27 infectious units of blood available for transfusion each year. This is about half the previous estimates. (The American Red Cross collects about half the donated blood in the United States annually.) With the widespread use of the recently FDA-approved Nucleic Acid Test (2002, see Chapter 12 for explanation) the numbers of infectious units of blood should be cut in half.

Autologous Transfusions—These are transfusions using **"self" (your own) blood.** It has been suggested that the two groups that benefit most from autologous blood storage are: expectant mothers facing caesarean section and certain elective surgical candidates where substantial loss of blood is anticipated. A person's blood is drawn and refrigerated for transfusion back into his or her own body when needed. (Stored whole blood remains usable for a maximum of 42 days.) This is the safest blood available.

INFECTION CONTROL PROCEDURES

With no cure or vaccine for HIV/AIDS, prevention of infection is of paramount importance. With the advent of the HIV/AIDS epidemic, health-care workers and others who are occupationally exposed to body fluids, especially blood, are understandably concerned about the risk of becoming HIV-infected. However, when precautions are observed, the risk is very small, even for those treating HIV/AIDS patients.

Two sets of infection control procedures are in use in hospitals, medical centers, physicians' offices, and units that deal with people in medical emergencies. One is called **universal precautions,** the other is **blood and body substance isolation.**

Universal Precautions

Universal precautions (Table 9-4) are standard practices that workers observe on the job to protect themselves from infections and injuries. These precautions or safety practices are called **universal** because they are used in all situations even if there seems to be no risk. Universal precautions had their beginnings in 1976 when barrier techniques were first recommended for the prevention of hepatitis B infection. Precautions required the use of protective eyewear, gloves, and gowns, and careful handling of needles and other sharp instruments. In 1977, hepatitis B immunoglobulin was recommended for those exposed to hepatitis B through needle sticks. In 1982, hepatitis B vaccine became commercially available and recommended for all health-care workers exposed to human blood.

In 1987, the CDC published "Recommendations for Prevention of HIV Transmission in Health-Care Settings." Blood and body fluid precautions should be consistently followed for all patients regardless of their blood-borne infection status. This extension of blood and body fluid precautions to all patients is referred to as **Universal Precautions (UP).**

Under universal precautions, the blood and certain body fluids of all patients are considered potentially infectious for HIV, hepatitis B virus (HBV), and other blood-borne pathogens.

Universal precautions are intended to prevent parenteral (introduction of a substance into the body by injection), mucous membrane, and broken skin exposure of health-care workers (HCWs), teachers, or any other person who may become exposed to blood-borne pathogens. In 1987, the CDC also published a report that got the immediate attention of most, if not all, informed health-care workers. The report stated that three health-care workers who were exposed to the blood of AIDS patients tested positive for HIV. What was so startling was that until that time, needle punctures and cuts were thought to be the only dangers in a clinical setting. These three cases appeared to involve only skin exposure to HIV-contaminated blood. One of the three cases involved a nurse whose chapped and ungloved hands were exposed to an AIDS patient's blood.

The second case involved a nurse who broke a vacuum tube during a routine phlebotomy on an outpatient. The blood splashed on her face and into her mouth. A blood splash was also involved in the third case. The worker's ungloved hands and forearms were exposed to HIV-contaminated blood (Ezzell, 1987).

These three cases of HIV infection informed health-care workers in the most dramatic way that they were all vulnerable. Perhaps these three cases produced a fear among health-care workers out of proportion to the actual risk of their becoming infected. Although calculations show that the risk of HIV infection after exposure to blood from an HIV/AIDS patient is about 1 in 200, if you are that one, probability is meaningless.

The universal precautions as published by the CDC currently apply to some 5.3 million health-care workers at 620,000 work sites across the United States and another 700,000 Americans who routinely come in contact with blood as part of their job, for example, people in law enforcement, education, fire fighting and rescue, corrections, laboratory research, and the funeral industry.

Table 9-4 Universal Precautions

DEFINITION

Universal precautions (UP) are a set of infection control practices developed by the Centers for Disease Control and Prevention in which health-care workers (HCWs) appropriately utilize barrier protection (gloves, gowns, masks, eyewear, etc.) for anticipated contact with blood and certain body fluids of *all* patients.

1. The hands and skin must be carefully washed when contaminated with blood or certain body fluids.
2. Particular care is taken to prevent injuries caused by sharp instruments.
3. Resuscitation devices should be available where the need is predictable.
4. HCWs with exudative lesions or weeping dermatitis should refrain from patient care until the condition resolves.

BLOOD AND BODY FLUIDS TO WHICH UP APPLY

Blood is the single most important risk source of HIV, HBV, and other blood-borne pathogens in the occupational setting. Prevention of transmission must focus on reducing the risk of exposure to blood and other body fluids or potentially infectious materials containing visible blood.

1. UP should be used when exposure to the following body fluids may be anticipated:
   a. Blood
   b. Serum plasma
   c. Semen
   d. Vaginal secretions
   e. Amniotic fluid
   f. Cerebrospinal fluid (CSF)
   g. Synovial fluid
   h. Pleural fluid
   i. Vitheous fluid
   j. Peritoneal fluid
   k. Pericardial fluid
   l. Wound exudates
   m. Any other body fluid containing visible blood (but not feces, urine, saliva, sputum, tears, nasal secretions, or sweat, unless they contain visible blood).
2. Note: Blood, semen, and vaginal secretions have been shown to transmit HIV. The others, with the exception of fluids containing visible blood, remain a theoretical risk.

RATIONALE

1. UP reduce the risk of parenteral, mucous membrane, and skin exposure to blood-borne pathogens such as, but not limited to, HIV and HBV.
2. For several reasons, focusing precautions only on diagnosed cases misses the vast majority of persons who are infected (many of whom are asymptomatic or subclinical) and who may be as infectious as the diagnosed cases. Persons who have seen a physician and have been diagnosed with acute or active disease represent only a small proportion of all persons with infection. Infectivity always precedes the diagnosis, which often is made once symptoms develop.

(Adapted from Mountain-Plains Regional AIDS Education Training Center. HIV/AIDS Curriculum, *Nov. 1994.)*

In July 1991, the CDC published an additional set of recommendations. These recommendations are also for preventing the transmission of HIV and hepatitis B virus to patients during exposure-prone procedures.

In summary, the concept of universal precautions assumes that all blood is infectious, no matter from whom and no matter whether a test is negative, positive, or not done at all. Rigorous adherence to universal precautions is the surest way of preventing accidental transmission of HIV and other blood-borne pathogens.

Blood and Body Substance Isolation (BBSI)

An alternative, and some believe superior, approach to the CDC's universal precautions in areas of high HIV prevalence is the system referred to as **body substance precautions or Blood Body Substance Isolation (BBSI)** (Gerberding, 1991).

In practice, these precautions are similar to universal precautions, in that prevention of needle stick injury and use of barrier methods of infection control are emphasized. Philosophically, however, the two are quite different. Whereas universal precautions place a clear emphasis on avoidance of blood-borne infection, body substance precautions take a more global view. Body substance isolation (BBSI) requires barrier precautions for all body substances (including feces, respiratory secretions, urine, vomit, etc.) and moist body surfaces (including mucous membranes and open wounds). BBSI is designed as a system to reduce the risk of transmission of all nosocomial (hospital-associated) pathogens, not just blood-borne pathogens. Gloves are worn for any anticipated or known contact with mucous membranes, nonintact skin, and moist body substances of all patients.

SEXUAL PARTNER NOTIFICATION

One of the most controversial issues in HIV prevention is **contact tracing** or **partner notification** of sexual contacts mostly because HIV/AIDS is considered an incurable disease with a great deal of stigma attached to the infected. Partner notification is the practice of identifying and treating people exposed to certain communicable diseases. The term "partner notification" is used by the CDC and some health-care providers because it more comprehensively describes the process by which the physician, other health-care workers such as Disease Intervention Specialists (DIS, someone who is specially trained in STD work), and the infected person may provide information to at-risk partners and sometimes to family, friends or care provider.

There are two very different approaches to informing unsuspecting third parties about their potential exposure to medical risk.

Each approach has its own history, including a unique set of practical problems in its implementation, and provokes its own ethical dilemmas. The first approach, involving the moral **duty to warn,** arose out of the clinical setting in which the physician knew the identity of the person deemed to be at risk. This approach provided a warrant for disclosure to endangered persons without the consent of the patient and could involve revealing the identity of the patient. The second approach—that of contact tracing—emerged from sexually transmitted disease control programs in which the clinician typically did not know the identity of those who might have been exposed. This approach was founded on the voluntary cooperation of the patient in providing the names of contacts. It never involved the disclosure of the identity of the patient. The entire process of notification was kept confidential (Bayer, 1992).

History of Sexual Partner Notification

The concept of partner notification was proposed in 1937 by Surgeon General Thomas Parran for the control of syphilis (Parran, 1937). By tracing and treating all known contacts of a syphilitic patient, the chain of transmission could be interrupted. According to George Rutherford (1988), contact tracing has been successfully used in a number of STDs beginning in the 1950s. It is still used in cases of syphilis, endemic gonorrhea, chlamydia, hepatitis B, STD enteric infections, and particularly in cases of antibiotic-resistant gonorrhea.

In 1985, when the HIV antibody was first used in screening the blood supply, notification of blood donors and other HIV-infected individuals and their contacts became possible. The strategy in HIV partner notification is the same as that used for the other STDs: to identify HIV-infected individuals, counsel them, and offer whatever treatment is available. In asymptomatic HIV-infected people only counseling is given. But, symptomatic HIV-infected patients receive counseling and treatment, if available, for their signs and symptoms. Partner notification depends on HIV-positive people to give the names of their partners; but they may be reluctant to do so fearing that their identification may result in physical abuse and loss of jobs and housing. For a review of partner notification read the article by Kevin Fenton et al., 1997. Those who oppose the use of partner notification call the investigators "sex police."

The Use of Sexual Partner Notification: Examples

In April 1993 an incarcerated male asked for an HIV test. The diagnosis was positive. Contact tracing turned up a network of 124 persons; all were linked by syringe sharing and syringe sharing with sex. One hundred twenty-one were contacted and offered an HIV test; 118 accepted the test; 44 were positive. One hundred thirteen of the 124 lived in the same county. The estimated cost for partner notification in this network was $13,969 (*MMWR*, 1995).

During a five-month period (February through June 1999) seven young people were diagnosed with HIV in a small rural town in Mississippi. The CDC, working with the Mississippi Health Department through partner contact investigation, defined a social network of 122 people. Persons in the network had a median age of 21 (range: 13 to 45 years). Of the 78 people tested for HIV infection, five women (median age, 16 years) and two men (median age, 25 years) were infected, all through heterosexual sex. Results of the interviews of the infected and noninfected people indicated a serious lack of HIV/AIDS prevention knowledge (*MMWR*, 2000).

At the 2004 Eleventh Conference of Retroviruses and Opportunistic Infections investigators said they found the number of new HIV infections in men from 37 southern colleges has risen rapidly in just a few years. In 2000 there were 6; in 2001, 19; in 2002, 29; and in 2003, 30. Of that total of 84 new infections, 73 were in blacks and 11 in whites. Sixty-three percent reported having sex with men only, 33% with both men and women, and 4% with women only. They attended 33 colleges in North Carolina, two in South Carolina, one in Georgia and one in Florida. The outbreak was identified in time for the authorities to ask North Carolina's colleges to include safer sex messages during Fall 2004 orientation sessions. Stories about the outbreak, appearing in campus newspapers, contact tracing or partner notification, and free HIV testing helped reveal and tie this minor college epidemic together.

Opposition to Sexual Partner Notification for HIV Disease

Currently opposition to partner notification continues on the grounds that: (1) it is not cost effective; (2) there is little evidence that those who are informed of their infection will do anything about changing the high-risk behaviors that got them infected in the first place; (3) the threat of social discrimination undermines the intent of contact tracing; (4) the unintended consequences of partner notification may include violence and even death from an abusive partner (Rothenberg, 1995); and (5) in 24 states homosexuality is a crime, and HIV-infected homosexuals fear prosecution if they acknowledge same-sex contacts because their sexual partner(s) may be prosecuted.

In Support of Sexual Partner Notification

The following discussion is from Stephen J. Joseph's article, "The Once and Future AIDS Epidemic," *Medical Doctor*, 37:92–104. Mr. Joseph is the past commissioner of health of New York City, 1986–1990.

The New York City medical examiner looked at a large sample of dead persons who were tested postmortem and found to be HIV positive, but who had no documentation in their medical records of having been diagnosed as infected. Over 35% of those persons had a readily identifiable spouse or steady sexual partner. Josephs asks,

> How can one justify, on clinical, public health or humanitarian grounds, *not* notifying that surviving partner, who might be the source of the infection in the deceased, or the recipient of infection? Arguments against this procedure border on the absurd; one has to start with the premise that increased medical knowledge is more dangerous than helpful to the individual, and that the rights of the uninfected count for nothing against the rights of the infected.

In January 2003, two separate studies, one by Patricia Kissinger and colleague, and the other by Tamara Hoxworth, reveal that previously held conceptions on the negative aspects of partner notifications are wrong. Their findings reveal that partner notification rarely damages relationships or promotes violence and that, if anything, exposure to partner notification contributes to safe behaviors. Partner notification did not lead to more relationship break-ups among HIV-

POINT OF INFORMATION 9.3

USING THE INTERNET FOR PARTNER NOTIFICATION OF HIV AND SEXUALLY TRANSMITTED DISEASES

An estimated one-third of internet visits by persons aged 18 years and over are to sexually oriented websites, chat rooms, and new groups that enable users to view sexual images or participate in online discussions of a sexual nature. Although "virtual sex" carries no risk for transmission of HIV or other sexually transmitted diseases (STDs), use of the internet to find partners for actual sexual activity does carry such risk. In Los Angeles County during 2001 through 2003, of 759 men who have sex with men (MSM) and who had early syphilis, 172 (23%) reported using the internet to meet sex partners. Because the internet enables sex partners to maintain anonymity by withholding identifying information (e.g., full name, address, and place of employment), it poses challenges for public health authorities. Use of the internet by public health authorities to notify sex partners of persons with HIV or STDs has been reported by the CDC in 2002. More recently in 2003, health department officials of Los Angeles and San Francisco have used email to notify anonymous partners of their health risk. The bottom line is that health officials are attempting to use the same technology that facilitates the transmission of HIV and other STDs to prevent and control their spread. The success in using email to warn anonymous partners will depend on legal challenges and nationwide acceptance.

positive individuals, in comparison to syphilis-infected persons.

DISCUSSION QUESTION: With current life-sustaining antiretroviral treatments available, is there an overwhelming excuse not to use partner notification—especially when so many HIV/AIDS experts promote the "hit early"—approach to therapy?

What is your response to the reasons for partner notification and to the opposition's point of view?

Legality of Sexual Partner Notification

Legal Obligations To Warn Third Parties at Risk—Many state laws permit, but do not require, disclosure by physicians to third parties known to be at significant future risk of HIV transmission from patients known to be infected. Thus, if a physician reasonably believes that a patient will share drug-injection equipment or have unprotected sex without informing a partner of the risk, the physician has discretion to inform the partner. Under some disclosure laws, the physician is required to first counsel the patient to refrain from the high-risk behavior, and, in providing the third party warning, the physician is prohibited from disclosing the patient's identity. In the absence of state laws permitting such disclosure, physicians may be held liable for breach of confidentiality for disclosing patient information to sex partners.

POINT TO PONDER 9.1

In 2004, a jury in Chicago ordered the mother of a man who died of AIDS to pay her son's former fiancée $2 million because she did not tell the woman about her son's condition. The attorney for the defense argued that the mother would have broken the state's HIV/AIDS confidentiality law if she had revealed her son's condition. The verdict is on appeal. **Your opinion is?**

The **duty to warn** may extend to nonpatient third parties in other contexts, based on the provider's primary duty to the patient. Thus, health-care professionals have a duty to inform patients that they have been transfused with HIV-contaminated blood and this duty may extend to third parties. A physician in one case failed to inform a teenager or her parents that she had been transfused with HIV-contaminated blood. When the young woman's sexual partner tested positive for HIV, the court upheld his claim against the physician based on the physician's failure to inform the patient. Similarly, courts have upheld that a health-care professional's duty to inform the patient of his or her HIV infection may extend to those the patient foreseeably puts at risk, such as a spouse or family member caregiver. On the other hand, courts have ruled that disclosure is wrongful in cases in which the third party, such as a family member, is not at actual risk of infection, or the physician has no knowledge that the patient has failed to disclose to the partner (Gostin et al., 1998).

At the beginning of 2005, at least 39 states in America have enacted partner notification laws that provide for penalties that range from a misdemeanor (a crime less serious than a felony, which is a serious crime) to attempted murder for anyone who does not reveal to a sexual partner that he or she is HIV positive (see Box 8.6). At least 44 states passed laws requiring or permitting workers (mostly health-care workers or public safety employees) to be notified of potential exposure of HIV. In some cases, the laws allow testing of the source patient. To date, Arkansas and Missouri are the only states that require patients to notify health-care providers of their HIV status before receiving care. All 50 states are now somewhere in the process of establishing the capacity for contact tracing at the request of a patient.

DISCUSSION QUESTION: If a law were passed that made persons who practiced high-risk behaviors and who contracted HIV/AIDS pay for their care and treatment or forego medical help—do you think these people would continue to engage in high-risk behaviors? Would this law be an effective means of HIV transmission prevention? Present examples to support your position. (Can you relate this scenario to those who smoke and develop cancer?)

For a detailed report on partner notification published by the WHO and UNAIDS see: http://www.who.int/asd/knowledge/rptngdiscl.html.

VACCINE DEVELOPMENT

It is always better to prevent disease than to treat it. Vaccines prevent diseases in those who receive them and protect those who come into contact with unvaccinated individuals who carry the disease.

Historically, vaccines have provided a safe, cost-effective, and efficient means of preventing illness, disability, and death from infectious diseases through the use of vaccinations.

It is now abundantly clear that no pharmacologic agent, no educational efforts directed to safer sex (regardless of how vigorously implemented), and no nutritional modification will stop this pandemic. Halting the spread of HIV requires an effective vaccine. Ending 2005, an estimated 75 million people will either be living with HIV or will have died from it.

POINT OF INFORMATION 9.4

THE GOAL OF DEVELOPING AN HIV VACCINE

PAST PRESIDENT CLINTON SETS A VACCINE GOAL

Speaking to the graduating class at Morgan State University on **May 18, 1997,** President Clinton invoked the legacy of John F. Kennedy's 1960s race to the moon and set a national target of developing an AIDS vaccine within the next 10 years (2007). This was **the first annual Vaccine Awareness Day in America.** The president said, "We dare not be complacent in meeting the challenge of HIV, the virus that causes AIDS." He then announced the creation of a research center at the National Institutes of Health in Bethesda, Md., to complete the task. That center, the Dale and Betty Bumpers Vaccine Research Center, opened in September 2000, and a year later began a DNA vaccine Phase I trial.

Robert Gallo stated, after the president's announcement, that "it is a serious possibility that we may never develop a vaccine for HIV."

In June 2001, Health and Human Services Secretary Tommy Thompson told scientists at a Geneva gathering that an HIV vaccine will be available within three to five years. There was a collective audible groan from his audience as they must have recalled a similar comment by Margaret Heckler on April 23, 1984.

May 18, 2004—**The seventh anniversary**—has passed, but only one preventive vaccine is in Phase III trials in Thailand, the same vaccine that failed U.S. tests in 2003. At this rate, Clinton's goal for a preventive HIV vaccine by year 2007 will not be reached.

May 18, 2004, was also the **seventh annual HIV Vaccine Awareness Day** honoring volunteers in 60 cities in America and some 25,000 people worldwide who have received one of 50 different experimental vaccines to prevent HIV infections. For this occasion people were asked to wear their red AIDS ribbon upside down to bring attention to this day.

One benefit from all of the HIV-vaccine research to date is that scientists found out that they know far less than they thought they knew about producing specific prevention vaccines. For example, scientists believed that vaccines work simply by producing antibodies, right? Well, probably not. This misconception coupled with basic ignorance of how they do work is stalling the urgent quest for an HIV/AIDS vaccine. No one yet has found out how highly successful vaccines like polio, measles, and hepatitis B actually protect people from disease. Phillippe Kourilsky, director of the Pasteur Institute, said, "We've had many successful vaccines over the past decades but we've missed a chance to see how these vaccines work. Each time a vaccine works the scientific community wanders off and leaves it to the public health workers to use it—and fails to invest in the research. If we had done that we would have been in a much better position to tackle the AIDS vaccine problem." Scientists are, for the first time, learning about the mechanisms of viral-host pathologies necessary to produce preventive vaccines.

What Is a Vaccine?

A vaccine is a suspension of whole microorganisms, or viruses, or a suspension of some structural component or product of them that will elicit an immune response after entering a host. In brief, vaccines mimic the organisms or virus that cause disease, alerting the immune system to be aware of certain viruses or bacteria. Because of this advance warning system, when the real organism or virus invades the body, the immune system marshals a response before the disease has time to develop.

Ideally, the body will make neutralizing **antibodies** that bind to and disable the foreign invader (**humoral immunity**) and trigger white blood cells called T cells to organize **attack cells** in the body to destroy those cells that have been infected by viruses (**cellular immunity**). Once the immune system's T cells and B cells, which make antibodies, are activated, some of them turn into **memory cells**. The more memory cells the body forms, the faster its response to a future infection. (See Chapter 5 for a discussion of the human immune system.) To date all successful vaccines prevent disease through the production of neutralizing antibodies.

The Use of Weakened or Inactivated Agents to Trigger Humoral and Cellular Immunity

Some vaccines, such as those against smallpox, polio (Sabins), measles, mumps, and tuberculosis, contain genetically altered or weakened organisms or viruses that are reproduced in the body after being administered but do not generally produce disease. Yet since the virus or bacterium is still alive, there is a small risk of developing the disease.

Whooping cough, cholera, and influenza vaccines are made of inactivated whole organisms and viruses or pieces of them. Because killed organisms and inactivated virus do not replicate inside the recipient, the vaccines confer only humoral immunity (the production of antibody), which may be short-lived.

Types of HIV Vaccines

Scientists are attempting to design three types of HIV vaccines: (1) a **preventive** or **prophylactic vaccine** to protect people from becoming HIV-infected (chance of success—doubtful). Historically, primary prevention most often referred to sterilizing immunity, in which a vaccine is given to those who have not yet been exposed to the infectious agent. This was the focus of many earlier vaccine searches and remains a target even though researchers will now accept far less than a 100% effective HIV vaccine and still feel successful. (2) A **therapeutic vaccine** (this is not a true vaccination, but a postinfection therapy to stimulate the immune system; the term "vaccination" is reserved for preventive strategies) for those who are already infected with HIV to prevent them from progressing to AIDS (chance of success—good); and (3) a **perinatal vaccine** for administration to pregnant HIV-infected women to prevent transmission of the virus to the fetus (chance of success—good).

What Is an Effective Viral Preventive Vaccine?

An effective preventive viral vaccine usually blocks viral entry into a cell, but vaccines are generally *not* 100% effective. For example, measles vaccine is 95% effective, tetanus 90%, hepatitis B 85%, and influenza 70%. Vaccine researchers attending the Seventh Conference on Retroviruses and Opportunistic Infections in February 2000 said that they may have to lower their sights regarding an HIV vaccine and settle for one that does not completely prevent HIV infection. Based on recent calculations, it has been estimated that a vaccine that is 30% effective against HIV can begin to eradicate the virus if it is widely administered and accompanied by prevention education. That's why the FDA has indicated that it will approve an HIV/AIDS vaccine at this level of efficacy. David Satcher, former U.S. surgeon general, said that without an effective HIV vaccine by year 2020, AIDS will kill 6.5 million Africans annually!

SIDEBAR 9.3

HIV VACCINES

The goal of any vaccine is to teach the immune system new, and hopefully better, ways to win the battle against the intruder. There are different types of immune responses, those we were born with (innate immunity) and those the immune system learns (acquired immunity). HIV vaccines are being created to exploit the side of the immune system that is learned by providing information to cells in new ways in hopes of enhancing their learning and making them more effective against the virus.

Progress to Date on an HIV Vaccine

A. There are no proven effective preventive or therapeutic HIV vaccines.

B. We don't yet know if the ability of a vaccine to induce HIV-specific immune responses tells us, in and of itself, if the vaccine (or the immune responses) is useful in treating or preventing HIV.

C. HIV vaccines researched to date have had minimal side effects, primarily pain, redness and swelling at the injection site, and sometimes transient fever, fatigue, and joint stiffness.

D. New vaccines, including the Merck DNA vaccine, have garnered much interest among activists, researchers, and people living with HIV. Only results of human study will tell us if this enthusiasm is warranted. Initial studies are underway.

E. Therapeutic HIV vaccine research is still in its infancy.

F. In previous studies, therapeutic vaccines were delivered monthly, by injection.

G. In the short term, it's likely anti-HIV therapy (drugs) will be required in therapeutic HIV vaccine studies.

Because HIV, once inside the cell, is capable of integrating itself into the genetic material of infected cells, a vaccine would have to produce a constant state of immune protection, which not only would have to block viral entry to most cells, but also would continue to block newly produced viruses over the lifetime of the infected person. Such complete and constant protection has never before been accomplished in humans, but it has been accomplished to some degree in cats that are vaccinated against the feline leukemia virus, also a retrovirus (Voelker, 1995). Perhaps more pertinent explanations for why there is still no HIV vaccine nor is one likely to be available soon are the facts that scientists lack sufficient understanding of HIV infection and the biology of HIV disease/AIDS is very complex.

Scientists know that the body defends itself against HIV in the early years of infection. But the great mystery has always been why it cannot neutralize HIV completely. One possibility is that the body has trouble seeing all the variant viruses. Like a Stealth fighter plane, some HIV may have hidden parts that do not show up on the immune system scanner. As a result, the immune system may not produce the right kind of antibody to neutralize all the variant HIVs.

The Ideal HIV Vaccine

An ideal HIV vaccine would be cheap to produce, stable at room temperature, easy to transport and administer without special equipment, completely safe, and would need only one dose to provide complete lifelong protection against all routes of transmission and all variants of HIV. All current vaccine candidates are likely to fall short of these criteria, although even an imperfect vaccine could deliver public health benefits and provide further insights for prevention and treatment strategies.

Requirements for an HIV-Specific Vaccine

Scientists agree that blocking an infection requires the production of neutralizing antibodies. This is how standard vaccines work: They show the immune system a protein that is unique to the germ. If the germ ever gets into the body, the defenses will quickly make antibodies that latch onto that protein, blocking the germ and destroying it. HIV, however, presents a changing target. It mutates so fast that it constantly changes the proteins on its surface. So a vaccine that triggers an attack against one strain of HIV may be powerless against another. Furthermore, the virus covers its surface with sugar, which hides its proteins from antibodies.

Is There Reason for Hope?

However long it takes only an effective preventive vaccine has any reasonable prospect of eliminating HIV as a global public health problem.

There are scientific reasons why there is hope that an HIV vaccine will ultimately be developed. **First,** studies of nonhuman primates that were given candidate vaccines based on HIV or SIV (simian immunodeficiency virus) have shown either complete or partial protection against infection with the wild type virus. **Second,** successful vaccines have been developed against the Feline Immunodeficiency Virus (FIV), a retrovirus. **Third,** almost all humans develop some form of immune responses that are protective or that are able to control the viral infection over a long time period. Some people have remained free of disease for over 25 years, often with undetectable viral loads. A group of sex workers from Nairobi and South Africa have remained HIV negative despite continuing high-risk exposure; resistance to HIV infection in these people is thought to be due to their ability to mount protective immune responses to HIV. This group and long-term survivors in America have provided insights into strategies for developing a vaccine.

Vaccine Expectations

Vaccines rarely prevent infection per se; rather they prevent or modify disease. Most vaccines currently in use—for example those for polio, tetanus, diphtheria, measles, hepatitis B, and influenza—prevent disease without actually preventing infection. They reduce the number of invading microorganisms, increase the rate of clearance of the infection, prevent the secondary consequences of infection, and prevent transmission. Similarly, few of the candidate HIV vaccines appear promising for preventing infection, and current expectations that HIV vaccines will prevent infection have yielded in the scientific community to the *hope* that they can produce a vaccine that can prevent disease.

TYPES OF EXPERIMENTAL HIV VACCINES

To make vaccines, scientists use either **dead microorganisms** and **inactivated viruses** or **attenuated viruses** and **microorganisms.** Attenuated (at-ten-u-ate-ed) means that viruses and other microorganisms are modified; they are capable of reproducing and invoking the immune response but lack the ability to cause a disease.

Use of Attenuated HIV Vaccine—In 1997 data from several labs revealed that a vaccine made from weakened or attenuated SIV-HIV's simian analog can cause AIDS-like symptoms in adult monkeys. These findings have worried some investigators about attempting to use an attenuated HIV vaccine in humans. Robert Gallo, director of the Institute for Human Virology, believes that a live HIV vaccine is too dangerous. He said that "live, low-replicating retroviruses almost always cause disease; that's been our experience in all animal systems. If those vaccinated do not get the disease in three years, it will not tell you what will happen in 10 years or 30 years."

Use of Whole Inactivated Viruses

To inactivate viruses for use in vaccines, the viruses are treated with formalin (for-mah-lin, a strong disinfectant) or another chemical. There is a danger in using inactivated viruses—**they may not all be inactivated.** Inactivated virus vaccines have been made against hepatitis B, rabies, influenza, and polio (Salk vaccine). Salk's first vaccine killed a number of recipients in the late 1950s because not all the polio viruses were destroyed; that is, some could still replicate.

Subunit Vaccines

Subunit vaccines are made from antigenic fragments of an organism or virus most suitable for evoking a strong immune response. Specific subunits can be mass produced and used in pure form to make a specific vaccine. Vaccine against hepatitis B is made from a subunit of the hepatitis B virus and produced in quantity in yeast.

In the United States, researchers are currently basing their vaccine strategies on the use of subunit proteins (gp 160 or gp 120) found in the envelope of HIV. (See Chapter 3 for glycoprotein, gp, discussion.)

DNA VACCINE

What Is a DNA Vaccine?

To make a DNA vaccine, a gene (or length of DNA) that is responsible for making a protein in the infectious virus or organism is inserted into a bacterial plasmid (a circular length of DNA that can replicate by itself inside a bacterial cell). Plasmids carrying the gene of choice are replicated in trillions of bacteria; the bacteria are then broken open and the trillions of plasmids, each carrying a copy of the gene, are purified. The purified genes/DNA are then given to a patient. Cells of the person take up the DNA and begin to make the exact protein the gene made while it was in the virus or microorganism from which it was taken. Such a protein is considered an antigen by the body and the immune system mounts a defense against it. Entering 2005, there were at least nine DNA vaccines in human trials.

Advantages of a DNA Vaccine

DNA vaccines have several potential advantages over the traditional methods. It is inexpensive and there's no risk of infection, as there is for inactivated or attenuated vaccines. DNA vaccines are also superior to protein-based vaccines because proteins, while being isolated to use an antigen in a vaccine, are easily destroyed. DNA vaccine produces the protein right in the host cell. Also, DNA is a very stable, cheap chemical, even at temperatures close to boiling, a decided advantage in developing countries where refrigeration is scarce.

In March 1996, the FDA approved the first human testing of a vaccine made with pure DNA. Apollon Inc. and the National Institutes of Health tested the experimental DNA-HIV vaccine. However, these experiments were not successful. New DNA vaccines are now in study.

Although scientists are mixing the old traditional ideas of vaccine production with the new, there remains the fear that the production of an effective HIV-preventive vaccine may not be possible. But, if it is, it won't be available before the year 2008 and even then it may only be 30% effective in prevention.

What the World Needs NOW Is a Vaccine to Prevent HIV Infection—Why Isn't There One?

On April 24, 1984, Margaret M. Heckler, who was then secretary of the Department of Health and Human Services, announced the discovery of the AIDS virus. She predicted an AIDS vaccine within two years. Even though the prediction proved wrong, very wrong, research was guided by the idea that finding the virus was the hard part, and vaccines could be made by simply injecting people with crucial viral proteins. Her optimism was most likely based on the success of the polio, measles, and flu vaccines. The approach to combating these diseases was: Isolate the virus, develop a vaccine, and prevent the disease! Since then, in the rush to develop new vaccines, scientists have only belatedly understood that their technical ability to mass produce vaccines has failed to match their knowledge about the cellular and molecular processes used by the body to protect itself from invading pathogens.

Vaccines are designed to provoke the immune system into making antibodies against a disease-causing agent. Most are made of inactivated or attenuated (genetically weakened) viruses and, in the case of some newer vaccines, extracts of viral coat proteins. In some cases, vaccination may result in worsening the disease. The distinction between protective, useless, and dangerous responses is essential for vaccine design.

Vaccines that work well are the most cost effective medical invention known to prevent disease.

PROBLEMS IN THE SEARCH FOR HIV VACCINE

In retrospect, in the movie *Rocky*, Rocky Balboa had it easy. Downing raw eggs at 5 a.m., sprinting through the streets of Philadelphia, pummeling sides of raw beef and pumping out one-armed push-ups prepared him to go the distance against world heavyweight champion Apollo Creed. Brute force was what it took. He eventually won! With regard to developing an HIV/AIDS vaccine, brute force is just one of the ingredients essential to winning.

HIV poses some unique problems for making a human vaccine. **First,** scientists have not established what immune responses are crucial for protecting the body against HIV infection. Studies over the last couple of years have shown that the cell-mediated arm of the immune system may be more important than the HIV antibody response. If this turns out to be true, investigators will have to regroup with respect to producing an HIV vaccine—most vaccines in field trials are geared toward producing sustained HIV antibody responses. Without this information, they cannot tailor the vaccine to produce the most essential immune response.

POINT OF INFORMATION 9.5

HIV VACCINE: PREVENTION AND THERAPY

A LOOK BACK AT THE FIRST VACCINE

The year 1996 marked the 200th anniversary of the first vaccine, which was developed against smallpox. As vaccine researchers launch a new century of challenging disease, they might find inspiration in the early beginnings of immunology, Edward Jenner's discovery.

According to lore, Jenner was a country doctor who heard a rumor that the cowpox virus could provide immunity to smallpox. Investigating the theory, Jenner endured the ridicule of his colleagues before proving his point.

In a now-famous 1796 experiment, Jenner extracted fluid from milkmaid Sarah Nelmes's hand and used it to inoculate the arm of 8-year-old James Phipps, infecting the boy with cowpox pus. Two months later, he again inoculated James but this time he added some smallpox residue. Phipps had been successfully immunized. Prominent physicians confirmed his work, and word of Jenner's method of vaccination spread quickly throughout the world. By 1801, an estimated 100,000 people in England had been vaccinated. A hundred years would pass before scientists understood that the cowpox virus stimulates the immune system to produce antibodies, which neutralize the deadly smallpox viruses as well as the mild cowpox viruses. Hailed as one of the great medical-social advances in history, Jenner's immunization method was among the first attempts to control disease on a national scale. It was also the first effort to protect the community as a whole, rather than the individual. A smallpox vaccine—and the field of vaccinology—was launched.

No viral epidemic has ever been conquered by drug therapy—prevention is the key, and primary prevention via vaccine inoculation is the cornerstone. But, there are enormous problems inherent in constructing an HIV vaccine. First, the virus has extraordinary diversity, leaving little chance that just a few subtypes could induce broadly protective immunity. Second, the attenuated (weakened) virus strategy that has been so successful for many infections, including smallpox, polio, and measles, will be very difficult if not impossible to implement for HIV. Third, HIV attacks the very immune cells that are essential in an immunization procedure, and the immune cell activation that accompanies any immunization will activate HIV replication.

Immunologists are still driven to produce an effective HIV-preventive vaccine. Since the first injections of an experimental HIV vaccine were given in the United States in 1987, over 15,000 uninfected adults have received about 50 experimental vaccines; 18 people have become HIV-infected from the vaccines (Bolognesi et al., 1998 updated). Only a handful of HIV vaccines have made it to the second of the three stages of trials that are needed before any vaccine can be marketed. In February 2003, it was announced that the only vaccine in Phase III trials in the United States, AIDSVAX, failed to protect against HIV infection. Worldwide, beginning 2005, there are some 76 experimental vaccines in basic research or animal testing. To date most HIV vaccines being developed target surface proteins of HIV. The reason for this approach is that the immune system first sees the outside boundaries of HIV—the surface proteins. There is a problem, however: Different clades or strains of HIV carry surface proteins that differ from each other because their genes have changed to make the different strains. Thus a vaccine made against one strain's surface proteins, say type B (United States), may not work against type C (Africa or India).

The hope is essentially, that the use of a therapeutic vaccine can reduce the burden of the disease and of its treatments. For example, if it prevented or delayed the emergence of drug-resistant viruses, it would ensure that people would not have to switch treatment for any reason other than side effects. This would, on balance, be a benefit for people's quality of life. Similarly, if it meant that people could continue to take less expensive first-line treatments rather than having to switch to more expensive drugs, this could be a major benefit to treatment programs, especially in countries with limited resources.

Second, it is too risky to use uninactivated or weakened HIV.

Third, HIV undergoes a high rate of mutation as it replicates, and strains from different parts of the world vary by as much as 35% in terms of the proteins that comprise the outer coat of the virus. Even within an infected individual, over a period of years, the virus may change its proteins by as much as 10%. This degree of antigenic drift or variation means that a vaccine made from one strain of HIV may not protect against a different strain. To prove effectiveness, vaccines may have to be tested in geographic areas where the prevalent strains are the same as the strain used in the vaccine.

Fourth, the immune response raised by the vaccine may be protective for only a short period of time. In such cases, booster vaccinations would be required too frequently to be practical.

Fifth, it is possible that the vaccine may make people more susceptible to HIV infection, a vaccine-induced enhancement of infection. That is, vaccine producers are concerned that if they are unable to make a completely HIV-neutralizing vaccine, the vaccine when taken might make the disease worse by boosting the number of antibodies that might enhance entry of HIV into cells. For example, enhancing antibodies are thought to be important in a few unusual viral infections such as dengue fever, Rift Valley fever, and yellow fever, in which the antibody binds to the virus and helps its entry into cells. Thus, the more of a certain kind of antibody you have, the worse the infection can become (Homsy et al., 1989; Levy, 1989).

Sixth, entering year 2005, no vaccine trial to date has been able to stimulate the cellular side of the immune system in the manner necessary to destroy HIV. For example, vaccines involved in current studies do not generate significant numbers of cytotoxic lymphocytes nor do they produce significant numbers of memory T cells that are necessary to recall the initial response against HIV.

Seventh, predictably, money—or rather, lack of it is an important obstacle. Even though vaccines are among the most cost-effective medical interventions ever devised, they are not big money-makers. **Drug companies are traditionally reluctant to invest in any form of vaccine development that carries high costs, low profits, and big risks of costly legal suits should accidents occur.** Their current analysis of the state of HIV/AIDS vaccine research is particularly bleak. Of the estimated **$30 billion to $40 billion spent globally each year** on HIV/AIDS research, care, and prevention, less than **$500 million** goes into vaccine research.

Eighth, the science is very tough. Animal models used to test HIV vaccines have severe limitations; no researcher has successfully demonstrated which immune responses correlate with protection from HIV.

In March 2002, Anthony Fauci told the Presidential Advisory Council on HIV and AIDS that a "broadly effective AIDS vaccine could be a decade or more away." In 2004, UNAIDS Executive Director Peter Piot said, "an HIV vaccine is not likely to be developed in the next decade."

In 2002 Bruce Walker of Harvard Medical School presented, at the Fourteeth International AIDS Conference, perhaps the most threatening information yet on why a vaccine may not be possible reasonably soon, if ever! One of Walker's HIV/AIDS patient's (codename ACO6) immune system was holding HIV in check without antiretroviral therapy. But then the man had unprotected sex and became superinfected—he became infected with a second genetically different strain of HIV. Even though his immune system was controlling the original HIV strain, it was unable to control the second strain, which ran rampant in his body. Many HIV/AIDS researchers fear that what happened to patient ACO6 could happen to someone who gets vaccinated against one strain of HIV but gets exposed to another strain. The chance of encountering a strain different from the one found in a vaccine is very high. Ronald Desrosiers, a well-known and respected veteran HIV vaccine researcher who is also from Harvard, described Walker's research as a "huge blow to vaccine development." (For additional information on HIV superinfection, see Ramos et al., 2002 and Chapter 4: Coinfection and Superinfection).

VACCINE THERAPY

The term "therapeutic vaccine" has become the popular terminology when it should be called therapeutic immunization because the administration of the vaccine occurs **after** infection has occurred. Thus, it is a postinfection treatment. A vaccine, by definition, means to prevent infection. Traditional vaccines are used to prime a person's immune system before a possible infection. Therapeutic vaccines attempt to teach a person's immune system to fight a virus long after it has infected them. With Epstein-Barr or other herpes viruses, for instance, the viruses cannot be eradicated from the body, but immunocompetent individuals are able to control the viruses so that they remain latent and are not constantly replicating. With HIV-infection, the body does a relatively poor job of controlling replication. The major principle underlying immune-based HIV therapies is to induce the immune system—through either immunization or vaccination—to do a more effective job at controlling replication.

Currently, the main growth area is in therapeutic vaccine research. Rather than using vaccines in an attempt to prevent initial infection, the idea is to vaccinate HIV-positive people being treated with HAART, then halt the drug therapy. Hopefully, the vaccine will stimulate their immune systems sufficiently to bring the infection under long-term control.

UPDATE 2004—At the June G8 summit meetings in Sea Island, Georgia, G8 officials from the United States, Japan, Germany, France, Britain, Italy, Canada and Russia announced the formation of the Global HIV Vaccine Enterprise to speed the development of an HIV/AIDS vaccine and streamline research and development efforts. The plan calls for the establishment of HIV vaccine development centers throughout the world, the expansion of manufacturing capabilities, the creation of standardized measurement systems, the construction of clinics for trials and the creation of rules allowing regulatory authorities in different countries to recognize the results of foreign clinical trials.

The bottom line is, without access to antiretroviral drugs and therapeutic immunization, most of the 41 million people living with HIV infection at the end of 2005 will die from AIDS.

Morality of Testing HIV Vaccines

Designers of the vaccine trials are confronted with a paradox unique to HIV/AIDS. If safer-sex and safe-needle practices are not taught, volunteers who believe they are protected by the unproven vaccine (which may actually be a placebo) could take more risks and increase their chances of becoming infected. If such practices are taught, and by ethical standards, they must be, volunteers could cut their risk so effectively that they are never exposed to HIV, leaving the vaccine with nothing to fight. In a way, the trial depends on the failure of education. To get around this potential conflict of interest, Phase III trials of an AIDS vaccine will have to rely on populations in which the number of infections remains high regardless of education, like young high-risk gay men and injection-drug users. There is also the additional problem of creating vaccine-induced HIV-positive persons who cannot be distinguished from those who are HIV-infected. They, too, will be subjected to adverse social, employment, and other discrimination following a positive antibody test.

HIV Vaccine Costs

Perhaps the most difficult moral question is the cost of the vaccine. A successful vaccine that sells for a high price will be of little use to poor and uninsured Americans and most people of developing nations, who have no more than a few dollars a year to spend on health care. Fifteen years have passed since the discovery of a vaccine for hepatitis B, a viral disease that is also spread by sexual contact and the sharing of hypodermic needles. But the product has yet to reach many poor people in the United States and Third World countries largely because it costs about $120 for a series of three injections.

Current Costs For Vaccine Development

The pharmaceutical sector estimates a cost of $50 million to $100 million, just to get to the point of identifying an effective vaccine. To build a production plant could cost between $100 and $200 million. How much it then costs to make the vaccine and deliver it is probably going to vary, depending on the size of the manufacturing plant and the type of vaccine that's made. Before a successful vaccine is found, produced in mass quantity, and made available at the market, the price will be several billion dollars.

HUMAN HIV VACCINE TRIALS

In the early 1990s, the biotechnology company Genentech, proceeded with a large study of its therapeutic HIV vaccine, rgp160. Results suggested that the vaccine made no impact on HIV disease progression and there was some indication that people who received the vaccine did slightly worse than those on the placebo. Genentech stopped the study and abandoned efforts in this arena. (Note: This vaccine was later sold to VaxGen, who modified it and is researching it as a preventive vaccine called AIDSVAX).

VaxGen's AIDSVAX Fails

On February 24, 2003, VaxGen announced the results of its three-year Phase III American study using this vaccine. Phase III data showed no significant difference in HIV prevention between the control or placebo group and those who received the vaccine injections. This vaccine will not protect the general public in the United States. Phase III AIDSVAX studies continued in Thailand but also failed.

Vaccine studies in humans are now in progress in the United Kingdom involving a combination of a DNA vaccine with a modified vaccina vaccine. The other, at Sydney's University of New South Wales, is under the direction of David Cooper. This vaccine contains a DNA vaccine with a bacterial protein known to enhance immune responses and a foulpox vector used to make several poultry vaccines.

UPDATE 2004—Researchers from the United States and Thailand announced plans to conduct a large-scale, U.S.-sponsored phase III trial of an HIV-1 vaccine comprised of two parts—the ALVAC canarypox vector expressing gp120 (Aventis-Pasteur) and monomeric gp120 (VaxGen). The vaccine will be administered to 16,000 HIV seronegative Thai subjects at an estimated total cost of U.S. $119 million. However, 22 prominent HIV/AIDS researchers argue in a letter to the journal *Science* that the proposed vaccine trial is unlikely to yield positive results and, therefore, is an inappropriate use of resources. They claim that prior studies of the vaccine components showed poor immunogenicity and no protective efficacy against HIV when given separately; that is, they failed, and that there is insufficient evidence to suggest that the combination vaccine will be any better. Stay tuned.

Vaccine Testing Confidentiality

Confidentiality must be maintained for the duration of a vaccine trial because people immunized with candidate vaccines who mount effective immune responses will appear positive for HIV antibody. They may be subject to the social stigma and discrimination associated with being truly HIV positive.

Vaccine-induced seroconversion may lead to difficulties in donating blood, obtaining insurance, traveling internationally, or entering the military. Vaccine-induced antibodies may be long-lived, thus volunteers in vaccine trials must be given some form of documentation that certifies that their antibody status is due to vaccination and not HIV infection. The National Institute of Allergy and Infectious Disease has provided a tamper-proof identification card to help uninfected participants in vaccine trials. The seroconversion issue may play a major role in recruitment efforts and in the future welfare of vaccine trial participants.

So the question for anyone is: Should I enter an HIV vaccine trial when I know safeguards are limited? I know why I ought to do it . . . but! (Class Discussion)

Vaccine Effectiveness and Infrastructure in Developing Nations

Both vaccine effectiveness and global infrastructure are necessary for HIV transmission to die out eventually. For example, some estimates indicate that to stem the HIV pandemic, the world would need 50% coverage with a 75% effective vaccine. Until a highly effective vaccine is in wide distribution, anti-HIV/AIDS programs aimed at encouraging behavioral change will remain essential in controlling the spread of HIV/AIDS.

Infrastructure—While making, purchasing, and bringing vaccines to developing countries is a daunting task, many experts say that the real challenges start at the airport, after the vials have been unloaded for distribution throughout the country. It's here that the issue of uptake—demand with a reality check—comes into play. Are there sufficient refrigerators, trucks, syringes, and syringe disposal facilities for the vaccine dispensaries? Are there trained personnel at these dispensaries? Is there capacity for community outreach and for follow-up to individuals who do not complete their immunization course? In short, is the necessary infrastructure in place? If the answer is no, then it will not matter whether a country has an explosive epidemic or an early one that could be stopped with a relatively small-scale immunization campaign. Without infrastructure, there is an unbridgeable gap between need and demand. The scientific consensus is that an effective HIV vaccine can be developed, this is a challenge that must be confronted and cannot be allowed to be repeated.

The Future Preventative Vaccine

Vaccine programs in the United States and in other nations are led by some of the most talented and dedicated scientists in the world. One must admire them and the dedicated public and private teams developing each vaccine candidate. Together they recognize that a preventative vaccine is our best long-term hope to control the pandemic, although it will not be a magic bullet replacing other preventive interventions. In February 2002, AIDS researchers Robert Gallo and Luc Montagnier, who fought a long and bitter battle over credit for the discovery of HIV and the resultant blood test, announced plans to collaborate on developing AIDS vaccines for Africa and other impoverished regions. The goal is to run clinical trials on at least five potential vaccines to prevent the spread of the disease and treat those who already suffer from AIDS. The new partnership, the Gallo/Montagnier Program for International Viral Collaboration, will first work to raise $4 million to start the organization and organize the research. They will work with the United Nations to develop research programs in areas with high populations of AIDS patients. The targeted areas include cities or regions in the United States, Italy, Canada, Africa, Central America, and Asia.

Disclaimer: The author of this book cannot be held responsible for any inaccuracies found in the inclusion of information by any organization, treatment, therapy, or clinical trial. The use of their information is not an endorsement of their facts or data. Any information found within this textbook should always be used in conjunction with professional medical advice.

For additional information about the search for an HIV/AIDS vaccine, the following literature is recommended.

AIDS Vaccine Research
Flossie Wong-Staal and Robert C. Gallo, Eds.
Marcel Dekker, 2002

Shots in the Dark: The Wayward Search for an AIDS Vaccine
by Jon Cohen
W.W. Norton, 2001

HIV and Molecular Immunity: Prospects for the AIDS Vaccine
by Omar Bagasra
Eaton Publishing, 1999

The Search for an AIDS Vaccine: Ethical Issues in the Development and Testing of a Preventive HIV Vaccine (Medical Ethics Series)
by Christine Grady
Indiana University Press, 1995

AIDS Vaccine Resources on the Internet

www.avac.org
www.niaid.nih.gov/daids/vaccine/default.htm
www.fhi.org/en/aids/hivnet/hivnet.html
www.vaccinealliance.org
www.vaccineadvocates.org/avacsite/inde.htm

These addresses will list most of the important vaccine internet addresses and serve as linkage to others.

SUMMARY

The key to stopping HIV transmission lies with the behavior of the individual. That behavior, if the experience of the past 24 years can be used as an indicator, has proven to be very difficult to change.

Changing sexual behavior and using a condom is referred to as **safer sex.** The latex condom is the only condom believed to stop the passage of HIV, and a spermicide should be used with the condom. Oil-based lubricants must not be used because they weaken the condom, allowing it to leak or break under stress. Water-based lubricants are available and should be used. There is at least one female condom, called a vaginal pouch, approved by the FDA and sold worldwide. It is inserted like a diaphragm. It offers protection to both sexual partners.

Over 8000 cases of HIV infection have come from contaminated blood transfusions. A test developed in 1985 to screen all donated blood in the United States has reduced the risk of HIV transfusion infection. But blood bank screening has reduced the size of the blood donor pool. Many hospitals are encouraging people who know they might need an operation to donate their own blood for later use—autologous transfusion.

At the end of 2004 the only FDA-approved vaccine in Phase III trials failed. Many top HIV/AIDS scientists have ruled out the use of an attenuated HIV vaccine. Inactivated whole virus vaccines are also being held back because there is no 100% guarantee that all HIV used in the vaccine will be inactivated.

Even if a vaccine does well in Phase III trials, will it be effective against all the HIV mutants in the HIV gene pool? Can the threat of vaccine-induced enhancement of HIV infection be overcome? How are vaccine testing agencies going to handle the ethical question of vaccine seroconverting normal subjects to positive antibody status? The social repercussions may be devastating for those who, when tested, test HIV positive even though they are HIV-free.

There are some 5.3 million health-care workers in the United States. It is crucial that they adhere to the Universal Protection Guidelines set down by the CDC, as a significant number of them are exposed to HIV annually. The risk of HIV infection after exposure to HIV-contaminated blood is about 1 in 200.

A few states have implemented HIV partner notification; many other states are beginning to experiment with HIV partner notification or contact tracing programs. It is too early to tell how successful locating and testing high behavioral risk partners will be, or the cost-to-benefit ratio. If these programs are to be successful, they will have to ensure confidentiality to those who are traced. Partner notification or contact tracing continues to work well for other sexually transmitted diseases.

REVIEW QUESTIONS

(Answers to the Review Questions are on page 438.)

1. Which is the better condom for protection from STDs, one made from lamb intestine or one made from latex rubber? Explain.
2. Which lubricant is best suited for condom use? Explain.
3. Briefly explain safer sex.
4. True or False: If a person has unprotected intercourse with an HIV-infected partner, he or she will become HIV-infected. Explain.
5. Yes or No: If injection-drug users (IDUs) were given free equipment—no questions asked—would that stop the transmission of HIV among them? Explain.
6. What is the current risk of being transfused with HIV-contaminated blood in the United States?
7. What do you think should happen in cases where a person who knows he or she is HIV positive lies at a donor interview, and donates blood?
8. Why do most scientists wish to avoid using an attenuated HIV vaccine or an inactivated HIV vaccine?
9. What is the advantage of using recombinant HIV subunits in making a vaccine?
10. Explain vaccine-induced enhancement. How does it occur?
11. Why is it necessary to practice strict confidentiality with respect to volunteers for AIDS vaccine tests?
12. What are universal precautions? Who formulated them?
13. True or False: Research continues to show that AIDS prevention messages are effective in causing teens to change their sexual behaviors.
14. True or False: Latex condoms eliminate the risk of HIV transmission.
15. True or False: Partner notification is usually performed by the infected individual or a trained and authorized health department official.
16. True or False: The Centers for Disease Control and Prevention estimates that as many as 1 in 100,000 units of blood in the blood supply may be contaminated with HIV.

17. True or False: The three types of vaccines that scientists are interested in developing are preventive, therapeutic, and perinatal vaccines.

18. True or False: Used disposable needles should be recapped by hand before disposal.

19. True or False: Prompt washing of a needle stick injury with soap and water is sufficient to prevent HIV infection.

20. True or False: The FDA approved the first vaccine for broad-scale testing in the United States in 1997.

21. HIV/AIDS is not curable, but it is preventable. Write a short essay on the best methods of prevention.

22. HIV/AIDS Word Search.

H	L	C	V	V	A	L	Y	L	L	L	E	C	T
H	C	A	E	L	B	D	O	K	R	S	A	R	A
R	T	N	G	R	O	S	N	S	D	I	A	Y	C
I	N	C	U	B	A	T	I	O	N	N	I	D	Q
N	E	G	I	T	N	A	O	M	S	N	M	D	U
R	D	T	S	M	F	C	I	M	Y	Y	M	U	I
O	N	E	M	I	X	A	I	R	N	X	U	V	R
A	E	P	O	C	E	S	S	N	D	O	N	U	E
W	X	P	D	T	S	U	D	B	R	A	E	N	D
P	P	S	N	I	E	A	N	P	O	A	S	R	C
N	O	H	O	N	F	L	T	H	M	T	Y	D	C
M	S	N	C	F	A	C	M	C	E	O	S	I	D
A	U	M	I	E	S	O	O	D	O	M	T	R	C
T	R	R	M	C	N	N	H	I	V	S	E	T	R
R	E	N	M	T	M	T	I	T	I	O	M	R	D
O	M	E	U	I	N	A	C	N	H	N	H	N	T
D	H	G	N	O	M	C	U	O	C	S	I	M	I
A	G	O	O	N	A	T	A	A	D	E	N	I	I
H	M	H	S	C	R	D	V	T	R	R	H	S	T
E	O	T	U	O	N	I	C	O	M	O	O	I	R
M	C	A	P	A	D	T	S	A	A	P	C	D	I
O	M	P	P	N	I	H	I	K	A	O	I	M	I
P	O	G	R	A	D	M	N	K	H	S	I	T	A
H	R	H	E	S	N	T	O	M	E	I	C	I	V
I	R	C	S	U	O	R	I	A	I	T	I	D	A
L	G	I	S	R	T	C	S	I	C	I	R	T	R
I	L	O	E	I	D	E	N	R	D	V	O	V	O
A	T	G	D	V	A	C	C	I	N	E	X	D	R

ACQUIRED
AIDS
ANTIBODY
ANTIGEN
BLEACH
CASUAL CONTACT
CDC
CONDOMS
DISEASE
EXPOSURE
HEMOPHILIA
HIV
IMMUNE SYSTEM
INCUBATION
IMMUNOSUPPRESSED
INFECTION
IV
KAPOSI
OPPORTUNISTIC
PATHOGEN
PWA
RISK
SAFE SEX
SEROPOSITIVE
STD
SYNDROME
T-CELL
TRANSMISSION
VACCINE
VIRUS

10

Prevalence of HIV Infections, AIDS Cases, and Deaths Among Select Groups in the United States and AIDS in Other Countries

Chapter Concepts

- AIDS is a new plague.
- Worldwide, ending 2005, heterosexuals will make up about 95% of people living with HIV/AIDS.
- Worldwide, 50% of new HIV infections are in people under age 25.
- Worldwide, about 14,000 new HIV infections occur daily.
- Worldwide, women represent 50% of all HIV-infected adults and 50% of AIDS deaths.
- In sub-Saharan Africa, 58% of the HIV-infected are women.
- AIDS is the world's leading cause of death by an infectious disease.
- AIDS is ranked fourth in causes of death worldwide.
- In the United States, men make up 77% of all AIDS cases; 23% are women.
- The majority of people with HIV/AIDS can be associated with certain lifestyle risks.
- HIV/AIDS can be associated with single or multiple exposure risks.
- HIV/AIDS cases can be separated by sex, age group, race, ethnicity, and sexual preference.
- Risk is strongly tied to social behavior.
- At-risk groups include homosexual and bisexual men, injection-drug users (IDUs), hemophiliacs, transfusion patients, and the sex partners of these people.
- HIV infection is strongly associated with injection-drug use.
- All military personnel are tested for HIV.
- Two per 1000 college students are HIV-infected.
- Thirty-seven states report the HIV-infected by name, thirteen by name to code or code.
- High rates of HIV infection have been found among prisoners.
- The greatest HIV threat to health-care workers is needle stick (syringe) injuries.
- All 50 states and U.S. territories must report all HIV and AIDS cases.
- January 2006: *reported* AIDS cases in the United States will reach 970,000, of which 541,000 will have ended in death.
- HIV reporting.
- People do not always tell the truth when completing questionnaires, especially with regard to sexual behavior.
- By the end of 2005, an estimated 65 million people worldwide will have been HIV-infected and about 24 million of them will have died of AIDS.
- HIV/AIDS crossword is presented.

The World Trade Center and the Pentagon were attacked on September 11, 2001. No one expected to see commercial airliners hijacked and flown into crowded skyscrapers and government buildings. Likewise, in the early 1980s, no one expected the sudden appearance of a deadly new disease spreading across the cities of America. Only a few years earlier some prominent scientists had declared that the fight against infectious disease was over. The horrible damage caused by fuel-laden airliners crashing into buildings and exploding into a fiery inferno was all but beyond our imagination. Equally beyond our imagination was this new disease that appeared 24 years ago. It was a disease that appeared without warning and seemed to lead to a painful, agonizing death in just a few weeks for some, a few months for others. The depth and scope of human destruction was so unprecedented that only a few people were quick to recognize the horror that was to come. On September 11, while the image of the jetliners with their passengers exploding into the World Trade Center was still painfully fresh, we were further stunned to see these seemingly invincible structures collapse, crushing almost 3000 men, women, and children in a vast cloud of toxic dust, rubble, and fire. A week earlier, no one would have believed that such pillars of concrete and steel could possibly collapse, let alone from the top down. In the tragedy that began to unfold in the beginning of the 1980s, scientists were puzzled and bewildered as they watched a disease that led to the collapse of the human immune system. Working from the inside out, here was a diabolically clever virus that destroyed the very system that was otherwise designed to defeat it. The beginning of AIDS in America in 1981 is the greatest plague in history to cross the globe, and the 2001 attack on the WTC and the Pentagon, the largest attack on the United States, caused the greatest loss of life since the Japanese attack on Pearl Harbor. The two have much in common. Both tragedies have appeared over and over again on television. Both require the best in people to set them right—to save lives. The United States and the world must not let terrorists or HIV determine our way of life. Both require acts of heroism and research. Both require sacrifice, resolve, and the determination to overcome. Both have placed America and the world in a race against time.

The News Media: The Result of Omission

When AIDS isn't visible in the media, it doesn't exist for many communities that rely on media to tell them what matters most. This is true not only in the West, but in Africa and in other developing nations, where stigma and discrimination against AIDS sufferers is deeply entrenched, and where silence and denial still drive many governments to cover up their frightening levels of HIV infection. Often the young people who most need to know how to protect themselves have few programs directed their way. At the same time, there are many stirring and effective responses led by unsung young heroes whose stories could inspire a greater mobilization. But who is going to tell these stories if the media doesn't? AIDS will surpass the bubonic plague or "Black Death" of Asia and Europe in the fourteenth century unless a vaccine is found. That plague killed about 40 million people. By the end of 2005, about 24 million will have died from AIDS. Each day AIDS kills about three times the number of people who died September 11, 2001, in New York City and at the Pentagon. No terrorist attack, no war or natural force of nature in our lifetime has ever killed 24 million people and threatens 40 million more with premature deaths in less than a quarter of a century! HIV is a virus of mass destruction.

A WORD ABOUT DATA

All HIV/AIDS data, including data from the Centers for Disease Control and Prevention (CDC), Joint United Nations Program on HIV/AIDS (UNAIDS), World Health Organization (WHO), and other organizations should be treated as broadly indicative of trends rather than accurate measures of HIV/AIDS prevalence. A large number of HIV/AIDS cases in developing countries, in particular, are underreported due to a lack of adequate medical and administrative personnel, the stigma associated with the disease, or the reluctance of countries to incur the loss of trade, tourism, and other losses that such revelations might produce.

Because morbidity and mortality of HIV/AIDS cases are multicausal, diagnosis and reporting can vary significantly, thereby distorting comparisons. The WHO and other international entities are dependent on such data despite its weaknesses and are often forced to extrapolate or build models based on relatively small samples, as in the case of HIV/AIDS. Changes in methodologies, moreover, can produce differing results; for example, the ranking of AIDS mortality ahead of TB mortality. This is partly due to the fact that HIV-positive individuals dying of TB were included in the AIDS mortality category in the most recent WHO survey. Another example is the global data on those living with HIV/AIDS. The UNAIDS global data say there are 38 million such people at the end of 2003, but this estimate results from a low estimate

BOX 10.1

REFLECTIONS FROM THE XV INTERNATIONAL AIDS CONFERENCE—BANGKOK, THAILAND, JULY 11–16, 2004

Every two years, the scientific community working on HIV/AIDS is joined by activists, health officials, and government leaders to take stock of the AIDS pandemic and to share the latest findings. Since the last International AIDS Conference in Barcelona, some 10 million people have been infected with HIV and about 6 million people have died from AIDS.

The 2004 XV International AIDS Conference was different from the previous 14 because the major point of the conference was clearly social and not scientific. Science took a back seat to the political rhetoric faulting (a) failed leadership in countries being devastated by the pandemic and (b) the developed nations, mainly the United States, for underfunding the global financial needs of the pandemic. Throughout the conference, Randall Tobias, the global AIDS coordinator for the U.S. government, dismissed the protests, saying the United States is doing more than all other countries put together.

This conference marks the first time a secretary general of the United Nations has attended. Kofi Annan spoke about the importance of leadership at every level. He said, "We need leaders everywhere to demonstrate that speaking up about AIDS is a point of pride, not a source of shame. There must be no more sticking heads in the sand, no more embarrassment, and no more hiding behind a veil of apathy. Leaders must not shirk their responsibilities, leaders must lead."

Delegates to the Convention

Between 19,000 and 20,000 delegates from 160 countries attended the six-day conference, the largest number of delegates ever. The theme for this conference was "Access For All"—access to HIV/AIDS medications; access to unbiased information and education about HIV/AIDS; access to effective prevention tools; access to comprehensive medical care; access to resources; and access to those things that will minimize the impact HIV/AIDS has on human lives.

During the six days of this conference about 40,000 people died of AIDS and about 82,000 became HIV infected.

AIDS Activists Protest

Before the conference even began protesters marched to the conference center demanding access to HIV/AIDS medications for all 6 million people who need them as well as access to condoms and clean needles (Figure 10-1). They called for world leaders to address the inequities in treatment availability. The protesters were especially eager to garner support for fully funding the Global Fund to Fight AIDS, Tuberculosis, and Malaria. Once the conference began there were daily protests against the United States and other G8 countries over drug pricing and funding policies as well as a perceived single-minded reliance on abstinence as the main weapon against AIDS. Protestors stormed the global AIDS

FIGURE 10-1 AIDS Activists Protesting the Cost of Antiretroviral Therapy Drugs at the XV International AIDS Conference in Bangkok, Thailand. *(Photograph courtesy of AP/Wide World Photos)*

summit throwing mock blood over posters for world leaders in protest over a shortfall in funding to tackle the epidemic. The protesters held a mock trial of the heads of the industrialized nations in the foyer of a conference center in the Thai capital. A young South African AIDS activist shouted out a list of charges against leaders such as U.S. President George W. Bush, British Prime Minister Tony Blair, French President Jacques Chirac, and Italian Prime Minister Silvio Berlusconi. The demonstrators shouted "shame" and defaced the large portraits of the leaders with wanted banners and red paint, saying they had reneged on promises to contribute $10 billion dollars to the fight.

How Far We Have Come—How Little Has Been Learned

The four-star Prince Palace Hotel in Bangkok separated its HIV-infected guests from the rest of its clientele. A hotel employee said that the separation was being done for reasons of hygiene. "It was for hygiene reasons. If they stay in separate rooms, it's easy for our maids to separate and wash their bedsheets and everything." The participants at an AIDS workshop at the hotel were kept to one floor and asked to eat in a separate dining area.

Antiretroviral Therapy

Representatives from the United Nations said that, of the six million people in need of drug therapy, about 400,000 to 500,000 were receiving them. The goal of 3 million people receiving drugs by 2005 (UN's 3 × 5 plan) will most likely not happen. The UN stated that, by 2007, $20 billion will be needed to fight AIDS, but it expected only $10 billion in donations.

Complicating the equations is lack of inexpensive backup drugs for patients who eventually will fail to respond to—or no longer tolerate—the current crop of lower-cost medications. That will happen to 10 to 20 percent of the patients who will end up needing the so-called second-line therapies that can cost as much as 20 times the price of the generics, threatening treatment programs down the road.

Prevention

Ugandan President Yoweri Museveni stirred up controversy at the conference for his strong stance that abstinence and faithfulness in marriage are the first line of defense against HIV. He said, "I look at the condom as an improvisation—not a solution—an improvisation." However, many researchers and advocates believe condoms must be an equal part in stopping HIV/AIDS. The Ugandan model is similar to the message championed by the U.S. government with its "ABC"—abstinence, be faithful, condoms when appropriate—approach at home and in the developing world, to the irritation of activists who argue Washington is dallying in ideology rather than science.

The World Is Slowly Crawling toward a Vaccine, but Only a Vaccine Can End This Pandemic

Seth Berkley, president of the International AIDS Vaccine Initiative (IAVI) said, "The world is inching toward a vaccine, when we should be making strides. The single biggest obstacle is that vaccine development is not a top scientific, political, and economic priority." Berkley noted that in the 24-year history of AIDS, "We've only had one vaccine that has been fully tested on humans. This is a global disgrace." That vaccine failed. Berkley said, "Annual spending on vaccine research is about $650 million, less than 1% of total spending on all health product development in the fight against AIDS. That level has to be doubled to $1.3 billion a year to help get more prototypes into testing and resolve tenacious puzzles as to exactly why HIV is such a stealthy foe. In funding terms, vaccines have always been the poor cousin of research for treatments, which are far more profitable for the pharmaceutical giants. And in scientific terms, no one yet knows what is the genetic recipe of antibodies or immune cells that can be primed for destroying the virus. In a planetary population of over 6 billion, not a single person has even been found whose immune system eradicated an infection by HIV.

Currently there are some 30 candidate vaccines, all focused on one approach. If they fail, it is back to square one, and at the moment the chance of success is not promising.

Testing for HIV, a Crucial Missing Link in the Fight against HIV/AIDS

Cuts of more than 90% in drug prices have made widespread access to antiretrovirals feasible in even the poorest countries, but testing remains a significant barrier. Over 90% of HIV infected people do not know they have HIV and can continue to transmit HIV. Also, most are severely ill before they are diagnosed, making them harder and more expensive to treat.

Will It Be Possible to Test All Who Need Testing?

The number of tests needed is daunting. For the World Health Organization to meet its target of treating 3 million people with HIV/AIDS in the developing world by 2005, some 5,000 people a day will need to be put on medication. That implies 500,000 tests a day will be needed, assuming that 50,000 will test positive in high prevalence countries and that 10% of these will need drugs immediately. Testing really is the missing link to HIV/AIDS care and prevention. Previously, the World Health Organization held that testing should come only at the initiative of a patient, who might dread the stigma and hopelessness a positive result might bring. Now, available AIDS drugs mean that the benefits of diagnosing the virus early outweigh potential concerns about discrimination.

Summary to the Conference

This conference was not about scientific inroads into the biology of HIV, its transmission or about its prevention. Perhaps this is because scientists have learned and shared their knowledge in past conferences. This conference was highly political, looking for the world's money and its leaders to make the difference that science has previously achieved. Many of the conference delegates felt there were too many opinions and too little scientific fact. In general terms this conference, after two years in the making, offered relatively little that was new or different. Ideas on treatment and prevention ruled the meetings but we knew that going in. So, was this conference necessary? The answer depends on who is asked. Many politicians and scientists attending this meeting felt disappointed. "The large sum of money could have been better spent elsewhere. There was insufficient value for the buck."

The conference ended with delegates committing to work towards progress in both treatment and prevention. The experts and others then headed home—the rich to one pandemic, the poor to another.

of 32 million and a high estimate of 44 million—recognizably a large margin for error. Such data come from using a relatively simple equation that calculated low- and high-risk populations with the crude adult death rate and the HIV-infected survival rate, along with variables including the start date of HIV for that country and the demand for risky sex. The bottom line is that estimates are just that—estimates. However, estimates prove to be very useful in predicting what is occurring now and in what lies ahead. Caution must be exercised, however, because the definition of AIDS differs between countries. The criteria used to diagnose a person with AIDS varies significantly from developed nations, where an HIV test is administered, to underdeveloped nations like Africa where HIV tests are generally not available and the diagnosis is based on certain signs and symptoms.

THE MILLENNIUM: YEAR 2005

As we begin this fifth year of the new millennium and enter into the early years of the third decade of AIDS, it is evident that the small epidemic recognized among a handful of homosexual men in 1981 is quite different from the global pandemic of today. For developed countries, current antiretroviral regimens are allowing HIV-infected individuals to live longer, healthier lives. However, the use of these regimens may be associated with complacency and a relapse to unsafe sexual practices resulting in sustaining the HIV epidemic, as will be discussed. For developing countries the problem is far more complex. The infrastructure to support the use of antiretroviral drugs is not in place, the cost of the drugs is too high, and the vast majority of HIV-infected people do not even know they are infected. Education and condom distribution campaigns have had limited success, but they remain the primary avenues of prevention.

Developed Countries: The AIDS Crisis Is Over?

In the developed nations, too many people think the AIDS crisis is over. Think again. UNAIDS estimated that, on average, in 2000 through 2005, each year about 2.5 million people died of AIDS and about 4 million a year became HIV-infected. It is estimated that by the end of 2005, 65 million people will have been HIV-infected, with about 41 million of them alive. Ninety-five percent of infections are in developing nations that hold 10% of the world's wealth. There is still neither a cure nor an effective vaccine.

Entering year 2005 there does not appear to be an immediate end to either the spread of HIV infection or the devastation caused by AIDS. Looking back over the 1980s and the 1990s, it is clear that scientists and the public have consistently underestimated the magnitude and the potential of the HIV/AIDS pandemic. Who would have imagined in the mid-1980s that HIV would eventually spread to every country of the world, infecting 65 million people and resulting in about 24 million deaths by the end of 2005?

Devastation by the Numbers

In 2002, UNAIDS warned that contrary to earlier expectations, the incidence of AIDS is still climbing even in the worst affected countries. Over 100 million people will die because of AIDS in the 45 most affected countries over the next 20 years, more than 7 times the number claimed by AIDS in those same countries during the first 20 years. In some countries, AIDS will kill half the women who became mothers in recent years, thereby creating countless more orphans. In Africa, women currently make up 70% of the HIV-infected population.

Estimates by the U.S. National Intelligence Agency

The National Intelligence Council predicted that by 2010 there will be 75 million cases of HIV infection in just five nations it deemed of strategic importance to the United States, none of them in Southern Africa—India with 25 million, China with 15 million, Russia with 10 million, Ethiopia with 10 million, and Nigeria with 15 million. These five countries contain 40% of the world's population. That total is four times the estimate of 25 million cases an international team of experts projected for those countries as part of a study published in 2001. The new intelligence estimate states Nigeria and Ethiopia will be the worst affected because AIDS will take a heavy economic toll and discourage foreign investment. Quarter-century, 2000 to 2025, projections based on a "mild" scenario estimate that 40 million people will die of AIDS, collectively, in India, China, and Russia! The infections are overwhelmingly occurring in heterosexuals. Worldwide the HIV/AIDS pandemic is not expected to peak for another 40 to 50 years.

Translating the Numbers

The numbers become so large that the individuals suffering and the personal, societal, and economic losses become impossible to measure or to even attempt to estimate. While there have been successes in slowing the epidemic in some communities and dramatic advances in survival in developed countries due to combination antiretroviral therapy, it is very important that people do not become complacent and focus on false beliefs that the pandemic is declining, that individuals are becoming less infectious, or that HIV control will be much better in the twenty-first century. If anything, the limited successes in prevention should encourage a continued effort to work harder at educating more people about how best to prevent further transmission through safer sex practices, antiretroviral therapy during pregnancy, treatment of STDs, provision of condoms, screening of the blood supply, the use of sterile needles, or many of the other avenues that can help to slow the pandemic until a vaccine can be developed.

Estimates by UNAIDS

Currently, UNAIDS reports that each day about 14,000 people become newly infected with HIV, or 10 men, women, and children per minute. Eleven percent of the newly infected people are under age 15. Over 50% of new infections are now occurring in people between ages 15 and 24, primarily due to sexual transmission. Worldwide women now represent 50% of all people over age 15 living with HIV infection. At the end of 1998, UNAIDS reported that AIDS had become the world's most deadly infectious disease. It reached this level of human devastation in just 18 years. Of all causes of death worldwide, AIDS has moved up to fourth place. These data are a bit ironic because as Peter Piot, executive director of UNAIDS said, "The pandemic is out of control at the very time when we know what to do to prevent its spread."

As shocking as these numbers are, they do not begin to adequately reflect the physical and emotional devastation to individuals, families and communities coping with HIV/AIDS, nor do they capture the huge deleterious impact of HIV/AIDS on the economies and security of nations, and entire regions.

HIV Infections: Overwhelming Numbers in the United States

It is easy to be overwhelmed by statistics in reporting on HIV infections and AIDS cases and to lose track of the human faces of the pandemic. But certain numbers, like the first half-million documented AIDS cases reported in October 1995 and about 540,000 dead ending 2005, take on a compelling quality of their own. Therefore, within this chapter there are many statistics presented on all facets of the HIV/AIDS pandemic. **After reading this chapter one will have gained a deeper insight into the spread of HIV and those who are affected by HIV: those who have it, and those who don't—those who will suffer and those who won't. In reality we are all impacted by this disease in one way or another.**

Prevalence/Incidence: How Many People Are HIV Positive?

The **prevalence** of a disease refers to the percentage of a population that is affected by it at a given time. **Prevalence = total number of AIDS cases at a given time divided by total population at the same time.** The **incidence** means the number of times an event occurs in a given time frame, for example, the number of new AIDS cases each month or new HIV infections each week (events that occur within a specified period of time). The two terms are similar. Much has been learned about the prevalence of HIV infection, HIV disease, and its terminal stage called AIDS since the 1981 CDC report that awakened the world to this new pandemic.

Although cases of AIDS appear retrospectively to have occurred in the United States as early as 1952, the **AIDS pandemic** in the United States is considered to have begun with the initial report in June 1981. Since then, the HIV/AIDS pandemic has become the most serious pandemic to occur worldwide since the Spanish flu of 1918, which killed between 30 million and 50 million people but lasted less than a year.

FORMULA FOR ESTIMATING HIV INFECTIONS

A newer formula proposed by the CDC for use in determining the number of HIV-infected persons in a given city is as follows:

National Number of Persons Living With AIDS (1997)		Number of PLWA in Your City (1997)

$$\frac{258{,}000}{900{,}000} \times \frac{\text{(e.g.,) } 1{,}000}{x} = 258{,}000$$

(Estimated national number of HIV-infected persons)

$$x = 900{,}000{,}000$$

$$x = \frac{900{,}000{,}000}{258{,}000}$$

$$= 3{,}488$$

(About 3488 persons in this sample city are HIV-infected.)

Single or Multiple Exposure Categories

In Table 10-1, the number of AIDS cases estimated ending 2005 is presented with respect to adult/adolescent, single or multiple exposure categories. For example, under *Single Mode of Exposure,* heterosexual contact accounts for 10% of all AIDS cases. Under *Multiple Modes of Exposure,* 5% of AIDS cases occurred among injection-drug users who also had heterosexual contact. Table 10-1 lists the numbers and percentages of all reported AIDS cases broken down into seven categories of people who contracted HIV/AIDS from a single risk mode of exposure and 26 categories of people who contracted HIV/AIDS from multiple risk modes of exposure. Note that of the total number of adult/adolescent AIDS cases, 77% occurred from single risk modes of exposure. Of this 77%, 44% occurred among men who had sex with men and about 20% among injection-drug users.

A composite representation of all AIDS cases by exposure category estimated beginning 2005 is shown in Chapter 8, Figure 8-1.

BEHAVIORAL RISK GROUPS AND STATISTICAL EVALUATION

Behavioral Risk Groups and AIDS Cases

As the pool of AIDS patients grew in number during 1981–1983, individual case histories were separated into **behavioral risk groups.** The early case histories of AIDS patients clearly separated people according to their social behavior and medical needs. AIDS patients were placed into the following six risk behavior categories: (1) homosexual and bisexual men; (2) injection-drug users; (3) hemophiliacs; (4) blood transfusion recipients; (5) heterosexuals; and (6) children whose parents are at risk. Each of these groups is considered to be at risk of HIV infection based on some common behavioral denominator. That is, those within these groups represented a higher rate of AIDS cases than people whose needs or behaviors excluded them from these groups. However, because there is some mixing between individuals in behavioral risk groups, HIV infection has gradually spread to lower-risk behavioral groups. Over time the behavioral risk groups have been aligned and defined according to age, exposure category, and sex (see Table 10-1).

A review of AIDS cases by sex/age at diagnosis, and race/ethnicity reported through December 2004 in the United States, shows that white, black, and Hispanic males between the ages of 20 and 44 make up 79% of all male AIDS cases. Between ages 20 and 59, they make up 97% of all male AIDS cases. People between the ages of 25 and 44 make up over half the nation's 138 million workers. In 2003 in the United States, 1 in 6 worksites with over 50 employees and 1 in 16 small businesses have/had an employee with HIV/AIDS.

Statistical Evaluation of Selected Risk Behavioral Group AIDS Cases

Adult/Adolescent AIDS Cases—On October 31, 1995, the United States reached a half-million (501,310) reported AIDS cases. Ending 2006, an estimated 1,013,000 AIDS cases and about 560,000 AIDS-related deaths will be reported to the CDC. Fifty-three percent of all AIDS cases reported occurred from 1995 through 2004. Cumulative through 2004, about 12% of AIDS cases have occurred among the heterosexual population, 25% occurred among injection-drug users, and 51% occurred in the male homosexual/bisexual IDU population. For 1996, the first time since the United States AIDS pandemic began, more blacks were diagnosed with AIDS (41%) than whites (38%). Table 10-2 presents the total estimated number of AIDS cases for 2005 and their distribution based on race, sex, and exposure group.

Figure 10-2 shows that the percentage of AIDS cases for ethnic-related adult/adolescent groups is in striking contrast to the population percentages of each group. Through year 2005 whites made up 71% of the population and represented 43% of adult/adolescent AIDS cases. Blacks made up 12.6% of the population but represented 38% of the adult/adolescent AIDS cases. Hispanics made up about 13% of the population but represented 18% of the adult/adolescent AIDS cases.

The CDC reported that in 2001, after seven straight years of declining AIDS cases, new AIDS cases increased nationally by about 5%. With a lapse in the practice of prevention, new AIDS cases are expected to increase through 2005.

According to 1993 data analyzed by Philip Rosenberg (1995 updated) of the National Cancer Institute, 1 in every 92 American men between ages 27 and 39 may be HIV-infected. The findings were especially dismal for black-American men, with 1 in 50 estimated to be HIV-infected. The estimate was 1 in 60 for Hispanic men. The statistics were equally high for women of color. One in 130 black-American women and 1 in 200 Hispanic women are estimated to be infected with HIV. By comparison, the number of white women infected with HIV was 1 in 3000. If the trends continue, Rosenberg noted, HIV/AIDS in young people and minorities must be considered "endemic in the United States."

According to estimates, ending 2005, black Americans will make up at least 49% of all new AIDS cases and Latinos 20%. Together blacks and Latinos will represent about 70% of all new AIDS cases but they are only a quarter of the population. The two populations also represent 85% of all pediatric AIDS cases. In 2005, black women will make up about 65% of new AIDS cases reported among females. White women will make up 15% and Latino women 20% of new AIDS cases. There are myriad factors contributing to the spread of HIV among blacks. Information about the threat of AIDS has not been disseminated widely or effectively enough, particularly among those under 21 who feel they are invulnerable. An official with the New York AIDS Coalition tells a story about a 15-year-old girl who said: "Don't tell me nothin' about no AIDS because that won't impact me. And if I was to get it, all I'd have to do is take a pill in the morning and I'll be O.K."

Table 10-1 Total Adult/Adolescent AIDS Cases by Single and Multiple Exposure Categories, Estimated Beginning 2006, United States

Exposure Category	AIDS Cases No.	(%)
Single Mode of Exposure		
1. Men who have sex with men	422,488	(44)
2. Injection-drug use 192,040	(20)	
3. Hemophilia/coagulation disorder	9602	(1)
4. Heterosexual contact	105,622	(11)
5. Receipt of blood transfusion	9602	(1)
6. Receipt of transplant of tissues/organs	20	(0)
7. Other/undetermined	110	(0)
Single Mode of Exposure Subtotal	739,354	(77)
Multiple Modes of Exposure		
1. Men who have sex with men; Injection-drug use	48,000	(5)
2. Men who have sex with men; hemophilia	174	(0)
3. Men who have sex with men; heterosexual contact	19,200	(2)
4. Men who have sex with men; receipt of transfusion/transplant	3700	(0)
5. Injection-drug use; hemophilia	180	(0)
6. Injection-drug use; heterosexual contact	48,000	(5)
7. Injection-drug use; receipt of transfusion	1600	(0)
8. Hemophilia; heterosexual contact	110	(0)
9. Hemophilia; receipt of transfusion/transplant	816	(0)
10. Heterosexual contact; receipt of transfusion/transplant	1605	(0)
11. Men who have sex with men; Injection-drug use; hemophilia	54	(0)
12. Men who have sex with men; Injection-drug use; heterosexual contact	9600	(1)
13. Men who have sex with men; Injection-drug use; receipt of transfusion/transplant	600	(0)
14. Men who have sex with men; hemophilia; heterosexual contact	25	(0)
15. Men who have sex with men; hemophilia; receipt of transfusion/transplant	42	(0)
16. Men who have sex with men; heterosexual contact; receipt of transfusion/transplant	259	(0)
17. Injection-drug use; hemophilia; heterosexual contact	85	(0)
18. Injection-drug use; hemophilia; receipt of transfusion/transplant	42	(0)
19. Injection-drug use; heterosexual contact; receipt of transfusion/transplant	1120	(0)
20. Hemophilia; heterosexual contact; receipt of transfusion/transplant	40	(0)
21. Men who have sex with men; Injection-drug use; hemophilia; heterosexual contact	15	(0)
22. Men who have sex with men; Injection-drug use; hemophilia; receipt of transfusion/transplant	16	(0)
23. Men who have sex with men; Injection-drug use; heterosexual contact; receipt of transfusion/transplant	160	(0)
24. Men who have sex with men; hemophilia; heterosexual contact; receipt of transfusion/transplant	8	(0)
25. Injection-drug use; hemophilia; heterosexual contact; receipt of transfusion/transplant	27	(0)
26. Men who have sex with men; Injection-drug use; hemophilia; heterosexual contact; receipt of transfusion/transplant	8	(0)
Multiple Modes of Exposure Subtotal	135,641	(14)
Risk Not Reported or Identified	86,418	(9)
Total AIDS Cases	960,200	(100)

(Source: For exposure categories. CDC HIV/AIDS Surveillance Report, *through 2003, 11:1–45, updated)*
Pediatric AIDS cases = 9800 (Total AIDS cases about 970,000)

Table 10-2 Adult/Adolescent Behavioral Risk Groups, Race and Sex: Percent of Total AIDS Cases—United States, 2005

HIV/AIDS	No. of Cases[a]	% of Cases
Exposure Group		
Men who have sex with men	19,927	46
Injection-drug user (IDU)	10,830	25
Homosexual/IDU	2166	6
Hemophiliac	87	0.2
Heterosexual contact	4765	12
Transfusion related	173	0.6
None of the above	4765	10
Total		100
Race/Ethnicity (all cases)		
White (non-Hispanic)	13,429	31
Black (non-Hispanic)	20,360	47
Hispanic	8664	20
Other	866	2
Sex (adults only)		
Male	33,356	77
Female	9964	23
Age Group (yrs)		
13–19	216	0.5
20–24	1560	3.6
25–29	6151	14.2
30–39	19,841	45.8
40–49	11,263	26
50–59	3162	7.3
60 and above	1126	2.6

[a]Cumulative estimated total = 43,320 adult/adolescent plus 180 pediatric = 43,500 = 100% of cases for 2005.

(Adapted from AIDS Surveillance Report, December 2003, updated)

Figure 10-3 is a U.S. map of estimated AIDS cases through 2005. Note that the highest incidence of AIDS cases occurs along the coastal regions.

Behavioral Risk Groups and Percentages of HIV-Infected People

Ending 2005, investigators found that HIV infection still remains largely confined to the populations at recognized behavioral risks: homosexual men, injection-drug users, heterosexual partners of injection-drug users, hemophiliacs, and children of HIV-infected mothers. In the general population, rates for HIV infection include 0.04% for first-time blood donors, 0.14% for military applicants, 0.33% for Job Corps entrants, 0.19% to 0.87% for child-bearing women, and 0.30% for hospital patients. Data reported by the CDC in the 1990s indicated that while the number of new AIDS cases increased by 5% in cities, it had increased by 37% in rural areas. This trend continues.

BOX 10.2

THE HIV/AIDS SCENARIO

HIV/AIDS is unstoppable in the short term. Because it takes an HIV infection so long to develop into AIDS, virtually all the AIDS cases that occur during the next 5 to 20 years will be the result of existing infections. Therefore, the epidemic cannot be materially reduced in this time frame by any reduction in new HIV cases. Worldwide, millions of HIV infections will progress into AIDS into the next decade. Will some of your friends and associates *still* regard HIV/AIDS as someone else's problem? HIV/AIDS has invaded *all* segments of society worldwide. **It is *everyone's* problem!**

Comments on a Variety of Individual Behavioral Risk Groups

Keep in mind that because a group of people is at risk for HIV does not mean that these people are predestined to become infected. People are placed within these groups because of their social behavior, a behavior that has been associated with a high, medium, or low risk of becoming HIV-infected. Essentially, there is no zero-risk group because a scenario can always be formulated to show that under certain circumstances one or more members of that group could become HIV-infected.

The point of placing people in behavioral risk groups is not to offend them but to provide a warning that certain behavior might make them more vulnerable to HIV infection. It is not race or ethnic group that places people at high or low risk for infection, it is their behavior.

The fact that AIDS was first identified in 1981 in seemingly well-defined behavioral groups (homosexual men, injection-drug users, hemophiliacs, Haitian immigrants) probably contributed to a false sense of security among people who did not belong to any of these groups. However, as information about HIV and AIDS accumulated, it

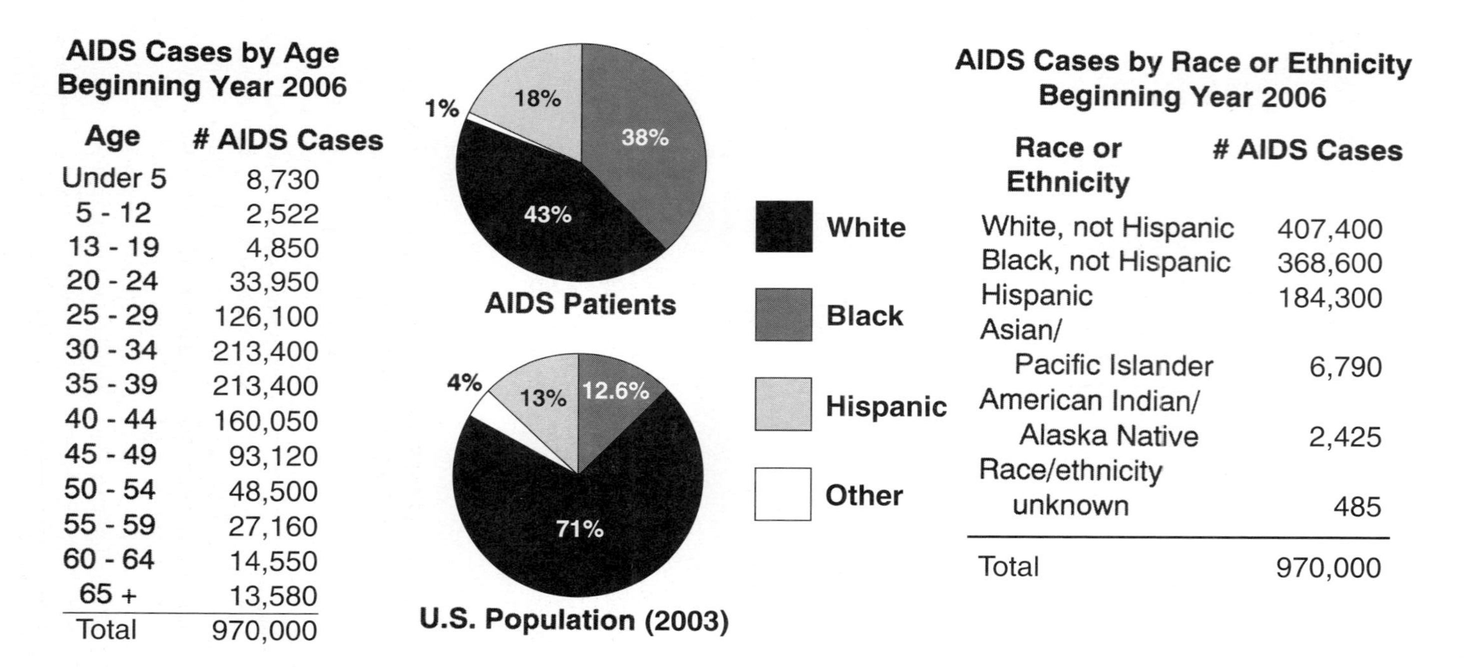

AIDS Cases by Age Beginning Year 2006

Age	# AIDS Cases
Under 5	8,730
5 - 12	2,522
13 - 19	4,850
20 - 24	33,950
25 - 29	126,100
30 - 34	213,400
35 - 39	213,400
40 - 44	160,050
45 - 49	93,120
50 - 54	48,500
55 - 59	27,160
60 - 64	14,550
65 +	13,580
Total	970,000

AIDS Cases by Race or Ethnicity Beginning Year 2006

Race or Ethnicity	# AIDS Cases
White, not Hispanic	407,400
Black, not Hispanic	368,600
Hispanic	184,300
Asian/ Pacific Islander	6,790
American Indian/ Alaska Native	2,425
Race/ethnicity unknown	485
Total	970,000

FIGURE 10-2 Estimated AIDS Cases by Age and Racial and Ethnic Classification. Adult AIDS cases show a disproportionate percentage among blacks and Hispanics. Fifty-seven percent of reported AIDS cases occur among racial and ethnic minorities. The figures reflect higher rates of AIDS in blacks and Hispanic injection-drug users and their sex partners. Percentages of the population are based on the numbers of AIDS cases in the United States estimated at the beginning of 2006. U.S. population is about 300 mil-

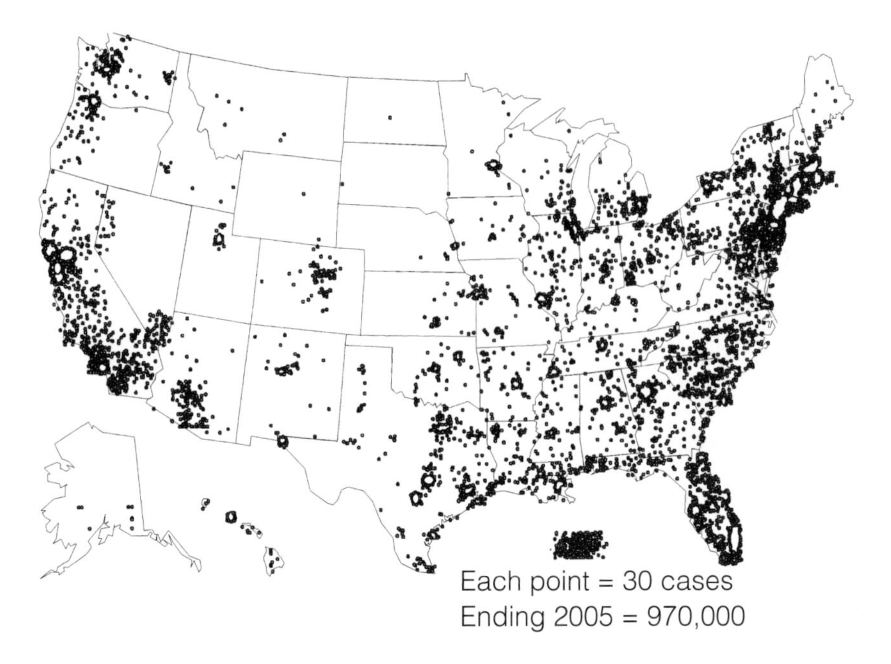

FIGURE 10-3 United States: Estimated Cumulative AIDS Cases and Approximate Location Ending 2005. *(CDC, Atlanta Surveillance Branch)*

became clear that HIV was transmitted in body fluids. This had grave implications for all social groups. On reflection, the people in the original high-risk groups simply had the bad luck of being in the way of a newly emerging infectious agent as it first began to spread. It is highly probable that in the different behavioral risk groups there are lifestyle or medical history factors that increase the efficiency with which the virus is transmitted.

Men Who Have Sex With Men (MSM)—In 1981, 100% of all AIDS cases reported to the CDC occurred in homosexual males. In the 1980s, the infection rate among gay men in San Francisco was over 60%. For the 1990s it dropped to about 40% per year. These findings underscore the need for community planning groups to consider culturally appropriate prevention services when addressing the HIV prevention needs of racial/ethnic

POINT OF INFORMATION 10.1

A REPORT FROM THE 11TH ANNUAL RETROVIRUS CONFERENCE BRINGS SHOCKING NEWS TO NEW YORK CITY

New York City has 3% of the U.S. population but has about 13% of the nation's AIDS cases. Approximately 44% are black and 32% are Latino. The 2004 conference report stated that over 75,000 people are living with AIDS or in some lesser state of HIV infection. In addition, about 25% of the infected do **NOT** know they are infected. This brings the number of HIV infected to about 93,750 and raises the infection rate for the population of New York City to 1.25%. According to the World Health Organization, an infection rate of one percent is a threshold at which government and civil services begin to feel the pinch of the health crisis caused by an epidemic. That puts New York City's epidemic on a par with half a dozen nations in Central and South America, as well as some countries in Africa and Asia.

About 2.8% of all the men in New York City have HIV infection or have been diagnosed with AIDS, about 4% of all men between the ages of 40 to 49 years have HIV infection or AIDS. And about 30% of infections are in women. In addition, about two-thirds of new infections were among heterosexuals, disproportionately among black women.

minorities. Also, there are indicators of a second wave in the HIV/AIDS epidemic among gay males.

Recent studies show that a significant percentage of gay men who had originally adopted safer sex behaviors are relapsing to unsafe sex (Stall et al., 1990; Lemp et al., 1994).

Size of the Homosexual Community—Estimating the size of the homosexual population continues to be a problem for the CDC. Because of the lack of available information on sexual practices, the CDC has relied on the 1948 Kinsey report, *Sexual Behavior in the Human Male,* for its estimate that 2.5 million (10%) American males are exclusively homosexual, while another 2.5 to 7.5 million have occasional homosexual contacts. These numbers are refuted by some recent surveys like the 1992 National Opinion Research Center that reports that among males, 2.8% are exclusively gay, and among women, 2.5% are exclusively lesbian. Judith Reisman argues in her 1990 book, Kinsey, Sex and Fraud, that male homosexuals make up about 1% of the population. The group, Coloradans for Family Values and the Washington-based Family Research Institute believe about 3% of the male population is exclusively gay.

Using data collected from HIV testing conducted in 1986 and 1987, the CDC estimates that 20% of exclusively homosexual men are HIV-infected. This means (depending on whose numbers are used) that about 1 million gay males in the United States have been HIV-infected. For bisexuals and men with infrequent homosexual encounters, the CDC tabulates a prevalence rate of 5%, meaning that about 100,000 of this population have been HIV-infected (Booth, 1988 updated). Approximately 6% of all IDU-associated AIDS cases occur in the homosexual/bisexual male population. These cases may reflect HIV transmission either by contaminated syringes or sexual activity.

POINT TO PONDER 10.1

ON THE DOWN LOW

On the down low refers to men who have sex with men but do not identify themselves as gay or bisexual. They often live secretive double lives, outwardly heterosexual, sometimes even married with children, but also having sexual relations with men. The phenomenon is more common among black men than whites. In 2000, in a study of 8780 HIV-positive men who said they were infected by having sex with a man, the CDC found that one-fourth of African-American respondents identified themselves as heterosexual, compared to only 6% of white men. As an example, at the National Conference of African-Americans and AIDS in February 2000, a Columbus, Ohio, publishing executive told several hundred health-care professionals, "I sleep with men, but I am not bisexual, and I am certainly not gay. I assure you that none of the brothers on the down low like me are paying the least bit of attention to anything you have to say."

AIDS experts are worried that regardless of race, men who call themselves heterosexual but are involved in secret sexual relations with men are fueling the rising incidence of HIV infection among women. A study published in the *Journal of the American Medical Association* called young bisexual men a bridge for HIV transmission to women. In that study, one in six men who had sex with men recently had sex with women. Nearly 25% of those men had recently had unprotected sex with both men and women.

Demographic Changes of HIV-Infected Men in the Gay Community

Data compiled by the CDC and reported in June 2001 reveals a striking change in the disease's demographics in America (HIV/AIDS Surveillance Report, June 2001). Year-end 2002 data shows that 74% of AIDS cases in white Americans, 37% in black Americans, and 42% in Latino Americans resulted from homosexual contacts. At the beginning of 2006, of the estimated 450,000 people living with AIDS in America, 36% were white, 42% were black and 20% were Hispanic. Entering 2006, AIDS continues to kill more black Americans under age 55 than heart disease, cancer, or homicide!

The national level among gay men is 1%. About a third of the gay men surveyed knew they were HIV-infected. Ending year 2005, gay males accounted for about 52% of 970,000 adult/adolescent AIDS cases in America and 65% of them have died. An estimated 177,000 gay men are living with AIDS and about 300,000 are progressing to AIDS (Wolitski et al., 2001; Catania et al., 2001 updated). Overall, of the

24 million AIDS deaths ending year 2005, less than 1 million will be gay men. These data overwhelmingly make this global pandemic a heterosexual pandemic.

Injection-Drug Users—According to the CDC, as many as 33% of the nation's 1.2 million injection-drug users may be HIV-infected. This behavioral risk group contains the nation's second largest group of HIV-infected and AIDS patients. An association between injection-drug use and AIDS was recognized in 1981, about two years before the virus was identified. AIDS in IDUs and hemophiliacs offered the first evidence that whatever caused AIDS was being carried in and transmitted by human blood. From the reported IDU AIDS cases in 1981 through 2005, 28% of all adult/adolescent AIDS cases were associated with IDUs. Of IDUs, 74% listed IDU as their only risk factor for HIV infection; 26% were also homosexual/bisexual. It is estimated that through 2005 40% of infected women and 22% of infected men became HIV-infected through IDUs.

IDU AIDS cases have been reported in all 50 states and the District of Columbia. Among the adult/adolescent heterosexual AIDS cases over half had sexual partners who are/were IDUs.

About 55% of all IDU-associated cases were reported in the Northeast, which represents about 20% of the population of the United States and its territories. The South reported 20% of IDU-associated AIDS cases, 5% from the Midwest, and the West reported the remaining 20%.

The rate of IDU-associated AIDS continues to be higher for blacks and Hispanics than for whites. Except for the West, where rates for whites and Hispanics were similar, this difference by race/ethnicity was observed in all regions of the country and was greatest in the Northeast. By 2005, overall IDU-associated male AIDS cases represented 9% of all AIDS cases in whites, 33% in blacks, 35% in Hispanics, 5% in Asians/Pacific Islanders, and 16% in American Indians/Alaskan Natives. For women it was: white, 42%; black, 40%; Hispanic, 39%; Asian/Pacific Islanders, 15%; and American Indian/Alaskan Natives, 45%.

Heterosexuals—The spread of HIV in the general population is relatively slow, yet potentially it is the source of the greatest numbers of HIV/AIDS cases. The CDC estimates there are about 150 million Americans without an identified at-risk behavior, the general population.

Data from the CDC for years 2004 and 2005 indicate that about 15% of all the AIDS cases and 35% of all new HIV infections in the United States occurred through heterosexual contact. Most of the heterosexual AIDS cases occurred in persons or the sexual partners of individuals with an identified behavioral risk. Relative to the general adult population, the number of heterosexual AIDS cases is only a fraction of 1%.

Global HIV/AIDS Cases and Heterosexuality

Worldwide beginning year 2006, there will be an estimated 11 million people living with AIDS and about 24 million will have died of AIDS. According to a June 1999 Worldwatch Institute report, South Africa's HIV pandemic is perhaps the worst on the globe. It has engulfed the country, and "barring a medical miracle, one of every five adults will die of AIDS over the next 10 years." This unprecedented social tragedy is also translating into an economic disaster. The working-age population is being lost to AIDS. In Zimbabwe, state morgues are extending their hours to cope with the soaring death rate, mostly as a result of AIDS. An estimated 3000 people now die every week in the southern African country, nearly 70% of them from AIDS-related illnesses. The main hospital in Harare has opened its morgue around the clock and other hospital and mortuary facilities have extended closing time by four hours. At the University of Durban-Westville in KwaZulu-Natal, 255 students recently tested HIV positive. In 1997, 14% of the population was HIV positive—in 2001 it was about 40%.

Worldwide, ending year 2005, **heterosexuals** will make up over 80% of the estimated 41 million living HIV-infected people. About 70% of these infected people live in sub-Saharan Africa. North America, Latin America, South and Southeast Asia, and Africa account for 93% of global HIV infections (Figure 10-4).

Of the 41 million living HIV-infected people worldwide, over 90% live in nonindustrial nations. It is estimated that through year 2005, a total of 35 million people will have/had AIDS worldwide, 77% of those cases will have occurred in Africa,

SIDEBAR 10.1

THE LATINO COMMUNITY

The **third annual National Latino AIDS Awareness Day,** an observance sponsored by the Latino Commission on AIDS, will be held on October 15, 2005, the last day of Hispanic Heritage Month, in cities throughout the United States. Latino leaders will sponsor activities that respond to the state of AIDS among Latinos in their communities. In recognition of the surging number of new infections among Latinas and young Latinos, organizers will use the day to promote and sponsor prevention activities, alert religious leaders and public officials to the need to reduce new infections, and care for Latinos infected with the virus.

"In 1983, my brother died of AIDS. It infected his wife, and a year later she died. And now we had the first set of orphans in our family. Despite everything, I saw the love and unity of my family—how we pulled together and took care of my brother and loved him. My sister left another set of orphans. My uncle, who is my godfather, lost children to HIV. His son died in a prison hospital of AIDS. His daughter died at the age of 33 and left another orphan child. My uncle wasn't around to see his daughter die because he died of suffering from watching. Last year we found out my 70-year-old uncle is infected with HIV. And that's just my family. I watch the news. I watch people talk about wars in other countries and I identify with the feelings of those people. I know they have bombs thrown at them. I know they have weapons pointed at them. But we have a weapon that is killing my community. We have to silence this weapon of AIDS, which is killing us."

This testimony was given by Marina Alvarez to 60 Latino community leaders from across the United States and Puerto Rico who convened at Harvard University in the spring of 1998. The rate of infection among Latinos in the United States reflects an increasingly dire situation. Currently, 19% of all AIDS cases in the United States occur among Latinos, even though Latinos represent about 14 % of the total population. For each of the years 1998 through 2005, Latinos represented an estimated 19% of new AIDS cases in America. About 72,000 Latinos are currently living with AIDS and between 140,000 to 190,000 are HIV-infected. The rate of HIV infection among Latino men is three times greater than the infection rate found among white, non-Hispanic men. Similarly, wo-men and children in the Latino community have rates of infection that are seven times greater than the rates found among white women and children. If these trends continue, the Harvard AIDS Institute projects that beginning 2006 the percentage of total annual AIDS attributable to the Latino population will eclipse the percentage of such cases attributable to the white, non-Hispanic population.

AIDS is the fourth leading cause of death for Latinos ages 25 to 34, and third among those ages 35–44. There are about 37 million Latino/Hispanic people in America. The rate of new AIDS cases among Latinos is about four times the rate among white Americans but about three times lower than the rate for black Americans.

Snapshot—New York Latino Population. New York has 12% of America's Latino population but 32% (56,000) of all Latino AIDS cases. Over 55% of HIV infection occurred through IDU. Forty percent of Latino women have become infected through sex with a male IDU. Latinos make up 33% of New York's inmate population and 50% of prison AIDS cases.

The Future Threat for Latinos—The AIDS epidemic threatens not only the present generation of Latinos but also future generations. At the end of the year 2000, over one-third of the members of the Latino community were under the age of 18. With rising rates of HIV infection among all youth, the implications for the Latino community are ominous. Rafael Campo, a physician at Beth Deaconess Medical Center said, "As overwhelming as this epidemic might seem to us right now, even greater devastation faces our next generation of irreplaceable young people whose future should be to lead this country into the next millennium; to take our places in medicine, business, in government; and to share with the world the tremendous brilliance and creativity of our culture" (Kao, 1999).

14% in Asia, 2.5% in North America, and 2% in Europe (UNAIDS, 1998 updated).

UNAIDS updated estimates are that, on average 4 million new HIV infections occurred annually from 1997 through 2005. The AIDS death toll for these nine years is estimated at 18 million of which 7 million were women and 4.5 million were under age 15 (Table 10-3). The United

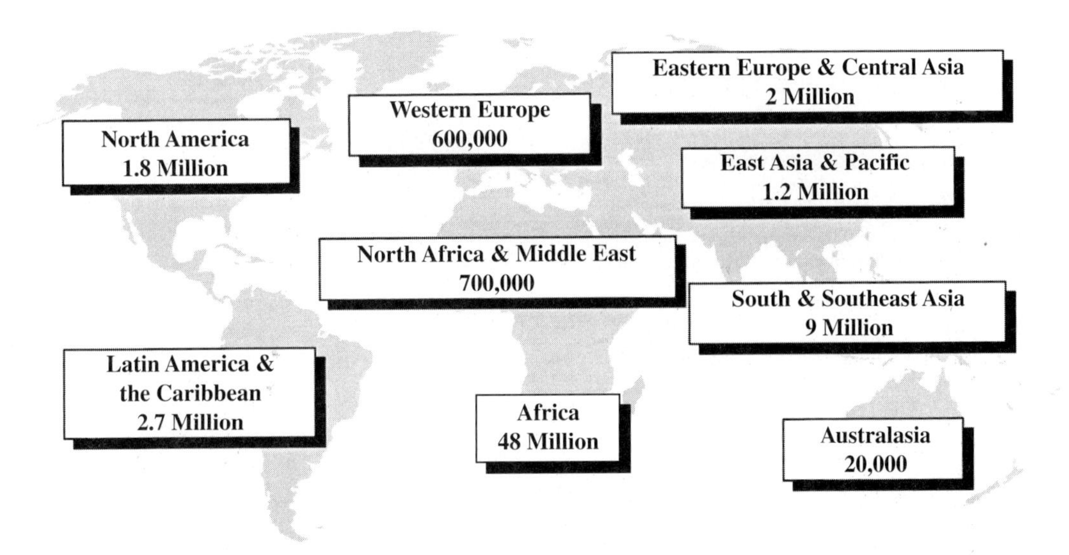

Estimated Global Total: 65 Million Ending 2005

FIGURE 10-4 Estimated Number of Global HIV Infections Projected by United Nations AIDS Program and World Health Organization ending year 2005. Of the estimated 75 million HIV-infected persons, about 41 million **are living** in some state of HIV/AIDS illness. Over 4.5 million of these are children. Of 209 countries reporting to the WHO and UNAIDS, 194 have reported AIDS cases. World population reached 6 billion in October 1999. *(Source: Pan American Health Organization Quarterly Report, September 1996 updated; UNAIDS Report on the Global HIV/AIDS Epidemic, December 1999 updated)*

Table 10-3 Leading Cause of Death Worldwide from 1998 estimated through 2005

Infectious Diseases	All Diseases
1. **HIV/AIDS**	1. Heart diseases
2. Diarrheal disease	2. Cerebrovascular diseases
3. Childhood diseases	3. Lower respiratory diseases
4. Tuberculosis	4. **HIV/AIDS**
5. Malaria	5. Obstructive pulmonary diseases
6. STDs excluding HIV/AIDS	6. Diarrheal diseases
7. Meningitis	7. Perinatal conditions
8. Tropical diseases	8. Tuberculosis

(*Data Courtesy of the World Health Organization, 1998 updated.*)

States will have about 1 million AIDS cases ending 2006—worldwide there will be over 34 times that number (Figure 10-5).

Worldwide, it is estimated that 8 in every 100 sexually active people aged 15 to 49 is HIV-infected. Overall, ending year 2006, AIDS will have orphaned over 15 million children.

Ending 2006, there will be an estimated 43 million people living with HIV infection. There are over a dozen countries whose population is equal or close to 43 million people!

Hemophiliacs—There are about 16,000 to 20,000 hemophiliacs in the United States. At least

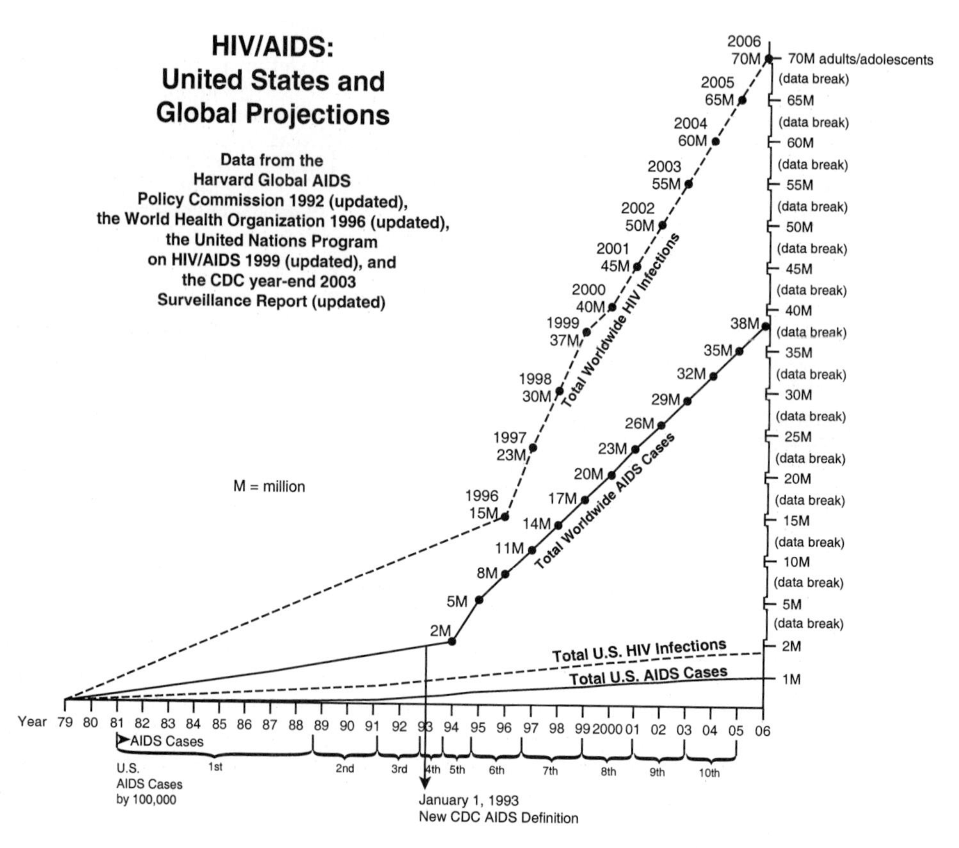

FIGURE 10-5 Estimated U.S. and Global Projections for Total Number of HIV and AIDS Cases. Ending year 2006, over 1 million U.S. AIDS cases will be reported to the CDC. Worldwide, the total number of AIDS cases is expected to reach about 42 million ending 2006. Through the year 2005, about 2% of all HIV infections and 2.5% of total AIDS cases will have occurred in the United States. William Hazeltine, AIDS researcher formerly at the Dana Farber Institute, and now with Human Genome Sciences, Inc., said that if prevention does not work or a cure is not found, some *1 billion* people could be HIV positive by the year 2025. Global population through 2006 is projected to be about 6.8 billion people. *(Source: Harvard Global AIDS Policy Commission 1992 and World Health Organization 1993, 1994, National Census Bureau 1998, and projected data)*

half are HIV positive and through 2003, about 3000 of them have died. Of those infected, over 98% received HIV in blood products that were essential to their survival. By mid-1985, an HIV blood screening test was put into effect nationally. From that point, the risk of HIV infection from the blood supply has been significantly lowered, but a small risk still exists. **Scandals** of knowingly selling HIV-positive blood for transfusions and in production of the blood factor essential for hemophiliacs, surfaced in France, Germany, the United States, Canada, and Japan in the 1990s. (For discussion, see Chapter 8.)

American Military—The incidence of HIV infection can be measured best in groups that undergo routine serial testing. Because active duty military personnel and civilian applicants for the service are routinely tested for HIV antibody, there is a unique opportunity to measure the incidence of HIV infection in a large, demographically varied subset of the general population.

The military began annual HIV testing of all service personnel in October 1985. Beginning 2004 the military began HIV testing of all personnel every two years. The current infection rate among all military personnel is about 0.02% or two per 10,000, much lower than for the general population.

College Students—Between April 1988 and February 1989, the blood of 16,863 students enrolled at 16 large universities and 3 private colleges was tested for HIV antibodies. Thirty students, 28 males and 2 females, tested HIV positive for an overall rate of 2 per 1000 or 0.2%. With 12.5 million students enrolled in American colleges and universities, the rate of 2 in 1000 means there are about 25,000 HIV-infected college students.

The study was conducted by the CDC with the American College Health Association (Gayle et al., 1988). Because the samples were not identified, students who tested positive could not be informed. All blood samples from 10 of the 19 campuses were HIV negative. At five campuses the rate of HIV-positive blood ranged from 4 to 9 samples per 1000. In contrast, the rate of HIV in the general heterosexual population is about 0.02% and the rate for military personnel in about the same age group tested over a similar time period was 0.14%.

The question is why college students have a rate of HIV infection 10 times higher than the general heterosexual population. Are college students less informed than the general heterosexual population?

Surveys indicate that college students are well educated about HIV/AIDS. Why, then, the higher rate of infection? Perhaps it's the age-old dilemma of information versus behavior. **They know what to do but they don't do it.** College students have always had information on drug use, alcoholism, pregnancies out of wedlock, and sexually transmitted diseases; but this knowledge has not appreciably reduced at-risk behaviors. STDs are at an all-time high in teenage and college students.

Prisoners—Ending 2005, the nation's prison population is at an all-time high, over 2 million adults contained in 1300 state and 71 federal prisons. Ninety percent are men.

According to the U.S. Bureau of Justice Statistics, 2.1% of these people or 42,000 are HIV positive and of these, 17% have AIDS (Rubel et al., 1997 updated). The total number of **prisoners, parolees,** and **probationers** in 2004 was about 6.5 million with an HIV-positive rate of about 0.8% or about 50,000 being HIV positive (DeGroot et al., 1996 updated). By 2005, over 6000 adult inmates in U.S. state and federal prisons and jails had died of AIDS. About 6000 people with AIDS were still in prisons and jails (*MMWR*, 1996c). In 2004, four states—New York, Florida, Texas, and California—had more than half the known cases of HIV in prison. Only two states and five prison systems in the United States distribute or sell condoms to male inmates: Mississippi and Vermont, New York City, Philadelphia, San Francisco, Los Angeles, and the District of Columbia (Hooker, 1996 updated) and only two of the six distribute dental dams and condoms to female inmates. **No U.S. correctional facility distributes bleach or syringes to inmates** (Mahon, 1996 updated).

Blood testing of inmates at 10 selected prisons indicated that 1 in every 24 prisoners is infected with HIV. This study also found that incarcerated women under age 25 had a higher HIV seroprevalence (5.2%) than incarcerated young men (2.3%). The rate was comparable among older women (5.3%) and older men (5.6%). Nonwhites were almost twice as likely to be infected as whites. Rates varied widely among the 10 institutions from 2.1% to 7.6% for men, and 2.7% to 14.7% for women (Vlahov et al., 1991). By 1999, 17 states tested all new inmates for HIV infection. Most (about 75%) prison systems will provide an HIV test on request. The House of Representatives passed legislation requiring all federal inmates serving at least six months to be HIV-tested within four months of entering the prison system.

In general, prisoners diagnosed with AIDS were infected prior to their incarceration—most are IDUs (Francis, 1987). Infected inmates most often transmit the virus to others through homosexual and drug-related activities. For that reason 21 states segregated HIV-infected and AIDS patient prisoners from the other prisoners. In September 1989, a court order in the state of Connecticut ended their prison segregation policy. Beginning 2004 only one Alabama womens prison continues to segregate its prisoners.

Large numbers of at-risk individuals cycle through prisons and jails every year. The turnover rate for some prisons is as high as 50% per year. On release from prison, the infected can easily spread HIV to others. For example, in 2004, the New York Commission of Corrections reported that about 67,000 inmates are spread out across 70 prisons in New York. About 9% are HIV-infected. About 29,000 or 43% of the inmates are released to the community each year. On average this means that about 3000 HIV-infected are released each year. Because of confidentiality laws no one will know who they are or where they go. In Florida, 4% of its prison population of 74,000 is HIV positive.

In April 2002, in order to stop the virus's spread to the inmates' partners outside prison, Florida passed legislation that makes HIV testing mandatory for inmates leaving Florida prisons. The purpose of the bill is to educate the inmates about the virus and their health status and to protect the public at large. Florida released about 78,000 inmates in 2003 through 2005. Now they will know which of the inmates are HIV positive and guide them into outside care facilities.

The Elderly—The face of AIDS has changed dramatically from the 1980s to the 1990s. AIDS cases among men, in particular gay men, and IDUs predominated the 1980s. But in the 1990s the numbers of AIDS cases among the elderly, women, their children, and people of color increased significantly. **The senior citizens of today did not grow up in the age of AIDS—they did not have sex and HIV in the same thought.**

Nationally about 15% of the total AIDS cases occur in people over age 50. By 2006, this represented about 146,000 people. Three percent of AIDS occurs in people over age 60 and 0.7% occurs in people over 70. One AIDS patient was 90 years old when diagnosed. He became HIV-infected from a blood transfusion during surgery at age 85. He became symptomatic after four years and died one year later of *Pneumocystis carinii* pneumonia.

In Florida, a state with a large retired population, the number of recorded AIDS cases among those over age 50 has risen from 6 in 1984 to 1341 in 1993 to about 15,000 by 2006, or about 14% of the state's cases. In Dade County, Fl., a popular destination among retirees, almost 20% of people with AIDS are seniors—one of the highest rates in the country for the 50-plus group. Forty-four percent of senior women in Florida with AIDS, and 18% of Florida men over 50 with AIDS are known to have become infected through heterosexual sex. These numbers are predicted to grow as the more sexually liberal baby boom generation ages.

Many seniors feel that HIV infection is just a problem for young people, homosexuals, and injection-drug addicts. Now, the use of Viagra compounds the problem. With the use of Viagra, seniors are much more sexually active and few practice safer sex. The manufacturer of Viagra is trying to incorporate safer sex messages in their advertisements.

Needle Stick Injuries—Health-care workers are in a quandary about the possibility of becoming HIV-infected via needle sticks. Articles such as "Needlestick Risks Higher Than Reports Indicate" or "The Risk of HIV Transmission via Needlesticks Is Low" convey conflicting impressions.

The kinds of needle sticks most likely to occur in hospital or health-care settings come from disposable syringes, IV line/needle assemblies, prefilled cartridge injection syringes, winged steel needle IV sets, vacuum tube phlebotomy assemblies, and IV catheters, in that order. Needle sticks and penetration of sharp objects account for about 80% of all health-care workers' exposures to blood and blood products. There are about 800,000 needle stick injuries from contaminated devices in health care settings each year. Of these, 16,000 devices are contaminated with HIV (Miller et al., 1997).

Ruthanne Marcus and colleagues (1988) reported that, across the board, health-care workers exposed to HIV-contaminated blood have about a 1 in 300 chance of becoming infected. Other more recent reports place the risk of HIV infection at 1 in 250.

ESTIMATES OF HIV INFECTION AND FUTURE AIDS CASES

In 1987, Otis Bowen, then secretary of health and human services, said "AIDS would make Black Death pale by comparison."

As long as the number of newly infected people each year exceeds the number who die, the pandemic will continue to build.

Who Reports AIDS Cases and to Whom? United States

AIDS cases are reported to the CDC through the SOUNDEX system, which involves translating names into specific sets of numbers and letters. While the resulting codes are not unique, when they are combined with other information, such as birthdates, individual cases can be followed without revealing names.

AIDS Reporting Systems

Reporting AIDS cases reveals past infections. When an AIDS case is reported, it is like looking at a 10- to 12-year-old photograph of the infection date (average time, without drug therapy, from infection to AIDS diagnosis is 10 to 12 years). AIDS became reportable in all 50 states, the District of Columbia, and U.S. territories to the CDC in Atlanta in 1986.

By the end of 1993, all 50 states, the District of Columbia, and four territories (Guam, Pacific Islands, Puerto Rico, and the Virgin Islands) reported adult/adolescent cases. The CDC also reported the numbers of adult/adolescent/pediatric AIDS cases per 100,000 population by state. For the 10 leading metropolitan areas of at least 500,000 population for AIDS beginning year 2006, see Table 10-4. Beginning year 2006, the 10 states and territories reporting the highest incidence of AIDS cases for adult/adolescents can be seen in Table 10-5.

The seven U.S. cities with the highest incidence of new AIDS cases for year 2005 were, highest to lowest: New York, Ft. Lauderdale, San Francisco, West Palm Beach, Jersey City, Newark, and Columbia, S. C. The American HIV/AIDS epidemic is now disproportionately affecting the South. Only 36% of the U. S. population lives in the South but the region is home to 40% of all people living with AIDS and 46% of newly identified cases. And, while the incidence of AIDS has been increasing in the South in recent years, in other regions of the country it has either remained constant or decreased.

With the widespread use of antiretroviral drugs beginning in 1996 there was a decline in AIDS cases but not in new HIV infections. Thus, it became clear that the total numbers of AIDS cases no longer accurately represent the pandemic. To track this pandemic now, it will be essential to have some form of reporting new HIV infections.

Table 10-4 Ten Metropolitan Areas Reporting Highest Number of AIDS Cases Through Year 2005

Metropolitan Area	Number of AIDS Cases
New York City	150,120
Los Angeles	54,210
San Francisco	41,700
Miami	29,190
Washington, D.C.	27,939
Chicago	26,271
Houston	23,977
Philadelphia	23,352
Newark	21,684
Atlanta	20,016
	417,000 or 43% of all AIDS cases in America.

Table 10-5 Ten States/Territories Reporting Highest Number of AIDS Cases Through Year 2005

State/Territory	Number of AIDS Cases
New York	177,292
California	153,560
Florida	99,116
Texas	67,008
New Jersey	53,746
Puerto Rico	32,806
Illinois	30,014
Pennsylvania	29,874
Georgia	27,920
Maryland	26,175
	698,000 or 72% of all AIDS cases in America.

Reporting HIV-Positive Cases

Reporting HIV cases reveals current infections. Beginning January 1, 2004, the last state, Georgia, began reporting newly identified HIV infections (Table 10-6). It is expected that HIV reporting will provide a more accurate view of recent transmission trends. The information will also help direct money to the most effective programs and could affect allocations for HIV patient care. The decision of states to report HIV cases was most influenced by the development of successful treatment, which allows many people with HIV to live healthier longer. But that progress has made it hard for statisticians to calculate backward to estimate time of infection. AIDS case reporting now has become much more of an indicator of who is getting treated, who is not getting tested early, and how effective their therapy is. Harold Jaffe, director of CDC's National Center for HIV, STD, and TB Prevention, said that compared with AIDS reporting, HIV case reporting puts a younger face on the epidemic. He added that initial reports demonstrate that there are higher proportions of women and minorities represented among HIV cases, which gives an indication of where infections are occurring. Focusing on HIV cases will

Table 10-6 Status of HIV Infection Reporting—United States, 2004

Confidential HIV Reporting Required		
Name (Adults/Adolescents)		Name to Code/Coded[1]
Alabama	New Jersey	California
Alaska	New York	Connecticut
Arizona	New Mexico	Delaware
Colorado	North Carolina	District of Columbia
Florida	North Dakota	Hawaii
Georgia	Ohio	Illinois
Guam	Oklahoma	Maine
Idaho	Pennsylvania	Maryland
Indiana	Puerto Rico	Massachusetts
Iowa	South Carolina	Montana
Kansas	South Dakota	New Hampshire[2]
Kentucky	Tennessee	Oregon
Louisiana	Texas	Rhode Island
Michigan	Utah	Vermont
Minnesota	Virgin Islands	
Mississippi	Virginia	
Missouri	West Virginia	
Nebraska	Wisconsin	
Nevada	Wyoming	

[1]As of January 1, 2004 all states must report the newly infected by name, name to code, or code.

[2]Other

also help health officials learn how well the prevention measures are working.

NEWLY INFECTED

For the 45,000 new HIV infections occurring in 1999 through 2005 annually, the CDC estimated, on average, that 67% of the infected were men. Of these men, 42% were infected via homosexual sex, 25% through IDU and 33% through heterosexual sex. Of these men, 54% are black, 26% are white, and 20% are Hispanics. Black American men make up 75% of new HIV infections among heterosexual cases. A small percentage are members of other racial/ethnic groups. Of the 15,000 plus new infections among women in the United States for 2001 through 2005, the CDC estimated that approximately 75% of women are infected through heterosexual sex and 25% through injection-drug use. Of newly infected women, approximately 64% are black, 16% are white, 20% are Hispanics, and a small percentage are members of other racial/ethnic groups. The CDC estimates that over the next four years, over 100,000 people under the age of 25 will become HIV-infected if current trends continue. The current increase in HIV infections is a reversal of prevention program successes. The overtly sick, the emaciated and those with visible Kaposi's sarcoma are rarely seen on the streets; antiretroviral drugs have the dying going back to work. The fear of infection and death has subsided. According to the CDC, new AIDS cases have increased between 1% and 5% annually nationwide in 2001 through 2005.

How Many People in the United States Have Been HIV-Infected?

How many people in the United States are HIV-infected? And just how many cases of AIDS are expected to occur and when? Although projections have been made, the numbers are in question. Why? They come from surveys and incidence data that lack rigorous scientific documentation. Researchers now believe that the current number of HIV-infected people in the United States is about 0.6% or about 1.5 million people.

ISSUES OF CREDIBILITY IN U.S. ESTIMATES

In 1992, a Harvard research group of 40 HIV/AIDS experts estimated that by 2000 there would be between 38 and 110 million HIV-infected adult/adolescents and 10 million children worldwide. There would be 24 million adult and several million pediatric AIDS cases. The estimates of HIV infections by the Harvard group is over twice that of the World Health Organization. In retrospect they were more correct. It is clear that the HIV/AIDS pandemic of the 1990s, in terms of new AIDS cases, was worse than for the 1980s.

According to the CDC projections in Figure 10-6, and adding in the 15% to 20% underreported AIDS cases, the estimates of the CDC fit relatively well. For 1993, before the implementation of the new AIDS definition, 65,211 cases were projected. Due to the 1993 change in the definition of AIDS, a total of 106,618 adult/adolescent cases were reported, a 37% increase over the expected and a 111% increase over the number of cases reported in 1992. After the backlog of AIDS cases based on the new definition were reported, new AIDS cases per year decreased as expected.

Rise in HIV/AIDS Cases Among Heterosexuals

In 1989, while the number of new AIDS cases rose by 11% among gay males, it increased by 36% or more among heterosexuals and newborns. In 1993, due to the new AIDS definition, heterosexual contact AIDS cases increased 130% over 1992, from 4045 to 9288.

The groups most affected by the expanded 1993 definition were women, blacks, heterosexual injection-drug users, and hemophiliacs. The increase was greater among women (151%) than among men (105%), and greater among blacks and Hispanics than whites. Young adults ages 13 to 29 accounted for 27% of the heterosexual contact cases. On average, women accounted for at least 35% of the heterosexual contact cases from 1993 through 2005.

Will There Be a Heterosexual AIDS Epidemic in the United States?

A heterosexual epidemic on a scale similar to any developing nation is very unlikely now or in the future. Heterosexual AIDS epidemics in the developing nations result largely from conditions that do not exist in America or in other industrialized countries. These include large-scale population shifts, little information about HIV/AIDS prevention, many migrant workers, widespread prostitution, deep reluctance to use condoms, and frequent, untreated sexually transmitted diseases. Still, according to the CDC, heterosexual transmission will slowly increase in the United States largely due to low-income black and Hispanic women. Among the things that put them at increased risk: exchanging sex for crack cocaine and having sex with injection-drug users and bisexual men. Also, HIV infection is a sexually transmitted disease and most people in America and other nations are heterosexual.

SHAPE OF THE HIV PANDEMIC: UNITED STATES

There is a common misconception that the AIDS epidemic is under control in the developed world. While the mortality rate associated with HIV has been sharply reduced, thanks to behavioral changes and antiretroviral drugs, they are imperfect solutions. Drug therapies are expensive, often toxic, and not a cure. AIDS education programs have impeded but have not stopped the epidemic. The United States has the highest rate of HIV infection among the world's most highly industrialized countries, in spite of the fact that it serves as a leader in AIDS education and prevention. About 1 in 200 people in the United States has been HIV-infected. The ratio in some sub-Saharan countries is 1 in 5 and in some African villages the ratio is 1 in 2 in women of childbearing age.

Reviewing Figure 10-7 one can extrapolate backward and see that there were about 19,000 people HIV-infected in 1977; and ending 1982, 190,000. Prior to 1981, 32 had died of AIDS.

Based on the HIV presentation of HIV infections in Figure 10-7, HIV infections reached their peak in 1982 and then rapidly declined. No one can say exactly which factors have been responsible for bringing the annual new HIV infection rate down from some 190,000 people in 1982 to 40,000 a year

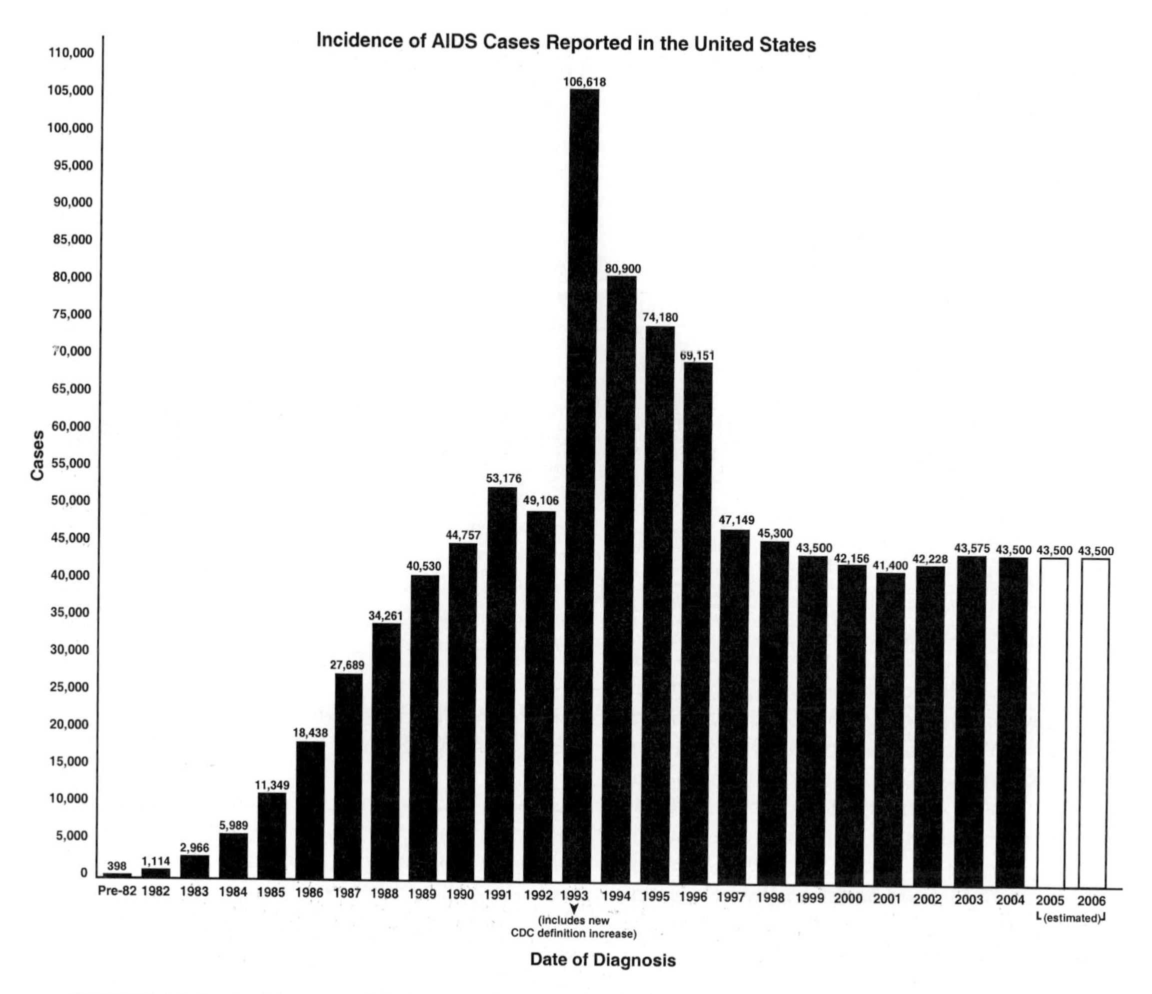

FIGURE 10-6 Incidence and Estimates of AIDS in Persons 13 and Older by Year of Diagnosis—United States, pre-1982 through 2006. The total for 1995 reflects a 9% reduction in AIDS cases from 1994 due to a drop in the backlog of persons to be identified as AIDS cases according to the 1993 definition of AIDS. AIDS cases reported in 2002 increased for the first time in 10 years, about a 2.2% rise over 2001. Data for years 2004 through 2006 reflect expected increases in AIDS cases resulting from therapy failure. Throughout the American pandemic, about 85% of persons with AIDS were/are ages 20–49.

each year from 1988 to 1998. From 1999 through 2005 there was an increase of about 5,000 new infections per year. For a breakdown of new HIV infections, see Figure 10-8. These figures are estimates with new cases of HIV infection now reportable by federal law, these data may change over the coming years.

The populations that are encompassed in the numbers of new infections are different from the past: HIV is now reaching younger people, it's reaching more women, it's reaching more communities of color. Forty-two percent of new infections are occurring in gay men, 35% in heterosexuals infected during sex, and 23% in injecting-drug users. Infections in heterosexual women are increasing more rapidly than in any other group. Unless a preventive vaccine is found, the rate of new HIV infections in the United States is expected to remain at about 45,000 annually. This may mean the new prevention campaigns, without a vaccine, will not sig-

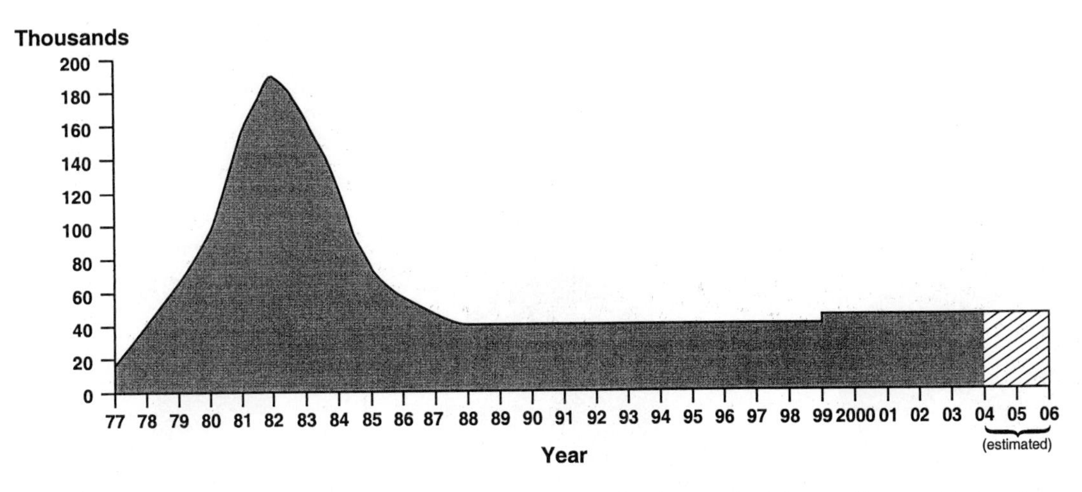

FIGURE 10-7 Estimated Rate of HIV Infection 1977 through 2006; United States. *(Source: Centers for Disease Control and Prevention)*

nificantly reduce new infections. Society has, in a sense, reduced new HIV infections to its lower limit.

DISCUSSION QUESTION: Is the last statement plausible—have we reached a point at which further expenditures will not significantly reduce the rate of infection—that all society can do now is wait on a preventive vaccine? What would be your solution given that we have actually reached this point and no vaccine is forthcoming?

SHAPE OF AIDS PANDEMIC: UNITED STATES

There has been a decline in new AIDS cases in every region of the United States. The decline began in 1996 and ended in 2001.

Although the noticeable drop in new AIDS cases was a renewed cause for hope, the downside is that the slowing progression to AIDS and to AIDS deaths (discussed next) means more HIV-infected people with better health are available to spread HIV. According to the CDC, beginning in 2005 there were about 427,000 adult/adolescent people and 4200 children reported to be living with AIDS in the United States. The estimated number of people currently living with HIV infection is over 1 million. The slowing of AIDS diagnosis makes tracking the epidemic harder. The ability to monitor the epidemic based on HIV infections does not, at the moment, compare with the CDC's ability to track the pandemic through the reporting of AIDS cases, but it will in the next year or so.

Changing U. S. HIV/AIDS Demographics Leads to Southern Discomfort

In January 2003 HIV/AIDS directors from 13 southern states and the District of Columbia produced a "Southern States Manifesto." The major complaint within the manifesto is that the South now leads the country in new HIV infections and overall AIDS cases but receives fewer federal dollars when compared to other HIV/AIDS regions. For example, over 130,000 people in the South have AIDS, compared to about 100,000 in the Northeast, 36,000 in the Midwest, and some 62,000 in the West. Still, the region is behind other areas of the nation in federal funding for HIV/AIDS programs. To illustrate the problem, Florida, with 12% of all AIDS cases, receives about 5% of federal prevention funds. In short, federal funds for HIV/AIDS are being misallocated. The South, with 38% of the population, has 40% of all people living with HIV/AIDS and

New HIV Infections

There are an estimated 950,000 people currently living with HIV in the United States, with about 45,000 new HIV infections occurring in the United States annually since 1999.

By gender, 70% of new HIV infections each year occur among men, although women are also significantly affected.

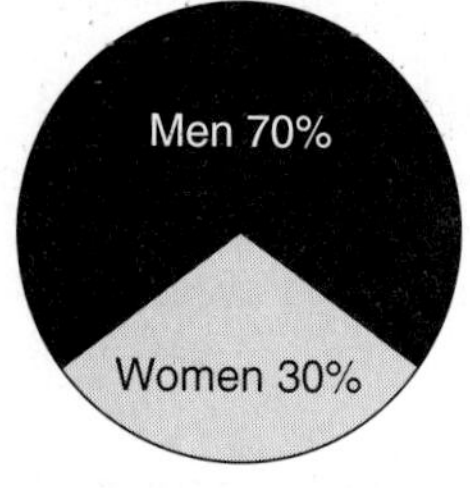

Estimates of annual new infections by gender (N ≅ 45,000)

By risk, men who have sex with men (MSM) represent the largest proportion of new infections, followed by men and women infected through heterosexual sex and injection-drug use.

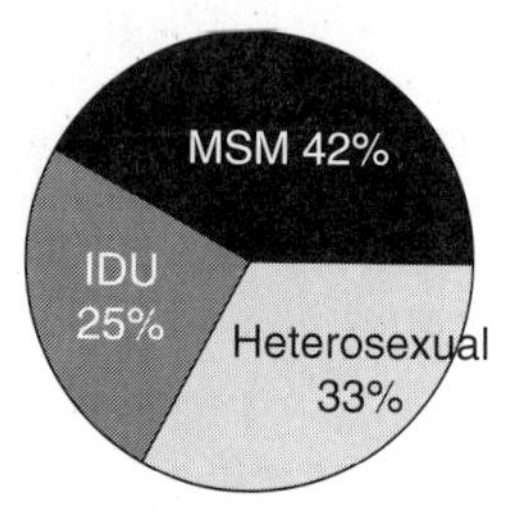

Estimates of annual new infections by risk (N ≅ 45,000)

By race, more than half of the new HIV infections occur among blacks, though they represent 13% of the U.S. population. Hispanics, who make up about 12% of the U.S. population, are also disproportionately affected.

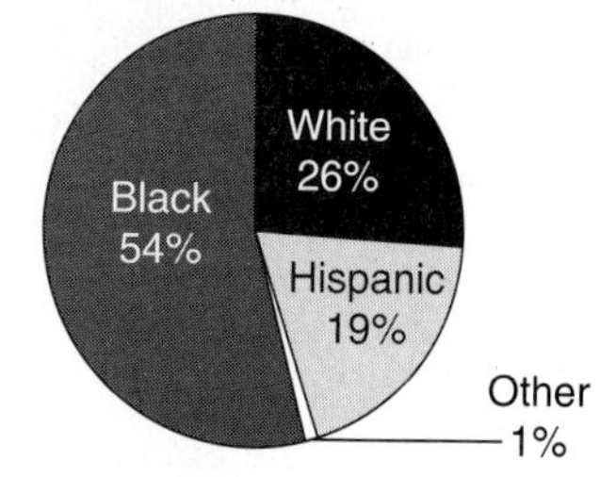

Estimates of annual new infections by race (N ≅ 45,000)

To better understand how the HIV/AIDS epidemic is affecting men and women, it is critical we look at race and risk by gender.

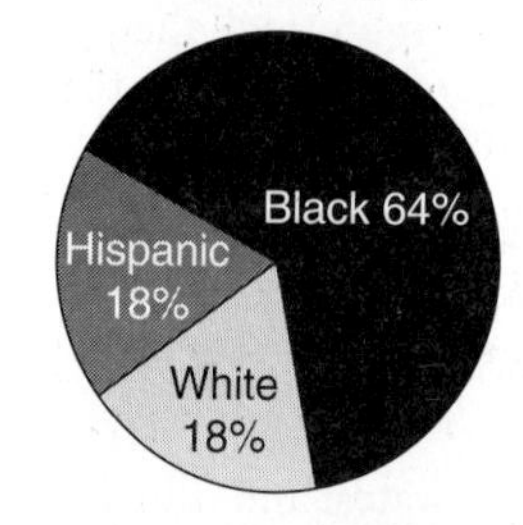

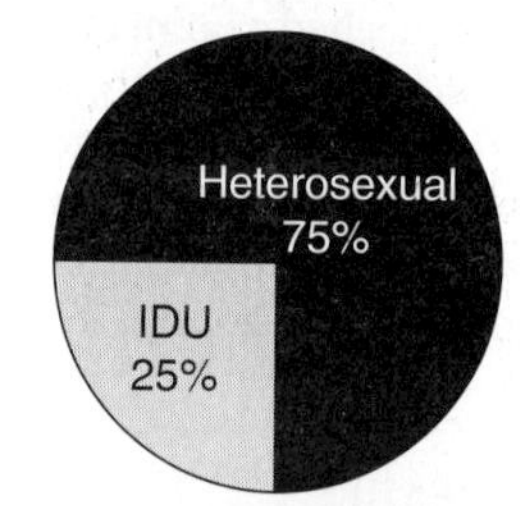

Estimates of annual new infections in women, U.S., by race and risk

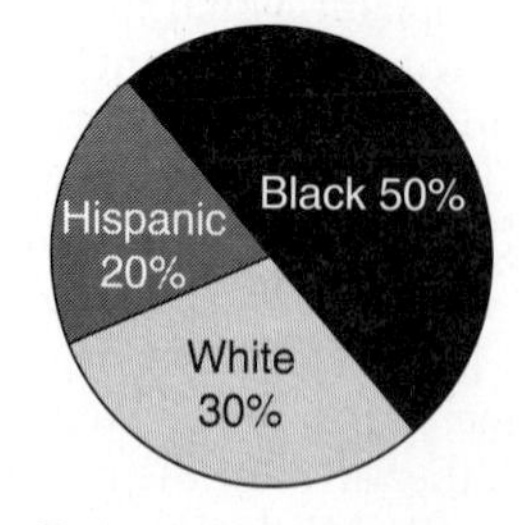

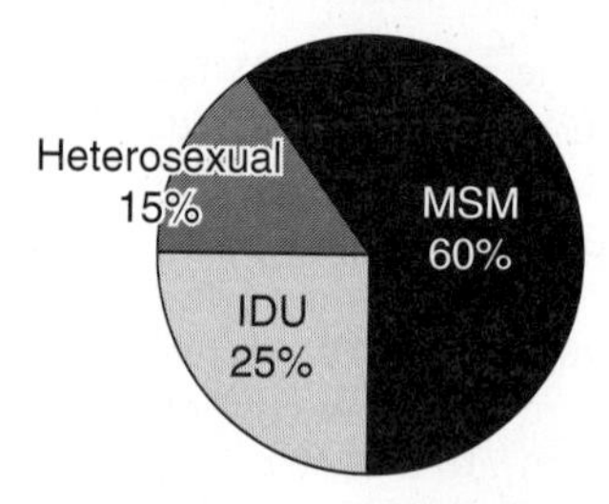

Estimates of annual new infections in men, U.S., by race and risk

FIGURE 10-8 A Glance at the HIV Epidemic in America. *(Source: Centers for Disease Control and Prevention updated)*

46% of newly identified HIV infections. Blacks make up 19% of the population of the South but 53% of the region's AIDS cases. The manifesto was sent to Congress for redress.

ESTIMATES OF DEATHS AND YEARS OF POTENTIAL LIFE LOST DUE TO AIDS IN THE UNITED STATES

"All my friends are dead." This expression is unique in a lifetime and is symbolic of reaching old age—except in a time of war. Too many young people worldwide have said it over the past 24 years because of AIDS.

Deaths Due to AIDS

Each year in the United States there are about 2,300,000 deaths. AIDS, from 1991 through 1995, caused at least 40,000 of these deaths each year and accounted for about 1.8% of all deaths for each of those years. That is, 2 people of each 100 who died, died of AIDS.

The good news is that between the end of 1995 and the end of 1996, AIDS deaths dropped in the United States for the first time since the pandemic began—25% nationwide (about 12,600 fewer deaths). Between 1996 and 1998 deaths had dropped by about 75% (Table 10-7). Similar data were reported from Europe. The sudden drop in

Table 10-7 New HIV Infections and AIDS Deaths in the United States

Year	New HIV Infections	AIDS Deaths	Accumulative 5-Year Intervals
1977	19,000	—	—
1978	40,000	—	—
1979	60,000	—	—
1980	90,000	30 (before 1981)	30
1981	160,000	128	
1982	190,000	463	
1983	160,000	1508	
1984	120,000	3505	
1985	70,000	6972	12,576
1986	60,000	12,110	
1987	50,000	16,412	
1988	40,000	21,119	
1989	40,000	27,791	
1990	40,000	31,538	121,546
1991	40,000	35,616	
1992	40,000	41,094	
1993	40,000	45,850	
1994	40,000	50,842	
1995	40,000	54,670	349,648
1996	40,000	38,296	
1997	40,000	22,245	
1998	40,000	18,823	
1999	45,000	18,249	
2000	45,000	17,672	463,906
2001	45,000	17,354	
2002	45,000	16,371	
2003	45,000	15,950	
2004	45,000	15,900	
2005	45,000	15,900	543,381
2006	45,000	15,900	559,281

Cumulative Deaths from AIDS in the United States from 1977 through 2003. Data for years 2004 through 2006 are estimated. AIDS deaths dropped by 75% from 1995 through year 2000. A similar or greater drop in AIDS deaths occurred in Europe over the same time period. It took from 1981 to about 1988, about 7 years, to reach the first 100,000 AIDS deaths. Over the next 18 years (2004) there were an additional 459,229 AIDS deaths.

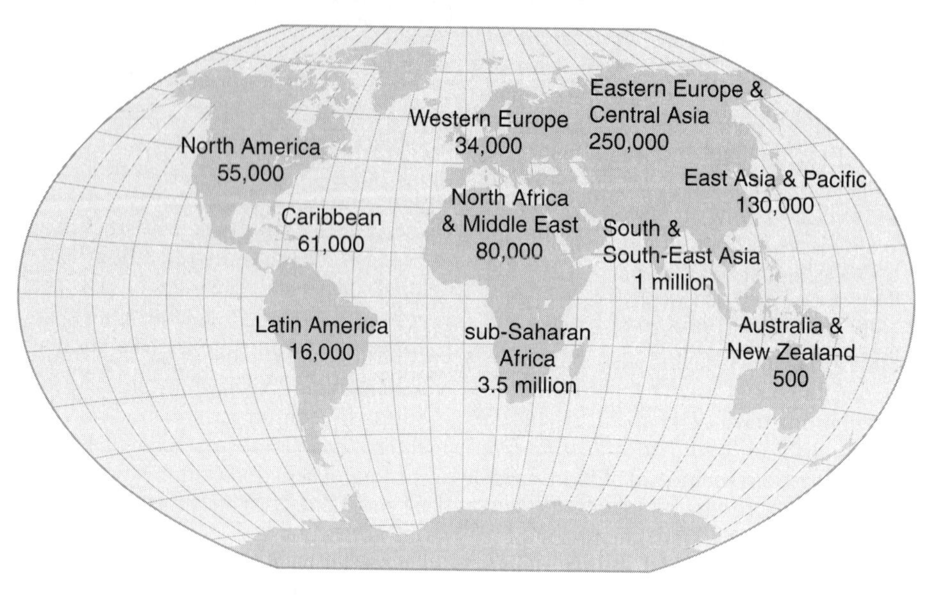

FIGURE 10-9 End-of-Year 2005 Estimate for New HIV Infections. *(UNAIDS, 2004 updated)*

expected deaths due to AIDS is believed to be associated with the use of combination drug therapy.

AIDS As a Cause of Death in the United States and Worldwide at the Beginning of 2005

United States

- AIDS is the fifth leading cause of death among people ages 25 to 44.
- AIDS is the leading cause of death of black American men ages 25 to 44.
- AIDS is currently the fourth leading cause of death among all U.S. women ages 25 to 44, and the second cause of death among black women ages 25 to 44.
- AIDS is the seventh leading cause of death among children ages 1 to 14.
- In some cities in the northeastern United States, AIDS is the leading cause of death among children ages 2 to 5.
- In the United States between five and six people per hour become HIV-infected 365 days a year (45,000) and every hour about two people die (16,000). Before 1996 one person died every 13 minutes.
- Annually, from 2002 through 2005, there have been about 16,000 AIDS deaths and 45,000 new HIV infections. Thus, the number of HIV-positive people has increased each year by about 30,000.
- Of those living with HIV infections, about 1 million at the start of 2005, 41% are white, 38% are black and 20% are Hispanic. Men make up 79% and women 21%.
- Through year 2010, an estimated 620,000 people will have died of AIDS.

Worldwide 2005 (See Figures 10-9, 10-10, and 10-11).

- Worldwide, 570 people become HIV-infected every hour, 365 days a year (5 million) and every hour 342 people die (3 million).
- Ending year 2005, about 65 million people will have been HIV-infected, about 95% in developing countries.
- Nine out of ten HIV-positive people are unaware they are infected.

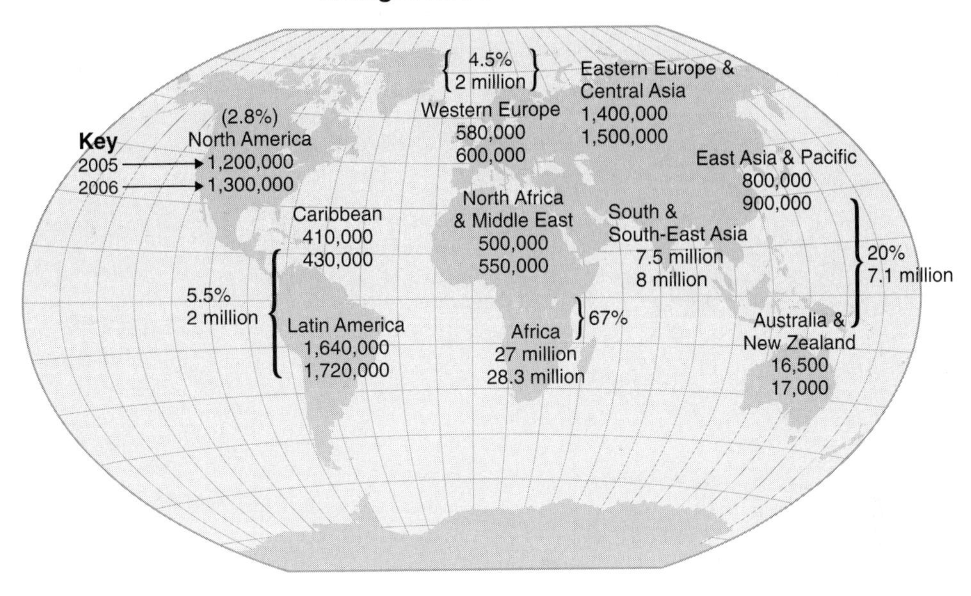

FIGURE 10-10 End-of-Year 2005 and 2006 Estimates of People Living with HIV Infection. Of the 41 million alive ending 2005, 96.4% are adult/adolescent, 50% are men, 50% women, with 3.6% under age 15. *(UNAIDS, updated)*

- Ending year 2005, about 24 million people will have died of AIDS. Less than 1 million will be gay men.
- 80% of those dying from AIDS are between ages 20 and 50. These are the people who enforce laws, harvest food, work the factories, heal the sick, keep students, and raise children.
- By the end of 2010 about 55 million people will have died of AIDS. Ending 2020 about 100 million people will have died of AIDS.
- If nothing changes, there will be an estimated 75 million to 100 million new HIV infections by 2020.
- It is estimated that new HIV infections will peak in 2060.

Former Surgeon General C. Everett Koop has said on a number of occasions that "AIDS is virtually 100% fatal." Looking back over the number of AIDS cases diagnosed and comparing them to the number of AIDS patients who have died would indicate that a diagnosis of AIDS is a death sentence. Table 10-7 presents a sobering look at the numbers of AIDS patients who have died in America since those first CDC-reported cases in 1981. People with HIV/AIDS were, through 1995, dying at the rate of about 3000 a month. Over 95% of those diagnosed with AIDS in 1981 have now died. And over 60% of AIDS cases diagnosed up through 2003 have died.

AIDS Deaths Postponed

Many of the deaths that would have occurred in the United States in 1996 through 2004 are being postponed through the use of combination cocktail

> **LOOKING BACK 10.1**
>
> In 1987, TV talk show host Oprah Winfrey said that 50 million heterosexual Americans will have died from AIDS by 1990.

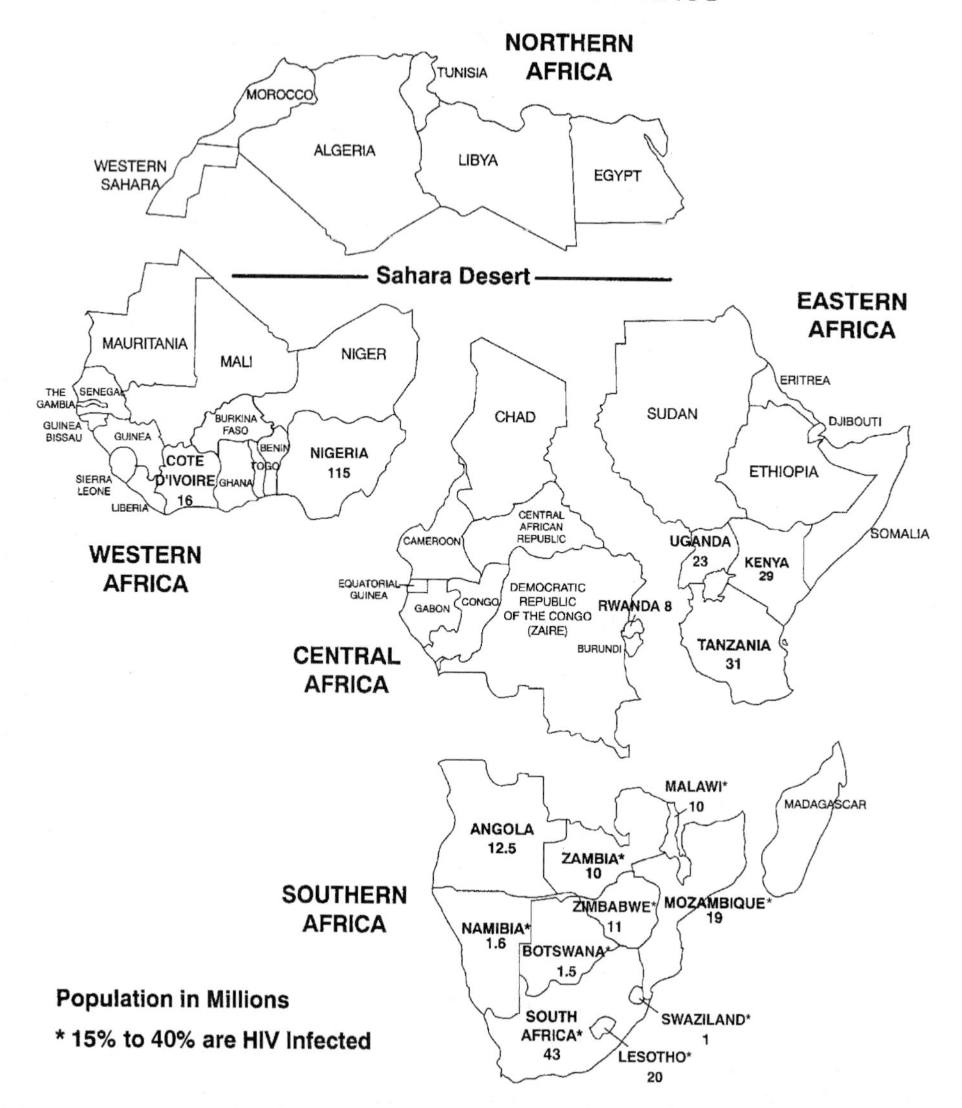

FIGURE 10-11 Map of Africa. Population of Africa = 800 million. Population of sub-Saharan Africa = 657 million. Prevalence of HIV infections and AIDS cases is highest in Southern Africa. South Africa with 45 million people has 5.3 million HIV-infected people, the highest incidence of HIV infection in the world. In seven African countries over 20% of the 15- to 19-year-old population is HIV-infected, from 20% in South Africa to about 40% among adults in Botswana. The lifetime risk of dying from AIDS for a boy who is currently 15 is 65% in South Africa and near 90% in Botswana. Karen Stanecki of the U.S. Census Bureau reported that in KwaZulu Natal, a province of South Africa, HIV infection has reached 46%. In Francistown, Botswana, HIV infection among adults is at 42%. Stanecki projected that AIDS deaths will peak in sub-Saharan Africa in 2020 at 6.5 million deaths/year; South Africa in 2010 at 940,000/year; Haiti in 2017 at 39,000/year; Thailand in 2000 at 66,000/year; Myanmar (Burma) in 2015 at 97,000/year; Brazil in 2002 at 623,000/year. *(Map courtesy of Centers for Disease Control and Prevention)*

drug therapy and in particular the use of at least one protease inhibitor within the drug cocktail.

Eight years of declining AIDS deaths means more Americans are living with HIV disease, which also means there are that many more people capable of spreading HIV.

The majority of AIDS deaths have occurred among homosexual/bisexual men (59%) and among women and heterosexual men who were injection-drug users (21%). Most (75%) of AIDS deaths have occurred among people 25 to 44 years of age and about 24% occurred in those over age 45. In 1992, HIV/AIDS became the number one cause of death among black and Hispanic men ages 25 to 44 and second among black women ages 25 to 44. Between January 1993 and December 1995 AIDS was the leading cause of death in all Americans aged 25 to 44 (*MMWR*, 1995a). For 1996 it dropped to second. In 2000, AIDS dropped to the eighteenth leading cause of death in the United States.

It now appears that for the 25–44 age group, AIDS is the leading cause of death among black American men, third for Hispanics and sixth for whites. After rapidly increasing in the 1980s, the annual rate of death from AIDS peaked in 1995 at about 50,000 deaths, then decreased annually through 2004. According to the CDC, people now dying from AIDS increasingly consist of females, blacks, and residents of the South.

According to the CDC, deaths from HIV disease/AIDS are currently underreported by 10% to 15%. Although most deaths occurred among whites, proportional rates have been highest for blacks and Hispanics.

The HIV/AIDS pandemic has taken on different forms in various parts of the world. In some areas, HIV infection has spread rapidly to the general population; in others, the spread has remained among higher-risk subpopulations, including sex workers and their customers, men who have sex with men, and injection-drug users. Worldwide, the adult prevalence rate is 1.07% of the population, and 50% of infections occur among women. AIDS is the fourth leading cause of death worldwide and the leading cause of death in sub-Saharan Africa. Because the number of people infected with HIV continues to expand, the annual number of deaths worldwide can be expected to increase. Since 1999, one in five HIV related deaths were among children and over half the adults who died of HIV-related causes were women. Without effective measures to prevent its spread, experts estimate that AIDS could take 100 million lives by 2010 and exceed 200 million by 2020. These estimates worry epidemiologists because with so many people with compromised immune systems, where and when will a new virus or bacterium evolve to take advantage of this huge pool of endangered people before moving into healthy people?

UNITED STATES AND GLOBAL: WHO IS DYING OF AIDS?

United States

The American AIDS epidemic once seen as a disease of middle-class gay white men has become a killer of minorities and the poor including African-American and Hispanic homosexual men, injection-drug users, and inner-city women. Year 2001 through 2004 government statistics show that minorities account for about 7 in 10 new AIDS cases.

In proportion, since 1985, the rate of death was higher for women than for men (*MMWR*, 1996d). **Through the year 1990, over 120,000 people died from AIDS. By the end of 2006, about 560,000 people will have died from AIDS.** To place the death rate due to AIDS in perspective, each day in America in 2004, about 6000 people died from assorted causes, while about 40 died each day from AIDS.

In 2004, the National Registry of Artists with AIDS listed 1803 writers, actors, directors, designers, composers, musicians, visual artists, film/video makers, dancers, architects, and interior designers who have died of AIDS.

Global

Worldwide, an estimated 3 million people died of AIDS each year from 2002 through 2005. Beginning 2002, half of these deaths were women and about 8% were children. People are dying at the rate of about 8,500 a day!

Prevalence and Impact of HIV/AIDS

However the impact of the disease is measured—by deaths, AIDS cases, or monetary losses—it is just beginning. The worldwide impact during the 1990s was 5 to 10 times that of the 1980s. As the pandemic continues, the United States will find itself progressively more involved with prevention/treatment programs and with the political changes that HIV/AIDS will bring about in countries with a high incidence of HIV/AIDS.

LIFE EXPECTANCY

In 1995, the Metropolitan Life Insurance Company published data that show that life expectancy reached its peak in the United States in 1992 at 75.8 years. For the years 1993 and 1994 it dropped to 75.5 years. However, with the use of anti-HIV drug therapies beginning in 1996 life expectancy has begun to rise again. In 2003 it was 74.4 years for men and 79.8 years for women for an average of 77 years. But, it continues to drop for those living in developing countries, especially in Africa.

A mortality analysis of homosexual males indicates that gay men in Vancouver, British Columbia, have a life expectancy 8 to 20 years shorter than nongay men in the same area (Hog et al., 1997). The researchers note that if the trend continues, "We estimate that nearly half of gay and bisexual men currently aged 20 years will not reach their 65th birthday." The study collected vital statistics from 1987 to 1992 and assessed mortality rates based on 3%, 6%, and 9% prevalence rates of homosexual and bisexual men. "Under even the most liberal assumptions, gay and bisexual men in this urban center are now experiencing a life expectancy similar to that experienced by all men in Canada in the year 1871."

CONNECTED BUT SEPARATE

Although all nations on Earth are connected in so many ways, there remains a separation of the haves and have-nots—the developed and developing nations. Two worlds: one with hope and one with 90% of all AIDS cases and orphans.

Worlds Apart

For the last several years, the joint United Nations Program on HIV/AIDS (UNAIDS) reported that each day, an estimated 14,000 men, women, and children in developing countries become infected with HIV. Fifty percent of these infections now occur in people ages 10 to 24. They are being infected at a rate of about 10 per minute, 600/hour, 14,400/day or 5 million/year. By the year 2020, there will be over 40 million orphans under the age of 15 in the 23 countries most affected by HIV. Most of these children will have lost their parents to AIDS. In contrast, in the industrialized world, new infections spread at the rate of 500 per day, less than one every 30 minutes. This disparity between rich and poor nations has increased dramatically over the course of the pandemic.

Africa, the HIV/AIDS Time Bomb: Ticking out of Control

HIV/AIDS is almost too large a problem in Africa to fully grasp. Like the bubonic plague in Europe in the 14th century and the flu epidemic of 1918 in the United States, the scale of the loss is staggering. HIV/AIDS is the worst human disease disaster the world has even seen.

Africa, with about 10% of the world's population, now accounts for about 90% of all new HIV infections. Beginning in 2005, about 83% of all AIDS deaths have occurred in Africa. The United Nations in 2002 reported that in Africa's 25 worst affected countries, 7 million agricultural workers have died from AIDS since 1985 and 16 million more could die by 2020. Also, unless there is some immediate relief, by 2006 most Africans will not live to see their forty-eighth birthday. Stefano Vella, president of the International AIDS Society, said in 2002, "This pandemic is not stabilizing, it is just beginning, it's just a baby beginning to grow." As a comparison, Vella said that if the United States had in proportion the same HIV infection rate as Botswana, between 35%–40%, there would be 40 million HIV-infected in America. The world and especially Africa has been living with a biological terrorist for the past 24 years with no end in sight.

Deepening the Misery

To add to the HIV/AIDS problem in Africa, in May 2002, reports out of Southern Africa, where as many as 20 million people are afflicted by hunger or malnutrition, reveal that this region of Africa is experiencing the worst food shortage in nearly 60 years, and HIV/AIDS is deepening the misery. Hunger is accelerating the onset of AIDS-related illnesses and death among household breadwinners. As a result, the AIDS epidemic is compounding the food problems by leaving the elderly and children to care for the sick, plant and harvest the crops, and take odd jobs for extra income. Seven countries facing the devastating

POINT OF VIEW 10.1

HIV STALKING AFRICA

A professor at South Africa's Pretoria University explained how HIV stalks the people of her country. "It is presumed if you get AIDS you have done something wrong. We have no language to talk candidly about sex, so we have no civil languages to talk about AIDS. As a result, the consequences of the silence march on: infection soars, stigma hardens, denial hastens death, and the chasm between knowledge and behavior widens." Because HIV is spread through sexual intercourse, migrant truck drivers and prostitutes commonly spread the virus.

Botswana, a country about the size of Texas, with a population of about 1.6 million, has the world's highest rate of HIV infection. In Francistown, where international highways and trade routes converge, at least 43% of adults are HIV positive. Prostitutes often trade sex for the transport of fruit, toilet paper, and toys to families. In Gaborone, the capital of Botswana, 44% of those between ages 15 and 49 are HIV-infected. Among women ages 25 to 30, about 50% are infected. Throughout the country there's premarital sex, sex as recreation, obligatory sex and its abusive counterpart, coercive sex, transactional sex, sex as a gift, sugar-daddy sex, extramarital sex, and multiple partners. Men rarely agree to use condoms and women often engage in dry sex to please their partners, both practices significantly increasing the risk of contracting or transmitting HIV. The question asked but not answered is: How did HIV infection become so advanced in one of the most democratic, wealthiest, best educated and least corrupt nations in Africa?

Children orphaned by AIDS add another complex dimension to Africa's epidemic because communities are becoming saturated with them. "Most orphans drop out of school, suffer malnutrition, ostracism, psychic distress . . . girls fall into prostitution and older boys migrate illegally to South Africa, leaving the younger ones to go on the streets." The death toll from AIDS in Africa threatens to wreck the region's frail economies, break down civil societies, and incite political instability.

drought have HIV prevalence rates over 20%: Botswana (40%), Lesotho (31%), Namibia (23%), South Africa (20%), Swaziland (34%), Zambia (22%), and Zimbabwe (34%).

By any and every measure, AIDS is a plague of biblical proportion. It is claiming more lives in Africa than in all the wars waging on the continent combined. AIDS is now the leading cause of death among all people of all ages in Africa and the progression of this pandemic has outpaced all projections (Figure 10-11). For example, in 1991, the World Health Organization predicted that by 1999 there would be 9 million infected and nearly 5 million deaths in Africa due to AIDS. Ending 2005, the numbers are about 48 million infected and 21 million deaths. Nearly all those currently HIV-infected in Africa, without drug therapy, will die around year 2010. In Botswana and Swaziland it is estimated that 80% of the population will die of AIDS over the next 20 years. It is estimated that there are 11,000 new HIV infections each day in Africa. From 1998 through 2005, AIDS killed, on average, 2.5 million annually, or 7000 per day. With such overwhelming numbers, it is important to remember that the data are about people, not numbers, and not facts and figures, but faces and families (see Figures 10-12 and 10-13).

Of all reported AIDS cases, 91% are estimated to have been heterosexually acquired. Infections through blood transfusions and perinatal transmission from mother to child have also contributed to the spread of HIV across the continent. Tuberculosis (TB) is the most common opportunistic infection among AIDS patients in Africa. The contribution of the factors that facilitate transmission—high prevalence of STD, low rate of male circumcision, the unequal status of women, migration, poverty, and patterns of social mixing—differs by country. The HIV/AIDS pandemic in Africa is like an explosion in slow motion, a slowly moving chain reaction—no sound, no blinding flash, no intense heat, no mushroom cloud, no buildings destroyed—just one silent death after another with no end in sight. Whole generations of people are in jeopardy with so little hope to go around in the developing nations.

FIGURE 10-12 The Long Journey Home. Friends chant while carrying the body of their friend to his mother's home in Haiti. It is many miles up through the mountains to a small rural village where he grew up. He left home at the age of 15 to find work. He is now returned home for burial at age 28. *(Photo by Mike Stocker/South Florida* Sun-Sentinel*)*

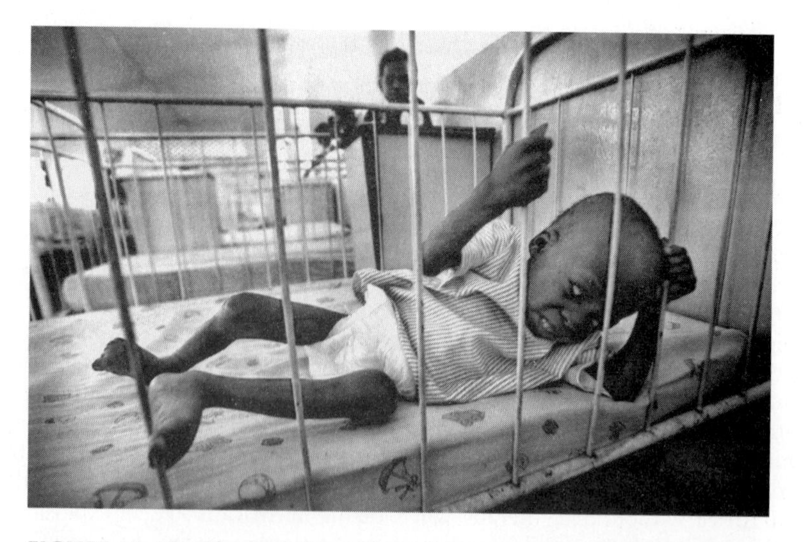

FIGURE 10-13 Charles is 3 years old. He is HIV positive and has tuberculosis. He was admitted to Grace Children's Hospital in Port-au-Prince in January 2001 suffering from seizures, chronic fever, skin lesions, loss of appetite, and weight loss. He and his mother have now died of AIDS. *(Photo by Mike Stocker/South Florida* Sun-Sentinel*)*

UNAIDS, 2000

"HIV will kill at least a third of the young men and women of countries where it has its firmest hold, and in some places up to two-thirds. Despite millennia of epidemics, war and famine, never before in history have death rates of this magnitude been seen among young adults of both sexes and from all walks of life." If death rates continue, even for a few more years, AIDS will kill more humans on the African continent than the 50 million who died on every front and in death camps in World War II.

Life Expectancy

AIDS already has sharply reduced life expectancy in many southern African countries. For instance, in Botswana, where more than one-third of adults are infected with HIV, life expectancy is now 39 instead of 74, as it would have been without the disease. It is projected that by 2010, life expectancy will be 27 in Botswana, 30 in Swaziland, 33 in Namibia and Zimbabwe, and 36 in South Africa, Malawi, and Rwanda. Without AIDS, it would have been around 70 in many of those countries. In the Central African Republic, Lesotho, Mozambique, Swaziland, Malawi, Zambia and Zimbabwe, a child born in 2004 would not be expected to see his or her 40th birthday. More people died of AIDS each year from 1999 through 2005 in Africa than in all the wars on the continent during those years. About 1 in 30 people is HIV-infected. Beginning year 2006, over 90% of all African children with perinatal HIV infection and 95% of all African AIDS orphans reside in sub-Saharan Africa.

Sub-Saharan Africa

The sheer number of Africans infected by this epidemic in this region is overwhelming. By the end of 2005, there will be an estimated 27 million HIV infected people living in sub-Saharan Africa. Seventy percent of all people living with HIV live here, as well as 90% of all cases of mother to child HIV transmission cases. In Rwanda, it is estimated that 80% of women are HIV positive resulting from rape by Rwandan soldiers during the genocide of a million Tutsi and Hutu. Overall, 58% of all new HIV infections in sub-Saharan Africa occur in women. In Botswana, Namibia, Swaziland, and Zimbabwe, current estimates show that between 20% and 40% of people aged 15 to 49 are living with HIV. In Zimbabwe, 30% to 50% of all pregnant women are now found to be infected, and at least one-third of these women will pass the infection on to their babies. South Africa, which escaped much of the epidemic in the 1980s, is now being hit particularly hard.

South Africa

By 2006 to 2007 about 7.7 million South Africans, roughly one in five people and one in three adults will be living with HIV/AIDS—and dying from it. Estimates are that by the end of 2010, about 7 million South Africans will have died from AIDS.

One in seven new infections on the African continent is occurring in South Africa, about 2000 infections per day. Ten percent or 200 are babies.

Military

In South Africa, according to *Johannesburg Mail and Guardian* newspaper, between 30% and 70% of the South African National Defense Force may be infected with HIV. In one unit in KwaZulu Natal, 90% of troops are infected. Some military units near Pietermaritzburg and on the South African Mozambique border also had HIV infection rates higher than 70%. South Africa's military infection rate is similar to neighboring countries. In Malawi, 75% of the military is HIV positive, and in Mozambique, 80% have tested positive. Forces in the Democratic Republic of the Congo and Angola also have high rates of HIV infection.

WORKFORCE—SOUTH AFRICA: SOME EXAMPLES

In April 2002, NMG-Levy, Agence France-Presse reported that by 2005 about 30% of South Africa's labor force will be HIV positive. The pandemic is taking dramatic toll on the most productive members of the population, those in their 20s, 30s, and 40s.

Mining

AngloGold, a gold mining company in South Africa, reported that about 30% of its 44,000 employees are already HIV positive.

Agriculture

The International Labor Organization stated that by 2010 over 30% of the farmers in South Africa would be HIV positive.

Nurses

An estimated 20% or 35,000 South African nurses are HIV positive. A Netcare group nursing manager recently told delegates that half of first-year students at one of the province's four nursing colleges are HIV positive. At another of the colleges, 70% of students are attending a local HIV clinic. At another, 21% of the students have volunteered information that they are HIV positive. In addition, 200 nurses a month are going abroad. The nursing manager said, "In our organization we are losing registered nurses. We are sitting with nurses who are dying now, and the students are even worse off."

EDUCATION: TEACHERS AND STUDENTS—SOUTH AFRICA AND SOME COUNTRIES IN SOUTHERN AFRICA

Senteza Kajubi, an education official in Africa said that about 30 million girls in sub-Saharan Africa are out of school. And that out of 100 million children in the world who do not attend school, 44 million are from Africa, the majority being girls from the sub-Saharan region.

The Impact of HIV/AIDS on Teachers, Students, and School Systems

As many teachers die every year of HIV/AIDS as qualify to teach. School districts are exhausting their annual budgets within two months by transporting deceased teachers to their homes. There is widespread closure of schools because HIV/AIDS has stripped them of their teachers. These nightmare scenarios afflict Kenya, Zambia, Botswana, Mozambique, Uganda, the Central African Republic, and South Africa. The devastating impact of HIV/AIDS on teachers and learners was a repeated concern at a national policy conference on teacher training and development convened by the Department of Education in Midrand, Gauteng, South Africa. The conference also heard that there has been an incredible 85% decrease in the number of students in pre-service teacher education programs between 1994 and 2002. These alarming trends are being identified at a time when the need for effective and widespread teacher training has never been greater. The country is poised to implement Curriculum 2005, the radically new teaching and learning methodology that has been on the drawing board since the mid-1990s. Nearly 400,000 in-service teachers have to be trained in the new curriculum. Wherc will they come from?

Nelson Mandela, former president of South Africa, said in September 2000 that in South Africa, 10 teachers die every month from AIDS and 1 student dies of AIDS each week in each of the country's 74 colleges and universities—roughly 120 teachers and 3600 students a year. About one in eight teachers is HIV positive. In Durban Westville University in South Africa, about 25% of students are HIV positive.

Kenya lost 1000 teachers to AIDS in 1996, 1200 in 1997, 1500 in 1998, and over 2000 in 1999. The death rate is now up to 6205 teachers in 2002 or 17 a day or about 170/month, and the trend continues. Also, one in five secondary students and one in five college students is HIV positive in Kenya.

Zambia and Swaziland are losing about 130 teachers/month. Two teachers die for one graduate. So the teachers are not being replaced. The pool of uneducated becomes larger as does the number of students in the classrooms that continue. In Tanzania, it is estimated that by 2010, between 15,000 and 30,000 teachers will die of AIDS. To date, in sub-Saharan Africa, about 3 million students lost their teachers to AIDS over the past several years. Many schools have closed because of teacher shortage.

In part of Malawi, Uganda, Botswana, and Zambia, over 30% of teachers are HIV positive. In the Central African Republic, 85% of teachers who died between 1996 and 1998 were HIV positive and died approximately 10 years before they were due to retire. In Mozambique 17% of teachers are HIV positive.

Burying the Dead

AIDS is the leading cause of death in South Africa. In 2002 through 2005, over 40% of all adult deaths were due to AIDS. They are dying at the rate of 4,200 a week!

In Soweto, South Africa, as in many South African cities and towns, burying those dying from AIDS has become a daily chore. They used to bury the dead on Saturdays. Too many are dying. "There are so many funerals, it is chaos. We don't know which funeral to attend. On some days we attend three funerals. Which neighbor or which family members' funeral should we attend? We must choose."

In Kenya, about 700 people die from AIDS daily. The country is sacrificing its forest for wood to build coffins. The disappearance of the forest is becoming an environmental disaster. The solution, said the environmental minister, is to use biodegradable plastic or synthetic coffins. For now, relatives are cementing wooden coffins into the ground to prevent thieves from stealing them and reselling them; the use of less valuable plastic coffins would discourage such theft.

As AIDS continues to claim the country's young, it has transformed black neighborhoods into open-air funeral parlors and neighbors into widows, orphans, and grieving relatives.

OTHER HIV/AIDS TIME BOMBS: ASIA, INDIA, CHINA, AND RUSSIA

Asia, with 60% of the world's population, is now home to an estimated 7.4 million people living with HIV/AIDS, and possibly as many as 10.5 million. About half a million people died of AIDS in the region in 2004 and about twice as many became newly infected. The epidemic is indeed surging ahead in Asia. The most worrying aspect of the pandemic in Asia is the sharp increase in HIV infections in China, Indonesia and Vietnam, which together have nearly 50% of Asia's population. China and India, with 2.3 billion people between them still have low national HIV prevalence rates—0.1% in China and 0.4% to 1.3% in India—but they have extremely serious epidemics in a number of provinces, territories, and states.

India

If by 2010 just 2% of India's one billion plus population is HIV-infected, that would be 20 million people (over twice the population of New York City). UNAIDS reports that about 7 to 8 million Indians are already infected and the virus is spreading rapidly in that country, primarily through heterosexual activities and injection-drug use. During the June 5, 2001, twentieth anniversary of AIDS, a physician from India said on the *Jim Lehrer News Hour*, that he believes the real number of HIV-infected in India is between 8 million and 12 million.

The Center for Strategic and International Studies in Washington estimates India will have 25 million HIV infected by 2010. The first case of AIDS was described in India in 1986. India is a vast heterogeneous country. Eighty percent of the people live in the countryside where many have never heard of HIV or AIDS. Some local authorities even challenge the existence of HIV or AIDS. Although it is believed that 83% of HIV infections occur through heterosexual intercourse with an infected partner, it is estimated that there will be between 4 million and 7 million HIV infections among female sex workers, their clients, and their families by year 2006.

Peter Piot, the executive director of UNAIDS, said in February 2002, "In the next few years, the number of HIV-positive people in India could surpass the number of HIV-positive South Africans if India does not step up prevention efforts. India currently has over 4 million HIV-positive residents compared with 5.3 million HIV-positive South Africans. However, in India, HIV-positive adults account for only 0.7% of the national population, whereas South Africa's HIV-positive residents account for 20% of the adult population." Piot estimated that in 10 years, tens of millions of Indians could have HIV. Piot said he is particularly concerned about the epidemic in the western state of Maharashtra where 2% of the adult population is HIV positive, and in the northeast where 30% to 40% of injection-drug users have the virus. Also, the large migrant populations in Uttar Pradesh and Bihar make those states extremely vulnerable to the HIV/AIDS epidemic. The Indian government has launched a national program to combat the spread of HIV, but it is hindered by social and cultural beliefs. There's still a perception that HIV is something for the very poor, homosexual men, and drug users. The shame, stigma, and discrimination associated with AIDS make it very hard to have open discussions and deliver explicit messages about HIV/AIDS.

SIDEBAR 10.2

VARIOUS ASPECTS OF DEATH IN SUB-SAHARAN AFRICA

BURIAL CHANGES BECAUSE OF TOO MANY AIDS DEATHS IN ZAMBIA

AIDS has changed the way people live. Now it is changing the way they are buried. As the AIDS toll rises, Zambia's local government authorities complain that burial ground is being filled up almost as soon as it is designated and predict a serious shortage soon. Zambians are being encouraged to look at cremation as a burial option. This has elicited serious debate. Zambians are by nature a very superstitious people who fear changes in cultural practices. University of Zambia's Department of History Professor Yizenge Chondoka says to shift peoples' thinking from burial to cremation will be hard. There are certain rites that can only be performed at graveyards to complete the burial process and ensure that the spirit of the buried is at peace. He asks, "If we begin to cremate, where will these rites be performed?" There is a psychological process that people need to go through to accept cremation. Otherwise, any misfortune that befalls a family will be attributed to the dead person whose spirit is wandering because it has not gone through the burials rites and is not settled on some place. Chondoka says Zambians have shrines at burial sites where they consult the spirits of the dead in times of need. "If we begin to cremate and throw or keep ashes in small clusters in our houses, where will the shrines be put up?" Gertrude Chamabantu, a marriage counselor, says as a Christian nation, the dead should be buried with the proper funeral rites just as Jesus was buried. Lusaka has three designated cemeteries (besides unofficial ones), which can take about 10,000 graves, but these are already full and people are now squeezing their dead on what used to be thoroughfares. Chandoka says Zambia has adequate land for graveyards. However, because the sites in the immediate vicinity have been filled up, people will travel farther to get to the graveyards.

Then there is the problem of economics. A teacher at the Zambian School of Education asks, "Do we really want valuable land to be taken up by graves?" AIDS is the leading cause of death in Africa. About 7600 Africans die daily from AIDS. Death in Africa is a very public affair. Funerals are big social events at which grief is expressed openly and lavishly. Families take large spaces in newspapers to announce the death of a loved one, complete with a picture, lists of achievements, and names of children. Now, about half the faces staring out of the public death notices are young. They have died of AIDS. But in contrast to the public rites of death, this increasingly frequent cause goes whispered or unmentioned. AIDS is a taboo subject in Africa. She said, "At the rate we are dying (from AIDS), three-quarters of Zambia will be a graveyard. There must be another way of disposing of our dead."

THE FUNERAL INDUSTRY—SOUTH AFRICA

Rising death rates in South Africa due to HIV/AIDS have led to the creation of a makeshift funeral industry. Many fly-by-night undertakers who are unlicensed and operate out of storefronts, compete to make funeral arrangements and leave bodies to decompose while they search for the cheapest means of disposal, creating a health hazard and raising costs to the government. The problem is greatest in Durban, capital of the hard-hit KwaZulu-Natal province. Morgues and cemeteries have run out of room and the unlicensed undertakers are tempted to cut corners by mishandling bodies—burying them in mass graves or abandoning them in mortuaries. The government has not regulated the new undertakers (who are mainly black) because they were previously disadvantaged, but established funeral directors (mostly white and Indian) complain that the new undertakers should be subject to the same regulations. The newcomers said they are subcontractors for licensed morticians—they sell coffins and transport the body for burial while a licensed mortician washes, dresses, and stores the body. Sometimes licensed morticians front for the newcomers by picking up bodies at morgues for a fee, a violation of health regulations. This corpse shell game often results in bodies being moved several times or left to decompose. As AIDS deaths rise, the problem will only worsen. Currently, AIDS-related deaths account for nearly a third of the country's 500 daily deaths and 40% of the deaths in KwaZulu-Natal. Deaths in South Africa could reach 16,000 a day by 2006, according to insurance experts. But talk of the rising death rate is all but taboo among government officials who, taking their cue from President Thabo Mbeki, barely acknowledge the extent of the HIV/AIDS epidemic or the rising death toll. Because of the stigma surrounding HIV/AIDS, many families do not claim bodies, leaving the government to dispose of them at a cost of $150 each. The government has rejected the idea of cremating the bodies because African tradition stipulates that a person cannot enter the spirit world if his or her body is not buried intact. Because morgues and cemeteries are out of space, a black market has grown up in stolen burial equipment. Crooked morgue workers sell corpses to favored undertakers, or to the highest bidder, sometimes even before bereaved families arrive to claim a body—leaving the relatives no choice but to pay the undertaker who collected the remains. Rival hearse drivers carry pistols and exchange death threats. Most ominous of all, government hospitals are being left with more and more unclaimed bodies—mostly AIDS victims—to bury at unbudgeted public expense. Undertakers who bid for this work can be tempted to cut corners by mishandling bodies. This growing demand for mandatory pauper burials has already led to scandal: bodies allegedly dumped in mass graves or left unburied in mortuaries for months at a time. What to do with bodies in Africa is becoming a massive problem!

THE CASKET/BURIAL PLOT INDUSTRY—ZIMBABWE

Griffin Shea reported on the demand for caskets and burial plots in Zimbabwe. Deep in the shadow of Harare's office high-rises, Luck Street is mostly islands of pavement in a river of mud and potholes. Despite its name, this side street in Zimbabwe's capital is where the city's least fortunate residents make their most-lasting purchases, a casket or coffin. With an economy in free fall and over 3000 people dying of AIDS every week, coffin-making has become one of the country's few reliable sources of income. In outlying townships, vendors line up caskets for sale next to tables of fruits and vegetables on the dusty roadside. But if you're on a budget—and almost everyone in Zimbabwe is—Luck Street is where you go for a bargain. At Sunshine Funeral Service, a darkened room behind a motorcycle repair shop, the owner and salesperson shows off his company's entire line of caskets, from a pressed-wood model that sells for about $15 to polished hardwood with shiny brass handles, $130. The owner said, "Last year, there were more people dying than in previous years. It is good business. People are dying." One thing that AIDS hasn't changed is the important place the funeral rites hold as part of the life cycle. Traditionalists say that the soul can take as long as a year to leave the body and join its ancestors in the spirit world—perhaps the reason for the traditional coffin burial. It is said that if people are not buried properly and given complete funeral rites by a medium, their souls will haunt their living relatives. After death, a number of rituals must be performed in order to assist the spirit. As funerals become an increasingly common part of life, Zimbabweans might have to find ways of blending their customs with modern realities. For about 30 cents per mile, the enterprising owners of Sunshine Funeral will send corpses back to their hometowns for burial, because most people prefer to be buried where they grew up. But skyrocketing inflation rates and a chronic fuel shortage have made that difficult for many urban dwellers, who are forced to buy a plot in cramped city cemeteries. According to Harare's director of cemeteries, 8 of the city's 10 cemeteries are full. So the city is clearing 5 square miles of land to expand one of its cemeteries on the outskirts, where most of the graves will be dusty plots in the bare earth. The city does not maintain the gravesite. The effect will be something like Arlington National Cemetery without the grass or trees—acres of land covered with grave plots. The new space should give Harare enough room to keep burying its dead for the next few years. But with 5 square miles expected to fill quickly, Sunshine Funeral Services won't have to worry about closing its doors anytime soon.

China

In China, HIV is prevalent in all 31 provinces, autonomous regions, and municipalities but epidemic patterns are different in different parts of the country. For example in Xinjiang, HIV among injection-drug users is between 35% and 80%, while in Anhui, Henan, and Shandong there is high HIV prevalence in rural people selling their HIV contaminated blood.

In August 2001, the deputy health minister held a first-ever news conference on the HIV/AIDS problem in China. A 2001 United Nations report estimated that over 1 million Chinese had HIV at the start of 2001, and that if current trends continue there could be 20 million by the end of 2010.

The first AIDS case was in an IDU discovered in the Yunnan province near the Burmese border in 1985. Since then HIV has spread rapidly among IDUs across China. Currently about 70% of China's IDUs are HIV positive. Eighty percent of reported AIDS cases are among IDU and prostitutes. There are at least eight HIV subtypes and several HIV recombinants now circulating in China.

HIV/AIDS: China's Titanic Peril—A current survey finds that most of China's population does not know what causes AIDS or how to prevent it—and 17% or 1 in 6 respondents had never heard of HIV or AIDS! Of respondents who had heard of HIV and AIDS, 73% did not know it was a virus, and 89% did not know how it can be detected. While 91% of these respondents knew that HIV can be transmitted, 22% could not identify even one route of transmission. Sixty-eight percent knew that HIV can be spread through sex. While 74% thought that HIV is preventable, 77% did not know that condoms offer protection, and 83% did not know infection could be avoided by not sharing injection needles. Eighty-four percent favored teaching prevention in schools. Among those least likely to be knowledgeable about HIV/AIDS were the poorest and least educated, women and farmers.

In 2004, Chinese officials predicted that over the next few years the major source of HIV infection would come from heterosexual contact through rising sexual freedoms and widespread HIV infected sex workers.

(The survey, "Current HIV/AIDS-Related Knowledge, Attitudes, and Practices Among the General Population in China: Implications for Action," was published on AIDScience.org (http://www.aidscience.org/articles/aidscience028.asp), a website run by the journal *Science.*)

Russia

HIV quietly crept into Russia in 1987 through sex between gay men—a practice that was then illegal, making the disease an unspeakable taboo. But even as the numbers grew rapidly in the mid-1990s, primarily through intravenous drug use, few people paid attention. In his January 2002 address to the nation, Russian President Vladimir Putin spoke about the nation's overall health crisis but made no specific mention of HIV or AIDS. Peter Piot, executive director of the United Nations AIDS Programs (UNAIDS) said in April 2002 that "Russia now has the fastest growing epidemic in the world, and I believe that the situation in Russia and the CIS (group of 12 ex-Soviet republics) is rapidly getting out of control." In year 2000 there were an estimated 80,000 HIV-infected. At the end of 2001 there were 250,000. According to the United Nations the actual number of HIV-infected in Russia is about 800,000, while Russia's director of anti-AIDS Center, Vadim Pokrovsky, has suggested about 1.5 million Russians are HIV-infected. If the current rate of infection continues, estimates are that over 5 million Russians will be infected by 2007. By 2015 between 5 million and 10 million will have died of AIDS. The major route of HIV transmission in Russia is through IDU, but its spread through heterosexual relations is growing rapidly. The age group most affected, as it is globally, is between ages 15 and 30. Half are under age 20. A major problem for Russia is that with this pandemic affecting the younger generations, Russia's shrinking population of 144 million people will continue to fall. Russia is faced with African-style depopulation, but in Africa its high birth rate moderates the effect of a large loss of life and will allow for population recovery.

Impact of HIV/AIDS on the World's Population

In 2000, the United Nations predicted the global population would grow from its current 6.3 billion to 9.3 billion in the year 2050. But population experts looking into the global HIV/AIDS disaster now believe the spread of the disease makes that figure too high. They believe the figure will be close to 8.8 billion, a discrepancy of about 500 million lives lost mainly due to the AIDS pandemic. The UN's World Population Prospects report estimates AIDS deaths in the 53 worst affected countries at 46 million by 2010 and rising to 200 million by 2050.

SUMMARY

In 1981, the CDC reported the first case of AIDS in the United States, and, from that time onward, has constantly tracked the prevalence of AIDS cases in different geographical areas and within different behavioral risk groups. In all behavioral risk groups, the common denominator is the exchange of body fluids, in particular blood or semen. The heterosexual population at large is considered to be at low risk for HIV infection in the United States. By 1993, all states and the District of Columbia, Puerto Rico, and the Virgin Islands have reported AIDS cases in people who have had heterosexual contact with an at-risk partner.

A major problem exists in attempting to determine the number of HIV-infected people. Several different approaches have been used by the CDC to estimate the total number of HIV infections. These estimates can be evaluated by examining their compatibility with available prevalence data.

With respect to race and ethnicity, the cumulative incidence of AIDS cases is disproportionately higher in blacks and Hispanics than in whites. The ratio of black to white case incidence is 3.2:1 and the Hispanic to white ratio 2.8:1. This racial/ethnic disproportion is also observed in HIV-positive blood donors and in applicants for military service. Even among homosexual and bisexual men and IDUs, where race/ethnicity-specific data are available, blacks appear to have higher seroprevalence rates than whites.

With regard to prostitution, in a large multicenter study of female prostitutes, black and Hispanic prostitutes had a higher rate of HIV infection than white and other prostitutes. This disproportion existed for both prostitutes who used injection drugs and for those who did not acknowledge injection-drug use.

The risk of new HIV infections in hemophiliacs and in people who receive blood transfusions has declined dramatically from 1985 because of the screening of donated blood and heat treatment of clotting factor concentrates. Evidence also indicates an appreciable decline in the incidence of new infections in homosexual men. However, the risk of new infections appears to remain high in IDUs and in their heterosexual partners.

There are some 5.3 million health-care workers in the United States. Even though they are supposed to adhere to Universal Protection Guidelines set down by the CDC for their protection, a significant number are exposed to HIV annually. A relatively small number of those infected have progressed to AIDS.

Estimating the number of HIV-infected people in the United States continues to be a numbers game. Various agencies and private industries have, for different reasons, attempted to determine the number of HIV-infected people. The numbers from the different groups vary widely. However, the 2005 estimated numbers of 1 to 1.5 million HIV-infected people may be too low.

Beginning year 2006, with over 1 million people living with HIV infection in the United States and with 45,000 new infections occurring each year from 1999 through the year 2005, the face of AIDS in America is changing. It's a younger and older face than it used to be. It's more likely to be a face of color than it used to be. And it is more likely to be female than it used to be. More people with HIV and AIDS are from areas outside major cities. Overall, the number of new AIDS cases appears to be leveling. But the HIV epidemic should be viewed as many different epidemics in different stages that vary according to age, race, gender, and locality. Although gay and bisexual men continue to make up the largest portion of new HIV infections, the epidemic is increasing more rapidly among people who become infected through heterosexual contact and through sharing syringes to inject drugs.

REVIEW QUESTIONS

(Answers to the Review Questions are on page 438.)

1. How did the CDC estimate the numbers of HIV-infected people in the United States? In what year was this done? In retrospect, how accurate are their estimates?
2. Why are people placed in potential HIV risk groups?
3. True or False: The time it takes for HIV-infected people to become AIDS patients is different for

each ethnic group, risk group, and exposure route. Explain.

4. What percentage of all U.S. HIV-infected IDUs are in the New York–New Jersey region?

5. What percentage of newborns from HIV-infected mothers are HIV-infected?

6. Eventually, what percentage of HIV-infected children will result solely from HIV-infected mothers?

7. What is the rate of college students currently HIV-infected? Is this more or less than the rate for military personnel? Explain.

8. Compare the college student rate of HIV infection with the rate of HIV infection for the general U.S. population.

9. What is the risk of a health-care worker converting to seropositivity after exposure to HIV-contaminated blood?

10. What single job-related event causes the greatest risk of HIV infection among health-care workers?

11. What percentage of the total number of AIDS cases in the United States represents health-care workers?

12. Are health-care workers more apt to become infected with the hepatitis B virus or the AIDS virus?

13. Do you think people of the United States openly and truthfully discuss their sexual habits with survey personnel? What does the text say?

14. Worldwide, how many AIDS patients are estimated to die by the end of 2006? In the United States?

15. Data on AIDS deaths indicated that of AIDS patients diagnosed between 1981 through 2005, in the United States,______% had died.

16. HIV/AIDS Crossword.

ACROSS

1. Surveillance agency
6. Invasion of microbes
9. Destroys microbe
10. Poisonous to cells
13. HIV test
15. Injection drug user
16. Antibodies detected

DOWN

1. Long duration
2. Communicable
3. Caused by HIV
4. Responds to antigen
5. Weakened
7. Phagocyte
8. Absence of symptom
11. Viral infection
12. Short duration
14. Medication

11 Prevalence of HIV Infection and AIDS Cases Among Women, Children, and Teenagers in the United States

Chapter Concepts

- Annual International Women's Day—March 8th.
- A global estimate of HIV-positive women is presented.
- In proportion there are significantly more black and Hispanic women with AIDS than white women.
- Injection-drug use is the major route of HIV infection for women.
- From 1994 through 2005 at least 38% of women infected in the United States contracted HIV from men through sexual intercourse.
- Injection-drug use and prostitution are strongly associated with HIV infection.
- AIDS is the leading cause of death for women ages 25 to 34.
- First 100,000 AIDS cases in women in the United States were documented in December 1997.
- About 30% of all HIV-infected adults/adolescents in the United States are women.
- About 40% of all new HIV infections in the United States are in women.
- Women make up about 25% of all U.S. AIDS cases.
- Black American women make up 13% of the female population but made up about 65% of new AIDS cases in 2000 through 2005.
- In the United States, about 70% of new AIDS cases among women now occur among those ages 30 to 49, 18% among those ages 20 to 29, and 12% among women over age 50.
- Women make up 50% of worldwide AIDS cases and AIDS deaths.
- Women make up 50% of worldwide HIV positives.
- Worldwide women account for 58% of AIDS cases among people ages 13 to 19.
- Of an estimated 3 million adult/adolescent AIDS deaths worldwide in 2004, 1.3 million were women.
- Pediatric means under age 13 in the United States and under age 15 in Canada and most underdeveloped nations.
- About 99% of new pediatric AIDS cases received the virus from their HIV-infected mothers.
- In proportion there are significantly more black and Hispanic pediatric AIDS cases than whites.
- Perinatal HIV infection without anti-HIV drug intervention in the United States is about 25%; with drug intervention it is about 8%. With drugs and caesarean section it is about 2%.
- Not all newborns who test HIV positive are HIV-infected.
- About 600,000 children under age 15 became HIV infected in 2002 and in each year through 2005.
- By year 2015, there will be an estimated 44 million HIV/AIDS-related orphans under age 15 in 23 underdeveloped countries.
- Seven of 10 people are sexually active by age 19.
- Eighty-six percent of all sexually transmitted diseases occur in the 15 to 29 age group.
- The total number of HIV-infected teenagers is unknown.
- Black and Hispanic teenagers account for a disproportionate number of AIDS cases compared to whites.
- Teenagers are being exposed to quality HIV prevention but they choose to ignore it.
- Sex thrills but AIDS kills.
- Worldwide people ages 14 to 24 account for at least 50% of all new HIV infections.
- HIV/AIDS crossword is presented.

INTERNATIONAL WOMEN'S DAY

Women play a central role in society and the benefits are apparent: families are healthier and better fed, savings and income rise, and a supportive environment is created. Take away women's ability to fulfill these roles and the entire societies fall apart. March 8th dates the annual celebration of International Women's Day. The Charter of the United Nations, signed in San Francisco in 1945, was the first international agreement to proclaim gender equality as a fundamental human right. Since then, International Women's Day has assumed a global dimension for women in developed and developing countries alike. International Women's Day is used as a time to reflect on progress made, to call for change, and to celebrate acts of courage and determination by ordinary women who have played extraordinary roles in the history of women's rights.

The global AIDS pandemic will be over 25 years old ending 2005. And at this time, there will be an estimated 65 million HIV infected people, about half of whom will be women. The face of AIDS has changed rapidly over this time span such that some say there has been a feminization of AIDS. It has been said that "a woman's work is invisible until it is not done." In 2005 over a million families realized that truth of this quotation after losing a female relative to AIDS. How many more women will be lost and families traumatized in 2006, into 2010 and beyond?

WOMEN: AIDS AND HIV INFECTIONS WORLDWIDE

The Gender Dimension of HIV/AIDS

In 2002 it was determined that worldwide half of AIDS cases are women.

Globally through 1997, women, children, and teenagers seemed to be on the periphery of the HIV/AIDS pandemic. In 2002 however, they became the center.

Women over age 15 make up over 50% of the 41 million people worldwide living with HIV/AIDS ending year 2005. The heightened vulnerability that gender inequality has created is now evident across the world, with over 50% of all new adult infections occurring in women. In sub-Saharan Africa, 58% of the HIV-positive adults are women. In parts of Latin America and the Caribbean, the proportion has reached as high as 45%, and this figure is on the rise. This alarming development must be recognized by concerned parties as an absolutely central

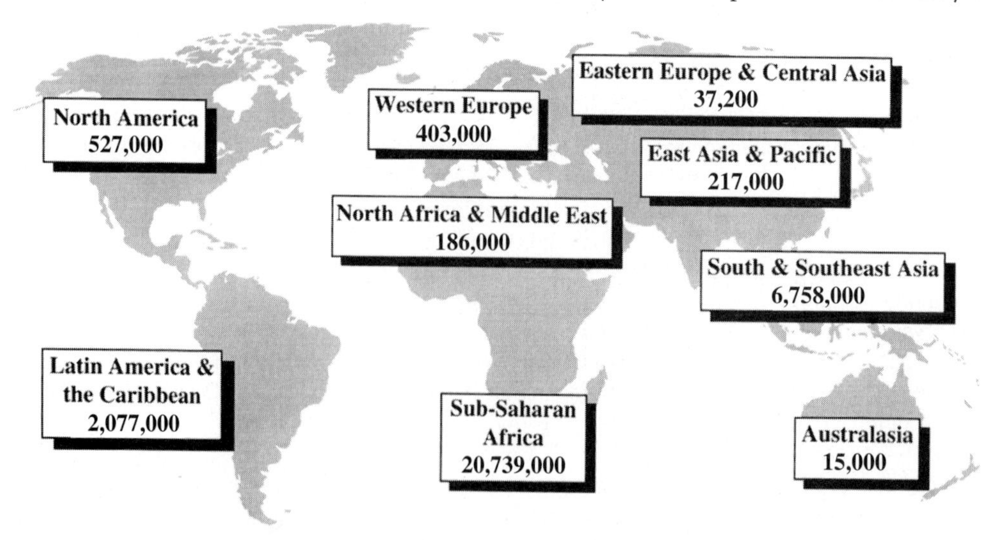

Global Total: 31 Million HIV-Positive Women

FIGURE 11-1 Estimate of HIV-Positive Women Worldwide Ending Year 2005. Ending 1993, women made up an estimated 8 million HIV cases. Ending year 2003, 48% (or 37.5 million) of all HIV positives worldwide were women. There will be about 33 million HIV-positive women globally ending year 2006 and about 11 million will have died.

consideration in the design of strategies to halt the spread of the disease worldwide.

Figures from UNAIDS show that the risk of infection is increasing for women everywhere—in developed and developing countries alike.

In sub-Saharan Africa, the **ratio of women to men** infected with HIV/AIDS is currently 7:5. In Rakai and Masaka, rural districts of Uganda, of women ages 15 to 25, 70% were HIV-infected. When compared with HIV-infected young men of the same villages and age group, there was a female-to-male ratio of 6:1. In Rwanda and Tanzania, women under age 25 to men under 25 is 2:1. The female-to-male ratio of AIDS cases among Ethiopian teenagers is 3:1; in Zimbabwe, it is 5:1. In Brazil, the ratio of female-to-male infection in Sao Paulo changed from 1:42 in 1985 to 1:2 in 1995. In Northern Thailand, 72% of sex workers are HIV-infected. By the year 2004, over 90% of HIV transmission worldwide was associated with heterosexual (vaginal) intercourse. Women are biologically more vulnerable to HIV infection than men because HIV in semen is in higher concentration than in vaginal and cervical secretions and because the vaginal area has a much larger mucosal area for exposure to HIV than the penis.

Table 11-1 Women as Percent of Adults Living with HIV/AIDS by Region, End of 2005

Region	Percent
Global	50%
Sub-Saharan Africa	58%
Caribbean	50%
North Africa/Middle East	50%
Latin America	37%
Eastern Europe/Central Asia	34%
South/Southeast Asia	30%
North America	25%
Western Europe	25%
East Asia	22%
Oceania	19%

(Source: UNAIDS, 2004 Report on the Global AIDS Epidemic, updated).

Gender Transmission of HIV

Transmission of HIV from male to female is 2 to 10 times more effective than from female to male

POINT OF INFORMATION 11.1

AFRICA: POVERTY AND RITUALS LEAD TO THE SPREAD OF HIV

A woman from Zimbabwe explained her use of dry sex to please her man, but it is not by choice. She used herbs from the Mugugudhu tree. After grinding the stem and leaf, she mixed a pinch of the powder with water, wrapped it in a bit of nylon stocking, and inserted it into her vagina for 10 to 15 minutes. The herbs swell the soft tissues of the vagina and dry it out. That makes sex very painful. But she adds, "Our African husbands enjoy sex with a dry vagina."

Many African women concur that dry sex hurts, but it is common throughout Southern Africa where the AIDS pandemic is out of control. Researchers who conducted a study in Zimbabwe had trouble finding a control group of women who did not engage in some form of the practice. Some women dry out their vaginas with *mutendo wegudo*—soil with baboon urine—that they obtain from traditional healers, while others use detergents, salt, cotton, or shredded newspaper. Research shows that dry sex causes vaginal lacerations and suppresses the vagina's natural bacteria, both of which increase the likelihood of HIV infection. Some AIDS workers believe the extra friction makes condoms tear more easily.

Africa contains thousands of cultures, some of which have strict sexual codes. But common to many sub-Saharan societies are the gender roles represented by dry sex: Women are unable to negotiate sex and so must risk infection to please the man. There are very few female checks and balances on male behavior. This stark inequality is part of the reason why HIV is spreading in countries with strict sexual codes, such as in Africa.

Lack of Authority

African women lack authority. Zimbabwe's Supreme Court ruled in 1999 that women have no more status of rights in the family than that of a "junior male"—usually an adolescent. In most sub-Saharan traditional cultures men pay for their wives, which gives them license to dominate the relationship. The very concept of marital rape doesn't exist in most of Africa and even the aunties, traditional marriage counselors for many young African wives, tell women that they cannot refuse sex with their husbands. Once a man has paid *lobola*—the word for dowry in several southern African languages, they are not forcing their wife to have sex. It's their right to have sex when they choose.

Many cultures especially in eastern and southern Africa provide **home guardianship/widow inheritance.** When a husband dies, one of his brothers or cousins marries the widow. This tradition guarantees that the children will remain in the late husband's clan and it also ensures that the widow and her children are provided for. When the guardian takes the widow, sexual intercourse is believed to cleanse her of the devils of death. A woman who refuses to take a guardian brings down *chira*—ill fortune on the entire clan. If her husband died of AIDS she might very well pass on the virus to her guardian. A Luo public health worker with the Red Cross said, "We have homes where all the males have died of AIDS because of this widow inheritance."

In some rural African villages, tradition holds that widows must sleep with the ritual "cleanser" —men who sleep with women after their husbands die—in order to be allowed to attend their husband's funerals or be inherited by their husband's brother or relative (a custom presented above). Unmarried women who lose a parent or child must also sleep with cleansers. Village elders in Gangre, Kenya, say the custom must be carried out or the community will be cursed with bad crops. Areas that still practice the tradition have the highest rates of HIV/AIDS. The cleansing job, held by hundreds of thousands of men across rural Africa, is seen as low class but essential to "purifying women." Cleansers are paid in cows and crops, as well as cash. They can be found in some rural parts of Uganda, Tanzania, and Congo. They are also a staple in Angola and across West Africa, specifically in Ghana, Senegal, Ivory Coast, and Nigeria. The tradition dates back centuries and is rooted in a belief that spirits haunt a woman after her husband dies. She is also thought to be unholy and "disturbed" if she is unmarried and abstains from sex.

Poverty: Black, Poor, and Female

Poverty means a day-to-day struggle for life in which individuals may be unable to afford the luxury of worrying about HIV/AIDS. The greatest inequality for women is poverty. Poverty is not a uniquely African phenomenon. Of the world's 1.3 billion people living in abject poverty, 70% are women—and most of them face the same basic problems as African women. In developing countries women are trapped in their reproductive roles. In numerous studies on HIV, women from Latin America, Asia, and Africa report that they dare not insist on safer sex or object to painful sex for fear of being abandoned by their men. In a 19-country study, the International Center for Research on Women found that the lower a woman's status, the greater her chance of becoming HIV-infected.

(World Health Organization, 1994). Paradoxically, since time immemorial women have been blamed for the spread of sexually transmitted diseases. Among certain peoples in Thailand and Uganda, STDs are known as "women's diseases." In Swahili, the language of much of East Africa, the word for STD means, literally, "disease of woman." Also, it is not coincidental that the countries in which HIV is now spreading fastest heterosexually are generally those in which women's status is low.

The World Health Organization reported that globally, during 2002 through 2005 there were 3 million AIDS deaths annually, on average 1.5 million of these deaths each year were women. That is about three female deaths per minute. Five women become HIV-infected every minute.

WOMEN: HIV-POSITIVE AND AIDS CASES—UNITED STATES

Women most at risk are ethnic minorities and the economically disadvantaged. Among sexually active teenagers, college students, and health-care workers nationwide, nearly 60% of the heterosexual spread of HIV is among women (Pfeiffer, 1991).

HIV/AIDS in women is a tragedy, but in addition, women are the major source of infection in infants. From 1993 through 2005, over 97% of HIV-infected children aged 0 to 4 years got the virus from their mothers.

At the end of 1988, women made up 6964 or 9% of the total adult AIDS cases in the United States. By the end of 2006, it is estimated that women will account for over 200,000 AIDS cases (Table 11-21).

Table 11-2 Reported AIDS Cases for Women, United States

Year	Number		Total
1981 (From June)	6		
1982	47		
1983	144		
1984	285		
1985	534		
1986	980		
1987	1701		
1988	3263		
1989	3639		
1990	4890		Through 1990
1991	5732	% Increase 1992–1994 151[b]	(15,489)
1992	6571		
1993	16,824[a]		
1994	14,379		
1995	14,100		Through 1995
1996	13,820		(73,095)
1997	11,651		
1998	10,500		
1999	10,800		
2000	11,000		Through 2000
2001	11,117		(130,989)
2002	11,300		
2003	12,000		
2004	12,900		Through 2006
2005	13,000		(estimated)
2006	13,200		(203,600)
Men	780,000		
Total:[d]	1,013,000[c]		

[a]The large increase in women's AIDS cases for 1993, over previous years was due to the January 1, 1993, implementation of the new definition of AIDS. Ending 2006, about 336,000 women will be living with HIV infection. Of these about 203,000 will be living with AIDS and about 99,000 will have died from AIDS.

[b]Reported male AIDS cases for 1993 were up 113%, in women, 128% (Hirschhorn, 1995).

[c]Total AIDS cases include 9800 pediatric.

[d]End of year 2006.

About 71% of all female AIDS cases have been reported in the 14 years between 1993 and the end of 2005. They have been reported from all 50 states and territories. About 67% of these females are between the ages of 13 and 39 (HIV/AIDS Surveillance, 1997 updated). Between 1989 and the end of 2005, female AIDS cases were and continue to be twice as frequent among black women as among white, and almost three times higher in black women than in Hispanic women (Figure 11-2).

In the United States, ending 2006, about three-quarters of the estimated 336,000 women living with HIV/AIDS will be black or Hispanic.

Among the most alarming HIV/AIDS statistics to emerge is that of HIV transmission through heterosexual sexual contact. Of new HIV infections in women, acquired through heterosexual contact, in 2002 through 2005 in the United States about 40% are women. Of AIDS cases that occur in women ages 13 to 24, about 76% are due to heterosexual contact.

Figure 11-3 presents the cumulative **source** of U.S. female AIDS cases through 2005. The median age for women reported with AIDS is 35 years, and women ages 25 to 44 account for 85% of female AIDS cases. Beginning year 2006, women

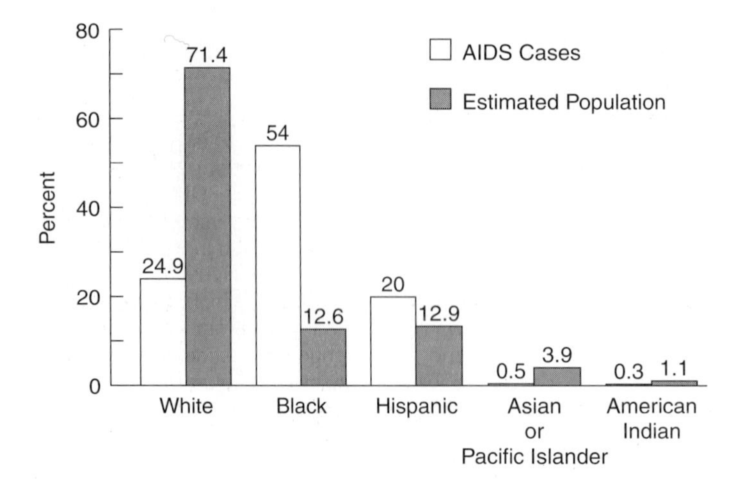

FIGURE 11-2 Incidence of AIDS Cases Among Women of Different Ethnic Groups, United States. Worldwide, every minute five women become HIV-infected, and every minute three women die of AIDS. *(Source: Global summary of HIV/AIDS Epidemic, 2003 updated)*

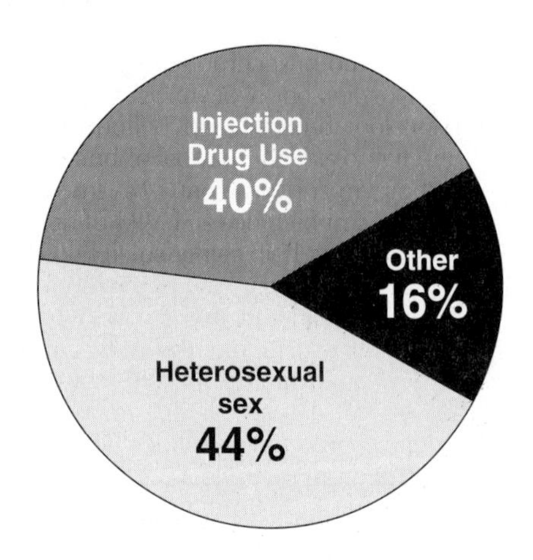

FIGURE 11-3 Cumulative Major Sources of HIV Infections in U.S. Women Through 2005. *(Source: U.S. Centers for Disease Control and Prevention, Surveillance Report Year 2003 updated)*

50 years and older accounted for 9% of all female AIDS cases. Most of these women became infected through heterosexual sexual activities (Schable et al., 1996 updated).

Beginning in 2001, the South accounted for the largest percentage of AIDS cases reported among women (46%), followed by the Northeast (36%), West (8%), Midwest (6%), and Puerto Rico and U.S. territories (4%). In the Northeast, 1.4% of women with AIDS resided outside metropolitan areas compared with 10.2% of women who resided outside metropolitan areas in the South. Of all AIDS cases among women, 62% were reported from five states: New York (26%), Florida (14%), New Jersey (10%), California (7%), and Texas (5%).

Of all new female AIDS cases, 50% are associated with heterosexual transmission. The second most frequent cause was injection-drug use at 31%. The AIDS rate for Hispanic women was eight times higher than for white women. The AIDS rate for black women was three times higher than for Hispanic women and 23 times higher than for white women.

As the frequency of HIV/AIDS increases in women, the question of whether AIDS will explode in the heterosexual community of the United States becomes more a question of when will the number of female AIDS cases equal male cases. For the first time, in 1997 American women made up over 20% (22%) of AIDS cases for the year. For 2006 the estimate is 30%.

Artificial Insemination

To date there are six cases of transmission through **artificial insemination** reported in the United States, and six other cases are known to have occurred (Joseph, 1993 updated). Four of eight Australian women who received semen from a single infected donor became infected. HIV-contaminated semen had been injected into the uterus through a catheter. In 2003, a woman in Tokyo became infected after being artificially inseminated with sperm from her HIV-positive husband. According to the report, several procedures involved in obtaining HIV-free sperm were not performed. Children have been born free of

BOX 11.1

THE FEMALE GENDER AND THE IMPACT OF HIV/AIDS

AIDS NOW HAS A WOMAN'S FACE

In the early 1990s women were on the periphery of the HIV/AIDS pandemic. In 2003–2005 women are at the epicenter. Worldwide, for the first time, ending 2002, women accounted for 50% of all living HIV/AIDS cases or 20 plus million cases ending 2005. But this figure hides the large variations in the world. (See Table 11-1.) Across the world, more women than men are now becoming HIV-infected and dying of this disease. The biggest factor appears to be how HIV is transmitted. Women are most at risk in countries where heterosexual sex is the main mode of transmission. This is the case in Africa, the Middle East, and the Caribbean. By comparison, HIV is mostly transmitted by men who have sex with men in Western Europe, Australia, New Zealand, and in North America.

QUESTIONS ABOUT THE FEMALE FACE OF AIDS

In 2002 it became clear that women equaled men in the global number of HIV/AIDS cases. But this alarming finding did not just happen overnight. The phenomenon of women and HIV/AIDS has grown relentlessly for over 21 years of this pandemic. What shocks our senses is how long it has taken to focus the world on the fact that this was happening. Why wasn't the trend identified much earlier? Why, when it emerged in cold statistical print did emergency alarm bells not ring out? Why has the continuing pattern of sexual carnage among young women, so grave as to lose an entire generation of women gone on unrecognized? And why did it take until 2003 for the UN to form a Task Force on the plight of women in Africa or until 2004 to put in place a Global Coalition on Women and AIDS?

HETEROSEXUAL TRANSMISSION

Why does heterosexual transmission strike more women than men? The answer has to do with physiology, economics, and culture. Women are physiologically more vulnerable to infection: If an HIV-positive man has unprotected sex just once with an HIV-negative woman, her chance of infection is around 1 in 300. Reverse the gender and the odds fall to around 1 in 1000. This means women have more incentive to insist on safer sex than men. But here's where economics comes in: In many cultures worldwide, women are denied equal access to education, income, ownership of land or other productive assets, even to credit. Too many women are left heavily dependent on men, and on exchanging sexual access to their bodies for the means of survival—both for themselves and their children. This situation makes negotiation of safer sex very difficult. Women are vulnerable for many reasons: They face domestic violence, at times worsened by conflict or insecurity; girls are the first to be pulled from school and put to work when HIV/AIDS strikes home; women lack the power and economic independence to negotiate sexual safety; women face the full brunt of the stigma and discrimination associated with HIV, which fuels their fear of getting tested and prevents them from seeking care if they are infected; and there are inequalities between the sexes and women have a lack of power to challenge these inequalities.

Global Impact: The Gender Imbalance Grows

It has been said that men are driving the HIV pandemic, but that women will ultimately be its main victims. According to Peter Piot, executive director of UNAIDS, higher deaths among young women due to HIV will lead to a hole in the age pyramid that has only been seen before in times of war, when more men die. According to UNAIDS, the situation is so badly tilted against women that they are now two to three times more likely to contract HIV than men. For example, so many more women will die from AIDS in the next 10 years than men that it will culminate in an unprecedented gender imbalance and will change reproductive choices dramatically in years to come.

SUMMARY: Peter Piot said, "Women may be vulnerable, but we must distinguish between vulnerability and weakness. Women have shown great courage and resourcefulness in facing the epidemic. They have practiced safer sex when it was dangerous to do so; they have successfully pushed through legal reforms protecting their rights; they have consistently provided care, both at home and in health care settings. Wherever we look, we see the hope women have generated by their actions."

POINT OF INFORMATION 11.2

PROFILE: WOMEN WITH HIV/AIDS IN THE UNITED STATES

At the end of 2005, there will be about 150 million American women in the United States.

The majority of women with AIDS in the United States reside in the Northeast and the South, are unemployed, and 83% live in households with income less than $10,000 per year. Only 14% are currently married, compared to 50% of all women in the United States ages 15 to 44 years. Twenty-three percent of HIV-infected women live alone, 2% live in various facilities, and 1% are homeless. Approximately 50% have at least one child younger than 15 years old. Similar to other population groups with AIDS in the United States, the majority of women with AIDS are from minority racial and ethnic groups, with 61% of all AIDS cases diagnosed in blacks, 19% in Latinos, and 18% in caucasians (HIV/AIDS Surveillance, 1998 updated).

AIDS in Older Women

Ending year 2005, an estimated 7.6% or about 15,000 AIDS cases will occur in women ages 55 and older: white 5250, Black 6900, Hispanic 2550, all others 300.

Two Typical Senior HIV/AIDS Anecdotes

A soft-spoken, 63-year old, self-described "churchy" woman from rural South Dakota, who has been living with HIV since her late 50s said, "I got very sick. I kept going to the doctor and he kept giving me antibiotics, and he would just kind of hit-and-run. He thought it was menopause, arthritis, high blood pressure, possibly a heart problem. Suddenly I found myself in the emergency room after two years of misdiagnosed infections, monthly visits to my primary care doctor, and an allergic reaction to daily doses of penicillin. That's where I was tested for HIV. When the doctor came into my room, I pulled on his sleeve and told him to sit down a minute. He said, 'What you have is really nasty, and you don't want to know.'"

In a second case, a 66-year-old woman said, "He was 10 years older than I was and he said, 'I've never worn a condom in my life and I'm not going to start now.' " Most older men say that. Older women won't insist that the man wear a condom. They don't ask where he's been. Older women especially don't pry. They can no longer have children, so why bother with condoms? Usually they tend not to be very assertive until they get HIV or an STD. Most seniors are first diagnosed with HIV in the hospital after they've already progressed to AIDS.

These two stories are typical of HIV-positive older women. The majority become infected through sexual contact with a husband or boyfriend. After months and sometimes years of being sick, they're tested for HIV as a last resort. Health-care providers routinely fail these women by not readily screening them for HIV and other sexually transmitted diseases. They also neglect to ask about their sexual and drug-using behaviors. Ailments that accompany HIV often masquerade as signs of aging, throwing physicians even further off track. Fevers, night sweats, tuberculosis, chronic fungal infections, shingles, decreased vision, *Pheumocystis carinii* pneumonia, and cervical cancer can signal HIV, AIDS, or aging.

HIV when the procedure for stripping HIV from sperm was used properly. For other cases of men infecting women see Box 8.6.

ANECDOTE

I was referred to an eye doctor who is a specialist in retinal detachment in HIV/AIDS patients. After looking at my chart, the first thing he said to me was "Sex or drugs?" I knew he was trying to pigeonhole me. But I said, "Excuse me?" and he repeated, "Sex or drugs?" I told him, "I don't do drugs anymore and sex with you, I don't think so!"

Commentary: Women with HIV/AIDS have to deal with a lot of attitude from people, including health-care providers.

Prostitutes

The term **"prostitute"** is used here in preference to the more recently coined **"sex worker."** No single term can adequately encompass the range of sex for money/drugs/friendship/accommodation transactions that undoubtedly occurs worldwide. However, the term prostitute is at least relatively clear in referring to those who are directly involved in trading sex for money or drugs. In the United States, prostitutes represent a diverse group of people with various lifestyles. About 31% of female IDUs admit to engaging in prostitution. They need money to support their drug habit, pay numbers of IDUs

SIDEBAR 11.1

REBEKKA'S STORY

At the young age of eighteen, Rebekka aspired to become a Playboy Playmate like she had seen in the magazines found under her grandfather's bed. Her pictures were sent to Hugh Hefner, and against all odds Rebekka was chosen to become a Playboy centerfold, Miss September '86. Rebekka's sunny disposition and enthusiastic spirit brought her all she could desire. It seemed a time when nothing could harm her. Rebekka said, "I led an exciting life as a Playmate. Traveling, meeting tons of people, parties, etc. However as time passed, I began to suspect something was wrong. I became fatigued easily and was plagued with a general feeling of malaise." She went to her doctor for tests and, as an afterthought, asked to be tested for HIV. It came back positive. Rebekka said, "Surprisingly I did not become infected living the wildlife in the light of Hollywood; no, it happened years earlier, as a teenager having unprotected sex with a young man I met on a summer beach vacation." She was diagnosed positive in 1989.

When her doctor called with her results, she became confused. "I thought my doctor was telling me I was pregnant, but I was positive for HIV, and the only thing I knew about HIV was death." Attempting to mask her anguish, she began using speed and partying heavily, and began taking 18 anti-HIV pills a day in hopes of curing her illness. Her hands and feet began tingling with the sensations of pins and needles. She had drug toxicity from the medication and was diagnosed with neuropathy, which is a severe form of nerve damage. She was rushed to the emergency room in Los Angeles and had seven spinal taps in four days and was diagnosed with two brain infections. She was given a second set of new medications for everything the doctors could find wrong. She began regaining her strength and she began partying again and her drugs failed again.

When placed on a third set of new HIV medications her pancreas ruptured. Broke and about homeless she thought about suicide. She mixed numerous pain pills with tequila and drove into a brick wall. She was in a coma for three and a half days. When she awoke she was transferred to the psychiatric ward of the hospital. After being released she decided to reveal her HIV infection. She went public with her illness in 1994 by speaking to family, friends, and the media. She was put on a fourth set of drugs, this time a three drug combination. It caused severe diarrhea. She had to wear diapers for 18 months. For the fourth time, her drugs failed. Her doctor prescribed a new drug regime, which she said was "pure evil." She did not have a bowel movement for eight days, and realized the medications were ruining her digestive system. Her medications were switched yet again, but this time it worked. She takes about 10 pills a day and is now into her 20th year of HIV infection. About five years ago Rebekka found her way into nutrition and bodybuilding (Figure 11-4). Her coach, now her husband and also HIV positive, trained her intensely. On May 31, 2004, Rebekka won first place in her first attempt in a women's bodybuilding competition.

FIGURE 11-4 Rebekka Armstrong: 20 years HIV positive and counting. After suffering antiretroviral failures, wasting and near death at 90 pounds, she switched her lifestyle to accept better nutrition, new anti-HIV drugs, exercise, weight lifting, and bodybuilding. *(Photograph by Joanne Greenstone, used with permission)*

Rebekka currently lives in Los Angeles with her trainer husband Oliver. Although infected with HIV for nearly 20 years and diagnosed with AIDS, she is in excellent health. Medications and a super healthy lifestyle that includes mornings of cardio and weight training have helped Rebekka to maintain a normal T-cell count (over 500) and undetectable viral load. She says she has never felt better.

subsequently have large numbers of prostitutes. Evidence is overwhelming—IDU, prostitution, and HIV infection are strongly associated.

However, non-injection-drug-using prostitutes in the United States play a small role in HIV transmission. This is believed to be because of the low incidence of HIV infection among their male clients and on the insistence by many prostitutes that their clients use condoms. In Africa, Asia, and other underdeveloped nations where the "Customer is King," prostitution plays a major role in HIV transmission.

Women Who Have Sex with Women (Lesbians)

Research on female-to-female transmission remains inconclusive. But the large numbers of HIV-positive women and women with AIDS should alert women who have sex with women that they cannot assume their partners are uninfected because they are lesbians. It has been reported that 80% of lesbian women have had sex with men during their lifetime. Also, certain sexual behaviors common among lesbians probably put them at risk for transmitting and receiving HIV through vaginal fluid, menstrual blood, sex toys, and cuts in the vagina, mouth, and on the hands.

The evidence is clear: HIV infection is present among lesbians, and lesbians engage in behaviors that put them at risk for HIV infection. Whether or not lesbians put their female sexual partners at risk is less clear. In fact, there is a great deal of controversy about this question. Some HIV/AIDS investigators believe the risk of sexual transmission increases as the number of HIV-positive lesbian partners increases. Others believe that lesbians are not getting infected through lesbian sex, but only through unsafe behaviors like IDU. Female-to-female transmission has been reported in one case and suggested in another (Curran et al., 1988; Chu et al., 1994).

George Lemp and colleagues (1995) reported on HIV infection in 498 lesbian and bisexual women. Six were HIV positive, but there was *no* evidence of women-to-women transmission. The Women's AIDS Network (1988) of San Francisco states in their publication *Lesbians and AIDS* that if there is a possibility that either woman is carrying the virus, she should not allow her menstrual blood, vaginal secretions, urine, feces, or breast milk to enter her partner's body through the mouth, rectum, vagina, or broken skin.

In 2003 Helena Kwakwa and colleagues reported that based on genetic evidence of the strain of HIV found in two women, sexual contact between the two resulted in the transmission of HIV. A 20-year-old female had exclusive sexual activity with an openly bisexual HIV-infected woman for two years before testing HIV positive. Their sexual relations involved oral contact and the use of sex toys. The investigators ruled out other possible means of HIV transmission in this case (Kwakwa et al., 2003).

Special Concerns of HIV/AIDS Women

First, HIV/AIDS has a profound impact on women, both as an illness and as a social and economic challenge. Women play a crucial role in preventing infection by insisting on safer sexual practices and caring for people with HIV disease and people with AIDS. The stigma attached to HIV/AIDS can subject women to discrimination, social rejection, and other violations of their rights. A study by Solley Zierler and colleagues (2000) estimates that about 21% of HIV-infected American women were assaulted by a partner or another relation after becoming HIV positive. The percentage of women receiving physical harm after receiving an HIV diagnosis in a developing nation is at least twice that found in America. In South Africa, for example, Gugu Dlamini was stoned to death by her neighbors after she revealed that she was HIV positive on World AIDS Day, 1999.

Women need to know that they can protect themselves against HIV infection. Women have a traditionally passive role in sexual decision making in many countries. They need knowledge about HIV and AIDS, self-confidence, the skills necessary to insist that partners use safer sex methods, and good medical care.

Efforts to influence women to practice safer sex must also be joined by efforts to address men and their responsibility in practicing safer sex.

Second, women become pregnant. Women who are ill and discover they are pregnant need

POINT OF INFORMATION 11.3

HIV INFECTION AMONG WOMEN

There are over 6 million women ages 18 to 40 in the United States who are unmarried and having sexual relationships. Those most at risk for HIV infection are: (1) those who have multiple sexual partners (defined as having more than four different partners/year), and (2) those women who do not insist on the use of a condom.

Three of the nation's top five metropolitan areas with the highest incidence of AIDS in women are located in Florida within a 70-mile radius of each other (West Palm Beach, Ft. Lauderdale, and Miami). This area is an epicenter of HIV infection for Florida women. The other areas of highest incidence of AIDS in women are Puerto Rico, followed by New Jersey, New York, the District of Columbia, Florida, Connecticut, Maryland, Delaware, Massachusetts, Rhode Island, Georgia, and South Carolina. In Florida's Palm Beach County the rate is 24%; in Broward and Dade Counties, 18%. The epidemiologist for the state of Florida stated that what is happening in Florida is happening in inner cities nationwide. What may distinguish the AIDS epidemic in women is that it hinges on the low self-esteem and lack of personal power experienced by women in many walks of life.

Most of Florida's women with AIDS are poor and receive their medical care through the public health system. Among women in South Florida, HIV transmission is associated with crack cocaine. Pam Whittington, director of the Boynton Community Life Center, a family support facility in southern Palm Beach County, said, "If you have 10 women on crack, probably 8 of them are HIV-infected."

Crack cocaine is cheap and readily available. Its use contributes to anonymous, high-risk sex with multiple partners. Those who cannot afford crack exchange sex for it. In isolated communities of crack users, there is a high degree of sharing sex partners, many of whom are HIV positive.

information about both the potential impact of pregnancy on their own health and maternal–fetal HIV transmission.

Third, women have the role of mothering. From this role come two important consequences. First, when a woman becomes ill with HIV disease or AIDS, her role as caretaker of the child or children or other adults in the household is immediately affected. The family is severely disrupted and each family member has to make adjustments. Second, the mother must cope with her own life-threatening illness while she also deals with the impact of the disease on her family. Demographic studies show that many women who are HIV-infected or have AIDS have young children; and these women are often the sole support of these children.

Fourth, a woman's illness may be complicated further by incarceration and the threat of foster care proceedings. If the mother is healthy enough to care for her child, she must still cope with the complex issues of medical and home care, school access, friends, and family stress.

Biology and the Clinical Course of AIDS Among Women and Men

According to Birgit van Benthem and colleagues (2002), sex differences with respect to response to HIV infection do exist. CD4 cell counts are higher in women than in men throughout infection and viral loads are lower initially in women than in men, although this difference eventually disappears. Despite these disparities, women have not been found to have a more rapid clinical progression to AIDS than men.

According to Arlene Bardeguez (1995) and other similar reports (Cohen, 1995; Garcia, 1995), biology does not influence the prevalence of AIDS-defining illnesses, with the exception of invasive carcinoma of the cervix and possibly Kaposi's sarcoma. Access to HIV-related care and therapies is the dominant factor influencing the prevalence of AIDS-defining illnesses among women. Injection-drug use in HIV-infected women leads to a higher incidence of certain diseases, particularly esophageal candidiasis, herpes

POINT OF VIEW 11.1

WOMEN + SEXUAL PARTNERS + DECEPTION = AIDS

Across the world, women in support groups or with a close friend have been telling their stories of trusting their sexual partners and ending up with AIDS. Their trust was violated—their lives forfeited. Here are a few examples of the thousands of similar cases worldwide.

1. She is 48 years old with curly red hair and bags beneath her eyes. She slouches slightly in the office chair, stretching out her feet. From her eye shadow to her sneakers, everything is blue. Married to one husband for 28 years, she has children and grandchildren. She also has AIDS. She did not use drugs or have multiple sexual partners. She did have sex with her husband without a condom!
2. One 23-year old had a boyfriend with hemophilia; he never used condoms and never mentioned HIV, even though he had already infected another woman.
3. A divorced man with two children did not tell his 46-year-old girlfriend he had AIDS, even when he was hospitalized with an AIDS-related infection.
4. A seven-year live-in partner of a woman denied infecting her, even though he tested positive for HIV; she did not know he was having sex outside their relationship.
5. Because she had only two boyfriends, because "we were perfectly ordinary," they did not use condoms.
6. This woman with a baby did not know "my man was shooting up drugs and sharing needles." Not until he died of AIDS.
7. She never dreamed her partner had used a needle. When the doctor said she had AIDS, she replied, "You have made a mistake. I cannot have AIDS. How could I have that?"

All these women discovered their HIV status only after they became seriously ill with infections they should not have had. Heterosexual transmission is rising dramatically. A seldom-mentioned fact is a large percentage of infected women are married or in committed relationships.

simplex virus, and cytomegalovirus. Once an initial diagnosis of AIDS has been made, several major AIDS-defining illnesses appear more frequently in women: toxoplasmosis, herpes genital ulcerations, and esophageal candidiasis.

The currently proposed female-specific markers of **HIV disease** include **cervical dysplasia** and **neoplasia** (tumor), **vulvovaginal candidiasis,** and **pelvic inflammatory disease** (PID). HIV-infected women will have a higher incidence of cervical abnormalities on routine screening.

No biology-related differences in the survival of HIV-positive persons have been documented when equal access to medical care is considered. The shorter observed survival of women in some studies is thought to occur because of the lack of access to physicians who are knowledgeable about HIV-related care and therapies. Shannon Hader and colleagues (2001) found no evidence between women and men infected with the virus in terms of natural history, progression, survival, and HIV-associated illnesses. Drug use, high-risk sexual behaviors, depression, and unmet social needs among infected women contributed to their underuse of HIV resources. Also in 2001, Timothy and colleagues reported that although viral loads were lower in women than men, the rates of progression to AIDS were similar.

In July 2000, a document entitled "Guide to the Clinical Care of Women with HIV: 2000 Preliminary Edition" was posted on the Web (hab.hrsa.gov/womencare.htm). The Ryan White CARE Act program of the Health Resources and Services Administration made this resource available to clinicians caring for women living with HIV/AIDS.

Female HIV/AIDS Deaths

Women's deaths in the United States rose from 18 cases in 1981 to an estimated 99,000 ending year 2006. That is, about 50% of all women with AIDS will have died. AIDS is the leading cause of death for all women between the ages of 25 and 34, the fourth leading cause of death for all women between the ages of 25 and 44, and the eighth leading cause of death in white women. It is the second leading cause of death for black women and the third leading cause of death for Hispanic women between the ages of 25 and 44 (*MMWR*, 1996b updated).

Identifying and Preventing HIV Infection

Currently, women make up 58% of adult/adolescent HIV infections in sub-Saharan Africa, 30% in Southeast Asia, and 23% in Europe and the United States.

Identification of HIV-Positive Women—At age 26, a woman and her physicians were baffled when she began suffering from a variety of strange medical conditions: fevers, throat sores, unexplained vaginal bleeding, and fatigue. **It took a variety of doctors and seven years to find out what was wrong. She tested HIV positive!**

This woman's difficulty in getting diagnosed points out the extent to which women still are invisible when it comes to AIDS. After more than 24 years into the epidemic, the message still hasn't reached primary care physicians: Their female patients may be at risk. This young woman said, **"I went into doctors' offices and all they saw was a white, middle-class woman, not someone at risk for HIV."**

Early identification of women with HIV infection is a pressing problem. Risk-based screening at a Johns Hopkins perinatal clinic showed that 43% of HIV-positive women were not identified as at risk on the basis of such screening, with infection being found in 20 (9.5%) of 211 women admitting to at-risk behaviors and in 15 (1.6%) of 949 who were not at risk according to their response to screening questions (Garcia, 1995).

Prevention: United States—To prevent HIV infection, women have been told to reduce their number of sexual partners, to be monogamous, and to protect themselves by using condoms. **But these goals, generally speaking, do not fit the realities of women's lives or may not be under their control.**

Women do not wear the condom. (A female condom is now available but not yet in heavy demand. See Chapter 9.) For women to protect themselves from HIV infection, they must not only rely on their own skills, attitudes, and behaviors regarding condom use, but also on their ability to convince their partner to use a condom. Gender, culture, and power may be barriers to maintaining safer sex practices.

Women who have more than one sexual partner in their lifetime often practice serial monogamy, remaining with one partner at a time. People living as couples reduce the number of their sexual partners. Still, in many phases of life, sex is practiced with new partners in new relationships. American women, on average, are single for many years before their first marriage; they might be single again after a divorce; they might marry again; and, in later phases especially, they might be widowed. For some women, multiple partners throughout life is an economic necessity; urging them to reduce the number of partners is meaningless unless the

BOX 11.2

ONE WOMAN'S COMMENTS ON HER HIV DISEASE TREATMENTS

"I was fired from my job when they found out I was HIV-positive. The boss said, 'You have a modern problem—and this is an old-fashioned business.'"

This woman, in 1995, was 29 years old and HIV positive since age 22. When her T4 cell count dropped to 250, her doctor put her on Zidovudine (ZDV, also called AZT). She was not given any literature or verbal explanation of how this drug would affect her. In 12 months she became anemic; she could not sleep or hold down food, and her menstrual cycle became erratic. Without explanation, she next received two other HIV replication inhibitors, ddI and ddC. The side effects were very bad: consistent premenstrual symptoms; mood swings; increased cravings for certain foods, alcohol, or drugs; breast tenderness; and bloating. But there was no literature for her to read and her doctor said, "I knew how the drugs affected men, but I knew nothing of what to expect when I gave these drugs to women." "I stopped taking these drugs—they were killing me faster than the virus! I have severe yeast infections, shingles, sinus infections, and a host of other infections. My doctor is dealing with me like I'm some kind of experiment."

economic situation for these women is improved (Ehrhardt, 1992). In addition, public health strategies, not necessarily targeted to women, can also play an important role for women. Syringe exchange and drug treatment are important strategies because almost half of all HIV infections in women are due to injection-drug use. Because women are now more likely to be infected by men through heterosexual contact, programs that specifically target men, especially IDUs, will have a beneficial impact on women's programs.

Prevention: Africa

Although they are exceptionally vulnerable to the epidemic, millions of young African women are dangerously uninformed about HIV/AIDS. According to UNICEF, over 70% of adolescent girls (ages 15–19) in Somalia and more than 40% in Guinea Bissau and Sierra Leone have never heard of HIV or AIDS. In countries such as Kenya and the United Republic of Tanzania, more than 40% of adolescent girls harbor serious misconceptions about how the virus is transmitted. One of the targets fixed at the UN General Assembly Special Session on HIV/AIDS in June 2001 was to ensure that at least 90% of young men and women should, by 2005, have the information, education, and services they need in order to defend themselves against HIV infection. This goal was not achieved. The vast majority of African women living with HIV still do not know they have been infected. One study found that 50% of adult Tanzanian women know where they could be tested for HIV, yet only 66% of these have been tested. In Zimbabwe, only 11% of adult women have been tested for the virus. Moreover, many people who agree to be tested prefer not to return and learn the outcome of those tests. An additional problem is that in many African countries where pregnant women agree to undergo HIV testing, most have no access to drug therapy to prevent mother-to-child transmission of HIV. Over half the HIV-infected women who were surveyed by Kenya's Population Council said they had not disclosed their HIV status to their partners because they feared it would expose them to violence or abandonment. Not only are voluntary counseling and testing services in short supply across the region, but stigma and discrimination continue to discourage people from discovering or disclosing their HIV status.

Childbearing Women

The extent of HIV infection among pregnant women is often used as an indicator of HIV penetration into the population at large. By this yardstick, several Asian countries have serious epidemics. In some of India's HIV/AIDS surveillance sites, more than 2% of pregnant women are infected, with some sites as high as 6%. Myanmar recorded prevalence rates of up to 5% among pregnant women in some areas in the country. In Thailand, HIV infection prevalence among pregnant women peaked at 2% nationally.

Worldwide each year, of an estimated 200 million women who became pregnant, 2.5 million are HIV positive. Annually, from 1999 through 2005, about 800,000 children each year were born HIV positive, and 600,000 died. Of the 4 million women who become pregnant each year in the United States, 6000 to 7000 are estimated to be HIV-infected. These pregnancies now result in about 200 HIV-infected children each year.

Women, in general, have two children before they find out they are infected (Thomas, 1989 updated). The birth of an infected child may serve as a **miner's canary**—the first indication of HIV infection in the mother.

Pregnancy and HIV Disease—Early findings in pregnant women indicated that those with T4 or CD4+ cell counts of less than **300/μL** of blood were more likely to experience HIV-associated illness during pregnancy. Pregnant HIV-infected women exhibit a greater T4 cell count decline during pregnancy than do women without HIV infection. T4 cell counts in the HIV-infected do not return to prepregnancy levels. However, the overall declines in HIV-infected women likely represent declines that would have occurred in the absence of pregnancy and suggest that pregnancy does not accelerate disease progression (Newell et al., 1997; Bessinger et al., 1998).

Over the last 10 years, HIV-positive pregnant women have been attracting more attention from the medical establishment, **first,** because there are better medications for the HIV-positive mother and fetus, and **second,** because of the relatively high incidence of HIV births.

As an aside, it should be mentioned that protease inhibitors reduce blood levels of the estrogen component in oral contraceptive pills so women taking both the pill and PIs may need to use back-up methods of contraception.

DISCUSSION QUESTION: Nationwide, approximately 2 of 1000 pregnant women are HIV-infected, an incidence much higher than that of fetal neural tube defects, for which pregnant women are screened routinely. Should all pregnancies be screened for HIV?

Reproductive Rights—These are central to a woman's right to control her body. The choices of becoming pregnant or terminating the pregnancy are continually disputed.

However, as more women become HIV-infected and give birth to HIV-infected children, childbearing may come under the surveillance of the state. Women of childbearing age may be among the first groups to undergo mandatory testing as part of an attempt to control the birth of HIV-infected newborns.

Reproductive rights take on new meaning with HIV-infected pregnancies. The state has traditionally expressed an interest in protecting the rights of the fetus. This interest was transcended in the 1973 ***Roe* v. *Wade*** decision when the Supreme Court recognized a woman's right to choose an abortion. The court ruled that a woman's right to privacy must prevail against the state's interest in protecting the future life of the fetus. (See Chapter 7 for recent information on preventing HIV infection of the fetus.)

INTERNET

Among the hundreds of websites containing information on HIV, there are key addresses that provide comprehensive information, including links to a vast array of other resources. The key sites listed here include special areas focused on **women and HIV.**

American Medical Association home page: http://www.ama-assn.org

Centers for Disease Control and Prevention (CDC) Home Page: http://www.cdc.gov

HIV Insite: Gateway to AIDS Knowledge: http://hivinsite.ucsf.edu

BOX 11.3

RUTH'S POEM

This poem was written by a mother living with HIV, to her daughter.

Here I am, a few years on
It was hard enough explaining where
Daddy had gone.
She knows there's something wrong with me
How can I explain about HIV?
I see the sadness, the fear in her eyes
Each time I have to be hospitalised,
And I see the way she looks at me
Each time I take my AZT.
"I thought these pills were to make you well,
so why are you still sick then? Go on mum,
tell.
Quick mum, here, be sick in this basin."
I can see within her tiny mind racing.
"I'll help you mum, watch you don't fall."
And she treats me like a china doll.
Then the moment for something, I was very
scared,
The question that I had not prepared.
"Why do I get a jag mum, why do I get
blood taken?"
The moment of truth now, there can be no
faking.
"Well, it's the doctors special way of
knowing
that inside you there's no bugs growing;
tiny bugs that can make you unwell."
I had no choice but the truth to tell.
"You don't have these bugs darling, so
there's no need to worry."
"But you have, mummy," she says in a hurry.
"Yes, my lamb, in me these bugs grow."
"When you die mummy, where will I go?"
I don't know the answer to that question,
Instead I make this stupid suggestion.
"Shall we go to the fridge and get some ice
cream?"
Then we had a cuddle and I said, "I'm here."
I wish I could take away all her fear.
Her blood just now may not be infected,
But by HIV she's most definitely affected.
I don't fear dying any more,
Just for a special little girl who's only four.

With Permission

BOX 11.4

HIV-INFECTED WOMEN: DIFFICULT CHOICES DURING PREGNANCY

She was 19 years old, a nursing student, pregnant, and HIV positive. She spent $4\frac{1}{2}$ months of pregnancy in constant fear for herself and for her baby. She waited, her health began to falter, then she decided to have an abortion.

Several studies have reported that HIV-positive women who perceived their risk of infecting their fetus to be greater than 50% were more likely to abort than those who perceived a lower risk. HIV-positive women who chose to continue their pregnancy cited the desire to have a child, strong religious beliefs, and family pressure (Selwyn et al., 1989).

For women who are HIV positive, pregnancy poses difficult choices. First, pregnancy may mask the presence of HIV disease symptoms and having a child poses other questions such as: Can the mother cope with a normal or infected child? Who will care for the child if the mother becomes too ill or dies? Such questions bring up a moral issue.

DISCUSSION QUESTION: Do couples have a right to have children when one of the partners is known to be HIV positive? If the woman is HIV positive? If both are HIV positive? Is there any stage of HIV disease/AIDS when you think a woman should lose the right to become pregnant? (See Chapter 7 for the four stages of HIV disease.)

The Body: www.thebody.com/women.html

AIDS Community Research Initiative of America (ACRIA): www.criany.org/treatment/treatment_edu_women.html

National Women's Health Information Center: www.4woman.gov/

Women Organized to Respond to Life-Threatening Disease (WORLD): www.womenhiv.org

PEDIATRIC HIV-POSITIVE AND AIDS CASES—UNITED STATES

The Pediatric HIV Conundrum

One of the biggest puzzles in understanding mother-to-child transmission of HIV is why the majority of babies born to HIV-infected women remain uninfected in utero, at birth, and—perhaps most remarkably—during breast-feeding. It's even more remarkable in view of studies suggesting that cell-free viral load in breast milk can vary from undetectable to more than 200,000 copies per mL, meaning that a breast-feeding infant may ingest up to millions of viral copies each day. This apparent resistance puts infants in the category of exposed, seronegative individuals who can repel or effectively control HIV despite repeated exposures. Katharine Lazuriaga and Sarah Rowland-Jones have both documented cases of infants apparently clearing a transient HIV infection. It is these immune defenses that vaccine researchers seek to boost, or mimic, with a neonatal vaccine. But there are little hard data on just what they are and how this apparent protection works.

Pediatric AIDS in the United States affects two age groups: (1) infants and young children who became infected through perinatal (vertical) transmission, and (2) school-age children, the majority of whom acquired HIV through blood transfusions (mostly hemophiliacs).

Ending 2005, about 9600 pediatric AIDS cases were reported and about 5280 (55%) have died from AIDS. Pediatric AIDS cases represent about 1% of the total number of AIDS cases to date. It is estimated that for each pediatric AIDS case reported there are three to four other HIV-infected children. Thus, an estimated 20,000 to 30,000 children in the United States have HIV. Of the pediatric AIDS cases, 3% were/are hemophilic children who received HIV-contaminated blood transfusions or blood products (pooled and concentrated blood factor VIII injections). About 5% of pediatric AIDS cases occurred in nonhemophilic children who were transfused with HIV-contaminated blood. From 1995 on, virtually all HIV-infected newborns contracted HIV vertically. The Pediatric AIDS Foundation reported in April 1995 hospital costs for each HIV-infected newborn were $35,000 per year. Worldwide beginning year 2005, about 3.5 million children will be living with HIV disease and about 6 million will have died from AIDS (Figures 11-5 and 11-6).

Ethnic Prevalence of Pediatric AIDS Cases

Children of color make up 14% of all children in the United States but account for 57% of pediatric AIDS cases. Whites make up 70% of children and account for 18% of pediatric AIDS cases. Hispanics make up 12% of children and account for 23% of pediatric AIDS cases.

ORPHANED CHILDREN DUE TO HIV INFECTION AND AIDS

"Orphan" is an English word that does not have an exact translation in many languages. The concept of orphan is a social construct, so the meaning assigned to it varies from one society to another. In the United States and Africa, typically the term is understood to mean a child who has lost either or both parents. UNAIDS reports orphans as children who have lost their mothers or both parents. They limit their estimates to children below age 15. The UN Convention on the Rights of the Child defines children as being below age 18, unless the age of majority (adulthood) is reached under national law. Who is a child and who is an adult is defined differently in different countries and among different cultures. This further complicates the meaning of the word because orphans are generally considered to be children.

Beginning year 2006, of the estimated 16 million AIDS-related orphans, 79% live or have lived in Africa. The problem of AIDS-related orphans will become much greater over the next 10 to 15 years.

AIDS orphans present a chilling illustration of the far-reaching effects of the AIDS pandemic.

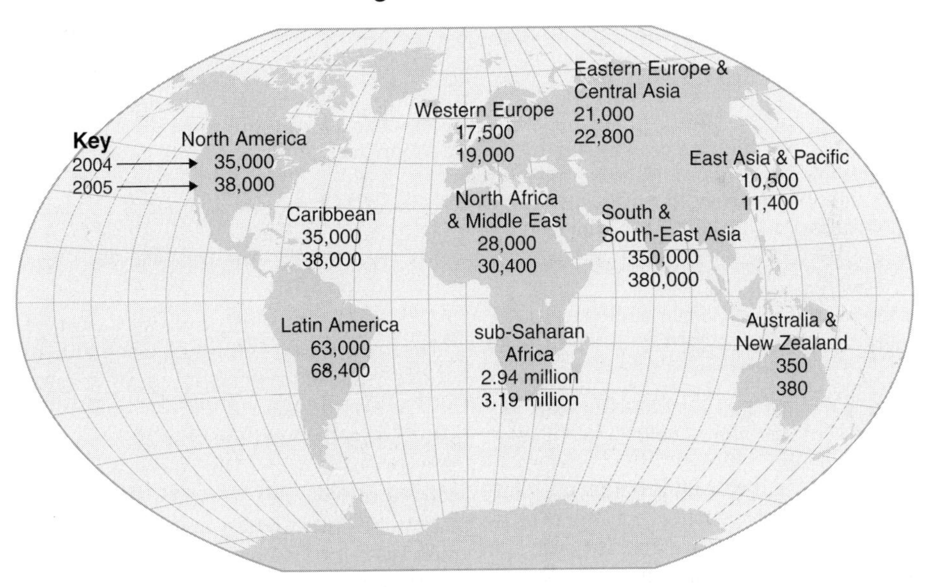

FIGURE 11-5 Year 2004 and 2005 Global Estimates of Children Living With HIV Infection. Note that sub-Saharan Africa and South and Southeast Asia have the highest burden of HIV-infected children. *(Courtesy of UNAIDS updated)*

Estimated Deaths in Children (<15 years)
Due to HIV/AIDS from 1981

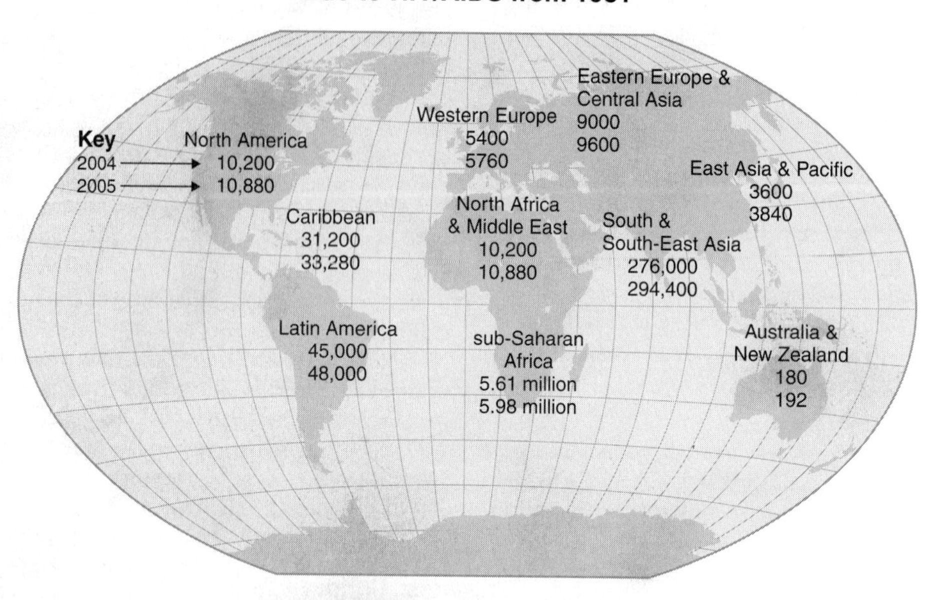

Totals are estimates for the entire globe.
Total: 2004—6 million
2005—6.4 million

FIGURE 11-6 Year 2004 and 2005 Global Estimates of AIDS Deaths in Children. (*Courtesy of UNAIDS 2001 updated*). As tragic as the data appear, these data on children dying of AIDS must be kept in perspective. Each year, inadequate drinking water and sanitation, indoor air pollution, accidents, injuries, and poisonings are just a few of the more preventable causes of 3 million deaths per year of children under age 5 and an estimated 12 million under age 15—about 24 times the number of children under age 15 who died from AIDS in 2004 and in 2005.

His mother was young, single, and HIV positive. When she went to the hospital to give birth, she checked in under a false name and address and then slipped out of the hospital leaving her baby who was only a few hours old. Today the boy is 4 years old. No one has yet to offer him a home.

An increasing number of HIV-infected children are being left in hospitals because their HIV-infected mothers and fathers are unable to care for them and no one else wants them. The hospital becomes their home.

As HIV continues to spread across the United States and HIV-infected women continue to become pregnant, the question is: What will happen to their HIV-infected babies? For one young woman who passed the HIV to her baby two years ago, the decision has been made. The baby has AIDS and is in foster care. The mother is very ill. The courts are now deciding whether her six other children should also be put in foster care.

Unless the course of the epidemic changes drastically, through the year 2005, the cumulative number of U.S. children, teens, and young adults left motherless due to AIDS will exceed 150,000. About 80,000 children under age 15 have already lost their mother or both parents to AIDS. The great majority of these children, uninfected by the virus—will begin to affect already burdened social services in major American cities. Things will get immediately worse in such places, where children already spend years going from foster home to foster home and caseworkers are overwhelmed by long lists of families needing everything from housing to medical care.

The Silent Legacy

Orphans are often referred to as the silent legacy of AIDS. It is expected that about a third of the children orphaned in America will be from New York City, which has the nation's largest number of AIDS cases. Other cities expected to be hit hard are Miami, Los Angeles, Washington, Newark, and San Juan, Puerto Rico. Most of these orphans will be the children of poor black or Hispanic women whose families are already dealing with stresses like drug addiction, inadequate housing, and health care. Relatives who might in other circumstances be called upon to care for the children often shun them because of the stigma attached to AIDS.

THE PHENOMENON OF AIDS ORPHANS

Sub-Saharan Africa

The HIV/AIDS orphan crisis is one of the greatest humanitarian and development challenges facing the global community. The orphan epidemic is still in its infancy. In the years and decades ahead, the impacts of HIV/AIDS on children, their families, and their communities will grow far worse—expanding to dimensions difficult to imagine at present. To date, over 90% of all children infected through mother-to-child transmission have been in sub-Saharan Africa. During 1995 through 2005, on average 600,000 children were born with HIV infection annually (about 1600 per day); of these children, 87% were in sub-Saharan Africa, 11% in Southeast Asia, 2% in Latin America and the Caribbean.

In Lesotho, Malawi, Mozambique, Swaziland, Zambia, and Zimbabwe people are battling a lethal mix of food shortages and HIV/AIDS. About one in four adults in the six countries now live with HIV or AIDS; increasing deaths and sickness have ground social safety nets way below the reach of poor households. The outlook for children is particularly bleak: The six countries are home to 2.5 million children who have lost one or both parents to AIDS, and 800,000 children under 15 who are HIV positive (Figure 11-7). In Africa,

FIGURE 11-7 AIDS Orphans Gathering to Get Food, Care, and Shelter in Malawi, Africa. (*Photograph courtesy of Ellen McCurley–The Pendulum Project*)

SIDEBAR 11.2

A GLIMPSE AT THE ORPHAN PROBLEM IN SOUTH AFRICA

To comprehend the AIDS pandemic ravaging Southern Africa/South Africa, one needs only to visit the village of Ingwavuma. Ingwavuma, a hilly community of eucalyptus trees and blooming orange aloe plants, is quietly suffering the same social meltdown that has devastated towns across KwaZulu-Natal, the South African province hardest hit by AIDS. Between 35% and 40% of Ingwavuma's population carries the virus. Many of the schools around Ingwavuma are struggling as teachers die and students miss classes to attend funerals of relatives. Annual school fees of about $5 a year have been slashed in half in many districts. Enrollment has nonetheless fallen by 30% over the last three years, and 75% of the kids are behind on their payments. The explosion in orphans around Ingwavuma is closing avenues to a better life even for the students who do manage to finish school.

Sthembile Nyawo, a 21-year-old recent graduate of Mthembu's school, had always hoped to find a job so she could help her family, particularly after her father's death from AIDS in 1998. In June, however, her mother died of AIDS, too. Now Nyawo spends her days looking after six of her younger brothers and sisters and trying to scrape together enough food with the help of her grandmother, whose $60 a month government pension is the only source of income for 16 family members.

Of the children playing in the dust outside Busisiwe Nhleko's hut, perched on a parched hillside near town, three are her brother's kids, orphaned after he and his wife died last year. Another four were dropped off by her dying sister. Four more arrived in the arms of various ill cousins, and her neighbors left another four when they died of AIDS; one baby has since died. Nhleko, age 38, who has 10 children of her own, now struggles to feed, bathe, and care for 24 kids each day, and without much help; her younger brother, her parents, and her husband have died of AIDS. "I don't have much time to tend to all of them. I've accepted all of this and tried to move on, but I still cry when the youngest ones call the older kids 'father' because they have no one else to call that."

Nhleko won't take time away from the 24 children she is raising to have an HIV test. "I've seen the dying and the pain of not being able to cure them. I couldn't stand to know. What would I do? What would they do?" she says, sweeping a hand toward the kids sitting in the dirt. "Life is not what it used to be," covering her face with her sleeve to hide the tears. "Every single person is a terrible loss," she said.

AIDS orphans are occurring at the rate of one every 15 seconds.

In April 2004, the International AIDS Trust and Children Affected by AIDS Foundation released a report stating that worldwide, every 14 seconds a child is orphaned by AIDS, and that by 2010 worldwide, there will be 25 million AIDS orphans (lost one or both parents to AIDS-related illness). The report also states that of the current 41 million people living with HIV/AIDS, about 3 million are less than age 15 and 12 million are between ages 15 and 24.

Mother-to-Child Transmission (MTCT) or Vertical HIV Transmission

Vertical transmission means that HIV passes directly from the infected mother into the fetus or infant. Data released in mid-2003 on 5000 mother-infant pairs showed that 40% of all HIV transmission within this group occured during breastfeeding, at least four weeks after delivery.

Stigma—Stigma is particularly strong surrounding mother-to-child transmission. The very phrase "mother-to-child" itself may be stigmatizing as it puts all the responsibility of transmission on the mother and none on the father of the child. Stigma stops women coming forward to get themselves tested. It reduces their choices when it comes to health care and family life once they are diagnosed as HIV positive and has a negative effect on their quality of life. Equally troubling is the lack of sympathy or respect given to pregnant women with HIV, especially in the developing nations where they are open to blame, ridicule, and rejection. For example, in rural Zambia a man stated, "If a pregnant woman is sick and has a sick and premature baby who dies before 3 months, then we know she is affected [infected with HIV] and turn away from her. This is our [HIV] test!"

Timing of HIV Transmission—Worldwide, over the past five years, 4 million HIV-infected children were born. Over 95% of these children were born in underdeveloped countries.

The exact time of HIV transmission to the fetus during pregnancy is unknown. It has been shown to occur as early as the fifteenth week of gestation, at or near the time of delivery and through breast-feeding.

Breast-feeding by mothers with HIV infection established *before* pregnancy increases the risk of vertical transmission by 16%. When a mother develops primary HIV infection while breast-feeding, the risk of transmission rises to 29%. In general, it is believed that 50% of HIV-positive babies are infected during the last two months of pregnancy and about 50% are infected during the birthing process or through the early months of breast-feeding (Miotti et al., 1999).

A working definition of the timing of maternal HIV transmission has been established to differentiate infants infected **in utero (in the uterus)** from those infected near the time of or **during delivery (perinatally).** In utero infection occurs in approximately 30% of HIV-infected infants. Children who are infected in utero have a more rapid progression to AIDS and generally become symptomatic during the first year of life. Those infected perinatally have no detectable HIV at birth but demonstrate HIV in the blood by 4 to 6 months of age. These children constitute the majority of HIV-infected infants and have a slower progression to AIDS, about 8% per year (Zijenah et al., 2004).

Rate of Transmission—The worldwide rate of HIV transmission from mother, without drug therapy, to child varies geographically. In Africa, maternal transmission is as high as 50%, producing about 1600 infected babies a day. In Europe and the United States, without the use of antiretroviral drugs the overall rate is 25% to 30%, producing less than 500 infected babies a year. The U.S. Public Health Service and 16 other national health organizations have recommended that HIV testing be offered to all women at risk prior to or at the time of pregnancy. Through year 2004, only two states—New York and Connecticut—mandate HIV testing for pregnant women, and seven states—Texas, New Mexico, Illinois, Tennessee, Michigan, Arkansas, and Florida—have pregnant women sign a test refusal form if they do not want an HIV test. The rate of vertical HIV transmission in New York state has decreased dramatically from 25% to about 3% since September 2003, when New York state mandated that all babies born to mothers not tested will be tested within 12 hours of birth.

Perinatal Transmission—Many factors that influence perinatal HIV transmission are not known; but, influencing factors do exist because one mother gave birth to an HIV-infected child, followed by an uninfected child, who was then followed by an infected child (Dickinson, 1988).

There are multiple factors involved in HIV transmission risk, including maternal immunity and viral load, placental conditions, route of delivery, duration of membrane rupture, and fetal factors (birth order, gestational age). This knowledge has led to trials of various interventions, such as drug therapy to reduce viral load and caesarean section to reduce HIV exposure during delivery.

Viral RNA Load Associated With Perinatal HIV Transmission—Although the close association between stage of HIV infection in a pregnant woman and likelihood of perinatal transmission has been established, there are no precise numerical criteria for pregnancies at high and low risk of transmission. Viral load measurements are now helping to quantitate this risk.

Through 2004, data continued to accumulate that suggest that the use of HAART and achievement of optimal viral load suppression is associated with the greatest reduction of vertical transmission. There is increasing concern about resistance using a single dose of Nevirapine for both mother and child. The identification of HIV infection in pregnant women is the biggest hurdle in reducing vertical transmission. Currently, there is a bill in the U.S. Congress that, if approved, would mandate HIV testing on any newborn whose mother refuses an HIV test.

States and Territories Most Affected by Pediatric Cases

There is evidence that HIV was present in female IDUs as early as 1977 because their babies developed AIDS (Thomas, 1988). As the number of

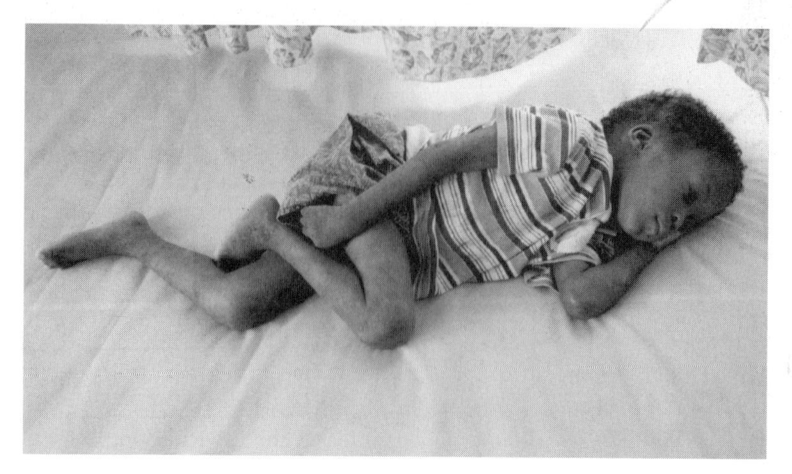

FIGURE 11-8 Josh, Age 3, Has Advanced-Stage AIDS. His older brother died from AIDS. Their mother died of AIDS, leaving them as AIDS orphans. *(Photo by Hilda M. Perez/South Florida Sun-Sentinel)*

HIV-infected women of childbearing age rises, so does the number of HIV-infected babies.

In the United States pediatric AIDS is most widespread among blacks, Hispanics, and the poor of the inner cities (Figure 11-8). Ending year 2005, New York will continue to have the highest incidence of pediatric AIDS cases, followed by Florida, New Jersey, California, Puerto Rico, and Texas. Combined, these cases accounted for 66% of all AIDS cases reported among children.

AIDS Cases Among Children Declining

There are many reasons to believe that with continued HIV counseling of HIV-positive pregnant women and use of AIDS drug cocktails that lower viral loads, fewer children will become infected and this will translate into a continued decrease in children with AIDS.

This possibility was inconceivable to many clinicians, families, and patients not too long ago. The rapid decline has many physicians suggesting that a goal of eliminating perinatal transmission may be attainable in the United States. However, Patricia Fleming of the CDC said at the 2002 International AIDS Conference, that it would be difficult to get the numbers much lower because the treatment isn't always effective. An estimated 130 HIV-positive babies still would have been born even if all the pregnant women had received treatment. The number of infected babies is likely to grow in the future because of the rise in new infections among women. Fleming said, "The simple fact is that the best way to prevent new infections in babies is to prevent infections in women."

Breast-feeding, Drug Therapy, and HIV Transmission

Without question, one of the most significant accomplishments of the HIV/AIDS era has been the dramatic reduction in transmission of HIV between mother and child. In America, from 1992 until the present time, perinatal transmission has declined over 80% and it is now possible, using combination anti-HIV drug therapy, to achieve transmission rates as low as 1%–2%, compared to 25%–30% a decade ago. (Figure 11-9.)

Conflicting Reports on Breast-feeding and Drug Therapy—Whether and Which Drugs Should Be Used Leads to Confusion

In reviewing the recent literature on breast-feeding, a variety of conflicting data have been reported by the respected authorities in this field of HIV/AIDS research. For example, it is not understood how AZT, Nevirapine, or other drugs lower the vertical HIV transmission rate, yet the following statements have been made:

1. If you give ZDV or Nevirapine to 100% of pregnant women, the mother should not breast-feed for fear of HIV transmission.
2. If you give ZDV or Nevirapine, safe and healthy formula or water to make formula is not available

FIGURE 11-9 Drops of Danger. This woman nurses her 4-month-old son. She, her husband, and three other children have been diagnosed with AIDS. She did not know she could pass HIV to her newborn via breast-feeding. She said that her husband was her only sexual partner. After her health began to decline and she tested HIV positive, her husband refused to be tested and he abandoned the family for another woman. *(Photo by Hilda M. Perez/South Florida* Sun-Sentinel*)*

or obtainable in most developing nations where vertical transmission rates are highest.

3. If you don't breast-feed, the babies will die of other diseases because they will lose the immunological benefits against the diseases that breast-feeding provides.
4. More babies will die because they are not breast-fed than will die from the small percentage that become HIV-infected perinatally.
5. A study published in 1999 and followed up in 2001 found that exclusive breast-feeding (with the addition of no other foods and liquids) leads to no additional risk of transmission of the virus than when there is no breast-feeding at all. And a review by the WHO Collaborative Study Team in 2000 noted that babies in developing countries who were not breast-fed were placed at a much higher overall risk of death from other infections.
6. On a scale of competing risks, a baby born in sub-Saharan Africa has a much greater chance of survival if breast-fed even by an HIV-infected mother than if breast-feeding is withheld.
7. Perhaps the most sobering comment concerning breast-feeding and the use of therapy is the "active use of drugs for millions of pregnancies may prevent hundreds of thousands of pediatric infections per year. However, these same women and their infected children exposed to short course Zidovudine plus Lamivudine or single-dose Nevirapine will be at substantial risk of treatment failure when antiretroviral therapy becomes available."

The World Health Organization and the United Nations continue to recommend the following: "When replacement feeding is acceptable, feasible, affordable, sustainable and safe, avoidance of all breastfeeding by HIV-infected mothers is recommended. Otherwise, exclusive breastfeeding is recommended during the first months of life."

Life Span of Untreated HIV-Infected Newborns in Sub-Saharan Africa

According to Taha Taha and colleagues of Johns Hopkins University (2000) over 50% of HIV-infected newborns die within 12 months. Eighty-nine percent of HIV-infected children alive at 6 months died by age 3 years. In comparison, the authors note that in a European study, only 18% of HIV-infected children died by age 3 years, and a U.S. study found that 75% lived to age 5 years.

HIV-Infected Newborns Now Having Children—a Third Generation of HIV/AIDS in America

Michelle McConnell of the CDC said, "It's a landmark in the HIV epidemic at least in the United States. Survival has increased to such an extent that not only are HIV-infected babies surviving but they're healthy enough to get pregnant and have healthy kids." McConnell was referring to the eight women living in Puerto Rico who contracted HIV from their mothers and who reported 10 pregnancies between August 1998 and May 2002. Five of the eight became pregnant accidentally; only two reported using condoms when they

POINT TO PONDER 11.1

DISTURBING THOUGHTS

Much has been said over the past year, by politicians and the press, about getting cheap antiretroviral drugs into developing countries in order to extend lives and reduce the transmission of HIV from the HIV-infected mother into her infant. However, Karen Beckerman of San Francisco General Hospital (2002) says that treatment programs that use antiretrovirals to prevent mother-to-child transmission are likely to succeed only in creating a generation of orphans and ruining the treatment chances of mothers. Beckerman raised the question "Is it justifiable to visit the antiretroviral mistakes of the industrialized world on regions that have been devastated by the HIV epidemic but are at least antiretrovirally naïve?" She states that active deployment of AZT/3TC during delivery for millions of pregnancies may prevent hundreds of thousands of pediatric infections per year. However, these same women and their infected children exposed to short-course Zidovudine plus Lamivudine or single-dose Nevirapine will be at substantial risk of treatment failure when antiretroviral therapy becomes available. Instead of inducing resistance to AIDS drug therapies, prophylaxis against mother-to-child transmission must be linked to preventing the creation of orphans.

Saving the Orphans, For What?

On average, about 600,000 HIV-infected babies have been born each year over the past seven years. But, save these children for what? To become orphans who then become street urchins who are poorly fed, undisciplined, unsheltered, and uncared for? These children form gangs and add to the growing crime wave in all orphan-populated cities. About 30% of these orphans are HIV positive and require care they cannot get. They are left in already crowded hospitals and homes that offer little more than a place to die. Regardless of the town, city, or country's economic plight, the growing mass of orphans tends to further decrease the economy and care available to healthy adults and school-age children. It is not universally agreed that saving an infant born to an HIV-infected mother is a sensible use of public funds in poor countries. It has been written that President Mbeki of South Africa follows this reasoning. He believes the cost of drugs and care would ruin the economy of South Africa. He is not alone in his beliefs. The infant will be born into extremely difficult circumstances. Death is predicted at least for its mother and probably for its father. Should public dollars be invested in preventing infant deaths—an investment that will translate into a rising tide of orphans? In hard-hit countries such as Zimbabwe, Botswana, Malawi, Namibia, Swaziland, and South Africa, it is predicted that ending 2005, one in three children younger than 15 years will be orphaned. Currently, about 40% of South Africa's population of 44 million people are under the age of 15. A report from the 2002 International AIDS Conference from UNAIDS states that 50% of South African new mothers could die because of HIV, and that mortality among 15- to 34-year olds will be 17 times higher because of AIDS.

African children infected and affected by HIV/AIDS are the ultimate development nightmare for a continent grappling with major socioeconomic problems. According to UNAIDS, every day 2000 infants contract HIV through their mothers throughout the world. At least 95% of these infants are born in Africa. Every day, worldwide, 6000 children lose one or both parents to AIDS. More than 90% of these children are Africans. Today in Africa, 95% of pregnant mothers do not have access to health programs that can significantly reduce the incidence of mother-to-child transmission of HIV. But, the worst is yet to come. According to UNAIDS, UNICEF, and USAID, by 2010 at least 20 million of the global 25 million AIDS orphans will live in Africa. This is in a continent where children face the deadly combination of high rates of infant deaths, vaccine-preventable deaths, under 5 mortality, diarrhea related deaths, and death from malaria. It is also a continent where children face major challenges of going to school, staying in school, eating nutritious meals, and having access to adequate sanitation.

DISCUSSION QUESTION: Take a stand, for or against saving the lives of some 42 million AIDS orphans who will populate the underdeveloped nations by year 2015. Support your presentation with facts, not just emotions. Answer the question, "Is the social cost of so many young people being raised without parental or adult guidance too high to tolerate?"

conceived. None of the babies born to the women, all of whom were teenagers when they conceived, were infected with the virus. All the mothers had received antiretroviral AIDS drugs consistently during pregnancy. The data showed that some women in the study reported becoming sexually active at around the same age that they learned of their HIV-positive status. That finding could indicate that teens and young adults infected with AIDS at birth are just as likely to engage in risky sex later in life as their peers who were not infected with the virus. It may also mean that the decision by many parents to shield their children from knowledge of HIV disease until later in adolescence may be too late. Since this report at least 15 similar cases have been reported by the CDC. In addition to these 23 cases, one has to consider the relatively large number of HIV-positive babies who have reached their teenage years and have become sexually active. As these numbers increase, the means of heterosexual/homosexual transmission increases among the young. Clearly, the use of highly active antiretroviral therapy is allowing an increasing number of young women who were born with HIV infection to live long enough to become sexually active and become pregnant (*MMWR*, 2003 updated).

GLOBAL HIV INFECTIONS IN YOUNG ADULTS

Over the summer of 1999, especially in the United States, young people ages 10 to 24, flocked to theaters to experience the **Dark Side of the Force.** Many thousands saw the STAR WARS movie repeatedly. But the real **Dark Side** of their lives is the threat of HIV infection and the **Force** should be their education to prevent their infection.

About half of all new HIV infections in the United States and worldwide are occurring in this age group. Hopefully, reading this section will encourage the young to stay on the **Light Side with the Force.**

In April 1998 the Joint United Nations Program on HIV/AIDS (UNAIDS) released a report showing that five young people ages 10 to 24 are now infected with HIV every minute, one every 12 seconds (it is now one every 10 seconds), and drew attention to the fact that Eastern Europe has emerged as a particular trouble spot for young people and HIV.

The UNAIDS report was released to mark the launch of a year-long initiative, "Force for Change: World AIDS Campaign With Young People," which aims to promote the participation of young people, and strengthen support for young people in their effort to fight AIDS. This report found that overall, children and especially young people still carry the heaviest burden of new infections. Among the key findings of the report (updated) are:

- More than half of all new HIV infections acquired after infancy occur among young people.
- Of the 41 million persons living with HIV/AIDS ending year 2005, at least a third will be young people.
- Every day, 7500 young people worldwide acquire the virus. This translates into 2.7 million infections each year in this group (ages 10 to 24).
- In the worst affected countries, HIV is spreading fastest among young people below the age of 24 years. And in places where the virus is spread predominantly through heterosexual intercourse—notably sub-Saharan Africa—67% of newly HIV-infected are young women. In South Asia it is 62%. In East and Southeast Asia it is 50%. In the Middle East and North Africa it is 41%.
- Overall, young people ages 14 to 24 make up about 20% of the world's population and account for about 50% of the 65 million HIV-infected ending 2005.

Young People and HIV/AIDS

Young people are central to any discussion of HIV/AIDS because there are so many! There are about 2 billion people ages 12 to 24 in the world. Half the world's population is under age 25. Young people are and will continue to be the sector of the population most affected by HIV/AIDS. As today's children grow up, the proportion of 12- to 24-year olds will continue to increase, particularly in developing countries.

POINT OF VIEW 11.2

PERILS OF UNSAFE SEX

As one teenager put it, "When you're a teenager, your hormones are raging and you think you're indestructible. But sex is how I got AIDS."

One of the hardest decisions facing a young person is whether to keep the disease a secret.

MTV Youth Poll—In 2000, less than 25% of young people surveyed in 16 countries across Asia, Europe, and the Americas consider themselves well informed about HIV/AIDS. Among the 16- to 24-year olds surveyed by the worldwide television network, 60% are concerned about becoming infected by HIV but just 24% said they know a lot about the disease. Some 25% of poll respondents believed only promiscuous people contracted HIV and 17% said it was improbable they would ever be exposed to the virus. (This means that 83% do not consider themselves to be at risk.) Seventy-nine percent of the poll respondents said they relied mostly on television as their news source for information about HIV/AIDS.

Why the High HIV Infection Rates Among Young People?

Of the world's young people, 85% live in developing countries, and this is where over 90% of the pandemic is now concentrated. But population percentages tell only part of the story. There are special reasons why young people are exposed to infection. Remember, that above all HIV is a sexually transmitted virus. Adolescence and youth are times of discovery, emerging feelings of independence, and the exploration of new behavior and relationships. It is also a time of examination, rebellion, and change. By definition adolescence is about taking risks and experimentation. Sexual behavior, an important part of this, can involve risks; the same is true of experimentation with drugs. At the time, young people get mixed messages. They are often faced with double standards calling for virginity in girls but early and active sexual behavior in boys. They have been told **"Just say no"** since the early 1980s, yet abuse continues, perhaps because they are confronted with hundreds of millions of dollars worth of media images of sex, smoking, and drinking as glamorous and risk-free. They are told to be abstinent, but exposed to a barrage of advertisements using sex to sell goods. Compounding the challenge, in the name of morality, culture, or religion, young people are often denied their right to education about the health risks of sexual behavior, and to important tools and services for protection. Among the world's young people, some are more exposed to HIV than others. Those living in what UNICEF terms "especially difficult circumstances" include young people who are out of school, who live on the streets, who share needles with other injecting-drug users, engage in commercial sex, or are sexually and physically abused. Young men who have sex with men are disadvantaged by the lack of information and services available to them and directed to their needs.

As the HIV/AIDS epidemic spreads, younger and younger age groups are becoming exposed to the risk of HIV. Infection spreads to younger age groups as men choose increasingly younger sexual partners. Many men believe, perhaps correctly, that younger girls are less likely to be infected with HIV, while others hold the mistaken belief that having sex with a virgin can cure AIDS.

HOW LARGE IS THE TEENAGE AND YOUNG ADULT POPULATION IN THE UNITED STATES?

In the United States there are about 28 million teenagers between the ages of 13 and 19 (73% white, 11% black, 16% Hispanic). This number is expected to increase to 43 million over the next 24 years (Sells et al., 1996 updated 2003). There are 18 million people ages 20 to 24, and about 22 million between the ages of 25 and 29. That's 68 million people between the ages of 13 and 29. Over 86 percent of all STDs occur in this age group. It has been estimated that 34% of all heterosexual adults with AIDS were infected with HIV as teenagers.

Young Women at Greatest Risk

In the United States, 25% of all new HIV infections are estimated to occur in young people between the ages of 13 and 20 and 50% in people under age 25. While more research needs to be done on this topic, several factors associated with young women are clear: (1) they tend to be partnered with **older men** who have had more sexual partners and have a greater chance of being infected with HIV and other sexually transmitted diseases; (2) their risk of HIV infection is greater because of their immature cervix and relatively low vaginal mucous production presents less of a barrier to HIV; and (3) many lack the education, social status, economic resources, and power in sexual partner relationships to make informed choices.

Safer Sex: Sex Thrills but AIDS Kills

There was a time when safer sex meant not getting caught by your parents. With time, sexually transmitted diseases and in particular HIV/AIDS have changed the meaning of safer sex. Today, over half of teenagers (ages 13 to 19) in the United States have had sex by the time they reach 16, and 7 in 10 are sexually active by 19. Many enter the sexual arena unprepared for the responsibility of their actions. About 1 million teenage women become pregnant outside of marriage each year.

Whether or not society openly discusses it, teens are having sex. Many women—and most men—have their first sexual relations prior to marriage, usually during their teens, and most often those first encounters are unprotected. Research in family planning has revealed that the quality of reproductive health information is generally low among adolescents. This is a reflection in part of the lack of social acceptance of providing sex education and contraceptive services to teens in many countries. In the developing world, contraceptive services are often available only to married women, and in some situations, only to women who have already borne one or more children.

The guiding philosophy in dealing with sexuality in many cultures is **"If you don't talk about sex they won't do it."** This logic, however, is critically flawed. Teenagers are sexual beings at varying stages of self-awareness and understanding. Many teenagers continue to engage in sexual intercourse despite lack of access to any accurate information about sex and, in most cases, engage in unsafe sex. **Safer sex requires an ability to distinguish between risky and nonrisky sexual activities and the emotional security to choose safer sex.**

Teens, Sexual Partners, and Sexually Transmitted Diseases

In a nationwide survey, 19% of high school students have had four or more sex partners by their junior year and 29% had four or more by their senior year (*MMWR*, 1992). Teenagers are experiencing skyrocketing rates of sexually transmitted diseases. Every 10 seconds a teenager somewhere in the United States is infected with an STD. (That's 3 million teenagers.) People under age 25 account for 66% of all new STDs every year. One in four teens will contract an STD before finishing high school. Experts fear that if these diseases are being transmitted, then HIV is too.

In April 1992, the *MacNeil/Lehrer Report* estimated that 40,000 to 80,000 teenagers will become HIV-infected during the 1990s. According to recent CDC estimates, it happened. Yet AIDS cases are relatively rare among 13- to 19-year olds. This is because of the 9- to 15-year time average from HIV infection to AIDS diagnosis. In 1981, there was one reported teenage AIDS case; by 1991 there were 789 reported AIDS cases. Ending 2005, there were an estimated 6000 (0.6% of the total AIDS cases). From 1991 to 2003 the number of AIDS cases in 13- to 19-year olds increased by 83%.

HIV/AIDS WON'T AFFECT US!

It is estimated that at least two teenagers per hour become HIV-infected. Regardless of available information on prevention, meaning that teenagers do know how HIV is transmitted, there has been a continuing increase in HIV infections within the preteen and early teen population. They continue to engage in sexual intercourse without condoms.

Two groups, teenage gay men and teenage women infected via heterosexual sex, account for about 75% of teenage HIV infections. Race is an important factor with regard to who becomes infected. Sixty-one percent of AIDS cases that occur in people ages 20 to 24, occur in blacks and Latinos, but they were HIV-infected in their teens (Collins et al., 1997).

ESTIMATE OF HIV-INFECTED AND AIDS CASES AMONG TEENAGERS AND YOUNG ADULTS

The total number of HIV-infected teenagers is unknown. **Federal health agencies estimate that teenagers make up about 20% or 200,000 of the HIV-infected population.** They are a silent pool for eventual cases of AIDS. About 20% of the total number of AIDS cases in the United States, or 1 in 5, occur in people ages 20 to 29 (about 194,000 ending 2005). About 50% of HIV-infected teenagers come from seven locations: New York, New Jersey, Texas, California, Florida, Washington D.C., and Puerto Rico. The overall male-to-female ratio of AIDS cases in the United States is now

POINT OF INFORMATION 11.4

HIV-INFECTED CHILDREN: LIVING LONGER—PAYING A PRICE

AIDS, in America, is the seventh leading cause of death in 15- to 24-year olds.

Adolescents infected with HIV since birth are a new population in the ever-changing HIV/AIDS pandemic. Data from the CDC show that before 1996, HIV-infected children lived to an average age of 9. After 1996, with the use of antiretroviral drugs, the average age has risen to 15 and continues to climb.

Six years ago, an estimated three children died each week of AIDS at Detroit's Children's Hospital. Today, that rate has dropped to about one child per year. But as the children's life expectancy increases, so do the number of complex social issues they must face, especially during adolescence. One mother of an HIV-positive child said she dreads the issue of dating because it's going to be emotionally crippling. "Unless current attitudes about those infected with HIV change, my daughter will not be very popular and that will hurt. It will just crush her." (Recommended reading, see Parker, 2000.) Her daughter is one of Michigan's 123 children under age 14 living with HIV. They have had their car tires flattened and parents of classmates forbade their children from playing with her. In response to these issues and new issues such as dating, the hospital has hired social workers and psychologists to hold support groups for these children, bringing them together to help them understand they aren't alone.

COMING OUT FOR AN "AIDS BABY": A SHORT STORY

If ever there was a time to tell her big secret, this was it, the seventh-grader thought. She and a few friends at a sleep over birthday party have sequestered themselves in a storage closet under a basement stairwell. They sat in a circle and talked for hours, promising, "Whatever we say here stays here." One girl shared her fear that her parents were on the verge of divorce. Another said she felt pressure to live up to her brother's example. There was a silence for a moment. Then the girl who had kept quiet for so many years took a deep breath and blurted a few quick words: "I have something to say. I'm HIV positive." Her friends took the news in stride. Until then, her friends had simply known her as their fun-loving buddy, the honors student, the girl with sarcastic wit who was as likely to use a big word they didn't understand as to address her friends as "dude." Now her friends knew something more: she was born an "AIDS baby," a term only vaguely familiar to most people her age. Three years after her disclosure, she reflects and strongly feels that she did the right thing. Coming out for these children is a complicated and terrifying task.

about 5:1, but among 13- to 19-year olds, the ratio in 1997 became about 1:1.

Sixty percent of new HIV infections in women now occur during their adolescent years (ages 13 to 24). About 500,000 people in the United States under the age of 29 are HIV positive.

Teenagers and Incidence of AIDS Cases by Gender and Color

Overall, it is estimated that there are over 13,000 AIDS cases in people under age 20. Over half were diagnosed before age 5.

Teenagers of color account for 47%, whites 32%, and Hispanics 19% of reported teen AIDS cases. Among males, white teenagers account for 41% of reported teen AIDS cases, followed by blacks (36%) and Hispanics (21%). Among females, black teenagers account for 66%, white for 17%, and Hispanic for 17%. Teenage females, unlike their adult counterparts, are more likely to become infected with HIV through sexual exposure than through injection-drug use. A program that followed a large group of HIV-infected adolescents found that although 85% of females contracted HIV infection through heterosexual intercourse, very few were aware that their male partners had HIV infection at the time of their exposure (Futterman et al., 1992).

Categories of Teenage HIV Infection

From 1996 through 2005, the breakdown for HIV infection among teenagers was estimated as follows: males who have sex with males account for about 29%, patients with hemophilia and

POINT OF INFORMATION 11.5

FEDERAL PUBLIC HEALTH POLICY SUPPORTS ABSTINENCE-ONLY PROGRAM: SEX EDUCATION

IN ONE LESSON—NO SEX!

When it comes to teens and sex, abstinence forces say the message is simple—don't do it. The other side says give them all the facts—including how to use a condom. No wonder parents and their children are confused.

Jenny, a cartoon teenage virgin, is about to give in to her boyfriend and climb into the backseat of his car. Suddenly, the emergency brake gives out and his car rolls until it teeters from a cliff off lover's lane. Their lives hang in the balance. That is when Windy, the good witch in hightops, leaps to the rescue. "Paul loves me," Jenny protests. Windy asks, "Oh. Is that why he asked you to do something that could mess up your life forever?" Using her time machine, Windy shows Jenny how she would have awakened pregnant. Had the car's brake not failed her boyfriend's condom would have. The cartoon, shown to sixth-graders at Burbank Elementary School, is one weapon in an arsenal of films, celebrity rallies, and school classes pushing a message of chastity in classrooms around the nation. Currently, only 18 states and the District of Columbia require schools to provide sex education.

Will it work? Years of heated debate on the failure versus benefits of **sexual abstinence-only** education continues. However, in August 1996, with the passage of the **Welfare Reform Act** the United States Congress allocated $50 million annually for five years to states that would institute abstinence-only educational programs (no discussion of contraception). This legislation specifically required funded programs to teach the social, psychological, and health gains to be realized by abstaining from sex; that abstaining from sexual activity outside marriage is the expected standard for all school-age adolescents; that a mutually faithful monogamous relationship in the context of marriage is the expected standard of human sexual activity; and that **abstaining from sexual activity is the only certain way to avoid pregnancy, STDs, and other associated health problems.**

Clearly there are numerous health, economic, and social benefits in delaying sexual onset in teenagers like disease prevention, but scientists say they **cannot** identify a long-term advantage to abstinence programs relative to safer sex programs.

Prevention Messages on Sexual Activity That Teenagers Should Hear?

Perhaps young people should receive two messages: one, promoting abstinence and the delay of sexual activity, the other, warning against high-risk behaviors and teaching teens how to protect themselves. These messages are not contradictory, but they are complex. "Don't drink, but if you drink, don't drive" is a similar complex message that has saved many people from death on highways. Prevention scientists offer significant evidence that safer-sex interventions work. But there is no clear and compelling evidence that abstinence-only programs work (DiClemente, 1998). The work of John Jemmott and colleagues (1998) and Douglas Kirby and colleagues (1998) demonstrate that for those adolescents that were sexually experienced at the beginning of their study, there was no difference in the proportion of the adolescents in the abstinence program relative to those in a safer-sex program with regard to having sexual intercourse. For those not sexually experienced, abstinence education was effective for a short time (three months). After 12 months the adolescents in the two groups were each as likely to become sexually active and were similar in their pregnancy and sexually transmitted disease rates.

Abstinence Catching on in Africa

Kenya—where 2.3 million out of 30 million people are HIV-infected, where 700 people die each day from AIDS, President Daniel Arap Moi has urged Kenyans to abstain from sex for at least two years to try to curb the spread of AIDS. The government announced plans to import 300 million condoms.

Swaziland—where 25% of its people are HIV positive, King Mswati III declared a five-year sex ban on women under age 18 in order to help stop the spread of HIV. Any man breaking this rule will be liable to pay about $160 or an animal, such as a cow. The tradition of preserving maidens' chastity, known as Imabali YeMaswati or Flower of the Nation, will be policed by traditional chiefs who still rule over much of Swazi society to preserve virginity among girls and combat AIDS. Under the rite, the girls wear woolen "do not touch me" tassels or different colors depending on their ages. Women in relationships and older than 19 years would be expected to wear red with black tassels, and those still virgins will wear blue with yellow.

Irony: King Mswati, eight weeks after his declaration, broke his sex ban rule. He married a 17-year-old woman as his eighth wife. He was fined one cow.

Namibia—In Namibia, 22% of the people are HIV positive. In attempting to educate ages 15 to 25 to abstain from sexual experiences, a survey of this group revealed that two major terms, "abstinence" and "faithfulness," were not understood. To these young people abstinence means "to be absent," and faithfulness means "faith," a religious meaning. Seventy-five percent of this age group never heard the word "monogamy" (generally referring to being faithful to one person—dictionary meaning "marriage with one person at a time").

Teenagers' Perception of Abstinence

The 2001 Alan Guttmacher Institute survey revealed that those ages 15 to 19 considered oral sex to be a substitute for sex. They believe that oral sex is risk free! In addition to this survey are the data from other public health-care workers who have found that young people believe that neither anal nor oral sex places them at risk and that these activities maintain their virginity! In 2002 the Henry J. Kaiser Family Foundation with *Seventeen* magazine published the results of a teen survey on oral sexual behavior. The results were 23% of students questioned in 7th through 12th grade said they have had oral sex. In 11th and 12th grade the number increased to 42%. Thirty percent didn't know that a boy or girl could become infected with HIV by having oral sex. In a 2003 survey at Northern Kentucky University, of 600 teens who took the abstinence pledge, 61% broke that pledge within a year. Of the 39% who did not break that pledge, over half had oral sex! One in three schools nationwide teaches an abstinence-only curriculum that forbids talking about oral sex or safer sex. Clearly some of the students who are getting the abstinence-only message believe that they are engaging in abstinent behavior when they are having oral sex. In defense of these young people it can be said that health educators themselves are no more clear. And why not? In 1998, President Clinton, in testimony about an affair with a White House intern said, with great sincerity, that he had not had "sexual relations" but had engaged only in oral sex!

UPDATE 2005 For 2003, the federal government invested $120 million in abstinence education. In 2005 the government invested $270 million. There are now over 700 abstinence-only sex education programs, bolstered by millions of federal dollars, within the United States.

An interim report written by independent researchers, (Mathematica Policy Research, under contract to the Department of Health and Human Services), stated that there is no evidence that abstinence-only programs prevent teen sex, pregnancy, or disease. Despite claims by advocates, no reliable evidence exists on whether the programs work. Most studies of abstinence education programs have methodological flaws that prevent them from generating reliable estimates of program impacts.

DISCUSSION QUESTION: Is it good science, poor politics, or poor science and good politics, or some other combination of events that made it logical for the federal government to earmark tax dollars specifically for abstinence-only educational programs? Do you sense a religious involvement in the government policy? Explain.

other coagulation disorders account for about 31%, heterosexual contact represents about 14%, injection-drug use about 17%, and non-hemophilia-related transfusions and organ donations about 9%.

Runaway Teenagers

Each year since 1993, an estimated 3.4 million adolescents dropped out of high school. Youth dropouts have higher frequencies of behaviors that put them at risk for HIV/STDs, and are less accessible to prevention efforts. Many teenagers run away from home.

The Homeless

An estimated 1 to 2 million runaway teenagers are homeless each year. These youth may engage in behaviors such as injection-drug use, having multiple sexual partners, exchange of sex for money

BOX 11.5

HIV SPREADING RAPIDLY AMONG TEENAGE GAY MALES

Combined results of the first two national surveys of young gay and bisexual men showed that 7% in the surveys were HIV-infected (*Health*, 1996 updated). Researchers interviewed 5273 men ages 15 to 22 in six urban counties: Miami's Dade County, Dallas County, and Los Angeles, San Francisco, Santa Clara, and Alameda counties in California. Results show that 5% of the young men ages 15 to 19 and 9% of those ages 20 to 22 were HIV-infected. Four percent of whites, 7% of Hispanics, and 11% of blacks were infected. And, 38% reported unprotected anal sex within the previous six months. These 1996/1999 data, when compared to those data of a similar 1994/1995 survey, were about the same. Education and prevention control measures do not appear to be changing teenage gay sexual behaviors with regard to preventing HIV infection.

or drugs, and unprotected sexual intercourse that place them at risk for HIV infection. HIV seroprevalence studies among homeless youth have shown rates that are higher than adolescents in other settings (Rotheram-Borus et al., 1991).

In 9 of 16 STD testing sites, HIV seroprevalence rates among homeless youth were higher than the median rates of youth attending STD clinics within the same city. These studies can only hint at how far HIV has infiltrated specific groups of teenagers.

If you are a teenager or know of one who needs help or has HIV/AIDS questions, call:

National Teenagers AIDS Hotline: 1-800-234-8336.

Adolescent AIDS Program: Montefiore Medical Center, 111 E. 210th St., Bronx, NY 10467; 1-718-882-0023.

AIDS Community Alliance: Works with HIV-positive and HIV-affected individuals. 44 North Queens St., Lancaster, PA 17603; 1-717-394-3380.

Bay Area Young Positive: Youth-run, offers counseling, resources, newsletter. 518 Waller St., San Francisco, CA 94117; 1-415-487-1616; email: BAYPOZ@aol.com.

SIDEBAR 11.3

CAN LIFE GET WORSE THAN BEING YOUNG, HOMELESS, AND DYING OF AIDS?

Regardless of cause, to have AIDS and nowhere to go, no one to help, and no one who cares about you means that one has hit the bottom rung of life. Many young people are there now! There are 15,000 to 20,000 young homeless people just in New York City. About 40% of them sell themselves in order to survive. Too many are becoming HIV-infected because of the demands of their lifestyle.

Case 1. In the 10th grade he was kicked out of his home—over time he meandered to San Francisco's tenderloin area. He began working as a street prostitute and injecting drugs. **He wanted to become HIV-infected.** He said he actually cried when he tested HIV negative. At age 19 he tested positive. This time he cried because he now recognized the mistakes he made along the way. He had no money, no job skills, and no friends. Once, near death he said he had an epiphany (a sudden realization). He went to a clinic, received job training, gave up drugs, got a job and said, "I really want to live!"

Case 2. Her history begins in a North Carolina group home. She left and drifted to Georgia where she believes she became infected from a man who gave her $300 to have unprotected sex. She said she was a night child—she went to night clubs and adult bookstores to pick up her tricks. By her twenty-first birthday she said she was ready to jump off the Golden Gate Bridge—if she could find it. After finding help at a clinic she is now preparing to take her GED (graduate equivalence degree).

Case 3. He was raped by the son of his foster parents. His life was spent in a series of Milwaukee foster homes. He became HIV positive at age 13. He found his way to San Francisco. He said "I felt I would be accepted there, it's a city that understood AIDS and the people that have it." He goes to a clinic that services 3000 other young homeless people. He believes he has been sentenced to death. He has not been able to become comfortable with his HIV status. (Adapted from Russell Sabin, *San Francisco Chronicle,* Oct. 18, 1998.)

POINT OF VIEW 11.3

THOUGHTS AND COMMENTS FROM A GENERATION AT RISK

"I was only 13 when I started having sex. I knew what AIDS was, and how you get it, but I was more worried about something else: getting pregnant. In fact, it was a visit to the health department to get birth control injections in January 2002 that I discovered I had HIV. I couldn't believe it, the disease I read about in health class and heard about on television and in movies was now a part of my life. I never thought it would happen to me. Now at age 16, I am back in school and take anti-AIDS drugs twice daily. I still have sex, I don't tell my boyfriends, but I make them use a condom.—**From Virginia**

I became HIV-infected at the same time I lost my virginity—at age 16. My 28-year-old boyfriend was an injection-drug user. He knew he was HIV positive but did not tell me. He has since died of AIDS. At 16 my only concern was pregnancy, so I took the pill and had unsafe sex. Living in Spain does not help either as the HIV-infected are discriminated against—so be careful out there.—**A message from Spain**

"If you're going to educate kids about AIDS, you have to educate them about drugs as well. If you're a youth, you're going to experiment with drugs, especially if you live in a metropolitan area. Even though you get stupid with drugs, you still think about things you don't want to do, but you do it anyhow."—**16-year-old HIV-positive youth from San Francisco**

"We grow up hating ourselves like society teaches us to. If someone had been 'out' about their sexuality. If the teachers hadn't been afraid to stop the 'fag' and 'dyke' jokes. If my human sexuality class had even mentioned homosexuality. If the school counselors would have been open to a discussion of gay and lesbian issues. If any of those possibilities had existed, perhaps I would not have grown up hating what I was. And, just perhaps, I wouldn't have attempted suicide."—**Kyallee, 19**

"People say HIV is this or that group's problem, not mine. But for HIV, it's a matter of risk behaviors, not risk groups. Because if you say it's a risk group thing, I don't identify with that group, so I'm not at risk. That makes people feel invincible to HIV."—**HIV-positive youth**

"I was infected with HIV by my first partner when I was 16 years old. Now at 20 I have this virus that's taking my life because everything I heard when I was younger was sugar-coated. We need more complete information than what we are being given. Even the pamphlets concerning HIV/AIDS prevention are too basic and bland. We need to know real stuff."—**Ryan, age 20**

"We, the young people of this country, need a place where we can go to ask our questions, where we won't be teased or ridiculed. We need a place where we can ask about our mixed up feelings, about sex, and about AIDS."—**15-year-old high school student from Concord, N.H.**

"If I could talk to the president, or a senator, or anyone in the federal government who can make a difference, I'd tell them to take a look, learn a lesson from the youth that are currently dealing with the disease. Listen to them, hear their stories and then see that they have a future. If they don't have that future, then we don't have an America."—**Allan, San Francisco**

(Adapted and updated from a *Report to the President,* March 1996)

Comments: Unlike teens who were infected by their mother perinatally and have grown up with the virus, the newly infected teens have daunting issues dumped on them virtually overnight. Do they tell anyone? How do they handle dating? How do they tackle the emotions clouding future relationships? And more immediately, how do they take on a life-saving medical regimen when they have the willpower of a teenager? Teens also have characteristics that work against treatment. They lead chaotic lives. Shun authority. Keep secrets. Feel invincible. And wear defiance like a badge of courage. That "you can't tell me what to do" attitude can be deadly. Some teens may not even know they have the virus. Years can pass before their viral loads are high enough to produce symptoms.

DISCUSSION QUESTION: If you were an HIV positive teen, would you tell others? Choose one and discuss.

A. Yes

B. Only with family and close friends

C. Only once I reached adulthood

D. Only if I planned to have sex

E. No

F. An option not listed

SUMMARY

AIDS surveillance and HIV seroprevalence studies indicate that a significant proportion of HIV infection among women in the United States is acquired through heterosexual contact. Because more men than women are HIV carriers, a woman is more likely than a man to have an infected heterosexual partner. The predominance of heterosexually acquired HIV infection in women of reproductive age has important implications for vertical HIV transmission to their offspring: Nearly 30% of children with AIDS were infected by mothers who acquired infection through heterosexual contact. One of the greatest tragedies of the AIDS pandemic is orphaned children. They are left in hospitals because (1) their parents have died of AIDS or cannot care for them, or (2) no one wants them. AIDS at present is relatively rare among 13- to 19-year olds, but 20% of AIDS cases, due to a 9- to 15-year period before AIDS diagnosis, had to begin with HIV-infected teenagers.

For the time being, however, abstaining from sex, mutual monogamy between uninfected partners, and the correct and consistent use of condoms are the only options that can be presented to young people for avoiding the sexual transmission of HIV. In order to decrease their risk of HIV infection today, it is essential that youth receive education about HIV and have access to health and rehabilitative services.

INTERNET

1. **The Coalition for Positive Sexuality (CPS) website (http://www.positive.org/cps),** which provides information and advice on sexuality, is produced by and targets adolescents. The coalition is a grassroots volunteer group based in Chicago. Their self-described mission is "to give teens the information they need to take care of themselves and in doing so, affirm their decisions about sex, sexuality, and reproductive control; second, to facilitate dialogue, in and out of the public schools, on condom availability and sex education." Included among the topics is information about safe sex, birth control, STDs, pregnancy, and being gay. Homosexual relations are discussed in the same manner as heterosexual relations.
2. Although CPS is aimed at youth in general, **Oasis (http://www.oasismag.com)** targets and is written primarily by gay youth. Most of the columns written by contributors, who range in age from 14 to 22, read a lot like personal high school journals, an approach that undoubtedly makes readers feel comfortable—like hearing from a friend. A monthly advice column on sexual health is written by a physician and an epidemiologist, who are based in the San Francisco area.

OTHER USEFUL SOURCES

National Runaway Switchboard: 1-800-621-4000

National Network Runaway Youth Service: 1-202-783-7949

American Institute for Teen AIDS Prevention: 1-817-237-0230

Teen AIDS Student Coalition on AIDS, Washington, D.C.: 1-202-986-4310

Positive Pediatrics/AIDS: 1-518-798-8940

Teen AIDS CDC: 1-800-342-2437

Teen AIDS Hotline: 1-800-440-8336

National Gay/Lesbian Youth Hotline: 1-800-347-8336

REVIEW QUESTIONS

(Answers to the Review Questions are on page 438.)

1. By the end of year 2006, how many women are expected to be HIV positive worldwide?

2. Globally, what percent of *new* HIV infections occur in women?

3. What are the major routes of HIV transmission into women?

4. What is the most likely way a female prostitute in the United States becomes HIV-infected?

5. During year 2006, how many women worldwide will have become HIV positive and how many will have died from AIDS?
6. AIDS is now the _____ cause of death for all women between ages _____ and _____. It is the _____ leading cause of death in _____ women between the ages of _____ and _____ and the _____ cause of death for black women ages _____ to _____.
7. By year 2004, how many HIV-infected women are of childbearing age in the United States?
8. By year 2004, how many states have not reported a pediatric AIDS case?
9. Since 1995, what percentage per year of HIV-infected newborns received HIV from their mothers?
10. List three major factors that are associated with perinatal HIV transmission.
11. Where do most of the orphaned AIDS children come from? Why are they called AIDS orphans?
12. What percentage of teenagers have had sexual intercourse by the age of 16?
13. Currently, _____ teens per hour are being HIV-infected in the United States? Worldwide _____ per hour.
14. How many teenagers are there between the ages of 13 and 19 in the United States?
15. What two groups of teenagers in the United States (adolescents) account for the most HIV infections _____, _____? What is the percentage?
16. What percentage of all new HIV infections in the United States occur between ages 13 and 20?
17. About how many people in the United States under the age of 29 are estimated to be HIV positive?
18. Is race an important factor with regard to who becomes HIV-infected? What race or races have the largest number of HIV-infected young adults?
19. Of the 41 million people living with HIV disease at the end of 2005, how many are young adults _____?
20. HIV/AIDS Crossword

ACROSS	DOWN
2. Primary source of pediatric AIDS	1. Sex worker
6. Most severely impacted cultural group	3. Number of copies of HIV RNA in the blood
7. Make up 20% of the infected population	4. Biologically more vulnerable
8. Maternal infection	5. Women who have sex with women
	9. Barrier

12 Testing for Human Immunodeficiency Virus

Chapter Concepts

- ELISA means **e**nzyme **l**inked **i**mmuno**s**orbent **a**ssay; it is a large-scale screening test for HIV infection.
- Western Blot is a confirmatory HIV test. It confirms the results of ELISA.
- The ELISA test has been used to screen all blood supplies in the United States since March 1985.
- HIV screening tests can produce both false positives and false negatives.
- A positive ELISA test only predicts that a confirmatory test will also be positive.
- False-positive readings result from a test's lack of specificity.
- There is a relationship between the incidence of HIV in the population being tested and the number of false positives reported. The higher the incidence, the fewer the false positives.
- Several new screening and confirmatory HIV tests are now available.
- June 27 is national HIV Testing Day.
- Screening of the nation's blood supply has improved.
- Saliva and urine HIV tests are FDA-approved.
- The polymerase chain reaction test is the most sensitive HIV RNA test currently available.
- Other HIV RNA tests available are Amplicor and the branched DNA test.
- Currently there are six FDA approved rapid tests available in the United States.
- AIDS cases have been reported in all 50 states.
- All states now have a name or coded HIV reporting system.
- Competency and informed consent are necessary for most HIV testing.
- Mandatory HIV testing does not mean people can be forced to undergo testing.
- HIV testing, for the most part, is on a voluntary basis.
- Compulsory HIV testing is used in the military, prisons, and in certain federal agencies.
- FDA has approved two home HIV test kits; one remains on the market.
- U.S. Public Health Service guidelines for annual prenatal HIV counseling and voluntary HIV testing of all pregnant American women.
- New York is the first state to legislate mandatory HIV testing and disclosure of newborn HIV status to mother and physicians.
- American Medical Association endorses mandatory HIV testing of all pregnant women and newborns.

Let's begin this chapter by asking, **WHY DOES TESTING MATTER? ANSWER:** Basic epidemiology holds that early knowledge of where a virus is moving—into which populations—is essential to slowing its spread. Even if a disease cannot be cured, knowing who the infected people are may help prevent the transmission of the disease to other people.

As of the beginning of 2006, of the estimated 41 million living HIV-infected people worldwide, at least **half** will become infected **before age 25.** About 10% will know they are HIV positive. HIV testing is not readily available in many places in developing nations. Where HIV testing is available, testing is seen as an important part of prevention and care. This chapter presents HIV testing information and some of the important problems connected with whom, how, and where to test.

HIV antibody testing has been available since 1985. Testing technology has evolved considerably over the years, with a variety of new and improved tests coming into use in daily practice. Because determining one's HIV status is the first step in prevention and treatment decisions, it is important to understand the tests being used today, including their limitations.

DETERMINING THE PRESENCE OF ANTIBODY PRODUCED WHEN HIV IS PRESENT

HIV antibody testing is a readily available, inexpensive, reliable, and accurate method to identify whether a person is infected with HIV. HIV antibodies are found in the blood and in other body fluids. When properly performed, HIV antibody testing is highly sensitive and specific.

Currently there are at least nine tests that detect HIV antibodies, antigens, or the nucleic acid of HIV in a person's body fluids. They are the: enzyme linked immunosorbent assay (ELISA), Western Blot, polymerase chain reaction (PCR), saliva and urine tests, Aplicor's branched DNA test, immunofluorescent antibody assay, rapid HIV test kits, and at-home HIV test kits. They are discussed in the following pages.

REQUESTS FOR HIV TESTING

HIV testing is offered at some 11,600 CDC-publicly funded sites and in other public and private settings. These testing sites are becoming overwhelmed with requests for HIV testing. But, most of the requests are repeats. The majority of those people at risk who have not been tested includes most hardcore IDUs and sexual partners of IDUs and people who are image sensitive. Currently, about 50 million blood and plasma samples are HIV tested annually worldwide. In the United States about 2.5 million blood and plasma samples are HIV tested annually.

In the developing world, where the greatest number of HIV-infected people are concentrated, HIV testing is done mostly for purposes of surveillance, which involves very small population samples and is done anonymously. Few people have any hope of treatment, so they feel little incentive to get tested. But even those who would want to know may not be able to find out. In many countries, there are no voluntary testing and counseling facilities; people have no acceptable way of learning if they are HIV-infected. An ongoing study at a rural hospital in South Africa suggests that only 2% of people who are HIV positive know their status. The situation in urban Kenya is equally poor.

SIDEBAR 12.1

ANNUAL NATIONAL HIV TESTING DAY

National HIV testing day began June 27, 1995. It was initiated by the **National Association of People Living with AIDS** in conjunction with the CDC. June 27, 2005, is the eleventh anniversary. The purpose of the annual **HIV Testing Day Campaign** is to encourage those at risk for HIV to take personal control and responsibility for their own health. It is estimated that about 300,000 Americans are unaware they are HIV-infected. In 2004 over 11,000 national, state, and local organizations participated in the campaign. Through 2005 about half of U.S. adults have been HIV tested.

REASONS FOR HIV TESTING

HIV testing is done to monitor the pandemic; to determine how many people are infected, how many are being infected in a given time period (incidence), and their location. Testing is used to determine the impact of prevention efforts to slow the spread of HIV, to prompt behavior change and to provide entry into clinical care. Also, if necessary, to provide a starting point for partner notification and education, and to protect the nation's blood supply.

Take the Test: Take Control

There is an immediate need to change perception about being HIV positive so that people feel good about taking the test to protect themselves and others, rather than the discrimination that now exists against those who have taken the test.

The Need for Routine HIV Testing

After 24 years of educating people about HIV/AIDS, about 50% of adults in the United States have never been tested for HIV (The CDC reports that 50% of heterosexuals and 35% of gay, lesbian, bisexual, and transgender adults have never been tested). Eighty percent of heterosexual adults say their reason for not being tested is that they do not consider themselves to be at risk. Unless people are tested in greater numbers, it may be impossible to break the cycle of HIV transmission in America. The failure of educational programs to stem the rates of 40,000 to 45,000 new HIV infections annually may need to be replaced by new policies on HIV testing. Joseph Inungu, a Central Michigan health science professor, is recommending an HIV test as part of the routine tests performed on office patients. Inungu analyzed data on people ages 18 to 80 who participated in a national health interview survey conducted by the National Center for Health Statistics. He found the groups least likely to be tested for HIV were men, people over the age of 50 or between age 18 and 19, people with a low level of education, people in rural areas, and people in the Northeast or Midwest. While adolescent groups are among the fastest growing population with HIV infection, few are tested for HIV unless they come to a physician for a sexually transmitted disease. Populations over the age of 50 have tended to feel they were not at risk but their numbers of HIV infected continue to increase. The groups most likely to get tested include blacks, people with higher educational levels, and people who are separated, divorced, or widowed.

IMMUNODIAGNOSTIC TECHNIQUES FOR DETECTING ANTIBODIES TO HIV

Refinements in the field of immunological testing, serology, and the study of antigen–antibody reactions have produced test names that reflect the component parts of the test being used. In most cases, tests are based on the detection of antibodies present in the serum, in this case antibodies to HIV. One immunological test uses antibodies, which if present in the person's serum, form a complex with a given antigen. An enzyme is then connected to the antibody. The presence of the antibody can be determined by adding a reagent that will form a colored solution if antibody to HIV is present. This is called the **enzyme linked immunosorbent assay** (ELISA). The ELISA (E-liz-a) test was first used in 1983 to detect antibodies against HIV.

Because the ELISA test detects the presence of antibodies made against HIV, it is called an **indirect test.** This test suggests that HIV is or was present. The Western Blot test, presented next, is also an indirect test for the same reason. Indirect tests stand in contrast to those tests that directly test for the presence of HIV's nucleic acid—a **direct test.** These tests say that HIV is currently present in the system. They are also presented.

ELISA HIV ANTIBODY TEST

A Fable for the HIV/AIDS Era

It was a medieval mystery. Somehow, needles were finding their way into some of the kingdom's haystacks. Cows were eating the needles: not a good thing. The king sent out a proclamation offering a sack of gold to the first person able to find a needle in a haystack. After 20 days, the contestants were still trying to locate the needles hidden among 10 haystacks placed in the palace courtyard. The king was despondent. From his tower he could see thousands of haystacks in fields across the kingdom. "This is terrible," he said. "Either our cows go hungry while we look for the needles, or we let them eat the hay along with some needles. Either way it's not good for the cows. We need a way to tell which haystacks have the needles; then we can feed our cows the good hay while we figure out how to get the needles out of the bad haystacks." He was a logical king. In 1981 the needle had no name but its presence was known. The search for this needle—like those in the haystacks—was intense and even after the needle was found to be HIV, investigators had to find a way to distinguish between those who carried and did not carry HIV—which haystacks carried the needles. Because HIV was in human blood it was essential to protect the nation's blood supply, thereby preventing people from receiving contaminated blood and blood products. But, like the good king, it took time to find the means to find the needle in the blood supply—HIV.

It was not until 1985—nearly four years after the first cases of AIDS were announced—that an antibody test was developed that could indicate whether a person was infected with HIV. Even after this discovery, the fact that there were no effective treatments for those who tested HIV positive, coupled with the widespread perception of stigma associated with HIV infection, left many questioning the value of HIV testing. That equation shifted markedly with the availability of potent combination therapy, which significantly delays the progression of HIV disease in many people. There is now widespread consensus among public health officials and community leaders regarding the importance of HIV testing and counseling in order to link individuals who test positive with medical care and to counsel them on how to reduce the risk of further transmission.

Screening the Nation's Blood Supply

The initial application of the ELISA test outside the research laboratory was used primarily in large-scale screening of the nation's blood supply. ELISA **testing of the existing blood supply and all newly donated blood began in the United States in March 1985.** Very quickly, however, testing also became seen as an important aspect of HIV prevention. In recent years, the discovery of treatments for HIV and associated opportunistic infections has further increased the benefits of early detection.

The ELISA test is used as a screening test because of its low cost, standardized procedures, high reproducibility, and rapid results.

Whole viruses are disrupted into subunit antigens for use. The subunits of HIV are then bound to a solid support system.

Two different solid support systems are used in the seven ELISA screening test kits licensed in the United States. Some attach or fix the antigens onto small glass beads (Figure 12-1), while others fix the antigens onto the sides and bottoms of small wells (microwells) in a glass or plastic microtiter plate. The serum to be tested is separated from the blood and is diluted and applied to the HIV-coated solid support systems (Figure 12-2). The ELISA test takes from 2.5 to 4 hours to perform (Carlson et al., 1989) and costs between about $8 in state-sponsored virology laboratories and about $60 to $75 in private laboratories.

In accordance with FDA recommendations, effective June 1992, blood collection centers in the United States began HIV-2 testing on all donated blood and blood components. The CDC does not recommend routine testing for HIV-2 other than at blood collection centers (*MMWR*, 1992).

Understanding the ELISA Test

The ELISA test determines if a person's serum contains antibodies to one or more HIV antigens. Although there are some minor differences among the FDA-licensed kits, test procedures are similar.

Problems with the ELISA Test

Any HIV screening test must be able to distinguish those individuals who are infected from those who are not. **The underlying assumption of an ELISA test is that all HIV-infected people will produce detectable HIV antibodies.** There are, however, problems with this assumption. **First,** although rare, there are documented cases of individuals who are infected but remain antibody negative. **Second,** the HIV-infected population in general does not produce detectable antibodies for six weeks to one or more years after HIV infection. This is called the **window period.** Most often, HIV antibody is detectable within 6 to 18 weeks. Thus, HIV-infected people can test HIV negative. This is a **false negative result.** In some HIV-infected persons, the virus ties up the available antibody as their disease progresses. Testing at this time may also produce false negative results.

An Unusual Case—At the VA Medical Center in Salt Lake City in 1997, a man tested HIV negative 35 times over a four-year period. Because his wife was HIV positive and because he demonstrated symptoms of HIV disease, tests other than the ELISA showed that he was HIV positive. (See Point of Information 12.1.) This case is unusual because (1) he was falsely negative almost four years beyond the window period; (2) the strain of HIV is typical of that found in the United States; and (3) the strain of HIV is closely related to the strain infecting his wife (Reimer et al., 1997).

False positive reactions may also occur. This means that the person's serum does not contain antibodies to HIV but the test results indicate that it does. Christine Johnson (2000) has compiled a list of 66 conditions taken from HIV/AIDS scientific literature that can cause false positive results.

People may test false positive who have an underlying liver disease, have received a blood transfusion or gamma globulin within six weeks of the test, have had several children, have had rheumatological diseases, malaria, alcoholic hepatitis, autoimmune disorders, various cancers, acute cytomegalovirus infection, or DNA viral infections; are injection-drug users; or have received vaccines for influenza or hepatitis B (Fang et al.,

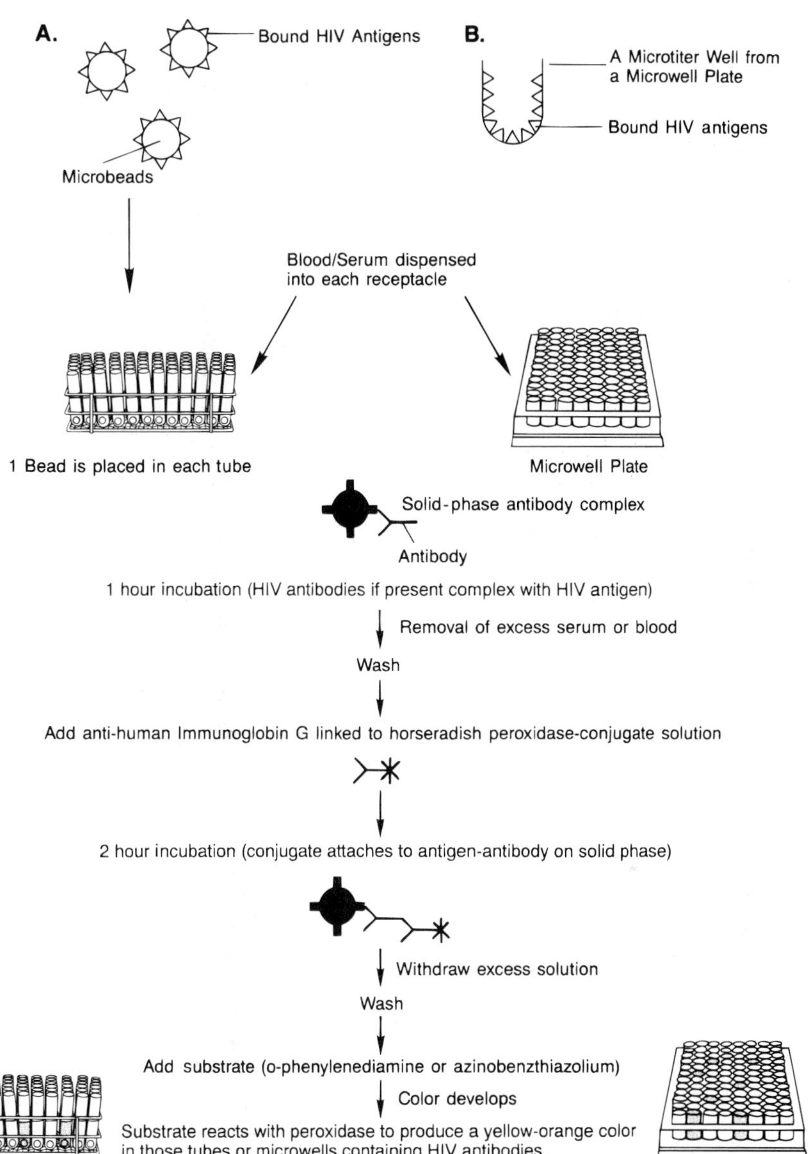

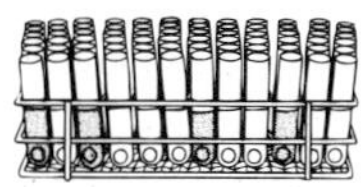

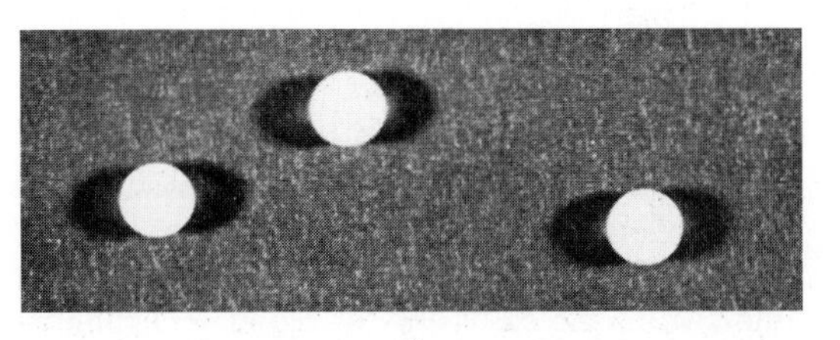

FIGURE 12-1 The ELISA Test. **A.** Microbeads with attached antigen in test tubes. **B.** Antigens bound to walls and bottom of microtiter wells. **C.** Microbeads are 7 mm in diameter. The test takes between 2.5 and 4 hours to perform.

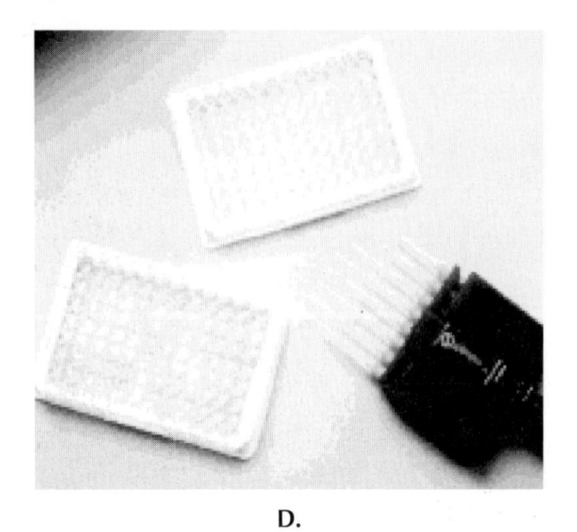

D.

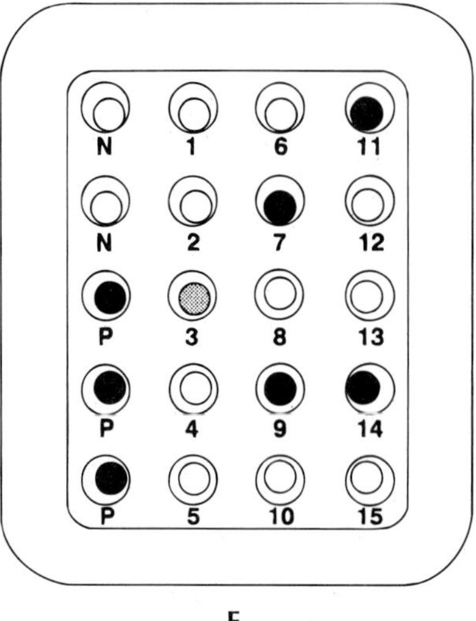

E.

FIGURE 12-1 *(continued)* **D.** Microwell plates showing positive (yellow) and negative (clear) test results. The microtiter dispenser handles eight microwells at a time. (**C.** and **D.**: *Photographs courtesy of the author*) **E.** Specimens positive for HIV antibody have a deeper color in this microwell tray. Serum specimens from 15 patients were tested for antibodies to HIV. Two negative and three positive control specimens are provided in the first column. In wells 7, 9, 11, and 14, the dark yellow color change, matching the color in the three positive control wells, indicates that the specimens are positive. Well 3 shows a weakly reactive result. The remaining specimens showed no color change and were interpreted as negative for HIV antibodies. *(Adapted from Fang et al., 1989)*

1989; MacKenzie et al., 1992). In each case, the person may have antibodies that will cross-react with the HIV antigens to give a false positive reaction. Other reasons for false positives are laboratory errors and mistakes made in reagent preparations for use in the test kits.

Why Is the ELISA Test Sensitivity and Specificity Set High?

Because the original purpose of the ELISA test was to screen blood, the sensitivity (ability to detect low-level color formation; see Figure 12-1) of the test was purposely set high. It was reasoned that it was better to have some false positives and throw away good blood rather than to take in any HIV-contaminated blood. Thus the ELISA test is a **positive predictive value** test. It only predicts that the serum tested will continue to test positive when a test with greater specificity, called a **confirmatory test,** is done.

In 1985, during the first month of donor screening, 1% of all blood tested HIV-antibody positive. On ELISA retesting of these samples, only 0.17% (17/10,000) were HIV-antibody positive. On subjecting these samples to a confirmatory test, only 0.038% (4/10,000) were actually HIV positive. These early tests produced about 24 false positives for every true positive result. The main reason for such a high false positive rate or **lack of specificity** was that something other than HIV produced an antibody or other substance that reacted with HIV antigen causing the HIV test to appear positive.

Although high sensitivity tests eliminate HIV-contaminated blood from the blood supply, there is a downside to high sensitivity testing when proper procedure is *not* used. People told that they have tested positive have become emotionally distraught. Former Senator Lawton Chiles of Florida, at an AIDS conference in 1987, told of a tragic example from the early days of blood screening in

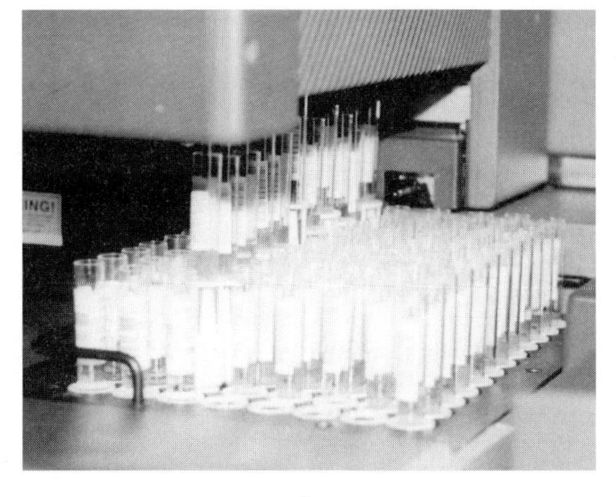

A.

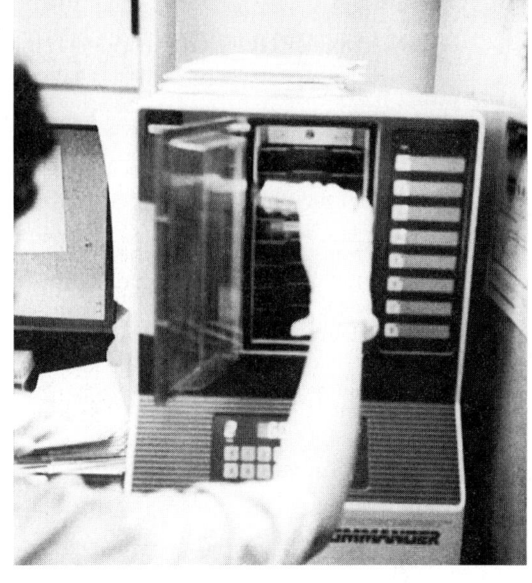

C.

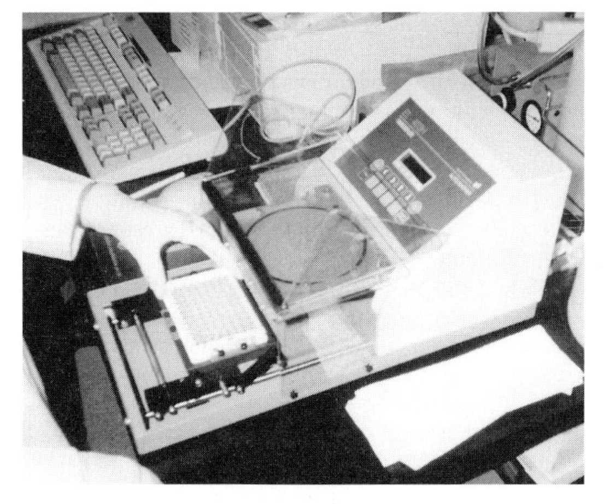

B.

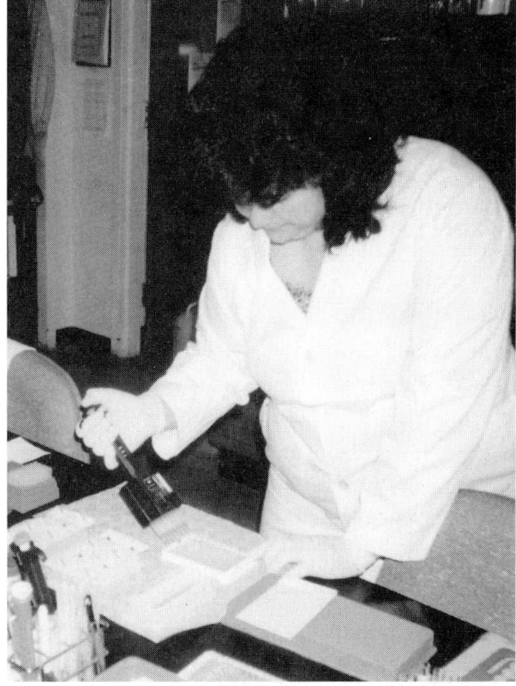

D.

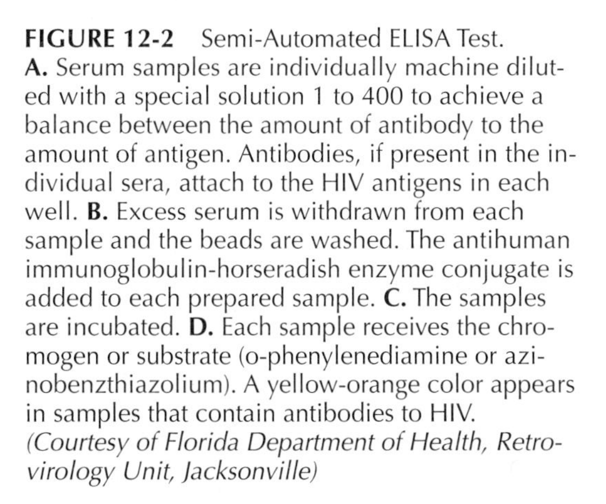

FIGURE 12-2 Semi-Automated ELISA Test. **A.** Serum samples are individually machine diluted with a special solution 1 to 400 to achieve a balance between the amount of antibody to the amount of antigen. Antibodies, if present in the individual sera, attach to the HIV antigens in each well. **B.** Excess serum is withdrawn from each sample and the beads are washed. The antihuman immunoglobulin-horseradish enzyme conjugate is added to each prepared sample. **C.** The samples are incubated. **D.** Each sample receives the chromogen or substrate (o-phenylenediamine or azinobenzthiazolium). A yellow-orange color appears in samples that contain antibodies to HIV. *(Courtesy of Florida Department of Health, Retrovirology Unit, Jacksonville)*

BOX 12.1

AN ASSUMPTION OF AIDS WITHOUT THE HIV TEST: IT SHATTERS LIVES

Case 1

San Francisco—For six years, a 53-year-old gay male lived in the world of AIDS. He stopped working, suffered the painful side effects of experimental drugs, and waited to die.

Now his doctors say he never had the disease.

His health shattered by AIDS treatment, his livelihood lost, he filed a $2 million claim against Kaiser Permanente health maintenance organization. He claims he underwent sustained treatment for full-fledged AIDS without receiving an HIV test.

His attorney said, "For six years he thought that the most he had was six months to live. So every day he'd wake up and think 'Is this the last day of my life?'"

To begin, this male says he checked into a San Jose hospital affiliated with Kaiser in 1986 with respiratory problems and doctors told him he had Pneumocystic pneumonia, considered a sure sign of AIDS at the time.

He underwent tests but *was not* given one to determine the presence of HIV, the virus associated with AIDS.

In 1986, he began taking the drug Zidovudine in high doses, which gave him a chronic headache, high blood pressure, and peripheral neuropathy—permanent pins and needles pains from his calves to his feet. He is battling an addiction to Darvon and other prescription drugs.

Under doctors' orders, he quit his job as a skin care technician and lives on government welfare and disability benefits of $600 a month.

(Associated Press, 1992)

Case 2

In 1980, she received a blood transfusion during surgery at a hospital in a Southeast Georgia town. During a checkup for a thyroid problem, a decade later, at a clinic in Hialeah, Fla., her blood was taken for testing.

On November 13, 1990, her telephone rang. She was asked to come down to the local health clinic where she was told she had AIDS. They were not sure how long she had to live. She was 45 years old. Her three sons were then teenagers; their father had died.

She kept the television on continually in a usually unsuccessful effort to block out the thought of AIDS.

The nights were the worst.

"I'd go to bed every night thinking about dying. What color do you want the casket to be? What dress do you want to be buried in? How are your kids going to take it? How will people treat them? I was afraid to go to sleep."

In 1992, her doctor put her on Didanosine, (ddI), which brought on side effects that included vomiting and fatigue.

"I had put my kids through hell. They were scared for me."

When she joined a local hospice group for AIDS patients, counselors heard her story and noted that her T cell counts had remained consistently high. At their suggestion, she was retested.

In November 1992, nearly two years to the day she was told she was HIV positive, another call came. She was greeted at the clinic with these words: **"Guess what? Your HIV test came out negative!"**

She sued the Florida Department of Health and Rehabilitative Services—the agency that performed the test—and the clinic and doctor who treated her.

A jury awarded her $600,000 for pain and suffering but cleared the clinic and said the bulk must be paid by the agency.

Case 3

Every day, four times each day, for six years Mark swallowed his antiretroviral drugs. Mark was told by a "fine physician" that he was HIV positive in July 1990. Regardless of the drugs, Mark felt sick and suffered further physical effects and depression. For reasons not given, Mark moved from Chicago to Ohio. His new physician was puzzled that Mark did not demonstrate signs or symptoms of HIV infection. His tests on Mark came back HIV negative. On investigation, the Chicago clinic could not produce any documents showing that Mark was ever HIV tested!

Florida. Of 22 blood donors who were told they were HIV positive by the ELISA test, seven committed suicide.

There continue to be false positive reactions among blood donors and low-level risk populations because of a low prevalence of HIV infec-

LOOKING BACK 12.1

THE *MIAMI HERALD* AND *CHICAGO TRIBUNE,* 1985

Miami Herald

In September 1985, the Metro Commissions of Dade County, Fl., voted to require all food service workers in Dade County be HIV tested. The commissioners' motivation is not in question. Aware of how the AIDS panic has hurt restaurant patronage in several other cities, they properly wanted to calm fears and protect food service jobs. Inadvertently, though, the commissioners may have sent the wrong signal. Medical experts say that AIDS is not spread by the kinds of contact that food handlers have with a restaurant's patrons or with their food. By ignoring this advice, commissioners implicitly seem to be questioning it. This may heighten the public's fears, not calm them. That is especially evident when one notes the test's inability to do what is expected of it. Richard Morgan, Dade's health director, says that testing all 80,000 of Dade's food service workers would take a year. So even if AIDS were spread through food—which it is not—the test would be of dubious value. A food service worker might test clean one day, become infected the next, continue working, and escape detection for months. Moreover, the high worker turnover in this seasonal business means that large numbers of these costly tests would continue to be given indefinitely. Thus the costs would not diminish much, even after all current workers had been tested. The AIDS-test requirement is part of a broader health-care ordinance that commissioners tentatively approved.

Chicago Tribune

In early 1985 Chicago's health commissioner protested federal guidelines requiring that blood donors be informed if their blood was HIV positive. In response the following article appeared on April 10, 1985: "Chicago Health Commissioner Lonnie Edwards is wrong to protest federal guidelines requiring that blood donors be informed if tests show that their blood contains antibodies to the virus linked to AIDS (Acquired Immune Deficiency Syndrome). The new test for AIDS antibodies is being phased into use in the nation's blood banks, after its approval early in March by the Department of Health and Human Services. Blood found to have AIDS antibodies will be discarded. By late April, all blood banks in the Chicago area will be doing such screening. But the antibody test is still controversial. It cannot show whether a blood donor currently has AIDS. A positive result only indicates that a person has been exposed at some time and in some way to the virus believed to cause the usually fatal disease—not that he is immune to it or that he will eventually become ill and die of it. Another problem is that the test is not totally accurate; both false negatives and false positives can occur. Some homosexual groups oppose the use of the AIDS screening test on the grounds it could violate individual privacy or that a third party might obtain test results and use them to discriminate against donors. But the presence of HIV in the nation's blood supply is becoming a serious problem. Dozens of recipients of blood or blood-based products have already died of AIDS presumably acquired from blood donors. In a few instances, these victims have passed the disease on to their spouses and young children."

tion in such populations. The American Red Cross Blood Services laboratories report that using current ELISA methodology, a specificity of 99.8% can be achieved.

Positive Predictive Value—The positive predictive value of the ELISA test indicates the percentage of true positives among total positives in a given population. To determine a positive predictive value:

Number of true positives ÷ (Number of true positives + false positives) × 100 = %

There is also a negative predictive test. A negative predictive value refers to the percentage of individuals who test truly negative; they *do not* have HIV. It is determined by:

Number of true negatives ÷ (Number of true negatives + false negatives) × 100 = %

To safeguard against false-positive tests, the CDC recommends that serum that tests positive be retested twice (in duplicate). If both tests are negative, the serum is considered HIV-antibody negative and further tests will only be done should signs or symptoms of HIV infection occur (Figure 12-3). If one or both of the tests is positive, the serum is subjected to a confirmatory test, usually a Western Blot (WB) (Figure 12-4). At blood banks, if the initial ELISA test is positive, the blood is discarded. If an individual's serum subjected to a confirmatory test is positive, the person is considered to be HIV-infected.

Although confirmatory tests can be used to determine true-positive results, they are too

FIGURE 12-3 Incoming Specimens. About 900 to 1000 blood specimens are received each day and are prepared for testing. From those samples, those testing negative are reported in 8 to 10 hours. Positive samples are reported in 3 to 4 days. *(Courtesy of Florida Department of Health, Retrovirology Unit, Jacksonville)*

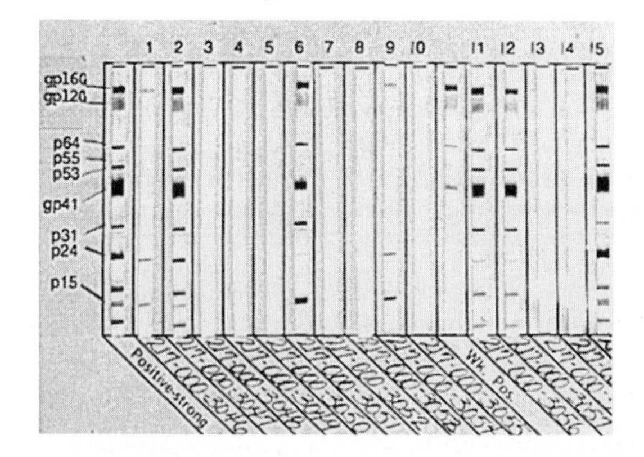

FIGURE 12-4 Western Blot Strips. Each WB strip contains nine separate antigenic proteins of HIV. Human serum or blood is applied directly to the strips. See text for details on reactions. Because false positives can sometimes occur with the ELISA test, additional testing is needed to evaluate specimens that are repeatedly reactive by ELISA. The Western Blot is more specific but less sensitive than the ELISA and is recommended for blood banks and organ donor centers. Its clinical usefulness in trials to aid in evaluating specimens that are questionably positive by other methods has been proven. It is not a screening test because it lacks a high level of sensitivity and is expensive. *(Courtesy of Roche Biomedical Laboratories, AIDS Testing Brochure)*

BOX 12.2

PRETEST PATIENT INFORMATION

It is important to read and understand the following information before having the HIV antibody test.

This test is not a test for AIDS, but a positive result does mean that there is a high probability an HIV-infected person will eventually develop AIDS.

Only anonymous test results remain absolutely confidential.

What Does the Test Reveal?

The test reveals whether a person has been exposed to HIV. The test detects the presence of HIV antibodies, an indication that infection has occurred. It does not detect the virus itself nor does it indicate the level of viral infection, only that one has been infected.

It is important to find out whether you are HIV positive or negative so you can prevent spreading the virus to others and seek early medical intervention.

What Tests Will Be Done?

Wherever the test is done, the procedures are similar. A blood sample is taken from the arm and analyzed in a laboratory using the ELISA test. If the first ELISA test is positive, the laboratory will run a second ELISA test on the same blood sample. If positive, another test, called Western Blot, is run on the same blood sample to confirm the ELISA result.

What Do the Test Results Mean?

See Table 12-1.

Table 12-1 The Meaning of Antibody Test Results

A Positive Results	B Negative Results
If you test positive, it does mean: 1. Your blood sample has been tested more than once and the tests indicate that it contained antibodies to HIV. 2. You have been infected with HIV and your body has produced antibodies. **If you test positive, it does not mean:** 1. That you have AIDS. 2. That you necessarily will get AIDS, but the probability is high. You can reduce your chance of progressing to AIDS by avoiding further contact with the virus and living a healthy lifestyle. 3. That you are immune to the virus. **Therefore, if you test positive, you should do the following:** 1. Protect yourself from any further infection. 2. Protect others from the virus by following AIDS precautions in sex, drug use, and general hygiene. 3. Consider seeing a physician for a complete evaluation and advice on health maintenance. 4. Avoid drugs and heavy alcohol use, maintain good nutrition, and avoid fatigue and stress. Such action may improve your chances of staying healthy.	**If you test negative, it does mean:** 1. No antibodies to HIV have been found in your serum at the time of the test. **Two possible explanations for a negative test result exist:** 1. You have not been infected with HIV. 2. You have been infected with HIV but have not yet produced antibodies. Research indicates that most people will produce antibodies within 6 to 18 weeks after infection. Some people will not produce antibodies for at least 3 years. A very small number of people may never produce antibodies. **If you test negative, it does not mean:** 1. That you have nothing to worry about. You may become infected; be careful. 2. That you are immune to HIV. 3. That you have not been infected with the virus. You may have been infected and not yet produced antibodies.

(*Adapted from the San Francisco AIDS Foundation*)

Positive Result—A positive result suggests that a person has HIV antibodies and may have been infected with the virus at some time. People with a positive result should assume that they have the virus and could therefore transmit it:

- by sex (anal, vaginal, or oral) where body fluids, especially semen or blood, get inside the partner's body;
- by sharing injection drug "works" (needles, syringes, etc.);
- by donating blood, sperm, or body organs; or
- to an unborn baby during pregnancy or to a newborn by breast-feeding.

Negative Result—If it has been four to six months since the last possible exposure to HIV, then a negative result suggests that a person is probably not infected with the virus. However, **false negatives** can occur. For example, if the test was too soon after HIV exposure, the body may not have had time to produce HIV antibodies. Collectively, studies show that 50% of HIV-infected persons seroconvert (demonstrate measurable HIV antibodies) by three months after HIV infection and 90% seroconvert by six months. A small percentage seroconverted at one year or later. Most important, a negative result does not mean that you are protected from getting the virus in the future—your future depends on your behavior.

Inconclusive Result—A small percentage of results are inconclusive. This means that the result is neither positive nor negative. This may be due to a number of factors that have nothing to do with HIV infection, or it can occur early in an infection when there are not enough HIV antibodies present to give a positive result. If this happens, another blood sample will be taken at a later time for a retest.

How Accurate or Error-Free is the Test Result?

The result is accurate. However, as with any laboratory test, there can be false positives and false negatives.

False Positive—A small percentage of all people tested may be told they have the HIV antibody when they do not. This can be due to laboratory error or certain medical conditions that have nothing to do with HIV infection.

False Negative—Some people are told that they are not HIV-infected when, in fact, they are. This can happen when the test is taken too soon after being infected—the body has not had time to produce HIV antibodies. This is called the **window period.** This is the most common cause for a false negative test.

Should the Test Be Performed?

Whether to have the test done is a personal decision. However, knowing your antibody status can help you address some very difficult questions. For example, if the test is negative, what lifestyle changes should be made to minimize risk of HIV infection? If the test is positive, how can others be protected from infection? What can be done to improve your chances of staying healthy?

(Source: Roche Biomedical Laboratories, Inc.)

labor intensive and expensive to be used in screening a large population. Thus the positive predictive value of an ELISA test is an important first step in large-scale screening. Recall, however, that **the predictive value depends on the prevalence of HIV infection in the population tested.** The higher the prevalence of HIV infection in a given population, the more likely a positive ELISA test is to be a true positive; and conversely, the lower the prevalence of infection, the less likely a positive ELISA test is to be a true positive.

Levels of Sensitivity and Specificity in Testing for HIV—A test's **sensitivity** is its capacity to identify all specimens that have HIV antibodies in them. A test's **specificity** is its capacity to identify all specimens that do not have HIV antibodies in them.

Sensitivity is determined as follows:

Number of true positives ÷ (Number of true positives + the false negatives) × 100 = %

If 100 persons are actually HIV-infected and the test identifies only 90 of them as such, then we can say that the test has 90% sensitivity.

Specificity is determined as follows:

Number of true negatives ÷ (Number of true negatives + false positives) × 100 = %

Assume that, in a group of 500 people being tested for HIV antibodies, 100 individuals are actually not infected. If test results show that only 90 out of the 100 are identified as not having the virus, then the test has 90% specificity.

WESTERN BLOT ASSAY

The gold standard for determining a true positive HIV-antibody test is the **Western Blot** (WB). This test is a method in which individual HIV proteins are used to react with HIV antibody in a person's serum. It should be understood that the WB test is not a true gold standard because it is not 100% certain, but it can come close to 100% if properly used.

Cells in which HIV is being cultured are lysed or broken open, and the mixture of cell components and HIV components (proteins) are separated from each other. The viral proteins are placed on a polyacrylamide gel, which then gets an electrical charge. The electrical current separates the viral proteins within the gel. This is called **gel electrophoresis.** The smallest HIV proteins will move quickly through the gel, separating from the next larger size, and so on.

Each different protein will arrive at a separate position on the gel. After proteins of similar molecular weight collect at a given site, they form a band; these bands are identified based on the distance they have run in the gel. Because each band is a protein produced as a product of a different HIV gene, the gel band patterns give a picture of the HIV genes that were functioning and the location of each gene's products on the gel (Figure 12-4). The protein or antigen bands within the gel are "blotted," that is, transferred directly, band for band and position for position, onto strips of nitrocellulose paper (Figure 12-4).

Once the antigen bands have been formed, serum believed to carry HIV antibodies is placed directly on them. That is, a test serum is added directly to antigen bands located on the nitrocellulose strip. If antibodies are present in the serum, they will form an antigen–antibody complex directly on the antigen band areas. (Figure 12-4). Positive test strips are then compared to two control test strips, one that has been reacted with known positive serum and one that has been reacted with known negative serum.

In contrast to the ELISA test, which indicates only the presence or absence of HIV antibodies, the WB strip qualitatively identifies which of the HIV antigens the antibodies are directed against. **The greatest disadvantage of the WB test is that reagents, testing methods, and test-interpretation criteria are not standardized.** The National Institutes of Health (NIH), the American Red Cross, DuPont company, the Association of State

SIDE ISSUES 12.1

A FALSE POSITIVE HIV REPORT

CASE 1

In January 2002, a man in Oklahoma City was awarded $1.4 million because he was wrongfully told he was HIV positive. At age 40 he received the news from a Health Maintenance Clinic. During the following four years, he became depressed, despondent, abused alcohol, and attempted suicide twice. Thinking he was HIV positive, he had unprotected sex with known HIV-positive partners. Reviewing his file to determine when he became infected he learned that his test four years earlier was HIV negative. He filed suit for negligence and won.

CASE 2

In November 2002, a Richland County, S.C., jury awarded $1.1 million to a woman who said Palmetto Health Richland Hospital misdiagnosed her with HIV. She said the diagnosis wrecked her life. "I was so depressed thinking I was going to die a horrible death. I gave up hope. Now I'm trying to get my life back together. I have lost so much of my life. I thank God everyday for helping me find the mistake." In her lawsuit, the hospital—then called Richland Memorial Hospital—diagnosed her as HIV-positive in February 1994. For several years she took anti-HIV medications, including the drug Zidovudine. She said she never had symptoms of the disease and that another test in 1998 by a different laboratory confirmed she had been misdiagnosed. The lawsuit said she suffered extreme depression, emotional distress, anxiety, fear, side effects from the drugs, and other related trauma because of the misdiagnosis.

and Territorial Health Officers (ASTHO), and the Department of Defense (DoD) each define a positive WB differently.

The WB procedure is labor intensive, takes longer to run (12 to 24 hours), and is therefore more costly than the ELISA test. **The WB is less sensitive than the ELISA but more specific.**

Because the WB lacks the sensitivity of the ELISA test, it is not used as a screening test. Despite the high specificity of the WB, false positives do occur, but they occur less frequently than with ELISA tests because the WB is only run on serum, blood, oral fluid, and urine already suspected of containing HIV antibodies.

Indeterminate WB

Western Blots may also turn out to be **indeterminate** in HIV infections—meaning, a person can be infected, but the blot is not conclusive—it looks positive but it may not be or is too poor to tell. The indeterminate WB results can occur either during the window period for HIV seroconversion, or during end-stage HIV disease. Indeterminate WBs have occurred in uninfected individuals because of cross-reacting autoantibodies related to recent immunization, prior blood transfusion, organ transplantation, autoimmune disorders, malignancy, infection with other retroviruses (for example, HIV-2), or pregnancy. Some patients have a persistent pattern of indeterminate reactivity that remains stable over several years in the absence of true HIV infection.

In general, most persons with an initial indeterminate Western Blot result who are infected with HIV will develop detectable HIV antibody within one month. Thus, clients with an initial indeterminate result should be retested for HIV infection after one month.

OTHER SCREENING AND CONFIRMATORY TESTS

There are a variety of HIV antibody and HIV antigen detection tests now on the market and others are on their way. A few of these tests have been singled out because they are currently in use or because of their potential to make a contribution in the field of HIV antibody–antigen testing methodology.

Immunofluorescent Antibody Assay

The **immunofluorescent antibody assay (IFA)** uses a known preparation of antibodies labeled with a fluorescent dye such as fluorescein isothiocyanate (FITC) to detect antigen or antibody. In the direct fluorescent antibody test, fluorescent antibodies detect specific antigens in cultures or smears. In the indirect fluorescent antibody test, specific antibody from serum is bound to antigen on a glass slide.

The indirect procedure is modified for use in detecting antibodies to HIV. Cells that are HIV-infected will have HIV antigens on their cell membranes and will later fluoresce when the antihuman fluorescent conjugate is added.

In late 1992, the FDA approved Fluorognost for marketing, the first assay for HIV-IFA confirmation and screening.

The assay allows doctors to do in-office tests for antibodies to HIV in human serum or plasma. As opposed to the Western Blot test, the current standard confirmation test, Fluorognost posts almost no indeterminate test results. In addition, the test takes only 90 minutes to complete, while the Western Blot takes from 12 to 24 hours to process. This FDA-approved test allows smaller health-care facilities, emergency rooms, and doctors' offices to conduct in-office HIV screening and confirmation with accuracy, ease, and low overhead.

Polymerase Chain Reaction

Interactions between HIV and its host cell extend across a wide spectrum, from latent to productive infection. The virus can persist in cells as unintegrated DNA, as integrated DNA with alternative states of viral gene expression, or as a defective DNA molecule. Determining the fraction of cells in the blood that are latently or productively infected is important for the understanding of viral pathogenesis and in the design and testing of effective therapies. Determining the number of infected cells in a heterogeneous cell population and the proportion of those cells that are carrying the virus but not producing new viruses requires the identification of the proviral DNA and viral mRNA in single cells.

The polymerase chain reaction (PCR) is a technique by which any DNA fragment from a single cell can be exponentially multiplied to an amount large enough to be measured. Thus PCR could be an ideal diagnostic test for HIV infection, since it directly amplifies proviral HIV DNA and

POINT OF INFORMATION 12.1

IMPROVED ELISA TESTING, BLOOD DONOR INTERVIEWS, SCREENING FOR P24 ANTIGEN, AND NUCLEIC ACID TESTING LOWER HIV IN NATION'S BLOOD SUPPLY

In 1993, approximately 6 per 100,000 blood donations collected by the American Red Cross tested positive for HIV antibody. In addition, an estimated 1 in 450,000 to 1 in 660,000 donations per year (18–27 donations) were infectious for HIV but were not detected by current screening tests.

In February 2002, a Texas man became the first person to get HIV from a blood transfusion in the United States since new and improved screening tests began in 1999. The infected blood given to him came from a heterosexual donor who didn't think he was at risk for HIV. The man donated his blood four times during 2000. At the time of his last donation, tests showed that his blood carried HIV. This prompted intensive testing of samples from his previous donations—including the blood given to the Texas man.

Screening for p24 Antigen—In August 1995, the FDA mandated that all blood and plasma collection centers screen all blood for p24 antigen. The FDA recommended p24 screening as an additional safety measure because recent studies indicated that p24 screening reduces the infectious window period (the FDA-approved Coulter p24 antigen blood test detects HIV as early as 16 days after infection). Among the 12 million plus annual blood donations in the United States, p24-antigen screening is expected to detect four to six infectious donations that would not be identified by other screening tests. FDA regards donor screening for p24 antigen as an interim measure pending the availability of technology that would further reduce the risk for HIV transmission from blood donated during the infectious window period (*MMWR*, 1996 updated).

Nucleic Acid Testing—Beginning Spring 1999, the American Red Cross and 16 member laboratories of the America's Blood Centers began testing donor blood for the Human Immunodeficiency Virus (HIV) type 1 and the hepatitis C virus with a new research testing method known as **nucleic acid amplification testing** (NAT). The power of NAT is its ability to detect the presence of infection by directly testing for viral nucleic acids rather than by indirectly testing for the presence of antibodies. The efficacy of such screening depends on the prevalence of the infection in the population and the duration of the window period. In most Western countries, HCV shows a higher prevalence and longer window period (80 days) than HBV (56 days) and HIV(16 days).

Pitfalls to the Use of NAT

There are some serious drawbacks to the use of NAT. For example, a cost-effective, automated system does not yet exist to perform the test on individual donor blood samples. The requirement of specially trained lab technicians, space constraints, a minimum 12 hours required to perform NAT, and the cost of a NAT test kit, which is currently 10 times higher than the ELISA test kit, will be difficult to overcome. Also, as NAT currently exists there is a 1% chance of obtaining false-negative results. Regardless, NAT testing continues at Red Cross blood collection centers.

Conclusion

In order to pursue its stated policy of seeking innovative means to increase the safety of the blood supply and to allow compliance with the European requirements, the FDA, in September 2001, licensed the first NAT for use in the screening of blood plasma donors for HIV and cytomegalovirus. FDA approval of NAT for broad commercial use came in February 2002.

To speed the transportation of blood products to areas affected by terrorist attacks in New York City on Sept. 11, 2001, the FDA released new emergency guidelines allowing donated blood to be shipped to crisis areas before HIV and hepatitis C testing is complete. An FDA statement says that due to the "recognized need for rapid and high volume blood collections under non-routine circumstances," some blood may have to be shipped before all testing is complete. The agency normally requires all blood donations be tested for such diseases before the blood is shipped. The emergency guidelines state that if the blood products are shipped before FDA testing has been completed, the

POINT OF INFORMATION 12.1 *(continued)*

product must be labeled "For Emergency Use Only" and must list the tests that have not yet been completed. The blood must indicate if screenings for HIV-1, HIV-2, and hepatitis C are not complete. The tests for the viruses should be completed "as soon as possible" and "appropriate actions taken in the event of a reactive screening test result."

POINT OF INFORMATION 12.2

NEW ELISA ANTIBODY TESTING AND VIRAL LOAD PROCEDURES REVEAL EARLY VS. LATE HIV INFECTION

STARHS—A Sensitive/Less Sensitive HIV Test

The best data for understanding recent changes in HIV transmission are measurements of the number of new infections in a defined time period (incidence of infection). But this has been difficult because ELISA testing simply gave a positive or negative response to the presence of HIV antibody without regard to the actual time of infection. However, in mid-1998 (Jensen et al.) through 1999 (McFarland et al.) a new testing strategy provided a means to detect new or early HIV infections versus older HIV infections. The new testing technique is called **STARHS (the Serologic Testing Algorithm for Recent HIV Seroconversions).** The test uses two different ELISAs to test a single blood sample to tell if an infection is old or new. By more accurately pinpointing the time of infection, STARHS may help patients identify when and from whom infection took place. Because people can live symptom-free with HIV for over a decade, there was no way previously to tell when they might have become infected. But the new technology changes that. For example, while it's hard for many sexually active people to recall the names and addresses of all their partners, spanning years of activity, it is typically a simple matter to make a list for the past four months. Armed with this new information, public health authorities can track down individuals who appear to be spreading HIV, and—for the first time—interrupt the chain of transmission on a large-scale basis.

HOW IT WORKS

The standard ELISA blood test is very sensitive and measures the presence of antibodies against HIV. The very sensitive or **highly tuned ELISA tests** can pick up even minute numbers of antibodies present in the first days of infection before the immune system has mounted a full response to the virus. Conversely, a less sensitive diluted or **detuned ELISA does the reverse.** It detects only the presence of antibodies at higher levels that typically appear three to six months after infection. By administering both the highly tuned and detuned ELISA tests at the same time, technicians can tell an individual's stage of infection. In brief, if someone tests positive on the sensitive test and negative on the detuned test, they likely have a recent infection. Positive results on both tests indicate the infection is more than four to six months old.

VIRAL LOAD RELATED TO STAGE OF INFECTION

As described in Chapter 4, a viral load test measures the number of viruses in the blood. Before the immune system produces antibodies to fight it, HIV multiplies rapidly. Therefore, this test will show a high viral load during the acute stage of HIV infection. Thus, a negative HIV antibody test and a high viral load indicates a recent HIV infection, most likely within the past two months. If both tests are positive, then HIV infection probably occurred a few months or more before the tests.

does not require antibody formation by the host. It is already used in settings where antibody production is unpredictable or difficult to interpret, such as in acute HIV infection or in the perinatal/postnatal period. The PCR is so sensitive that it can detect and amplify as few as 6 molecules of proviral DNA in 150,000 cells or 1 molecule of viral DNA in 10 μL of blood.

Now that there are some good anti-HIV therapies available to help slow the onset of AIDS, the diagnosis of individuals who carry the provirus is critical because they may benefit from early treatment. The PCR test will become even more important with the advent of an HIV vaccine. Vaccinated people will become HIV-antibody positive. The PCR test will be used to identify those who are truly HIV-infected.

RAPID HIV TESTING

A rapid test for detecting antibody to HIV is defined as a screening test that reveals the presence or absence of antibodies for HIV during the length of the patient's visit to the clinic, which is usually between 10 and 30 minutes. In the United States, 10% to 20% of pregnant women do not know their HIV status at the time of delivery. In Africa and Asian countries, it is estimated that over 90% of pregnant women do not know their HIV status at delivery. But the HIV status of pregnant women is essential to prevent mother-to-child HIV transmission.

Rapid HIV test results are also necessary for deciding whether to initiate treatment for health-care workers after accidental exposures to patient body fluids, and there is a need for rapid HIV tests to assist with diagnosis and appropriate treatment of persons who may have opportunistic infections due to AIDS in urban emergency departments, which have been shown to have high rates of undiagnosed HIV infection among their patient populations. In short, the value of rapid HIV tests in public health has been well established.

There are over 30 different rapid HIV tests currently marketed worldwide. The first FDA-approved rapid HIV antibody test occurred in 1992. It is called the Single Use Diagnostic System (SUDS). Because of an unacceptable level of false positive results, it is no longer used.

FDA Rapid Tests Now in Use

In February 2004, the FDA approved Trinity Biotech's 10-minute Uni-Gold rapid serum/plasma/whole blood HIV test. In November 2002 the FDA approved OraSure's 20-minute whole blood HIV-1 and HIV-2 OraQuick Rapid Antibody Test, and in March 2004 the FDA approved their 20-minute OraQuick Saliva Antibody Test, in which a treated cotton swab, run along the gum line, is placed in a testing liquid and read in about 20 minutes (Figure 12-5). Either test costs about $15.00. OraQuick's rapid test is now available in clinics and physicians' offices nationwide. Each year about 750,000 or 30% of 2.5 million tested do not return a week later to receive their test results. With this new test, in less than 20 minutes, they can learn preliminary information about their HIV status, allowing them to get the care they need to slow the progression of their disease and to take precautionary measures to help prevent the spread of this deadly virus. Unlike other antibody tests for HIV, this test can be stored at room temperature, requires no specialized equipment, and can be used outside of traditional laboratory or clinical settings.

POINT OF VIEW 12.1

SOME RELATIVE DRAWBACKS TO THE CURRENT HIV SCREENING TEST—ELISA/WESTERN BLOT

Regardless of the high sensitivity and high specificity or overall test performance, there are a few test-related problems. For example, although only 24 hours are required to complete the ELISA/Western Blot testing procedures, most labs batch specimens for processing, causing a one- to two-week wait for definitive results. The results of several studies have shown that at least 35% of those who test positive and 42% of those who test negative never returned for their test results. Perhaps because of concern about maintaining anonymity or other factors, many individuals with HIV are not tested until they develop symptoms. Up to one-third of patients receive their HIV diagnosis within two months of an AIDS diagnosis.

Cost and time invested in the two-step process (pre- and post-test counseling) have also impeded testing. Even when tested without charge at a publicly funded clinic, clients must take time out of work or caring for children for waiting time, counseling, or travel. The true cost of this process has been estimated at $41 per test. And, individuals tested in private offices may incur not only the cost of the test by commercial or hospital labs, but also the physician's office fee (Sax et al., 1997).

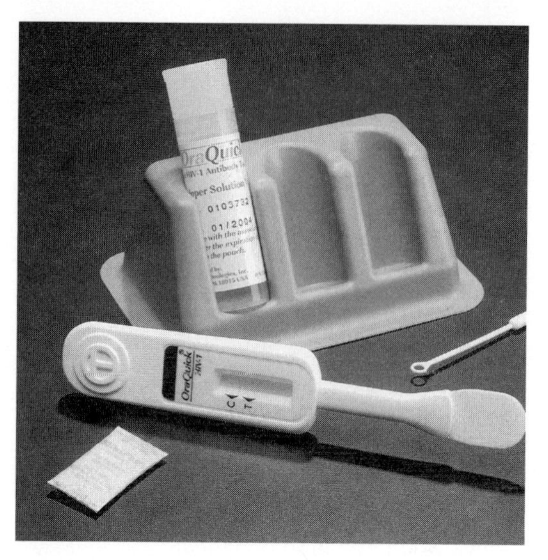

FIGURE 12-5 HIV Antibody Testing and OraSure Technology. OraSure is a simple lollipop-like device that is an easy-to-use oral HIV antibody-testing technology now available to consumers through health professionals. The product does not require the use of needles or blood and is about as effective as traditional HIV blood tests. The patient places the specially treated pad attached to the handle between the lower cheek and gum and swabs the area. The device is then placed in a vial with preservative. If HIV antibodies are present, the device will display two reddish-purple lines in about 20 minutes. *(Copyright OraSure Technologies, Inc. 2004.)*

The 3-Minute Test—On April 17, 2003, the FDA approved the MedMira's Reveal Rapid HIV-1 Antibody Test, the first FDA-approved rapid, point-of-care test designed to detect HIV-1 antibodies within three minutes. With the FDA approval, MedMira can market the test for the detection of HIV-1 antibodies in serum or plasma. It is designed for point-of-care diagnostic purposes in medical settings such as hospitals and clinics.

The 1-Minute Test—In March 2004, Hedley Technologies delivered their Ultra Rapid 60 Second HIV Test Kit to the Uganda AIDS Commission. At this time only trained healthcare workers will be able to administer the test.

The Testing Counseling Requirement—Although CDC officials still say counseling is important, they do not want the education component to deter people from getting tested. They concluded the counseling requirement might have kept busy doctors from providing tests and can make getting tested a tiring process for patients. Persons whose rapid test results are reactive will be counseled about their likelihood of being infected with HIV and precautions to prevent HIV transmission. They will be told to return for definitive test results before medical referrals or partner counseling is initiated. A simple message to convey this information could be a statement that "Your preliminary test result was positive but we won't know for sure if you are HIV-infected until we get the results from your confirmatory test. In the meantime, you should take precautions to avoid possible transmission of the virus."

Summary

Through 2004 the FDA has approved at least six rapid HIV screen tests for use in the United States. All have at least a year long shelf life stored at room temperature and have sensitivity and specificity values equal to the ELISA test. In 2004, New York City and the state of California, by law, made an HIV rapid test available to anyone who wanted to take it.

HIV Gene Probes

Gene probes or genetic probes are an idea borrowed from methodologies used in recombinant DNA research. The idea is to isolate a DNA segment, make many copies of it, and label these copies with a radioisotope or other tag compound. If the DNA sequence copied is contained in any of the HIV genes, the labeled copies of this DNA sequence can be used to hybridize or attach to DNA of cells that contain HIV DNA. This method of DNA probe analysis eliminates the need of searching for HIV gene products or antibodies to these products to prove that a person is HIV-infected.

At least two HIV-specific probes are on the market. One uses a radioactive sulfur label on the DNA for detection (^{35}S). This probe hybridizes to about 50% of the entire HIV genome and most specifically hybridizes to the HIV polymerase region. A second probe, also using ^{35}S, is an **RNA**

probe. It is being used to detect HIV RNA in peripheral blood or tissue samples. **RNA probe hybridization** allows detection of 1 HIV-infected cell out of 400,000 uninfected cells. Specifically, the assay detects the presence of HIV in whole white blood cells as soon as the virus begins to replicate. The ^{35}S-labeled probes enter the white blood cells and combine with HIV; the procedure does not require DNA extraction and results are obtained in just over one day (Kramer et al., 1989).

VIRAL LOAD: MEASURING HIV RNA

Recent studies have shown that a very rapid turnover of HIV RNA occurs in the plasma of infected patients, with approximately 30% of the total virus population in the plasma being replenished daily. HIV RNA viral load measurements are used in the management of HIV-infected patients, both in predicting rate of progression and monitoring response to antiretroviral therapy (see Chapters 4, 7, and 11 for additional information on HIV RNA load). Until June 1996, reliable measures of HIV RNA were available only in research laboratories. And, the quantitation of HIV from a clinical specimen required very expensive, labor-intensive, and difficult-to-reproduce culture techniques. Recently, however, quantitation of HIV viral load in plasma specimens has been accomplished with a variety of techniques that measure HIV RNA and are less expensive and easier to perform. Currently, three commercial assays are available: (1) FDA-Approved Amplicor Ultra-Sensitive HIV Monitor Test, which couples reverse transcription to quantitative polymerase chain reaction (PCR) (Roche Molecular Systems); (2) FDA-approved Bayer Quantiplex HIV RNA assay, a branched DNA (bDNA) technology; and (3) nucleic acid sequence-based amplification. The Ultra-Sensitive test, and the branched DNA test are briefly reviewed. Because there is a lot of variability with respect to viral load determinations, depending on which test is used, the same test or assay should always be used for the same patient or person.

Amplicor Ultra-Sensitive HIV Monitor Test

This multistep test includes specimen preparation, reverse transcription and PCR, and nonradioactive detection.

The new Amplicor HIV Monitor Ultra-Sensitive Test can measure about 50 copies of HIV RNA/mL.

Branched DNA Testing

Another new assay for measuring HIV levels directly, developed by Chiron, uses a branched DNA (bDNA) technology to amplify a signal to indicate the presence of HIV RNA.

Using the bDNA assay, nucleic acids can be detected directly in clinical samples. Signal amplification of HIV RNA, a luminescence, or spot of light, means that HIV RNA is present and the brighter the light (or signal) the more HIV RNA is present in the test sample. Chiron's current assay can measure down to 40 copies of HIV RNA/ml of sample.

FDA APPROVES TWO HOME HIV ANTIBODY TEST KITS

On May 14, 1996, the FDA, which for years opposed home-based HIV testing kits because of the lack of face-to-face counseling, reversed its stance by saying the benefits of early detection of HIV infection outweigh any risks posed by the test. FDA Commissioner David Kessler said, "We are confident that this new home system can provide accurate results while assuring patient anonymity and appropriate counseling."

The FDA approved the home test because in a 1994 study by the CDC of people at increased risk of infection, like injection-drug users and sexually active homosexual men, 42% indicated that they would use a home test.

Test Kit Operation

A person who buys the kit uses an enclosed lancet to prick his or her finger and places three drops of blood on a test card with an identification number. The card is mailed to a laboratory for HIV testing, and samples that test positive are retested to ensure reliability. People who use the home system do not submit names, addresses, or phone numbers with the specimen sent in on filter paper. Therefore, the HIV test results are anonymous. To get results, the individual calls three days later and punches into the phone his or her identification number.

If the caller's test results are positive or inconclusive, he or she will be connected to a counselor who will explain the results, urge medical treatment, and, if necessary, make a referral to a local doctor or health clinic. If the person's results are negative, he or she will be connected to a recording that will note that it is possible to be infected with HIV and still test negative if the antibodies to HIV haven't yet developed. A counselor is available for anyone who tests negative and wants to discuss the results.

The FDA said the kit is as reliable as tests conducted in doctors offices and clinics.

Test Kit Availability

The first FDA-approved HIV test kit, called Confide HIV Testing Service, was made available in June 1996. It was withdrawn from the marketplace in June 1997 due to poor sales. A second FDA-approved HIV home test kit went on sale nationwide in July 1996. This kit, called **Home Access Express HIV Test** (Figure 12-6) (1-800-448-8378), lets people take a blood sample at home, mail it to a laboratory, and, three days to a month later, learn by phone their results. The two tests are very similar to each other with regard to use and performance.

In May 2003, the administrators at Home Access reported they had processed over 450,000 HIV tests since FDA approval in July 1996. The overall HIV-positive rate for these tests was 0.08% (8 of 10,000 were positive). The general population rate in America is 0.03%. These data suggest that this form of testing appeals to an at-risk population.

WHO SHOULD BE TESTED FOR HIV INFECTION?

People report getting tested for many reasons, including wanting to learn their HIV status, feeling at risk, illness, doctor or sexual partner referral, pregnancy, and because the test was offered. The main reason given for not getting tested is not feeling at risk. A quarter of U.S. adults say they want

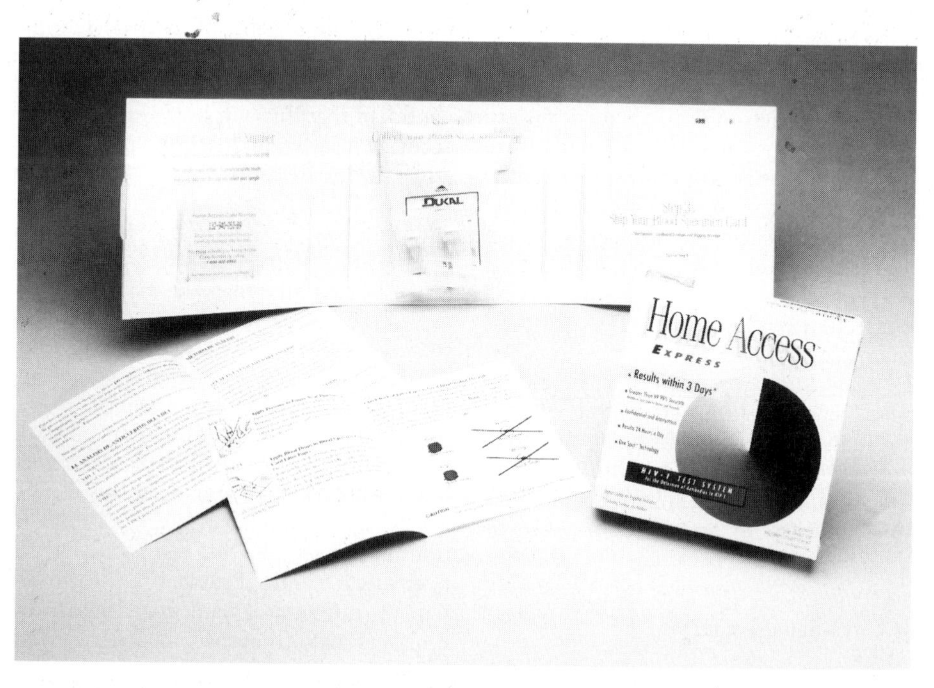

FIGURE 12-6 Home Access Anonymous HIV Test Kit. This kit was FDA-approved in July 1996 and provides access to professional counseling and medical/social service referrals 24 hours a day, 7 days a week. *(Photograph courtesy of Home Access Health Corp., Illinois)*

more information about HIV testing, including the different types of tests available, what test results mean, how much tests cost, and where to get tested. A recent survey of teens found that over two-thirds would not know where to go to get tested.

The First Step

Testing for antibodies to HIV is an important first step in establishing a diagnosis of HIV infection. However, attempts to isolate or publicly identify people with HIV/AIDS can actually fuel the spread of HIV. People outside the tested group may feel invulnerable, and then fail to make necessary changes in their behavior.

False Sense of Security

The effects of stimulating a false sense of security are well illustrated in Germany where, in certain towns, prostitutes are required to be checked for certain STDs every week and are given health inspection cards. Many customers think these cards guarantee against disease and so refuse to use condoms.

The problem is not confined to Germany. A European shipbroker whose work frequently takes him to Southeast Asia, where he has a regular sex partner, said: "I tell her when I'm due to arrive and she has an HIV test just before. If she has an up-to-date health card I know I'm safe."

In fact, a negative HIV test is no guarantee the tested person is truly HIV negative—he or she may be in the window period. The decision to test must be based on people's risk behaviors and/or symptoms.

Who Should Be Tested

As to who needs to be HIV tested, a complete history and physical examination will give the best answer to this question. Decisions based on individual indications are often more appropriate than decisions based on one's classification (for example, all pregnant women or all single men between 20 and 49 years of age). Initial assessment for current or past behaviors (within the past 10 years) should include:

1. Persons with **risk behaviors** such as:
   a. Anal sexual activity, male or female
   b. Injection-drug use
   c. Frequent casual heterosexual activity
   d. Encounters with prostitutes
   e. Previous treatment for sexually transmitted diseases (*Condyloma acuminata* (genital warts), herpes simplex virus, gonorrhea, syphilis, Chlamydia)
   f. Blood transfusions, especially before 1985
   g. Sexual activity with partners having any of the above
   h. Infants born to women involved in any of the above
2. Persons with symptoms such as:
   a. Fever, weight loss (unexplained)
   b. Night sweats
   c. Severe fatigue
   d. Recent infections, especially thrush and shingles (varicella-zoster)
3. Persons with **signs** (based on physical exam) such as:
   a. Weight loss
   b. Enlarged lymph nodes and/or tonsilar enlargement
   c. Oral exam (candidiasis, oral hairy leukoplakia)
   d. Skin lesions (Kaposi's sarcoma, varicella-zoster, psoriasis)
   e. Hepatosplenomegaly (enlarged liver)
   f. Mental status examination showing changes
4. Pregnant women who have demonstrated high-risk behavior.
5. Prisoners. A number of states (about 30 ending 2005) now have legislation to HIV test all incoming and/or outgoing (released) inmates of state prisons (see Chapter 10).
6. Gay and bisexual men. Hoping to head off a new surge of infections, the government recommended for the first time, in May 2002, that sexually active gay and bisexual men get tested at least once a year for HIV.

Addressing Barriers to HIV Testing

Knowledge of HIV infection status can benefit the health of individual persons and the community. Thus, HIV testing should be as convenient as possible to promote client knowledge of HIV infection status. Efforts should be made to remove or lower barriers to HIV testing by ensuring that:

- Testing is accessible, available, and responsive to client and community needs and priorities;

- Anonymous and confidential HIV testing are available;
- The testing process considers the client's culture, language, sex, sexual orientation, age, and developmental level; and
- Confidentiality is maintained. (In some places, like rural communities, **confidentiality** can't be assured, so people may decide not to get tested.

WHY IS HIV TEST INFORMATION NECESSARY?

A recent CDC random-digit-dial telephone survey of Americans ages 18 to 65 revealed that **42% of those surveyed** said they had been HIV tested. Geographic differences were great. For example, 26% of people in South Dakota said they had been tested versus 60% in Washington, D.C. (*MMWR*, 1999). The CDC also reported that about 25 million Americans are HIV tested or retested each year. Some 10,000 publicly funded counseling and testing programs conduct about 2.5 million of these tests annually.

Thirty percent of adults who seek HIV testing do so to find out their HIV status; 12% are tested because of hospitalization or surgery; 16% for application for insurance; and 7% to enter the military. Another 1% are referred by their doctor, the health department, or sexual partner, and 4% are tested for HIV for immigration reasons (Hooker, 1996).

IMMIGRATION INTO THE UNITED STATES

Having HIV can make it very difficult to immigrate into the United States. Everyone 15 years of age and older who applies for a green card has to take an HIV test. If you test positive you cannot get a green card unless you get an HIV waiver. An HIV waiver is a special permission from the Immigration and Naturalization Service (INS) that allows an HIV-positive person to obtain permanent residency. A green card applicant who is the spouse, unmarried child, or parent of a U.S. citizen or legal permanent resident may be able to get a waiver if he or she meets the other requirements. Before you can get a waiver, you must show the INS that you are able to pay your medical expenses if you get sick. In other words, you must show that you will not need government benefits such as Medicare or free county services. You must also show that you are not a danger to the public health—that is, you must demonstrate

SIDEBAR 12.2

BEST POLICY FOR PRENATAL TESTING?

Given the efficacy of antiretroviral drugs, if administered to the mother while the fetus is in the womb or to the baby within 48 hours after birth, in reducing perinatal HIV transmission, it has become a priority to maximize the number of pregnant women who consent to prenatal HIV testing. Three distinct approaches to obtaining consent for prenatal HIV testing are used: (1) the "**opt-in**" or voluntary policy (the test, after counseling, is offered to the woman, she may refuse); (2) the "**opt-out**" policy (the woman is not counseled about HIV/AIDS but is informed that the HIV test is part of a battery of prenatal tests, which are automatic but she may refuse HIV testing by signing a form rejecting the test); and 3) the mandatory newborn HIV testing approach (the mother is informed that the newborn will be tested, with or without her consent, if her HIV status is unknown at delivery). At press time, four states—New York, Connecticut, Arkansas, and Tennessee—mandate HIV testing of the newborn. Connecticut also mandates HIV testing of all pregnant women. Five states—Texas, New Mexico, Illinois, Michigan, and Florida—use the "opt out" policy. The CDC recently evaluated the efficacy of these three approaches in obtaining prenatal HIV tests in the United States and Canada. As a result of these evaluations, the CDC in April 2003 unveiled their HIV testing strategy for pregnant women. The new strategy specifically urges the testing of all pregnant women rather than relying upon patients to volunteer for testing. The guidelines also make HIV testing a routine part of care in doctors' offices and clinics, rather than waiting for patients to specifically request it. The strategy is advisory but has some authority: CDC will ask state and local governments to adhere to it in exchange for federal funding.

knowledge of how HIV is spread and what you can do to minimize the chances of transmission.

Section 212(a)(1)(A)(I) of the Immigration and Nationality Act renders inadmissible any applicant for a visa or admission who is found, according to the regulations published by the secretary of health and human services (HHS), to have a communicable disease of public health significance, which includes HIV infection. However, in view of humanitarian and family unity concerns, the law also provides waivers to inadmissibility, which are discretionary and granted on a case-by-case basis. Two specific waiver policies have been implemented for applicants seeking admission as **nonimmigrants** (for a temporary period of time) who are inadmissible due to HIV infection:

Routine HIV Waiver Policy—Nonimmigrants may be granted a waiver for admission to the United States for 30 days or less to attend conferences, receive medical treatment, visit close family members, or conduct business. The applicant must demonstrate that he or she is not currently afflicted with symptoms of the disease; there are sufficient assets, such as insurance, that would cover any medical care that might be required in the event of illness while in the United States; the proposed visit to the United States is for 30 days or less; and that the visit will not pose a danger to public health in the United States.

The Designated Event Policy—This policy facilitates the admission of HIV-positive persons to attend certain designated events, which are considered to be in the public interest, such as academic and educational conferences and international sports events. To initiate the process, HHS writes a letter to the Department of State (DOS) regarding a specific event. DOS recommends the event to the attorney general who designates the event and authorizes a blanket waiver. This blanket waiver allows HIV-positive applicants seeking admission to the United States specifically to participate in the designated event to be admitted for the duration of the event without being questioned about their HIV status.

If the INS discovers a noncitizen is HIV positive, he or she can be prevented from entering the country even if they have a green card. The INS does not require an HIV test to reenter. However, they are allowed to ask you if you have HIV. If you reenter the United States with AIDS-related materials such as pamphlets or medicine such as AZT in your luggage, or if you are clearly sick, the INS may ask you if you are HIV positive and detain you for medical examination. However, tourists and other nonimmigrants such as business people, temporary workers, or those seeking medical treatment are generally not asked to take an HIV test. There is no HIV test when applying for citizenship. Only persons applying for permanent residency are required to be tested by the INS. Always consult with an immigration/naturalization attorney.

BOX 12.3

THE PHYSICIAN'S DILEMMA

While on vacation, a physician came upon a motorcycle accident. He administered cardiopulmonary resuscitation for 45 minutes to one of the riders, who was bleeding from the mouth. After the man died, the physician asked the emergency department doctor in charge, who was a personal friend, to test the deceased for HIV.

The doctor said he needed consent. The physician said, "The guy died. I have a wife and kids."

The incident prompted this physician to raise the issue of testing without informed consent. This brings up the pro and con issues on informed consent. After reading the Pro and Con, decide where you stand on the issue.

Pro—*Physician*: "Consent for every test we do is a nice luxury, but we don't do it for syphilis or hepatitis B. Only with this disease have we departed from public health policy. The majority with HIV are not aware they're infected. Not in the history of medicine have we chosen a policy to protect the infected."

Con—*Physician*: "It's extremely paternalistic. We should be trying to get patients involved in their care and give them informed choices on what is possible. I favor consent on any test. I am afraid that liberal HIV testing laws and policies are being fostered out of physician self-interest rather than patient need.

Doctors don't like HIV particularly, and they don't like people who have HIV infection. They want a way to be in control. They want to know the patient's status, and their reasons are less medical or for the patient's best interest than for the physician's."

DISCUSSION QUESTION: What is your position?

POINT OF INFORMATION 12.3

BLACK MINISTERS LEAD BY EXAMPLE

FEBRUARY 7, 2000—FIRST ANNUAL BLACK HIV/AIDS AWARENESS DAY: GET TESTED, EDUCATED, INVOLVED

Five national organizations founded National Black HIV/AIDS Awareness and Information Day to mobilize Black-American communities in the fight against the epidemic.

During June 4 and 5, 2000, at least 50 black ministers in Chicago were HIV tested. Jessie Jackson, who led the movement, said, "There's a certain taboo and even a certain shame to AIDS." Balm in Gilead, a national organization established 12 years ago to help African-American churches deal with AIDS began airing television commercials about how the clergy can help. While similar mass testing has taken place in Atlanta, this is the first movement in Chicago and the first time Jackson's Rainbow/PUSH coalition has been involved. In the United States, AIDS is the number one cause of death of African-American men and women 25 to 44 years old. Statistics show that in Chicago, two-thirds of all new AIDS cases from 1999 through 2004 were among black Americans, compared to 37% one decade before.

Illegal Immigrants and HIV

According to immigration authorities, an unknown number of the 8 million undocumented immigrants in the United States are infected with HIV and the secretive existence they live out of fear of deportation is adding to their health risks, according to medical experts. In the worst cases, immigrants are dying because they cannot access health care or they are spreading the disease because they do not know they are infected. Undocumented persons with HIV may be getting a patchwork of inconsistent care, turning to emergency rooms for treatment of opportunistic infections and then returning to the shadows, afraid that deportation to a homeland without medicines or health care is a certain death sentence.

Immigration and Entry into Foreign Countries

An increasing number of foreign countries require that foreigners be tested for HIV prior to entry. This is particularly true for students or long-term visitors. Information available from 193 countries at the beginning of 2005 reveals that 62 countries require an HIV test prior to entry, on arrival, or on application for residency. Entering 2004, a total of 105 countries had entry or residence restrictions against HIV-infected people. There are 89 countries that have no restrictions. Before traveling abroad, check with the embassy of the country to be visited to learn entry requirements and specifically whether or not HIV testing is a requirement. If the foreign country indicates that U.S. test results are acceptable "under certain conditions," prospective travelers should inquire at the embassy of that country for details (which laboratories in the United States may perform tests and where to have results certified and authenticated) before departing the United States. For a copy of HIV Testing Requirements for Entry into Foreign Countries, send a self-addressed, stamped, business-size envelope to: Bureau of Consular Affairs, Room 5807, Department of State, Washington, D.C. 20520.

TESTING, COMPETENCY, AND INFORMED CONSENT

Competency is often used interchangeably with capacity; it refers to a person's ability to make an informed decision. For example, to consent to medical treatment, a person must be mentally capable of comprehending the risks and benefits of a proposed procedure and its alternatives. While a health-care provider can assess competence, a legal finding of competency is often required based on the testimony of a mental health professional. Mental illness by itself does not indicate that a person is incompetent to make medical decisions. Various degrees of mental incapacity may occur with HIV infection, requiring an assessment of competency. AIDS Dementia Complex (ADC) occurs in approximately 70% of HIVinfected patients at some point in HIV disease/ AIDS and may interfere with the patient's capacity to provide an informed consent.

POINT OF INFORMATION 12.4

U.S. PUBLIC HEALTH SERVICE (USPHS) RECOMMENDATIONS FOR HIV COUNSELING AND VOLUNTARY TESTING FOR ALL PREGNANT WOMEN

Knowledge of HIV status is important for several reasons. **First,** women who know their serostatus can gain access to HIV-related care and therapies (such as *Pneumocystis carinii* pneumonia prophylaxis, TB screening, and antiretroviral therapy) during pregnancy and postpartum as needed. **Second,** HIV-infected women can be offered antiretroviral therapy to potentially block maternal–fetal transmission of HIV. **Third,** an obstetrician would postpone rupture of amniotic membranes and avoid scalp electrodes or other potentially invasive procedures, all of which may be cofactors for enhanced transmission of HIV. **Fourth,** antiretroviral drugs can be offered to infants of HIV-seropositive women who have recently delivered.

This report (*MMWR,* 2001) contains the recommendations of a 10-member USPHS task force on the use of Zidovudine to reduce perinatal transmission of HIV. In summary, the recommendations are:

1. Health-care providers should encourage all pregnant women to be tested for HIV infection, both for their own health and to reduce the risk for perinatal HIV transmission. Four million women become pregnant each year in the United States.
2. HIV testing of pregnant women and their infants should be voluntary. In voluntary testing, the reason for the test, how it is administered, and the person's right to privacy and confidentiality must be explained. This allows the person the choice of taking or refusing the test and giving or not giving demographic data.
3. Uninfected pregnant women who continue to practice high-risk behaviors (such as IV drug use and unprotected sexual contact with an HIV-infected or high-risk partner) should be encouraged to avoid further exposure to HIV and to be retested in the third trimester of pregnancy.
4. For women who are first identified as being HIV-infected during labor and delivery, health-care providers should consider offering intrapartum and neonatal antiretroviral drugs.

In mid-1995 the CDC reported that routine HIV counseling and voluntary testing for all pregnant women has already proved effective in several communities nationwide. In one innercity hospital in Atlanta, for example, 96% of women chose to be tested after being provided HIV counseling.

According to the Pediatric AIDS Foundation, about $350 million a year could be saved by testing pregnant women for HIV. The average hospital bill for a baby born infected with HIV is $35,000 annually for the 8 to 10 years the child lives.

Informed Consent

Informed consent is not just signing a form but is a process of education and the opportunity to have questions answered. The concept of informed consent includes the following components: full disclosure of information, patient competency, patient understanding, voluntariness, and decision making. The process of obtaining informed consent involves appropriate facts being provided to a competent patient who understands the information and voluntarily makes a choice to accept or refuse the recommended procedure or treatment.

When the concept of informed consent is applied clinically, complexities arise regarding both

POINT OF INFORMATION 12.5

A WEAK LINK IN HIV TESTING

The movement from a **positive** test result to treatment is a weak link in the overall care of the HIV-infected. Jeffrey Samet and colleagues (1998) reported that although a majority of patients (61%) sought medical care in the first year after their diagnosis, 39% delayed treatment for longer than one year, 32% for longer than two years, and 18% for longer than five years. Their report did not include those persons using home testing kits. It is believed that these people are even further removed from testing to medical care follow-up.

BOX 12.4

BOXER STRIPPED OF FEATHERWEIGHT TITLE AFTER POSITIVE HIV TEST

It took Ruben Palacio 12 years to win a world title. On the eve of his first defense, he became the first champion to test positive for HIV. The British Boxing Board of Control said, "We can't risk the life of another boxer by letting him fight. It's a kind of disease that can be spread via blood contact, and boxing is a sport where that is likely to happen."

Palacio is the first active world title holder known to have tested positive for the AIDS-causing virus. Esteban DeJesus, who held the WBC lightweight boxing title in the 1970s, contracted AIDS after his retirement and died in 1989.

HIV testing has been a routine part of the pre-fight medical examination in Britain for years. In February 1990, African heavyweight champion Proud Kilimanjaro of Zimbabwe was barred from a fight with Britain's Lennox Lewis because he refused to give details of an HIV test to the British Boxing Board of Control.

His manager said, "This brings the HIV thing into perspective. Instead of going home with the largest paycheck of his life, he is going home with an HIV test result that means he will die."

In March 1996, Tommy Morrison, a former U.S. heavyweight contender disclosed that he is HIV positive. Entering 2005, at least 10 states require professional fighters and kickboxers, licensed in those states, to be HIV tested.

the content and the process. The concept contains ambiguous requisites such as "appropriate" facts, "full" disclosure, and "substantial" understanding. The process is affected by many variables including the communication skill and range of practice style of the physician; the maturity, intelligence, and coping strategies of the patient; and the interaction between the physician and the patient (Hartlaub et al., 1993).

Testing Without Consent

By 2003, at least 29 states had laws that allowed HIV testing without informed consent under certain conditions. The required conditions vary and include:

1. Patient or other authorized person is unable to give or withhold consent.
2. Test result will help determine treatment.
3. Patient is unable to give consent, and physician can document that a medical emergency exists and that the test is needed for diagnosis and treatment.
4. Test is needed to protect the health of other patients, health workers, or emergency or law enforcement personnel.
5. Several states require post-test counseling.
6. In 27 states, teenagers must have signed parental consent to be HIV tested. In 23 states (Alabama, Arizona, California, Colorado, Connecticut, Delaware, Florida, Georgia, Hawaii, Illinois, Iowa, Michigan, Nebraska, Nevada, New Mexico, North Carolina, North Dakota, Ohio, Rhode Island, Tennessee, Utah, Washington, Wyoming) minors can consent to HIV testing and treatment.

Generally, HIV antibody testing without consent is legally considered battery. Legal liability for "unlawful touching" may result from performing an HIV antibody test without consent. Such a procedure may also constitute an illegal search.

DISCUSSION QUESTION: Are federal and state governments overemphasizing personal privacy at the expense of prevention? (Defend your answer with examples/situations.)

Voluntary Named HIV Testing

In **voluntary named** HIV testing, the individual freely provides his or her name. In this type of testing, an individual voluntarily seeks to learn his or her HIV status and receives a result that is known both to the individual and the test provider/testing agency. An advantage to named testing is that health-care providers can contact the person tested if he or she does not return for the results.

Voluntary Unnamed HIV Testing

HIV testing is voluntary, but the identity of the person being tested is not placed on the blood sample or the testing form. As a result, the only person who can link the test result with an individual is the person being tested. This form of testing may encourage people concerned about HIV infection status to obtain testing as it eliminates risk of discrimination or stigmatization. However, it places the exclusive responsibility for seeking counseling, support, and preventive measures on the individual who is infected. Unnamed testing permits reporting of test data to public health authorities without the risk of breaching confidentiality.

Testing that is voluntary may miss populations that disproportionately need to be reached. The people least likely to have the virus, it appears, are the most likely to say yes to a test, and the people most likely to be infected are the most likely to say no. In one study, infection rates were 5.3 times as high among people who refused HIV testing as among people who consented to it. In voluntary anonymous, the downside is that such testing reduces the probability that they will return for post-test counseling and linkage to follow-up services, and a substantial reduction in partner notifications (Moser, 1998).

Mandatory HIV Testing

HIV testing is mandatory if it is required to participate in a process or activity that is not itself required. For example, if an HIV test is required for travel to some foreign countries, or to donate blood, this is considered mandatory testing because, while the test is required, one is not required to travel or donate blood. In mandatory testing, care must be taken to ensure that people are not in fact forced to undergo testing. At least in theory, mandatory testing is a form of voluntary testing: People can decide not to participate in the process or activity for which testing is required. In practice, however, the degree of voluntary consent is in some cases questionable. For example, in a situation where employment is not possible unless one agrees to be tested, and one needs that job, the voluntary nature of the test appears to have vanished (*AIDS, Health and Human Rights,* 1995).

Why Mandatory Testing?

Mandatory testing is for the protection of a certain group or the public at large. Although it is not anonymous, results are kept confidential on a need-to-know basis. Mandatory testing for HIV continues to be angrily debated primarily because of the possibility of error when running large numbers of test samples, inadvertent loss of confidentiality, and lack of overall benefit to those who are found to be HIV positive. Mandatory HIV testing is routine for blood donors and military and Job Corps personnel.

Mandatory HIV Testing of Newborns and Disclosure of Test Results to Mothers and Physicians

Perhaps no call for mandatory HIV testing has caused as much recent controversy as those requiring all pregnant women to take the test or their newborns will be tested. On June 26, 1996, New York became the first state in the nation to mandate and disclose the HIV status of newborns to mothers and physicians. Governor George E. Pataki signed into law legislation known as the **"Baby AIDS" bill,** which authorizes the state health commissioner to establish a comprehensive program of HIV testing of newborns. Over the next eight years, there was an 80% decline in HIV-infected babies born to infected mothers.

The passage of this bill was immediately followed by the American Medical Association announcement endorsing mandatory testing of all pregnant women and newborns for the AIDS virus. Entering year 2004 only the states of Connecticut, Arkansas, Tennessee, and New York require health-care providers to offer an HIV test to *every* pregnant woman as early in her pregnancy as possible—if she refuses the test, the baby is tested.

Compulsory HIV Testing

In **compulsory testing,** a person cannot refuse to be tested. Compulsory testing may be forced onto an individual, groups, communities, or even entire populations. A court may order an individual to be tested, or a government may decree or legislate that, for example, commercial sex workers, homosexuals, prisoners, hospital patients, or persons seeking immigration must be tested.

In Colorado, Florida, Georgia, Kentucky, Illinois, Michigan, Nevada, Rhode Island, Utah, and West Virginia, HIV testing is compulsory for people convicted of prostitution. However, many prostitutes are back on the streets before their test results are in. In many cases, the prostitutes could not be found for follow-up counseling. In Duval County, Fl., county judges agreed to impose a 30-day jail term for convicted prostitutes, a time period long enough to get their test results and provide counseling. Prostitutes have to sign the test results sheet. They are released as soon as they do.

Under Florida law, a prostitute who knows he or she is carrying HIV but continues to offer sexual favors can be jailed for one year.

In June 1999, the state of Oregon passed legislation that allows a judge to order a person accused of a crime to be tested for HIV. The person also would be tested for other communicable diseases if he or she transmitted bodily fluids to a victim. The results would not become public record.

In mid-2004, the governor of Wisconsin signed into law a bill that allows teachers to require students to be tested if a teacher is exposed to students' blood. The teachers in this state were added to a list of professionals who are permitted to require HIV tests on people whose blood they are exposed to. The list includes firefighters, police, emergency medical technicians, and other health care workers.

At least 45 states and the District of Columbia authorize HIV testing for charged or convicted sex offenders (Hooker, 1996 updated).

Fear of Compulsory Testing—If a massive compulsory screening test program were implemented, would it be possible to keep results confidential? What would be done with the information? For example, would the state prevent an uninfected person from marrying an infected one? Officials fear that mandatory testing will drive many people who might have volunteered for anonymous testing underground and away from health care. These people will be lost to the counseling and education that would benefit them and others. The reason for going underground would be fear of discrimination and social ostracism if found to be HIV-infected.

A case can be made that a compulsory program could maintain strict confidentiality even with large numbers of people being tested. But, it would appear that the political powers and public in general are not ready for broad-scale compulsory testing in the United States.

Confidential, Anonymous, and Blinded Testing

Both confidential and anonymous testing involve the use of informed consent forms that are, to date, with exception of the U.S. military, Job Corps workers, and certain criminals, done on a voluntary basis. Blinded testing does not, because of procedure, require informed consent.

Confidential HIV Testing—The person's name and test results are recorded. A consent to HIV testing must be given freely and without coercion. The volunteer does not have to provide any information unless he or she wants to.

The following example demonstrates one of the problems with confidential testing: A young homosexual male with signs of oral thrush agreed to an HIV test. Later that day, he called and asked that

BOX 12.5

CONFIDENTIALITY AND SEXUAL PARTNER BETRAYAL

This story took place in an HIV/AIDS clinic in the South. A husband and wife came into the clinic for an HIV test. They said the *only* reason for requesting the test was that they wanted to begin a family and hoped that nothing in their past would have led to either of them being HIV positive. The tests were completed.

The husband came in on a Monday; the wife came in that Friday.

Monday A.M.

Counselor: Mr. X, your test came back HIV positive.

Reactions and counseling were similar to those presented in this chapter. Then Mr. X said he wanted to be the one to tell his wife; *he insisted on it.* The counselor agreed, Mr. X left the clinic agreeing to come back for a follow-up counseling session.

Friday A.M.

Counselor: Mrs. X, your HIV test was negative.

Mrs. X: That's wonderful news. I can't wait to tell my husband. We've been waiting for my results. I want to get pregnant immediately.

Mrs. X received HIV-negative counseling and left the clinic very happy.

Clearly, the husband did not tell his wife the truth about his test results. A follow-up phone call to the husband went unanswered; so did a letter from the clinic. Several months later, Mrs. X called the counselor to tell her that she was pregnant!

What do you think the counselor should do now?

1. Inform the woman about her husband.
2. Take no action.
3. Call the husband and discuss the situation.
4. Threaten the husband with legal action if he does not tell his wife.
5. Your position?

Discuss the moral, ethical, and legal responsibilities of each participant.

his blood *not* be sent to the lab. He was a teacher in a parochial school and feared the results would be revealed. His sample was set aside, but the laboratory courier mistakenly took it for testing. The result was positive, yet no one could tell the patient. A malpractice attorney said to make certain all records of the test were deleted and to send the patient a letter urging him to return for a blood test. He never appeared (Wake, 1989).

Many national public health agencies and committees favor a confidential screening and counseling program that includes all individuals whose behavior places them at high risk of HIV exposure. These agencies recommend that the following eight groups seriously consider volunteering for periodic HIV antibody testing:

1. Homosexual and bisexual men
2. Present or past injection-drug users
3. People with signs or symptoms of HIV infection
4. Male and female prostitutes
5. Sexual partners of people either known to be HIV-infected or at increased risk of HIV infection
6. Hemophiliacs who received blood clotting products prior to 1985
7. Newborn children of HIV-infected mothers
8. Immigrants from Haiti and Central Africa since 1977

Anonymous HIV Testing—No name is ever given. This is also a form of voluntary testing. It differs from confidential testing only in that those who request anonymity receive a bar-coded identification number. They provide no personal information and they come back at a predetermined time to find out if their test number is positive or negative. No follow-up occurs. Forty-one states have anonymous test sites providing over 2 million tests a year (Nash et al., 1998 updated). Things change when the individual seeks treatment: If one goes to a doctor and the doctor does a viral load test, he reports the person and their viral load test to the health department. There is no way to keep treatment for HIV anonymous.

Blinded HIV Testing—This occurs when blood or serum is available for HIV testing as a result of another medical procedure wherein the patient's blood has been drawn for analysis. In this case, the demographic data have been recorded and can be used for epidemiological studies even if the name of the individual is withheld and a bar code is used. In 1988, the CDC asked for a blinded study of all 1989 newborn blood samples taken in certain cities in 45 states for metabolic studies. The name and other demographics of each newborn were recorded on the label of each tube. After the metabolic tests were completed, the name was changed into a bar code and leftover blood was sent to a state HIV testing center.

DISCUSSION QUESTION: What are reasons for and against blinded testing?

SUMMARY

HIV infection can be detected in three ways: first, by HIV-antibody or antigen testing prior to the signs and symptoms of AIDS; second, by detecting the presence of HIV nucleic acid; and third, by physical examination after symptoms occur.

The test most often used to screen donor blood at blood banks and individuals referred to testing centers is the ELISA test. ELISA (enzyme linked immunosorbent assay) is a highly sensitive and specific test that determines the presence of HIV antibodies in a person's blood or serum. The ELISA test was first used in 1985 to reduce the number of HIV-infected blood units for blood transfusions.

Because the ELISA test is only a predictive test that gives the percentage chance that a person is truly positive or truly negative, serum from those who test positive is retested in duplicate. If still positive, the serum is then subjected to a Western Blot (WB) test. The WB is a confirmatory test. If it is also positive, the person is said to be HIV-infected.

Other screening and confirmatory tests are available. The indirect immunofluorescent antibody assay (IFA) is relatively quick and easy to perform. Although it can be used as a screening test, it is generally used as a confirmatory test. The test is similar to the ELISA test except that the analysis is made by looking for a fluorescent color, indicating the presence of HIV antibodies, with a dark field light microscope. In 2003 the FDA approved 20-minute (OraQuick) and 3-minute (MedMira) rapid HIV tests for physicians' offices, medical labs, and hospitals, respectively.

The polymerase chain reaction (PCR) is a process wherein a few molecules of HIV proviral DNA can be amplified into a sufficient mass of DNA to be detected by current testing methods. It can determine if newborns of HIV-infected mothers are truly HIV positive. The branched-DNA assay offers a means of detecting the presence of HIV RNA directly within the clinical sample.

Gene probes are also being used to detect small HIV proviral DNA sequences in cells of people who are HIV-infected but not yet making antibodies.

REVIEW QUESTIONS

(Answers to the Review Questions are on page 438.).

1. What is the acronym for the most commonly used HIV-antibody test, and what does each letter stand for?
2. What basic immunological assumption is this test based on?
3. Does a single positive HIV antibody result mean the person is HIV-infected? Explain.
4. Is there a specific test for AIDS? Explain.
5. What is currently the most frequently used HIV confirmatory test in the United States?
6. What is the name of one additional confirmatory test in use in the United States?
7. How is HIV antibody detected in the ELISA test?
8. True or False: All newborns who are antibody positive are HIV-infected and all go on to develop AIDS. Explain.
9. What is the greatest shortcoming of the ELISA and WB tests?
10. What are the two major problems in interpreting ELISA test results?
11. What two factors may account for false-positive and false-negative results?
12. What is the relationship between false-positive results and prevalence of HIV in the population?
13. In an HIV screening test, what is a positive predictive value? Why is it called a predictive value?
14. What is the current gold standard of confirmatory tests in the United States?
15. What is the major problem in using this test?
16. Why is the polymerase chain reaction (PCR) considered so useful in HIV testing? Name two situations when PCR can be significant in HIV testing.
17. How quickly can one get reliable HIV test results using the OraQuick HIV test?
18. How quickly can one get reliable HIV test results using the MedMira HIV test?
19. True or False: Two home-use HIV antibody test kits are now available in the United States.
20. Using the ELISA test, when are HIV antibodies first detectable?
21. How early are HIV antigens detectable in human serum?
22. What are three benefits of early identification of HIV-infected people?
23. Name the four kinds of testing privacy available to people who want to take an HIV test.
24. What is the major difference between an anonymous and a blind HIV test?
25. Why would someone want an anonymous test?
26. True or False: The ELISA serological test is adequate to confirm HIV infection.
27. True or False: Pre- and post-HIV-antibody test counseling is recommended any time an HIV antibody test is performed.

13 AIDS and Society: Knowledge, Attitudes, and Behavior

In the end, they will say, we died not at the hands of our enemies, but in the silence of our friends.

Martin Luther King

Chapter Concepts

- HIV infections keep on going and going . . .
- HIV/AIDS is here to stay.
- The HIV/AIDS devastation is now.
- Inaccurate journalism leads to public hysteria.
- Vignettes on AIDS.
- It's 24 years later, and what do we know about HIV/AIDS?
- Use of explicit sexual language on TV and in journalism.
- Goal of sex education: to interrupt HIV transmission.
- Education, Just Say Know.
- What does the red ribbon mean?
- Education is not stopping HIV transmission.
- Students still have misconceptions about HIV transmission.
- The general public, homophobia, and HIV transmission.
- Employees are not well informed and fear working with HIV/AIDS-infected co-workers.
- Teenagers are not changing sexual behaviors that place them at risk for HIV infection.
- Physician–patient relationships in the HIV/AIDS era.
- U.S. Supreme Court renders its first ever ruling in the HIV/AIDS pandemic.
- Educating employees about HIV/AIDS.
- Placing the risk of HIV infection in perspective.
- Federal response to the AIDS pandemic—create an AIDS industry.
- Global AIDS, Tuberculosis, and Malaria fund begins, June 2001.

THE NEW MILLENNIUM AND HIV/AIDS

September 11 and June 27, 2001—The first is the date of the terrorist attack on the U.S. in New York. The second is the date on which 189 nations signed up to a UN Declaration of Commitment on HIV/ AIDS in the same city. One day changed the world, the other apparently changed very little.

HIV/AIDS is becoming more of a global disaster with each passing year of the new millennium. This disaster will be with us for many years to come, perhaps lifetimes or generations to come—like smallpox, the bubonic plague, cholera, and other diseases from ancient times. It does not appear that there are any new surprises that scientists are about to spring on this virus—but this virus continues to surprise our best scientists. HIV just keeps spinning its genetic building blocks looking for its next jackpot—how to overcome the next onslaught of antiretroviral drugs. Those jackpots keep coming up and the virus is winning the drug war. In developed nations, too many people think the AIDS crisis is over—they have been fooled by erroneous television and press coverage. Globally in 2002 through 2005, between 2 million and 3 million people died of AIDS annually

LOOKING BACK 13.1

AIDS COALITION TO UNLEASH POWER: ACT UP

The gay, lesbian, and transgender communities and their heterosexual supporters must be acknowledged for their immediate, aggressive, and humanitarian response to AIDS. Years before government was ready to accept its rightful role, these communities were caring for the sick, fighting for treatment and research, and making remarkable changes in personal and organizational behavior to curb the risk of AIDS. Their efforts and accomplishments are without precedent in modern medical history. In years past, an AIDS action group called ACT UP (AIDS to Unleash Power) under its mantra, "Silence = Death," urged people to get out there and raise hell because nobody else was doing anything about this disease. In order to promote prevention, they draped a 35-foot condom over Republican Jesse Helms's house, scattered the ashes of people who died of AIDS on the White House lawn, and tossed a coffin in front of a San Francisco hospital after an AIDS patient was denied a liver transplant. And these were some of the milder forms of protest. But it took these actions to get the attention of mainstream America and its politicians. Today's more recently infected people may not recognize the names of those who demonstrated in the streets, stopped meetings, defied government policies, and demanded adequate research funding and better medicines for AIDS, but they should know that without the efforts of these early warriors in the AIDS battle, the nationwide infrastructure of care, prevention, and treatment education would be much less than it is. The successes achieved by gay-empowered ACT UP later empowered women and heterosexuals to demand action for better drugs and a government commitment to finding an HIV vaccine, which served as model for international AIDS activism that has recently achieved great price reductions in HIV/AIDS drugs for their countries. ACT UP has also served as a model for people with other diseases, like breast cancer, to form action groups to demand more from their government.

Whatever weakness and failings exist in the current structures, they do provide structures to build and improve upon. Those who have died have left a legacy that can serve as a guide in the future as the epidemic cuts its way around the world.

and in each year between 4 million and 5 million new infections occurred. But estimated numbers of AIDS deaths and new infections for 2006 and 2007 are even higher!

AIDS in the new millennium is like a train heading toward a horrific wreck. As the train gains speed, the global rate of HIV infections increases. As T4 or CD4+ cells drop, the distance to the wreck becomes shorter; on impact millions more will have died of AIDS. Like the **Energizer bunny,** HIV infections keep on going and going and . . .

The impact of AIDS is so monumental on societies as a whole, and on communities and families in particular, that there is no precedent in human history. There's nothing from the Black Death (a European plague in the fourteenth century) to the world wars of the twentieth century that even approximates it. That we've never had such numbers or seen the focus on a single gender, or ever had so many orphans, so many social breakdowns in various sectors, has became an overwhelming linkage of events of which there are no modern parallels, and therefore, we have to respond in ways that are unprecedented. AIDS is going to leave a fossil-like imprint on civilization that we can't yet begin to imagine.

HIV/AIDS IS AN UNUSUAL SOCIAL DISEASE

AIDS, like other severe epidemics in America, has a powerful social force that has allowed it to become established and that has promoted its rapid spread. But this disease is unique in the sense that it seems to track the fault lines in our society and profits from the social flaws and weaknesses inherent in societies in transition. The spread of AIDS is determined by very powerful social and economic factors.

The epidemic is not afflicting us because of chance or bad luck. HIV was destined to spread because of specific social and economic factors. HIV is mainly sexually transmitted and therefore closely associated with and intertwined in people's relationships. The forces in our society that promote lasting and loyal relationships, unity, social order, and stable family and community life inhibit epidemics. The forces inherent in a disordered social environment—in which there is a high degree of family and community instability, where individual and community stress is high, and where significant numbers of people are dislocated from their homes and families—serve to aid the spread of this disease. AIDS is a schizophrenic condition. It is a pandemic worldwide, but it has not yet fully emerged. It often promotes stigma and rejection at the very time when people need comfort, compassion, and support. Stigma and discrimination further serve to propagate the disease because in this climate individuals do not feel confident or free to disclose their status, often not even to their sexual partners or their spouses. The disease remains silent and spreads relentlessly. HIV is transmitted through our most creative and spiritual potential, our ability to create life. It is also passed in the breast milk of mothers to their newborn infants; the very food of life can sow seeds of death. These spiritual connections can trigger negative and harmful perceptions and thoughts, promoting the concept that this epidemic is God's punishment to the wicked and to the sexually immoral. These can easily be translated into negative actions, blame, and rejection, further serving to keep the epidemic hidden and invisible, and fostering continued denial of the problem and its spread. Unlike most other terminal or life-threatening diseases, this one affects mainly the young and middle-aged adults, the adults on whom we all depend; they drive the economy, they parent and teach the children, they care for the sick, and keep the planes and trains running on time.

AIDS IS HERE TO STAY

The past 24 years of the global AIDS pandemic have taught us that HIV disease is a permanent part of life on planet Earth. HIV is simultaneously a virus and a phenomenon. When it is viewed only as a virus, it is hard to see why HIV prevention is a problem. People normally want to avoid harming themselves and others. When viewed as a phenomenon, HIV points to the many personal and societal causes of disease transmission. It points to the difficulties people have in making their intentions match their behavior. It points to the inequalities in relationships, which help to spread HIV. It points to societies' reluctance to prepare young people to manage their intimate relations, to admit to sexual diversity, to responsibly manage complex health and social problems such as drug abuse, and to provide access to health care. It points to the world's failure to care about improving living conditions in poor countries.

These are indeed the dark days of the global AIDS pandemic. Ninety-five percent of people who need treatment are dying without it. Necessary prevention and vaccine programs go unfunded. By the time you read this, another 3 million men, women, and children will have died since Summer 2004. About 24 million people will have died of AIDS ending 2005. And yet, the darkest days of this global pandemic are ahead. Many millions more will become HIV-infected and many millions more will die from AIDS. Nicholas Everstadt, a fellow at the American Enterprise Institute, a neoconservative think tank, estimated that up to 155 million people could become HIV-infected in just three countries (China, Russia, and India) between 2000 and 2025.

AIDS

AIDS was first described in the United States in 1981. Who would have thought then that 25 years later about 65 million people or about 1 in every 90 people on Earth would be infected with the virus that causes AIDS? And that this virus, re-

SIDEBAR 13.1

A MOTHER'S NIGHTMARE

Recently, at a bus stop, a woman saw a man walking toward her. The man was so thin she could see his face, leg, and arm bones. His eyes were sunken and sad. As he got closer, she recognized it was her son! He was living in the streets, an injection-drug user; he told her he was dying of AIDS. She was too stunned to be hurt, she just put her arms around him and let him cry.

There are some things mothers can fix—this wasn't one of them. But she set out trying to find him a place to live and a doctor to care for him. At age 35, her son is watching himself shrink into an old man. His mother, age 61, wonders how she will pay for his burial.

gardless of the involvement of the world's governments, the best scientists, and the expenditure of over $300 billion just in the United States, continues to spread out of control in many nations of the world. There is no way at the present time to prevent the 5 to 6 million new HIV infections each year. No drugs offer a cure, and most of the 41 million living with HIV-infected, beginning 2006, cannot afford and will not receive those drugs that offer some temporary improved quality of life. An HIV-preventative vaccine now is wishful thinking. But, there is one commodity in plentiful supply—blame—enough for everyone, everywhere.

BLAME SOMEONE, DÉJÀ VU

The greater **hostility** and greatest **stigma** tends to be assigned to diseases in which individuals are seen as responsible for having the disease; in which the disease's course is fatal; in which fear of transmission is a major issue; and in which the disease leads to highly visible and frightening physical expressions. All these conditions are associated with HIV disease and AIDS. With AIDS more than any other disease in history, people have found verbal mechanisms for distancing themselves from thoughts of personal infection. Worldwide, from the onset of this pandemic, people have learned in a relatively short time to **categorize, rationalize, stigmatize,** and **persecutize** those with HIV disease and AIDS. AIDS statistics are published in categories to identify how many gay or bisexual men, injection-drug users, persons with hemophilia, and so on, have developed AIDS. Also listed are the countries, states, and cities with the highest incidence of the disease, along with which racial and ethnic groups are highest among reported AIDS cases. **By focusing on categories of people, have we made it possible for society to rationalize that AIDS belongs to somebody else? Have we made the thought of HIV/AIDS somewhat impersonal? HAVE WE FOUND A WAY TO BLAME SOMEONE ELSE?**

Placing blame does not always require reason and tends to focus on people who are not considered normal by the majority. Thus, minorities and foreigners are often singled out to blame for something, sometimes anything. Epidemics of plague, smallpox, leprosy, syphilis, cholera, tuberculosis, and influenza have historically focused social blame onto specific groups of people for spreading the diseases by their "deviant" behavior. Blaming others leads to their stigmatization and persecution.

While the Black Death, a pandemic of bubonic plague, swept across Europe in the fourteenth century, blame was variously attached to Jews and witches, followed by the massacre and burning of the alleged culprits. In Massachusetts between 1692 and 1693, some 20 people were hanged or burned at the stake after being accused of having the powers of the devil. Eighty percent of those accused were women. When Hitler blamed Jews, communists, homosexuals, and other undesirables for the economic stagnation of Germany in the 1930s, the result was death camps and ultimately World War II. Now there is a new plague—HIV/AIDS. What blame comes packaged with this new disease?

Jonathan Mann, former head of the World Health Organization's Global Program on AIDS, said in 1998 that there are really three HIV/AIDS epidemics, which are phases in the invasion of a community by the AIDS virus.

First is the epidemic of silent infection by HIV, often completely unnoticed. **Second,** after a period of incubation/clinical latency that may last for years, is the epidemic of the disease itself.

Third, and perhaps equally important as the disease itself, is the epidemic of social, cultural, economic, and political reaction to HIV/AIDS. The willingness of each generation to place blame on others when believable explanations are not readily available simply recycles history. We have been there before; we have placed blame on others and it will continue. With respect to the HIV/AIDS pandemic, blame has been disseminated among nations. There is no shortage of political, economic, social, or ethical issues associated with this new disease.

FEAR: PANIC AND HYSTERIA OVER THE SPREAD OF HIV/AIDS IN THE UNITED STATES

With the 1981 announcement by the U.S. Public Health Service and the CDC that there was a new disease, AIDS quickly became a symbol for our darkest fears. Responsible public officials gave out conflicting messages: **reassurance** on one hand and **alarm** on the other. **Public panic and hysteria began.**

People with HIV disease and AIDS are still abused, ridiculed, and maligned. Some people believe that AIDS is divine retribution for immoral lifestyles. People who have not indulged in high-risk lifestyles (for example, newborns and recipients of blood products) continue to be labeled as **innocent victims,** implying perhaps that other HIV-infected individuals are guilty for their behavior that led to their infection and therefore deserve their illness.

Families and communities continue to be divided on their beliefs and acceptance of HIV/AIDS patients. Federal and state agencies stand accused of a lack of commitment and compassion in the war against AIDS. The bottom line is that **value judgments** are associated with HIV/AIDS because the disease involves the most private areas of people's lives—**sex, pregnancy, drug use, and finances.**

The Fear Factor

Worldwide, the political, medical, and legal communities used the media or vice-versa, to scare people about a new disease called AIDS. The result has been to scare people into fearing other people rather than the disease. For example, a 1990 survey of 1000 black American church members in five cities found that more than one-third of them believed the AIDS virus was produced in a germ warfare laboratory as a form of genocide against blacks.

Another third said they were unsure whether the virus was created to kill blacks. That left only one-third who criticized the theory.

These findings held firm even among educated individuals. Rumors that AIDS was created to kill blacks have circulated in the black community for years, and the belief is still endorsed by some black leaders.

A poster in the Swiss STOP AIDS Campaign focuses on how we think about people with AIDS:

> For the doctors, I am HIV-positive; for some neighbors, I am AIDS-contaminated; for my friends, I am Claude-Eric.

Soon after young homosexual men began dying in large numbers, a barrage of frightening rhetoric began filling the airwaves, television, the popular press, and even the most reputable scientific journals. The AIDS disaster was here. One health-care administrator stated, "We have not seen anything of this magnitude that we can't control except nuclear bombs."

In 1986, Myron Essex of the Department of Cancer Biology at the Harvard School of Public Health noted,

> The Centers for Disease Control and Prevention (CDC) has been trying to inform the public without overly alarming them, but we outside the government are freer to speak. The fact is that the dire predictions of those who have cried doom ever since AIDS appeared haven't been far off the mark . . . The effects of the virus are far wider than most people realize. It has shown up not just in blood and semen but in brain tissue, vaginal secretions, and even saliva and tears, although there's no evidence that it's transmitted by the last two.

In 1987, columnist Jack Anderson reported that the Central Intelligence Agency (CIA) concluded that in just a few years heterosexual AIDS cases would *outnumber* homosexual cases in the United States. Also in 1987, Otis Bowen, former secretary of health and human services, said AIDS would make the Black Death that wiped out one-third of Europe's population in the Middle Ages pale by comparison. In 1988, sex researchers William Masters, Virginia Johnson, and Robert Kolodny stated in their book, *New Directions in the AIDS Crisis: The Heterosexual Community,* that there was a possibility of HIV infection via casual transmission—from toilet seats, handling of contact lenses from an AIDS patient, eating a salad in a restaurant prepared by a person with AIDS, or from instruments in a physician's office used to examine AIDS patients.

In contrast to these reports is the 1988 article by Robert Gould in *Cosmopolitan* reassuring women that there is practically no risk of becoming HIV-infected through ordinary vaginal or oral sex even with an HIV-infected male. The vaginal secretions produced during sexual arousal keep the virus from penetrating the vaginal walls. His explanation was: "Nature has arranged this so that sex will feel good and be good for you."

Reaction Based on Fear

In December 1997, a male entered a bar and held a syringe full of blood against a female patron's throat. He said the blood contained HIV. He demanded money. He was caught and charged with robbery and attempted murder, pending the outcome of HIV tests of the blood from the syringe. The man thought the fear of AIDS would be sufficient to rob the bar.

In Taiwan, people with HIV have been hired to work as debt collectors. The widespread fear of HIV/AIDS is being used to force people to pay their debts. The HIV-infected need a job and the loan agency needed a clever way to get their money back. The loan agency believes it's a proper business agreement.

In rural Zimbabwe to be a widow and old is very dangerous. Self-appointed witch hunters backed by community leaders are accusing widows of bewitching people with AIDS. If the widow is lucky, she is banished from the village. If not, she undergoes a brutal cleansing ritual. Witchcraft dates back to the communities' forefathers and continues. Elderly women and widows are often referred to as witches. The increased number of deaths in rural Zimbabwe is believed to be caused by possessing a powerful charm that is being exposed to the young people. The prevalent belief is that HIV/AIDS can only attack a person if he or she has been bewitched or made unfortunate with the use of charms. One woman whose husband and two children died of AIDS was given one hour to leave the village and was not allowed to take her possessions—she lost everything. In another case, in an exorcism ceremony, the woman was made to crouch over a large bucket of boiling water with a blanket over her head. After 10 minutes the blanket was removed—her face and arms were scalded and disfigured.

While causing physical and mental suffering to those widows identified as witches, witch hunters are profiteering by cheating villagers in exorcising ceremonies. For one to be exorcised of witchcraft, one has to pay a large price. Some families have had to give their livestock in order to pay the witch hunters. When a village calls in witch hunters it also has to pay for their keep.

Fear of AIDS is understandable, given that AIDS is fatal and is communicable. AIDS appeared suddenly and spread quickly—it took a number of years to identify the virus that causes it and the mechanisms for spread. Yet, today the routes of transmission are well established and widely known, as are the precautionary measures that can be taken to prevent its spread. Early on, fear of AIDS took an unhealthy turn, anxieties were projected onto those who were hit the hardest by the disease, and the fear of AIDS became an irrational fear of *people* with AIDS.

WHOM IS THE GENERAL PUBLIC TO BELIEVE?

Because of the complexity of HIV disease, a great deal of press coverage of AIDS issues reflects what scientists say to journalists. A journalist's responsibility is to check that the facts are accurate, but not necessarily to judge their overall merit. Why should a good story be spiked just because other scientists disagree with the data interpretation? When scientists say contradictory things to the public, how can the public assess whom to believe? Science has a duty to inform and educate the public, but it must neither frighten people unnecessarily nor give them unjustified expectations. Claims of **"AIDS cures"** in the popular press need to be based on much more than just test tube data. Whatever the need to attract research funding, is five minutes of fame ever worth a day of fear or weeks of false hopes for many? The popular press has provided HIV-infected persons with a roller coaster ride between hopelessness and fantasies of imminent cure.

As a result of journalistic promises, there was and still is a range of emotions that run from real hope of a cure to public panic and hysteria. In at least five states, children with AIDS were barred from attending local public schools. The case of 12-year-old Ryan White of Kokomo, Ind., was made into a TV movie, *The Ryan White Story,* in 1989.

In some localities, police officers and health care workers put rubber gloves on before apprehending a drug user or wear full cover protective suits when called to the scene of an accident. In other communities, church members, out of fear of HIV infection, have declined communion wine from the common cup.

Since the epic announcement in 1981, HIV/AIDS has refashioned America. HIV/AIDS is a disease molded to the times, one that strikes hardest at the outcasts—gay men, injection-drug users, prostitutes, and impoverished whites, blacks, and Hispanics. HIV/AIDS has brought forth uncomfortable questions about sex, sex education, homosexuality, the poor, and minorities. The disease has inevitably polarized the people, accentuating both the best and worst worldwide. Many churches, schools, and communities have responded to the new disease with compassion and tolerance; others have displayed hate and reprisals of the worst kind.

As Camus wrote in *The Plague*, "The first thing (the epidemic) brought . . . was exile." Anyone who carried the disease could inspire terror. They became pariahs in society.

People with hate in their hearts torched the house of the Ray family and their three HIV-positive hemophilic children in Arcadia, Fla. Someone shot a bullet through the window of Ryan White's home to let the teenager know he should not attend the local high school. After he died, his 6-foot, 8-inch gravestone was overturned four times, and a car ran over his grave!

Former President Ronald Reagan: His Policy on AIDS

Early on, the federal government and its public health apparatus showed little interest in the HIV/AIDS epidemic. Former President Ronald Reagan never once met with former Surgeon General C. Everett Koop to talk about AIDS despite Koop's pleas. Koop said, "If AIDS had struck legionnaires or Boy Scouts, there's no question the response would have been very different."

By the time Reagan delivered his first speech on the AIDS crisis in 1987, over 40,000 men, women, and children had been diagnosed with AIDS and over 28,000 Americans had died of AIDS. Recently released documents from the Reagan administration show that the president's delay in addressing the AIDS issue was based on the perceived "political risk" in doing so. It took nine years and over 115,000 AIDS deaths before the Congress and former President George Bush enacted the nation's first comprehensive AIDS-care funding package—the Ryan White CARE Act (1990).

The primary purpose of the act is to provide emergency financial assistance to localities that are disproportionately affected by the human immuno-deficiency virus epidemic. An eligible metropolitan area (EMA) is any metropolitan area for which there have been reported to the CDC a cumulative total of more than 2000 Acquired Immune Deficiency Syndrome (AIDS) cases for the most recent five years for which data are available. The amended act requires formula grants based on the estimated number of persons living with AIDS in the EMA. Estimates were derived by using methods specified in the act. The amount of funds received by each EMA (under Title I) or state (under Title II) is determined by the locality's proportion of the total estimated number of living persons with AIDS.

Misconceptions About HIV/AIDS Linger

Despite widespread reports that casual contact does not spread the virus, families have walked out of restaurants that employed gay waiters and hospital workers have quit rather than treat HIV/AIDS patients.

In 1989, a man was barred in Anderson, Ind., from coaching his daughter's intramural basketball team because he had been diagnosed with AIDS. The 37-year-old father received the virus in a blood transfusion during surgery in 1984.

Each example points out that regardless of education, the public assumes the virus can be casually transmitted. **FEAR IS BEING TRANSMITTED BY CASUAL CONTACT—NOT THE VIRUS.** How would you react if a good friend, classmate, or coworker told you he or she was HIV positive? What if you found out that your child's schoolmate, a hemophiliac, had AIDS? What if you were told this child had emotional problems or a biting habit? What if your work put you in direct physical contact with people who might be HIV positive?

An AIDS diagnosis for one person resulted in his physician's refusal to treat him, his roommate left him, his friends no longer visited him, his attorney advised him to find another attorney, and his clergyman failed to support him. They were all afraid of "catching" AIDS. In another case, a mother whose young son has AIDS sent cupcakes to his classmates on his birthday. School officials would not permit the children to eat the cupcakes. The elementary school principal said the school had a policy against homemade food because it could spread diseases such as AIDS.

In Jacksonville, Fla., Leanza Cornett (Miss America 1993), using her reign as a national platform to teach about AIDS, was told by public school officials not to use the word "condom" while addressing student groups. In Bradford County, Fla., she was told she could not mention the name of the disease (AIDS) in three elementary schools she planned to visit.

In Hinton, W. Va., one woman was killed by three bullets and her body dumped along a

BOX 13.1

EXAMPLES OF UNCONTROLLED FEAR AND HOSTILITY

In Cleveland, **November 1998,** a black American sex worker was forced, as part of his plea bargain, to appear on an NBC-affiliated TV station WKYC to identify himself as a person living with HIV. The judge said her goal with the unusual sentencing was to protect the public. In **December 1998, Gugu Dlamini of South Africa was beaten to death the day after she revealed that she was HIV positive during a commemoration of World AIDS Day, December 1, 1998.** She was one of many unsung heroes of the daily struggle against HIV. Her death reminds us how stigmatizing the disease AIDS still is, and how much courage it takes for people with HIV to be open about their condition because fear still controls people's behavior.

In **Colonial Heights, Va., February 1999,** the 4th U.S. Circuit Court of Appeals found that although the Americans With Disabilities Act prohibits discrimination against people with AIDS, the law does not require U.S.A. Bushidokan, in Colonial Heights, to accept Michael Montalvo, then 14, for its Japanese-style sparring classes. Michael was 12 when he attempted to join the karate school. The owners barred him from class after learning about his HIV infection. The decision drew criticism from AIDS groups, which said it reflects fear rather than legitimate health concerns. An AIDS Action leader said, "Communities are right to become concerned about HIV, but cases like this detract from giving young people the information they need and give a false sense of how it is transmitted." The Montalvos, have since moved out of Virginia because they believe public disclosure of Michael's infection led to discrimination.

In December 1999, Mercy Makhalemele of Kwamashu, South Africa, told her husband that she was HIV positive. He shoved her into a pot of boiling water, scalding her. But she still had to go to her job selling shoes. Her husband came to her job and told her to go home, get her things, and leave him. He could not live with someone with HIV. That was 10 A.M. By 3 P.M. she was fired from her job. Adding to all this, her only child, a toddler, was diagnosed with AIDS. She was unable to find treatment for her. Her child died thereafter.

In February 2000, The *Los Angeles Times* reported on the death of Israeli pop star Ofra Haza. She died of AIDS. Her condition and death were a closely guarded family secret because of the stigma associated with an AIDS death in Israel.

In March 2003, in the southern Uganda city of Mbarara, a woman was battered and killed after she refused to have sex with her HIV-positive husband. In another case, a woman died after her husband, without warning, poured acid on her because she refused to sleep with him without a condom. Both men were HIV positive. These are but two recent examples of what is happening in Africa because of HIV-associated fears.

In February 2004, an Arizona surgeon who refused to operate on an HIV-positive patient agreed to pay $160,000 in fines for violating the American with Disabilities Act. And in India, a Catholic priest refused to bury a parishioner who died of AIDS. In April 2004, in Port of Spain, the family of a young man suffering from the HIV disease cleans him by donning boots and hosing down the bed, with him in it. They feed him by pushing food along the floor to him. And in Indonesia, a 24-year-old man was handcuffed by his family and guarded by a security guard in a room in his own house. Another man was locked up in a room built separate from the family's house. His food was delivered to him under the door.

In spite of the fact that 99% plus Americans are knowledgeable about HIV/AIDS, the *unthinkable* continues. A complaint was filed in April 2003 against a personal care home in Lycoming County, Penn., because the adminsitration refused admission to a man with AIDS on the basis that his illness would make staffers "uncomfortable." In the complaint, AIDS Law Project of Pennsylvania alleges that the 56-year old man—who is also legally blind and has heart disease—is illegally being denied care. The man's social worker, in an attempt to schedule an introductory admissions appointment for the man, offered to train the staff about HIV/AIDS and prevention/infection control but she was refused. The outcome of the case is pending.

The way people with HIV are treated is likely to continue to be based more on fear, prejudice, and judgmentalism than on the facts of science. Supreme Court rulings are not going to change that. What the courts do, however, is make it absolutely clear that such prejudicial behavior will not be legally tolerated.

remote road. Another was beaten to death, run over by a car, and left in the gutter. Each woman had AIDS and told people. And each, authorities say, was killed because she had AIDS. Lawyers and advocates for AIDS patients say the similar slayings, two counties and six months apart, illustrate the arrival of AIDS in the American countryside and the fear and ignorance it can unearth.

In Calcutta, India, October 2000, the body of a 45-year-old male who died of AIDS laid in the morgue of a hospital for 45 days because undertakers refused to take it for cremation. The "doms"—low-caste Hindus who take bodies for cremation—refused to remove the corpse after learning that he died from AIDS. Some of the doms threatened to stop work unless the body was removed immediately by someone else. India accounts for 60% of HIV cases in Asia and 20% of the world's HIV infections. But AIDS awareness is still extremely limited and the HIV-infected are more often than not ostracized by their families and local communities. The fear and stigma associated with HIV/AIDS has no boundaries.

What Do We Know?—Beginning year 2005 and over 24 years of the AIDS pandemic, it is clear that the scare headlines and tactics lack substance. From what has been learned about the biology of HIV, it appears the virus is not casually nor easily spread but it has reached the magnitude of the great plagues and a vaccine has not yet been found!

However, a survey of 3500 American adults was conducted by the National Institute of Allergy and Infectious Diseases on their state of knowledge concerning the availability of an HIV/AIDS vaccine. The results were reported in May 2003. One in five white Americans, or 20%, believe that an HIV vaccine already exists but is being kept secret from patients and the general public. Twenty-eight percent of Hispanics and 48% of blacks held this belief. Also, 42% of those surveyed did not know that vaccines require testing on human volunteers before being made available to the public. About 33% believe that vaccines in study could cause an HIV infection in humans receiving a test vaccine.

Fallout from AIDS: The Spector of Discrimination—Like most complex problems, the AIDS epidemic poses special problems. One of the most disturbing is discrimination against the HIV-infected. People worldwide have from the very

POINT OF VIEW 13.1

UNIVERSITY STUDENTS' EXPERIENCES

A university student taking a senior semester course on HIV/AIDS in 1999 experienced first-hand the fear-related ignorance that still exists.

This student had accompanied her friend to the office of a physician in a nearby small town. While the student was in the waiting room, she was using the time to read the given class assignment in her textbook, ***AIDS Update 1999.*** People coming into the office looked at the empty chairs on each side of her—then looked at her and sat elsewhere. Even as the office filled and fewer and fewer seats were available, no one chose to sit near her. She said, "People took one look at the book and chose another seat some distance from me." She said, "So much for education on nontransmission of HIV by casual contact!"

In November 2004, as I stood in line at my local CompUSA store, my cell phone rang. I answered the call and had a brief conversation with my wife. During the course of this conversation my wife asked what are the chances of me having free time that evening. As I have been extremely busy with my course load, without hesitation I replied, "I can't do anything tonight, I really have to take that AIDS test." This statement referred to the take home test assigned to me for my Biology of AIDS class and had nothing to do with an HIV test. Immediately after making this statement, I noted a woman behind me leave the line I was in and move to another much longer line to await checkout. It did not immediately occur to me why she had done this, but as time passed, I caught her staring at me on three separate occasions as I waited to be served. On exiting the store I looked back and saw her still glaring at me. It was clear to me that she moved lines in reaction to my AIDS statement to my wife. This woman had no reason to fear me even if I was HIV infected. Following her rude eavesdropping on my conversation, ignorance and fear guided her actions as they do so many people that do not understand the transmission of HIV.

Since placing these events in this book, students in each of my HIV/AIDS courses through 2005 have experienced similar discrimination while reading their AIDS textbook in a beauty salon, and in the office waiting rooms of a physician, veterinarian, optometrist, and a car wash.

beginning of the AIDS pandemic learned to categorize, rationalize, stigmatize, and "persecutize" those with HIV disease and AIDS. Perhaps the worst display of stigmatization and discrimination occurs against children, especially those of school age. Guidelines from the U.S. surgeon general, federal and state health officials, and the medical community have not calmed the fears of misinformed parents. Many stories have made headlines and television news concerning children who have been barred from attending school. While the courts can order admission, they cannot assure peer and adult acceptance. Persuasion must come through a better understanding of the disease. Parents can be reassured through reminders that HIV/AIDS is not transmitted by casual contact. Students must also be educated about HIV/AIDS.

The U.S. Department of Health and Human Services report in their "Healthy People 2000 Review—1998–1999" that 41.4% of America's college students received information on HIV/AIDS at their respective college campuses. The year 2000 target was to raise the number of college students receiving HIV/AIDS information on their campuses to 90%. *It failed.* Entering 2005 the target has still not been achieved.

The biggest difference between HIV/ AIDS and other diseases is the larger amount of social discrimination. Society does not reject those with a variety of sexually transmitted diseases or with cancer, diabetes, heart disease, or any other health problems to the degree that it rejects people with HIV disease or AIDS.

The AIDS pandemic has taught people about risk behavioral groups, homosexuals in particular. In some, this has promoted tolerance and understanding; in others, it has reinforced feelings of hatred. Information on HIV disease and AIDS, how it is spread, and how to prevent becoming infected, has, over the past 24 years, become a part of TV talk shows, movies, TV advertisements, and newspaper and magazine articles.

Phil Donahue, host of a former popular TV show, said in 1990, "On *Donahue,* we're discussing body cavities and membranes and anal sex and vaginal lesions. We've discussed the consequences of a woman's swallowing her partner's semen. No way would we have brought that up five years ago. It's the kind of thing that makes a lot of people gag."

The language, photography, and artwork used by the media are explicit and have upset certain religious groups. They believe that open use of language about condoms, homosexuality, anal sex, oral sex, vaginal sex, and so on promotes promiscuity.

DISCUSSION QUESTION: How can people learn to prevent HIV infection and AIDS without talking about sexual behavior and injection-drug use? Does it seem at times as if opponents of sex education would rather have people suffer with AIDS than have them learn about sex?

Regardless of who is correct, few could have predicted in 1980 the casualness with which these topics are now presented in the media. If the AIDS pandemic has done nothing else, it surely has affected the nature of public discourse. In 1987, prior to the TV broadcast of the **National AIDS Awareness Test,** viewers were warned of objectionable material. By 1990, few if any such viewer warnings were given.

How Can Information Help?—There is great hope that information will lead the nation past its social prejudice and forward to compassion for those who are HIV-infected or have AIDS. More than any disease before, AIDS has proved that ignorance leads to fear and information can lead to compassion. **The need for compassion is great.**

Admiral James Watkins, chairman of the 1988 Presidential Commission on the HIV Epidemic, reported that "33% to 50% of physicians in some of our major hospitals would not touch an AIDS patient with a ten-foot pole."

A friend of a dying AIDS patient who was in the hospital with Pneumocystic pneumonia went to visit him. As he was leaving, his dying friend said, "Thank you for coming—thank you for touching me." He said, "I can't even imagine being at a point in my life that I would be so grateful for someone touching me, that I would have to say thank you."

It would appear that although biotechnology has provided methods of HIV detection, new drugs, and hope for a vaccine, human emotional responses have not changed much from those demonstrated during previous epidemics.

AIDS EDUCATION AND BEHAVIOR

> I said education was our **"basic weapon."** Actually it's our **only** weapon. We've got to educate everyone about the disease so that each person can take responsibility for seeing that it is spread no further.
>
> C. Everett Koop
> Former U.S. Surgeon General

BOX 13.2

HOW SOME PEOPLE RESPONDED AFTER LEARNING THAT SOMEONE HAD AIDS

VIGNETTES ON COMMUNITY BEHAVIOR AND AIDS

In **Colorado Springs, Colo.,** Scott Allen's wife, Lydia, had contracted HIV from a blood transfusion hours before their son Matthew was born. A second son, Bryan, was also born before Lydia learned of her HIV infection. Scott wasn't infected, but was dismissed as minister of education at First Christian Church in Colorado Springs when he sought his pastor's consolation. Matt was kicked out of the church's day care center and the family was told to find another church.

When the family moved to Dallas and moved in with Scott's father, Allen, and his wife, church after church refused to enroll Matt in Sunday school. Allen, a former president of the Southern Baptist Convention wrote in his book, *Burden of a Secret: A Story of Truth and Mercy in the Face of AIDS,* "Good churches. Great churches. Wonderful people. Churches pastored by fine men of God, many of whom I had mentored. Nobody had room for a boy with AIDS."

Bryan, an infant, died in 1986, Lydia died in 1992, and Matt died in 1995.

In **Florida** in 1987, Mrs. Ray, the mother of three hemophilic HIV-infected sons (Ricky, 14; Robert, 13; and Randy, 12) turned to her pastor for confidential counseling. He responded by expelling the family from the congregation and announcing that the boys were infected. As a result, the boys were not allowed to go to church, school, stores, or restaurants. Barbers refused to cut their hair. Some townspeople interviewed said they were terrified at having the boys in the community. They had to move to another town. The Rays sued the DeSoto County School District. They agreed to pay a $1.1 million settlement in 1988. Ricky Ray died of AIDS on December 13, 1992, at age 15. Robert died at age 22 in 2000. Randy, age 24, was diagnosed with AIDS in May 1993 and is still living.

In **Duval County, Fla.,** (1990), the foster parents of a 3-year-old AIDS child who was infected by his mother, were forced to leave their church because other parents insisted that the child not be allowed to attend the church nursery. The pastor went along with the majority. When presented with CDC findings that HIV is not casually transmitted, one parent scoffed, "I called the CDC for information and they asked me what I was going to use it for." He then asked the congregation, "How can you believe anyone like that?"

In **California,** a young man arrived home one evening to find that the locks had been changed. A few days later he discovered that everything he had ever touched had been thrown out—clothes, books, bed sheets, toothbrush, curtains, and carpeting. Even the wallpaper had been stripped from the walls and trashed. The day before, he had told his friends he had AIDS. "Overnight, I had no friends. I slept on park benches. I stole food. I passed bad checks. No one would come near me. I was told that I had 14 weeks to live."

In a second California incident, volunteer firefighters refused to help a 1-year-old baby with AIDS at a monastery that cares for unwanted infants. The baby was reported to be choking. Although the fire department has agreed to respond to such calls in the future, one firefighter quit, saying he was frightened because he had not been trained to deal with AIDS victims.

In **Charleston, W. Va.,** in July 1995, a 10-year postal mail carrier who refused to deliver mail to a couple with AIDS was indefinitely suspended with pay after an educational class failed to change his mind. The mailman said he was afraid of cutting himself on the home's metal mail slot and becoming infected from envelopes or stamps the couple had licked.

Almost daily, similar senseless acts of violence and cruelty occur across the United States as a response to AIDS. Such episodes of panic, hysteria, and prejudice are perpetuated by the very people society uses as role models: clergy, physicians, teachers, lawyers, dentists, and so on. Philosopher Jonathan Moreno said, "Plagues and epidemics like AIDS bring out the best and worst of society. Face to face with disaster and death, people are stripped down to their basic human character, to good and evil. AIDS can be a litmus test of humanity."

THE LIFE OF RYAN WHITE

In Kokomo, Ind., Ryan White was socially unacceptable. He was not gay, a drug user, black, or Hispanic. He was a hemophiliac; he had AIDS. His fight to become socially acceptable, to attend school, and to have the freedom to leave his home for a walk without ridicule made him a national hero (Figure 13-1).

Ryan's short life was a profile in courage and understanding. Like many other people with

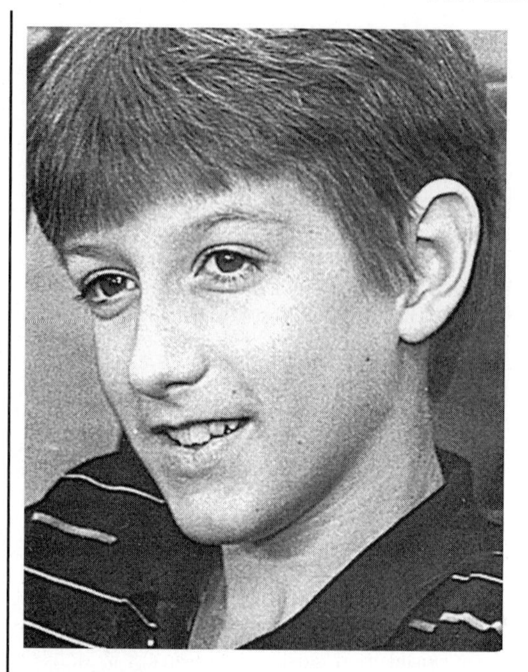

FIGURE 13-1 Ryan White Was Diagnosed with AIDS in 1984 and Died on April 8, 1990. This young male became another teenage AIDS tragedy. He gained the respect of millions across the United States before he died of an AIDS-related lung infection. *(Courtesy AP/Wide World Photos)*

AIDS, Ryan tried to change the public's misconception of how HIV is transmitted. Ryan suffered most from the indignities, lies, and meanness of his classmates and his classmates' parents. They accused him of being a "fag," of spitting on them to infect them with the virus, and other fabrications. Ryan said he understood that this discrimination was a response of fear and ignorance. Ryan got the virus from blood and blood products essential to his survival. Ryan's wish was to be treated like any other boy, to attend school, to study, to play, to laugh, to cry, and to live each day as fully as possible. But AIDS was an integral part of his life. AIDS may not have compromised the quality of his life as much as the residents of his community did. One day, at age 16, as Ryan talked about AIDS to students in Nebraska, another boy asked Ryan how it felt knowing he was going to die. Showing the maturity that endeared him to all, Ryan replied "It's how you live your life that counts." Ryan White died, a hero of the AIDS pandemic, at 7:11 A.M. on April 8, 1990. He was 18 years old.

For some, the occurrence of recently estimated 45,000 new HIV infections in the United States each year is evidence that HIV education and prevention efforts have failed. If HIV prevention programs are held to a standard of perfection and are expected to protect 100% of the people from disease 100% of the time, the efforts are by definition doomed to failure. No intervention aimed at changing behaviors to promote health has been or can be 100% successful, whether for smoking, diet, exercise, or drinking and driving. For example, even though warnings regarding the health effects of smoking were issued in 1964, warning labels on cigarettes were not mandated until 1984, and smoking-related illness still remains a major cause of death.

Because some of the behaviors and activities that need to change in order to avert HIV infection are pleasurable, it should be no surprise if short-term interventions do not lead to immediate and permanent behavior changes. An important difference between HIV infection and other life-threatening diseases is that HIV can be contracted by a single episode of risk-taking behavior. **Once HIV-infected there is no second chance—no giving up the behavior, like drinking alcohol or smoking, that will make any difference;** HIV disease progresses to AIDS.

After 24 years of experience with HIV, it has demonstrated that lasting changes in behavior needed to avoid infection can occur as a result of carefully tailored, targeted, credible, and persistent HIV risk-education efforts. Given experience in other health behavior change endeavors, no interventions are likely to reduce the incidence of HIV infection to zero; indeed, insisting on too high a standard for HIV risk-reduction programs may actually undermine their effectiveness. A number of social, cultural, and attitudinal barriers continue to prevent the implementation of promising HIV risk-reduction programs. The remote prospects for a successful vaccine for HIV and the difficulty in finding long-lasting effective drug treatments have underscored the importance of sustained attention to HIV prevention and education.

This sounds so easy: Educate people and they will do the right thing. **Wrong.** Knowledge does not guarantee sufficient motivation to change sexual behavior or stop the biological urge to have sex. Education has not stopped teenage pregnancy, nor has the knowledge about cigarettes causing lung cancer stopped people from smoking.

Perhaps the reason education is not as effective as it could be is because the public receives its education by daily doses from the mass media. With so much going on in the world, people have become more or less dependent on the media for information essential to their well-being. Gordon Nary (1990) said, "The public wants to know what's right or wrong in five three-second images or 25 words or less. It wants simple problems with simple solutions. It wants *Star Wars* with good and evil absolutely defined. The media often respond to these demands."

Mathew Lefkowitz (1990) says that the word "AIDS" has been infused with an irrational fear that has nothing to do with the illness. He states that the word has been politicized in such a way that it can and has been used as a weapon. Lefkowitz relates the parallel between today's use of the word AIDS with Eugene Ionesco's classic absurdist play *The Lesson*. In the play, a professor stabs a girl to death with the word *knife*—not with a knife but with the word knife. It appears that today the word *AIDS* is being used to stab those whom we fear; namely the HIV-infected. How long will it take the educational process to work? The virus is not as much our enemy as we are. For example, in Washington, D.C., 1995, because a national public radio commentator was furious at U.S. Senator Jesse Helms of South Carolina for having the audacity to suggest that the government spends too much money on AIDS research, he said, "I think he ought to be worried about what's going on in the good Lord's mind because if there is retributive justice, he'll get AIDS from a transfusion—or one of his grandchildren will get it."

POINT OF VIEW 13.2

HATRED EXPRESSED

CASE 1

In April 1996, a youth traffic safety program coordinator with the Florida Department of Transportation received a pledge form requesting a donation for the Florida AIDS Ride, an Orlando to Miami Beach bike race to raise money for AIDS programs. The person responded to the pledge form by writing to the race organizers that AIDS "was created as a punishment to the gay and lesbian communities across the world. As far as the gay and lesbians of this world . . . let them suffer the consequences!"

CASE 2

In May 1998 a St. Charles, Mo., man was arrested and charged with assault for allegedly injecting his 11-month-old son, in 1992, with blood containing HIV that police believe the man stole from a hospital where he worked as a phlebotomist. He allegedly injected his son with HIV because he hated paying child support. The boy was diagnosed with AIDS in 1996. In January 1999, the man was sentenced to life in prison.

Perhaps zealots need to realize that the purpose of civil exchange is to arrive at wisdom through reasoned debate, not to verbally intimidate those who differ into silence.

DISCUSSION QUESTION: What is your opinion?

Public AIDS Education Programs

Over the years billions of dollars have been spent by federal and state health departments and private industry to inform the public about cardiovascular risks, health risks associated with sexually transmitted diseases (STDs), smoking and lung cancer, chewing tobacco and oral cancer, drug addiction, alcohol consumption and driving drunk, and seat belt use to name just a few. In some cases these campaigns were eventually supported by specific state and federal legislation. Tobacco advertisements were outlawed on TV, and drivers in some states who are not buckled up must pay a fine. But even with laws to support these educational

POINT OF VIEW 13.3

NEWSPAPER HEADLINE SHOCKS READERS

On June 7, 1996, a large bold print headline of the Jacksonville, *Florida Times Union* newspaper, a very conservative newspaper, read:

STUDY: ORAL SEX POSES HIV INFECTION RISK

This headline in this newspaper would never have happened in this author's lifetime without 15-plus years of HIV/AIDS history. During this time, declarations involving all types of sexual behavior and drug use appeared during prime time TV programming and voluminous publication of such topics in all types of regional, state, and federal pamphlets, magazines, journals, and public reading materials. Yet, this newspaper received 124 phone calls from people angry about this headline in their newspaper. Their main objection concerned the use of the words "Oral Sex." Many others objected but did not call—some dropped their subscription.

DISCUSSION QUESTION: How would you feel about this headline appearing in your hometown newspaper?

programs, many adults have failed to change established behavior patterns. In the larger cities, educators must combat the fear that AIDS is a government conspiracy to eliminate society's "undesirables"—minorities, drug addicts, and homosexuals. They must overcome cultural and religious barriers that prevent people from using condoms to protect themselves.

It might also be added that there have always been educational programs against crime, but from 1990 through 2005 more new jails were built in the United States than ever before. In short, educational programs on TV, radio, in newspapers, and in the popular press have achieved only limited success in changing peoples' behavior.

It is not that education is unimportant; it is essential for those who will use it. That is the catch. Although education must be available for those who will use it, too few, relatively speaking, are using the available education for their maximum benefit. In general, people, especially young adults, do not do what they know. They sometimes do what they see, but most often do what they feel. In short, knowledge in itself may be necessary but it is insufficient for behavioral change. A variety of studies have failed to show a consistent link between knowledge and preventive behaviors (Fisher, 1992; Phillips, 1993).

Costs Related to Education/Prevention—The assertion that spending more money on educational programs will ensure disease prevention for the masses, as the examples given suggest, may not be the case. In particular, peoples' behaviors regarding the prevention of HIV infection do not appear to be changing significantly despite the billions of dollars used to produce, distribute, and promote HIV/AIDS education. The major educational thrust is directed at how not to become HIV-infected. Most of this information is being given out to people ages 13 and older.

The problem with AIDS education is that communicating the information is relatively easy but changing behavior, particularly addictive and/or pleasurable behavior, is quite difficult. The mass media have provided near saturation coverage of key AIDS issues and it is very unlikely that significant numbers of future HIV infections in the United States will occur in individuals who did not know the virus was transmitted through sexual contact and IV drug use. Yet new infections over the past four years occurred at an incidence of about 45,000/year.

Although humans are capable of dramatic behavioral changes, it is not known what really initiates the change or how to speed up the process.

Public School AIDS Education: Just Say Know

Some information relevant to AIDS education can be learned from educational programs that have been designed to reduce pregnancy and the spread of STDs among teenagers. However, data from a variety of high school sex education classes offered across the country indicate that teenagers are learning the essential facts but they are not practicing what they learn. **They do not do what they know.**

Risky sexual behavior is widespread among teenagers and has resulted in high rates of STDs. Over 25% of the 15 million STD cases per year occur among teenagers. One in six teens has been infected with an STD. Over 50% of sexually active teenagers (11 million) report having had

BOX 13.3

WHEN ONE WITH AIDS COMES FORTH

Father Paul made the decision to preach on AIDS because of a phone call he had received informing him that a former parishioner was coming to Jacksonville. "George has AIDS. He will be in Church on Sunday. With your permission, he will be receiving Communion."

Father Paul granted permission and welcomed George's attendance. He sensed, however, that some might not agree to have George in church or receive Communion. In his parish, a number of people refuse to believe that HIV is not transmitted by saliva from the lavitha (Communion cup).

By Sunday morning's sermon, over 60% of the congregation had learned of George.

Father Paul began his sermon, "Today's Gospel lesson, Luke 10:25–37, tells us the Parable of the Good Samaritan . . . The Parable challenges us to take stock of who our neighbors are who have needs that we can meet. . . . is it not also true that our neighbors are being harmed by AIDS? . . . Many Orthodox Christians are good about reaching out to the needy and indigent. But we are not so willing to reach out to those with AIDS."

Father Paul reminded everyone about the faith of the church. "To believe that one can contract sickness and AIDS from Holy Communion is blasphemy against the Holy Spirit. It is also to render everything that the Bible and the Church teaches about Communion meaningless."

As he spoke, he saw George near the back of the Church. Though only 47 years old, George looked 60. George was weak, abnormally thin, spoke with a rasp, and walked with a cane.

Before Sunday's Liturgy, Father Paul had discussed with George his pastoral concerns about the people's anxieties. George proposed a solution, he would receive Communion last.

Father Paul introduced George to the congregation and announced that he would be receiving last.

At Communion time, several of the congregants assisted George to the front of the church. Then, the same individuals and several others lined up behind him.

About 10 people received Communion after George. Father Paul asked one why he did it."Father, did you not tell us in your sermon that we had to be Good Samaritans? It would have been a very unloving and discriminatory act to allow George to go last."

George died several weeks later in New York City. The news was received with sadness.

George, thank you for coming to Jacksonville. God brought you to us to help us grow. May God remember you in his Kingdom.

(Adapted with permission from Father Paul Costopoulos, Jacksonville, Fla.)

FREEDOM FOR COMPASSION: CHILDREN AND AIDS

Charlie the doll was pressed against the antique glass case. Jeff, a blue-eyed, blond-haired boy of 10, looked at him closely. "Someday Charlie will leave this cage," he pondered, "and someday he will be free."

A few visits later, Jeff devised a plan to buy Charlie. He began his task by seeking employment as a leaf raker, a car washer, and the best of panhandlers among friends. After a while his hope faded and his energy waned, but he did not despair. He had met adversity before; in fact, for most of his life. When Jeff was 4 years old, he had contracted AIDS.

The family of John Calvin Presbyterian Church met Jeff because other congregations had turned him away, telling his family that Jeff's illness was a punishment from God. Jeff planned his own memorial service, but five different congregations ignored him by making excuses that the songs he had chosen from the play, "Peter Pan," and the balloons he had requested would not be appropriate.

One Sunday morning as I was beginning my sermon, Jeff's mother wheeled him down the main aisle to a front pew. I wondered what would happen if this church rejected him, too? How would the other children treat him? But my fears were relieved after church at the coffee fellowship. Parents introduced themselves to Jeff's mother and the children included Jeff within their circle of games. People earnestly gathered to accompany Jeff and his family on their special journey: members ran errands, provided transportation, and brought in food. This outpouring of help came at a crucial time: Jeff's mother had given up her own business to take care of Jeff, emotional pressures contributed to Jeff's parents' divorce, the family lost their home to bankruptcy court, and Jeff's brother and sister suffered from the prejudice of schoolmates and others.

BOX 13.3 *(continued)*

A few weeks prior to Jeff's death, I bought Charlie the doll. As Jeff's fragile hands began to untie the shiny silver ribbon that secured the purple box, large tears began to trickle down his sunken cheeks. When he discovered what was inside, Jeff smiled and said, "Now he's free . . . and someday I will be, too." Jeff's freedom arrived on March 2, 1988.

Those of us who knew Jeff have gained freedom as well. Jeff, and others like him, have introduced us to a new appreciation of life. Through them we have been reminded of how fragile we are and of our precious responsibility to live each moment fully.

Jeff's memorial service was just as he had planned it. The sanctuary of the church was filled with the nurses and doctors who had worked with him, the many hospice volunteers who had given him solace, his buddies from the Tampa AIDS Network who had held his hand, the many other friends he had made on his journey. Hundreds of brightly colored helium balloons were released into the sky at the end of the service. We celebrated Jeff's life—and ours as well. Our celebration continues as we minister with others who are traveling this very difficult path. Our strength and hope are renewed with each encounter, for we have been given the freedom for compassion.

Adapted from Rev. Jim Hedges, pastor of John Calvin Presbyterian Church, Tampa, Fla., in Church and Society, *Vol. 79, No. 3 (January/February 1989). Reprinted with permission.*

BOX 13.4

THE WAY WE ARE

The young man worked alongside a 35-year-old woman helping to teach the disabled to function. She was very attractive with a wonderful sense of humor. He watched her every movement—he was in love from a distance; she inspired him to new heights in his work and thoughts about his future. She understood his emotions so she was not surprised when, after work one evening, he offered dinner and a moonlight walk around the lake. During the walk, she could tell from the conversation that his young hormones were flowing—as were her own. In the awkwardness of saying goodnight she said, "I know you want me—I would like that very much." As his broad face glowed she said, "I must share something with you. I have been diagnosed with AIDS but there are no outward signs yet." The love that moments before surged through his body crashed down around him; he felt ill yet sympathetic—his urge for sex vanished. He could not bring himself to look at her as he said, "Please forgive me, I just can't." She nodded her understanding and explained following an auto accident in 1984 it was either a blood transfusion or death. "I made the choice for life and whatever comes with it." The young man cried as he walked home. He left the job the next morning.

two or more sexual partners; and fewer than half say they used a condom the first time they had intercourse.

Teenage Perceptions About AIDS and HIV Infection in the United States—A recent survey by *People* magazine indicated that 96% of high school students and 99% of college students knew that HIV is spreading through the heterosexual population; but the majority of these students stated that they continued to practice unsafe sex and that 26% of American teenagers practice anal intercourse. Data such as these have prompted a number of medical and research people to express concern for the next generation. If HIV becomes widespread among today's teenagers, there is a real danger of losing tomorrow's adults. Available data suggest that teenagers have not appreciably changed their sexual behaviors in response to HIV/AIDS information presented in their schools or from other sources.

Teenagers at high risk include some 200,000 who become prostitutes each year and others who become IDUs. About 1% of high school seniors have used heroin and many from junior high on up have tried cocaine (Kirby, 1988). A large number of children ages 10 and up consume alcohol. Is it possible that too much hope is being placed on education to prevent the spread of HIV? Teenagers must be convinced that they are vulnerable to HIV infection and death. Until then, it only happens to someone else. The World Health Organization estimates that worldwide, beginning 2006, there will be 16 million HIV-infected teenagers.

College Students—Everyone must know and act on the fact that a wrong decision about having

POINT OF VIEW 13.4

THE RED RIBBON

Frank Moore II, a Manhattan painter who was instrumental in launching the overlapping red ribbon as a symbol of AIDS awareness in 1991, died April 22, 2002, from AIDS-related complications at the age of 48. Moore, who said that his paintings represented a journal of his long battle with HIV, was a board member of Visual AIDS, a Manhattan-based group that raises money to fund artists with HIV/AIDS and helps maintain the art of people with the disease. The red ribbon became an international symbol of AIDS awareness and has been used in other colors by groups to represent different causes.

The color red was chosen because the disease was a blood-borne disease. The shape of the ribbon was meant to signify the connectedness of all of us with or without the disease, who wanted to make a statement of visible support for those who were infected or had AIDS.

THE EVENT

We were sitting in a small Italian restaurant. I had just come back to town from a presentation on AIDS. The jacket I wore still had the red ribbon on the lapel. As we enjoyed our meal, I noticed a woman at the next table who appeared to be glaring at me and making statements to her companion. At one point her voice became loud enough for us to hear her say, "I am sick and tired of those people trying to push the lifestyle of homosexuals down our throats" as she was looking right at me. She then said, "That red ribbon is a sign of a sick person trying to make all of us sick too. That ribbon and all that it stands for ruins my day." With that she and her companion left the restaurant.

That outburst left my family and me embarrassed and confused. My children deserved an explanation. I don't think I have ever explained the idea of the red ribbon to anyone before. Like so many things we observe in life, after a while they become understood by each in his or her own way. This woman expressed her way rather forcefully. To my children I said that the ribbon is a symbol to call attention to a social problem that needs a solution. I went on to say, "Do you recall the song 'Tie A Yellow Ribbon Round the Old Oak Tree' in 1973 and what that meant? And, do you recall the ribbons tied around trees, on car antennas, mailboxes and so on while our 56 servicemen were held captive in Iran in 1980 and again for our captives in the 1991 Persian Gulf War? Remember the first lady Nancy Reagan's campaign using red ribbons for 'Just say no to drugs' and more recently the pink ribbons for women against breast cancer and most recently the purple ribbons for stopping violence in our schools? These are all symbolic gestures to show support for those enduring suffering and pain. All the ribbons then and now serve to connect people emotionally, to help unite people in a common cause, to help people feel less isolated in a crisis."

I explained to my children that the problem with the red ribbon now is similar to what occurred over the long time period our soldiers were in captivity—people begin to wear the ribbon as an accessory.

The Author

sexual intercourse can take away the future. For example, a young college student had a three-year nonsexual friendship with a local bartender. She was bright, well educated, and acutely aware of AIDS. After drinks one evening, as their friendship progressed toward sexual intercourse, she asked him if he was "straight" (a true heterosexual) and he said yes. But he was a bisexual. It was a single sexual encounter. She graduated and left town. She found out that the bartender died of AIDS three years later. She did not think much of it until she was diagnosed with AIDS five years after their affair. This young, talented, bright, and personable girl has since died, she lost her future. It is difficult to change something as complex as personal sexual behavior regardless of knowledge. It also brings up at least one other important point in personal relationships: **telling the truth.**

Nothing but the Truth?

During the 1988 Psychological Association Convention, the following facts were presented with respect to telling the truth or lying in order to have sex. The data came from a survey of 482 sexually experienced southern California college students:

FIGURE 13-2 Dawn Beckhols at age 23, about the time of her infection. She believed that she became infected via one sexual encounter while on vacation. Dawn died of AIDS on July 13, 1997, at age 33. *(Photo courtesy of L. Schwitters)*

SIDEBAR 13.2

RECOMMENDED FILM: DAWN'S GIFT

She relates her story of becoming HIV-infected. Running time, 38 minutes. Cost, $5.50 plus about $2 for shipping.

Make check out to L. Schwitters—Dawn's Gift and send to:

L. Schwitters—Dawn's Gift
1745 Brookside Dr. SE
Issaquah, WA 98027
1-425-392-9161 fax 1-425-837-9971
email dawnsgift@hotmail.com
Purchase orders accepted

1. 35% of the men and 10% of the women said they had lied in order to have sex.
2. 47% of the men and 60% of the women reported they had been told a lie in order to have sex.
3. 20% of the men and 4% of the women said they would say they had a negative HIV test in order to have sex.
4. 42% of the men and 33% of the women said they would never admit a one-time sexual affair to their long-term partner.

Researchers at the French National Research Institute reported in 2002 that men and teenage boys are far less likely than females to tell their main sexual partners they have been diagnosed with HIV or other sexually transmitted diseases. Researchers found that 14% of men diagnosed with an STD in the past five years had not told their main partners, compared with just 2% of women. Similarly, 51% of boys who had been diagnosed with an STD had not talked about it with their partner at the time, in contrast to 9% of girls. In a 2002 online poll Gay.com/PlanetOut.com Network wanted to know what people thought about the fact that a San Francisco court awarded $5 million in damages to a man who claims he was infected with HIV from his ex-lover, a former city health commissioner, who lied about his HIV status. When asked, "Should lying about one's HIV status to a sexual partner be a crime?" 69% of respondents said yes, 8% said no. Another 20% answered, "only if someone is infected as a result."

Adult Perceptions About HIV Infection and AIDS in the United States—A 1996 survey by the Kaiser Foundation: *Survey on Americans and AIDS/HIV* found that with respect to having confidence in AIDS sources, 6% did not believe the public health officials, 9% did not believe the U.S. surgeon general, 11% did not believe the newspapers, 16% did not believe Magic Johnson, 17% did not believe church/religious organizations, 25% did not believe the media, and 34% did not believe the U.S. government. Thirty-three percent of people did not know there are drugs and other treatments for HIV disease or AIDS and 29% did not know there were ways to prevent or reduce the spread of AIDS. Less than 5% could name a single person they thought was a national leader on AIDS. These data still hold true today!

One aspect about what adults now think about the AIDS pandemic has changed dramatically. The proportion of Americans naming HIV/AIDS as the nation's number one health problem has been steadily declining over time. In 1987, seven in ten Americans (70%) named HIV/AIDS as the most urgent health problem facing the nation. In 2004,

LOOKING BACK 13.2

AMERICAN HIGH SCHOOL GRADUATIONS AND AIDS

High school graduates of 1998–1999 were the first to graduate from U.S. public high schools without the possibility of experiencing an HIV/AIDS-free world. The recognizable spread of HIV on a global scale began in the mid-1970s but was first reported in the United States in 1981. The graduates of 1998–1999 entered a world where hundreds of thousands had already died of AIDS in America and millions had died of the disease worldwide.

15% named HIV/AIDS as the nation's number one health problem. Americans are now more likely to name HIV/AIDS as the most urgent health problem facing the world than facing the nation. In 2004, the Associated Press ran a telephone survey about public attitudes on AIDS. When asked "How concerned are you personally about becoming HIV infected?", 48% were somewhat concerned, 52% were not. But, 51% were concerned about their sons and daughters becoming HIV infected. To, "do you personally know anyone who has died from AIDS or tested positive for HIV?" Yes, 39%; No, 61%. Fifty percent favored teaching safer sex as the major focus in prevention while 40% favored teaching abstinence only.

POINT OF VIEW 13.5

THE RIGHT TO PRIVACY

Arthur Ashe (Figure I-3) died of AIDS in February 1993; he was *forced* to go public about having AIDS. Did that affect his life? Yes. He had to alter a lifestyle he was trying to live with his family. He could no longer live a normal day with his family.

The immediate defense from the press, which forced Arthur Ashe to go public, is that it serves the public interest to identify prominent people with AIDS. The news media says that it increases awareness—and therefore, theoretically, action—to fight back against AIDS. **DO YOU AGREE?**

Organizations can write and broadcast about HIV/AIDS any time, about anybody they want to. **IS THIS MORALLY RIGHT?**

Some defenders of the news media say that a public figure cannot have it both ways—cannot deliberately keep himself in the limelight with sports equipment endorsements and so forth, and then turn around and say, "I want to be a private person." **But what is public and what is private?**

As Ellen Hume asks, "Is there no moment of a public figure's life that is not open to prurient exposure? Does being a political officeholder or, as in Ashe's case, a sports champion, mean that the public owns all of your life, including your life in the bedroom, the doctor's office, the church confessional, or the psychiatrist's couch?" **WHAT IS YOUR REPLY TO HER?**

Ms. Hume asks, "Should journalists end their scrutiny of public figures?" She thinks not. But she says, "Not every revelation serves the public interest. No American president and few athletes, astronauts, journalists or other heroes could have survived this Spanish Inquisition. Isn't it time for the press to develop a more sophisticated sense of priorities and ethics to go along with its extraordinary new power?" **What is your reply**?

The Workplace—If a person is not working near or beside someone who is HIV positive, they will be relatively soon. But they may not know it because a person's right to privacy prevails over an employee's right to know.

Two thousand adult employees were phone interviewed during a national survey concerning their attitudes about AIDS. The survey (Hooker, 1996) revealed that:

- AIDS is the **chief health concern** among 20% of U. S. employees (cancer was the primary concern of 32%; heart disease of 7%).
- 67% of employees predicted that their co-workers would be uncomfortable working with someone with HIV or AIDS.
- 32% thought an HIV-positive employee would be fired or put on disability at the first sign of illness.
- 24% said that an HIV-positive employee should be fired or put on disability at the first sign of illness.
- 75% of employees said they wanted their employers to offer a formal AIDS education program.

AIDS can have a variety of impacts in the workplace. The obvious one, of course, is on the individual employee who is diagnosed with HIV. The probability of sickness and death obviously affects the individual and his or her ability to continue to contribute to the organization's activities and goals.

Employees' fear of AIDS can create a widespread loss of teamwork and productivity, and create an environment that is inhumane and insensitive toward the infected employee.

As the incidence of HIV and AIDS increases, the impact on organizations will obviously increase as well. While there are important logical and moral reasons for ensuring that infected people are not discriminated against, there are also practical reasons for addressing the employee HIV/AIDS problem.

An educated workforce, aware of the facts regarding diagnosis, testing, treatment, and transmission of HIV, can have a positive impact on the overall health of all employees. People are more inclined to openly acknowledge their HIV status and to seek treatment when assured of a supportive workplace environment. This increases productivity.

Whether additional public education will change these attitudes over time is unknown.

American Perceptions About HIV/AIDS in Developing Nations Based on the Henry J. Kaiser Telephone Poll of 1042 Randomly Selected Adults in 2002

- 74% supported President Bush's proposal to spend $500 million over three years to stop mother-to-child HIV transmission.
- 31% said the United States was not spending enough on HIV/AIDS in developing countries.
- 80% said that HIV/AIDS in other countries would affect the quality of life in America.
- 51% said that America has no more obligation to fund HIV/AIDS in developing countries than other wealthy countries.
- 47% said additional funding for HIV/AIDS in Africa would not lead to meaningful progress.
- 80% said that Africans were unwilling to change their unsafe sexual practices to prevent HIV infection.
- 75% said the African government was not doing enough to help fight HIV/AIDS.
- 51% said HIV/AIDS prevention and treatment is a high priority for U.S. spending on health care in Africa.
- 67% said that the worst is yet to come for Africa.

The survey also found that many Americans think the HIV/AIDS crisis in the United States is over.

- 15% of white Americans believe a cure for AIDS exists.
- 25% of black Americans believe a cure exists.
- 18% of Latinos believe there is a cure.

THE CHARACTER OF SOCIETY

Key HIV/AIDS-Related Laws Enacted Through 2005

The following laws were enacted specifically to help and protect those living with HIV.

The Americans with Disabilities Act of 1990 (Public Law 101–336) prohibits discrimination against any qualified individual with a disability—including people living with HIV/AIDS—in employment, public services, telecommunications, and public accommodations.

The Ryan White Comprehensive AIDS Resources Emergency (CARE) Act of 1990 (Public Law 101–381) authorizes funds to "improve the quality and availability of care for individuals and families with HIV diseases." The primary purpose of the act is to provide emergency assistance to localities that are disproportionately affected by HIV/AIDS. The CARE Act provides funding to cities, states, and other public and private nonprofit entities to develop, coordinate, and operate systems of care. Congress has twice reauthorized the act, which is currently authorized through September 2005.

The Housing Opportunities for People With AIDS (HOPWA) Act of 1991 (Public Law 101–625) provides housing assistance to low-income people living with AIDS. Funds are allocated to states and cities.

The National Institute of Health Revitalization Act of 1993 (Public Law 103–43) provides authority for a permanent, independent Office of AIDS Research at NIH and requires the director of that office to "act as the primary Federal official with responsibility for overseeing all AIDS research conducted or supported by NIH."

The Ricky Ray Hemophilia Relief Fund Act of 1998 (Public Law 105–369) mandates a single payment of $100,000 to any individual infected with HIV if the individual has any blood-clotting disorder and was treated with blood-clotting agents between July 1, 1982, and December 31, 1987.

The Ticket to Work/Work Incentives Improvement Act of 1999 (Public Law 106–170) enables states to create new Medicaid buy-in programs for working individuals with disabilities and authorizes state demonstration programs to provide Medicaid to workers with potentially severe disabilities, including HIV/AIDS, who are not yet disabled.

The Global AIDS, Tuberculosis and Malaria Relief Act of 2000 (Public Law 106–264) authorizes funds for U.S. participation in the global response to HIV/AIDS, TB, and malaria.

Rumors of Destruction

Today, friends are asked on street corners, at social gatherings, or over telephones: "Did you hear that he/she has AIDS?" Or: "Do you believe that he/she might be infected? You never know with the life they lead!" Some of the famous people rumored to have HIV/AIDS are Madonna, Elizabeth Taylor, Burt Reynolds, and Richard Pryor.

Rumors ruin lives. People suddenly subtly lose services; the lawn boy quits, no reason given. Quietly, job applications are turned down or car and homeowner insurance policies are canceled, and so on. In one case, after rumors of HIV infection spread in a small town, a man, if he was served in local bars at all, received his drinks in plastic cups. A health club refunded his membership dues. His apartment manager asked him to leave and when his toilet backed up the maintenance man came in wearing a hat over a World War II gas mask, deep water fishing boots, a raincoat, and rubber gloves. In frustration, he had an HIV test. The results were negative and he gave copies of the test to every "joint" in town, his physician, dentist, theater manager, grocery store manager . . . He felt this approach was better than running. Do you agree?

In another case in Brantly County, Ga., population 11,077, the 22-year-old mother of a 2-year-old son was the subject of a rumor that she was HIV-infected. The rumor also stated that she had had intercourse with 200 men in the past year. To convince the townspeople, she took the HIV test and was not HIV-infected. This young woman had to circulate the results of her blood test around town, but it was still not enough to stop the rumor. A newspaper in nearby Waycross quoted an unnamed source saying that this woman was HIV positive.

There was the case of a compassionate person who opened a home for helping AIDS patients. Rumor quickly spread that the entire neighborhood was in danger, especially after the mail carrier refused to deliver mail and was ordered to wear rubber gloves and return to the post office for disinfection. To help the neighborhood understand AIDS and stop unfounded fears, a seminar was held at the AIDS home, but no one would enter the house.

A young person with AIDS reluctantly returned home—it meant revealing that he was gay to his family—and to the community. He said to his parents, "I have good news and bad. The bad is that I'm gay. The good is I have AIDS. I won't be around long enough to interfere with anyone." Once the word got out, a catering service refused to do the annual family Christmas party. They could not hire a practical nurse. Family and friends who used to drop in stayed away. People whispered that the son had gay cancer . . . that it was lethal . . . that it could be caught from dishes, linens, a handshake, and breathing in the same air he breathed out.

In Athens, Ala., the headline of the town's newspaper read, "Athens doctor: 'I don't have AIDS.' " This doctor, a prominent pediatrician in town for 18 years, had to produce a public defense to dispel the gossip that he had AIDS. The doctor offered a $2000 reward for any information about who began the rumor. To date no one has collected the money, but the townspeople have gathered to support him.

In Nebraska, a man sued a prominent woman for starting a rumor that he had AIDS. The Nebraska Supreme Court upheld a lower court ruling that the man was slandered and he received $25,350 in damages.

In Atlanta in March 1999, the CDC said it had received many inquiries about reports that drug users infected with HIV had left contaminated needles in public places, on seats in movie theaters and under gas pump handles. Some reports have falsely indicated that CDC confirmed the presence of HIV in the needles. But the CDC said in a statement, "CDC has not tested such needles. Nor has the CDC confirmed the presence or absence of HIV in any samples related to these rumors. The majority of these reports and warnings appear to have no foundation in fact. CDC is not aware of any cases where HIV has been transmitted by a needle-stick injury outside a health-care setting."

Good News, Bad News, and Late News

Good News—Good news is thinking and believing you're HIV-infected and you're not. There must be a reason to think you're infected, so not being infected is, as some would say, a new lease on life. All too often that feeling is soon forgotten and many people continue lifestyles that place them at risk for infection.

Bad News—Bad news is thinking you're HIV-infected and you are. It is difficult to predict what a person sees, hears, or does after being told he or she is HIV positive. For example, some have contemplated suicide, some have committed suicide, others have become completely fatalistic and proceeded to live a reckless and careless lifestyle that endangered others. Some have said it's like death before you're dead.

You are never prepared to hear the bad news regardless of how sure you are that you're infected. For example, one man who had suffered from night sweats, fevers, weight loss, and other classic symptoms knew he was infected. Yet when told of the positive test results he said "I got so angry, I ripped a shower out of the wall." Being told you are infected is totally devastating, said another infected person. "You feel that everyone is looking at you, everyone can tell you're dirty." Another person said, "The fact that I'm HIV-positive completely dominates my life. There is not a waking hour that I do not think about it. I feel like a leper. I live between hope and despair." Still another said, "I did not leave the house for two days after being told I was HIV-positive. The initial shock was that I was contaminated—unhealthy, soiled, unclean. I carried this burden in isolation for over two years. After all, I had met people who had AIDS but never a person who said—I'm HIV-infected."

The bad news is not confined to those hearing they are HIV-infected; it touches everyone they know—lovers, family, friends, and employers. Nothing remains the same. The more symptomatic one becomes, the greater the social and human loss. One symptomatic mother said, "Whenever I tell my four-year-old I am going to the doctor, he screams because he knows I could be gone for weeks. I try to put him in another room playing with his sister when they come for me." This woman died. Relatives care for her children.

Late News—Late news is remaining **asymptomatic** after infection. Asymptomatic can be defined as when no clinically recognizable symptoms appear that would indicate HIV infection. During this time period, the virus can be transmitted to sexual partners. By the time either antibodies and/or clinical symptoms appear, the news is too late for those who might have been spared infection had their sexual partner tested positive or demonstrated clinical symptoms early on.

Munchausen's Syndrome: When People Pretend to be HIV Positive or to Have AIDS

As Hector Gavin wrote in 1843 in his 400-page history *Of Feigned and Factitious Diseases,* "The monarch, the mendicant, the unhappy slave, the proud warrior, the lofty statesman, even the minister of religion . . . have sought to disguise their purposes, or to obtain their desires, by feigning mental or bodily infirmities."

Donald Craven (1994) reported that a growing number of people may pretend to have AIDS either because of emotional disorders or because they want to gain access to free housing, medical care, and disability income. In one case, seven patients with self-reported HIV infection were treated for an average of 9.2 months in a clinical AIDS program before their sero-negative status was discovered at the hospital (in general, hospitals in the United States do not require a written copy of the HIV test results proving that someone is HIV positive). Craven noted that "because patients with AIDS often have preferred access to drug treatment, prescription drugs, social security disability insurance, housing and comprehensive medical care, the rate of malingering may increase and reach extremes."

Other doctors who have treated AIDS patients said that they have seen many patients who repeatedly come to their offices fearing that they have AIDS, even though multiple tests have shown that they do not (Zuger, 1995, Mileno, 2001).

An Oklahoma physician has written about the AIDS **Munchausen's syndrome**—an emotional disorder in which people pretend to have the disease simply to get attention from doctors. Confidentiality requirements make AIDS a perfect illness for people suffering from Munchausen syndrome because the laws shield them from being discovered.

Physicians' Public Duty: An Historical Perspective of Professional Obligation

Physicians enjoy a virtual monopoly on medical care, social status, and generous financial remu-

BOX 13.5

TELLING STORIES ABOUT AIDS-RELATED EVENTS

At the International AIDS Society-USA course in New York in March 2001, Mary Fisher, a mother, author, and AIDS activist who was diagnosed with AIDS in July 1991, reminded those in attendance of the importance of telling stories by and to the community of people affected by HIV and AIDS. She said that a community is defined by its stories; stories of victory and loss, of heroes and scapegoats, of tragedy and triumph. For the American AIDS community to become a community again, we must find ways to tell the stories again, to let others know that each story has a name, each with a purpose, each with a life. Here are two of her favorite HIV/AIDS stories.

During Bill Clinton's last run for the presidency—1996—I was invited to speak at an AIDS-related event in Little Rock, Arkansas. It was an awards night for regional folk who'd made significant contributions to the fight against HIV. The room was packed with social workers, people with AIDS, family members, religious leaders, a few politicians, and journalists—in other words, the room was packed with Democrats. Out of deference to me, every speaker had been very discreet never to mention politics or Republicans, until the community awards were being handed out, and the last recipient wanted to talk. She was a wonderful, elderly public health nurse: bright, quick, tiny, 77 years old, and feisty. And you could hear every politically correct person in that room stop breathing when she reached up, grabbed the microphone and said, "I've had it with them dumb Republicans. For 15 years, I've talked to them dumb Republicans. Over and over, I've explained there ain't but three ways you can get AIDS: swap needles or blood, have sex, or get born with it. And, for 15 years them dumb Republicans been askin', 'But can't you get it from mosquitoes?'" She paused for a moment, and then she said, "I'm telling y'all tonight that, from now on, I'm gonna tell 'em, 'Yep, you can get it from mosquitoes—but only in three states: Florida, Louisiana, and Arkansas. Cause them's the only places mosquitoes grow so big Republicans can have sex with 'em.' "

Her second story is taken from the book *I'll Go Quietly*.

> Billy Cox came out of his hospital bed in Birmingham, Alabama, to bring me a hug in Montgomery. I'd first met him a year earlier at the University of Alabama at Birmingham where I was visiting Michael Saag (Michael is known for his work with antiretroviral drug therapies). Michael wanted me to meet Billy, to see his spunk and spirit. "Billy is the boxer in the ring," Michael once observed. "The doctors and nurses and medical staff, we're just the trainers in his corner. His friends and family are his fans, cheering him on." Now, a year later, I'd come to Montgomery to speak. But what I said there was not as eloquent as the events that soon played out in the life of Boxer Billy and Cousin Michael. Six weeks after he'd brought his hug to Montgomery—7 years, 4 months, and 3 days after testing positive of the AIDS virus—Billy Cox died on November 23, 1994. On Billy's last day, Michael Saag was leaving town for a few days and stopped in just to say good-bye. When he heard Billy's labored breathing, he called the family together and told them the end was near. And then—as nurses and old friends and Billy's family crowded into the room forming a remarkable community bound only by love for the boxer—Michael rested his head on Billy's chest and, unashamed, before the crowd, sobbed, "I'm sorry, I'm sorry." Science has limits. Even community has bounds. But no one will ever know what love might do.

Susan Paxton—Lifting the Burden of Secrecy

I have been living with HIV for over thirteen years. Like most people, my self-esteem was shattered after my diagnosis and it was a long, slow and often painful journey to get where I am today. For almost a decade I lived a double life. I became increasingly active in the global response to AIDS, yet was unable to come out as a positive woman in my local area (Figure 13-3).

Six months after my diagnosis, I started speaking to health workers and school students about living with HIV. I was well received and experienced a great sense of relief after sharing my secret. I considered going public in the media to raise awareness about HIV infection in women in Australia.

FIGURE 13-3 Susan Paxton Is an AIDS Activist, Facilitator, Trainer, Writer and Community-Based Researcher. She has been living with HIV for over a decade. She is the corecipient of the Australian Government's inagural Jonathan Mann Memorial Scholarship. Her postdoctoral research, based at the Australian Research Center for Sex, Health, and Society, La Trobe University, investigates the barriers to involving HIV-positive people in the global response to AIDS. *(Photo courtesy of Susan Paxton)*

Then, a year after my diagnosis, I fell ill, fainted in my bathroom and split my head open. A doctor was called to my home, where I lay in a pool of blood. I immediately told her of my HIV status and she backed away, suggested I come to her surgery later and left without examining me. Ten minutes later, her secretary telephoned to say the doctor could not tend my wound because she had no way of disposing of the dressings—yet universal precautions to prevent infection were supposedly in place within all health care facilities. It took five hours before I received several stitches at a public hospital. The following morning, my son and I were surprised and confused by the arrival of the police squad who proceeded to carry out a three-hour search of our home. An employee in the doctor's office heard that I was HIV-positive, assumed I was an injecting-drug user and telephoned the police. Because of the trauma my son and I experienced over these two days, it took me almost a decade before I made the decision to go public in the media.

I continued doing HIV education in closed groups and addressing international AIDS conferences, yet I lived in intense fear that my son would face discrimination if I disclosed my status in the media.

My health remained good for several years. By 1999 however, I had a viral load >750,000 copies/ml and a CD4 cell count of 50/μL and my major coronary artery was in spasm daily.

When I started the drugs, I was doing research for my doctoral thesis, examining the role of people living with HIV in AIDS education. One afternoon, after listening to a paper I was preparing, my son suddenly announced the he didn't mind how open I was. The following month, I was nominated to carry the Olympic Torch in the lead up to the Sydney 2000 Olympics.

Coming out in the media was a huge relief and extremely frightening. I completed my PhD and continue with research on HIV and human rights.

Nkosi Johnson

Nkosi died at 6 A.M. on June 1st in his Melville, Johannesburg home with his foster mother, Gail Johnson, at his bedside. He had turned 12 years old on February 4, 2001.

Danny Schechter, executive editor of Mediachannel.org writes of Nkosi. "I've met Nkosi and I love him, so can't profess any objectivity here. On the other hand, you probably haven't heard about him or, for that matter, heard much about the larger cause that he embodies and championed. It is not your fault if you don't know about Nkosi or the growing army of other Nkosis in Africa, because no one is telling you about them. Not with any regularity. Not with any context or explanation. Not in a way that will encourage you to care. Alas, when an issue like this is not on TV regularly in the United

States, it doesn't exist for millions of us." Sadly, Schechter shows that it matters not who or where you are, violence is an everyday commodity available to all. On April 27, 2001, South African Freedom Day: Some armed robbers slipped into his Johannesburg home at one o'clock in the morning, pointing a gun at the woman named Grace who was taking care of him. They took the TV and the VCR and whatever else they could grab. She was traumatized. Nkosi, who cannot speak, saw what happened and had seizures the next day." Nkosi lost his parents to AIDS, and now AIDS claimed his 12-year-old existence.

April 2004. Sister Priscilla Dlamini, a 55-year old nurse of Gingindlovu, South Africa, clutched a corner of her billowing black wimple as she pointed down the muddy dirt road that runs past the Holy Cross AIDS Hospice, where she works. She tells this story: "The first house, there, the white one, you see it on the right," her thick finger tracing the path of the road to a thatched roof barely visible above the cane. "The father and the mother died of AIDS, so did the boy and two girls. That pink house over there, seven died. And there. All eight dead." Her hand swept back toward the horizon to cloud-shrouded mountains. "Everywhere between here and there are empty houses. In the mountains, it is even worse. And where there are people in the houses, there are graves beside them." From one wardroom comes a shrill sound: a 7-year-old girl lets out a whooping cry, punctuated by a dry, congested cough; she is thirsty. Her hollow eyes are round with fear and overflow with tears. In Zulu she cries, "I want to go home. Why do you keep me? Why? Why?" Sister Priscilla offers water as she leans over and whispers a few words of comfort, but she does not tell her the truth. The girl will never go home, even when she dies. Her father and mother are dead. There is no home to go to. Sister Priscilla opened the hospice because so many people dying from AIDS were being left in the sugar cane fields by their families for the clinic workers to find. She said, "People come home from Durban and the other cities to die. But relatives do not accept them. They chase them away or dump them on the edge of the sugar cane plantations and we go around picking them up and bring them here." Some of the dying children arrive at the hospice with nothing, not even identification documents. We give them a stone to hold before they die, and tell the children, "Your mother held this stone."

The Sister knows that death is never far away in the heart of what people here call the AIDS belt, a region in rural KwaZulu-Natal Province that stretches along the Indian Ocean from Richards Bay 80 miles southwest toward the port city of Durban. This is where South Africa faces the full fury of the AIDS pandemic and its social, economic and political devastation. It is also here that the South African government confronts an awful truth: There is too much to do and too little to do it with. Deaths from AIDS complications will continue to rise for many years to come.

neration. Thus the medical profession is uniquely entrusted with the knowledge to care for those with HIV disease/AIDS or any other contagious disease and has a clear responsibility to do so, absent compelling considerations to the contrary. A fair and reasonable share of medical risk just naturally goes with the professional territory.

Robert Fulton and Greg Owen (1988) stated that throughout history, plagues and pestilences have challenged humankind. In his book, *Plagues and People*, William McNeill (1976) cited the many death-dealing epidemics in Europe. He wrote that one advantage the West had over the East in the face of deadly epidemics was that caring for the sick was a recognized duty among Christians. The effect of a prolonged epidemic more often than not strengthened the church when other social institutions were discredited for not providing needed services. McNeill further observed that the teachings of Christianity made life meaningful, even in the immediate face of death: Not only would survivors find spiritual consolation in the vision of heavenly reunion with their dead relatives or friends, but God's hand was also seen in the work of the life-risking caregivers.

The United States has also had its share of plagues and epidemics; one of the most notable was the outbreak of yellow fever in Philadelphia in 1793. Thousands of citizens perished. William Powell (1965), in his book *Bring Out Your Dead*, describes Philadelphia at the time of the yellow fever plague: The dying were abandoned, the dead left unburied, orphaned children and the elderly wandered the streets in search of food and shelter. Nearly all who could fled the city, including the president, leaving the victims of the fever to their fate. Among those who remained, however, were Benjamin Rush, M.D., the mayor, a handful of medical colleagues and their assistants, and a number of clergy. With the help of a small group of laborers and craftsmen, they undertook the enormous tasks of maintaining law and order, providing medical care, food, and shelter to the sick and helpless, as well as gathering up and burying the dead.

Dr. Rush and the others remained at their posts because of their overriding sense of professional obligation along with a conviction inspired by the precept of the New Testament "Blessed are the merciful, for they shall obtain mercy" (Matthew 5:9).

But this vision, shared by Christians for centuries, along with a sense of professional commitment, may not be sufficient to persuade contemporary health-care workers to stay at their posts.

The baby boom generation, educated and self-oriented, has learned to blame AIDS on groups society defines as deviant: homosexuals, prostitutes, and drug abusers. There is a significant probability, therefore, that today's young health-care practitioner may turn away from HIV/AIDS patients.

Medical Moral Issues

HIV/AIDS represents a new era in medicine, one in which physicians are faced with complex moral issues. When the American Medical Association (AMA) issued a statement to the effect that it is unethical to refuse to treat HIV/AIDS patients, that statement reflected a deep concern in the medical community about the possibility of their becoming infected by treating patients with HIV/AIDS.

The AMA statement for an ethical call to arms is *unprecedented* in this century. It is the result of a spreading fear that HIV/AIDS is too contagious to tolerate in spite of the knowledge that the virus is not transmitted via casual contact. Emotions, not education, are in control of those whose fears exceed reality. But these emotions are real and they are having an impact on the medical community.

There is an ongoing dilemma concerning the rights of the physician and other health-care workers to practice medicine in a safe environment as opposed to the rights of HIV-infected and AIDS patients to receive care and medical support. Although the risk of HIV transmission through medical occupational exposure appears to be quite low, the fact that it is possible at all, coupled with the uniformly fatal prognosis associated with AIDS, suggests that physicians, nurses, and other health-care workers have legitimate concerns about health risks.

In a report by C.E. Lewis and colleagues (1992), almost half the primary care physicians in the Los Angeles area had refused to treat HIV-infected patients or had planned not to accept them as regular patients.

In a more recent survey of U.S. doctors in residency training, 39% said that a surgeon or other specialist had refused to treat a patient with AIDS in the resident's care. In Canada, only 13% reported a specialist had refused to treat a patient with AIDS; in France, only 8% said a specialist had rejected care.

Further, 23% of U.S. doctors would not care for AIDS patients if they had a choice as compared with 14% of Canadian physicians and 4% of French doctors.

The survey results may be a disturbing indicator of how U.S. physicians view their work and a reflection of cultural and political attitudes here that view those with AIDS with veiled hostility.

To combat that fear, medical schools are now providing their students with disability insurance that covers AIDS. Hospitals have adopted policies that require physicians to treat AIDS patients or face dismissal.

Risk Protection of the Patient: Political Dimensions—The American Medical Association's position is that HIV/AIDS doctors "should consult colleagues as to which activities the physician can pursue without creating a risk to patients."

There must be a rational relationship between the degree of risk, the morbidity and mortality of the disease, and the consequences of policies and procedures implemented to reduce the risk. Invasive surgery involves many risks. There is an overriding ethical obligation to reduce these risks in ways that do not create more harm than good. The justification of policies to prevent life-threatening illnesses with a risk factor of 1 in 10,000 is significantly greater than those with a risk factor of 1 in 100,000. More lives can be

POINT OF INFORMATION 13.1

DEALING WITH DISCRIMINATION: THE AMERICANS WITH DISABLILITIES ACT

Although incidents of discrimination are disheartening, they are only a part of the story. Another part of the story is how discrimination has been fought and how courts, legislatures, and other social institutions have responded with attempts to reduce HIV/AIDS discrimination and to minimize its impact. To this end a brief synopsis of the Americans with Disabilities Act is presented along with the first U.S. Supreme Court ruling based on this act.

AMERICANS WITH DISABILITIES ACT

The primary federal nondiscrimination statute that prohibits discrimination on the basis of a person's disability or health status is the Americans with Disabilities Act (ADA) of 1990. The ADA provides that no individual "shall be discriminated against on the basis of disability in the full and equal enjoyment of the goods, services, facilities, privileges, advantages or accommodations of any place of public accommodation." The ADA's definition of "public accommodation" specifically includes hospitals and professional offices of health-care providers. A critically important issue under the ADA is whether persons with **asymptomatic** HIV infection have a disability and thus are protected under the ADA. Disability is defined as a physical or mental impairment that substantially limits one or more of the major life activities of the individual, a record of such impairment, or being regarded as having an impairment. In the past, many courts have ruled or assumed as undisputed that HIV infection, as the underlying cause of a life-threatening illness, is a disability. However, several recent court decisions have held that HIV does not automatically qualify as a disability, and in each case there must be an individualized determination as to whether the infection actually limits, in a substantial way, a major life activity. The ADA's legislative history, however, indicates that Congress intended to include HIV infection within the definition of disability, and the Equal Employment Opportunity Commission's regulations embody that view. In its first AIDS case ever, the Supreme Court had to decide whether and to what extent persons with HIV infection are protected under the ADA.

THE CASE OF *BRAGDON* V. *ABBOTT*: UNITED STATES SUPREME COURT

1998

In the 17-year history of the HIV/AIDS pandemic, the U.S. Supreme Court had never considered a case directly involving HIV or AIDS until March 30, 1998, when oral arguments began in the case of *Bragdon* v. *Abbott*. The *Bragdon* case is also the first time the court has ever heard a case involving the ADA. On September 16, 1994, Sidney Abbott, age 37, went to her dentist in Bangor, Maine, to get a cavity filled. Dr. Randon Bragdon refused to fill a gum-line cavity in his office when he read on her medical form that she was HIV positive. He told her he could do the procedure in a hospital, a change of venue that would have added approximately $150 to the bill. According to a cover story about the case in the *American Bar Association Journal*, Dr. Bragdon did not have privileges to practice in any area hospitals, nor had he applied for them. Dr. Bragdon maintains that he could have sought and received permission to perform occasional procedures without having been granted full privileges.

Abbott's Lawsuit

Her lawsuit argues that in refusing to treat her in his office, Bragdon violated the ADA and the Maine Human Rights Act. Federal district and appeals courts both agreed with her.

United States Supreme Court Decision June 25, 1998

The Supreme Court ruled 5 to 4, upholding the District Court and the First Circuit Court of Appeals, finding that Bragdon violated Abbott's rights to treatment under the provisions of the ADA of 1990. Abbott's HIV infection constituted a disability under the ADA in that her HIV infection "substantially limits" a major life activity—her ability to reproduce and bear children (to have a child she places her husband at risk for HIV infection and risks infecting her child). Justice Kennedy, in delivering the opinion of the court, held that from the moment of infection and throughout every stage of the disease, HIV infection satisfies the statutory and regulatory definition of a "physical impairment." Applicable Rehabilitation Act regulations define "physical or mental impairment" to mean "any physiological disorder or condition affecting the body['s] hemic and lymphatic [systems]." HIV infection falls well within that definition. The medical literature reveals that the disease follows a predictable and unalterable course from infection to inevitable death. It causes immediate abnormalities in a person's blood, and the infected person's white cell count continues to drop throughout the course of the disease, even during the intermediate stage when its attack is concentrated in the lymph nodes. Thus, HIV infection must be regarded as a physiological disorder with an immediate, constant, and detrimental effect on the hemic and lymphatic systems.

DISSENT IN PART

Justice O'Connor stated that Abbott's claim of a disability should be evaluated on an individual basis and that she has not proven that her symptomatic HIV status substantially limited one or more of her major life activities. "In my view, the act of giving birth to a child, while a very important part of the lives of many women, is not generally the same as the representative major life activities of all persons—caring for one's self, performing manual tasks, walking, seeing, hearing, speaking, breathing, learning, and working"—listed in regulations relevant to the Americans with Disabilities Act. Based on that conclusion, there is no need to address whether other aspects of intimate or family relationships not raised in this case could constitute major life activities; nor is there reason to consider whether HIV status would impose a substantial limitation on one's ability to reproduce if reproduction were a major life activity.

U.S. SUPREME COURT RULES ON INSURANCE COVERAGE CAP FOR HIV/AIDS TREATMENT

In January 2000, the U.S. Supreme Court let stand a ruling that allowed an insurance company to provide less coverage for AIDS-related illnesses than for other conditions under the same policy. The high court, without comment, refused to hear the appeal brought by two HIV-positive Chicago men who claimed that their insurance company's policies violate the Americans with Disabilities Act. The two policies in question were issued by Mutual of Omaha. One policy set a $25,000 lifetime coverage limit for AIDS-related illnesses and the other contained a $100,000 cap, while both allowed a $1 million cap for other illnesses. Attorneys for Mutual of Omaha argued that the insurance company had not discriminated because it offered the men the same coverage offered to other customers. In 1998, the federal judge in Chicago ruled in favor of the two men, but the 7th U.S. Circuit Court of Appeals reversed that ruling. The appeals court said the ADA guarantees access to insurance but does not regulate the content of coverage.

NATIONAL SURVEY OF DENTISTS IN CANADA: REFUSAL TO TREAT HIV-INFECTED PATIENTS

According to the report by Gillian McCarthy and colleagues (1999) there are about 15,232 dentists in Canada. A random sample of 6444 answered a survey on whether they would refuse to treat HIV-infected patients. Some 4281 dentists responded. The conclusion presented by the investigators was that 1 in 6 (or 17%) would refuse to treat an HIV-infected patient. The refusal was associated with the respondent's lack of belief in an ethical responsibility to treat patients with HIV and his or her fears of becoming infected from their patients.

DISCUSSION QUESTION: Clearly a 5 to 4 ruling is not an overwhelming mandate to support Abbott's lawsuit. Is a simple majority, 55% in this case, sufficient or because this case has vast implications, should it require a two-thirds majority, 6 in favor, 3 against (67%)? What are the legal and moral issues in accepting a simple majority versus a two-thirds ruling?

Bragdon raised a question for you to consider as you research the question above. Looking at **Magic Johnson**—he asks if it really makes sense to consider someone "disabled" who can earn millions of dollars playing professional basketball, or go to work, or otherwise perform the tasks of daily living. Your response is?

saved by innovative approaches to preventing the kind of injuries that have accounted for 75% of all cases of occupationally related HIV infection than by preventing HIV-infected health-care workers from performing invasive surgery.

The risk of HIV infection by an HIV-infected

POINT OF INFORMATION 13.2

HIV TRANSMITTED BY AUSTRALIAN AND FRENCH SURGEONS

A breakdown in infection control procedures is being blamed for the transmission of HIV to five patients of an Australian surgeon. It's believed the virus was transmitted from one patient to four others during minor skin surgery on a single day in November 1989. Health officials say one patient, a gay man, is believed to have been the infection source. **The CDC calls the case the first known patient-to-patient transmissions of HIV in a health-care setting** (*American Medical News*, 1994 37:2.) In October 1995, after the HIV seropositive status of an orthopedic surgeon in Saint Germain en Laye (suburb of Paris) was announced in the medical press, the French director of general health decided to inform and offer testing to the patients operated on by this surgeon. Review of the medical history of the surgeon suggests that he most probably became infected with HIV in May 1983. The diagnosis of HIV infection and AIDS were made simultaneously in May 1994. This investigation identified 3004 patients who had undergone at least one invasive procedure by the surgeon; 2458 patients were able to be contacted by mail. The serologic status of 968 patients was ascertained; 967 are negative. Only one patient, who was negative before a prolonged operation performed by the surgeon in 1992 (10 hours) is HIV positive. Typing of the viral strains of the patient and the surgeon comparing nucleotide sequences of two viruses showed they are closely related. In February 2000, the surgeon was ordered to pay $107,000 to the infected patient. Similar to the more recent dentist-Bergalis case, the source of HIV in these cases will most likely never be documented to everyone's satisfaction.

In only the third known case of HIV being transmitted to a patient by a health-care worker, a 61-year-old female patient became infected. At the Free University of Brussels, in July 1996 the woman developed severe symptoms of HIV infection. Viral RNA sequence studies showed the virus came from her nurse who at the time had advanced HIV disease but did not know she was infected (Goujan, 2000).

surgeon is only one-tenth the chance of being killed by lightning, one-fourth the chance of being killed by a bee, and half the chance of being hit by a falling aircraft. The 1 in 100,000 risk of transmission from an infected surgeon equals the probability of death we face bicycling 2 miles each way to school for a month, or commuting 15 miles round-trip by car for a year.

Similarly, our chance of dying from anesthesia during an operation is roughly 10 times our chance of being infected by a surgeon known to be infected with HIV. The risk to a person undergoing invasive surgery by a surgeon of unknown HIV status is about 1 in 20 million of becoming HIV-infected (Daniels, 1992). The CDC estimates the risk to a patient by an HIV-infected surgeon to be between 1 in 42,000 and 1 in 420,000 (Lo et al., 1992).

Finally, to place the risk of the patient's HIV infection/death in perspective, the National Academy of Science reported that each year as many as 98,000 Americans die from medical mistakes made by physicians, pharmacists, and other health-care professionals. More Americans die from medical mistakes than from breast cancer, highway accidents, or AIDS (Weiss, 1999). According to a study released in January 2000 by the *Canadian Public Health Journal*, Tobacco Control, 10,000 people die worldwide every day due to smoking. Also, 50% of young people who continue to smoke will die from tobacco-related causes. Of everyone alive today, 500 million will eventually be killed by tobacco. And, smoking is responsible for 90% of all lung cancers and 75% of chronic bronchitis and emphysema. In the United States, obesity now kills about 400,000 a year, alcohol abuse 110,000 each year, and automobile and firearm-associated deaths, 43,000 and 34,000 annually.

Patients' Right to Know If Their Physician Has HIV/AIDS—In a recent Gallup Poll, 86% of those polled felt that they had the right to know if a health-care worker treating them was HIV-infected. Many lawyers also take this position. The courts appear to be moving toward an interpretation of the doctrine of informed patient consent as "what a reasonable patient would want to know," rather than "what a reasonable physician would disclose." Because it is so difficult for surgeons to avoid occasionally cutting themselves during surgery, it has been suggested that the best solution is not to have HIV-infected surgeons perform surgery at all.

Public anxiety on HIV/AIDS and medical care is becoming increasingly tinged with hysteria. A recent national Gallup Poll undertaken for *Newsweek* asked a representative sample of 618 adults, "Which of the following kinds of health care workers should be required to tell patients if they are infected with the AIDS virus?"

The answers were: surgeons 95%; all physicians 94%; dentists 94%; all health-care workers 90%.

Clearly, people do not differentiate between doctors who perform invasive procedures and those who do not. However, the patient could ask what the probability is of a single dentist (Acer) infecting six of his patients (Bergalis, Web, and four others). Extremely low, yet it did happen! The lowest of probabilities and best of guidelines and precautions do not stop the fire of fear. It must also be mentioned that the same *Newsweek* poll found that 97% of those interviewed felt that HIV-infected patients should tell their health-care workers that they are infected.

There is one important aspect related to this poll that needs to be addressed. That is, many of the people interviewed stated that if they knew their surgeon was HIV-infected, they would "get another surgeon." This switching dilemma may, at some point, have the majority of the population needing surgery standing in line for the uninfected surgeons. Services provided by the reduced number of surgeons, it could be claimed, at some point, result in increased costs and diminished quality.

HIV-Infected Health-Care Professionals' Duty To Disclose—Several courts have held that health-care professionals have a duty to disclose their HIV status to patients or health authorities, assuming that their professional activities pose a risk of transmission to patients. The Maryland Court of Appeals ruled that a surgeon has a duty to inform his patients of his infection; even if the patient has not actually been exposed and tests HIV negative, the contact with the surgeon may subsequently give rise to a claim for their infliction of mental distress due to fear of transmission. Courts justify orders to disclose based on a duty to protect patients and on the doctrine of informed consent. Requiring disclosure to patients, of course, can severely jeopardize a health-care professional's career. To avoid this result, some states allow the professional to continue practicing, with appropriate restrictions and supervision, but without disclosing his or her HIV status (Gostin et al., 1998).

In February 2004, the Quebec Medical Association (QMA), a division of the Canadian Medical Association, has adopted a policy that requires doctors to disclose their HIV positive status to their employers, yet protects the physicians' confidentiality. The association adopted the policy after Ste-Justine hospital revealed that one of its surgeons had operated on more than 2614 children without the hospital administration's knowing she had HIV. All the children were HIV tested, and all were HIV negative.

Physician–Patient Relationships

In most states, physicians may not test a patient for HIV antibodies without written permission from the patient. Physicians can run tests for any other infectious diseases without written permission.

DISCUSSION QUESTION: Do you think this is fair and equal medical practice?

Because the majority of HIV-infected patients are asymptomatic, there is no way of telling who is or is not infected. Therefore, for the protection of health-care workers, everyone must be treated as though they were HIV-infected.

HIV has been a silent medical threat for over 24 years and continues into the new millennium. Wherever blood is drawn, a wound is examined, a dressing is changed, or anything that involves blood, needles, or surgery is done, there is an unspoken fear that HIV might be present. The patient sees this preventive attitude of physicians and other health-care workers by the new look: wraparound smocks, gloved hands, and masks. The patient wonders whether his or her physician is an HIV carrier and the physician assumes that the patient may be a carrier. A recent Associated Press article presented some examples of how the AIDS pandemic has changed doctor–patient relationships:

1. In many operating rooms, doctors and nurses wear wraparound glasses in case of blood splashes.
2. In many emergency rooms, health-care workers cover themselves with caps, goggles, masks, gowns, gloves, shoe covers, and blood-proof aprons.
3. Infection control specialists must spend time convincing hospital workers it is safe to enter an

HIV/AIDS patient's room to perform normal duties ranging from picking up dinner trays to fixing the plumbing.

4. At some hospitals, all emergency room trash, no matter how innocuous, is treated as hazardous waste.
5. Many doctors and nurses routinely pull on gloves whenever they give an injection or draw blood.
6. Mouth-to-mouth resuscitation is simply not done at many hospitals. Instead a mask and valve device is used to avoid direct contact with the patient's mouth.
7. A surgeon at San Francisco General Hospital takes the AIDS drug Zidovudine whenever he operates on people he thinks are infected.
8. Physicians are increasing life insurance policies to provide for their families in case they become HIV-infected.

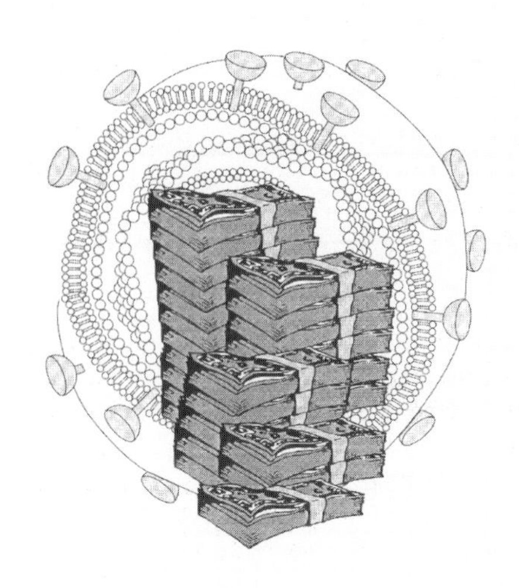

FIGURE 13-4 The Enormous Sum of Federal, State, and Private Sector Monies Spent on HIV/AIDS Has Made This Disease an Industry unto Itself.

The health-care professionals' fear of getting AIDS will persist as long as there is a risk that HIV can be transmitted in the workplace. **The goal is not to eradicate that fear, but to prevent it from compromising the quality of patient care and from threatening the health professional's own well-being** (Gerbert et al., 1988).

In early 1990, Lorraine Day quit her post as chairperson of the orthopedics department at San Francisco General Hospital and abandoned her surgical practice because of her fear of exposure to HIV. "I have two children to think about, and operating was too dangerous. If I was a skydiver, people would say I was an irresponsible mother, but to me, surgery now, it was just as risky."

Dr. Day said during a TV interview, "Our risk is one in 200 per single (needle) stick with AIDS blood and it can be the first one—it doesn't take 200. And I ask you, if you came to work every day and flipped the light switch on in your office and only one out of 200 times you were electrocuted would you consider that low-risk?"

While Dr. Day has been called a scaremonger, many surgeons in private conversations call her a "hero" for raising the risk issue.

FEDERAL AND PRIVATE SECTOR FINANCING: CREATION OF AN AIDS INDUSTRY

Twenty-four years ago, when the public was just learning about a new disease that would be called AIDS, the scientists tracking down the cause were already thinking about how their research could be marketed. French and American groups eventually claimed to have codiscovered HIV independently and in different ways. In one respect their approach was the same: Shortly before announcing their discoveries, both rushed to file patents that described how to determine whether a person's blood harbored the virus. By doing this, they gave birth to the HIV/AIDS industry. (Figure 13-4)

Financing the AIDS Industry: A Quilt with Many Holes

Federal Government—In 1990, the U.S. Congress did something quite rare: It allocated money specifically for the treatment of one disease—HIV/AIDS. In some ways the increased commitment of federal and state government to cancer research and treatment in the early 1970s is similar to what happened in the war on AIDS in the 1980s. In both decades, there was a major funding surge to stimulate research, therapy, and prevention. A major difference, however, is that dollars for cancer came more slowly over a longer time period that began well before the 1970s. With AIDS, federal funding began in 1981 (Figure 13-5) and has increased at an unprecedented rate.

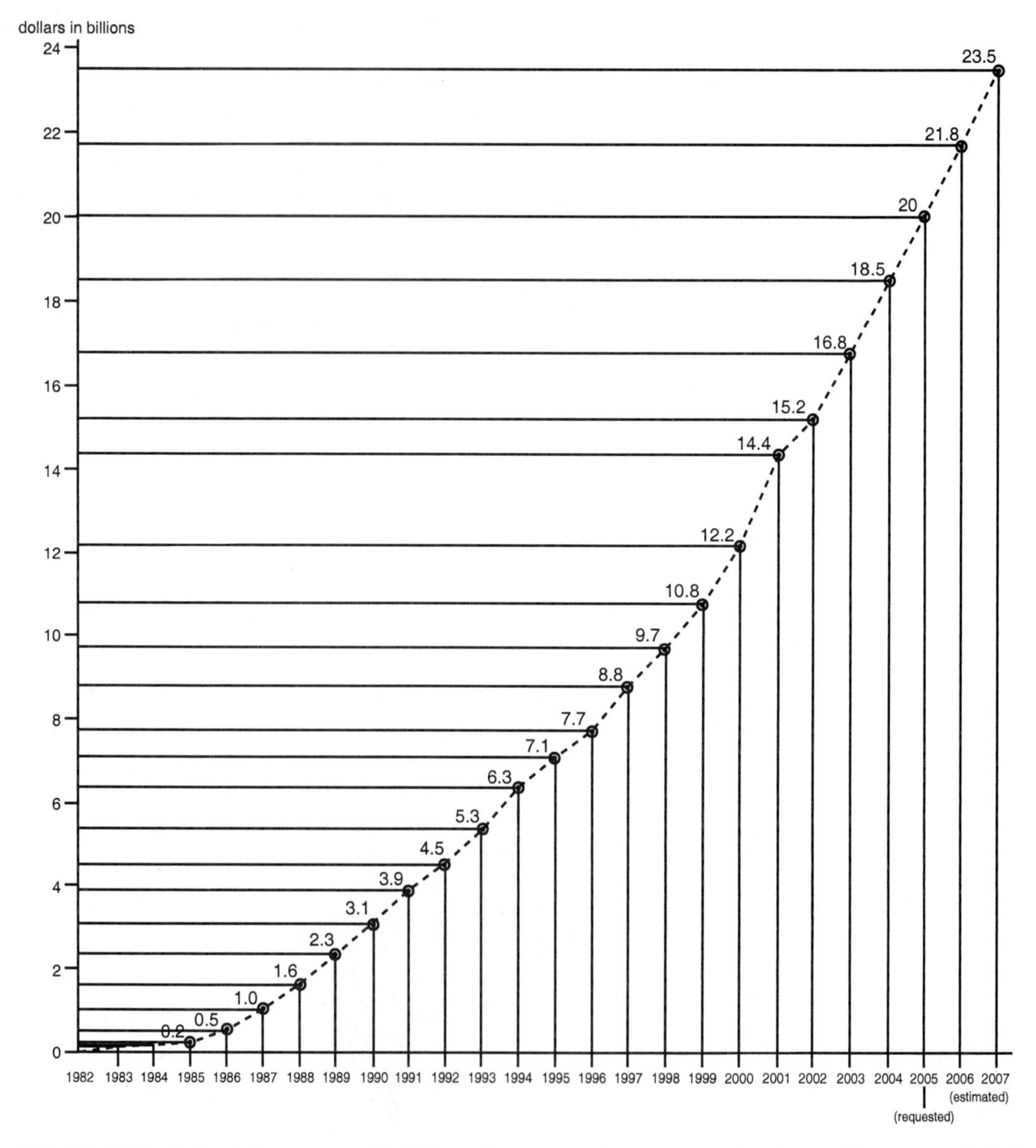

FIGURE 13-5 Federal Government AIDS Expenditures to the Year 2004. Federal expenditures were $1 billion in 1987, and $16 billion for 2004. Bernhard Schwartlander of UNAIDS said in 2002 that the United States is now spending over $20 billion a year on prevention, care, and research. By the end of 2006 the U.S. federal government will have spent over $190 billion over 25 years on this disease.

No Cheap Way Out: HIV/AIDS Is a Very Expensive Disease

During the 24 years from 1982 through 2005, federal spending on AIDS-related projects increased from $8 million in 1982 to $20 billion for 2005. The Presidential Advisory Council on HIV/ AIDS provided the president with six AIDS goals that will be funded with the federal budget for AIDS. The goals are to: (1) develop a cure, (2) reduce/ eliminate new infections, (3) guarantee care/service for the HIV-infected, (4) fight against HIV/

AIDS discrimination, (5) quickly translate scientific advances into improved care/prevention, and (6) provide support for international AIDS efforts. (See Figure 13-6).

Private Sector Funding: United States—In addition to the money spent by the federal government, collectively the states also spend between $6 billion and $8 billion each year. The private sector spends about $5 billion a year. By the end of 2005, the federal government spent about $170 billion for HIV/AIDS-related work, and the states and private sector spent about the same. In reality, adding in unspecified federal dollars that went to AIDS-related projects and the dollars spent in the state and private sector would most likely bring the total AIDS-related expenditures to over $300 billion. Yet, in spite of this massive expenditure on HIV/AIDS, the United States still does not have the underpinning of a uniform health care system to provide an organized, controlled use of AIDS funds. The current funding is **heterogeneous** and provides **unequal access** to HIV/AIDS care.

AIDS Costs as a Percentage of the National Federal Budget

The $20 billion for 2005 represents about 0.7% of the $3 trillion federal budget.

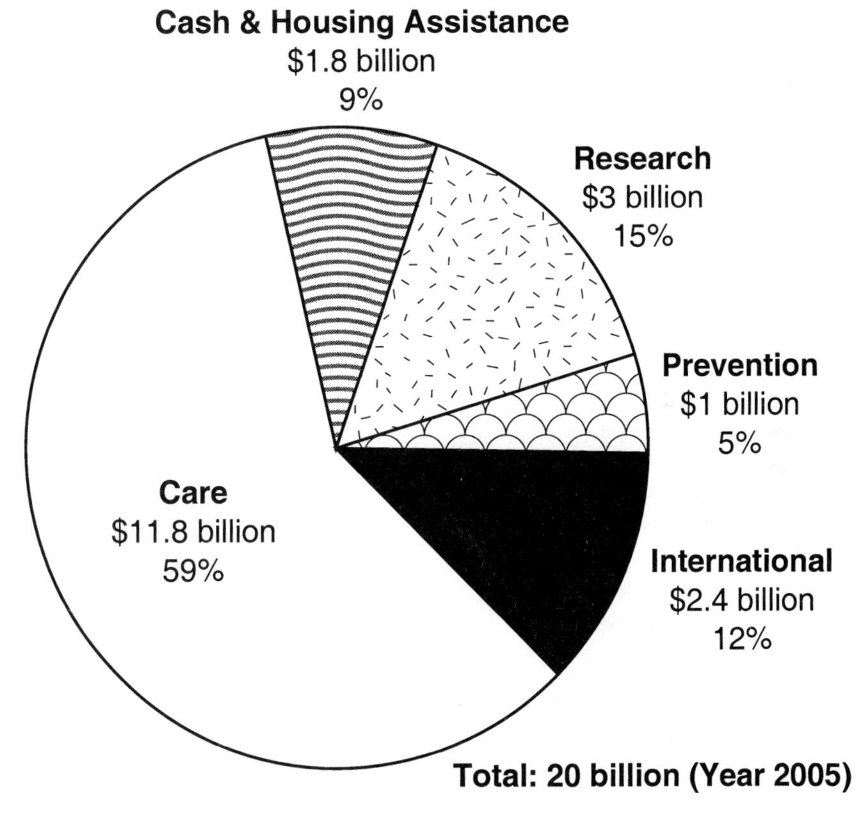

FIGURE 13-6 Total Federal HIV/AIDS Spending by Category FY 2005. Federal HIV/AIDS spending is divided generally into four categories: Care and Assistance, Research, Prevention, and International.

Programs under **Care and Cash/Housing Assistance** are those that deliver health-care services, support services, and disability assistance to individuals with HIV/AIDS. Major health-care programs in this category include Medicaid, Medicare, Supplemental Security Income (SSI), Social Security Disability Income (SSDI), Ryan White CARE Act, Ricky Ray Hemophilia Relief Fund, and Housing Opportunities for People With AIDS (HOPWA).

POINT OF INFORMATION 13.3

A VIEW OF HIV/AIDS FUNDING

It should be noted that only about half of all federal HIV/AIDS funding is specifically designated for HIV/AIDS programs. Most of the federal help comes from programs such as Medicaid or Medicare that are intended to provide care and services to any qualifying person, based on his or her medical, functional, and income status. People with HIV/AIDS qualify for these programs in much the same manner as people with cancer or heart disease do. The level of support that is identified as "for HIV/AIDS" is an estimate of what portion of the overall national spending goes for HIV/AIDS care. In addition, there are other federal programs that are not specifically designed for HIV/AIDS services and that serve people with HIV/AIDS, but for which HIV/AIDS-related spending estimates cannot be provided. For example, federally supported tuberculosis control programs are undoubtedly used by people with HIV/AIDS, but estimates of the proportionate spending are not available. Estimates of funding for these programs that may be used by many people with HIV/AIDS are not included. Spending by the federal government as an employer providing health insurance to its workers with HIV/AIDS through the Federal Employee Health Benefits Program (FEHBP) is also not included in total federal spending data. In addition, federal funding is not the only funding for HIV/AIDS programs. States supply a substantial portion of overall Medicaid spending, ranging from a maximum of 50% of the total Medicaid funding in comparatively affluent states to a minimum of 17% in the poorest states. States are also required to make matching payments for some federal programs including the Ryan White CARE Act. In the Supplemental Security Income (SSI) program, many states make the supplemental payments to raise disability cash assistance above the federal minimum. Many states also operate HIV/AIDS programs of their own. Beyond federal and state funding, many local governments, foundations, and charities also provide HIV/AIDS services. Public hospitals, for example, are usually supported by counties or cities and are often the sites of much clinical HIV/AIDS care. Likewise, free clinics and specialized HIV/AIDS service organizations provide both care and prevention services. Foundations, such as the Clinton and Gates, have made substantial contributions to HIV/AIDS programs. Federal funds have, nonetheless, remained predominant during the HIV/AIDS pandemic.

It is difficult to estimate overall HIV/AIDS spending. Just trying to work with federal dollars for any individual fiscal year is difficult. For example, the federal budget for any individual fiscal year takes several years to prepare, carry out, and audit. For instance, on preparing this textbook (mid-2004), final accounts of what was spent in FY 2003 have closed; Congress has finished appropriations for FY 2004; the president had delivered his budget for FY 2005 to Congress; and the Executive Branch agencies are preparing their recommended budget proposals for FY 2006.

AIDS Expenditures per Death Compared to Other Major Diseases Causing Death

Federal spending for AIDS research, educational programs, counselor training, testing, and prevention programs has been compared with federal spending for other diseases. There is a discrepancy between the total federal dollars spent on certain diseases and the number of deaths they cause. In 2004, with about 16,000 deaths, HIV/AIDS received $18.5 billion. That's equivalent to about $1,156,000 per AIDS-related death! In 2004, cancer caused about 560,000 deaths (about 1500 people/day) and received $3.3 billion or about $5900/death. During the same time, there were 1,220,000 new cases of cancer in the United States; 183,000 were new cases of breast cancer and there were over 42,000 breast cancer-associated deaths. One in seven to nine women will get breast cancer. Heart disease, which caused some 750,000 deaths in 2004 (43% of all deaths for the year), received about $1.5 billion or about $2000/death. Some 2 million Americans require insulin injections for type I diabetes and perhaps as many as 14 million have type II diabetes (diet controlled). Diabetes, a severe progressive disorder, received $541 million, or about 5% of the AIDS budget in 2000 or about $2847/death. It is estimated that 190,000 people will die annually because of diabetes complications. Stroke, which affects millions of Americans and is involved in about 160,000 deaths per year, received $212 million or $1325/death. Overall, between 1981 and the beginning of 2006, while 543,000 people died of AIDS, 19 million died of heart disease and 10.3 million died of cancer. For the year 2005 20 billion federal dollars will be allocated for HIV/ AIDS programs. There will be about 15,000 AIDS-related deaths or over $1.3 million/death! This is the most expensive disease in history relative to time, 1981 through 2005.

U.S. GOVERNMENT BELIEVES HIV/AIDS IS A THREAT TO NATIONAL SECURITY: THE UNITED NATIONS BELIEVES IT IS A WEAPON OF MASS DESTRUCTION

In addition to spending enormous sums of money on HIV/AIDS, the federal government declared in May 2000 that "AIDS is a threat to our national security." U.S. Secretary of Health Donna Shalala said, "We know that infectious diseases know no borders, that they can affect this country, and in this case it is both in our economic interest and in our national security interest to work on these infectious diseases abroad. The high rates of AIDS in Africa are putting security and stability at risk by disabling national armies, disrupting economies and killing off people who might become the next generation of leaders. Basically AIDS is uncoupling the economic gains in Africa as African countries are forced to shift more resources to their health care systems from their economic investments." Shalala also said that countries in other parts of the world face a growing HIV/AIDS crisis. "Eastern Europe has an AIDS problem, Russia has it, India has it, every country in the world that we do business with, but more importantly our need to be politically stable, is suffering from this huge onslaught of AIDS. And that makes the relationship economically and from the security point of view relevant to America's national security."

DISCUSSION QUESTION: Present reasons that agree or disagree with the federal government's declaration that HIV/AIDS is a threat to America's national security.

GLOBAL HIV/AIDS FUNDING FOR UNDERDEVELOPED NATIONS

In many ways, we are of one world. Africa's destiny is our destiny. There is hope on the horizon, but that hope will only be realized if the developed nations take constructive action together. As South Africa's Archbishop Desmond Tutu said: "If we wage this holy war together—we will win."

The bottom line is this: There is no vaccine or cure for AIDS in sight, and the world is at the beginning of a global pandemic, not the end. What is happening in Africa now is just the tip of the iceberg. As goes Africa, so will go India and the Newly Independent States of the former Soviet Union. There must be a sense of urgency for the developed world to work together; to learn from their failures and successes, and to share their experience with those countries that now stand on the brink of disaster. Millions of lives—perhaps hundreds of millions of lives—hang in the balance. AIDS is a devastating human tragedy that requires global help.

HINDSIGHT Fifteen years ago, when the AIDS death toll in the United States crossed 100,000, few paid heed to a grim prediction by the World Health Organization (WHO) that "by the year 2000, 40 million persons may be infected with HIV." In the developed world, AIDS was seen as a serious but small disease, restricted to gay men, drug users, hemophiliacs, and their infants. In the developing world, just a few courageous voices were warning about the silent spread of a deadly new plague. In retrospect, WHO's grim prediction was 20 million too low! Africa is in crisis. In some countries, about 40% of the adult population is infected. Many millions have died, and millions more will follow, leaving their societies trapped in poverty, burdened with a generation of orphans, and facing demographic catastrophe. The grim statistics are not confined to Africa. Asia and the Caribbean face explosive HIV epidemics, while the nations of the former Soviet empire are looking at an overwhelming increase of drug addiction, untreated sexual diseases, and the unchecked spread of HIV. Finally the world has begun to take notice. In January 2000, the United Nations Security Council held a precedent-setting special session, in which for the first time it identified a disease—AIDS—as a global security threat. A second UN HIV/AIDS special session of the General Assembly was held June 25–27, 2001.

BOX 13.6

AIDS PROGRAMS: AN EPIDEMIC OF WASTE?

Through 2002 there has been constant pressure on politicians and leaders of private industry to contribute higher levels of money in the fight against HIV/AIDS. In 2002, AIDS activists marched in Washington in an attempt to get Congress to allocate $2.5 billion to the Global AIDS, Tuberculosis, and Malaria Fund. Their belief is that America is not spending enough money at home or globally on HIV/AIDS. Others, however, protest this notion and feel that if graft and corruption were eliminated there would be sufficient money such that yearly increases in the federal HIV/AIDS program would not be called for. To support this belief, a group called Citizens Against Government Waste (CAGW) released a report in February 2002 on **"AIDS Programs: An Epidemic of Waste"** (www.cagw.org). Citizens Against Government Waste is the nation's largest nonpartisan, nonprofit organization dedicated to eliminating waste, fraud, abuse, and mismanagement in government. Some of the fraud found by this group is listed exactly as it is presented in their report:

- CAGW has obtained a copy of a $20,000 grant from the Vermont Department of Public Health to the Twin State Women's Network (TSWN) to be used for a weekend retreat. Topics for the weekend included "Toys 4 Us" and "Self Loving/Self Healing: Discussing the Role of Masturbation as a Tool for Healing." TSWN also received: $1,500 for long distance phone calls; $1,000 for books, including "The New Good Vibrations for Sex" manual; and $250 for videos, choices of which included "Fire in the Valley: A Guide to Masturbation for Women" and "Fire in the Valley: A Guide to Masturbation for Men." Each participant received a welcome bag filled with mints and chocolate and each room was equipped with welcome packets containing condoms, lubricant, candles, massage lotion, and lip balm. TSWN received 86% of its funds from government sources, including the Centers for Disease Control (CDC).
- Positive Force in San Francisco receives $1 million a year from the CDC. The group offers flirting classes and, last July, hosted a workshop on how to have anal intercourse if you suffer from diarrhea. (Diarrhea is a common side effect of AIDS.)
- On February 28, 2002, the Stop AIDS Project of San Francisco, which received nearly $700,000 from the CDC in fiscal 2001, will sponsor "GUYWATCH: Blow by Blow." The advertisement for the seminar reads, in part: "What tricks do you want to share to make your man tremble with delight?"
- A Central Florida AIDS Unified Resources (CENTAUR) staffer spent $600,000 in Ryan White CARE Act money on tickets to Disney World, hotels, and restaurants.
- In April 2001, *The New York Post* revealed New York City was spending nearly $180,000 a week ($9 million a year) on hotel rooms for HIV and AIDS patients. That month, the city had reserved 20 rooms at the Sofitel Hotel in Midtown Manhattan at $329 a piece. Advocates say DASIS must use the expensive hotels because it has ruined its relationship with lower-cost hotels by not paying bills on time. New York city received $52.6 million in Housing Opportunities for People With AIDS (HOPWA) program funding in fiscal 2001.
- The University of California-San Francisco AIDS Health Project (AHP), which received a $633,765 grant from the CDC prevention in fiscal 2001 and continually receives nearly 85% of its funding from government sources, sponsored a workshop in November in physical intimacy, focusing on "holding, kissing, licking, sucking, and . . ."
- A doctor in Puerto Rico used $2.2 million in federal funds to buy luxury items like cars and jet skis, while severely neglecting the AIDS patients in his care.
- More than $20 million in grant money intended to help house AIDS patients was collected—but never spent—in Los Angeles.
- AID Atlanta, Inc., which received more than $3.5 million from the government in fiscal 2000 and only $1.2 million in private contributions, sponsors "Deeper Love: A Workshop for Gay and Bisexual Men of African Descent" that addresses such subjects as dating, relationships, and erotica. The program lists the following topics of discussion: "Dirty talk: what makes it good; Tossing salad; Strollin' in the park, through the trails; The art of latex; Safety versus trust." AID Atlanta, Inc. also sponsors "Slipping and Sliding" where men can explore their needs and desires and learn how to fulfill them.
- FBI investigation into the South Dallas Health Clinic revealed that more than $60,000 in the

Title I funds had been spent on calls to psychic hotlines and on shopping trips to Neiman Marcus.

- The nonprofit Tampa Hillsborough Action Plan (THAP) gives its top executives plenty of perks despite its financial woes. THAP boss and THAP chief executive officer rang up nearly $1000 in meal charges in a three-week period and were also afforded the use of sport utility vehicles. THAP boss received up to $45,000 a year annually for the maintenance of his vehicles. THAP's top executives also received four season tickets for Tampa Bay Buccaneers games and two season tickets for both the Tampa Bay Devil Rays and the Tampa Bay Lightning. Meanwhile, THAP owed nearly $25,000 in delinquent payroll taxes. THAP receives $450,000 a year from the federal government to provide housing to people with AIDS.
- Not in the CAGW investigation is the Los Angeles County Auditor Controller's Office report that states "officials in the county's Office of AIDS Programs and Policy cannot account for $83 million it spent in 2001." That is more money than the budget for many American cities.

CONCLUSION OF CAGW

Before new resources are added to the $13 billion in federal money currently allotted for AIDS-related programs, the Departments of Health and Human Services and Housing and Urban Development should conduct extensive audits of the Ryan White CARE Act Title I and the HOPWA program. Such audits will give Congress more incentive to reform or eliminate these antiquated and duplicated social programs. Congress should redirect many CDC prevention grants to international AIDS relief efforts or increased funds for researching an AIDS cure. Many CARE Act programs, including all of Title I, should be phased out and incorporated into existing federal safety net programs such as Medicaid and Medicare. This would ensure necessary, life-saving medical care to those with HIV and AIDS who are low-income or uninsured, while also eliminating nonessential AIDS services. It would also save money to bolster the AIDS Drug Assistance Program.

In May 2002, U.S. Treasury Secretary Paul O'Neill took a 10-day tour of Africa. On his return, he called for increased access to HIV/AIDS treatment and greater accountability for assistance programs. O'Neill went to Chris Hani Baragwanath Hospital in Soweto where he met with HIV-positive women whose children had been treated at birth with the antiretroviral drug Nevirapine to reduce the risk of vertical HIV transmission. O'Neill, who has been critical of foreign aid in the past, asked why the South African government was not providing treatment to all HIV-positive pregnant women. O'Neill said, "This whole business about having so much money . . . and it not going primarily to treatment is just a stunning revelation." In an interview with ABC's *This Week,* on May 26, O'Neill echoed his call for greater accountability. "My problem is that it isn't clear why we aren't getting better choices about the priority use of the money that is already there. For me, this is about getting real results on the ground."

Since the CAGW report in 2003, several hundred cases of financial fraud and corruption have been uncovered in the United States and worldwide—far too many to report here. One is the recent case of antiretroviral drugs provided, at cost, to various countries in Africa. The drugs were smuggled out of Africa into Britain, France, The Netherlands, Germany, and other developed countries and sold for millions of dollars in profit.

DISCUSSION QUESTION: Do you think in Africa, there is at least an equal amount of fraud and corruption as found in America with regard to the allocation and spending of HIV/AIDS dollars donated for the prevention and treatment of HIV/AIDS? Support your opinions with examples.

United Nations Secretary-General Kofi Annan Calls for Large-Scale Mobilization in Fight Against AIDS, Tuberculosis, and Malaria

In mid-2001, calling the battle his personal priority, the secretary-general outlined five priority areas for the global campaign.

1. Preventing further spread of the epidemic, especially by giving young people the knowledge and power to protect themselves.
2. Reducing HIV transmission from mother to child, which he called "the cruelest, most unjust" infections of all.
3. Ensuring that care and treatment is within reach of all.

4. Delivering scientific breakthroughs. Finding a cure and vaccine for HIV/AIDS must be given increased priority in scientific budgets.
5. Protecting those made most vulnerable by the epidemic, especially orphans.

To achieve these five goals, Annan called world leaders to help finance the campaign against AIDS, tuberculosis, and malaria in Africa. In April 2001, he said, "a war chest of 7 billion to 10 billion U.S. dollars is needed annually, over an extended period of time, to wage an effective global campaign against AIDS. In July 2001 the Group of Eight (G8) industralized countries (Britian, Canada, France, Germany, Italy, Japan, Russia, and the United States) proposed the creation of a new Global Fund to obtain $10 billion a year dedicated to the battle against HIV/AIDS, tuberculosis, and malaria. Current spending on AIDS in developing countries totals around $1 billion annually." This is many billions of dollars less than they spend on their military. Although Annan's figure of $10 billion for combating AIDS, tuberculosis, and malaria seems like a large sum, it is equivalent to:

- Four days of global military spending.
- Ten days of running the Organization for Economic Cooperation and Development (OECD).
- The cost of 100 Eurofighters (jet fighter planes).

A contribution of $10 billion a year would be equivalent to the amount of money spent in 60 days in the United States on soft drinks or in 35 days on fast foods.

Annan also wants all antiretroviral drugs to be sold to underdeveloped nations at 5% of the cost to people in the developed world.

How Far Will Ten Billion Dollars Go In Africa?

Should Africa alone receive $10 billion annually from the developed world, it would average out that each living HIV-infected African would receive $312 per year ($10 billion divided by 32 million HIV-infected) for as long as the money was provided. This would occur only if the money was actually given to the people. This money, along with free or very low drug costs could be of considerable help, especially since many African countries spend less than $5 to $10 per person a year on public health. But, used in this way, this large sum of money would not be available to build needed medical facilities or import the thousands of doctors necessary to treat the HIV-infected. That will take a few hundred billion more dollars.

An article on "Estimating the Cost of Expanded AIDS Treatment in Africa," that appeared in the June 2001 issue of *Topics in HIV Medicine,* states that it would cost $1.12 billion each year to treat 1 million HIV-infected Africans. Ending 2005 there will be an estimated 32 million Africans living with HIV. The cost, based on information presented in the article, would be about $36 billion a year if all were to be treated equally.

POINT OF INFORMATION 13.4

THE GLOBAL FUND TO FIGHT HIV/AIDS, TB, AND MALARIA

HIV/AIDS, TB, and malaria kill over 6 million people annually with an additional 350 million people suffering from these diseases. This fund was set up in response to widespread public criticism of governments' apathy to the health crisis in developing countries, especially concerning HIV/AIDS. The health status of the poor, women, men, and children is deteriorating in many parts of the world, and the fund is a unique opportunity to mobilize international political will and resources to address this crisis in a new way, rather than continuing with business as usual.

The Global Fund was initiated through the United Nations General Assembly Special Session on AIDS (UNGASS) in June 2001. The year 2001 has been recorded in AIDS history as the year when political commitment to the disease moved to center stage. UNGASS demanded real political commitment, and activists had to think on their feet as they adjusted to the opaque UN negotiations.

Comments Relative to the War Chest

The Global Fund: Which Countries Owe How Much?—proposed that contributions to the Fund should be made according to an *Equitable Contributions Framework,* in which donor countries contribute in relation to the sizes of their economies. But, many nations refuse to pledge money because there is no mechanism in place to handle corruption and who besides Africa will receive how much of the fund, and what the money will be spent for. For example, there remains a large division among the 184 countries of the United Nations General Assembly on whether most of the money should be spent on prevention rather than antiretroviral drugs. Many members of the General Assembly said that about half the $10 billion should be spent on drugs for Africa and the other half be spent on prevention programs in Asia and the former Soviet nations where the epidemic is expanding out of control. Clearly there are no easy choices, but choices must be made. The United States was the first and only country by mid-2001 to offer Annan $200 million toward his global AIDS fund. Twenty-nine other nations have followed. Also making financial commitments are three foundations and two agencies. Total financial pledges from all countries and private foundations, etc., through 2008 is about $6 billion. But, will the pledges be fulfilled? The financial goal for 2005 is $3.6 billion but the fund is $2.7 billion short. No single nation has committed more money to this fund toward the African HIV/AIDS problem than the United States.

FORMS OF U.S. FOREIGN MONETARY AND OTHER ASSISTANCE FOR HIV/AIDS

A U.S. Congressman once said, "A billion dollars here, a billion dollars there, it begins to add up." With respect to U.S. HIV/AIDS support, those hundreds of millions and billions of dollars have indeed added up. The United States has provided and continues to provide more money globally to foreign governments and nongovernmental organizations (NGOs) than any country on earth. The U.S. federal government has donated billions of dollars, through a variety of organizations, directly to people in need and to the Global AIDS, TB, and Malaria Fund. Direct spending for international HIV/AIDS activities by the U.S. began in 1986 with a $1.1 million investment, through several U.S. agencies that had already started international HIV/AIDS projects. Spending increased steadily and reached $2.8 billion.

Direct Assistance through Government-to-Government Agreements and Bilateral Aid

A key strategy is to form bilateral working relationships with governments around the world to cooperatively conduct prevention, care, and research programs. This approach to development assistance is based on promoting broad-based economic growth through policy analyses, technical assistance, and training. In addition, emergency relief is structured to help nations make the transition to sustainable development. In addition to developing working relationships, the U.S. also provides bilateral support directly to governments to address HIV/AIDS.

Loans to Developing Countries—Loans are another mechanism used by the United States and others to provide international assistance. Last year, the U.S. Export-Import (Ex-Im) Bank announced plans to provide $1 billion in loans per year for five years to support the purchase of HIV/AIDS medications made in the U.S. The Ex-Im Bank also announced that it would increase the standard repayment terms for HIV/AIDS pharmaceutical sales in these countries from six months to five years.

Debt Relief—According to the World Bank, 33 of the world's 41 most heavily indebted countries are in Africa with a total debt of $230 billion. There is a growing movement on the part of developed nations to forgive all or part of that debt. Debt relief can enable developing nations to spend more of the resources on health programs addressing HIV/AIDS. Last year, Congress approved nearly $450 million in debt relief for developing nations, primarily in sub-Saharan Africa. Japan cancelled $3 billion in debts owed by African nations in 2003–2004.

International Global Fund—In 2002, the G8 countries (United States, Russia, Great Britain, France, Germany, Italy, Japan, and Canada) organized a new International Global Fund, which became operational in January 2002. To this Fund through 2006, the U.S. will have contributed an estimated $7.2 billion (if budget requests for 2005 and 2006 are approved by Congress). These international global funds are to be used to support programs for HIV/AIDS (60%), TB, and malaria (40%). This Fund should not be confused with the Global AIDS, TB, and Malaria Fund.

President George W. Bush's Emergency Plan for AIDS Relief through the International Global Fund: January 28, 2003

President Bush announced in the State of the Union address the Emergency Plan for AIDS Relief, a five-year, $15 billion initiative to turn the tide in combating the global HIV/AIDS pandemic. "To meet an urgent crisis abroad, tonight I propose the Emergency Plan for AIDS Relief—a work for mercy beyond all current international efforts to help the people of Africa. I ask the Congress to commit $15 billion over the next five years, including nearly $10 billion in new money to turn the tide against AIDS in the most afflicted nations of Africa and the Caribbean." This commitment of resources will help the 14 most afflicted countries in Africa and the Caribbean (Haiti and Guyana). Specifically, the initiative is intended to:

- Prevent 7 million new infections (60% of the projected new cases in the target countries): The initiative will involve large-scale prevention efforts including voluntary testing and counseling. The availability of treatment will help prevention efforts by providing an incentive for individuals to be tested (20% of funds).
- Treat 2 million HIV-infected people: Capitalizing on recent advances in AIDS treatment, the president's Emergency Plan for AIDS Relief will be the first to provide advanced antiretroviral treatment on a large scale in the poorest afflicted countries (55% of funds).
- Care for 10 million HIV-infected individuals and AIDS orphans: To provide a range of care, including support for AIDS orphans (25% of funds).
- Bush's plan calls for treating two million people in 15 countries over five years (2 x 5) and should not be confused with the World Health Organization's plan to treat three million HIV infected in 38 countries by 2005 (3 x 5).

The $15 billion in funding for this initiative virtually triples the U.S. commitment to international AIDS assistance. Funding began with $2 billion in fiscal year 2004, and will increase thereafter. The $15 billion includes $1 billion for the Global Fund To Fight HIV/AIDs, Tuberculosis, and Malaria, conditioned on whether or not the fund shows results.

An important aspect of Bush's Emergency Plan is that it issues a challenge to every other member of the G8 countries to follow suit. In a sense he has placed a moral burden on these countries. On May 26, 2003, President Bush signed into law the $15 billion program that he called "a great mission to rescue." This program is the largest monetary commitment for an international public health initiative involving a specific disease in history!

In early 2004, United States AIDS activists have asked President Bush to allocate $5.4 billion for 2005 and $30 billion by 2008 "in order to pay a fair share of what is needed by the Global AIDS, TB, and Malaria fund to meet grant requests."

Summary—A report from the London-based Panos Institute entitled "Missing The Message – Twenty Years Of Learning From HIV/AIDS," says "After years of neglect, more money and political interest are being directed towards AIDS than ever before." However, it says spending large sums of money in hopes of achieving rapid results has often brought "disappointing or short lived" results. Implying that lessons haven't been learned means that the total world response to HIV/AIDS isn't panning out in the way that one would possibly hope that it could. That is, money in and of itself will not solve the AIDS crises in Africa or in any other developing nation. Whether it is $5 billion or $50 billion—pick a number—the money is useless if those funds do not reach the vulnerable, the HIV infected, the dying and whole communities that need them. Institutions and policies must be firmly in place to assure that the allocated funds can be well spent. To demand less would be irresponsible. We can all agree that the AIDS pandemic is a horror of unimaginable proportions, but spending vast sums of money without accountability never, ever works. It does not bring about solutions, but it does bring about disorder and discord. Stephen Lewis, a United Nations Special Envoy, HIV/AIDS for Africa, said in his presentation at the Fourteenth International AIDS Conference in Durban, South Africa, "What is wrong with the world? People are dying in numbers that are the stuff of science fiction. Millions of human beings are at risk. Communities, families, mothers, fathers, children are like shards of humanity caught in a maelstrom of destruction. They're flesh and blood human beings, for God's sake; is that not enough to ignite the conscience of the world? Why should we have to produce all these tortured rationales to drive home such an obvious point? This pandemic has done something dreadful to the instinct for compassion. I don't really understand what's happening; I don't really understand why the simple act of saving or prolonging human life isn't sufficient anymore. It's irrational to need a balance sheet of geometric calculation and economic architecture. It's sick."

The Problem in Perspective: Wealth, Poverty, and AIDS

The relationship between poverty and HIV transmission is not simple. If it were, South Africa might not have Africa's largest epidemic, for South Africa is rich by African standards. Botswana is also relatively rich, yet this country has the highest levels of infection in the world. While most people with HIV/AIDS are poor, many of the infected are not poor. Undernourishment; lack of clean water, sanitation, and hygienic living conditions; generally low levels of health, compromised immune systems, high incidence of other infections including genital infections and exposure to diseases such as tuberculosis and malaria; inadequate public health services; illiteracy and ignorance; pressures encouraging high-risk behavior, from labor migration to alcohol abuse and gender violence; an inadequate leadership response to either HIV/AIDS or the problems of the poor; and finally, lack of confidence or hope for the future—all companions of poverty promote HIV infection!

The cycle of poverty intensifies as individuals, households, and communities living with HIV/AIDS find that lost earnings, lost crops, and missing treatment make them weaker, make their poverty deeper, and push the vulnerable into poverty. Inequality sharpens the impact of poverty, and a mixture of poverty and inequality may be driving the epidemic. A South African truck driver is not well paid compared to the executives who run his company, but he is rich in comparison to the people in the rural areas he drives through. For the woman at a truck stop, a man with 50 rand ($10) is wealthy; her desperate need for money to feed her family may buy him unprotected sex, even though she knows the risks.

According to the UNAIDS chief epidemiologist, by the end of 2006 there will be 43 million living HIV-infected people globally. About 40 million of these people live on less than $2 a day. In many of the high HIV incidence countries in Southern Africa, like Kenya, Botswana, Zambia, Malawi, Nigeria, Swaziland, and Uganda, about 50% of their populations live on a dollar or less per day! An HIV-positive American can focus on his T cell counts; an HIV-positive African in a rural village still has to focus on finding clean water and food.

Politicians, economists, and AIDS specialists rarely say this bluntly, but the truth is that most of those 48 million people have simply been written off because the first priority for the first few billion dollars is prevention, not treatment. An economist who studies AIDS in South Africa said, "You can't give up on the infected because of the message it sends. But if he had $1 billion to spend most wisely, I would spend it on giving women more power, caring for orphans, and getting them education."

FINALLY, THE QUESTION: How much would it cost to contain the global AIDS pandemic? The answer is: How much have you got? How much would it cost to banish ignorance, to deaden lust, to shame rape, to stop war, to enrich the poor, to empower women, to defend children, to make decent medical care as globally ubiquitous as Coca-Cola—in short, to get rid of all the underlying causes of the pandemic in the developing nations. Much of the world at risk for HIV/AIDS can't read. Most of the world at risk has never used a condom. Most of the world at risk has never heard of ACT UP and wouldn't dare heckle a president. And most of the world with AIDS thinks it doesn't have the disease and doesn't know anyone who does, because 95% of those infected in the developing nations have never been tested. (Most estimates come from anonymous testing at prenatal clinics.) Most of the world cannot afford the antiretroviral drugs.

SUMMARY

In 1981, the CDC announced a new disease affecting the homosexual population. This disease was later called AIDS. Many religious people believed this was a sign that homosexuality should be punished. The few facts available at that time gave rise to a great deal of fantasy and fear. Affected people were either seen as innocent victims or it was felt that they deserved the disease. Contracting AIDS labeled a person as less than desirable, a homosexual, or one who practiced deviant forms of sexual behavior. But even the so-called innocent victims, the children, the hemophiliacs, and the recipients of blood transfusions were not spared social ostracism. If you had AIDS, you were twice the victim—first of the virus and second of the social behavior.

Children were barred from attending school, adults from their jobs, and both from adequate medical care. For example, there are still relatively few dentists who will treat AIDS patients and a significant number of surgeons refuse to operate on AIDS patients. Years have passed, but many misconceptions about HIV/ AIDS linger on.

Fear is being casually transmitted rather than the virus. A significant number of people, after years of broad-scale education, still believe that the AIDS virus can be casually transmitted from toilet seats, drinking glasses, and even by donating blood.

The fallout from the fear of the AIDS pandemic has been a major change in sexual language in TV advertisements, magazines, and radio. Condoms once spoken about only in hushed tones and kept under the counter in most drug stores, are now spoken of everywhere as a means of safer sex. AIDS, perhaps more than any other disease, has demonstrated that ignorance leads to fear and knowledge can lead to compassion.

To achieve understanding and compassion, people must be educated as to their HIV risk status and how they can keep it low. Many hundreds of millions of dollars have been spent to inform the public of the kinds of behavior that either place them at risk or reduce their risk for HIV infection. The problem is that although people are getting the information, too many refuse to act on it. Former Surgeon General C. Everett Koop's office mailed 107 million copies of the brochure "Understanding AIDS" to households in the United States. Fifty-one percent of those who received it said they never read it. Even among those who read the brochure are those who refuse to change their sexual behavior. Old habits are difficult to break.

To date, the hard evidence shows that only the homosexual population has significantly modified their sexual behavior as evidenced by the drop in the number of new cases of AIDS among them from 1988 through 2004.

A major problem looming on the horizon is the prospect of HIV being spread in the teenage population. Large numbers of teens use drugs and alcohol, have multiple sex partners, and believe they are invulnerable to infection.

The AMA stated in 1988 that physicians may not refuse to care for patients with AIDS because of actual risk or fear of contracting the disease. Some physicians get around this through referral to other physicians who will treat AIDS patients. There is one area of medicine that takes issue at having to treat AIDS patients: surgery. Because it is difficult not to accidently get cut during surgery, surgeons have been the leading advocates for HIV testing of all surgical patients so they will know their risks before performing surgery.

On the other hand, patients say they have a right to know if their physician, especially a surgeon, is HIV-infected. Surveys indicate that most people would not want to be treated by an HIV-infected physician.

On June 25, 1998, the U.S. Supreme Court ruled that the Americans with Disabilities Act protected HIV-infected people. Even though they experienced no symptoms, they are to be treated as handicapped.

REVIEW QUESTIONS

(Answers to the Review Questions are on page 438.)

1. Name three major sources of information that contributed to the early panic and hysteria about the spread of AIDS.
2. Give three examples of unfounded public fears to AIDS infection.
3. Fear of the casual transmission of AIDS parallels what other earlier STD epidemic?
4. What evidence is there that it is difficult to get people to change their behavior even though they know it is harmful to their well-being?
5. What is the major thrust of AIDS education in the United States?

6. If education is the key to preventing HIV infection and new cases of AIDS, and most people interviewed say they have been educated, why is it not working?
7. Why are today's teenagers in danger of contracting and spreading HIV?
8. Yes or No: Do physicians have a right to refuse to treat AIDS patients? Support your answer.
9. Do patients have a right to know if their physician is HIV-infected or has AIDS?
10. What is the primary means of offsetting the bias toward people with AIDS in the workplace?
11. Who is the current secretary of the United Nations and how much money does he believe is needed to fight AIDS globally each year?
12. About how much money has been pledged to the Global AIDS Fund through 2004?
13. Compared with the money the federal government spends on research and treatment to combat other health and medical problems such as heart disease and cancer, do you think federal spending on AIDS research and treatment is too high, too low, or about right? Support your choice with credible evidence.

Answers to Review Questions

CHAPTER 1

1. Acquired Immune Deficiency Syndrome
2. No. AIDS is a syndrome. A syndrome is made up of a collection of signs and symptoms of one or more diseases. AIDS patients have a collection of opportunistic infections and cancers. Collectively they are mistakenly referred to as the AIDS disease.
3. In 1983 by Luc Montagnier
4. 1981
5. LAV
6. Five; 1982, 1983, 1985, 1987, and 1993
7. The ARC or AIDS Related Complex definition was a middle ground used before AIDS was better understood. It became meaningless after the 1985 expanded definition listed organisms and symptoms that indicated that two states existed: HIV infection and progression to AIDS.
8. It allows HIV-infected persons earlier access into federal and state medical and social programs.

CHAPTER 2

1. The unbroken transmission of infection from an infected person to an uninfected person.
2. The answer to both questions is unknown at this time.
3. HIV/AIDS Word Search Solution

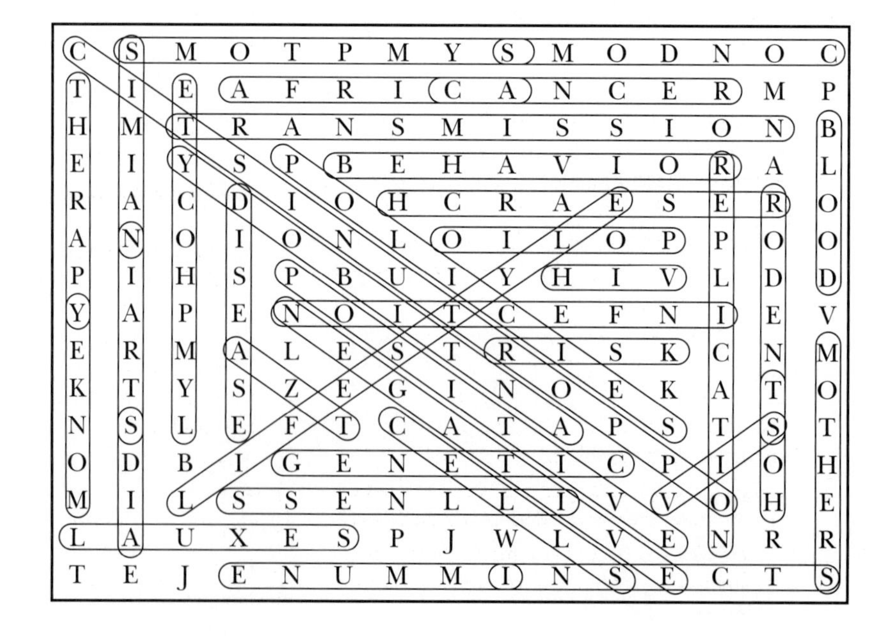

AIDS
AZT
AFRICA
ANTIBODY
BEHAVIORAL
BLOOD
CANCER
CELLS
CONDOMS
DISEASE
GENETIC
HIV
HOST
ILLNESS
IMMUNE
INFECTION
INSECTS
LIFESTYLE
LYMPHOCYTE
MONKEY
MOTHERS
NEGATIVE
OPPORTUNISTIC
POLICIES
POLIO
POSITIVE
REPLICATION
RESEARCH
RISK
RODENTS
SEXUAL
SIMIAN
SIV
STRAIN
SYMPTOMS
THERAPY
TRANSMISSION

CHAPTER 3

1. Because it contains RNA as its genetic message and a reverse transcriptase enzyme to make DNA from RNA.
2. GAG-POL-ENV; at least six
3. Because HIV has demonstrated an unusually high rate of genetic mutations; (1) the reverse transcriptase enzyme in HIV is highly error prone (makes transcription errors), and (2) a variety of HIV mutants have been found within a single HIV-infected individual.
4. The reverse transcriptase enzyme is highly error prone, making at least one, and in many cases more than one, deletion, addition, or substitution per round of proviral replication.

CHAPTER 4

1. Not really, because there is no way as yet to remove the provirus from the cell's DNA.
2. A physiological measurement that serves as a substitute for a major clinical event.
3. March; July, 25
4. 8
5. Zidovudine, Didanosine, Zalcitabine, Stavudine, Lamivudine.
6. Becoming incorporated into DNA as it is being synthesized, thereby stopping reverse transcriptase from attaching the next nucleotide.
7. (a) Clinical biological side effects; (b) the selection of drug-resistant HIV mutants.
8. (a) The number of copies of HIV RNA present in the plasma; (b) this number indicates the reproductive activity of HIV at the time and, if therapy is being used, the effect of the therapy on the reproductive ability of the virus.
9. Saquinavir mesylate, saquinavir (Fortovase) ritonavir, indinavir, nelfinavir, amprenavir, atazanavir.
10. Indinavir, because either as a monotherapy or in combination, it is the most successful at dropping the viral load.
11. They physically interact with the reverse transcriptase enzyme and interfere with its function.
12. To suppress HIV replication, thereby reducing the number of mutant RNA strands produced.
13. A
14. B
15. To be determined by the instructor.
16. Answer depends on the credible facts the student presents.
17. D
18. D
19. True
20. C
21. D
22. False
23. A
24. False
25. D
26. D

CHAPTER 5

1. T4 helper cells; because T4 cells are crucial for the production of antibodies, a depletion of T4 cells results in immunosuppression, which results in OIs.
2. CD4 is a receptor protein (antigen) secreted by certain cells of the immune system, for example, monocytes, macrophages, and T4 helper cells. It becomes located on the exterior of the cellular membrane and happens to be a compatible receptor for the HIV to attach and infect the CD4-carrying cell.
3. The question of true latency after HIV infection has not been settled. Most HIV/AIDS investigators currently believe there is a latent period, a time of few if any clinical symptoms and low levels of HIV in the blood. Other scientists, currently the minority, believe there is no true latency. The virus hides out in the lymph nodes, slowly reproducing, and slowly killing off the T4 cells. The virus is always present, increasing slowly in numbers over time.
4. True
5. False
6. True
7. False
8. False
9. True
10. False
11. True
12. True
13. True
14. B, assessing risk of disease progression.

CHAPTER 6

1. OI is caused by organisms that are normally within the body and held in check by an active immune system. When the immune system becomes suppressed, for whatever reason, these agents can multiply and produce disease.
2. *Pneumocystis carinii;* lungs, pneumonia
3. *Isospora belli*
4. *Mycobacterium avium intracellulare*
5. False. HIV has not been found in KS tissue. KS is believed to develop as a result of a suppressed immune system and not the virus *per se.*
6. Classic KS, as described by Moritz Kaposi; and KS associated with AIDS
7. False. KS normally affects gay males. It is highly unusual to find KS in hemophiliacs, injection-drug users, and female AIDS patients.
8. True
9. True
10. True (Answer provided in POI 6.1)
11. Opportunistic Infections Crossword Solution

CHAPTER 7

1. The 6-stage Walter Reed System and the 4-group CDC system
2. About 30%; about 90%
3. AIDS Dementia Complex
4. Skin—Kaposi's sarcoma
   Eyes—CMV retinitis
   Mouth—thrush or hairy leukoplakia
   Lungs—Pneumocystis pneumonia
   Intestines—diarrhea
5. True
6. False. The average time is 6 to 18 weeks.
7. False. HIV infection leads to HIV disease. AIDS is the result of a weakened immune system that allows opportunistic infections to occur.
8. False. The average length of time is about 10 to 11 years.
9. Instructor's evaluation
10. E, all of the above.
11. Word Search Solution

OPPORTUNISTIC INFECTIONS CROSSWORD SOLUTION

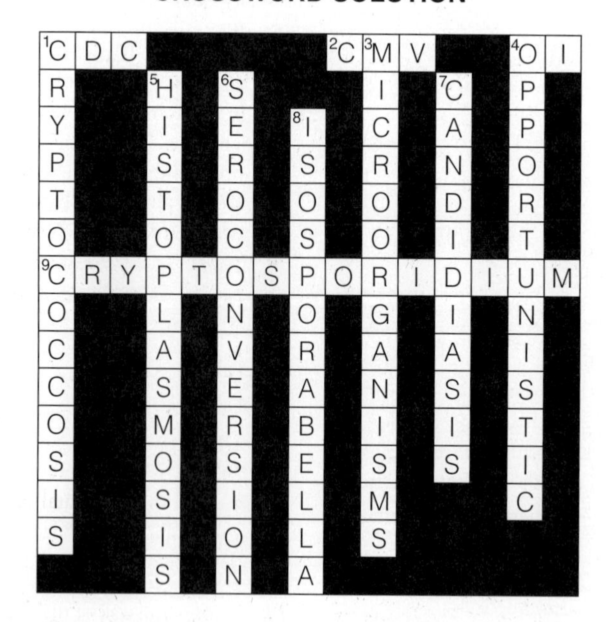

WORD SEARCH SOLUTION

C	A	L	C	A	D	V	A	N	C	E	D	H	I	I
I	P	H	H	S	E	P	T	I	C	E	M	I	A	O
T	J	C	R	P	T	D	L	N	X	T	D	V	M	N
A	E	I	O	S	A	I	U	R	C	U	E	U	U	E
M	L	N	N	U	U	S	D	E	N	C	T	S	R	G
O	B	E	I	B	T	E	A	P	O	A	E	N	T	I
T	A	G	C	G	N	A	A	L	T	R	C	K	C	T
P	R	O	G	R	E	S	S	I	O	N	T	L	E	N
M	U	H	S	O	T	E	O	C	D	N	A	L	P	A
Y	S	T	T	U	T	N	O	A	U	S	B	F	S	T
S	A	A	A	P	A	N	K	T	S	G	L	S	N	S
A	E	P	G	A	V	N	O	I	T	C	E	F	N	I
U	M	H	E	E	Y	E	F	O	M	E	T	S	Y	S
T	H	E	R	A	P	Y	C	N	E	T	A	L	M	E
B	N	T	N	O	I	T	A	B	U	C	N	I	G	R

ACUTE
ADULT
ADVANCED
AIDS
ANTIGEN
ASYMPTOMATIC
ATTENUATED
CHRONIC
CLASSIFY
DETECTABLE
DISEASE
HIV
INCUBATION
INFANT
INFECTION
LATENCY
MEASURABLE
MUTATION
PATHOGENIC
PROGRESSION
REPLICATION
RESISTANT
SEPTICEMIA
SEROCONVERT
SPECTRUM
STAGE
SUBGROUP
SYSTEM
THERAPY

CHAPTER 8

1. False. The United States currently *reports* most of the world's AIDS cases.
2. Cases of AIDS-related death, according to the CDC definition, can be traced back to 1952 in the United States and to the mid-1950s in Africa.
3. HIV-1 and HIV-2 show a 40% to 50% genetic relationship to each other.
4. False. HIV-1 and HIV-2 are both transmitted via the same routes. HIV-2 is spreading globally in similar fashion to HIV-1.
5. True. All scientific and empirical evidence to date indicates that HIV is *not* casually transmitted.
6. Through sexual activities: exchange of certain body fluids—blood and blood products, semen and vaginal secretions; and from mother to fetus or newborn by breast milk.
7. False. There is only one documented case of HIV infection caused by deep kissing. HIV has been found in the saliva of infected people in very low concentration, and saliva has been shown to have anti-HIV properties.
8. True; but this assertion has been proven to be untrue. Insects, in particular mosquitoes, have not been shown to transmit HIV successfully.
9. False. According to studies involving the sexual partners of injection-drug users and hemophiliacs, HIV transmission from male to female is the more efficient route. This is believed to be due to a greater concentration of HIV found in semen than in vaginal fluid.
10. The answer may be true or false. There have been cases in which a single act of intercourse has resulted in HIV infection. However, the majority of surveys on the sexual partners of injection-drug users and hemophiliacs indicate that the number of sexual encounters may increase the risk of HIV infection but does not guarantee infection. Sexual partners of infected people have remained

HIV-free after years of unprotected penis-vagina or penis-anus intercourse.

11. The percentage of fetal risk varies widely in a number of hospital studies. At the moment, the risk as reported without Zidovudine therapy varies from less than 30%. For Africa the figures most commonly used are 30% to 50%. With the use of Zidovudine therapy, the risk has been cut to about 8%. Using Zidovudine and a caesarean section reduces HIV transmission to about 2%.
12. E, all of the above.
13. True
14. True
15. True
16. True
17. True
18. False
19. True
20. True
21. False
22. True
23. True
24. True
25. B, mosquito bites.
26. HIV/AIDS Word Unscramble

ABMETARUTS	=	MASTURBATE
ACIPMDEN	=	PANDEMIC
AERIBRR	=	BARRIER
AGSEASM	=	MASSAGE
ANLA	=	ANAL
ATOUNITM	=	MUTATION
BENSTCENIA	=	ABSTINENCE
BLAUEXSI	=	BISEXUAL
BLMSKRITAE	=	BREASTMILK
BYUDILSFOD	=	BODYFLUIDS
CDRLSAIIMEP	=	SPERMICIDAL
DBOOL	=	BLOOD
DEIPCIME	=	EPIDEMIC
EGEIVTAN	=	NEGATIVE
FOCTNEINI	=	INFECTION
HEHALIIMPO	=	HEMOPHILIA
HREUSALXEOET	=	HETEROSEXUAL
IAVLNAG	=	VAGINAL
IDSA	=	AIDS
IVH	=	HIV
LSNCMIAOOO	=	NOSOCOMIAL
MEENS	=	SEMEN
MNTSNAOISISR	=	TRANSMISSION
MREPS	=	SPERM
NCEIALBCUMMO	=	COMMUNICABLE
NITSARS	=	STRAINS
NOEUTESRIRC	=	INTERCOURSE
OEULXMASOH	=	HOMOSEXUAL
OMODNC	=	CONDOM
PESREH	=	HERPES
PRLIATANE	=	PERINATAL
RSNIAUVONET	=	INTRAVENOUS
SAXELU	=	SEXUAL
SDT	=	STD
SILAAV	=	SALIVA
SIPEN	=	PENIS
SNRISEGY	=	SYRINGES
TFYELSEIL	=	LIFESTYLE
TLACER	=	RECTAL
UNOASNISFTR	=	TRANSFUSION
VPIETISO	=	POSITIVE
VSURI	=	VIRUS

CHAPTER 9

1. Latex condoms. They are known to stop the transmission of viruses. This may not be true for animal intestine condoms.
2. Water-based lubricants. Oil-based lubricants weaken the latex rubber, causing them to leak or break under stress.
3. Safer sex is having sexual intercourse with an *uninfected* partner while using a condom.
4. The answer may be true or false. There have been cases where a single act of intercourse resulted in HIV infection. However, the majority of surveys completed by sexual partners of injection-drug users and hemophiliacs indicate that the number of sexual encounters may increase the risk of HIV infection but does not guarantee infection. Sexual partners of infected persons have remained HIV-free after years of unprotected penis-vagina intercourse.
5. No. IDUs exist between "fixes." They lose things, they may not care to pick up new equipment—they need the "fix" now, it may be easier to share. Circumstances vary considerably among the IDU. Just giving them free equipment is no assurance that they will use it.
6. Between 1 in 39,000 and 1 in 200,000.
7. Have several students read their answers for promoting class discussion. Compare their response to that given in the text (that they should be punished).
8. Because attenuated HIV may mutate to a virulent form, causing an HIV infection; there is no absolute guarantee that 100% of HIV are inactivated.

9. Because at no time will a whole HIV be present in the vaccine. Only a specific subunit of the HIV will be present in pure form so the vaccine should be free of any contaminating proteins that might prove toxic to one or more persons receiving the vaccine.

10. It is a situation wherein HIV antibodies might predispose the host to become more easily HIV-infected. For whatever reason, it appears that the HIV antibody complex enters the cell more easily than HIV alone.

11. Because of the severe forms of social ostracism that occur when it is learned that someone is HIV positive or belongs to a high-risk group (gay, injection-drug user, bisexual).

12. Universal precautions are a list of rules and regulations provided by the CDC to help prevent HIV infection in health-care workers.

13. False
14. False
15. True
16. True
17. True
18. False
19. False
20. False, 1998
21. Instructor evaluation
22. HIV/AIDS Word Search Solution

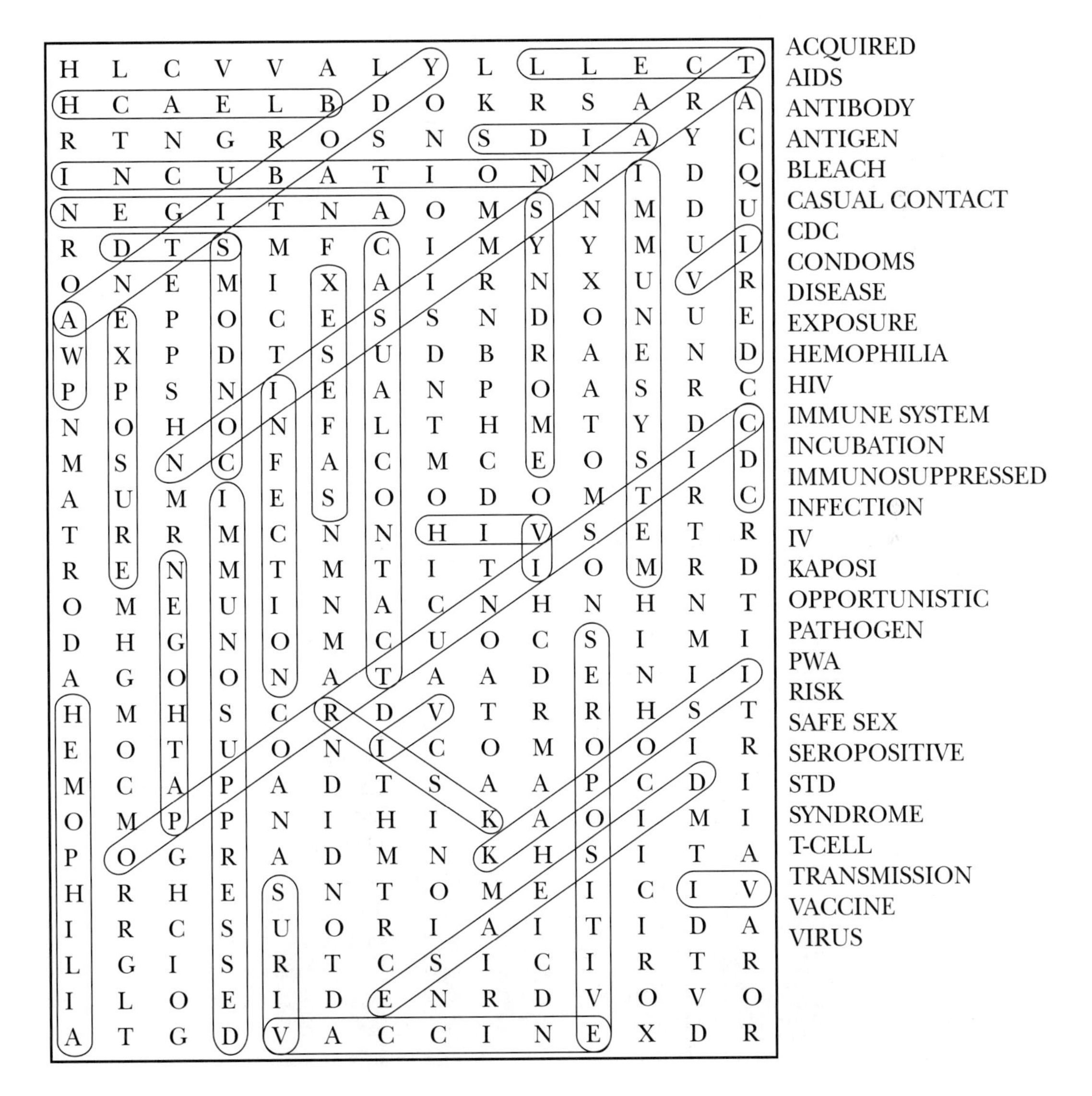

CHAPTER 10

1. The CDC said for each AIDS case there are from 50 to 100 HIV-infected people in the population. They got the 1 to 1.5 million figure using the 50 HIV-infected/AIDS cases; 1986; they are believed to be within plus or minus 10%.

2. Because their social and sexual behaviors and medical needs place these people at a greater risk for HIV exposure than those not practicing these behaviors or who do not need blood or blood products.

3. False. Studies show that the time for progression from HIV infection to AIDS is the same regardless of parameters.

4. 52%

5. 30% to 50%

6. 99%

7. Two per 1000 students; more: the rate for military personnel is 1.4 per 1000.

8. College students 2/1000, general population 0.2/1000; this means the rate of HIV infection on college campuses is about 10 times higher than in the general population.

9. One in 250 to 300

10. Needle stick injuries

11. 3.5%

12. Hepatitis B virus

13. GAO—300,000 to 485,000
CDC—270,000

14. Student's answer; text says no. People most often do not tell the truth—much depends on where, when, why, and who is doing the survey. There are just too many variables involved to believe sexual surveys.

15. Worldwide over 30 million, United States about 560,000.

16. 56% $\frac{543{,}000}{970{,}000}$

17. HIV/AIDS Crossword Solution

CHAPTER 11

1. Approximately 40 million women and 40 million men

2. About 50%

3. Injection-drug use, being a sexual partner of an IDU, and through heterosexual contact.

4. IDU

5. 2.5 million; 1.5 million

6. Leading; 25 and 34; fifth; white; 25 and 44; leading; 25 and 44

7. An estimated 100,000

8. None, all states now have reported pediatric cases.

9. Virtually all—100%

HIV/AIDS CROSSWORD SOLUTION

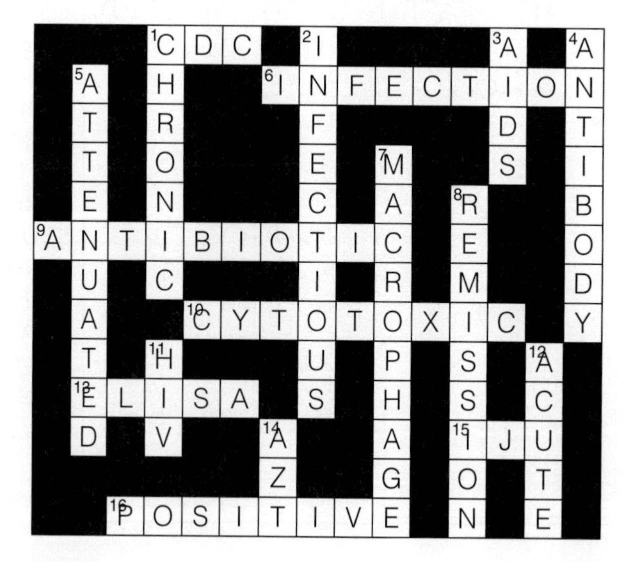

10. (1) Maternal viral load; (2) route of delivery, and (3) duration of early membrane rupture.
11. Most orphaned AIDS children have mothers who are IDUs and are themselves HIV-infected. They are AIDS orphans because (1) their parents abandon them due to illness or death; and (2) these children are HIV-infected or demonstrate AIDS and therefore no one wants them.
12. Over 50%
13. Two; 300
14. 28 million
15. Teenage gay men and teenage women; 75%
16. 25%
17. About 500,000
18. Yes; blacks and Latinos
19. One-third or about 15.8 million
20. HIV/AIDS Crossword Solution

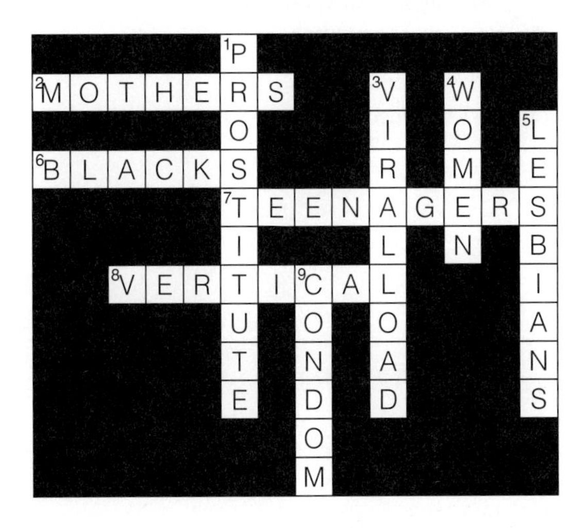

CHAPTER 12

1. ELISA; enzyme linked immunosorbent assay
2. That the body will produce antibody against antigenic components of the HIV virus after infection occurs.
3. No; a positive antibody result must be repeated in duplicate and if still positive, a confirmatory test is performed prior to telling people they are HIV-infected.
4. No; AIDS is medically diagnosed after certain signs and symptoms of specific diseases occur.
5. Western Blot
6. Indirect immunofluorescent assay
7. By a color change in the reaction tube; the peroxidase enzyme oxidizes a clear chromogen into color formation. This occurs if the HIV antibody–antigen enzyme complex is present in the reaction tube.
8. False. Some newborns receive the HIV antibody passively during pregnancy. About 30% to 50% of HIV-positive newborns are truly HIV positive; it is unknown whether all HIV-positive newborns go on to develop AIDS. Not all have been discovered and it has not been determined whether 100% of HIV-infected adults or babies will develop AIDS.
9. They are not 100% accurate.
10. Determining that positive and negative tests are truly positive and negative and not falsely positive or negative.
11. Using either too high or too low cut-off points in the spectrophotometer and the presence of cross-reacting antibodies.
12. The percentage of false positives will increase as the prevalence of HIV-infected people in a population decreases.
13. It is a screening test value that represents the probability that a positive HIV test is truly positive; because screening tests are not 100% accurate.
14. Western Blot
15. There is no standardized WB test interpretation. Different agencies use different WB results (reactive bands) to determine that the test sample is positive.
16. Because the PCR allows for the detection of proviral DNA in cells before the body produces detectable HIV antibody; PCR reactions can be used to determine if high-risk (or anybody), antibody-negative people are HIV-infected but not producing antibodies and whether newborns are truly HIV positive or passively HIV positive.
17. 20 minutes
18. 3 minutes
19. False. The FDA approved two home-use HIV antibody test kits in 1996, but one was withdrawn from the market.
20. Between 6 and 18 weeks after HIV infection.
21. As early as two weeks after infection.
22. (1) Changes in their lifestyles that reduce stress on their immune systems may delay the onset of illness.

    (2) They can practice safer sex and hopefully not transmit the virus to others.

(3) The earlier the detection, the earlier they can enter into preventive therapy.

23. Mandatory with confidentiality; voluntary with confidentiality, anonymous, and blinded.
24. For anonymous testing no personal information is given; in blind tests, the name is deleted but the demographic data remain.
25. Because there are many examples of breaches of confidence, which destroys trust and subjects people to social stigma.
26. False
27. True

CHAPTER 13

1. Newspapers, TV, radio, magazines, etc.
2. Barring children from public schools, police wearing rubber gloves during arrests, not going to a restaurant because someone who works there has AIDS, firing AIDS employees, etc.
3. Syphilis
4. Use of tobacco products, alcohol, drugs; nonuse of seat belts and motorcycle helmets, etc.
5. The ways by which one can become HIV-infected and how not to become HIV-infected.
6. Because most of the new cases of HIV infection and AIDS occur in high-risk groups that will not or cannot change sexual and drug practices.
7. Because a larger percentage of teenagers are sexually active with more than one partner, use drugs, use alcohol, and think they are invulnerable to infection and death.
8. According to the AMA, no. Physicians may not refuse to care for patients with AIDS because of actual risk or fear of contracting the disease.
9. The CDC and AMA state that a patient's right to that information should be determined on a case-by-case basis where surgery will be performed. There is no legal requirement for physicians to tell their patients of their HIV status.
10. Worker information sessions that explain how the virus can and cannot be transmitted.
11. Kofi Annan; $7 billion to 10 billion
12. 5 billion plus
13. Review students' evidence—share with class.

Glossary

ACRONYMS

ACTG AIDS Clinical Trial Group

ADA Americans with Disabilities Act

AIDS Acquired Immunodeficiency Syndrome

AZT Azathioprine (a misnomer for Zidovudine or Azidothymidine)

CD Cluster Differentiating Antigen

CD4 a protein imbedded on the surface of a T lymphocyte to which HIV most often binds—a CD4+ or T4 cell.

CD8 a protein imbedded on the surface of a T lymphocyte suppressor cell—a T8 cell.

CDC Centers for Disease Control and Prevention (part of PHS)

3TC Lamivudine; nucleoside analog

DHHS Department of Health and Human Services

DNA deoxyribonucleic acid

d4T stavudine; nucleoside analog

ddC dideoxycytosine; nucleoside analog

ddI dideoxyinosine; nucleoside analog

FDA Food and Drug Administration (part of PHS)

HAART Highly Active Antiretroviral Therapy

HIV Human Immunodeficiency Virus

IDU Injection-Drug User

LAV lymphadenopathy-associated virus

NCI National Cancer Institute (part of NIH)

NIAID National Institute of Allergy and Infectious Diseases (part of NIH)

NIH National Institutes of Health (part of PHS)

NNRTI Non-nucleoside reverse transcriptase inhibitor

PCR (polymerase chain reaction) a very sensitive test used to detect the presence of HIV

PHS Public Health Service (part of DHHS)

PLWA Person Living With AIDS

RNA ribonucleic acid

ZDV Zidovudine; major drug in treating HIV/AIDS; nucleoside analog

For the newest anti-HIV drugs, names, and use, see Chapter 4.

TERMS

Acquired Immunodeficiency Syndrome (AIDS): A life-threatening disease caused by a virus and characterized by the breakdown of the body's immune defenses. (See AIDS.)

Acute: Sudden onset, short-term with severe symptoms.

Acyclovir (Zovirax): Antiviral drug for herpes 1 and 2 and herpes zoster.

Adjuvant: The active ingredient in vaccines that improves the human immune system response by attracting immune cells into the region where the vaccine is injected.

AIDS (Acquired Immunodeficiency Syndrome): A disease caused by a retrovirus called HIV and characterized by a deficiency of the immune system. The primary defect in AIDS is an acquired, persistent, quantitative functional depression within the T4 subset of lymphocytes. This depression often leads to infections caused by opportunistic microorganisms in HIV-infected individuals. A rare type of cancer (Kaposi's sarcoma) usually seen in elderly men or in individuals who are severely immunocompromised may also occur.

AIDS dementia: Neurological complications affecting thinking and behavior; intellectual impairment.

AIDSVAX: Trade name for all formulations of VAXGEN's vaccine.

Analog (analogue): A chemical molecule that closely resembles another one but which may function differently, thus altering a natural process.

Anal sex: A type of sexual intercourse in which a man inserts his penis in his partner's anus. Anal sex can be insertive or receptive.

Anemia: Low number of red blood cells.

Antibiotic: A chemical substance capable of destroying bacteria and other microorganisms.

Antibody: A blood protein produced by mammals in response to a specific antigen.

Antigen: A large molecule, usually a protein or carbohydrate, which when introduced into the body stimulates the production of an antibody that will react specifically with that antigen.

Antigen-presenting cells: B cells, cells of the monocyte lineage (including macrophages and dentritic cells), and various other body cells that present antigen in a form that T cells can recognize.

Antigenemia: Presence of viral proteins (antigens).

Antiretroviral therapy: Treatment with drugs designed to prevent HIV from replicating in HIV-infected persons. Highly active antiretroviral therapy (HAART) is an antiretroviral regimen that includes multiple classifications of antiretroviral drugs.

Antiserum: Serum portion of the blood that carries the antibodies.

Antiviral: Means against virus; drugs that destroy or weaken virus.

Apoptosis: Cellular suicide, also known as programmed cell death. A possible mechanism used by HIV to suppress the immune system. HIV may cause apoptosis in both HIV-infected and HIV-uninfected immune system cells.

Asymptomatic carrier: A host that is infected by an organism but does not demonstrate clinical signs or symptoms of the disease.

Asymptomatic seropositive: HIV positive without signs or symptoms of HIV disease.

Attenuated: Weakened. See Live/attenuated vaccine.

Atypical: Irregular; not of typical character.

Autoimmunity: Antibodies made against self tissues.

B lymphocytes or B cells: Lymphocytes that produce antibodies. B lymphocytes proliferate under stimulation from factors released by T lymphocytes.

B and T cell lymphomas: Cancers caused by proliferation of the two principal types of white blood cells—B and T lymphocytes.

Bacterium: A microscopic organism composed of a single cell. Many but not all bacteria cause disease.

Blood count: A count of the number of red and white blood cells and platelets.

Bone marrow: Soft tissue located in the cavities of the bones. The bone marrow is the source of all blood cells.

Canarypox: A virus that infects birds and is used as a live vector for HIV vaccines. It can carry a large quantity of foreign genes. Canarypox virus cannot grow in human cells, an important safety feature.

Cancer: A large group of diseases characterized by uncontrolled growth and spread of abnormal cells.

Candida albicans: A fungus; the causative agent of vulvovaginal candidiasis or yeast infection.

Candidiasis: A fungal infection of the mucous membranes (commonly occurring in the mouth, where it is known as thrush) characterized by whitish spots and/or a burning or painful sensation. It may also occur in the esophagus. It can also cause a red and itchy rash in moist areas, for example, the vagina.

Capsid: The protein coat of a virus particle.

CC-CKR-5 (CKR-5): Receptor for human chemokines and a necessary receptor for HIV entrance into macrophage.

CD: Cluster differentiating-type antigens found on T lymphocytes. Each CD is assigned a number: CD1, CD2, etc.

CD4 (T4 cell): White blood cell with type 4 protein embedded in the cell surface—target cell for HIV infection.

CD8 cell: Suppressor white blood cell with type 8 protein embedded in the cell surface.

Cell-mediated immunity: The reaction to antigenic material by specific defensive cells (macrophages) rather than antibodies.

Cellular immunity: A collection of cell types that provide protection against various antigens.

Chain of infection: A series of infections that are directly or immediately connected to a particular source.

Chemokines: Chemicals released by T cell lymphocytes and other cells of the immune system to attract a variety of cell types to sites of inflammation.

Chemotherapy: The use of chemicals that have a specific and toxic effect upon a disease-causing pathogen.

Chlamydia: A species of bacterium, the causative organism of *Lymphogranuloma venereum,* chlamydial urethritis, and most cases of newborn conjunctivitis.

Chromosomes: Physical structures in the cell's nucleus that house the genes. Each human cell has 22 pairs of autosomes and two sex chromosomes.

Chronic: Having a long and relatively mild course.

Clade: Related HIV variants classified by degree of genetic similarity; nine are known for HIV.

Cleavage site: One of nine sites (peptide bond) within the gag-pol polyprotein (peptide precursor) that is cleaved by HIV-1 protease to form functional subunits of GAG (p17, p7, p24) and POL (protease, reverse transcriptase, integrase).

Clinical latency: Infectious agent developing in a host without producing clinical symptoms.

Clinical manifestations: The signs of a disease as they pertain to or are observed in patients.

CMV: See cytomegalovirus.

Cofactor: Factors or agents that are necessary or that increase the probability of the development of disease in the presence of the basic etiologic agent of that disease.

Cohort: A group of individuals with some characteristics in common.

Communicable: Able to spread from one diseased person or animal to another, either directly or indirectly.

Condylomata acuminatum (venereal warts): Viral warts of the genital and anogenital area.

Confidential HIV test: An HIV test for which a record of the test and the test results are recorded in the client's chart.

Confirmatory test: A highly specific test designed to confirm the results of an earlier (screening) test. For HIV testing, a Western Blot or, less commonly, an immunofluorescence assay (IFA) is used as a confirmatory test.

Congenital: Acquired by the newborn before or at the time of birth.

Core proteins: Proteins that make up the internal structure or core of a virus.

Cross-resistance: Development of resistance to one agent as an antibotic, that results in resistance to other, usually similar agents.

Cryptococcal meningitis: A fungal infection that affects the three membranes (meninges) surrounding the brain and spinal cord. Symptoms include severe headache, vertigo, nausea, anorexia, sight disorders, and mental deterioration.

Cryptococcosis: A fungal infectious disease often found in the lungs of AIDS patients. It characteristically spreads to the meninges and may also spread to the kidneys and skin. It is due to the fungus *Cryptococcus neoformans.*

Cryptosporidiosis: An infection caused by a protozoan parasite found in the intestines of animals. Acquired in some people by direct contact with the infected animal, it lodges in the intestines and causes severe diarrhea. It may be transmitted from person to person. This infection seems to be occurring more frequently in immunosuppressed people and can lead to prolonged symptoms that do not respond to medication.

Cutaneous: Having to do with the skin.

CXCR-4 (FUSIN): Receptor for human chemokines and a necessary receptor for HIV entrance into T4 cells.

Cytokines: Powerful chemical substances secreted by cells. Cytokines include lymphokines produced by lymphocytes and monokines produced by monocytes and macrophages.

Cytomegalovirus (CMV): One of a group of highly host-specific herpes viruses that affect humans and other animals. Generally produces mild flu-like symptoms but can be more severe. In the immunosuppressed, it may cause pneumonia.

Cytopathic: Pertaining to or characterized by abnormal changes in cells.

Cytotoxic: Poisonous to cells.

Cytotoxic T cells: A subset of T lymphocytes that carry the T8 marker and can kill body cells infected by viruses or transformed by cancer.

Dementia: Chronic mental deterioration sufficient to significantly impair social and/or occupational function. Usually patients have memory and abstract thinking loss.

Dendritic cells: White blood cells found in the spleen and other lymphoid organs. Dendritic cells typically use threadlike tentacles to "hold" the antigen, which they present to T cells.

Didanosine: Also known as videx; see ddI—inhibits HIV replication.

Dissemination: Spread of disease throughout the body.

DNA (deoxyribonucleic acid): A linear polymer, made up of deoxyribonucleotide repeating units. It is the carrier of genetic information in living organisms and some viruses.

DNA vaccine (nucleic acid vaccine): Direct injection of a gene(s) coding for a specific antigenic protein(s), resulting in direct production of such antigen(s) within the vaccine recipient in order to trigger an appropriate immune response.

DNA viruses: Contain DNA as their genetic material.

Dysentery: Inflammation of the intestines, especially the colon, producing pain in the abdomen and diarrhea containing blood and mucus.

Efficacy: Effectiveness.

ELISA test: A blood test that indicates the presence of antibodies to a given antigen. Various ELISA tests are used to detect a variety of infections. The HIV ELISA test does not detect AIDS but only indicates if viral infection has occurred.

Endemic: Prevalent in or peculiar to a community or group of people.

Enteric infections: Infections of the intestine.

ENV: HIV gene that codes for protein gp160.

Envelope proteins: Proteins that comprise the envelope or surface of a virus, gp120 and gp41.

Enzyme: A catalytic protein that is produced by living cells and promotes the chemical processes of life without itself being altered or destroyed.

Epidemiology: Science that deals with the incidence, distribution, and control of disease in a population.

Epidemic: Affecting many persons at once, outbreak or rapid, sudden growth or development.

Epitope: A specific site on an antigen that stimulates specific immune responses, such as the production of antibodies or activation of immune cells.

Epivir: See 3TC.

Epstein-Barr virus (EBV): A virus that causes infectious mononucleosis. It is spread by saliva. EBV lies dormant in the lymph glands and has been associated with Burkitt's lymphoma, a cancer of the lymph tissue.

Etiologic agent: The organism that causes a disease.

Etiology: The study of the cause of disease.

Extracellular: Found outside the cell wall.

Factor VIII: A naturally occurring protein in plasma that aids in the coagulation of blood. A congenital deficiency of Factor VIII results in the bleeding disorder known as hemophilia A.

Factor VIII concentrate: A concentrated preparation of Factor VIII that is used in the treatment of individuals with hemophilia A.

False negative: Failure of a test to demonstrate the disease or condition when present.

False positive: A positive test result caused by a disease or condition other than the disease for which the test is designed.

Fellatio: Oral sex involving the penis.

Fitness: The ability of an individual virus to replicate successfully under defined conditions.

Follicular dendritic cells: Found in germinal centers of lymphoid organs.

Fomite: An inanimate object that can hold infectious agents and transfers them from one individual to another.

Fortovase: A more easily assimilated form of saquinavir.

Fulminant: Rapid onset, severe.

Fungus: Member of a class of relatively primitive organisms. Fungi include mushrooms, yeasts, rusts, molds, and smuts.

FUSIN: See CXCR-4.

Gammaglobulin: The antibody component of the serum.

Ganciclovir (DHPG): An experimental antiviral drug used in the treatment of CMV retinitis.

Gene: The basic unit of heredity; an ordered sequence of nucleotides. A gene contains the information for the synthesis of one polypeptide chain (protein).

Gene expression: The production of RNA and cellular proteins.

Genitourinary: Pertaining to the urinary and reproductive structures; sometimes called the GU tract or system.

Genome: A complete set of genes in a cell or virus.

Genotype: The sequence of nucleotide bases that constitutes a gene.

GP41: Glycoprotein found in envelope of HIV.

GP120: Glycoprotein found in outer level of HIV envelope.

GP160: Precusor glycoprotein to forming gp41 and gp120.

Globulin: That portion of serum that contains the antibodies.

Glycoproteins: Proteins with carbohydrate groups attached at specific locations.

Gonococcus: The specific etiologic agent of gonorrhea discovered by Neisser and named *Neisseria gonorrhoeae.*

Granulocytes: Phagocytic white blood cells filled with granules containing potent chemicals that allow the cells to digest microorganisms. Neutrophils, eosinophils, basophils, and mast cells are examples of granulocytes.

Hemoglobin: The oxygen-carrying portion of red blood cells that gives them a red color.

Hemophilia: A hereditary bleeding disorder caused by a deficiency in the ability to synthesize one or more of the blood coagulation proteins, for example, Factor VIII (hemophilia A) or Factor IX (hemophilia B).

Hepatitis: Inflammation of the liver; due to many causes including viruses, several of which are transmissible through blood transfusions and sexual activities.

Hepatosplenomegaly: Enlargement of the liver and spleen.

Herpes simplex virus I (HSV-I): A virus that results in cold sores or fever blisters, most often on the mouth or around the eyes. Like all herpes viruses, it may lie dormant for months or years in nerve tissues and flare up in times of stress, trauma, infection, or immunosuppression. There is no cure for any of the herpes viruses.

Herpes simplex virus II (HSV-II): Causes painful sores on the genitals or anus. It is one of the most common sexually transmitted diseases in the United States.

Herpes varicella zoster virus (HVZ): The varicella virus causes chicken pox in children and may reappear in adulthood as herpes zoster. Herpes zoster, also called shingles, is characterized by small, painful blisters on the skin along nerve pathways.

Histoplasmosis: A disease caused by a fungal infection that can affect all the organs of the body. Symptoms usually include fever, shortness of breath, cough, weight loss, and physical exhaustion.

HIV (Human Immunodeficiency Virus): A newly discovered retrovirus that is said to cause AIDS. The target organ of HIV is the T4 or CD4 subset of T lymphocytes, which regulate the immune system.

HIV positive: Presence of the human immunodeficiency virus in the body.

Homophobia: Negative bias toward or fear of individuals who are homosexual.

Human leukocyte antigens (HLA): Protein markers of self used in histocompatibility testing. Some HLA types also correlate with certain autoimmune diseases.

Humoral immunity: The production of antibodies for defense against infection or disease.

Immunity: Resistance to a disease because of a functioning immune system.

Immune complex: A cluster of interlocking antigens and antibodies.

Immune response: The reaction of the immune system to foreign substances.

Immune status: The state of the body's natural defense to diseases. It is influenced by heredity, age, past illness history, diet, and physical and mental health. It includes production of circulating and local antibodies and their mechanism of action.

Immunoassay: The use of antibodies to identify and quantify substances. Often the antibody is linked to a marker such as a fluorescent molecule, a radioactive molecule, or an enzyme.

Immunocompetent: Capable of developing an immune response.

Immunoglobulins: A family of large protein molecules, also known as antibodies.

Immunostimulant: Any agent that will trigger a body's defenses.

Immunosuppression: When the immune system is not working normally. This can be the result of illness or certain drugs (commonly those used to fight cancer).

Incidence: The total number of new cases of a disease in a defined population within a specified time, usually one year.

Incubation period: The time between the actual entry of an infectious agent into the body and the onset of disease symptoms.

Indeterminate test result: A possible result of a Western Blot, which might represent a recent HIV infection or a false positive.

Indinavir: Crixivan, a protease inhibitor drug.

Infection: Invasion of the body by viruses or other organisms.

Infectious disease: A disease that is caused by microorganisms or viruses living in or on the body as parasites.

Inflammatory response: Redness, warmth, and swelling in response to infection; the result of increased blood flow and a gathering of immune cells and secretions.

Injection-drug use: Use of drugs injected by needle into a vein or muscle tissue.

Innate immunity: Inborn or hereditary immunity.

Inoculation: The entry of an infectious organism or virus into the body.

Integrase: HIV enzyme used to insert HIV DNA into host cell DNA.

Interferon: A class of glycoproteins important in immune function and thought to inhibit viral infection.

Interleukins: Chemical messengers that travel from leukocytes to other white blood cells. Some promote cell development, others promote rapid cell division.

Intracellular: Found within the cell wall.

In utero: In the uterus.

In vitro: "In glass"—pertains to a biological reaction in an artificial medium.

In vivo: "In the living"—pertains to a biological reaction in a living organism.

IV: Intravenous.

Kaposi's sarcoma: A multifocal, spreading cancer of connective tissue, principally involving the skin; it usually begins on the toes or the feet as reddish blue or brownish soft nodules and tumors.

Lamivudine: Nucleoside analog inhibits HIV replication.

Langerhans cells: Dendritic cells in the skin that pick up antigen and transport it to lymph nodes.

Latency: A period when a virus or other organism is in the body but in an inactive state.

Latent viral infection: The virion becomes part of the host cell's DNA.

Lentiviruses: Viruses that cause disease very slowly. HIV is believed to be this type of virus.

Lesion: Any abnormal change in tissue due to disease or injury.

Leukocyte: A white blood cell.

Leukopenia: A decrease in the white blood cell count.

Live or attenuated vaccine: A vaccine in which an active virus is weakened through chemical or physical processes in order to produce an immune response without causing the severe effects of the disease. Attenuated vaccines currently licensed in the United States include measles, mumps, rubella, polio, yellow fever, and varicella.

Live-vector vaccine: A vaccine that uses a non-disease-causing organism (virus or bacterium) to transport HIV or other foreign genes into the body, thereby stimulating an effective immune response to the foreign products. This type of vaccine is important because it is particularly capable of inducing cytotoxic leukocyte activity. Examples of organisms used as live vectors in HIV vaccines are canarypox and vaccinia.

Log: 10-fold difference.

Lymph: A transparent, slightly yellow fluid that carries lymphocytes, bathes the body tissues, and drains into the lymphatic vessels.

Lymph nodes: Gland-like structures in the lymphatic system that help to prevent spread of infection.

Lymphadenopathy: Enlargement of the lymph nodes.

Lymphadenopathy syndrome (LAS): A condition characterized by persistent, generalized, enlarged lymph nodes, sometimes with signs of minor illness such as fever and weight loss, which apparently represents a milder reaction to HIV infection.

Lymphatic system: A fluid system of vessels and glands that is important in controlling infections and limiting their spread.

Lymphocytes: Specialized white blood cells involved in the immune response.

Lymphoid organs: The organs of the immune system where lymphocytes develop and congregate. They include the bone marrow, thymus, lymph nodes, spleen, and other clusters of lymphoid tissue.

Lymphokines: Chemical messengers produced by T and B lymphocytes. They have a variety of protective functions.

Lymphoma: Tumor of lymphoid tissue, usually malignant.

Lymphosarcoma: A general term applied to malignant neoplastic disorders of lymphoid tissue, not including Hodgkin's disease.

Lytic infection: When a virus infects the cell, the cell produces new viruses and breaks open (lyse), releasing the viruses.

Macrophage: A large and versatile immune cell that acts as a microbe-devouring phagocyte, an antigen-presenting cell, and an important source of immune secretions.

Major histocompatibility complex (MHC): A group of genes that controls several aspects of the immune response. MHC genes code for self markers on all body cells.

Malaise: A general feeling of discomfort or fatigue.

Malignant tumor: A tumor made up of cancerous cells. The tumors grow and invade surrounding tissue, then the cells break away and grow elsewhere.

Messenger RNA (mRNA): RNA that serves as the template for protein synthesis; it carries the information from the DNA to the protein synthesizing complex to direct protein synthesis.

Microbes: Minute living organisms including bacteria, viruses, fungi, and protozoa.

Microorganisms: Microscopic plants or animals.

Molecule: The smallest amount of a specific chemical substance that can exist alone. To break a molecule down into its constituent atoms is to change its character. A molecule of water, for instance, reverts to oxygen and hydrogen.

Monoclonal antibody: Custom-made, identical antibody that recognizes only one epitope.

Monocyte: A large phagocytic white blood cell which, when it enters tissue, develops into a macrophage.

Monokines: Powerful chemical substances secreted by monocytes and macrophages. They help direct and regulate the immune response.

Morbidity: The proportion of people with a disease in a community.

Morphology: The study of the form and structure of organisms.

Mortality: The number of people who die as a result of a specific cause.

Mucosal immunity: Resistance to infection across mucous membranes.

Mucous membrane: The lining of the canals and cavities of the body that communicate with external air, such as the intestinal tract, respiratory tract, and the genitourinary tract.

Mucous patches: White, patchy growths, usually found in the mouth, that are symptoms of secondary syphilis and are highly infectious.

Mucus: A fluid secreted by membranes.

Mutant: A new strain of a virus or microorganism that arises as a result of change in the genes of an existing strain.

Natural killer cells (also called NK cells): Immune cells that kill infected cells directly within four hours of contact. NK cells differ from other killer cells, such as cytotoxic T lymphocytes, in that they do not require contact with antigen before they are activated.

Neisseria gonorrhoeae: The bacterium that causes gonorrhea.

Neonatal: Pertaining to the first four weeks of life.

Neoplasm: A new abnormal growth, such as a tumor.

Neuropathy: Group of nerve disorders—symptoms range from tingling sensation and numbness to paralysis.

Neutralizing antibody: The kind of antibody that prevents a virus from entering a cell. It is hoped that a vaccine will produce neutralizing antibody because if HIV is prevented from entering cells, it cannot replicate and dies in the bloodstream within a few hours.

Nevirapine: Non-nucleoside analog inhibits HIV replication.

Notifiable disease: A notifiable disease is one that, when diagnosed, health providers are required, usually by law, to report to state or local public health officials. Notifiable diseases are those of public interest by reason of their contagiousness, severity, or frequency.

Nucleic acids: Large, naturally occurring molecules composed of chemical building blocks known as nucleotides. There are two kinds of nucleic acid, DNA and RNA.

Nucleoside analog: Synthetic compounds generally similar to one of the bases of DNA.

Nucleotide of DNA: Made up of one of four nitrogen-containing bases (adenine, cytosine, guanine, or thymine), a sugar, and a phosphate molecule.

Oncogenic: Anything that may give rise to tumors, especially malignant ones.

Opportunistic disease: Disease caused by normally benign microorganisms or viruses that become pathogenic when the immune system is impaired.

P24 antigen: A protein fragment of HIV. The p24 antigen test measures this fragment. A positive test result suggests active HIV replication and may mean the individual has a chance of developing AIDS in the near future.

Pandemic: Occurring over a wide geographic area and affecting a high proportion of the population.

Parenteral: Not taken in through the digestive system or lungs (intravenous, intramuscular, subcutaneous).

Parasite: A plant or animal that lives, grows, and feeds on another living organism.

Pathogen: Any disease-producing microorganism or substance.

Pathogenic: Giving rise to disease or causing symptoms of an illness.

Pathogenicity: Capable of causing a disease.

Pathology: The science of the essential nature of diseases, especially of the structural and functional changes in tissues and organs caused by disease.

Perianal glands: Glands located around the anus.

Perinatal: Occurring in the period during or just after birth.

Pestilence: A virulent, devastating contagious disease that is caused by a bacterium, for example, *Yersina pestis*, which causes the plague.

Phagocytes: Large white blood cells that contribute to the immune defense by ingesting microbes or other cells and foreign particles.

Phenotype: A defined behavior; specifically drug susceptibility with regard to HIV drug resistance.

PID (pelvic inflammatory disease): Inflammation of the female pelvic organs; often the result of gonococcal or chlamydial infection.

Placebo: An inactive substance against which investigational treatments are compared to see how well the treatment worked.

Plague: A calamity; an epidemic of disease causing a high rate of mortality.

Plasma: The fluid portion of the blood that contains all the chemical constituents of whole blood except the cells.

Plasma cells: Derived from B cells, they produce antibodies.

Platelets: Small oval discs in blood that are necessary for blood to clot.

PLWA: Person Living With AIDS.

Polymerase chain reaction: Method to detect and amplify very small amounts of DNA in a sample.

Positive HIV test: A sample of blood that is reactive on an initial ELISA test, reactive on a second ELISA run of the same specimen, and reactive on Western Blot, if available.

***Pneumocystis carinii pneumonia* (PCP):** A rare type of pneumonia primarily found in infants and now common in patients with AIDS.

Prenatal: During pregnancy.

Prevalence: The total number or percentage of cases of a disease existing at any time in a given area.

Primary immune response: Production of antibodies about 7 to 10 days after an infection.

Prime-boost: In HIV vaccine research, administration of one type of vaccine, such as a live-vector vaccine, followed by or together with a second type of vaccine, such as a recombinant subunit vaccine. The intent of this combination regimen is to induce different types of immune responses and enhance the overall immune response, a result that may not occur if only one type of vaccine were to be given for all doses.

Prophylactic treatment: Medical treatment of patients exposed to a disease before the appearance of disease symptoms.

Protease: Enzyme that cuts proteins into peptides (breaks down proteins).

Protease inhibitors: Compounds that inhibit the action of protease.

Proteins: Organic compounds made up of amino acids. Proteins are one of the major constituents of plant and animal cells.

Protocol: Standardization of procedures so that results of treatment or experiments can be compared.

Protozoa: A group of one-celled animals, some of which cause human disease including malaria, sleeping sickness, and diarrhea.

Provirus: The genome of an animal virus integrated into the chromosome of the host cell, and thereby replicated in all the host's daughter cells.

Quasispecies: A complex mixture of genetic variants of an RNA virus.

Race: Beginning in 1976 the federal government's data systems classified individuals into the following racial groups: American Indian or Alaskan Native, Asian or Pacific Islander, black, and white.

Rapid HIV test: A test to detect antibodies to HIV that can be collected and processed within a short interval of time (approximately 3–30 minutes).

Rate: A rate is a measure of some event, disease, or condition in relation to a unit of population, along with some specification of time.

Receptors: Special molecules located on the surface membranes of cells that attract other molecules to attach to them. (For example, CD4, CD8, and CC-CKR-5).

Recombinant DNA: DNA produced by joining pieces of DNA from different sources.

Recombinant DNA techniques: Techniques that allow specific segments of DNA to be isolated and inserted into a bacterium or other host (like yeast or mammalian cells) in a form that will allow the DNA segment to be replicated and expressed as the cellular host multiplies.

Remission: The lessening of the severity of disease or the absence of symptoms over a period of time.

Retroviruses: Viruses that contain RNA and produce a DNA analog of their RNA using an enzyme known as reverse transcriptase.

Reverse transcriptase: An enzyme produced by retroviruses that allows them to produce a DNA analog of their RNA, which may then incorporate into the host cell.

Ritonavir: Novir, a protease inhibitor drug.

RNA (ribonucleic acid): Any of various nucleic acids that contain ribose and uracil as structural components and are associated with the control of cellular chemical activities.

RNA viruses: Contain RNA as their genetic material.

Sarcoma: A form of cancer that occurs in connective tissue, muscle, bone, and cartilage.

Saquinavir: Invirase, a protease inhibitor drug.

Secondary immune response: On repeat exposure to an antigen, there is an accelerated production of antibodies.

Sensitivity: The probability that a test will be positive when the infection is present.

Septicemia: A disease condition in which the infectious agent has spread throughout the lymphatic and blood systems, causing a general body infection.

Seroconversion: The point at which an individual exposed to HIV has detectable antibodies to HIV in their serum.

Serologic test: Laboratory test made on serum.

Serum: The clear portion of any animal liquid separated from its more solid elements, especially the clear liquid that separates in the clotting of blood (blood serum).

Shigella: A bacterium that can cause dysentery.

Specificity: The probability that a test will be negative when the infection is not present.

Spirochete: A corkscrew-shaped bacterium; for example, *Treponema pallidum.*

Spleen: A lymphoid organ in the abdominal cavity that is an important center for immune system activities.

Squamous: Scaly or plate-like; a type of cell.

Statistical significance: The probability that an event or difference occurred as the result of the intervention (vaccine) rather than by chance alone. This probability is determined by using statistical tests to evaluate collected data.

Stavudine: Also known as Zerit; See d4T—inhibits HIV replication.

STD (sexually transmitted disease): Any disease that is transmitted primarily through sexual practices.

Subclinical infections: Infections with minimal or no apparent symptoms.

Subtype: Also called a clade. With respect to HIV isolates, a classification scheme based on genetic differences.

Subunit vaccine: A vaccine that uses only one component of an infectious agent rather than the whole to stimulate an immune response.

Suppressor T cells: A subset of T cells that carry the T8 marker and turn off antibody production and other immune responses.

Surrogate marker: A substitute; a person or agent that replaces another, an alternate.

Surveillance: The process of accumulating information about the incidence and prevalence of disease in an area.

Susceptible: Inability to resist an infection or disease.

Syndrome: A set of symptoms that occur together.

Systemic: Affecting the body as a whole.

T cell growth factor (TCGF, also known as interleukin-2): A glycoprotein that is released by T lymphocytes on stimulation by antigens and that functions as a T cell growth factor by inducing proliferation of activated T cells.

T helper cells (also called T4 or CD4 cells): A subset of T cells that carry the CD4 marker and are essential for turning on antibody production, activating cytotoxic T cells, and initiating many other immune responses.

T lymphocytes or T cells: Lymphocytes that mature in the thymus and that mediate cellular immune reactions. T lymphocytes also release factors that induce proliferation of T lymphocytes and B lymphocytes.

T8 cells: A subset of T cells that may kill virus-infected cells and suppress immune function when the infection is over.

Therapeutic HIV vaccine: A vaccine designed to boost the immune response to HIV in a person already infected with the virus. Also referred to as an immunotherapeutic vaccine.

Thrush: A disease characterized by the formation of whitish spots in the mouth. It is caused by the fungus *Candida albicans* during times of immunosuppression.

Thymus: A primary lymphoid organ high in the chest where T lymphocytes proliferate and mature.

Titer: Level or amount.

Tolerance: A state of nonresponsiveness to a particular antigen or group of antigens.

Toxic reaction: A harmful side effect from a drug; it is dose dependent, that is, becomes more frequent and severe as the drug dose is increased. All drugs have toxic effects if given in a sufficiently large dose.

Toxoplasmosis: An infection with the protozoan *Taxoplasma gondii,* frequently causing focal encephalitis (inflammation of the brain). It may also involve the heart, lungs, adrenal glands, pancreas, and testes.

Transcription: The synthesis of messenger RNA on a DNA template; the resulting RNA sequence is complementary to the DNA sequence. This is the first step in gene expression.

Translation: The process by which the genetic code contained in a nucleotide sequence of messenger RNA directs the synthesis of a specific order of amino acids to produce a protein.

Treponema pallidum: The bacterial spirochete that causes syphilis.

Tropism: Involuntary turning, curving, or attraction to a source of stimulation.

Tumor: A swelling or enlargement; an abnormal mass that can be malignant or benign. It has no useful body function.

V3 loop: Section of the gp120 protein on the surface of HIV; appears to be important in stimulating neutralizing antibodies.

Vaccine: A preparation of dead organisms, attenuated live organisms, live virulent organisms, or parts of microorganisms that is administered to artificially increase immunity to a particular disease.

V.D.: Contagious disease usually acquired through sexual intercourse.

Vector: The means by which a disease is carried from one human to another.

Venereal: Venus = love, sexual desire; involves the sexual organs and related to sexual pleasure; comes through contact of sexual organs.

Venereal warts: Viral *Condylomata acuminata* on or near the anus or genitals.

Viral load: The total amount of virus in a person's blood.

Viremia: The presence of virus in the blood.

Virulence: The quality of expression or the expression of the disease.

Virus: Any of a large group of submicroscopic agents capable of infecting plants, animals, and bacteria; characterized by a total dependence on living cells for reproduction and by a lack of independent metabolism.

Western Blot: A blood test used to detect antibodies to a given antigen. Compared to the ELISA test, the Western Blot is more specific and more expensive. It can be used to confirm the results of the ELISA test.

Wild type: A genotype or phenotype circulating prior to selection of drug resistance.

X-ray: Radiant energy of extremely short wavelength used to diagnose and treat cancer.

Zalcitabine: Also known as HIVID; see ddC—inhibits HIV replication.

Zidovudine: Also known as Retrovir; see ZDV—inhibits HIV replication. Mistakenly referred to as AZT.

References

CHAPTER 1

Barre-Sinoussi, Francoise, et al. (1983). Isolation of a T-lymphocyte retrovirus from a patient at risk for acquired immune deficiency syndrome (AIDS). *Science*, 220:868–871.

Gallo, Robert C. (1987). The AIDS virus. *Sci. Am.*, 256:47–56.

Hahn, Beatrice, et al. (2000). AIDS as a zoonosis: Scientific and public health implications. *Science*, 287:607–614.

Hobson, Simon Wain, et al. (1991). LAV revisited: Origins of the early HIV-1 isolates from Institut Pasteur. *Science*, 252:961–965.

Marlink, Richard. (1996). Lessons from the second AIDS virus HIV-2. *AIDS*, 10:689–699.

Morbidity and Mortality Weekly Report. (1981). Pneumocystis pneumonia Los Angeles, 30:250–252.

Morbidity and Mortality Weekly Report. (1982). Update on acquired immune deficiency syndrome (AIDS) United States, 31:507–508, 513–514.

Morbidity and Mortality Weekly Report. (1990). Surveillance for HIV-2 infection in blood donors United States, 1987–1989, 39:829–831.

Morbidity and Mortality Weekly Report. (1993). 1993 revised classification system for HIV infection and expanded surveillance case definition for AIDS among adolescents and adults, 41:1–19.

Morbidity and Mortality Weekly Report. (1994). Update: Impact of the expanded AIDS surveillance case definition for adolescents and adults on case reporting—United States, 1993, 43:160–170.

Sprecher, Lorrie. (1991). Women with AIDS: Dead but not disabled. *The Positive Woman*, 1:4.

Stadtmauer, Gary, et al. (1997). Primary Immune Deficiency Disorders that mimic AIDS. *Infections in Medicine*, 4:899–905.

CHAPTER 2

Andrews, Charla. (1995). "The Duesberg Phenomenon." What does it mean? *Science*, 267:157.

Bailes, Elizabeth, et al. (2003). Hybrid origin of SIV in chimpanzees. *Science*, 300:1713.

Balter, Michael. (1998). Virus from 1959 sample marks early years of HIV. *Science*, 279:801.

Baum, Rudy. (1995). HIV link to AIDS strengthened by epidemiological study. *Chem. Eng. News*, 74:26.

CDC Weekly. (1988). Extremists seek to blame AIDS on Jews. July 11.

Cherry, Mike. (1999). AZT critics swayed South African president. *Nature*, 402:225.

Cohen, Jon. (1993). Keystone's blunt message: "It's the virus, Stupid". *Science*, 260:292–293.

Connor, Edward. (1994). Reduction of maternal infant transmission of HIV with zidovudine treatment (ACTG 076). *N. Engl. J. Med.*, 331:1173–1180.

Culliton, Barbara J. (1992). The mysterious virus called "Isn't." *Nature*, 358:619.

Darby, Sarah, et al. (1995). Mortality before and after HIV infection in the complete UK population of haemophiliacs. *Nature*, 377:79–82.

Duesberg, Peter H. (1990). Duesberg replies [to the charges of Weiss and Jaffe]. *Nature*, 346:788.

Duesberg, Peter H. (1993). HIV and AIDS. *Science*, 260:1705–1708.

Duesberg, Peter H. (1995a). The Duesberg Phenomenon: Duesberg and other voices. *Science*, 267:313.

Duesberg, Peter H. (1995b). Duesberg on AIDS causation: the culprit is noncontagious risk factors. *The Scientist*, 9:12.

Duesberg, Peter H., et al. (1998). The AIDS Dilemma: drug disneases blamed on a passenger virus. *Genetica*, 104:85–132.

Editorial. (1995). More conviction on HIV and AIDS. *Nature*, 377:1.

Gibbs, Wayt. (2001). Dissident or Don Quixote? *Scientific American*, 285:30–32.

Grmek, Mirko. (1990). *History of AIDS: Emergence and Origin of a Modern Pandemic.* Princeton, NJ: Princeton University Press.

Hahn, Beatrice, et al. (1999). Origin of HIV-1 in the chimpanzee *Pan troglodytes troglodytes. Nature*, 397:436–441.

Hahn, Beatrice, et al. (2000). AIDS as a zoonosis: scientific and public health implications. *Science*, 287: 607–614.

Harris, Steven. (1995). The AIDS heresies: A case study in skepticism taken too far. *Skeptic*, 3, no. 2:42–58.

Hirsch, Vanessa, et al. (1995). Phylogeny and natural history of the primate lentiviruses, SIV and HIV. *Current Opinion Genetic Development*, 5:798–806.

Holder, Constance. (1988). Curbing Soviet disinformation. *Science*, 242:665.

Koprowski, Hilary. (1992). AIDS and the polio vaccine. *Science*, 257:1024–1026.

Korber, Bette, et al. (1998). Limitations of a molecular clock applied to considerations of the origin of HIV-1. *Science*, 280:1868–1871.

KORBER, BETTE, et al. (2000). Timing the ancestor of the HIV-1 pandemic strains. *Science*, 288:1789–1796.

KORBER, BETTE, et al. (2000). Timing the origin of the HIV-1 pandemic. Programs and abstracts of the 7th Conference on Retroviruses and Opportunistic Infections, January 30–February 2, San Francisco. Abstract L5.

LEVY, JAY. (1995). *HIV and the Pathogenesis of AIDS.* Washington, D.C.: ASM Press.

MOORE, JOHN. (1996). À Duesberg, adieu! *Nature*, 380:293–294.

National Institute of Allergy and Infectious Diseases. (1995). The relationship between the human immunodeficiency virus and the acquired immunodeficiency syndrome. *National Institutes of Health*, 1–61.

PEETERS, MARTINE, et al. (2002). Risk to human health from a plethora of simian immunodeficiency viruses in primate bushmeat. *Emerging Infectious Diseases*, 8:451–457.

ROBBINS, KENNETH, et al. (2003). U.S. Human Immonodeficiency Virus type I epidemic: Date of origin, population history, and characterization of early strains. *Virology*, 77:6359–6366.

ROOT-BERNSTEIN, ROBERT. (1993). *Rethinking AIDS.* New York: Free Press.

SANTIGO, MARIO, et al. (2002). SIVcpz in wild chimpanzees. *Science*, 795:465.

SANTIAGO, MARIO, BEATRICE HAHN, et al. (2003). Amplification of a complete simian immunodeficiency virus gnome from fecal RNA of a wild chimpanzee. *Virology*, 77:2233–2242.

STONE, RICHARD. (1995). Congressman uncovers the HIV conspiracy. *Science*, 268:191.

SULLIVAN, JOHN, et al. (1995). HIV and AIDS. *Nature*, 378:10.

VANGROENWEGHE, DANIEL. (2001). The earliest cases of human immunodeficiency virus type 1 group M in Congo-Kinshasa, Rwanda and Burundi and the origin of acquired immune deficiency syndrome. *Philos. Trans. R. Soc. Lond. B. Biol. Sci.;* 356 (1410):923–925.

VIDAL, NICOLE, et al. (2000). Unprecented degree of HIV-1 Group M genetic diversity in the Democratic Republic of Congo suggests HIV-1 pandemic originated in Central Africa. *J. Virology* 74:10,498–10,507.

WEISS, ROBIN A., et al. (1990). Duesberg, HIV and AIDS. *Nature*, 345:659–660.

CHAPTER 3

BONHOEFFER, SEBASTIAN, et al. (1995). Causes of HIV diversity. *Nature*, 376:125.

BRIX, DEBORAH, et al. (1996). Summary of track A: Basic science. *AIDS*, 10 (suppl. 3): S85–S106.

BRODINE, STEPHANIE, et al. (1997). Genotypic variation and molecular epidemiology of HIV. *Infect. Med.*, 14:739–748.

COHEN, JON. (1997). Looking for leads in HIV's battle with immune system. *Science*, 276:1196–1197.

COHEN, MITCHELL, et al. (1994). When bugs outsmart drugs. *Patient Care*, 28:135–146.

COLLINS, KATHLEEN, et al. (1998). HIV-1 Nef protein protects infected primary cells against killing by cytotoxic T Lymphocytes. *Nature*, 391:397–401.

DELWART, ERIC, et al. (1993). Genetic relationships determined by a DNA heteroduplex mobility assay: Analysis of HIV envGenes. *Science*, 262:1257–1262.

DERDEYN, CYNTHIA, et al. (2004). Envelope-constrained naturalization-sensitive HIV-1 after heterosexual transmission. *Science*, 303:2019–2022.

DEVEREUX, HELEN, et al. (2002). In vitro HIV-1 compartmentalisation: drug resistance associated mutation distribution. *J. Med. Virology*, 66:8–12.

DIAZ, RICARDO, et al. (1997). Divergence of HIV quasispecies in an epidemiology cluster. *AIDS*, 11:415–422.

DIMMROCK, N.J., and S.B. PRIMROSE, (1987). *Introduction to Modern Virology*, 3rd ed. Oxford: Blackwell Scientific Publications.

ESSEX, MAX. (1996). Deciphering the mysteries of subtypes: communities of color, Spring/Summer: *Harvard AIDS Rev.*, 18–19.

FIELDS, BERNARD. (1994). AIDS: Time to turn to basic science. *Nature*, 369:95–96.

FISCHL, MARGARET. (1984). Combination retroviral therapy for HIV infection. *Hosp. Pract.*, 29:43–48.

FRITZ, CHRISTIAN, et al. (1995). A human nucleoprotein-like protein that specifically interacts with HIV-Rev. *Nature*, 376:530–533.

GAO FENG, et al. (1999). Origin of HIV-1 in the chimpanzee Pan troglodytes troglodytes. *Nature*, 397:436–441.

GAO FENG, et al. (1992). Human infection by genetically diverse SIVSM-related HIV-2 in west Africa. *Nature*; 358:495–499.

GARRUS, JENNIFER, et al. (2001). Tsg 101 and the vacuolar protein sorting pathway are essential for HIV-1 budding. *Cell*, 107:55–65.

GREENE, WARNER. (1993). AIDS and the immune system. *Sci. Am.*, 269:99–105.

HILDRETH, JAMES. (2001). Adhesion molecules, lipid rafts and HIV pathogenesis. HIV Pathogenesis Keystone Symposium, March 28–April 3, Presentation 038.

HU, DALE, et al. (1996). The emerging genetic diversity of HIV. *JAMA*, 275:210–216.

JETZT, AMANDA, et al. (2000). High rate of recombination throughout the HIV genome. *Journal of Virology*, 74:1234–1240.

KOHLEISEN, MARKUS, et al. (1992). Cellular localization of Nef expressed in persistently HIV-1 infected low-producer astrocytes. *AIDS*, 6:1427–1436.

LENNOX, JEFFREY. (1995). Approaches to gene therapy. *International AIDS Society—USA*, 3:13–16.

LEVIN, BRUCE, et al. (2001). Epidemology, evalution and future of the HIV/AIDS pandemic. *Emerging Infectious Diseases*, 7:505–511.

LI, CHIANG. (1997). Tat protein perpetuates HIV-1 infection. *Proc. Natl. Acad. Sci. USA*, 94:8116–8120.

LUM, JULIAN, et al. (2003). Vpr R77Q is associated with long-term non-progressive HIV and impaired induction of apoptosis. *J. Clin. Invest.*, 111:1547–1554.

MARX, PRESTON, et al. (2001). Serial human passage of simian immunodeficiency Virus by unsterile injections and the emergence of epidemic HIV in Africa. *Philos. Trans. R. Soc. Lond. B. Biol. Sci.*, 356:911–920.

MATSUYA, HIROAKI, et al. (1990). Molecular targets for AIDS therapy. *Science*, 249:1533–1543.

MATTHEWS, STEPHEN, et al. (1994). Structural similarity between p17 matrix protein of HIV and interferon-2. *Nature*, 370:666–668.

MOORE, JOHN, et al. (1994). The who and why of HIV vaccine trials. *Nature*, 372:313–314.

Morbidity and Mortality Weekly Report. (1993). Nosocomial enterococci resistant to vancomycin—United States, 1989–1993, 42:597–599.

NOWAK, MARTIN A. (1990). HIV mutation rate. *Nature*, 347:522.

NOWAK, MARTIN A., et al. (1991). Antigenic diversity thresholds and the development of AIDS. *Science*, 254:963–969.

NOWAK, RACHEL. (1995). How the parasite disguises itself. *Science*, 269:755.

PATRUSKY, BEN. (1992). The Intron story. *Mosaic*, 23:20–33.

POTASH, MARY JANE, et al. (1998). Peptid inhibitors of HIV-1 protease and viral infection of peripheral blood lymphocytes based on HIV-1 ViF. *Proc. National Academy of Science*, 95:13,865–13,868.

PRESTON, BRADLEY D., et al. (1988). Fidelity of HIV-1 reverse transcriptase. *Science*, 242:1168–1171.

ROBERTSON, DAVID, et al. (1995). Recombination in HIV 1. *Nature*, 374:124–126.

ROSEN, CRAIG A. (1991). Regulation of HIV gene expression by RNA-protein interactions. *Trends Genet.*, 7:9–14.

SAGG, MICHAEL S., et al. (1988). Extensive variation of human immunodeficiency virus Type-1 *in vivo. Nature*, 334:440–444.

SAGG, MICHAEL S., et al. (1995). Improving the management of HIV disease. *Advanced Causes in HIV Pathogenesis*, pp. 1–30. February 25, Swissotel, Atlanta (Michael Sagg, Program Chair).

SIMON, FRANCOIS, et al. (1998). Identification of a New human immunodeficiency virus Type I Distinct from Group M and Group O. *Nature Medicine*, 4:1032.

SOMASUNDARAN, M., et al. (1988). Unexpectedly high levels of HIV-1 RNA and protein synthesis in a cytocidal infection. *Science*, 242:1554–1557.

SOTO-RAMIREZ, LUIS, et al. (1996). HIV-Langerhans' cell tropism associated with heterosexual transmission of HIV. *Science*, 271:1291–1293.

STEVENSON, MARIO. (1998). Basic Science: Highlights of the 5th Retrovirus Conference. *Improving the Management of HIV Disease*, 6:4–10.

TORRES, YOLANDA, et al. (1996). Cytokine network and HIV syncytium-inducing phenotype shift. *AIDS*, 10:1053–1055.

UNAIDS. (1997). Implications of HIV variability for transmission: Scientific and policy issues. *AIDS*, 11:1–15.

VARTANIAN, JEAN-PIERRE, et al. (1992). High-resolution structure of an HIV-1 quasispecies: Identification of novel coding sequences. *AIDS*, 6:1095–1098.

WILLS, JOHN W., et al. (1991). Form, function and use of retroviral Gag protein. *AIDS*, 5:639–654.

Workshop Report from the European Commission/Joint United Nations Program on HIV/AIDS. (1997). HIV-1 subtypes: Implications for epidemiology, pathogenicity, vaccines, and diagnostics. *AIDS*, 11:17–36.

WU, YUNTAO, et al. (2001). Selective transcription and modulation of resting T cell activity by preintegrated HIV DNA. *Science*, 293:1503–1506.

CHAPTER 4

AIDS CLINICAL CARE. (1998). Report on the Fifth Conference on Retroviruses and Opportunistic Infections. 10:27–29.

AUTRAN, BRIGITTE, et al. (1997). Positive effects of combined antiretroviral therapy on CD4+ T cell homeostasis and function in advanced HIV disease. *Science*, 277:112–116.

BACK, DAVID. (2001). Pharmacology to the fore. *PRN Notebook*, 6:11–14.

BALTER, MICHAEL. (1997). HIV survives drug on-slaught by hiding out in T cells. *Science*, 278: 1227.

BARTHWELL, ANDREA. (1997). Substance use and the puzzle of adherence. *Focus*, 12:1–4.

BEEKER, AMANDA STEPHEN, et al. (2002). Young HIV-infected adults are at greater risk for medication nonadherence. *Medscape HIV/AIDS eJournal* B(4).

BLANKSTON, JOEL, et al. (2001). Structured therapeutic interruptions: A Review. *Hopkins HIV Report*, 13. 1, 8–9, 13.

BOZZETTE, SAMUEL, et al. (2001). Expenditures for the care of HIV infected patients in the era of HAART. *N. Eng. J. Med.*, 344: 817–823.

British Guidelines Coordinating Committee. (1997). British HIV Association guidelines for antiretroviral treatment of HIV seropositive individuals. *Lancet*, 349:1086–1091.

BURMAN, WILLIAM, et al. (1998). The Case for Conservative Management of Early HIV Disease. *JAMA*, 280:93–95.

CARPENTER, CHARLES, et al. (1997). Antiviral therapy for HIV infection in 1997. *JAMA*, 277:1962–1969.

CARPENTER, CHARLES, et al. (2000). Antiretroviral therapy for Adults: Updated recommendations of the International AIDS Society—USA Panel. *JAMA*, 283:381–390.

CARR, ANDREW, et al. (1998). Pathogenesis of HIV protease inhibitor-associated peripheral lipodystrophy, hyperlipidemia and insulin resistance. *Lancet*, 351:1881–1883.

CARR, ANDREW, et al. (1998a). A syndrome of peripheral lipodystrophy, hyperlipidemia and insulin resistance in patients receiving HIV protease inhibitors. *AIDS*, 12: F51–58.

CARR, ANDREW, et al. (1998b). Lipodystrophy associated with an HIV protease inhibitor. (Images in Clinical Medicine) *N. Engl. J. Med.*, 339:1296.

CARR, ANDREW, et al. (1998c). Abnormal fat distribution and use of protease inhibitors. *Lancet*, 351:1736 (letter).

CARR, ANDREW, et al. (1998d). Pathogenesis of HIV-1 protease inhibitor-associated peripheral lipodystrophy, hyperlipidaemia and insulin resistance. *Lancet*, 351:1881–1883.

CASCADE COLLABORATION (2000). Survival after the introduction of HAART in people with known duration of HIV-1 infection. 355:1158–1159.

CASPER, TONY. (2000). How to make billions out of the misery of millions. Online. Available: http://www.dispatch.co.2a [2000, May 18].

CHUN, TAE-WOOK, et al. (2000). Relationship between pre-existing viral reservoirs and the re-emergence of plasma viremia after discontinuation of highly active anti-retroviral therapy. *Nature Medicine*, 6:757–762.

CLARK, DAWN, et al. (1999). T cell renewal impaired in HIV-1 infected individuals. Presented at: Sixth Conference on Retroviruses and Opportunistic Infections; January 31– February 4, Chicago. Abstract 22.

CLOUGH, LISA, et al. (1999). Factors that predict incomplete virological response to protease inhibitors-based antiretroviral therapy. *Clinical Infections Diseases*. 29:75–81.

CLUMECK, NATHAN. (1995). Summary to the use of saquinavir for HIV therapy. *AIDS*, 9(Suppl. 2):533–534.

COFFIN, JOHN. (1995). HIV population dynamics *in vivo*: Implications for genetic variation, pathogenesis and therapy. *Science*, 267:483–489.

DANNER, SVEN, et al. (1995). Short term study of the safety, pharmacokinetics and efficacy of ritonavir. *N. Engl. J. Med.*, 333:1528–1533.

DEEKS, STEVEN, et al. (1997). Genotypic-resistance assays and antiretroviral therapy. *Lancet*, 349:1489–1490.

DEEKS, STEVEN, et al. (1999). HIV RNA and CD4 cell count response to protease inhibitor therapy in an urban AIDS clinic: Response to both initial and salvage therapy. *AIDS*, 13:F35–F43.

DEEKS, STEVEN, (2003). To switch or not to switch—when is it a question? *PRN Notebook*, 8:12–15.

DE MARTINO, MAURIZIO. (1995). Redox potential status in children with perinatal HIV-1 infection treated with zidovudine. *AIDS*, 9:1381–1383.

DICKOVER, RUTH, et al. (1996). Identification of levels of maternal HIV-RNA associated with risk of perinatal transmission. *JAMA*, 275:599–605.

DOBKIN, JAY. (1997). Fortovase: Son of Invirase. *Infections in Medicine*. 14:926, 934.

D'SOUZA, M. PATRICIA, et al. (2000). Current evidence and future directions for targeting HIV entry. *JAMA*, 284: 215–222.

DUBÉ, MICHAEL, et al. (1997). Protease associated hyperglycaemia. *Lancet*, 350:713–714.

DYBUL, MARK. (2002). HAART interruption strategies: How, when, and why? *PRN Notebook*, 7:10–14.

ERICKSON, JOHN, et al. (1990). Design, activity, and 2.8 angstrom crystal structure of a C_2 symmetric inhibitor complexed to HIV protease. *Science*, 249:527–533.

FAGARD, CATHERINE, et al. (2003). A prospective trial of structured treatment interruptions in human immunodeficiency virus infection. *Archives of Internal Medicine*, 163:1220–1226.

FINZI, DIANA, et al. (1999). Latent infection of CD4(+) T cells provides a mechanism for life-long persistence of HIV-1, even in patients on effective combination therapy. *Nature Medicine*, 5:512–517.

FLEXNER, CHARLES. (1996). Pharmacokinetics and pharmacodynamics of HIV protease inhibitors. *Infect. Med.*, 13:16–23.

FRATER, ALEXANDER, et al. (2002). Comparative response of African HIV-1 infected individuals to highly active antiretroviral therapy. *AIDS*, 16:1139–1146.

FREEDBERG, KENNETH, et al. (2001). The cost effectiveness of combination antiretroviral therapy for HIV disease. *N. Eng. J. Med.*, 344:824–831.

GERBER, JOHN. (1996). Drug interactions with HIV protease inhibitors. *Improv. Manage. HIV Dis.*, 4:20–23.

GERVASONI, CRISTINA, et al. (1999). Redistribution of body fat in HIV-infected women undergoing combined antiretroviral therapy. *AIDS*, 13:465–472.

GOLDSCHMIDT, RONALD, et al. (1995). Antiretroviral strategies revisited. *J. Am. Board Fam. Pract.*, 8:62–69.

GOLDSCHMIDT, RONALD, et al. (1998). Individualized strategies in the era of combination antiretroviral therapy. *J.A.M. Board Family Practice*, 11:158–164.

HELLERSTEIN, MARC, et al. (1999). Directly measured kinetics of circulating T lymphocytes in normal and HIV-infected humans. *Nat. Med.*, 5:83–89.

HENDERSON, DAVID, et al. (2001). HIV postexposure prophylaxis in the 21st Century. *Emerging Infectious Diseases*, 7:254–258.

HENRARD, DENIS, et al. (1995). Natural history of HIV cell-free viremia. *JAMA*, 274:554–558.

HIRSCHEL, BERNARD, et al. (1998). Progress and problems in the fight against AIDS. *N. Engl. J. Med.*, 338:906–908.

HO, DAVID, et al. (1995). Rapid turnover of plasma virions and CD4 lymphocytes in HIV-1 infection. *Nature*, 373:123–126.

HOGG, WOOD, et al. (2003). Effect of medication adherence on survival of HIV infected adults who start HAART when the CD4+ cell count is 200 to 350. *Ann. Intern. Med.*, 139:810–816.

HU, DALE, et al. (1996). The emerging genetic diversity of HIV. *JAMA*, 275:210–216.

JURRIAANS, SUZANNE, et al. (1994). The natural history of HIV infection: Virus load and virus phenotype independent determinants of viral course? *Virology*, 204:223–233.

KAISER FAMILY FOUNDATION. (1999). National monitoring project: *Annual Report* 1–45 and Appendix 1–14.

KATZ, MARK. (2000). Top 10 HIV/AIDS stories of 2000. *Being Alive Newsletter*, Feb 01. http://www.beingalivela.org.

KAUFMANN, DANIEL, et al. (1998). CD4-cell count in HIV-1 infected individuals remaining viraemic with HAART. (Swiss HW Cohort Study) *Lancet*, 351:723–724.

KENYON, GEORGE. (2001). Resistance study to re-evaluate HAART, *Nature*, 7:515.

LAURENCE, JEFFREY. (1996). The clinical promise of HIV protease inhibitors. *AIDS Reader*, 6:39–41, 71.

LAWRENCE, JOHN, et al. (2003). Structured treatment interruption in patients with multi-drug resistant HIV. *N. Engl. J. Med.*, 349: 837–846.

LEE HUANG, SYLVIA, et al. (1999). Lysozyme and RNases as anti-HIV components in beta-core preparations of human chorionic gonadotropin. *Proceedings of the National Academy of Sciences.* 96:2678–2681.

LIPSKY, JAMES. (1998). Abnormal fat accumulation in patients with HIV-I infection. *Lancet*, 351:847–848.

LO, JOAN, et al. (1998). "Buffalo Hump" in Men with HIV-1 infection. *Lancet*, 351:867–870.

MARKOWITZ, MARTIN, et al. (1995). A preliminary study of ritonavir, an inhibitor of HIV protease. *N. Engl. J. Med.*, 333:1534–1539.

MAYERS, DOUGLAS. (1996). Rational approaches to resistance: Nucleoside analogues. *AIDS*, 10 (Suppl. 1):S9–S13.

MELLORS, JOHN, et al. (1995). Quantitation of HIV-1 RNA in plasma predict outcome after seroconversion. *Ann. Intern. Med.*, 122:573–579.

MELLORS, JOHN. (1996). Clinical implications of resistance and cross-resistance to HIV protease inhibitors. *Infect. Med.*, 13:32–38.

MERRICK, SAMUEL. (1997). Managing antiretrovirals in HIV-infected patients. *AIDS Reader*, 7:16–27.

MILLER, KIRK, et al. (1998). Visceral abdominal-fat accumulation associated with use of indinavir. *Lancet*, 351:871–875.

Morbidity and Mortality Weekly Report. (1998). Guidelines for the use of antiretroviral agents in pediatric HIV infection, 47:1–38 No. RR-4.

Morbidity and Mortality Weekly Report. (1994). zidovudine for the prevention of HIV transmission from mother to infant. 43:285–287.

MOYLE, GRACME. (1995). Resistance to antiretroviral compounds: Implications for the clinical management of HIV infection. *Immunol. Infect. Dis.*, 5:170–182.

NIGHTINGALE, STUART. (1996). From the Food and Drug Administration: First protease inhibitor approved. *JAMA*, 275:273.

NOWAK, MARTIN. (1995). How HIV defeats the immune system. *Sci. Am.*, 273:58–64.

O'BRIEN, MEGAN, et al. (2003). Patterns and correlates of discontinuation of the initial HAART regimen in an urban outpatient cohort. *AIDS*, 34:407–414.

PALELLA, FRANK, et al. (1998). Declining morbidity and mortality among patients with advanced HIV infection. *N. Engl. J. Med.*, 338:853–860.

PERRIN, LUC, et al. (1998). HIV treatment failure: Testing for HIV resistance in clinical practice. *Science*, 280:1871–1873.

PIATAK, MICHAEL, et al. (1993). High levels of HIV-1 in plasma during all stages of infection determined by competitive PCR. *Science*, 259:1749–1754.

PINKERTON, STEVEN, et al. (2004). Cost effectiveness of post exposure prophylaxis after sexual or injection drug exposure to HIV. *Arch. Intern. Med.*, 164:46–54.

RICH, JOSIAH, et al. (1999). Misdiagnosis of HIV infection by HIV plasma viral load testing: A case series. *Annals of Internal Medicine Online* (1/5/99) 130:37.

ROBERTS, JOHN, et al. (1988). The accuracy of reverse transcriptase from HIV-1. *Science*, 242: 1171–1173.

SAKSELA, KALLE, et al. (1994). Human immunodeficiency virus type 1 mRNA expression in peripheral blood cells predicts disease progression independently of the number of CF4+ lymphocytes. *Proc. Natl. Acad. Sci. USA*, 91:1104–1108.

SAX, PAUL. (2004). Two new looks at the question of "When to Start?" re-emphasize 200 as the critical threshold. *AIDS Clinical Care*, 16:9–10.

SCHMIT, JEAN-CLAUDE, et al. (1996). Resistance-related mutations in the HIV protease gene of patients treated for 1 year with protease inhibitor ritonavir. *AIDS*, 10:995–999.

SCHRAGER, LEWIS, et al. (1998). Cellular and anatomical reservoirs of HIV in patients receiving potent antiretroviral combination therapy. *JAMA*, 280:67–71.

SIMBERKOFF, MICHAEL. (1996). Long-term follow-up of symptomatic HIV-infected patients originally randomized to early vs. later zidovudine treatment: Report of a Veterans Affairs cooperative study. *AIDS*, 11:142–150.

SIMONI, JANE, et al. (2003). Antiretroviral adherence interventions: A review of current literature and ongoing studies. *Topics in HIV Medicine*, 11:185–198.

STEPHENSON, JOAN. (1996). New anti-HIV drugs and treatment strategies buoy AIDS researchers. *JAMA*, 275:579–580.

TELENTI, AMALIO, et al. for the Swiss HIV Cohort Study. (1998). CD4 T cell counts in HIV-infected individuals remaining viraemic with highly active antiretroviral therapy followed in the Swiss HIV Cohort Study (SHCS). *Antiviral Ther.*, 3(suppl 1):53.

VALDEZ, HERNAN, et al. (1999). Human immunodeficiency virus 1 protease inhibitors in clinical practice. Archives of *Internal Medicine.* 159:1771–1776.

VELLA, STEFANO. (1995). Clinical experience with saquinavir. *AIDS*, 9(suppl 2):S21–S25.

VELLA, STEFANO. (1997). Clinical implications of resistance to antiretroviral drugs. *AIDS Clin. Care*, 9:45–47, 49.

VELLA, STEFANO, et al. (1996). HIV resistance to antiretroviral drugs. *Improv. Manage. HIV Dis.*, 4:15–18.

VIARD, JEAN-PAULA, et al. (2004). Impact of 5 years of maximally successful HAART on CD4 cell count and HIV-1 DNA level. *AIDS*, 18:45–49.

VIGAN, ALESSANDRA, et al. (2003). Increased lipodystrophy is associated with increased exposure to highly active antiretroviral therapy in HIV infected children. *J. Acq. Immune Def. Syndromes*, 15; 32:482–489.

WAIN-HOBSON, SIMON. (1995). Virologies mayhem [editorial]. *Nature*, 373:102.

WALKER, BRUCE, et al. (1998). Treat HIV infection like other infections—Treat it. *JAMA*, 208:91–93.

WEI, XIPING, et al. (1995). Viral dynamics in human immunodeficiency virus type 1 infection. *Nature*, 373:117–122.

WEI, XIPING, et al. (2002). Emergence of resistant human immunodeficiency virus type 1 in patients receiving fusion inhibitor (T-20) monotherapy. *Antimicrob. Agents. Chemother.*, 46(6):1896–1905.

WILLIAMS, CAROLYN, et al. (2004). Persistent GB virus C infection and survival in infected men. *N. Engl. J. Med.*, 350:981–990.

WONG, JOSEPH, et al. (1997). Recovery of replication-competent HIV despite prolonged suppression of plasma viremia. *Science*, 278:1291–1294.

CHAPTER 5

AMEISEN, JEAN CLAUDE. (1994). Programmed cell death apoptosis and cell survival regulation: relavance to cancer. *AIDS*, 8:1197–1213.

AMIGORENA, SEBASTIAN, et al. (1994). Transient accumulation of new class II MHC molecules in a novel endocytic compartment in B lymphocytes. *Nature*, 369:113–120.

BAKKER, LEENDERT J., et al. (1992). Antibodies and complement enhance binding and uptake of HIV-1 by human monocytes. *AIDS*, 6:35–41.

BARR, PHILIP, et al. (1994). Apoptosis and its role in human disease. *BioTechnology*, 12:487–494.

BAXENA, S., et al. (1985). Immunosuppression by human seminal plasma. *Immunol. Invest.*, 14: 255–269.

BERKMAN, S. (1984). Infectious complications of blood transfusions. *Sem. Oncol.*, 11:68–75.

BLEUL, CONRAD, et al. (1996). The lymphocyte chemoattractant SDF-1 is a ligand for LESTR/fusin and blocks HIV entry. *Nature*, 382:829–833.

BLUMBERG, N., et al. (1985). A retrospective study of transfusions. *Br. Med. J.*, 290:1037–1039.

BOLOGNESI, DANI P. (1989). Prospects for prevention of and early intervention against HIV. *JAMA*, 261:3007–3013.

BROWN, PHYLLIDA. (2001). Cinderella goes to the ball. *Nature*, 410:1018–1020.

CALLEBAUT, CHRISTIAN, et al. (1993). T cell activation antigen, CD26 as a cofactor for entry of HIV in CD4+ cells. *Science*, 262:2045–2050.

CHEN, SI-YI, et al. (1997). Second major HIV coreceptor inactivated by intrakine gene therapy. *Proc. Natl. Acad. Sci. USA*, 94:11,567–11,572.

COHEN, JON. (1993). Keystone's blunt message: "It's the virus, Stupid." *Science*, 260:292–293.

COHEN, JON. (1993). HIV cofactor comes in for more heavy fire. *Science*, 262:1971–1972.

COHEN, JON. (1997). Exploiting the HIV-chemokine nexus. *Science*, 275:1261–1264.

CONNER, RUTH, et al. (1994). human immunodeficiency virus type 1 variants with increased replicative capacity develop during the asymptomatic stage before disease progression. *J. Virol.*, 68: 4400–4408.

CORBEIL, JACQUES, et al. (2001). Temperal gene regulation during HIV-1 infection of human CD4+ T cells. *Genome Research*, 11:1198–1204.

DEAN, MICHAEL, et al. (1996). Genetic restriction of HIV infection and progression to AIDS by a deletion allele of the CKR-5 structural gene. *Science*, 273:1856–1861.

DENG, HONGKUI, et al. (1996). Identification of a major co-receptor for primary isolates of HIV-1. *Nature*, 381:661–666.

DOHERTY, PETER. (1995). The keys to cell-mediated immunity. *JAMA*, 274:1067–1068.

DRAGIC, TATJANA, et al. (1996). HIV-1 entry into CD4+ cells is mediated by the chemokine receptor CC-CKR-5. *Nature*, 381:667–673.

EDGINGTON, STEPHEN M. (1993). HIV no longer latent, says NIAID's Fauci. *BioTechnology*, 11:16–17.

EMBRETSON, JANET, et al. (1993). Massive covert infection of helper T lymphocytes and macrophages by HIV during the incubation period of AIDS. *Nature*, 362:359–362.

EUGEN-OLSEN, JESPER, et al. (1997). Heterozygosity for a deletion in the CKR-5 gene leads to prolonged AIDS-free survival and slower CD4 T-cell decline. *AIDS*, 11:305–310.

FAUCI, ANTHONY, et al. (1995). Trapped but still dangerous. *Nature*, 337:680–681.

FITZGERALD, LYNN. (1988). Exercise and the immune system. *Trends Genet.*, 2:1–12.

FOSTER, R., et al. (1985). Adverse effects of blood transfusions in lung cancer. *Cancer*, 55:11,951–12,202.

FOX, CECIL H., et al. (1991). Lymphoid germinal centers for reservoirs of HIV type I RNA. *J. Infect. Dis.*, 164:1051–1057.

FOX, CECIL. (1996). How HIV causes disease. *Carolina Tips*, 59:9–11.

Geijtenbeek, Teunis, et al. (2000). Identification of DC-SIGN, a novel dendritic cell specific ICAM-3 receptor that supports primary immune response. *Cell*, 100:575–585.

Gelderblom, H.R., et al. (1985). Loss of envelope antigens of HTLV III/LAV, a factor in AIDS pathogenesis. *Lancet*, 2:1016–1017.

Groenink, Martijin, et al. (1993). Relation of phenotype evolution of HIV to envelope V2 configuration. *Science*, 260:1513–1516.

Haase, Ashley, et al. (1996). Quantitative image analysis of HIV infection in lymphoid tissue. *Science*, 274:985–990.

Haase, Ashley. (1999). Population biology of HIV-1 infection: Viral and CD4+ T cell demographics and dynamics in lymphatic tissues. *Annu. Rev. Immunol.*, 17:625–656.

Heath, Sonya, et al. (1995). Follicular dendritic cells and HIV infectivity. *Nature*, 377:740–744.

Hengartner, Michael. (1995). Life and death decisions: Ced-9 programmed cell death in *C. elegans*. *Science*, 270:931.

Hill, Mark, et al. (1996). Natural resistance to HIV. *Nature*, 382:668–669.

Ho, David. (1995). Pathogenesis of HIV infection. *International AIDS Society–USA*, 3:9–12.

Ho, David, et al. (1995). Rapid turnover of plasma virons and CD4 lymphocytes in HIV infection. *Nature*, 373:123–126.

Kingsley, L., et al. (1987). Risk factors for seroconversion to HIV among male homosexuals. *Lancet*, 8529:345–348.

Kion, Tracy, et al. (1991). Anti-HIV and Anti-Anti-MHC antibodies in alloimmune and auto-immune mice. *Science*, 253:1138–1140.

Klein, Jan, et al. (2000a). The HLA System: Part I. *NEJM*, 343: 702–709.

Klein, Jan, et al. (2000b). The HLA System: Part II. *NEJM*, 343: 782–786.

Knight, Stella. (1996). Bone-marrow-derived dendritic cells and the pathogenesis of AIDS. *AIDS*, 10:807–817.

Kunal, Saha, et al. (2001). Isolation of primary HIV-1 that target CD8+ lymphocytes using CD8 as a receptor. *Nature Medicine*, 7: 65–72.

Kwong, Peter, et al. (1998). Structure of an HIV gp120 envelope glycoprotein in complex with the CD4 receptor and a neutralizing human antibody. *Nature:* 648–659.

LaPierre, A., et al. (1992). Exercise and health maintenance in AIDS. In Galantino, M.L. (ed.), *Clinical Assessment and Treatment in HIV: Rehabilitation of a Chronic Illness*, Chap. 7. Thorofare NJ: Slack, Inc.

Laurence, Jeffrey. (1996). Where do we go from here? *AIDS Reader*, 6:3–4, 36.

Lee, Sang Kyung. (2002). The functional CD8 T cell response to HIV becomes type–specific in progressive disease. *J. Clin. Invest.*, 110:1339–1347.

Liu, Rong, et al. (1996). Homozygous defect in HIV-1 coreceptor accounts for resistance of some multiple-exposed individuals to HIV-1 infection. *Cell*, 86:367–377.

Marmor, Michael, et al. (2001). Homozygous and heterozygous CCR-5-32 genotypes are associated with resistance to HIV infection. *J. Acq. Immune Def. Syndromes*, 27:472–481.

McDermott, David, et al. (2000). Chemokine promoter polymorphism affects risk of both HIV infection and disease progression in Multicenter AIDS Cohort Study. *AIDS*, 14:2671–2678.

McNicholl, Janet, et al. (1997). Host genes and HIV: The role of the chemokine receptor gene CCR-5 and its allele (Δ32CCR-5). *Emerg. Infect. Dis.*, 3:261–271.

Moir, Susan, et al. (2000). B cells of HIV-1 infected patients bind virons through CD21-complement interactions and transmit infectious virus to activated T cells. *J Exp Med*, 192:637–646. Published online August 28, at http:// www.jem.org.

Moir, Susan, et al. (2001). HIV induces phenotypic and functional perturbations of B cells in chronically infected individuals. *PNAS*, 98:10,362–10,367.

Moore, John. (1997). Coreceptors: Implications for HIV pathogenesis and therapy. *Science*, 276:51–52.

Mountain-Plains Regional HIV/AIDS Curriculum, 4th ed. (1992). Mountain-Plains Regional AIDS Office, University of Colorado Health Sciences Center, Denver, CO 80262.

Nowak, Martin, et al. (1995). HIV results in the frame: Results confirmed. *Nature*, 375:193.

O'Brien, Stephen. (1998). AIDS: A role for host genes. *Hospital Practice*, 33:53–79.

Olinger, Gene, et al. (2000). CD4-negative cells bind HIV-1 and efficiently transfer virus to T cells. *J. Virol.*, 74:8550–8557.

Pantaleo, G., et al. (1993). HIV infection is active and progressive in lymphoid tissue during the clinically latent stage of disease. *Nature*, 362: 355–358.

Pennisi, Elizabeth. (1994). A room of their own. *Science News*, 145:335.

Pianz, Oliver, et al. (1996). Specific cytotoxic T cells eliminate cells producing neutralizing antibodies. *Nature*, 382:726–729.

Pope, Melissa. (2002). Dendritic cells: Immune activators or virus facilitators? *PRN Notebook*, 7:8–10.

Rizzuto, Carlo, et al. (1998). A conserved HIV gp120 glycoprotein structure involved in chemokine receptor binding. *Science*, 280:1949–1953.

Samson, Michel, et al. (1996). Resistance to HIV-1 infection in caucasian individuals bearing mutant alleles of the CCR-5 chemokine gene. *Nature*, 382:722–725.

Schmid, Sandra, et al. (1994). Making class II presentable. *Nature*, 369:103–104.

Schnittman, Steven M., et al. (1989). The reservoir for HIV-1 in human peripheral blood is a T cell that maintains expression of CD4. *Science*, 245:305–308.

SINHA, ANIMESH, et al. (1990). Autoimmune diseases: The failure of self-tolerance. *Science*, 248:1380–1387.

SPRENT, JONATHAN, et al. (1994). Lymphocyte life-span and memory. *Science*, 265:1395–1400.

STEINMAN, RALPH. (2000). DC-SIGN: A guide to some mysteries of dendritic cells. *Cell*, 100:491–494.

STROMINGER, JACK, et al. (1995). The Class I and Class II proteins of the human major histocompatibility complex. *JAMA*, 274:1074–1076.

SUBBRAMANIAN, RAMU, et al. (2002). The presence of ADCC—but not NA—antibodies in serum was associated with viral neutralization in the presence of complement (Comparison of human immunodeficiency virus (HIV)-specific infection enhancing and inhibiting antibodies in AIDS patients. *J. of Clin. Micro.*, 40:2141–2146).

TULP, ABRAHAM, et al. (1994). Isolation and characterization of the intracellular MHC class II compartment. *Nature*, 369:120–126.

UNANUE, EMIL. (1995). The concept of antigen processing and presentation. *JAMA*, 274:1071–1073.

WEI, XIPING, et al. (1995). Viral dynamics in HIV type I infection. *Nature*, 373:117–122.

WEI, XIPING, et al. (2003). Antibody neutralization and escape by HIV–1. *Nature*, 422:307–312.

WEISS, ROBIN, et al. (1996). Hot fusion of HIV. *Nature*, 381:647–648.

WITKIN, S., and SONNABEND, J. (1983). Immune responses to spermatozoa in homosexual men. *Fertil. Steril.*, 39:337–341.

WYATT, RICHARD, et al. (1998). The antigenic structure of the HIV gp120 envelope glycoprotein. *Nature*, 705–711.

YEH, EDWARD. (1998). Life and death of a cell. *Hospital Practice*, 33:85–92.

ZHU, TOUFU. (1993). Genotypic and phenotypic characterization of HIV-1 patients with primary infection. *Science*, 261:1179–1181.

ZINKERNAGEL, ROLF. (1995). MHC-restricted T-cell recognition: The basis of immune surveillance. *JAMA*, 274:1069–1071.

CHAPTER 6

AMIN, NAVIN M. (1987). Acquired immunodeficiency syndrome, Part 2: The spectrum of disease. *Fam. Pract. Recert.*, 9:84–118.

AWE, ROBERT J. (1988). Benefits, promises and limitations of zidovudine (AZT). *Consultant*, 28:57–72.

BALFOUR, HENRY. (1995). Cytomegolovirus retinitis in persons with AIDS. *Postgrad. Med.*, 97:109–118.

BUCHANAN, KENT, et al. (1998). What makes *Cryptococcus neoformans* a pathogen? *Emerg. Infect. Dis.*, 4:71–83.

CANNON, MICHAEL, et al. (2003). Risk factors for Kaposi's sarcoma in men seropositive for both human herpes virus 8 and human immunodeficiency virus. *AIDS*, 17:215–222.

CHANG, YUAN, et al. (1994). Identification of Herpes virus-like DNA sequences in AIDS-Associated Kaposi's Sarcoma. *Science*, 266:1865–1869.

CHIN, DANIEL. (1992). Mycobacterium avium complex infection. *AIDS File: Clin. Notes*, 6:7–8.

Coalition News. (1993). Pet guidelines for people with HIV. 2:4–5.

CURRIER, JUDITH, et al. (1997). Pathogenesis, prevention and treatment of opportunistic complications. *Improv. Manage. HIV Dis.*, 4: S17–S18.

DALEY, CHARLES L. (1992). Epidemiology of tuberculosis in the AIDS era. *AIDS File: Clin. Notes*, 6:1–2.

DANNENBERG, ARTHUR M. (1993). Immunopathogenesis of pulmonary tuberculosis. *Hosp. Pract.*, 28:51–58.

DEWIT, STEPHANE, et al. (1991). Fungal infections in AIDS patients. *Clin. Adv. Treatment Fungal Infect.*, 2:1–11.

ENNIS, DAVID M., et al. (1993). Cryptococcal meningitis in AIDS. *Hosp. Pract.*, 28:99–112.

Emergency Medicine. (1989). Fighting opportunistic infections in AIDS. 21:24–38.

ERNST, JEROME. (1990). Recognize the early symptoms of PCP. *Med. Asp. Hum. Sexuality*, 24:45–47.

EZZELL, CAROL. (1993). Captain of the men of death. *Science News*, 143:90–92.

GANEM, DONALD. (1996) Kaposi's sarcoma-associated herpesvirus. *Improv. Manage. HIV Dis.*, 4(3):8–10.

GOTTLIEB, MICHAEL S., et al. (1987). Opportunistic viruses in AIDS. *Patient Care*, 23:139–154.

GROSS, DAVID J., et al. (1989). Update on AIDS. *Hosp. Pract.*, 25:19–47.

GROSSMAN, RONALD J., et al. (1989). PCP and other protozoal infections. *Patient Care*, 23:89–116.

GRULICH, ANDREW. (2000). Cancer risk in persons with HIV/AIDS in the era of combination antiretroviral therapy. *AIDS Reader*, 10:341–346.

GUARINO, M., et al. (1995). Progressive multifocal leucoencephalopathy in AIDS: Treatment with cytosine arabinoside. *AIDS*, 9:819–820.

Guidelines, U.S. Public Health Service. (1995). Preventing OIs in persons with HIV disease. *AIDS Reader*, 5:172–179.

HARDEN, C.L., et al. (1994). Diagnosis of central nervous system toxoplasmosis in AIDS patients confirmed by autopsy. *AIDS*, 8:1188–1189.

HARRIS, CHARLES. (1993). TB and HIV: The boundaries collide. *Medical World News*, 34:63.

Harvard AIDS Institute. (1994). *Special Report—Opportunistic Infections.* Fall issue:1–14.

HERNDIER, BRIAN, et al. (1994). Pathogenesis of AIDS lymphomas. *AIDS*, 8:1025–1049.

HESSOL, NANCY. (1998). The changing epidemology of HIV related cancers. *The AIDS Reader*, 8:45–49.

HUGHES, WALTER. (1994). Opportunistic infections in AIDS patients. *Postgrad. Med.*, 95:81–86.

JACOBSON, MARK A., et al. (1988). Serious cytomegalovirus disease in the acquired immunodeficiency syndrome (AIDS): Clinical findings, diagnosis, and treatment. *Ann. Intern. Med.*, 108:585–594.

JOINER, K.A., et al. (1990). Toxoplasma gondii: Fusion competence of parasitophorous vacuoles in Fe receptor-transfected fibroblasts. *J. Cell Biol.*, 109:2771.

KLEDAL, THOMAS, et al. (1997). A broad spectrum chemokine antagonist encoded by KS-associated herpesvirus. *Science*, 277:1656–1659.

KOEHLER, CHRISTOPHER. (2002). Consumption, the great killer. *Modern Drug Discovery*, 5:47–48.

LAURENCE, JEFFREY. (1995). Evolving management of OIs. *AIDS Reader*, 5:187–188, 208.

LAURENCE, JEFFREY. (1996). Where do we go from here? *AIDS Reader*, 6:3–4, 36.

LEDERGERBER, BRUNO, et al. (1999). AIDS-related OI occurring after initiation of potent antiretroviral therapy: A Swiss cohort study. *JAMA*, 282:2220–2226.

LOONEY, DAVID, (1996). Kaposi's sarcoma. *Improv. Manage. HIV Dis.*, 4:21–24.

LYNCH, JOSEPH P. (1989). When opportunistic viruses infiltrate the lung. *J. Resp. Dis.*, 10:25–30.

MCGRATH, MICHAEL, et al. (1994). Identification of a common clonal human immunodeficiency virus integration site in human immunodeficiency virus-associated lymphomas. *Cancer Res.*, 54:2069.

MEDOFF, GERALD, et al. (1991). Systemic fungal infections: An overview. *Hosp. Pract.*, 26:41–52.

MONETTE, PAUL. (1988). *Borrowed Time: An AIDS Memoir.* New York: Avon Books.

MONFORTE, ANTONELLA D'ARMINIO, et al. (1992). AIDS-defining diseases in 250 HIV-infected patients: A comparative study of clinical and autopsy diagnoses. *AIDS*, 6:1159–1164.

Morbidity and Mortality Weekly Report. (1993). Estimates of future global TB morbidity and mortality. 4:961–964.

Morbidity and Mortality Weekly Report. (1995). USPHS/IDSA guidelines for the prevention of opportunistic infections in persons infected with HIV: A summary. 44:1–34.

Morbidity and Mortality Weekly Report. (1995). 1995 revised guidelines for prophylaxis against PCP for children infected with or perinatally exposed to HIV. 44:1–10.

Morbidity and Mortality Weekly Report. (1997). 1997 USPHS/IDSA guidelines for the prevention of OIs in persons infected with HIV. 46:1–46.

Morbidity and Mortality Weekly Report. (1999). USPHS/IDSA guidelines for the prevention of opportunistic infections in persons infected with HIV. 48:1–59.

MURPHY, ROBERT. (1994). Opportunistic infection prophylaxis. *Int. AIDS Soc.–USA*, 2:7–8.

NEWTON, HERBERT. (1995). Common neurologic complications of HIV infection and AIDS. *Am. Fam. Phys.*, 51:387–398.

OSMOND, DENNIS, et al. (2002). Prevalance of Kaposi sarcoma-associated herpes virus infection in homosexual men at the beginning of and during the HIV epidemic. *JAMA*, 287:221–225.

PHAIR, JOHN, et al. (1990). The risk of *Pneumocystis carinii* among men infected with HIV-1. *N. Engl. J. Med.*, 322:161–165.

POWDERLY, WILLIAM G., et al. (1992). Molecular typing of Candida albicans isolated from oral lesions of HIV-infected individuals. *AIDS*, 6:81–84.

POWDERLY, WILLIAM, et al. (1998). Recovery of the immune system with antiretroviral therapy: the end of opportunism? *JAMA*, 280:72–77.

RINALD, CHARLES, et al. (2001). Primary human herpesvirus 8 infection generates a broadly specific CD8+ T cell response to viral lytic cycle proteins. *Blood*, 97:2366–2373.

ROSSITCH, EUGENE, et al. (1990). Cerebral toxoplasmosis in patients with AIDS. *Am. Fam. Pract.*, 41:867–873.

RUSSELL, JAMES. (1990). Study focuses on eyes and AIDS. *Baylor Med.*, 21:3.

SAID, JONATHAN, et al. (1997). KS-associated herpesvirus/ human herpesvirus type 8 encephalitis in HIV-positive and -negative individuals. *AIDS*, 11:1119–1122.

SCADDEN, DAVID. (2002). Lymphoma in the setting of HIV disease. *PRN Notebook*, 7:21–25.

SEPKOWITZ, KENT. (1998). Effect of HAART on natural history of AIDS-related opportunistic disorders. *Lancet*, 351:228–230.

SIBLEY, L. DAVID. (1992). Virulent strains of *Toxoplasma gondii* comprise single clonal linage. *Nature*, 359:82–85.

SMALL, PETER. (1996). Tuberculosis research: Balancing the portfolio. *JAMA*, 276:1512–1513.

SOLOWAY, BRUCE. (1998). Report on the Fifth Conference on Retroviruses and Opportunistic Infections. *AIDS Clinical Care*, 10:27–29.

STANFORD, J.L., et al. (1993). Old plague, new plague, and a treatment for both? *AIDS*, 7:1275–1276.

STRINGER, JAMES, et al. (1996). Molecular biology and epidemiology of Pneumocystis carinii infections in AIDS. *AIDS*, 10:561–571.

TUAZON, CARMELITA, et al. (1991). Diagnosing and treating opportunistic CNS infections in patients with AIDS. *Drug Therapy*, 21:43–53.

United States Public Health Service and Infectious Diseases Society of America. (2001). Guidelines for the prevention of opportunistic infections in persons infected with human immunodeficiency virus. *MMWR*, November 28, 1–65.

WALLACE, MARK R., et al. (1993). Cats and toxoplasmosis risk in HIV-infected adults. *JAMA*, 269:76–77.

WALZER, PETER D. (1993). *Pneumocystis carinii*: Recent advances in basic biology and their clinical application. *AIDS*, 7:1293–1305.

WHEAT, L. JOSEPH. (1992). Histoplasmosis in AIDS. *AIDS Clin. Care*, 4:1–4.

CHAPTER 7

ABOULKER, JEAN-PIERRE, et al. (1993). Preliminary analysis of the concorde trial. *Lancet*, 341:889–890.

ANDERSON, ROBERT E., et al. (1991). CD8 T lymphocytes and progression to AIDS in HIV-infected men: Some observations. *AIDS*, 5:213–215.

BALTER, MICHAEL. (1998). Chemokine mutation slows progression. *Science*, 279:327.

BALTIMORE, DAVID. (1995). Lessons from people with nonprogressive HIV infection. *N. Engl. J. Med.*, 332: 259–260.

BARNHART, HUIMAN, et al. (1995). *Abstracts of the 2nd National Conference on Human Retroviruses*, Washington, D.C., p. 161, abstr. 575.

BARTLETT, JOHN, et al. (1998). Primary care of HIV infection. *Hospital Practice*, 33:53–55.

BOLOGNESI, DANI P. (1989). Prospects for prevention of and early intervention against HIV. *JAMA*, 261:3007–3013.

BRYSON, YVONNE J., et al. (1995). Clearance of HIV infection in a prenatally infected infant. *N. Engl. J. Med.*, 332:833–838.

BUCHBINDER, SUSAN P., et al. (1992). Healthy long-term positives: Men infected with HIV for more than 10 years with CD4 counts of 500 cells. *Eighth International Conference on AIDS*, Amsterdam, July 1992, abstr. TUCO572.

BUCHBINDER, SUSAN P., et al. (1994). Long-term HIV infection without immunologic progression. *AIDS*, 8:1123–1128.

BUCY, R. PAT. (1999). Viral and cellular dynamics in HIV-1 disease. *Improving Management of HIV Disease*, 7:8–11.

BURCHAM, JOYCE, et al. (1991). CD4 % is the best predictor of development of AIDS in a cohort of HIV-infected homosexual men. *AIDS*, 5:365–372.

CAO, YUNZHEN, et al. (1995). Virologic and immunologic characterization of long-term survivors of human immunodeficiency virus type 1 infection. *N. Engl. J. Med.*, 332:201–208.

CATALAN, JOSÉ, et al. (2001). The changing picture of HIV: A chronic illness, again? *Focus*, 16:1–4.

Centers for Disease Control and Prevention. (2001). Report of NIH panel to define principles of therapy of HIV infection and guidelines for the use of antiretroviral agents in HIV infected adults and adolescents. Updated February 5, as a living document: www.hivatis.org.

CHUN, TAE-WOOK, et al. (1998). Early establishment of a pool of latently infected, resting CD4+ T cells during primary HIV-1 infection. *PNAS*, 95:8869–8873.

CLERICI, MARIO, et al. (2003). Mucosal and systemic HIV specific immunity in HIV exposed but uninfected heterosexual men. *AIDS*, 17:531–539.

COCCHI, FIORNZA, et al. (1995). Identification of RANTES, MIP-1a and MIP-1b as the major HIV-suppressive factors produced by CD8+ T cells. *Science*, 270:1811–1815.

COFFIN, JOHN M. (1995). HIV population dynamics in vivo: Implications for genetic variation, pathogenesis and therapy. *Science*, 267:483–489.

COHEN, JON. (1993). Keystone's blunt message: "It's the virus, Stupid." *Science*, 260:292–293.

COHEN, JON. (1995). High turnover of HIV in blood revealed by new studies. *Science*, 267:179.

COLLMAN, RONALD. (1997). Effect of CCR-2 and CCR-5 variants on HIV disease. *JAMA*, 278:2113–2114.

CONANT, MARCUS. (1995). The current face of the AIDS epidemic. *AIDS Newslink*, 6(Fall):1–9.

COULIS, PAUL A., et al. (1987). Peptide-based immunodiagnosis of retrovirus infections. *Am. Clin. Prod. Rev.*, 6:34–43.

DAVIS, SUSAN, et al. (1995). Prevalence and incidence of vertically acquired HIV infection in the U.S.A. *JAMA*, 274:952–955.

DEACON, NICHOLAS, et al. (1995). Genomic structure of an attenuated quasi species of HIV from a blood transfusion donor and recipients. *Science*, 270:988–991.

ENSOLI, F., et al. (1990). Proviral sequences detection of human immunodeficiency virus in seronegative subjects by polymerase chain reaction. *Mol. Cell Probes*, 4:153–161.

ENSOLI, F., et al. (1991). Plasma viraemia in seronegative HIV-1 infected individuals. *AIDS*, 5:1195–1199.

ESCAICH, SONIA, et al. (1991). Plasma viraemia as a marker of viral replication in HIV-infected individuals. *AIDS*, 5:1189–1194.

FAUCI, ANTHONY S. (1988). The scientific agenda for AIDS. *Issues Sci. Technol.*, 4:33–42.

FENG, YU, et al. (1996). HIV-1 entry cofactor: Functional cDNA cloning of a seven-transmembrane, G protein-coupled receptor. *Science*, 272:872–877.

FREED, ERIC, et al. (1994). HIV infection of non-dividing cells. *Nature*, 369:107–108.

FRENKEL, LISA, et al. (1998). Genetic evaluation of suspected cases of transient HIV infection of infants. *Science*, 280:1073–1077.

GEGNEY, THOMAS. (2000). Long term nonprogressors: The study of HIV infection without progression to AIDS. Research Initiative Treatment Alert (RITA), 6, no. 2. June 2000. http://www.aegis.org/pubs/rita/ 2000/RI00064.html

GOLDSCHMIDT, RONALD, et al. (1997). Treatment of AIDS and HIV-related conditions—1997. *J. Am. Board Fam. Pract.*, 10:144–167.

GRUBMAN, SAMUEL, et al. (1995). Older children and adolescents living with perinatally acquired HIV infection. *Pediatrics*, 95:657–663.

HAAS, DAVID, et al. (2000). Evidence of a source of HIV-1 within the central nervous system by ultraintensive sampling of cerebrospinal fluid and plasma. *AIDS Res. Hum. Retroviruses*, 16:1491–1502.

HANLEY, DANIEL F., et al. (1988). When to suspect viral encephalitis. *Patient Care*, 22:77–99.

HAY, CHRISTINE. (1998). Immunologic response to HIV. *AIDS Clinical Care*, 10:1–3.

HAYNES, BARTON, et al. (1996). Toward an understanding of the correlates of protective immunity to HIV infection. *Science*, 271:324–328.

HENRARD, DENIS, et al. (1995). Natural history of HIV cell-free viremia, *JAMA*, 274:554–558.

HO, DAVID, et al. (1995). Rapid turnover of plasma virons and CD4 lymphocytes in HIV infection. *Nature*, 373:123–126.

HO, DAVID. (1996). HIV pathogenesis. *Improv. Manage. HIV Dis.*, 4:4–6.

HORSBURG, C.R., et al. (1989). Duration of HIV infection before detection of antibody. *Lancet*, ii:637–639.

IMAGAWA, D.T., et al. (1989). Human immunodeficiency virus type I infection in homosexual men who remain seronegative for prolonged periods. *N. Engl. J. Med.*, 320:1458–1462.

Italian Register for HIV Infection in Children. (1994). *Lancet*, 343:191–195.

JOHNSON, RICHARD, et al. (1996). Quantitation of human immunodeficiency virus in brains of demented and nondemented patients with acquired immunodeficiency syndrome. *Ann. Neurol.* 39:392–395.

KATZENSTEIN, TERESE, et al. (1996). Longitudinal Serum HIV RNA quantification: Correlation to viral phenotype at seroconversion and clinical outcome. *AIDS*, 10:167–173.

KELLY, MAUREEN, et al. (1991). Oral manifestations of human immunodeficiency virus infection. *Cutis*, 47:44–49.

KIRCHHOFF, FRANK, et al. (1995). Brief report: Absence of intact NEF sequences in a long-term survivor with nonprogressive HIV-1 infection. *N. Engl. J. Med.*, 332:228–232.

KLINE, MARK. (1995). Long-term survival in vertically acquired HIV infection. *AIDS Reader*, 5:153.

LEARMONT, JENNIFER, et al. (1999). Immunological and virologic status after 14 to 18 years of infection with an attenuated strain of HIV-1. *N. Engl. J. Med.*, 340:1715–1722.

LEVY, JAY. (1995). HIV and long-term survival. *Int. AIDS Soc. USA*, 3:10–12.

LIPTON, STUART. (1997). Treating AIDS dementia. *Science*, 276:1629–1630.

MCARTHUR, JUSTIN. (1999). Declining incidence of neurologic complications of HIV disease. *Hopkins HIV Report*, 11:8.

MCGUIRE, DAWN. (1993). Pathogenesis of brain injury in HIV disease. *Clin. Notes*, 7:1–11.

MERIGAN, THOMAS, et al. (1996). The prognostic significance of viral load, codon 215-reverse transcriptase mutation and CD4+ T cells on HIV disease progression. *AIDS*, 10:159–165.

MIGUELES, STEPHEN, et al. (2000). HLA B5701 is highly associated with restriction of virus replication in a subgroup of HIV infected long term nonprogressors. *Pro. Natl. Acad. Sci USA*, 97: 2709–2714.

MIGUELES, STEPHEN, et al. (2002). HIV-specific CD8+ T cell proliferation is coupled to perforin expression and maintained in nonprogressors. *Nature Immunology*, 3:1061–1068.

MISHRAHI, MICHELINE, et al. (1998). CCR-5 chemokine receptor variant in HIV-1 mother to child transmission and disease progression. *JAMA*, 279:277–280.

Morbidity and Mortality Weekly Report. (1996). Persistent lack of detectable HIV-antibody in a person with HIV-infection—Utah, 1995. 45:181–185.

Morbidity and Mortality Weekly Report. (1997). Revised guidelines for performing CD4+ T-cell determinations in persons infected with HIV. 46:1–4.

National Institute of Allergy and Infectious Diseases. (1989). Tests confirm lack of mental impairment in asymptomatic HIV-infected homosexual men. June: 1–2.

NEWELL, MARIE, et al. (1996). Detection of virus in vertically exposed HIV-antibody-negative children. *Lancet*, 347:213–215.

NIELSON, CLAUS, et al. (1993). Biological properties of HIV isolates in primary HIV infection: Consequences for the subsequent course of infection. *AIDS*, 7:1035–1040.

NOWAK, M.A., et al. (1990). The evolutionary dynamics of HIV-1 quasispecies and the development of immunodeficiency disease. *AIDS*, 4:1095–1103.

PANTALEO, GUISEPPE, et al. (1993). The immunopathogenesis of HIV infection. *N. Engl. J. Med.*, 328:327–335.

PANTALEO, GUISEPPE, et al. (1995). Studies in subjects with long-term nonprogressive human immunodeficiency virus infection. *N. Engl. J. Med.*, 332:209–216.

PEDERSEN, C., et al. (1989). Clinical course of primary HIV infection: Consequences for subsequent course of infection. *Br. Med. J.*, 299:154–157.

PERELSON, ALAN, et al. (1996). HIV dynamics in vivo: Viron clearance rate, infected cell life-span and viral generation time. *Science*, 271:1582–1586.

PHILLIPS, ANDREW N., et al. (1991a). p24 Anti-genaemia, CD4 lymphocyte counts and the development of AIDS. *AIDS*, 5:1217–1222.

PHILLIPS, ANDREW N., et al. (1991b). Serial CD4 lymphocyte counts and development of AIDS. *Lancet*, 337:389–392.

PHILLIPS, ANDREW, et al. (1994). A sex comparison of rates of new AIDS-defining disease and death in 2554 AIDS cases. *AIDS*, 8:831–835.

PRICE, RICHARD W. (1988). The brain in AIDS: Central nervous system HIV infection and AIDS dementia complex. *Science*, 239:586–593.

QUINN, THOMAS. (1997). Acute primary HIV infection. *JAMA*, 278:58–62.

RANKI, A., et al. (1987). Long latency precedes overt seroconversion in sexually transmitted human immunodeficiency virus infection. *Lancet*, ii:589–593.

RANKI, ANNAMARI, et al. (1995). Abundant expression of HIV NEF and Rev proteins in brain astrocytes *in vivo* is associated with dementia. *AIDS*, 9:1001–1008.

REYNES, JACQUES, et al. (2001). CD4 T cell surface CCR-5 density as a host factor in HIV-1 disease progression. *AIDS*, 15:1627–1634.

RICHMAN, DOUGLAS. (1995). Antiretroviral resistance and HIV dynamics. *Int. AIDS Soc.*, 3:15–16.

ROQUES, PIERRE, et al. (1995). Clearance of HIV infection in 12 perinatally infected children: Clinical, virological and immunological data. *AIDS*, 9:F19–F26.

ROSENBERG, PHILIP, et al. (1994). Declining age at HIV infection in the United States. *N. Engl. J. Med.*, 330:789–790.

ROYCE, RACHEL A., et al. (1991). The natural history of HIV-1 infection: Staging classifications of disease. *AIDS*, 5:355–364.

SAX, PAUL, et al. (1995). Potential clinical implications of interlaboratory variability in CD4+ T-lymphocyte counts of patients infected with human immunodeficiency virus. *Clin. Infect. Dis.*, 21:1121–1125.

SCHONNING, KRISTIAN, et al. (1998). Chemokine receptor polymorphism and autologous neutralizing antibody response in long-term HIV infection. *J. acquired immune deficiency syndrome and Human Retrovirology*, 18:195–202.

SHERNOFF, MICHAEL. (1997). A history of hope: The HIV roller coaster. *Focus*, 12:5–7.

SKURNICK, JOAN, et al. (2002). Correlates of nontransmission in US women at high risk of HIV-1 infection through sexual exposure. *J. Infect. Dis.*, 185:428–438.

SOLOWAY, BRUCE, et al. (2000). Antiretroviral failure: A biopsychosocial approach. *AIDS Clinical Care*, 12:23–25, 30.

STRAMER, SUSAN L., et al. (1989). Markers of HIV infection prior to IgG antibody seropositivity. *JAMA*, 262:64–69.

The European Collaborative Study. (1994). *Pediatrics*, 94:815–819.

VANHEMS, PHILLIPPE, et al. (1999). Recognizing primary HIV infection. *Infections in Medicine*, 16:104–108, 110.

VOELKER, REBECCA. (1995). New studies say viral burden tops CD4 as a marker of HIV-disease progression. *JAMA*, 275:421–422.

WAIN-HOBSON, SIMON. (1995). Virological mayhem. *Nature*, 373:102.

WEI, XIPING, et al. (1995). Viral dynamics in HIV type 1 infection. *Nature*, 373:117–122.

WINKLER, CHERYL, et al. (1998). Genetic restriction of AIDS pathogenesis by an SDF-1 chemokine gene variant. *Science*, 279:389–393.

YAN XU, et al. (2004). HIV-1 mediated apoptosis of neuronal cells: Proximal molecular mechanisms of HIV-1 induced encephalopathy. *PNAS*, 101:7070–7075.

YU, KALVIN, et al. (2000). Primary HIV infection. *Postgraduate Medicine*, 107:114–122.

CHAPTER 8

ABBOTT, ALISON. (1995). Murder charges brought in German HIV blood products case. *Nature*, 376:628.

ABBOTT, ALISON. (1996). Japan agrees to pay HIV-blood victims. *Nature*, 380:278.

ALDHOUS, PETER. (1991). France will compensate. *Nature*, 353:425.

AMIRKHANIAN, YURI, et al. (2001). AIDS knowledge, attitudes and behavior in Russia: Results of a population based, random-digit telephone survey in St. Petersburg. *Int. J. STD AIDS*, 12:50–57.

ARCHIBALD, D.W., et al. (1990). *In vitro* inhibition of HIV-1 infectivity by human salivas. *AIDS Res. Human Viruses*, 6:1425–1431.

BAGASRA, OMAR, et al. (1994). Detection of HIV proviral DNA in sperm from HIV-infected men. *AIDS*, 8:1669–1674.

BARON, SAMUEL, et al. (1999). Why is HIV rarely transmitted by oral secretions? Saliva can disrupt orally shed, infected leukocytes. *Archives of Internal Medicine*, 159:303–310.

BENNETT, AMANDA, et al. (1996). AIDS fight is skewed by federal campaign exaggerating risks. *Wall Street Journal*, May 1, A1.

BOLLING, DAVID R. (1989). Anal intercourse between women and bisexual men. *Med. Asp. Human Sexuality*, 23:34.

BOYER, PAMELA J., et al. (1994). Factors predictive of maternal-fetal transmission of HIV. *JAMA*, 271:1925–1930.

BREO, DENNIS L. (1991). The two major scandals in France's AIDS-GATE. *JAMA*, 266:3477–3482.

BURKETT, ELINOR. (1995). The gravest show on earth: America in the age of AIDS. Boston: Houghton Miflin.

BUTLER, DECLAN. (1994a). Allain freed to face new charges? *Nature*, 370:404.

BUTLER, DECLAN. (1994b). Blood scandal raises spectre of Dreyfus case. *Nature*, 371:548.

CAMERON, WILLIAM D. (1989). Female to male transmission of human immunodeficiency virus type 1: risk factors for seroconversion in men. *Lancet*, 2:403–407.

CDC (Centers for Disease Control and Prevention). (1990). *HIV/AIDS Surveillance Report*, Oct.:1–18.

Centers for Disease Control and Prevention. AIDS among persons aged >50 years—US 1991–1996. *MMWR*, 47:21–27.

CHU, S.Y., et al. (1990). Epidemiology of reported cases of AIDS in lesbians, United States 1980–89. *Am. J. Public Health*, 80:1380.

COHEN, J.B., et al. (1989). Heterosexual transmission of HIV. *Immunol. Ser.*, 44:135–137.

COHEN, JON. (1995). Bringing AZT to poor countries. *Science*, 269:624–626.

CONANT, MARCUS. (1995). The current face of the AIDS epidemic. *AIDS Newslink*, 6:14–18.

CONNER, EDWARD, et al. (1994). Reduction of maternal-infant transmission of HIV with zidovudine treatment. *N. Engl. J. Med.*, 331:1173–1180.

COOMBS, ROBERT W., et al. (1989). Plasma viremia in HIV infection. *N. Engl. J. Med.*, 321:1526.

COTTONE, JAMES A., et al. (1990). The Kimberly Bergalis case: An analysis of the data suggesting the possible transmission of HIV infection from a dentist to his patient. *Phys. Assoc. AIDS Care*, 2:267–270.

COUTSOUDIS, ANNA, et al. (1999). Influence of infant-feeding patterns on early mother-to-child transmission of HIV-1 in Durban, South Africa: a prospective cohort study. *Lancet*, 354:9177, 9471.

COUTSOUDIS, ANNA, et al. (2001). Method of feeding and transmission of HIV-1 from mothers to children by 15 months of age: prospective cohort study from Durban, South Africa. *AIDS*, 15:379–387.

CURRAN, JAMES W., et al. (1988). Epidemiology of HIV infection and AIDS in the United States. *Science*, 239:610–616.

DEL ROMERO, JORGE et al. (2002). Evaluating the risk of HIV transmission through unprotected orogenital sex. *AIDS*, 16:1296–1297.

DEMARTINO, MAURIZIO, et al. (1992). HIV-1 transmission through breast-milk: Appraisal of risk according to duration of feeding. *AIDS*, 6:991–997.

DES JARLAIS, DON C., et al. (1989). AIDS and IV drug use. *Science*, 245:578.

DES JARLAIS, DON, et al. (1995). Maintaining low HIV-seroprevalence in populations of injecting drug users. *JAMA*, 274:1226–1231.

DES JARLAIS, DON, et al. (2000). HIV incidence among injection drug users in New York City, 1992–1997; Evidence for a declining epidemic. *Am J. Public Health*, 90:352–359.

DEVINCENZI, I., et al. (1989). Risk factors for male to female transmission of HIV. *Br. Med. J.*, 298:411–415.

DILLON, BETH, et al. (2000). Primary HIV infections associated with oral transmission. Program and abstracts of the Seventh Conference on Retroviruses and Opportunistic Infections; January 30–February 2; San Francisco, Calif. Abstract 473.

DORFMAN, ANDREA. (1991). Bad blood in France. *Time*, 138:48.

DROTMAN, PETER. (1996). Professional boxing, bleeding, and HIV testing. *JAMA*, 276:193.

EDWARDS, SARA, et al. (1998). Oral sex and the transmission of viral STD's. *J. Infect. Dis.*, 74:6–10.

ELSON, JOHN. (1991). The dangerous world of wannabes. *Time*, 138:77–80.

Emergency Cardiac Care Committee, American Heart Association. (1990). Risk of infection during CPR training and rescue: Supplemental guidelines. *JAMA*, 262:2714–2715.

European Mode of Delivery Collaboration, The. (1999). Elective Caesarean section versus vaginal delivery in prevention of vertical HIV-1 transmission: A randomized clinical trial. *Lancet*, 353:1035–1039.

FOX, PHILIP. (1991). Saliva and salivary gland alterations in HIV infection. *J. Am. Dental Assoc.*, 122:46–48.

FRIEDLAND, GERALD H. (1991). HIV transmission from health care workers. *AIDS Clin. Care*, 3:29–30.

GAUTHIER, DEANN et al. (1999). Bareback sex, bugchasers and the gift of death. *Deviant Behavior*, 20:85–100.

GIBBONS, MARY. (1994). Childhood sexual abuse. *Am. Fam. Phys.*, 49:125–136.

Global AIDSNEWS. (1994). A new approach to STD control and AIDS prevention. 4:13–14, 20.

GODDARD, JEROME. (1997). Why mosquitoes cannot transmit the AIDS virus. *Infect, Med.*, 14:353–354.

GOLDSCHMIDT, RONALD, et al. (1995). Antiretroviral strategies revisited. *J. Am. Board Fam. Pract.*, 8:62–69.

GOSTIN, LAWRENCE. (1994). HIV testing, counseling and prophylaxis after sexual assault. *JAMA*, 271:1436–1444.

GOTO, Y., et al. (1991). Detection of proviral sequences in saliva of patients infected with human immunodeficiency virus type 1. *AIDS Res. Hum. Retroviruses*, 7:343–347.

GROSSKURTH, HEINER, et al. (1995). Impact of improved treatment of sexually transmitted diseases on HIV infection in rural Tanzania. *Lancet*, 346:530–536.

GUINAN, MARY. (1995). Artificial insemination by donor: Safety and secrecy. *JAMA*, 273:890–891.

HARRISON, LEE, et al. (2000). Drugs cut HIV in semen: Safe sex still crucial. *Ann. Int. Med.*, 133: 280–284.

HAWTHORNE, PETER. (1999). An epidemic of rapes. *Time*, 154, no. 18: p 59.

HECHT, FREDERICK, et al. (1998). Sexual transmission of an HIV-1 variant resistant to multiple reverse-transcriptase and protease inhibitors. *N. Eng. J. Med.*, 339:307–311.

HELPERIN, DANIEL, et al. (1999). Viewpoint: male circumcision and HIV infection: 10 years and counting. *Lancet*, 354(9192):1813–1815.

HIV/AIDS Surveillance Report. (December 1996). 8:1–39.

HIV/AIDS Surveillance Report. (December 1997). 9:1–43.

HIV/AIDS Surveillance Report. (December 1999). 10:1–44.

HIV/AIDS Surveillance Report. (June 2001). 13:1–41.

HOLDEN, CONSTANCE. (1994). Switzerland has its own blood scandal. *Science*, 264:1254.

HOLMES, WILLIAM, et al. (2002). HIV-seropositive individuals optimistic beliefs about prognosis and relation to medication and safe sex adherence. *Journal of General Internal Medicine*, 17:9,677–683.

HOLMSTROM, PAUL, et al. (1992). HIV antigen detected in gingival fluid. *AIDS*, 6:738–739.

HOLTGRAVE, DAVID (2004). Estimation of annual HIV transmission rates in the United States, 1978–2000. *JAIDS*, 35:89–92.

HOOKER, TRACEY. (1996). HIV/AIDS: Facts to consider: 1996 National Conference of State Legislators, Denver, Colorado, February. 1–64.

HORN, TIM. (2001). Safety and efficacy of solid organ transplantation in HIV positive patients. *PRN Notebook*, 6:19–24.

HU, DALE J., et al. (1992). HIV infection and breastfeeding: Policy implications through a decision analysis model. *AIDS*, 6:1505–1513.

JACKSON, BROOKS. (1999). Progress in reducing mother-to-infant HIV transmission. *The Hopkins Report*, 11:2–3.

JAYARAMAN, KRISHNAMUNTHY. (1995). HIV scandal hits Bombay blood centre. *Nature*, 376:285.

JOSEPH, STEPHEN C. (1993). Dragon within the gates: The once and future AIDS epidemic. *Med. Doctor*, 37:92–104.

JOVAISAS, E., et al. (1985). LAV/HTLV III in 20-week fetus. *Lancet*, 2:1129.

KAHN, JAMES, et al. (1998). Acute HIV-1 Infection. *N. Eng. J. Med.*, 339:33–40.

KAISER, JOCELYN. (1996). Pasteur implicated in blood scandal? *Science*, 272:185.

KATNER, H.P., et al. (1987). Evidence for a Euro-American origin of human immunodeficiency virus. *J. Natl. Med. Assoc.*, 79:1068–1072.

KATZ, MITCHELL, et al. (2002). Impact of HAART on HIV seroincidence among men who have sex with men: San Francisco. *Am J. Public Health*, 92:388–394.

KIND, CHRISTIAN, et al. (1998). Prevention of vertical HIV transmission: Additive protective effect of elective caesarean section and zidovudine prophylaxis. *AIDS*, 12:205–210.

KINGSLEY, L.A., et al. (1990). Sexual transmission efficiency of hepatitis B virus and human immunodeficiency virus among homosexual men. *JAMA*, 264:230–234.

KLEIN, HARVEY. (2000). Will blood transfusion ever be safe enough? *JAMA*, 284:238–240.

KOSS, M.P., et al. (1991). Deleterious effects of criminal victimization on women's health and medical utilization. *Arch. Intern. Med.*, 151:342–347.

KRIEGER, JOHN, et al. (1998). Risk of sexual transmission of HIV unaffected by vasectomy. *J. of Urology*, 159:820–825.

KUHN, LOUISE, et al. (1994). Maternal-infant HIV transmission and circumstances of delivery. *Am. J. Public Health*, 84:1110–1115.

LACAYO, RICHARD. (1997). Assault with a deadly virus. *Time*, 149:82.

LAGA, MARIE. (1991). HIV infection and sexually transmitted diseases. *Sexually Transmitted Dis. Bull.*, 10:3–10.

LAGA, MARIE, et al. (1993). Non-ulcerative STDs as risk factors for HIV transmission in women: Results from a cohort study. *AIDS*, 7:95–102.

LEWIS, S.H., et al. (1990). HIV-1 introphoblastic villous Hofbauer cells and haematological precursors in eight-week fetuses. *Lancet*, 335:565.

LICHTMAN, STUART M., et al. (1991). Greater attention urged for HIV in older patients. *Infect. Dis. Update*, 2:5.

MARLINK, RICHARD, et al. (1994). Reduced rate of disease development after HIV infection as compared to HIV-1. *Science*, 265:1587–1590.

MCGOWAN, JOHN, et al. (2004). Risk behavior for transmission of human immunodeficiency virus (HIV) among HIV-seropositive individuals in an urban setting. *Clin. Infect. Dis.*, 38:122–127.

MIIKE, LAWRENCE. (1987). Do insects transmit AIDS? Office of Technological Assessment, Sept. 1:43.

MONZON, O.T., et al. (1987). Female to female transmission of HIV. *Lancet*, 2:40–41.

Morbidity and Mortality Weekly Report. (1988). Update: Universal precautions for prevention of transmission of human immunodeficiency virus, hepatitis B virus, and other blood-borne pathogens in healthcare settings. 37:377–382, 387–388.

Morbidity and Mortality Weekly Report. (1990a). Possible transmission of HIV to a patient during an invasive dental procedure. 39:489–493.

Morbidity and Mortality Weekly Report. (1990b). HIV infection and artificial insemination with processed semen. 39:249–256.

Morbidity and Mortality Weekly Report. (1991a). Update: Transmission of HIV infection during an invasive dental procedure—Florida. 40:21–27, 33.

Morbidity and Mortality Weekly Report. (1991b). Drug use and sexual behaviors among sex partners of injecting-drug users—U.S. 40:855–860.

Morbidity and Mortality Weekly Report. (1992). Child-bearing and contraceptive-use plans among women at high risk for HIV infection—Selected U.S. sites, 1989–1991. 41:135–144.

Morbidity and Mortality Weekly Report. (1994a). Human immunodeficiency virus transmission in household settings—United States. 43:347, 353–357.

Morbidity and Mortality Weekly Report. (1994b). Guidelines for preventing transmission of HIV through transplantation of human tissue and organs. 43:1–15.

Morbidity and Mortality Weekly Report. (1994c). Medical-care expenditures attributable to cigarette smoking—United States, 1993. 43:469–472.

Morbidity and Mortality Weekly Report. (1994d). zidovudine for the prevention of HIV transmission from mother to infant. 43:285–287.

Morbidity and Mortality Weekly Report. (1995). Use of AZT to prevent perinatal transmission (ACTG 076): Workshop on implications for treatment, counseling, and HIV testing. 44:1–12.

Morbidity and Mortality Weekly Report. (1997). Transmission of HIV possibly associated with exposure of mucous membrane to contaminated blood. 46:620–623.

Morbidity and Mortality Weekly Report. (1998). AIDS among people aged 50 years—United States, 1991–1996. 47:21–27.

MUNZER, ALFRED. (1994). The threat of secondhand smoke. *Menopause Manage.*, 3:14–17.

NAFTALIN, RICHARD J. (1992). Anal sex and AIDS. *Nature*, 360:10.

NEWELL, Marie-LOUISE, et al. (1990). HIV-1 infection in pregnancy: Implications for women and children. *AIDS*, 4:S111–S117.

NOWAK, RACHEL. (1995). Rockefeller's big prize for STD test. *Science*, 269:782.

OLESKE, JAMES M. (1994). The many needs of HIV-infected children. *Hosp. Pract.* 29:81–87.

OMETTO, LUCIA et al. (1995) Viral phenotype and host-cell susceptibility to HIV infection as risk factors for mother-to-child HIV transmission. *AIDS*, 9:427-434.

PADIAN, NANCY S., et al. (1991). Female to male transmission of HIV. *JAMA*, 266:1664–1667.

PADIAN, NANCY, et al. (1997). Heterosexual transmission of HIV in northern California: Results from a ten-year study. *Am. J. Epidemiol.*, 146:350–357.

PATTERSON, JULIE, et al. (1995). Basic and clinical considerations of HIV infection in the elderly. *Infect. Dis.*, 3:21–34.

PECKHAM, CATHERINE, et al. (1995). Mother-to-child transmission of HIV. *N. Engl. J. Med.*, 333:298–302.

PETERMAN, THOMAS A., et al. (1988). Risk of human immunodeficiency virus transmission from heterosexual adults with transfusion-associated infections. *JAMA*, 259:55–58.

PETO, RICHARD. (1992). Statistics of chronic disease control. *Nature*, 356:557–558.

POURTOIS, M., et al. (1991). Saliva can contribute in quick inhibition of HIV infectivity. *AIDS*, 5:598–599.

QUINN, THOMAS, et al. (2000). Viral load and heterosexual transmission of human HIV-1. *N. Engl. J. Med.*, 342: 921–929.

READ, JENNIFER. (1999). The mode of delivery and the risk of vertical transmission of human immunodeficiency virus type 1—A meta-analysis of 15 prospective cohort studies. *N. Eng. J. Med.*, 340:977–987.

REICHHARDT, TONY. (1995). Top aide to face charges in French HIV blood scandal. *Nature*, 375:349.

RILEY, LAURA, et al. (1999). Elective caesarean delivery to reduce the transmission of HIV. *N. Eng. J. Med.*, 13:1032–1033.

ROGERS, DAVID, et al. (1993). AIDS policy: Two divisive issues. *JAMA*, 270:494–495.

ROTHENBERG, RICHARD, et al. (1998). Oral Transmission of HIV. *AIDS*, 12:2095–2105.

ROZENBAUM, W. et al. (1988). HIV transmission by oral sex. *Lancet*, 1:1395.

SCHACKER, TIMOTHY, et al. (1998). Frequent recovery of HIV from genital herpes simplex virus lesions in HIV infected men. *JAMA*, 280:61–66.

Science in California. (1993). AIDS: I want a new drug. *Nature*, 362:396.

SCOTT, G.B., et al. (1985). Mothers of infants with the acquired immunodeficiency syndrome: Evidence for both symptomatic and asymptomatic carriers. *JAMA*, 253:363–366.

SEGARS, JAMES H. (1989). Heterosexual anal sex. *Med. Asp. Human Sexuality*, 23:6.

SELWYN, PETER A. (1986). AIDS: What is now known. *Hosp. Pract.*, 21:127–164.

SHERNOFF, MICHAEL. (1988). Integrating safer-sex counseling into social work practice. *Social Casework: J. Contemp. Social Work*, 69:334–339.

SPITZER, P.G., et al. (1989). Transmission of HIV infection from a woman to a man by oral sex. *N. Engl. J. Med.*, 320:251.

SPURGEON, DAVID. (1994). Canadian AIDS suit raises hope for HIV-blood victims. *Nature*, 281.

SPURGEON, DAVID. (1996). Canadian inquiry points the finger. *Nature*, 663.

ST. LOUIS, MICHAEL E., et al. (1993). Risk for perinatal HIV transmission according to maternal immunologic, virologic and placental factors. *JAMA*, 269:2853–2860.

STRYKER, JEFF, et al. (1993). AIDS policy: Two divisive issues. *JAMA*, 270:2436–2437.

SWENSON, ROBERT M. (1988). Plagues, History and AIDS. *Am. Scholar*, 57:183–200.

SWINBANKS, DAVID. (1993). American witnesses: Testify in Japan about AIDS risks. *Nature*, 364:181.

UNAIDS. (1997). Implications of HIV variability for transmission: Scientific and policy issues. *AIDS*, 11:S1–S15.

VAN DE PERRE, PHILIPPE, et al. (1993). Infective and anti-infective properties of breast milk from HIV-infected women. *Lancet*, 341:914–918.

VITTECOQ, D., et al. (1989). Acute HIV infection after acupuncture treatments. *N. Engl. J. Med.*, 320:250–251.

VOELKER, REBECCA. (1996). HIV guide for primary care physicians stresses patient-centered prevention. *JAMA*, 276:85–86.

WEBB, PATRICA, et al. (1989). Potential for insect transmission of HIV: Experimental exposure of *Cimex hemipterous* and *Toxorhynchites amboinensis* to human immunodeficiency virus. *J. Infect. Dis.*, 160:970–977.

WILL, GEORGE F. (1991). Foolish choices still jeopardize public health. *Private Pract.*, 24:46–48.

Women's AIDS Network. (1988). Lesbians and AIDS: What's the connection? San Francisco AIDS Foundation, 333 Valencia St., 4th Floor, P.O. Box 6182, San Francisco, CA 94101-6182.

WOOLLEY, ROBERT J. (1989). The biologic possibility of HIV transmission during passionate kissing. *JAMA*, 262:2230.

ZIELER, SALLY, et al. (2001). Violence and HIV: Strategies for primary and secondary prevention. *Focus*, 16: 1–4.

ZIGLER, J.B., et al. (1985). Postnatal transmission of AIDS-associated retrovirus from mother to infant. *Lancet*, 1:896–897.

CHAPTER 9

ABDALA, NADIA, et al. (1999). HIV-1 can survive in syringe for more than 4 weeks. *J. Acq. Imm. Def. Syndromes*, 20:73–80.

ABDALA, NADIA, et al. (2001). Use of bleach to disinfect HIV-1 contaminated syringes. *Am. Clinical Lab.*, 20:26–28.

ANDERSON, FRANK W.J. (1993). Condoms: A technical guide. *Female Patient*, 18:21–26.

ANDERSON, JOHN, et al. (1998). Needle hygiene and sources of needles for injection drug users: Data from a National Survey. *J. Acq. Imm. Def. Syndromes*, 18:S147

BARBER, HUGH R.K. (1990). Condoms (not diamonds) are a girl's best friend. *Female Patient*, 15:14–16.

BAYER, RONALD, et al. (1992). HIV Prevention and the two faces of partner notification. *Am. J. Public Health*, 82:1158–1164.

BOLOGNESI, DANIEL, et al. (1998). Viral envelope fails to deliver? *Nature*, 391:638–639.

BURNETT, JOSEPH. (1995). Fundamental basic science of HIV. *Cutis*, 55:84.

BURRIS, SCOTT, et al. (1996). Legal strategies used in operating syringe-exchange programs in the United States. *Am. J. Public Health*, 86:1161–1166.

BURRIS, SCOTT, et al. (2000). Physician prescribing of sterile injection equipment to prevent HIV infection: Time for action. *Ann. Intern. Med.*, 133:218–226.

CIVIC, DIANE, et al. (2002). Ineffective use of condoms among young women in managed care. *AIDS*, 14:779–788.

CLEMENT, MICHAEL J. (1993). HIV disease: Are we going anywhere? *Patient Care*, 27:13.

Consumer Reports. (1989). Can you rely on condoms? 54:135–141.

DRUCKER, LURIE, et al. (1997). An opportunity lost: HIV infection associated with lack of a national needle-exchange program in the USA. *Lancet*, 349:604–608.

ETZIONI, AMITAI. (1993). HIV sufferers have a responsibility. *Time*, 142:100.

EZZELL, CAROL. (1987). Hospital workers have AIDS virus. *Nature*, 227:261.

FARR, GASTON, et al. (1994). Contraceptive efficacy and acceptability of the female condom. *Am. J. Public Health*, 84:1960–1964.

FENTON, KEVIN, et al. (1997). HIV partner notification: Taking a new look. *AIDS*, 11:1535–1546.

FINDLAY, STEVEN. (1991). AIDS: The second decade. *U.S. News World Rep.*, 110:20–22.

FISHER, PETER. (1990). A report from the underground. *International Working Group on AIDS and IV Drug Use*, 5:15–17.

FRANCIS, DONALD, P., et al. (1987). The prevention of acquired immunodeficiency syndrome in the United States. *JAMA*, 257:1357–1366.

FREZIERES, RON, et al. (1999). Evaluation of the efficacy of a polyurethane condom: Results from a randomized, controlled clinical trial. *Family Planning Perspectives*, 31:81–87.

GERBERDING, JULIE LOUISE. (1991). Reducing occupational risk of HIV infection. *Hosp. Pract.*, 26:103–118.

GIBSON, DAVID, et al. (2001). Effectiveness of syringe exchange programs in reducing HIV risk behavior and HIV seroconversion among injecting drug users. *AIDS*, 15:1329–1341.

GOSTIN, LAWRENCE, et al. (1998). HIV infection and AIDS in the public health and health care systems: The role of law and litigation. *JAMA*, 279:1108–1113.

GRIMES, DAVID A.(1992). Contraception and the STD epidemic: Contraceptive methods for disease prevention. *The Contraception Report: The Role of Contraceptives in the Prevention of Sexually Transmitted Diseases*, III:1–15.

GUTTMACHER, SALLY, et al. (1997). Condom availability in New York City public high schools: Relationships to condom use and sexual behavior. *Am. J. Public Health*, 87:1427–1433.

HAGEN, HOLLY. (1991). Studies support syringe exchange. *Focus*, 6:5–6.

HEARST, NORMAN, et al. (2004). Condom promotion for AIDS prevention in the developing world: Is it working? *Studies in Family Planning*, 35:39– 47.

HEARST, NORMAN, et al. (1995). Collaborative AIDS prevention research in the developing world: The CAPS experience. *AIDS*, 9(suppl 1):51–55.

HOMSY, JACQUES, et al. (1989). The Fe and not CD4 receptor mediates antibody enhancement of HIV infection in human cells. *Science*, 244:1357–1359.

HOOKER, TRACEY. (1996). HIV/AIDS: Facts to consider, 1996. *National Conference of State Legislature*. Denver, February, 1–64.

HOTGRAVE, DAVID. (2002). Estimating the effectiveness and efficiency of U.S. HIV prevention efforts using scenario and cost-effectiveness analysis. *AIDS*, 16:2347–2350.

HOXWORTH, TAMARA, et al. (2003). Changes in partnerships and HIV risk behaviors after partner notification. *Sexually Transmitted Diseases*, 30:83–88.

JUDSON, FRANKLYN N. (1989). Condoms and spermicides for the prevention of sexually transmitted diseases. *Sexually Transmitted Dis. Bull.*, 9:3–11.

KAPLAN, EDWARD, et al. (1993). Let the needles do the talking! Evaluating the New Haven needle exchange. *Interfaces*, 23:7–26.

KIRBY, DOUGLAS, et al. (1998). The impact of condom distribution in Seattle schools on sexual behavior and condom use. *Am. J. Public Health*, 89:182–187.

KISSINGER, PATRICIA, et al. (2003). Partner notification for HIV and syphilis: Effects on sexual behaviors and relationship stability. *Sexually Transmitted Diseases*, 30:75–82.

LACKRITZ, EVE, et al. (1995). Estimated risk of transmission of HIV by screened blood in the United States. *N. Engl. J. Med.*, 333:1721–1725.

LEVY, JAY A. (1989). Human immunodeficiency virus and the pathogenesis of AIDS. *JAMA*, 261:2997– 3006.

LEWIS, DAVID. (1995). Resistance of microorganisms to disinfection in dental and medical devices. *Nature Med.*, 1:956–958.

LURIE, PETER, et al. (1998). A sterile syringe for every drug user injection: How many injections take place annually and how might pharmacists contribute to syringe distribution? *J. Acquired Immune Defic. Syndr. Hum. Retrovirol*, 18:545–551.

LURIE, PETER, et al. (1994). Ethical behavioral and social aspects of HIV vaccine trials in developing countries. *JAMA*, 271:295–302.

MANDELL, WALLACE, et al. (1994). Correlates of needle sharing among injection drug users. *Am. J. Public Health*, 84:920–923.

Morbidity and Mortality Weekly Report. (1988). Partner notification for preventing human immunodeficiency virus (HIV) infection—Colorado, Idaho, South Carolina, Virginia. 37:393–396; 401–402.

Morbidity and Mortality Weekly Report. (1989). Guideline for prevention of transmission of HIV and hepatitis B virus to health care workers. 38:3–17.

Morbidity and Mortality Weekly Report. (1992). Sexual behavior among high school students—United States, 1990. 40:885–888.

Morbidity and Mortality Weekly Report. (1993a). Update: Barrier protection against HIV infection and other sexually transmitted diseases. 42:589–591.

Morbidity and Mortality Weekly Report. (1993b). Impact of new legislation on needle and syringe purchases and possession—Connecticut 1992. 42:145–147.

Morbidity and Mortality Weekly Report. (1995). Notification of syringe-sharing and sex partners of HIV-infected persons—Pennsylvania, 1993–1994. 44:202–204.

Morbidity and Mortality Weekly Report. (1996). School-based HIV-prevention education—United States, 1994. 45:760–764.

Morbidity and Mortality Weekly Report. (1997). Update: Syringe-exchange programs—United States, 1996. 46:565–568.

Morbidity and Mortality Weekly Report. (2000a). Cluster of HIV-Infected adolescents and young adults—Mississippi, 1999. 49: 861–864.

Morbidity and Mortality Weekly Report. (2000b). Notice to readers: CDC statement on study results of product containing nonoxynol-9. 49:717.

Office Nurse. (1995). Contraception: how today's options stack up. 8:13–14.

PARRAN, THOMAS, P. (1937). *Shadow on the Land: Syphilis.* New York: Reynal and Hitchcock.

RAMOS, ARTUR, et al. (2002). Intersubtype human immunodeficiency virus Type 1 superinfection following seroconversion to primary infection in two injection drug users. *J. Virol.* 76:7444–7452.

RAYMOND, CHRIS ANNE. (1988). U.S. cities struggle to implement needle exchanges despite apparent success in European cities. *JAMA*, 260:2620–2621.

REZI, CRISTINA, et al. (2003). Safety and acceptability of the Reality condom for anal sex among men who have sex with men. *AIDS*, 17:727–731.

ROPER, WILLIAM L., et al. (1993). Commentary: Condoms and HIV/STD prevention—clarifying the message. *Am. J. Public Health*, 83:501–503.

ROTHENBERG, KAREN, et al. (1995). The risk of domestic violence and women with HIV infection: Implications for partner notification, public policy, and the law. *Am. J. Public Health*, 85: 1569–1576.

RUTHERFORD, GEORGE W. (1988). Contact tracing and the control of human immunodeficiency virus infection. *JAMA*, 259:3609–3670.

SATTAR, SYED, A., et al. (1991). Survival and disinfectant inactivation of HIV: A critical review. *Rev. of Infect. Dis.*, 13:430–447.

SELLERS, DEBORAH, et al. (1994). Does the promotion and distribution of condoms increase teen sexual activity? Evidence from an HIV prevention program for Latino youth. *Am. J. Public Health*, 84:1952–1958.

SPRUYT, ALAN, et al. (1998). Identifying condom users at risk for breakage and slippage; Findings from three International Sites. *Am. J. Public Health*, 88:239–240.

STIMSON, GERRY V., et al. (1989). Syringe exchange. *International Working Group on AIDS and IV Drug Use*, 4:15.

STRATHDEE, STEFFANIE, et al. (2002). Sex differences in risk factors for HIV seroconversion among injection drug users: A 10-year perspective. *Archives of Internal Medicine*, 161:1281–1288.

STRATTON, P., et al. (1993). Prevention of sexually transmitted infections: Physical and Chemical barrier methods. *Infect. Dis. Clin. North Am.*, 7(4):841–859.

Time. (1992). Closed: Needle Park. 139:53.

U.S. Public Health Service. (1994). Counseling to prevent unintended pregnancy. *Am. Fam. Phys.*, 50(5):971.

VOELKER, REBECCA. (1995). Lessons from cat virus. *JAMA*, 273:910.

WEINSTEIN, STEPHEN P., et al. (1990). AIDS and cocaine: A deadly combination facing the primary care physician. *J. Fam. Prac.*, 31:253–254.

WHITE, NICK, et al. (1988). Dangers of lubricants used with condoms. *Nature*, 335:19.

WODAK, ALEX. (1990). Australia smashes international needle and syringe exchange record. *International Working Group on AIDS and IV Drug Use*, 5:28–29.

CHAPTER 10

BOOTH, WILLIAM. (1988). CDC paints a picture of HIV infection in U.S. *Science*, 242:53.

BOOTH, WILLIAM. (1989). Asking America about its sex life. *Science*, 243:304.

CATANIA, JOSEPH, et al. (2001). The continuing HIV epidemic among gay men. *Am. J. Public Health*, 91:907:914.

CHIN, J., et al. (1990). Projections of HIV infections and AIDS cases to the year 2000. *Bull. WHO*, 68:1–11.

COUTINHO, ROEL, et al. (1996). Summary of Track C: Epidemiology and public health. *AIDS* 10(suppl. 3):S115–S121.

DE GROOT, ANNE, et al. (1996). Barriers to care of HIV-infected inmates: A public health concern. *AIDS Reader*, 6:78–87.

DOLAN, KATE, et al. (1995). AIDS behind bars: Preventing HIV spread among incarcerated drug infections. *AIDS*, 9:825–832.

EL-SADR, WAFFA, et al. (1994). *Managing Early HIV Infection: Quick Reference Guide for Clinicians.* Agency for Health Care Policy and Research. Publication #94-0573, Rockville, MD.

FAY, ROBERT E., et al. (1989). Prevalence and patterns of same-gender sexual contact among men. *Science*, 243:338–348.

FELDMAN, MITCHELL, et al. (1994). The growing risk of AIDS in older patients. *Patient Care*, 28:61–72.

FILLIT, HOWARD, et al. (1989). AIDS in the elderly: A case and its implications. *Geriatrics*, 44:65–70.

FRANCIS, DONALD P. (1987). The prevention of acquired immunodeficiency syndrome in the United States. *JAMA*, 257: 1357–1366.

GAYLE, HELENE. (1988). Demographic and sexual transmission differences between adolescent and adult AIDS patients, U.S.A. *Fourth International Conference on AIDS.*

GOSTIN, LAWRENCE, et al. (1997). National HIV case reporting for the United States. *N. Engl. J. Med.*, 337:1162–1167.

HENDERSON, DAVID, et al. (1990). Risk for occupational transmission of human immunodeficiency virus type 1 (HIV-1) associated with clinical exposures. *Ann. Intern. Med.*, 113:740.

HIRSCHHORN, LISA. (1995). HIV infection in women: Is it different? *AIDS Reader*, 5:99–105.

HIV/AIDS Surveillance Report. (December 1996). 8:1–36.

HIV/AIDS Surveillance Report. (December 1997). 9:1–43.

HIV/AIDS Surveillance Report. (December 1999). 11:1–45.

HOGG, ROBERT, et al. (1997). HIV disease shortens life expectancy of gay men by up to 20 years. *Inter. J. Epidemiol.* 26:657–661.

HOOKER, TRACEY. (1996). HIV/AIDS: Facts to consider—1996. National Conference of State Legislatures, Denver. pp. 1–64.

KAO, HELEN. (1999). Leaders unite on HIV in the Latino community. *Harvard AIDS Review Women and AIDS.* Spring:13–15.

LEMP, GEORGE. (1991). The young men's survey: Principal findings and results. A presentation to the San Francisco Health Commission, June 4.

MAHON, NANCY. (1996). New York inmates' HIV risk behaviors: The implications for prevention policy and programs. *Am. J. Public Health*, 86:1211–1215.

MARCUS, R., et al. (1988). AIDS: Health care workers exposed to it seldom contract it. *N. Engl. J. Med.*, 319:1118–1123.

MARTORELL, REYNALDO, et al. (1995). Vitamin A supplementation and morbidity in children born to HIV-infected women. *Am. J. Public Health*, 85:1049–1050.

MICHAELS, DAVID, et al. (1992). Estimates of the number of motherless youth orphaned by AIDS in the United States. *JAMA*, 268:3456–3461.

MILLER, PATTI, et al. (1997). Compensation for occupationally acquired HIV needs revamping. *Am. J. Public Health*, 87:1558–1562.

Morbidity and Mortality Weekly Report. (1990). HIV prevalence, projected AIDS case estimates. Workshop, October 31–November 1, 1989. 39:110–119.

Morbidity and Mortality Weekly Report. (1991). The HIV/AIDS epidemic: The first 10 years. 40:357–368.

Morbidity and Mortality Weekly Report. (1992). Surveillance for occupationally acquired HIV infection—United States, 1981–1992. 41:823–824.

Morbidity and Mortality Weekly Report. (1993). Update: Mortality attributable to HIV infection among persons aged 25–44 years—United States, 1991 and 1992. 42:869–873.

Morbidity and Mortality Weekly Report. (1994). Heterosexually acquired AIDS-United States, 1993. 43:155–160.

Morbidity and Mortality Weekly Report. (1995). First 500,000 AIDS cases—United States, 1995. 44:849–853.

Morbidity and Mortality Weekly Report. (1996a). AIDS associated with injection drug use—United States, 1995. 45:392–398.

Morbidity and Mortality Weekly Report. (1996b). HIV/AIDS education and prevention programs for adults in prisons and jails and juveniles in confinement facilities—United States, 1994. 45:268–271.

Morbidity and Mortality Weekly Report. (1996c). Update: Mortality attributable to HIV infection among persons aged 25–44 years—United States, 1994. 45:121–125.

Morbidity and Mortality Weekly Report. (2000). HIV/AIDS among racial/ethnic minority men who have sex with men—United States, 1989–1998.

Nations Health Report. (1995). Women learn of progress, share deep concerns on HIV/AIDS issues. XXV:10.

OSBORNE, JUNE E. (1993). AIDS policy advisor foresees a new age of activism. *Fam. Prac. News*, 23:1, 45.

REISMAN, JUDITH. (1990). *Kinsey, Sex and Fraud: The Indoctrination of a People.* Lafayette, LA: Huntington House Press.

ROSENBERG, PHILIP. (1995). Scope of the AIDS epidemic in the United States. *Science*, 270:1372–1376.

RUBEL, JOHN, et al. (1997). HIV-related mental health in correctional settings. *Focus*, 12:1–4.

RYDER, ROBERT, et al. (1994). AIDS orphans in Kinshasa, Zaire: Incidence and socioeconomic consequences. *AIDS*, 8:673–679.

UNAIDS. (1997). The HIV/AIDS situation in 1997: Global and regional highlights, Geneva 27, Switzerland, pp. 1–14.

VLAHOV, D., et al. (1991). Prevalence of antibody to HIV-1 among entrants to U.S. correctional facilities. *JAMA*, 265:1129.

WEISFUSE, ISAAC C., et al. (1991). HIV-1 infection among New York City inmates. *AIDS*, 5:1133–1138.

WILSON, J., et al. (1990). Keeping your cool in a time of fear. *Emergency Medical Services*, 19:30–32.

WOLITSKI, RICHARD, et al. (2001). Are we headed for a resurgence of the HIV epidemic among gay men? *Am. J. Pub. Health*, 91:883–888.

CHAPTER 11

Aggleton, Peter, et al. (1994). Risking everything? Risk behavior, behavior change, and AIDS. *Science*, 265:341–345.

Allen, J. R., et al. (1988). Prevention of AIDS and HIV infection: Needs and priorities for epidemiologic research. *Am. J. Public Health*, 78:381–386.

Bardeguez, Arlene. (1995). Managing HIV infection in women. *AIDS Reader, Suppl.*, Nov/Dec, pp. 2–3.

Bessinger, Ruth, et al. (1997). Pregnancy is not associated with the progression of HIV disease in women attending an HIV outpatient program. *Am J. Epidemiol.*, 147:434–440.

Bloomberg School of Public Health and Johns Hopkins University. (2001). Youth and HIV/AIDS: Can We Avoid Catastrophe? http://www.jhuccp.org/pr/112/112print.shtml

Butler, Declan. (1993). Whose side is focus of AIDS research? *Nature*, 366:293.

Chu, Susan, et al. (1994). Female-to-female sexual contact and HIV transmission. *JAMA*, 272:433.

Cohen, Jon. (1995). Women: Absent term in the AIDS research equation. *Science*, 269:777–780.

Collins, Chris, et al. (1997). Outside the prevention vacuum: Issues in HIV prevention for youth in the next decade. *AIDS Reader*, 7:149–154.

Cotton, Paul. (1994). U.S. sticks head in sand on AIDS prevention. *JAMA*, 272:756–757.

Cu-rin, Susan. (1999). Antiretroviral treatment during pregnancy. *Improving Management of HIV Disease*, 7:14–18.

Curran, James W., et al. (1988). Epidemiology of HIV infection and AIDS in the United States. *Science*, 239:610–616.

Dickinson, Gordon M. (1988). Epidemiology of AIDS. *Int. Ped.*, 3:30–32.

DiClemente, Ralph. (1998). Preventing sexually transmitted infections among adolescents: A clash of ideology and science. *JAMA*, 279:1574–1575.

Ehrhardt, Anke A. (1992). Trends in sexual behavior and the HIV pandemic. *Am. J. Public Health*, 82:1459–1464.

Farzadegan, Homayoon, et al. (1998). Sex differences in HIV-1 viral load and progression to AIDS. *Lancet*, 352:1510–1514.

Futterman, Donna, et al. (1992). Medical care of HIV-infected adolescents. *AIDS Clin. Care*, 4:95–98.

Hader, Shannon, et al. (2001). HIV infection in women in the United States. *JAMA*, 285: 1186–1192.

Health. (1996). AIDS still spreading rapidly among young gay men. *Am. Med. News*, 39:30.

HIV/AIDS Surveillance Report. (1995). Year end, 7:1–36.

HIV/AIDS Surveillance Report. (1996). Year end, 8:1–39.

HIV/AIDS Surveillance Report. (1997). December, 9:1–34.

Holden, Constance. (1998). World-AIDS The worst is yet to come. *Science*, 278:1715.

Ioannidis, John, et al. (1999). Maternal viral load and the risk of perinatal transmission of HIV-1. *NEJM*, 341:1698.

Jemmot, John, et al. (1998). Abstinence and safer sex high risk-reduction interventions for African American adolescents: A randomized controlled trial. *JAMA*, 279:1529–1536.

Johnson, Timothy, et al. (1995). Current issues in the primary care of women with HIV. *Female Patient*, 20:51–58.

Joseph, Stephen C. (1993). The once and future AIDS epidemic. *Med. Doctor*, 37:92–104.

Kirby, Douglas, et al. (1997). The impact of the Postponing Sexual Involvement Curriculum among youths in California. *Fam. Plann. Perspect.*, 29:100–108.

Kwakwa, Helena, et al. (2003). Female to female transmission of human immunodeficiency virus. *Clinical Infectious Diseases*, 36:e40–e41.

Lemp, George, et al. (1995). HIV seroprevalence and risk behaviors among lesbians and bisexual women in San Francisco and Berkley, California. *Am. J. Public Health*, 85:1549–1552.

Mandelbrot, Laurent, et al. (1997). Natural conception in HIV-negative women with HIV-infected partners. *Lancet*, 349:850–851.

Mandelbrot, Laurent, et al. (1998). Perinatal HIV-1 Transmission. *J. Am. Med. Assoc.*, 280:55–60.

Miotti, Paolo, et al. (1999). HIV transmission through breast-feeding. *JAMA*, 282:744–749.

Morbidity and Mortality Weekly Report. (1992). Selected behaviors that increase risk for HIV infection among high school students—United States, 1990. 41:236–240.

Morbidity and Mortality Weekly Report. (1993). Update: Acquired immunodeficiency syndrome—United States, 1992. 42:547–557.

Morbidity and Mortality Weekly Report. (1994). Update: Impact of the expanded AIDS surveillance case definition for adolescents/adults on case reporting—United States, 1993. 43:160–170.

Morbidity and Mortality Weekly Report. (1996a). HIV testing among women aged 18–44 years—United States, 1991 and 1993. 46:733–736.

Morbidity and Mortality Weekly Report. (1996b). Update: Mortality attributable to HIV infection among persons aged 25–44 years—United States, 1994. 45:121–125.

Morbidity and Mortality Report. (1999). Surveillance for AIDS-defining opportunistic illnesses, 1992–1997. 48:1–20.

Morbidity and Mortality Weekly Report. (2003). Pregnancy in perinatally HIV infected adolescents and young adults—Puerto Rico, 2002. 52:149–151.

Newell, Michael, et al. (1997). Immunological markers in HIV-infected pregnant women: The European Collaborative Study and the Swiss HIV Pregnancy Cohort. *AIDS*, 11:1859–1865.

Office of National AIDS Policy. (1996). Youth and HIV/AIDS: An American agenda. *Report to the President*, pp. 1–14.

Parker–Bush, Tracey. (2000). Perinatal HIV: Children with HIV grow up. *Focus*, 15:1–4.

Pfeiffer, Naomi. (1991). AIDS risk high for women; care is poor. *Infect. Dis. News*, 4:1, 18.

Rotheram-Borus, M., et al. (1991). Sexual risk behaviors, AIDS knowledge, and beliefs about AIDS among runaways. *Am. J. Public Health*, 81:208–210.

Schable, Barbara, et al. (1996). Characteristics of women 50 years of age or older with heterosexually acquired AIDS. *Am. J. Public Health*, 86:1616–1618.

Sells, Wayne, et al. (1996). Morbidity and mortality among US adolescents: An overview of data and trends. *Am. J. Public Health*, 86:513–519.

Selwyn, Peter A., et al. (1989). Knowledge of HIV antibody status and decisions to continue or terminate pregnancy among intravenous drug users. *JAMA*, 261:3567–3571.

Sterling, Timothy, et al. (2001). Initial plasma HIV-1 RNA levels and progression to AIDS in women and men. *N. Engl. J. Med.*, 344:720–725.

Taha & Taha, et al. (2000). Morbidity among HIV-1 infected and uninfected African children. *Pediatrics*, 106: http://www.pediatrics.org/cgi/content/full/106/6/e77

Thomas, Patricia. (1988). Official estimates of epidemic's scope are grist for political mill. *Med. World News*, 29:12–13.

Thomas, Patricia. (1989). The epidemic. *Med. World News*, 30:41–49.

UNAIDS, (1997). Global summary of the HIV/AIDS epidemic. *Report on the global HIV/AIDS epidemic.* December 1997:1–25.

Van Benthem, Birgit, et al. (2002). The impact of pregnancy and menopause on CD4 lymphocyte counts in HIV infected women. *AIDS*, 16:919–924.

Women's AIDS Network. (1988). Lesbians and AIDS: What's the connection? *San Francisco AIDS Foundation*, 333 Valencia St., 4th Floor, P.O. Box 6182, San Francisco, CA 94101-6182.

World Health Organization, Geneva. (1994). *Women's Health*, p. 18.

Wortley, Pascale, et al. (1997). AIDS in women in the United States. *JAMA*, 278:911–916.

Yao, Faustin K. (1992). Youth and AIDS: A priority for prevention education. *AIDS Health Promotion Exchange No. 2*, Royal Tropical Institute, The Netherlands: 1–3.

Zierler, Sally, et al. (2000). Violence victimization after HIV infection in a US probability sample of adult patients in primary care. *Am. J. Pub. Health*, 90:208–215.

Zijenah, Lynn, et al. (2004). Timing of mother to child transmission of HIV-1 and infant mortality in the first six months of life in Harare, Zimbabwe. *AIDS*, 18:273–280.

CHAPTER 12

AIDS, Health and Human Rights. (1995). Francois-Xavier Bagnoud Center for Health and Human Rights–Harvard School of Public Health, pp. 1–162.

Allen, Brady. (1991). The role of the primary care physician in HIV testing and early stage disease management. *Fam. Pract. Recert.*, 13:30–49.

Anderson, John, et al. (1992). HIV antibody testing and post-test counseling in the United States: Data from the 1989 National Health Interview Study. *Am. J. Public Health*, 82:1533–1535.

Angell, Marcia. (1991). A dual approach to the AIDS epidemic. *N. Engl. J. Med.*, 324:1498–1500.

Bayer, Ronald, et al. (1995). Testing for HIV infection at home (Sounding Board). *N. Engl. J. Med.*, 332:1296–1299.

Belongia, Edward A., et al. (1989). Premarital HIV screening. *JAMA*, 261:2198.

Carlson, Desiree A., et al. (1989). Testing for HIV risk from therapeutic blood products. *Pathology and Pathophysiology of AIDS and HIV Related Diseases* (Eds. Jami J. Harawi and Carl J. O'Hara), St. Louis: C.V. Mosby Co.

Cordes, Robert, et al. (1995). Pitfalls in HIV testing. *Postgrad. Med.*, 98:177–189.

El-Sadr, Wafaa, et al. (1994). Managing early HIV infection: Agency for Health Care Policy and Research. *Clinical Practice Guideline on Evaluation and Management of Early HIV Infection.* January, 7:1–37.

El-Sadr, W., et al. (1994). *Managing early HIV infection: quick reference guide for clinicians.* AHCPR Publication No. 94-0573. Rockville, MD.

Fang, Chyang T., et al. (1989). HIV testing and patient counseling. *Patient Care*, 23:19–44.

Gostin, Lawrence, et al. (1997). National HIV case surveillance is urged. *N. Engl. J. Med.*, 337:1162–1167.

Hartlaub, Paul, et al. (1993). Obtaining informed consent: It is not simply asking "do you understand?" *J. Fam. Pract.*, 36:383–384.

Hegarty, J.D., et al. (1988). The medical care costs of human immunodeficiency virus infected children in Harlem. *JAMA*, 260:1901–1905.

Hooker, Tracey. (1996). HIV/AIDS: Facts to consider—1996. Natural conference of State Legislatures, February, pp. 1–64.

Intergovernmental AIDS Report. (1989). Illinois court overrules mandatory HIV testing for prostitutes and sex offenders, 2:1–18.

Janssen, Robert, et al. (1998). New testing strategy to detect early HIV-1 infection for use in incidence estimates and for clinical and prevention purposes. *JAMA*, 280:42–48.

Jenny-Avital, Elizabeth, et al. (2001). Erroneously low or undetectable plasma HIV-1 RNA load, determined by PCR, in West African and American patients with non-B subtype HIV infection. *Clinical Infectious Diseases*, 32:1227–1230.

Johnson, Christine. (2000). Factors known to cause false positive HIV antibody test results. In *Alive and Well* [online]. Available: http://www.aliveandwell.org.

Keller, G.H., et al. (1988). Identification of HIV sequences using nucleic acid probes. *Am. Clin. Lab.*, 7:10–15.

KRAMER, F.R., et al. (1989). Replicatable RNA reporters. *Nature,* 339:401–402.

MACKENZIE, WILLIAM R., et al. (1992). Multiple false positive serologic tests for HIV, HTLV-1 and hepatitis C following influenza vaccination, 1991. *JAMA,* 268:1015–1017.

MCFARLAND, WILLIAM, et al. (1999). Detection of early HIV infection and estimation of incidence using a sensitive/less sensitive enzyme immunoassay testing strategy at anonymous counseling and testing sites in San Francisco. *JAIDS,* 22:484–489.

MERCOLA, JOSEPH M. (1989). Premarital HIV screening. *JAMA,* 261:2198.

MIIKE, LAWRENCE. (1987). *AIDS Antibody Testing.* Office of Technological Assessment Testimony to the U.S. Congress. October:1–21.

Morbidity and Mortality Weekly Report. (1992). Testing for antibodies to HIV-2 in the United States. 41:1–9.

Morbidity and Mortality Weekly Report. (1996). U.S. Public Health Service Guidelines for testing and counseling blood and plasma donors for HIV type I antigen. 45:1–9.

Morbidity and Mortality Weekly Report. (1998b). Update: Counseling and testing using rapid tests—United States, 1995, 47:211–215.

Morbidity and Mortality Weekly Report (1999). HIV Testing—United States, 1996. 48:52–55.

MOSER, MICHAEL, (1998). Anonymous HIV testing. *Am. J. Pub. Health,* 88:683.

MORE, DANIEL, et al. (2000). Utility of an HIV-1 RNA assay in the diagnosis of acute retroviral syndrome. *Southern Medical Journal,* 93:1004–1006.

MULLIS, KARY B., et al. (1987). Process for amplifying, detecting, and/or cloning nucleic acid sequences. (U.S. Patent No. 4,683,195). *Official Gazette of the U.S. Patient and Trademark Office,* vol. 1080, no. 4, July.

NASH, GRANT, et al. (1998). Health benefits and risks of reporting HIV-infected individuals by name. *Am. J. Pub. Health,* 88: 876–879.

PASSANNANTE, MARIAN R., et al. (1993). Responses of health care professionals to proposed mandatory HIV testing. *Arch. Fam. Med.,* 2:38–44.

PHILLIPS, KATHRYN, et al. (1995a). Potential use of home HIV testing. *N. Engl. J. Med.,* 332:1308–1310.

PHILLIPS, KATHRYN, et al. (1995b).Who plans to be tested for HIV or would get tested if no one could find out the results? *Am. J. Prevent. Med.,* 11(3):156.

REIMER, LARRY, et al. (1997). Undetectable antibody reported in a patient with typical HIV. *Clin. Infect. Dis.* 25:98–103.

RHAME, FRANK S., et al. (1989). The case for wider use of testing for HIV infection. *N. Engl. J. Med.,* 320:1242–1254.

RICH, JOSIAH, et al. (1999). Misdiagnosis of HIV infection by HIV-1 Plasma Viral load testing: A case series. *Annals of Internal Medicine,* 130:37–39.

SAMET, JEFFREY, et al. (1998). Trillion Niron delay: Time from testing positive for HIV to presentation for primary care. *Arch. Int. Med.,* 158:734–740.

SCHEFFEL, J.W. (1990). Retrocell HIV-1 passive haemagglutination assay for HIV-1 antibody screening. *J. Acquired Immune Deficiency Syndromes,* 3:540–545.

WAKE, WILLIAM T. (1989). How many patients will die because we fear AIDS? *Med. Econ.,* 66:24–30.

WOFOY, C. B. (1987). HIV infection in women. *JAMA,* 257:2074–2076.

CHAPTER 13

American Medical Association News. (1991). Ruling fuels debate over HIV-infected doctors. May:1, 41–43.

BURRIS, SCOTT. (1996). Human immunodeficiency virus-infected health care workers. *Arch. Fam. Med.,* 5:102–106.

CRAVEN, DONALD, et al. (1994). Fictitious HIV Infection. *Ann. Intern. Med.,* 121:763–766.

DANIELS, NORMAN. (1992). HIV-infected professionals, patient rights and the 'switching dilemma.' *JAMA,* 267:1368–1371.

EICKHOFF, THEODORE C. (1989). Public perceptions about AIDS and HIV infection. *Infect. Dis. News,* 2:6.

FISHER, J.D., et al. (1992). Changing AIDS risk behavior. *Psychol. Bull.,* 111:455–474.

FOURNIER, A.M., et al. (1989). Preoperative screening for HIV infection. *Arch. Surg.,* 124:1038–1040.

FULTON, ROBERT, et al. (1988). AIDS: Seventh rank absolute. In *AIDS: Principles, Practices and Politics.,* Inge B. Corliss, et al., eds. Bristol, PA: Hemisphere.

GERBERT, BARBARA, et al. (1988). Why fear persists: Health care professionals and AIDS. *JAMA,* 260:3481–3483.

GOSTIN, LAWRENCE, et al. (1998). HIV infection and AIDS in the public health and health care systems: The role of law and litigation. *JAMA,* 279:1108–1113.

GOUJAN, CHRISOPHE, et al. (2000). Phylogenetic analysis indicates an atypical nurse to patient transmission of HIV-1. *J. Virology,* 74:2525–2532.

HAGEN, M.D., et al. (1988). Routine preoperative screening for HIV: Does the risk to the surgeon outweigh the risk to the patient? *JAMA,* 259:1357–1359.

HEGARTY, JAMES D., et al. (1988). The medical care costs of HIV-infected children in Harlem. *JAMA,* 260:1901–1909.

HEREK, GREGORY M., et al. (1993). Public reaction to AIDS in the United States: A second decade of stigma. *Am. J. Public Health,* 83:574–577.

J. Virology, 74:2525–2534.

JAPENGA, ANN. (1992). The secret. *Health,* 6:43–52.

KIRBY, D. (1988). The effectiveness of educational programs to help prevent school-age youth from contracting AIDS: A review of relevant research. United States Congress.

LEFKOWITZ, MATHEW. (1990). A health care system in crisis: The possible restriction against HIV-infected health care workers. *PAACNotes,* 2:175–176.

LEWIS, C.E., et al. (1992). Primary care physicians' refusal to care for patients infected with HIV. *West. J. Med.*, 156:36–38.

LO, BERNARD et al. (1992). Health care workers infected with HIV. *JAMA*, 267:1100–1105.

MCCARTHY, GILLIAN, et al. (1999). Factors associated with refusal to treat HIV infected patients: The results of a National Survey of Dentists in Canada. *Am. J. of Public Health*, 89:541–545.

MCNEILL, WILLIAM H. (1976). *Plagues and People*, Garden City, NJ: Anchor Press.

MICHAELS, DAVID, et al. (1992). Estimates of the number of youth orphaned by AIDS in the United States. *JAMA*, 268:3456–3461.

MILENO, MARIA, et al. (2001). Factitious HIV syndrome in young women. *The AIDS Reader*, 11:263–268.

Morbidity and Mortality Weekly Report. (1990). HIV-related knowledge and behavior among high school students—Selected U.S. cities, 1989. 39: 385–396.

NARY, GORDON. (1990). An editorial. *PAACNotes*, 2:170.

PHILLIPS, KATHRYN A. (1993). Subjective knowledge of AIDS and Use of HIV testing. *Am. J. Public Health*, 83:1460–1462.

POLDER, JACQUELYN A., et al. (1989). AIDS precautions for your office. *Patient Care*, 23:161–171.

POWELL, JOHN H. (1965). *Bring Out Your Dead.* New York: Time-Life Inc.

RHAME, FRANK S., et al. (1989). The case for wider use of testing for HIV infection. *N. Engl. J. Med.*, 320:1242–1254.

ROWE, MONA, et al. (1987). *A Public Health Challenge: State Issues, Policies and Programs, Volume 2.* Intergovernmental Health Policy Project, George Washington University.

STRYKER, JEFF, et al. (1995). Prevention of HIV infection. *JAMA*, 273:1143–1148.

VOELKER, REBECCA. (1989). No uniform policy among states on HIV/AIDS education. *Am. Med. News*, September 3: 28–29.

WEISS, RICK. (1999). Thousands of deaths linked to medical errors. Online. Available: http://www.washingtonpost.com (November 30)

ZUGER, ABIGAIL. (1995). The high cost of living. *Sci. Am.* 273:108.

Index

Note: Italicized letters *f* and *t* following page numbers indicate figures and tables, respectively.

A
Abacavir (Ziagen), 74*t*, 80
Abbott, Sidney, 421
Abdala, Nadia, 247
Abraham, Yvonne, 213
ABT-378 (Kaletra), 74*t*, 80, 90
Accessory protein, 55
Acer, David, 202
Acquired Immune Deficiency Syndrome (AIDS). *See also* HIV disease; Human Immunodeficiency
  cases, 3*t*
    adult/adolescent cases, 297–299
    behavioral risk groups, 296–297, 299–301
    demographic changes, 302–308
    in developing countries, 1
    estimating, 296, 308–310, 311, 315
    first US reports, 23–24, 29, 41
    new AIDS cases, 30, 68, 211, 295, 310–328
    prevalence/incidence, 296
    reporting systems, 308–310
    single/multiple exposure categories, 296
    total number infected persons, 1, 16–17
    in USA, 193, 295–296, 310–315, 319
    young adults, 355–362
  causes
    alternative views, 32–37
    CDC reports, 26–27
    in general, 24–26, 32
  cures, 70–72, 101–102, 400
  data sources, 291–294
  deaths, 15, 29, 140, 315–328
    life expectancy, 320, 323
    sobering reports, 96–100
    women, 342–343
  definition, 29, 171, 294
    AIDS surveillance case definition, 29–30, 171*t*
    clinical description, 29
    impact, 30
    names for virus, 27–28
    1981–1992, 29–30
    1993, 30
  diagnosis, 11, 162, 167
    pediatric, 188–189
  discrimination, 403–404
  fear about, 398–400, 401, 402
  FRAID, 19
  in history, 11
  misconceptions about, 401–404, 412–414
  overview, 11–14, 23, 26–27, 397–398, 414–416
  pandemic, 296
  progression
    beta-2 microglobulin, 179–180
    in general, 179
    p24 antibody levels, 179
    p24 antigen levels, 179
    T4/T8 lymphocyte levels, 180
  remission, 101–102
  research, 18
  as syndrome, 23
  transmission, vectors, 23–24
ACT UP, 435
  discussed, 396
ACT UP/Golden Gate, 32
ACT UP/San Francisco, 32
Activism, 292–294
Acupuncture, 225
Acute retroviral syndrome, 167. *See also* Retrovirus
Acycolvir (Zovirax), 149
ADA. *See* Americans with Disabilities Act
ADC. *See* AIDS dementia complex
ADCC. *See* Antibody-dependent cell cytotoxicity
Adenovirus. *See also* Virus
  discussed, 121
Adherence, 97
Adolescent AIDS Program, 361
Adolescents. *See also* Children
  AIDS cases, 355–362
    categories, 358–360
  AIDS perception, 410–412, 413
  homeless, 360–361
  runaways, 360
  sexual activity, 408–412
Africa. *See also* South Africa
  abstinence, 359–360
  AIDS cases, 41–42, 319, 320–321, 429
    women, 295, 332–333, 344, 400
  AIDS funding, 432
  AIDS origin, 42–46, 65, 66
  AIDS orphans, 349–350, 354
  AIDS prevention, 245, 252, 255, 258, 264
  AIDS progression, 182–183
  AIDS testing, 366
  AIDS transmission, 23–24, 27, 195, 205, 207–208, 222, 230
  anti-HIV drugs, 98
  breast feeding, 235
  funeral industry, 326–327
  HIV-2 infection, 28–29

home guardianship/widow inheritance, 334
sexual assault, 228–229
AID Atlanta, 430
"AIDS: A Different View", 33
AIDS Clinical Trials Group (ACTG), 142–143
Protocol 076, 34
AIDS cocktails, 82, 84–85. *See also* Anti-HIV therapy
AIDS Community Alliance, 361
AIDS dementia complex (ADC), 186–188, 388
"AIDS Dilemma, The" (Ruesberg/Rasnick), 34
AIDS funding
"AIDS Programs: An Epidemic of Waste", 430–431
federal funding, 425–427, 428
global fund, 433–434
global funding
in general, 432–433
International Global Fund, 433–434
US contribution, 433–434
per death, compared to other diseases, 428–429
as percentage of national budget, 427
private sector funding, 427
UN call, 431–432
AIDS Health Project, 430
AIDS Healthcare Foundation, 217
AIDS Memorial Quilt, 14–16
Names Project, 16
AIDS pandemic, 296. *See also* Epidemic
"AIDS Programs: An Epidemic of Waste", 430–431
AIDS Project Los Angeles, 82
AIDS surveillance
AIDS definition, 29–30, 171*t*
discussed, 29
AIDS Treatment News, 100
AIDS Update 1999, 403
AIDSVAX, 285–286
Alan Guttmacher Institute, 360
Alcohol, 423
Allen, Scott, 405
Alvarez, Marina, 304
Alzheimer's disease, 7
American Association of Blood Banks, 268
American Bar Association Journal, 421
American College Health Association, 307
American Heart Association, 192
American Medical Association, 255, 391, 420
American Social Health Association, 257
Americans with Disabilities Act (ADA), 402, 414, 421–422
Amplicor HIV Monitor Ultra Sensitive test, 92, 383
Amplicor Q-PCR test, 92
Amprenavir (Agenerase), 74*t*, 79–80
Amyl nitrate, 27
Anderson, Allen, 236
Anderson, Jack, 399
Anemia, 131
AngloGold, 323
Angola, 64. *See also* Africa
Annan, Kofi, 431–432
Anorexia, 151
Anthrax, 9
Anti-HIV therapy. *See also* Human Immunodeficiency Virus; Prevention; Testing; Viral load; *specific drugs*
antiretroviral therapy, 178, 293
long-term non-progressors, 178, 181–184
rapid progressors, 178
slow progressors, 178
costs, 241–242
current state, 104
drug complications
drug holidays, 91, 100–105
in general, 90
mutant HIV, 175
side effects, 75–77
strategic drug interruption, 91
structured treatment interruption, 91, 100–105
drug resistance
coinfection, 85
combination therapy and, 84–86
in general, 6, 174
genotypic analysis, 87
genotypic/phenotypic susceptibility profiles, 87–88
HIV replication measurement, 88
nucleoside analog mutants, 80
phenotypic analysis, 87
protease inhibitor mutants, 80–84
replication capacity test, 88
resistance testing, 93
superinfection, 86
drugs, 38, 74*t*
AIDS cocktails, 82, 84–85
antisense drugs, 74*t*
assembly inhibitors, 74*t*
combination therapy, 101–102
drug cocktails, 80, 84–85
economics, 96
effectiveness, 93
entry inhibitors, 103–105
expanded access use, 73
with FDA approval, 72–75
integrase inhibitors, 74*t*
long-term dilemma, 104
monotherapy, 75, 81–84
non-nucleoside analog reverse transcriptase inhibitors, 74*t*, 77–79, 265
non-nucleoside reverse transcriptase inhibitors, 74*t*, 96
nucleoside analog reverse transcriptase inhibitors, 74t, 86–87, 96
nucleoside/non-nucleoside analog reverse transcriptase inhibitors, 74*t*, 75–77
post-exposure prevention, 103
protease inhibitors, 74*t*, 77–80, 86–87, 96, 101–102

reverse transcriptase inhibitors, 74–77
surrogate markers, 72
zinc finger, 74*t*
future for, 95–96
in general, 68–70
HAART, 83–84, 93, 96–100
side effects, 91
Lazarus effect, 82
management
ratio of physicians to population and patients, 95
treat early/treat late, 91–94
number of persons receiving, 82
for opportunistic infection, 142
pre-exposure prophylaxis, 87
problems with, 97–100
adherence, 97
compliance, 97
costs, 98, 99
duration, 97
side effects, 98–99
underdosing, 97
side effects, HAART, 91
standard therapy, 83
treatment
kitchen sink, 86–87
salvage therapy, 86
treatment failure
clinical failure, 86, 175
in general, 86
immunologic failure, 86, 175
virologic failure, 86, 175
Antibiotics, 63
Antibody, 25, 37, 279. *See also* Human Immuno-deficiency Virus
antibody test, 366
B cell production, 121
cellular immunity, 279
discussed, 114, 117, 120–121, 284
HIV disease and, 121–122
HIV protected from, 123–124
infection-enhancing antibody, 123
neutralizing antibody, 112
production, 176
Antibody-dependent cell cytotoxicity (ADCC), 131
Antigen. *See also* p24
CD4 antigen, 113
discussed, 117
superantigen, 132
Antigen-presenting cell (APC), 112, 118, 132
Antisense drugs, 74*t*
APC. *See* Antigen-presenting cell
APOBEC3G enzyme, 58–59
Apoptosis, 131
discussed, 132–133
Arkansas, 278, 351, 391
Armed Forces Institute of Pathology, 10
Armstrong, Rebekka, 339
Arthritis, 50
Artificial insemination, 201, 225, 336–338. *See also* Sexual activity; Women
Ashe, Arthur, 11, 413
Asia, 223–224
Associated Press, 413, 424
Atazanavir (Reyataz), 74*t*, 80
Atevirdine, 77
Atherosclerosis, 7
Athletes, 217–219
Australia, 336
Automatic, 254
Automobiles, 423
AZT treatment, 34

B
B cell
discussed, 110–112, 119
production, 121
B cell lymphoma, 7
Bacteria, Group A streptococcus, 2
Bagasra, Omar, 200
Bailes, Elizabeth, 44
Bailey, Robert, 203
Baltimore, David, 125, 184
Bardeguez, Arlene, 341
Barebacking, 212–214. *See also* Homosexual activity
Bay Area Reporter, 32
Bay Area Young Positive, 361
Bay Guardian, 32
Bay Times, 32
Bayer Quantiplex HIV RNA assay, 383
Bayer, Ronald, 210
BBSI. *See* Blood and body substance isolation
Becker, Stephen, 97
Behavioral risk groups, 296–297. *See also* Prevention
Bergalis, Kimberly, 202
Berger, Ed, 128
Bergeron, Michael, 260
Berkley, Seth, 293
β-Chemokine receptor, 128
BioTechnology Company, 285
Blacks, 40, 297, 302, 319, 342–343
HIV infection, 130, 352
Tuskegee experiments, 234
Blame, 7, 398
Bleul, Conrad, 128
Blood, 133. *See also* Blood bank; Hemophiliac; Human Immuno-deficiency Virus
blood safety, 271–273, 379–380
autologous transfusion, 273
Nucleic Acid Test, 272
blood screening, 271, 368
window period, 271
HIV transmission, 199–200, 268–275
Blood and body substance isolation (BBSI), 275

Blood bank, 24, 200, 268
contaminated blood, 209–210, 268–272, 306, 346, 379
Blood donor, 268
Blood Feuds (Feldman), 210
Blossom, 9
Bluette, Skip, 22, 69, 70
BMS805, 105
Boccachio, 4
Boehringer, 96
Bolognesi, Dani, 176
Bone loss, 91
Borrowed Time (Monette), 148, 150
Bovine spongiform encephalopathy, 5
Bowen, Otis, 308, 399
Bradbury, Ray, 204
Bragdon, Dr. Randon, 421, 422
Bragdon v. Abbott, 421
Brain, cerebral spinal fluid, 135
Branched DNA test, 366, 383. *See also* Testing
Brands, Martien, 33
Brazil, 8, 64, 245, 333
Breast cancer, 428
Breast feeding, 232, 235, 346, 351, 352–353
Bridges, Fabian, 226
Bring Out Your Dead (Powell), 419
Bristol-Myers Squibb, 96, 105
Buchbinder, Susan, 181–182, 208
Buffalo hump, 99, 99*f*
Burden of a Secret (Allen), 405
Burnet, Sir MacFarlane, 6
Burnett, Joseph, 247
Bush, George, 71, 271, 401, 433
Bush meat, 8, 42. *See also* Chimpanzee; Monkey

C
CAGW. *See* Citizens Against Government Waste
Calgary Herald, 219
California, 95, 336, 352, 357, 361
NEP, 267
Callen, Michael, 132
Campo, Rafael, 304
Camus, Albert, 4, 401
Canada, 242, 268, 320, 432
public confidence survey, 234–235
Canadian Public Health Journal, 423
Cancer, 144*t*, 428. *See also specific cancers*
in general, 154–155
viral associations, 7
Cancer Research, 33
Candida albicans, 155t
discussed, 145
Candidal esophagitis, 140, 142, 145, 341
Candlelight Memorial, 16
Cannon, Michael, 157
Caprine arthritis-encephalitis virus, 50
Cardiopulmonary resuscitation (CPR), 192
Casper, Toby, 96
Cat
feline immunodeficiency virus, 5, 40, 50, 281
feline leukemia virus, 280
Catalan, Jose, 173
Catholics for a Free Choice, 259
CC-CKR-5 (R-5), 128–129, 183
CCKR-2 (R-2) receptor, 129
CCKR-3 (R-3) receptor, 129
CD4 cell, 11, 68–69
CD4 antigen, 113
CD4+ cell, 25, 27, 56, 178
cell pool, 124
in general, 113, 118, 122, 124–127, 183–184
loss of, 114
profound cell depletion, 135
CD4+ cell count, 72, 83, 140, 185
complications and, 170*t*
use of, 116–120, 180–184
CD4+ cell disease, 25
coreceptor, 183
non-CD4 type cell, 129
T4 cell and, 131
CD41 cell, 131
CDC. *See* Centers for Disease Control and Prevention
Cell receptor, 123
Center for Strategic and International Studies, 325
Centers for Disease Control and Prevention (CDC), 1, 24, 26–27, 29, 32, 87, 103, 140, 171, 241, 245, 254, 291, 368, 398, 415
SAFE program, 253
Central Intelligence Agency (CIA), 399
Central nervous system (CNS)
AIDS infection
AIDS dementia complex, 186–188, 388
anti-HIV therapy, 186
in general, 185–186
Cervical cancer, 7
Cervical dysplasia, 342
Chamabantu, Gertrude, 326
Chemokine, 117, 128, 130. *See also specific chemokines*
Chemokine receptors, 183
Chicago Tribune, 373
Children. *See also* Adolescents; Women
"AIDS baby", 358
AIDS deaths, 13
AIDS diagnosis, 188–189
AIDS infection, 232–233
AIDS orphans
Africa, 349–350, 354
in general, 321, 323, 347–349
AIDS transmission, 183
child molestation, 227
pediatric AIDS
decline in, 352
ethnic prevalence, 347

perinatal transmission, 233, 351
states affected by, 351–352
transmission, 231–233, 346–348
sex trade, 223–224
testing
mandatory testing, 391
prenatal testing, 386
Chiles, Lawton, 370
Chimpanzee, 187–188. *See also* Monkey
AIDS origin, 42–46, 61
SIVcpz, 42–44, 65, 67
China
AIDS cases, 9, 64, 65, 195, 209–210, 222, 252, 295, 327–328
blood supply, 209–210
HIV survey, 236–237
SARS, 5
Chirac, Jacques, 25
Chiron, 89, 383
Cholera, 3*t*, 5, 6, 7, 40
Chondoka, Yizenge, 326
Chowdhry, Tanvir, 100
CIA, 40
CIA. *See* Central Intelligence Agency
Circumcision, 203–204
Citizens Against Government Waste (CAGW), 43
Cleanliness, 2
Clevici, Mario, 183
Clinton, Bill, 278, 360, 417
CNS. *See* Central nervous system
Coates, Tom, 250
Cocaine, 341. *See also* Drug use
Coccidioidomycosis, 143
Cohen, Jon, 33
Coinfection, 85
College students, AIDS infection, 307
CollegeClub.com, 244
Coloradans for Family Values, 302
Combivir, 74*t*, 96, 217
Community Programs for Clinical Research on AIDS (CPCRA), 96
Compassion, 404, 409
Compliance, 97
Conant, Marcus, 250
Condom. *See* Sexual activity
"Condom Hut", 255
Confide HIV Testing Service, 384
Congo, 41, 44, 64
Connecticut, 267, 307, 351, 391
Consumer Reports, 259
Cooper, David, 83, 286
Corbeil, Jacques, 131
Cornett, Leanza, 401
Cosmopolitan, 399
Costopoulos, Paul, 409
Coulis, Paul, 176
Coutsoudis, Anna, 235
Cox, Spencer, 105
CPCRA. *See* Community Programs for Clinical Research on AIDS
CPR. *See* Cardiopulmonary resuscitation
Craven, Donald, 416
Crixivan, 74*t*
discussed, 73
Cruise ships, 225
Cryptococcal meningitis, 150
Cryptococcus neoformans, 145
discussed, 145–147
Cryptosporidium, 150–151, 155*t*
Cuba, 64, 224–225, 249–250
CXCKR-4 (R-4), 128
Cytokine, 117
Cytomegalovirus, 24, 133, 144*t*, 155*t*, 342
discussed, 121, 148
retinitis, 150

D
Daneels, Godfreid, 259
Darrow, William, 214
Data sources, 291–294. *See also* Education
Davis, Julia, 244
Dawn's Gift, 412
Day, Lorraine, 425
Deacon, Nicholas, 182
Deadly Deception (Willner), 33
Dean, Michael, 130
Defoe, Daniel, 4
DeJesus, Esteban, 219
del Romero, Jorge, 208
Delacorte, Eric, 42
Delavirdine (Rescriptor), 74*t*, 77
Dendritic cell, 135, 203. *See also* Human Immunodeficiency Virus
discussed, 124–127
DC-SIGN, 127
dendritic trojan horse, 127
follicular dendritic cell, 136
Dengue fever, discussed, 5
Denial, 24, 166
Dental dam, 251
Dentist, 422
HIV transmission, 201–203
Desrosiers, Ronald, 284
Developed nations. *See also* Developing nations; *specific nations*
AIDS epidemic, 294
Developing nations. *See also* Africa; *specific nations*
AIDS transmission, 205
AIDS treatment, 140
condoms, 252, 258–259
drug use, 221–222
prostitution, 208–224
vaccines, 286–287
Devereux, Helen, 62

DHHS. *See* U.S. Department of Health and Human Services
Diabetes, 7, 428–429
Dickens, Charles, 1
Didanosine (Videx), 74*t*, 75
Diphtheria, 5
DIS. *See* Disease Intervention Specialist
Discovery, 33
Discrimination, 403–404, 406, 427
Disease. *See also* Epidemic; Opportunistic infection; Virus; *specific diseases*
  drug resistance, 6
  infectious disease
    blame and, 7
    as cause of death, 6
    discussed, 5–6
    disease paradigm, 4
    first recognition, 2
Disease Intervention Specialist (DIS), 275
"Dissident or Don Quixote" (Gibbs), 33
Dlamini, Gugu, 340, 402
Dlamini, Priscilla, 419
Donahue, 404
Donahue, Phil, 404
Driskill, Richard, 202
Drug holidays, 91. *See also* Anti-HIV therapy
  discussed, 100–105
Drug legalization, 270
Drug resistance. *See* Anti-HIV therapy
Drug use, 24, 27, 133, 212, 213, 224, 341. *See also* Anti-HIV therapy; Injection-drug user; *specific drugs*
Duesberg, Peter, 32–36, 219
Dupont, 104

E
Ebola virus, 4, 5, 46, 110
  discussed, 8
Education. *See also* Prevention
  about condom, 260
  AIDS impact, 324–325, 404
  behavior and, 404–407
  costs, 408
  prevention and, 242, 242–243, 244, 245, 272, 367
  public education programs, 407–408
  public schools education programs, 408–411
  UK survey, 235–236
Edwards, Sara, 208
Efavirenz (Sustiva), 74*t*
80/20 rule, 204
Eiseley, Loren, 19
ELISA. *See* Enzyme linked immunosorbent assay
Elson, John, 217
Employment, following remission, 82–83
Emtricitabine (Emtriva), 74*t*, 80
Encephalitis, 149
Entry inhibitors, 103–105. *See also* Anti-HIV therapy
ENV gene, 56, 62
Envelope, 51, 56
Enzyme linked immunosorbent assay (ELISA) test, 65, 366, 367. *See also* Testing
  detuned ELISA test, 380
  in general, 367–368, 379–380
  highly tuned ELISA test, 380
  problems with, 368–370
    false negative, 368
    false positive, 368–370
    window period, 368
  sensitivity and specificity, 370–377
    confirmatory test, 370
    positive predictive value, 370, 373–377
  STARHS, 380
  understanding, 368
Epidemic. *See also* Disease
  AIDS pandemic, 296, 311–315, 396–397
  defined, 191
  historical, 1–2
    Plagues in History, 3*t*
Epidemiology, 191
Epstein-Barr virus, 7, 133, 144*t*, 155*t*, 284
Equitable Contributions Framework, 432
Esquire, 105
Essex, Max, 65
Essex, Myron, 399
Ethiopia, 295, 333. *See also* Africa
Eugene-Olsen, Jesper, 130
Evans, Barry, 208
Everstadt, Nicholas, 397
Exercise, 134
Eye disease, 148–149
  retinitis, 150

F
Fallopius, Gabrielle, 253
Falwell, Jerry, 27
Family Research Institute, 302
Famvir, 149
Farr, Gaston, 261
Fauci, Anthony, 77, 124, 135, 284
FDA. *See* U.S Food and Drug Administration
Fear, 166, 196, 228, 398–400, 401, 402, 406, 420
Feinberg, Mark, 61
Feldman, Eric, 210
Feline immunodeficiency virus, 5, 281
Feline leukemia virus, 280
Female Health Company, 261
Fenton, Kevin, 276
Feshbach, Murray, 222
Fidelity, 251. *See also* Sexual activity
Findlay, Steven, 246
Finzi, Diana, 102
Firearms, 423
Fisher, Mary, 417

Fleming, Patricia, 352
Florida, 214, 225, 227, 276, 308, 309, 336, 341, 351, 352, 357, 361, 401
Florida Times Union, 408
Flu (Kolata), 10
Fluorognost, 378
"Force for Change", 355
Forstein, Marshall, 214
Fosamprenavir (Lexiva), 74*t*, 80
Foscavir, 149
Foster, Roland, 266
"4 H disease", 27
Fox, Cecil, 102, 135
FRAID, 19
France, 65, 261, 268, 420, 432
Frankfurt Clinic Cohort, 84
Frezieres, Ron, 260
Friis-Moller, N., 91
Frontline, 226
Fulton, Robert, 419
Funding. *See* AIDS funding
Fungal disease. *See also* Opportunistic infection
  discussed, 143–147
FUSIN, 128

G
GAG gene, 55, 62
Gallo/Montagnier Program for International Viral Collaboration, 287
Gallo, Robert, 25, 27, 105, 187, 278, 281, 287
Gallup Poll, 423
Gambian rat, 5
Ganciclovir, 150
Garrett, Laurie, 105
Garrus, Jennifer, 48
Gates Foundation, 87
Gastrointestinal problems, 79
Gaulthier, DeAnn, 213
Gavin, Hector, 416
Gay Men's Health Crisis, 212, 213
Gay-Related Immune Deficiency (GRID), 27
Gay.com/PlanetOut.com, 412
Gays. *See* Homosexual activity
Gender power, 245
Genentech, 285
Genetic revolution, 60
Genital ulcer disease (GUD), 230–231
Germany, 64, 432
Gibbs, Wayt, 33
Gibson, Steven, 250
"Gift, The", 214
Glaser, Paul & Elizabeth, 206–207
Glaxo Wellcombe, 96
Global AIDS, Tuberculosis and Malaria Relief Act, 415, 430
Glycan shield, 125
Glycoprotein (gp), 51, 125
  gp120, 125
Goat, 50
Goddard, Jerome, 195
Golden Triangle, 223–224
Goldsworthy, Bill, 219
Gonorrhea, 3*t*, 208, 231
Gordon, Alexander, 2
Gottlieb, Michael, 1, 24, 29
Gould, Robert, 399
gp. *See* Glycoprotein
Graham, Billy, 27
Great Britain, 432
GRID. *See* Gay-Related Immune Deficiency
Grmek, Mirko, 45
Group A streptococcus, 2
GUD. *See* Genital ulcer disease
"Guide to the Clinical Care of Women with HIV", 342
Guilt, 82, 166
Guinea Bissau, 28
Gutknecht, Gil, 34, 37

H
HAART. *See* Highly Active Anti-Retroviral Therapy
Hader, Shannon, 342
Hahn, Beatrice, 42, 45
Hairy leukoplakia, 179
Haiti, 37, 44, 45, 205, 225
  Kaposi's sarcoma, 26
Halperin, Daniel, 203
Hand washing, 2
Hantavirus, 2, 5
Hare, Peter, 217
Harrison, Lee, 200
Harvard AIDS Institute, 304
Hatred, 407
Hawaii, NEP, 267
Hawthorne, Peter, 228
Health care worker, needle stick injury, 308
Hearst, Norman, 216, 256
Heart disease, 91, 429
Heath, Sonya, 135
Heckler, Margaret H., 282
Helms, Jesse, 396, 407
Hemophiliac. *See also* Blood
  AIDS infection, 34, 37, 181, 209, 233, 305–306, 346, 405
  AIDS transmission, 23
Henry J. Kaiser Foundation, 260, 272, 360, 412, 414
Hepatitis B, 7, 133, 192, 221, 253, 268, 273, 279
  discussed, 10–11
Hepatitis C, 133, 144*t*, 221, 268
  discussed, 5
  as opportunistic disease, 144*t*, 147–148
Herbal medicine, 95, 166
Heroin, 224, 270

Herpes. *See* Human herpes virus
Heterosexual activity. *See also* Sexual activity
  AIDS transmission, 66, 183, 195, 203, 303, 333, 336
    adolescents, 361
    in general, 204, 311
    global transmission, 303–305
    non-USA, 205
    risk estimates, 216–220
    vaginal/anal intercourse, 205–206
    viral load and, 220
Highly Active Anti-Retroviral Therapy (HAART), 93, 96–100, 234. *See also* Anti-HIV therapy
  discussed, 83–84, 245–246
  effect of, 211–212
  future for, 95–96
  giga HAART therapy, 86–87
  for opportunistic infection, 142
  response to, 170
  side effects, 91
Hippocrates, 9
Hirsch, Vanessa, 41
Hispanics, 297
Histoplasmosis, 143, 144*t*
  discussed, 145
History of AIDS (Grmek), 45
Hitler, Adolph, 398
HIV. *See* Human Immunodeficiency Virus
HIV disease. *See also* Acquired Immune Deficiency Syndrome; Human Immunodeficiency Virus
  antibody production, 176
  defined, 162
  disease progression, 174–175, 181–184
    drug resistance, 174
    HIV clearance, 175
    treatment failure, 175
    viral pool, 174
  HIV/AIDS progression classification, 177–179
    persistent generalized lymphadenopathy, 177
    viremic, 177
  life expectancy
    case presentations, 172
    in general, 171–172
    symptoms and impairment, 173
    2003 federal clinical guide, 172–173
  neuropathy, peripheral neuropathy, 75, 188
  pregnancy and, 344–345
  primary infection route, 167–169
  spectrum, 162–164
  stages
    acute state, 167–169
    advanced stage, 170–171
    asymptomatic stage, 167, 169, 175–176, 416, 421
    in general, 164–167, 175
    symptomatic stage, 170
  transmission, 175
HIV reservoir, 123
HIV Testing Day Campaign, 366
HLA. *See* Human leukocyte antigen
Ho, David, 136, 169
Hodgkin's disease, 145
Hoffmann, Geoffrey, 133
Holmes, Oliver Wendell, 2
Holmes, William, 211
Holmstrom, Paul, 192
Holtgrave, David, 204
Home Express HIV Test, 384
Home HIV test, 366. *See also* Testing
  discussed, 383
Homeless, 360–361
Homosexual activity. *See also* Sexual activity
  AIDS risk, 133, 211–214, 320
    Miami's South Beach, 214
  AIDS transmission, 195, 297
    anal intercourse, 206
    barebacking, 212–214
    circuit parties, 213
    MSM, 301–302
    receptive/insertive partner, 205–206
  demographic changes, 302–303
  extant of, 302
  syphilis study, 212–213
Homozygote, 130
HOPWA. *See* Housing Opportunities for People With AIDS
Horton, Richard, 9
Hot Zone, The (Preston), 8
Housing Opportunities for People With AIDS (HOPWA), 414, 430
Hoxworth, Tamara, 276
Huff, Douglas, 40
Hulley, Stephen, 216
Human herpes virus, 110, 208, 253, 284
  genital herpes, 158, 231
  herpes simplex, 140, 144*t*, 148, 155*t*, 341–342
  herpes zoster virus, 149
  HHV 6, 2, 133
  HHV 7, 2, 133
  HHV 8, 2, 133, 156
  HSV 1 & 2, 148–149
  Kaposi's virus and, 156–158
Human Immunodeficiency Virus (HIV), 5. *See also* Acquired Immune Deficiency Syndrome; Anti-HIV therapy; HIV disease; Immune system; Virus
  biology
    HIV lifecycle, 50–55
    long terminal repeat, 52
    provirus, 51–52, 70–72, 159
    replication rate, 136–137
    transcription, 52
    viral RNA strand/RNA transcript, 52–55
    virus characteristics, 48–50
  as cause of AIDS, 32
    alternative views, 32–37
    evidence, 37–39

Koch's Postulates, 38–39
"origin of HIV", 39–46
discovery
in general, 27
HTLV III, 27
naming virus, 27–28
epidemiology, 191
in general, 192
HIV1/HIV2, 193
evolutionary impact, 61
in general, 1, 25, 32, 121
genetics
accessory protein, 55
antibody protection, 123–124
antigenic variation, 62–63
APOBEC3G enzyme, 58–59
clades/subtypes, 62–67
gene function, 57–58
gene sequence, 57
in general, 55
genetic instability, 60
genetic recombination, 61–63
genetic revolution, 60
mutations, 41, 59–60, 60–63
nef, 57, 58
nine HIV genes, 55–56
rev, 57
reverse transcriptase, 60, 62
six HIV genes, 56–57
tat, 57, 58
transcription error, 137
variants, 59–60
vif, 57, 58–59
vpr, 57, 58
vpu, 57, 59
HIV-2
compared to HIV-1, 61, 63, 193
first reported case, 27, 28–29
origin, 41, 44–46, 65
HIV disease, 25
HIV messenger RNA, 57
HIV reservoir, 123
infection
acute infection, 124, 167
acute retroviral syndrome, 167
aspects of, 176
asymptomatic infection, 167, 169, 175–176, 416, 421
barriers to, 251
biological latency, 169
chronic infection, 167
estimating, 296
HIVs required for, 199
how HIV enters the body, 192
incubation and latency, 164
lower set point, 168
monocytes/macrophages, 134–137
primary infection, 124
primary infection route, 167–169
rapid progressors to AIDS, 130
set point, 185
symptomatic AIDS, 164
variation of, 163
virulence, 231
window of infectivity before seroconversion, 168
with/without drugs, 178–179
mutations, 41, 59–60, 60–63
origin of HIV, 39–46
biological warfare, 40
cats, 40
chimpanzee, 42–46
CIA, 40
UFOs, 40
receptor swap, 128–129
replication capacity, 88
resistance
genetic resistance, 129–130
R-5 gene, 130
strain, macrophage-tropic HIV strain, 128
subtypes, 63–67
A, 64, 65
AB, 65
B, 65
C, 64, 65
D, 64, 65
E, 64, 65
international subtypes, 64–65
M, 63–64
mysteries of, 66
N, 63, 65–67
non-B, 64
"non-M non-O", 65–67
O, 63, 64, 65, 66
"outlier", 65
related to transmission, 65
syncytium-inducing HIV, 56, 128, 162
transmission
acupuncture, 225
artificial insemination, 201, 225, 336–338
behavioral/social factors, 192, 198–199
biological factors, 192
blood, 199, 268–275
body fluids, 199–200
breast milk, 201
caesarean section, 233
dentist, 201–203
drug use, 220–221
80/20 rule, 204
family settings, 195–197
global patterns, 195, 198*t*
human bite, 225
insects, 192, 194–195, 235
kissing, 192, 200–201, 235

noncasual transmission, 197–198
organ transplant, 229
other routes, 192, 195, 225, 246–247
pediatric transmission, 231–233
perinatal transmission, 233
saliva, 200
semen, 199–200
sexual assault USA, 225–228
sexual transmission, 23–24, 66, 192, 195, 197–198, 201, 203–231
sperm, 200
subtypes, 65
vertical transmission, 233
workplace transmission, 234
viral fitness, 88
virulence, 231
Human leukocyte antigen (HLA), discussed, 111–112, 119
Human papillomavirus, 7
Human rights, 1
Hume, Ellen, 413

I
IAS-USA. *See* International AIDS Society-USA
IDU. *See* Injection drug user
IFA. *See* Immunofluorescent antibody assay
Illinois, 351
Immigrants, 386–388
designated event policy, 387
foreign countries, 388
illegal immigrants, 388
routine HIV waiver policy, 387
Immigration and Nationality Act, 387
Immune system. *See also* Human Immunodeficiency Virus
AIDS/HIV affecting, 27, 69–70, 122, 142
branches of, 117–120
cellular immunity, 117
dysfunction, in general, 124
function of, 109–110
APC, 112
class I protein, 110, 111–112, 120
class II protein, 111–112, 120
HLA, 111
immune response, 110
MHC, 112
peptide, 112
"self/nonself", 110
self protein, 111
in general, 109, 117–120
humoral immunity, 114, 120
immune surveillance, 112
infection and, 115–116
lymphocytes, T cells/B cells, 110–112, 114
Immunofluorescent antibody assay (IFA), 366. *See also* Testing
discussed, 378
Immunoglobin, 114
Immunopathogenesis of HIV Infection (Pantaleo), 173
Incubation, clinical incubation, 164
India, 403
AIDS cases, 64, 182–183, 216, 295, 325
Indinavir (Crixivan), 74*t*, 79, 80
Indonesia, 195
Infection (Horton), 9
Influenza, 3*t*, 279
discussed, 9–10
1918 pandemic, 10
Injection drug user (IDU)
AIDS transmission, 23, 65, 195, 220–223, 265–266, 297, 303, 351–352
needle/syringe, 24, 45, 195, 197–198, 265–268, 307, 308
needle exchange program, 265–268, 269–271
Insects, HIV transmission, 192, 194–195, 235
Insurance, 422
Integrase, 51, 77
Integrase inhibitors, 74*t*, 76*f*
Interleukin, 134, 178
International AIDS Conference, 15*t*
"AIDS: A Different View", 33
Durban, 14, 35, 39
International AIDS Society-USA (IAS-USA), recommendations, 94
International Labor Organization, 19
International Women's Day, 332
Internet, 277, 287, 345
Inungu, Joseph, 367
Inventing the AIDS Virus (Duesberg), 34, 219
Ionosco, Eugene, 407
Ipsos-Reid, 234

J
Jackson, Jesse, 388
Jaffe, Harold, 26, 33, 309
Japan, 336, 432
Jenner, Edward, 9, 283
Jim Lehrer News Hour, 325
Johannesburg Mail and Guardian, 323
Johnson, Earvin "Magic", 134, 217–218, 247, 412, 422
Johnson, Nkosi, 418–419
Johnson, Sherry, 202
Johnson, Virginia, 399
Jones, Cleve, 14
Joseph, Stephen J., 276
Journal of the American Medical Association, 6, 302
Junin virus, 8
Justice, Dr. Barbara J., 153

K
Kajubi, Senteza, 324
Kaplan, Art, 229
Kaplan, Edward, 267
Kaposi, Moritz, 156

Kaposi's sarcoma (KS), 26, 27, 41, 144*t*, 155, 341. *See also* Cancer; Opportunistic infection
in general, 155–156
edema, 156
indolent, 156
herpes and, 156–158
Katz, Mitchell, 211
Kenya, 182, 324–325, 344, 359. *See also* Africa
Kessler, David, 383
Kidney disease, 91, 99, 140
Kilimanjaro, Proud, 390
Kinsey, Sex and Fraud (Reisman), 302
Kion, Tracy, 133
Kirby, Douglas, 257
Kissing, 192, 200–201, 235. *See also* Sexual activity
Kissinger, Patricia, 276
Kitchen sink, 86–87
Klemmens, Dr. Sallie, 153
Koch, Robert, 39
Koch's Postulates, 38–39
Kolata, Gina, 10
Kolodny, Robert, 399
Koop, C. Everett, 317, 401
Korber, Bette, 44
Kourilsky, Phillippe, 279
Kramer, Larry, 24, 99, 229
Krieger, John, 200
KS. *See* Kaposi's sarcoma
Kwakwa, Helena, 340
Kwong, Peter, 125

L
Lacayo, Richard, 226
Lackritz, Eve, 271
Lactic acidosis, 75–76, 91
Lamivudine (Epivir), 74*t*, 75, 80
Lancet, The, 203
Langerhans cells, 65, 203
Lassa virus, 5
discussed, 8
Latency
biological latency, 169
clinical latency, 164
Latin America, 205, 252, 261
AIDS transmission, 195, 205, 222–223, 225
Latinos, 297, 302, 319, 343
discussed, 304
LAV. *See* Lymphadenopathy-associated virus
Lavitt, Mara, 22, 69, 70
Lazarus effect, 82
Lazuriaga, Katharine, 346
LeCarre, John, 51
Lederberg, Joshua, 6
Lee, Jong Wook, 18
Lefkowitz, Mathew, 407
Legionnaires disease, 2
Legislation, 414–415
Lemp, George, 340
Lentivirus, 50
Lesbians and AIDS, 340
Lesson, The (Ionesco), 407
Leukemia, 50, 145
feline leukemia virus, 280
Leukocyte, discussed, 117
Lewis, C.E., 420
Lewis, Lennox, 390
Lewis, Stephen, 434
Lifestyle, 248
Ligand, 123
Lipodystrophy, 91, 99
Liu, Rong, 130
Liver disease, 5, 7, 75–76, 91, 99, 140, 148
Lopinavir/Novir, 80
Los Angeles Times, 40, 402
Louganis, Greg, 219
Loviridine, 77
Lupus, 50
Lyme disease, 2
Lymph node, 135
Lymphadenopathy, 27
Lymphadenopathy-associated virus (LAV), 27
Lymphocyte. *See also* CD4 cell; T4 cell
B lymphocyte, 119, 120
cytotoxic T lymphocyte, 119, 168, 183–184
discussed, 117, 118
helper T lymphocyte, 113, 118
killer T lymphocyte, 113
memory cell, 120
natural killer cell, 120
suppressor T cell, 120
T cell/B cell, 110–112, 114
T lymphocyte, 118, 180
Lymphocyte surface receptor, 117
Lymphokine, 118, 134*f*
Lymphoma, 144t. *See also* Opportunistic infection
discussed, 158

M
Maathal, Wangari, 40
MAC. *See* Mycobacterium avium complex
Machanick, Philip, 34
Machupo virus, 8
MacNeil/Lehrer Report, 357
Macrophage
role in infection, 134–137
trojan horse, 135
Maggiore, Christine, 32
Major histocompatibility complex (MHC), 58, 111, 112
self-major histocompatibility complex, 112
Makhalemele, Mercy, 402
Malaria, 2, 3*t*, 63, 95, 131, 140, 210, 431
discussed, 10–11
Mallon, Mary, 6

Mandela, Nelson, 324
Mann, Jonathan, 398
Mansergh, Gordon, 211
Marcus, Ruthanne, 308
Markoff, Niro Asistent, 179
Marlburg virus, 5
  discussed, 8
Marmor, Michael, 130
Marquez, Gabriel Garcia, 4
Marshal, Michael Leonard, 179
Marx, Preston, 45, 63
Massie, Robert, 181
Masters, William, 399
Matassa, Matthew, 16
Mbeki, Thabo, 34–36, 327, 354
McCarthy, Gillian, 422
McConnell, Michelle, 353
McGee, Darnell, 226
McGowan, John, 212
McNeil, Donald, 166
McNeill, William, 419
Measles, 3*t*, 7, 279
Media, 291
Medicaid, 264, 415
Medical Doctor, 276
Mellors, John, 80
Melton, George, 179
Merck & Co., 73, 96
Methamphetamine, 212. *See also* Drug use
Metropolitan Life Insurance Company, 320
Mexico, 143
Meyers, Gerald, 41
MHC. *See* Major histocompatibility complex
Miami. *See also* Florida
  South Beach, 214
Miami Herald, 373
Michigan, 351
Microbicides, 263–265
Midwest, 143, 145
Migueles, Stephen, 182
Military, 306–307, 323, 392
Milk, Harvey, 16
Miner's canary, 344
Mississippi, 276, 307
Missouri, 278
Mitochondria, 200
Mitochondriosis, 91
Modeste, Wendi Alexis, 152–154
Monette, Paul, 148, 150
Monkey, 187–188. *See also* Chimpanzee
  AIDS transmission, 28
    green monkey, 65–66
    Sooty Mangabey, 41, 61, 67
  bush meat, 8, 42
Monkeypox virus, discussed, 5
Monocyte, role in infection, 134–137
Monotherapy, 75. *See also* Anti-HIV therapy
  discussed, 81–84
Montagnier, Luc, 25, 27, 72, 227, 287
Montalvo, Michael, 402
Moore, Frank, II, 411
Morbidity/Mortality Weekly, 234
Moreno, Jonathan, 405
Morrison, Tommy, 219, 390
Morse, Stephen, 45
MTV youth poll, 356
Munchausen's Syndrome, 416
Murphy-O'Connor, Cormac, 259
Museveni, Yoweri, 293
Mutual of Omaha, 422
Mycobacterium avium complex (MAC), 140, 144*t*, 151, 155*t*

N

Naftalan, Richard, 206
Names Project, 16
Napoleon, 4
Nary, Gordon, 407
Nasopharyngeal carcinoma, 7
NAT. *See* Nucleic acid test
National Academy of Science, 423
National AIDS Awareness Test, 404
National Association of People Living with AIDS, 366
National Black HIV/AIDS Awareness and Information Day, 388
National Center for Health Statistics, 367
National Condom Day, 251
National Hemophilia Foundation, 147
National Institute of Allergy and Infectious Disease (NIAID), 114, 135, 286, 403
  SMART study, 95–96
National Institute of Health, 87, 282, 377
National Institute of Health Revitalization Act, 414
National Latino AIDS Awareness Day, 304
National Marriage Project, 244
National Opinion Research Center, 302
National Registry of Artists with AIDS, 319
National Teenagers AIDS Hotline, 361
Native Americans, 1–2, 7, 9
Necrotizing fasciitis, 2
Needle, 24, 45, 195, 197–198, 265–268, 307, 308. *See also* Injection drug user; Needle exchange program
  needle stick injury, 308
Needle exchange program (NEP). *See also* Injection-drug user
  California, 267
  controversy about, 269–271
  discussed, 265–268
  global, 267–268
  Hawaii, 267
  New Haven, 267
  New York, 267
  Tacoma, 267

Nef gene, 121
Nelfinavir (Viracept), 74*t*, 79
Nelme, Sarah, 283
Neoplasia, 342
Netherlands, 258, 267, 268
Nevirapine (Viramune), 74*t*, 76, 77, 352, 431
New Directions in the AIDS Crisis, 399
New England Journal of Medicine, The, 24
New Haven. *See also* Connecticut
  NEP, 267
New Haven Register, 22
New Jersey, 309, 336, 352, 357
New York, 212, 226–227, 301, 304, 307, 309, 336, 349, 351, 352, 357, 391, 430
  NEP, 267
New York Post, 430
New York Times, 26, 99
Newsline, 132
Newsweek, 34, 193, 424
Ngugi, Elizabeth, 182
Nigeria, 295
1984 (Orwell), 39
NNRTIs. *See* Non-nucleoside analog reverse transcriptase inhibitors
Non-Hodgkin's lymphoma, 155
Non-nucleoside analog reverse transcriptase inhibitors (NNRTIs), 74*t*, 75–79, 265. *See also* Anti-HIV therapy
  discussed, 77–79
North Carolina, 276
Norwegian sailor, 66
Nowak, Martin, 136
NRTIs. *See* Nucleoside analog reverse transcriptase inhibitors
Nucleic acid test (NAT), discussed, 379
Nucleoside analog reverse transcriptase inhibitors (NRTIs), 74*t*, 86–87, 96. *See also* Anti-HIV therapy
Nucleoside/non-nucleoside analog reverse transcriptase inhibitors, 75–77. *See also* Anti-HIV therapy
Nuke. *See* Nucleoside/non-nucleoside
Nyawo, Sthembile, 350

O
O-Brien, Stepen, 131
Obesity, 423
O'Brien, Megan, 79
O'Brien, Stephen, 61
Of Feigned and Factitious Diseases (Gavin), 416
Office of Technology Assessment (OTA), 195
OI. *See* Opportunistic infection
"On the Contagiousness of Puerperal Fever" (Holmes), 2
On the low down, 302
Onc gene, 55
Oncogene, 55, 159
Oncovirus, 50
O'Neil, Paul, 431
Opportunistic infection (OI). *See also* Disease
  anti-HIV therapy for, 142, 179
  bacterial disease, 151–154
  changing spectrum of, 142–143
  fungal disease, 143–147
  in general, 69–70, 139, 144*t*, 154, 155*t*, 321
  global variation, 143
  prophylaxis against, 140
  protozoal disease, 149–151
  socioeconomic factors, 143
  viral disease, 147–149
OraSure/OraQuick, 381
Oregon, 391
Organ transplant, 229
Orwell, George, 39
Osmond, Dennis, 158
Osterhaus, Albert, 41
OTA. *See* Office of Technology Assessment
Owen, Greg, 419

P
p24, 55. *See also* Antigen
  screening for, 379
Palacio, Ruben, 390
Pancreatis, 75
Panos Institute, 434
Papovirus, 144*t*
Parade, 179
Parcelsus, 104
Parran, Thomas, 275
Partner notification. *See also* Sexual activity
  examples, 276
  in general, 275
    duty to warn, 275, 277–278
  history of, 275–276
  Internet for, 277
  legality of, 277–278
  pros and cons, 276–277
Passkey, 125
Pataki, George E., 391
Paterson, Floyd, 219
Patterson, Bruce, 37
Paxton, Susan, 417
PCP. *See Pneumocystis carinii*
PCR. *See* Polymerase Chain Reaction
Pediatric AIDS Foundation, 346, 389
Pelvic inflammatory disease, 342
Pentafuside (enfuvirtide/fuzeon), 105
People Living With AIDS (PLWA), 91
PEP. *See* Post-exposure prevention; Pre-exposure prophylaxis
Peptide, 112
Perez, Jorge, 250
Peripheral neuropathy (PN), 75, 188
Persistent generalized lymphadenopathy (PGL), 177

PGL. *See* Persistent generalized lymphadenopathy
Phagocyte, 117–118
Pharmaceutical industry, 73
Pharmacia, 104
Philpott, Sean, 130
Phipps, James, 283
Physician
  AIDS transmission, 423–424
    duty to disclose, 424
    patients' right to know, 423–424
  medical moral issues, 420
  physician-patient relationships, 424–425
  public duty, 416–420
  risk protection of the patient, 420–423
Piot, Peter, 13, 98, 246, 295, 325, 328, 337
Plague, 3. *See also* Epidemic
  Black Plague, 5, 7, 11, 130–131, 291, 396, 398, 399
Plague, The (Camus), 401
Plagues and People (McNeill), 419
Plasma, 88
  viral RNA in plasma, 89–90
PLWA. *See* People Living With AIDS
PN. *See* Peripheral neuropathy
Pneumocystis carinii (PCP), 26–27, 133, 140, 144*t*, 155*t*, 179
  discussed, 145
Pneumonia, 26, 140, 148
Poe, Edgar Allan, 4
Pokrovsky, Vadim, 222, 328
POL gene, 55, 62
Poland, HIV survey, 236
Polio, 3*t*, 6
  discussed, 7–8
Polymerase Chain Reaction (PCR), 366. *See also* Testing
  discussed, 378–381
Pope, Melissa, 127
Pornography, 219
Positive Force, 430
Post-exposure prevention (PEP), 103
Poverty, 98, 192, 224–225, 333–334, 435
Powell, Colin, 11, 71
Powell, William, 419
Power, Lisa, 235
Prairie dog, 5
Pre-exposure prophylaxis (PEP), discussed, 87
Presidential Advisory Council on HIV/AIDS, 426
Preston, Richard, 8
Prevention. *See also* Anti-HIV therapy; Education; Testing; Vaccine
  behavioral change, in general, 247–248
  compared to cancer, 243
  compared to treatment, 240–241
  education and, 242, 242–243, 244, 245, 272
  gender power, 245
  in general, 240, 243–247
  global prevention, 241, 243, 246, 252
  goal of, 241
  guidelines, 248*t*–249*t*
  HAART, 245–246
  infection control
    blood and body substance isolation, 275
    in general, 273
    universal precautions, 273–275
  investing in, 241–242
  misinformation and, 245
  National Prevention Information Network, 254
  preventive behavior, 248
  quarantine and, 248–251
  safer sex, 242–243, 251, 252
  San Francisco, 250
  types of, 241
  vaccine
    costs, 285
    defined, 279
    in developing nations, 286–287
    expectations for, 281
    in general, 278–279, 283
    human trials, 285–286
    Internet resources, 287
    problems with, 282–284
    requirements for, 280–281
    testing, 285, 293–294
    testing confidentiality, 286
    types of, 279–282
    vaccine therapy, 284–285
Prisoners, AIDS infection, 307
Proceedings of the National Academy of Sciences, 33
Progressive multifocal leukoencephalo-pathy, 155. *See also* Opportunistic infection
  discussed, 159
Project Inform, 100
Prostitution. *See also* Sexual activity; Women
  AIDS and, 27, 38, 182–183, 208–224, 321, 338–339, 391
  developing countries, 208–224
  sex trade, 222–224
  USA, 208
Protease, 51, 77, 78
Protease inhibitors, 74*t*, 86–87, 96, 142
  discussed, 77–80, 101–102
  drug-resistant mutants, 80–84
Protease paunch, 98*f*, 99
Protein
  class I protein, 110, 111–112
  class II protein, 110, 111–112
Protein processing, 112
Protestantism, 7
Public confidence surveys, 234–237
Public opinion, 193
Puerperal fever, 2
Puerto Rico, 143, 222–223, 349, 352, 353, 357, 430
Purchase, Dave, 267
Putin, Vladimir, 328

Q
QMA. *See* Quebec Medical Association
Quality of life, 140
Quarantine, 248–251. *See also* Prevention
Quasispecies, 62
Quebec Medical Association (QMA), 424

R
R-4. *See* CXCKR-4
R-5. *See* CC-CKR-5
Ramses, 1, 9
RANTES, 128–129
Rapid HIV test, 366. *See also* Testing
  discussed, 381–383
  1-minute test, 382
  3-minute test, 382
Rasnick, David, 34–36
Ray, Ricky, Robert, Randy, 24, 405
Reagan, Ronald, 23, 24, 25, 26, 401
Receptor swap, 128–129
"Recommendations for Prevention Of HIV Transmission in Health-Care Settings" (CDC), 273
Red Crescent, 210
Red Cross, 210, 272, 373
Regulatory protein, 55
Reisman, Judith, 302
Renzi, Cristina, 264
Replication capacity, 88
Replication capacity test, 88. *See also* Testing
Research Triangle Institute, 234
Retrovirus. *See also* Human Immuno-deficiency Virus; Virus
  acute retroviral syndrome, 167
  discussed, 49–55
  lentivirus, 50
  oncovirus, 50
  spumavirus, 50
"Retro-viruses as Carcinogens and Pathogens" (Duesberg), 33
Reverse transcriptase (RT), 60, 62, 77
Reverse transcriptase inhibitors (RTIs), 74–77
Ribonuclease enzyme, 51
Richman, Douglas, 175
Richmond, Tim, 219
Ricky Ray Hemophilia Relief Fund Act, 414
Rinaldo, Charles, 157
Risk, 204
Ritonavir (Norvir), 74*t*, 79, 80
Rizzuto, Carlo, 125
RNA probe, 382–383
Robbins, Kenneth, 44
Roche, 96, 105
Rocky, 282
Roe v. Wade, 345
Rosenberg, Eric, 181
Rosenberg, Philip, 297
Ross, Judith Wilson, 234
Rous, Peyton, 55
Rous sarcoma virus, 55
Rowland-Jones, Sarah, 346
RT. *See* Reverse transcriptase
RTIs. *See* Reverse transcriptase inhibitors
Rumors, 415
Rush, Benjamin, M.D., 420
Russia, 432
  AIDS cases, 8, 40, 195, 221–222, 295, 328, 429
Rutherford, George, 275
Ruth's Poem, 345
Ryan White CARE Act, 342, 401, 414, 428
Ryan White Story, The, 400

S
Saag, Michael, 62, 241, 417
Sabĭa virus, 8
SAFE program, 253
Saint-Fleur, Henri-Claude, 37
Saliva, 192, 199, 200
Saliva test, 366, 381. *See also* Testing
Salmonella typhi, 6
Samet, Jeffrey, 389
Samson, Michel, 130
San Francisco, 211, 307, 309, 430
  ACT UP/San Francisco, 32
  Candlelight Memorial, 16
  prevention, 250
San Francisco Chronicle, 211
Saquinavir (Fortovase), 74*t*, 79, 80
Saquinavir mesylate (Invirase), 74*t*, 79
Satcher, David, 279
Scare tactics, 12. *See also* Fear
Scarlet Fever, 3*t*
SCH-C, 105
Schering Plough, 105
Schizophrenia, 7
Science, 33, 87
Scientific American, 33
Scotland, 267
SDF-1. *See* Stromal Derived Factor-1
Sea animal morbillivirus, 5
Seattle Times, 200
Segars, James, 206
Semen, 133, 199–200
Semmelweis, Ignaz Phillip, 2
Senior citizens, 236, 308
  women, 338
Septicemia, 189
Seroconversion, 162
  window of infectivity before seroconversion, 168
Serologic Testing Algorithm for Recent HIV Seroconversion (STARHS), 380. *See also* Enzyme linked immuno-sorbent assay test; Testing
Serum, 88
Seventeen, 360

Severe acute respiratory syndrome (SARS), 18, 49
discussed, 5
Sexual activity. See also Heterosexual activity; Homosexual activity; Prostitution
abstinence, 247, 251, 359–362
AIDS transmission, 23–24, 66, 192, 195, 197–198, 201, 203–231, 215t
activities affecting, 208
circumcision affecting, 203–204
number of partners, 206–208
orogenital sex, 208
sexual assault, 225–229
STD's affecting, 229–231
anal sex, 24
condom, 24, 211–214, 247, 251
advertising about, 254
buying, 255
choosing, 251, 253–254
contact dermatitis, 251
in developing nations, 258
education about, 260
female condom, 260–261, 264
in general, 251, 255–258
history, 253
lubricants, 261–263
as medical device, 251
plastic condom, 251, 254, 260
polymer gel condom, 260
polyurethane condom, 253, 254
quality, 259–260
religion and, 258–259
in schools, 255
size, 251–253
dental dam, 251
fidelity, 251
Kinsey study, 302
kissing, 192, 200–201, 235
microbicides, 263–265
partner notification, 219, 275–278
preventive behavior, 248
prostitution
developing countries, 208–224
USA, 208
safer sex, 242–243, 251, 359–360
spermicide, 251, 254
unprotected sex, 211, 355
extreme sex party, 211
Russian Roulette party, 211
unsafe sex, 242–243
Sexually transmitted disease (STD). See Sexual activity; *specific STD's*
SFV. See Simian foamy virus
Shalala, Donna, 270, 429
Sharp, Matt, 90
Shea, Griffin, 327
Sheep, 50
Shoemaker, Lisa, 202
Siliciano, Robert, 123
Simian foamy virus (SFV), 8
Simian immunodeficiency virus (SIV), 28, 41, 50, 67
SIVcpz, 42–44, 65
Simmons, Roy, 218
Simoni, Jane, 93
Single Use Diagnostic System (SUDS), 381. See also Testing
SIV. See Simian immunodeficiency virus
60 Minutes, 202
Smallpox, 1–2, 3t, 5, 110, 131
discussed, 7–9
SMART study, 95–96
Smith, Jerry, 218
Smith, Robert, 242
Sneezing, 192, 235
Snow, John, 40
Social Security Administration (SSA), 29
Society, 414–415
Sodroski, Joseph, 188
Sonnabend, Joseph, 80
SOUNDEX, 308
South Africa. See also Africa
AIDS infection, 65, 295, 303, 320–321, 323–325, 435
AIDS/Natal Province, 166
AIDS policies, 34–36, 340
Durban conference, 14, 235, 434
South African Freedom Day, 419
Sperm, 200
Spermicide, 251, 254
Sports, 217–219
Sports Illustrated, 217
Spumavirus, 50
Squamous cell carcinoma, 155
SSA. See Social Security Administration
SSI. See Supplemental Security Income
St. Louis Post-Dispatch, 226
Stallone, Sylvester, 219
STARHS. See Serologic Testing Algorithm for Recent HIV Seroconversion
Stavudine (Zerit), 74t, 75
STD. See Sexual activity; *specific STDs*
STI. See Structured treatment interruption
Stigma, 350–351, 397, 398
Stop AIDS Project, 430
Stramer, Susan, 176
Strategic drug interruption, 91
Strathdee, Steffanie, 266
Stress, 27
discussed, 133–134
Stromal Derived Factor-1 (SDF-1), 128, 183
Structural protein, 55
Structured treatment interruption (STI), 91
discussed, 100–105
Subbramanian, Ramu, 123
SUDS. See Single Use Diagnostic System
Sun-Sentinel, 37

Superinfection, 86
Supplemental Security Income (SSI), 428
Surrogate markers, 72
Survivors. *See also* Anti-HIV therapy; Prevention
  discussed, 82–83
  with therapy, 178–179
  without therapy, 178
Swartout, Judith, 153
Sweden, 267
Swine flu, 9
Switzerland, 270
Syncytia, 56
Syphilis, 7, 12, 208, 210, 211, 231, 253, 275–276
  Tuskegee experiments, 234

T
T cell. *See also* T4 cell
  attack cell, 279
  discussed, 110–112
  helper T cell, 118
  inducer T cell, 119
  memory cell, 279
  naive T cell, 124
T cell antigen receptor (TCR), 112
T4 cell, 27, 37, 56, 83
  cell count, 116–120, 140, 180–184, 185
  cell loss, 114
    apoptosis, 132–133
    autoimmune mechanisms, 133
    cofactors, 132, 133
    in general, 131–132
    HIV cellular transfer, 133
    impact, 134
    profound depletion, 135
    superantigens, 132
    syncytia formation, 132
  cell receptor
    cluster, 128
    FUSIN, 128
  helper cell, 25
  helper cell disease, 25
  HIV affecting, 113, 122, 124–127, 131, 183–184
  immune cell number, 72
  lymphocyte count, 11
  memory T4 cell, 124, 178
  T lymphocyte, 68–69
  T4 lymphocyte, 134
T8 cell, cell count, 180
T20, 105
Tacoma, NEP, 267
Taiwan, 400
Tale of Two Cities, A (Dickens), 1
Tampa Hillsborough Action Plan (THAP), 431
Tattoo, 225
TCR. *See* T cell antigen receptor
Tears, 192
Tennessee, 351, 391
Tenofovir (Viread), 74*t*, 80, 87
Terrorism, 1, 290, 395
Testing. *See also* Anti-HIV therapy; Prevention; Viral load test; *specific tests*
  anonymous testing, 393
  assumptions and, 372
  barriers to, 385–386
  blinded testing, 393
  Chicago Tribune, 373
  competency and informed consent, 388–390
    named/unnamed testing, 390–391
    testing without consent, 390
  compulsory testing, 391–392
  confidential testing, 392–393
  confidentiality/partner betrayal, 392
  counseling requirement, 382
  direct test, 367
  false positive/negative, 368, 376
  gene probes, 382–383
  in general, 365–366
  immigrants, 386–388
  immunodiagnostic techniques, 367
  indirect test, 367
  mandatory testing, 391
  Miami Herald, 373
  National AIDS Awareness Test, 404
  post-exposure prevention, 103
  prenatal, 386
  pretest patient information, 375–376
  reasons for, 366–367, 386
  replication capacity test, 88
  requests for, 366
  saliva test, 366, 381
  sensitivity/specificity, 376–377
  who should be tested?, 384–385
  window period, 368, 376
Tetanus, 279
Texas, 336, 351, 352, 357
Thailand, 64, 66, 143, 224, 286, 333
  Bangkok conference, 14, 292–294
THAP. *See* Tampa Hillsborough Action Plan
This Week, 431
Thompson, Tommy, 71, 278
3TC (Epivir), 80
Thrive Magazine, 217
Thrush, 145
Ticket to Work/Work Incentives Improvement Act, 414–415
Time, 217, 228
Tobacco, 423
Topics in HIV Medicine, 432
Toxic Shock Syndrome, 2
Toxoplasma gondii, 149–150
Toxoplasmosis, 26, 149–150
Traditional medicine, 95
"Treatise on the Epidemic Puerperal Fever of Aberdeen" (Gordon), 2

"Treatise on the Management of Pregnant and Lying-In Women" (White), 2
Trillion, 113
Trimeris, 105
Trizivir, 74*t*, 80
Trujillo, Alfonso Lopez, 258, 259
Truth, 411–412
Truvada, 80
Tsg101 cell, 48
Tshabalala-Msimang, Manto, 36
Tuberculosis, 3*t*, 5, 133, 140, 144*t*, 321, 431
  discussed, 10–11, 151–154
Turned Down, 254
Tutu, Desmond, 429
Typhoid, 6, 131
"Typhoid Mary", 6
Typhus, 3t, 4

U
UFOs, 40
Uganda. *See also* Africa
  AIDS transmission, 207–208
Ukraine, 222
UN Convention on the Rights of the Child, 347
UNAIDS. *See* United Nations Program on HIV/AIDS
UNDCP. *See* United Nations Drug Control Program
UNDP. *See* United Nations Development Program
UNESCO. *See* United Nations Educational, Scientific and Cultural Organization
Unexpected Universe, The (Eiseley), 19
UNFPA. *See* United Nations Population Fund
UNICEF. *See* United Nations Children's Fund
United Nations Children's Fund (UNICEF), 19, 223–224, 354
United Nations Development Program (UNDP), 19
United Nations Drug Control Program (UNDCP), 19
United Nations Educational, Scientific and Cultural Organization (UNESCO), 19, 44
United Nations Population Fund (UNFPA), 19
United Nations Program on HIV/AIDS (UNAIDS), 13, 19, 222, 242, 245, 254, 261, 291, 294, 304, 337, 347, 354, 355
Upjohn, 104
Urine test, 366. *See also* Testing
U.S. Department of Health and Human Services (DHHS), 96, 404
  guidelines, 94
U.S. Export-Import Bank, 433
U.S Food and Drug Administration (FDA), 200, 251, 282, 368
  approved drugs, 72–75
U.S. Public Health Service, 259, 351, 389, 398
USA Today, 33

V
Vaccination, 63. *See also* Vaccine
Vaccine. *See also* Anti-HIV therapy; Prevention
  costs, 285
  defined, 279
  in developing nations, 286–287
  expectations for, 281
  in general, 278–279, 280
  human trials, 285–286
  Internet resources, 287
  problems with, 282–284
  requirements for, 280–281
  testing, 285
  testing confidentiality, 286
  types of, 279–280
    dead/attenuated microorganisms, 281
    DNA vaccine, 281–282
    perinatal vaccine, 279
    preventive vaccine, 279
    prophylactic vaccine, 279
    subunit vaccine, 281
    therapeutic vaccine, 279
    whole inactivated virus, 281
  vaccine therapy, 284–285
Vaccine Awareness Day in America, 278
Valleroy, Linda, 211
Vandamme, Anne-Mieke, 28
Vangroenweghe, Daniel, 44
Varicella-zoster, 144*t*
Vasectomy, 199–200
Vaughan, Dr. Victor, 10
VaxGen AIDSVAX, 285–286
Vermont, 307
Viagra, 99, 211, 308
Viard, Jean-Paula, 83
Vidal, Nicole, 44
Vietnam, 216, 252
Viral fitness, 88
Viral load, 72, 82, 83. *See also* Anti-HIV therapy; Viral load testing
  determination, and questions people ask, 92–94
  in general, 88
  HIV RNA, 89–90, 184–185, 382–383
    detectable vs. undetectable levels, 93–94
    measurement, 383
  opportunistic infection and, 142–143
  perinatal transmission, 351
  relation to infection stage, 380
  sexual transmission, 220
Viral load assay. *See also* Viral load test
  bDNA, 89
  NASBA, 89
  PCR, 89
Viral load disconnect, 72
Viral load test, 88–89. *See also* Testing; Viral load testing
  available tests, 92
  results, 92–93
Viral load testing
  discussed, 88–89

questions people ask, 92–94
relevance, 93
Viral pool, 174
Viremic, 177
ViroLogic, 88
Virus. *See also* Disease; Human Immunodeficiency Virus; *specific viruses*
animal origins, 41
cancer associations, 7
characteristics, 48–50
long terminal repeat, 52
provirus, 51–52, 70–72, 124
cancer connection, 159
regulatory proteins, 55
retrovirus, 49–50
transcription, 52
viral RNA strand/RNA transcript, 52–55
viral specificity, 49
Visna virus, 50
Vulvovaginal candidiasis, 342

W
Walker, Bruce, 181, 284
War, 4
biological warfare, 40
Washington Post, 26
Wasting syndrome, 151, 171
Watkins, James, 404
WB. *See* Western Blot
Weapon of mass destruction, 11, 429
Webb, Barbara, 202
Weiss, Robin, 33
Welform Reform Act, 359
West Nile virus, 18
discussed, 8–9
Western Blot (WB) test, 366, 374. *See also* Testing
discussed, 377–378
gel electrophoresis, 377
indeterminate test, 378
What If Everything You Thought You Knew About AIDS Was Wrong? (Maggiore), 32
White, Charles, 2
White, Nick, 261
White, Ryan, 400–401, 405–406
WHO. *See* World Health Organization
Willner, Robert, 33
Wisconsin, 392
Witchcraft, 398, 400
Witek, James, 96
WLQY-AM, 37
Wolinsky, Steven, 37
Women. *See also* Children
AIDS infection, 30, 145, 198*t*, 295, 323
adolescent, 357–358
artificial insemination, 201, 225, 336–338
drug use, 221
in general, 332–334, 340–341
prevention, 343–344
prostitution, 27, 38, 182–183, 208–224, 321, 338–339, 391
USA, 334–336, 338
AIDS progression, 182–183
AIDS transmission, 38, 192
caesarean section, 233
in general, 341–342
groupies, 218–219
lesbians, 339–340
pediatric transmission, 231–234, 346–348, 351
perinatal transmission, 233, 351
pregnancy, 344–345, 346
vertical/MTCT, 350–351
deaths, 342–343
gender roles, 245
International Women's Day, 332
pregnant women risks, 76, 250
rape, 225–229
reproductive rights, 345
side effects, 99
testing, 389
Women Alive, 100
Women's AIDS Network, 340
Workplace, 413–414
World, 100
World AIDS Day, 16–17, 259
World Bank, 19, 433
World Health Organization (WHO), 5, 6, 9, 18–19, 26, 84, 154, 210, 222, 245, 259, 264, 291, 294, 301, 334, 429
Wu, Yuntao, 58
Wyatt, Richard, 125

X
Xiping Wei, 105, 125, 136, 169

Y
Yecs, John, 202
Yellow Fever, 3*t*, 5, 419–420
Yerkes Primate Center, 61
Yoder, Michael, 242

Z
Zalcitabine (Hivid), 74*t*, 75
Zambia, 326–327
Zidovudine (retrovir) (ZDV) (AZT), 34, 74*t*, 75, 77, 80, 81, 85, 150, 233, 235, 343, 352
Zierler, Solley, 340
Zinc finger, 74*t*